T0230028

Lecture Notes in Computer Science 508

Edited by G. Goos and J. Hartmanis

Advisory Board: W. Brauer   D. Gries   J. Stoer

S. Sakata (Ed.)

Applied Algebra, Algebraic Algorithms and Error-Correcting Codes

8th International Conference, AAECC-8
Tokyo, Japan, August 20-24, 1990
Proceedings

Springer-Verlag

Berlin Heidelberg New York
London Paris Tokyo
Hong Kong Barcelona
Budapest

Series Editors

Gerhard Goos
GMD Forschungsstelle
Universität Karlsruhe
Vincenz-Priessnitz-Straße 1
W-7500 Karlsruhe, FRG

Juris Hartmanis
Department of Computer Science
Cornell University
Upson Hall
Ithaca, NY 14853, USA

Volume Editor

Shojiro Sakata
Department of Knowledge-Based Information Engineering
Toyohashi University of Technology
Tempaku, Toyohashi, Aichi 441, Japan

CR Subject Classification (1991): E.4, I.1, G.2, F.2, E.3

ISBN 3-540-54195-0 Springer-Verlag Berlin Heidelberg New York
ISBN 0-387-54195-0 Springer-Verlag New York Berlin Heidelberg

Printing and binding: Druckhaus Beltz, Hemsbach/Bergstr.
2145/3140-543210 - Printed on acid-free paper

PREFACE

The present volume contains the Proceedings of the Eighth International Conference on Applied Algebra, Algebraic Algorithms and Error-Correcting Codes (AAECC-8) held in Tokyo, Japan, August 20–24, 1990. The annual international AAECC conferences cover a range of topics related to applied algebra, algebraic algorithms, error-correcting codes, cryptography, computational methods in algebra and geometry, and computer algebra in general. This volume contains papers on error-correcting codes (including some new aspects of their applications, e.g. coded modulation and sequences with good correlation properties), computational algebra and geometry, bases, algorithms and designs for symbolic and algebraic computation, and cryptography. The previous proceedings of AAECC conferences (AAECC-2, 3, 4, 5, 6) were also published by Springer-Verlag as Lecture Notes in Computer Science Volumes 228, 229, 307, 356, 357.

The 8th AAECC conference took place for the first time outside Europe. It was held in the buildings of Nihon University near the center of the giant metropolis of Japan as the Second International Joint Conference together with the International Symposium on Symbolic and Algebraic Computation (ISSAC'90). The extended abstracts submitted for AAECC-8 were each evaluated by two international referees. About 50 contributions were presented at the conference. Then the respective authors were required to submit a full paper version, again to be refereed by two international referees. After the screening, this volume contains two invited contributions and 31 full papers, including a paper presented at AAECC-7.

I would like to express my sincere thanks to the referees, whose help was essential to have this volume published in a short time. We thank all participants who helped make the conference so nice. Many thanks are due to Prof. E. Goto as the General Chairman of the conference and to the members of the Program Committee, in particular to Prof. J. Calmet, Prof. H. Imai, Prof. R. Kohno, Prof. H. F. Mattson, Jr., Prof. T. Mora and Prof. A. Poli for their kind encouragement and helpful suggestions in preparing this volume. We are also indebted to Prof. H. Imai, Dr. Y. Iwadare and Prof. R. Kohno for their efficient help as Local Organizers. Special thanks go to our sponsors: Matsushita Electric Industrial Company Ltd., Mitsubishi Electric Corporation, Advanced Company Ltd., Inoue Foundation for Science, and International Communications Foundation.

The next AAECC conference will be held in New Orleans, October 1991. We hope it will continue to succeed in making an important contribution in the world of computer science and mathematics.

Shojiro Sakata

March 1991

CONFERENCE OFFICERS

General Chairman of the Second International Joint Conference ISSAC'90 and AAECC-8:
E. Goto, University of Tokyo, Tokyo

AAECC Conference Co-Chairman:
H. Imai, Yokohama National University, Yokohama

AAECC Program Committee:
T. Beth, J. Calmet, A. C. Hearn, J. Heintz, L. Huguet, H.Imai (Co-Chairman), Y. Iwadare, M. Kasahara, R. Kohno, R. Loos, H. Lüneburg, H. F. Mattson, Jr., A. Miola, T. Mora, E. Olcayto, A. Poli (Co-Chairman), S. Sakata (Proceedings Editor), K. Yamaguchi

Local Organization:
H. Imai, Y. Iwadare, R. Kohno

CONTENTS

Computational Algebra and Geometry

†*presented at AAECC-7*

COVERING RADIUS AND WRITING ON MEMORIES

Gérard D. COHEN

ENST
46 rue Barrault
75634 Paris cedex 13 France

ABSTRACT

We investigate and survey some connections between coding theory and the problem of writing on binary memories subject to constraints on transitions between states. More precisely, the existence of good covering codes is used for two purposes :
- computing the capacity (maximum achievable rate) for some special classes of translation-invariant constraints
- constructing error-correcting WOM- codes.
 The talk is based mostly on two papers :
[A] "Writing on some binary memories with constraints(W*M's), presented at "Geometries, Codes and Cryptography", Udine, Italy, 19-23 June 1989,
[B] "Error-correcting WOM-codes", co-authored with G. ZEMOR, submitted to IEEE Trans. on Inform. Theory.

1.1- INTRODUCTION : W*M's

We consider a binary storage medium consisting of n cells on which we want to store and update information. These operations must be performed under some constraints, dictated by technology, cost, efficiency, speed,... In the past few years, many models have been studied, which we list here in more or less chronological order : write-once memory or WOM ([18], [22], [5]), write-unidirectional memory or WUM ([3], [6], [20]) write-efficient memory or WEM [1] and write-isolated memory or WIM [7], [11].We also consider the related problems of reluctant memories, WRM [9] and defective memories WDM [10], [14]. The initial model (WOM) representing the first generation of optical disks differs fondamentally from the others in that writing is irreversible. In this paper, * stands for U, E, I, R or D. Accent will be put on general methods rather than specific constructions and some already published proofs will be only sketched.

We assume we have a W∗M with n positions, which we use for writing one message among M possible. We want to be able to continue the process indefinitely, under a specific constraint depending on ∗. The problem is : what is aymptotically the maximum achievable rate R of the W∗M, defined as (log being to the base 2) $R := (1/n) \log M$?

We assume that the encoder (writer) is informed about the previous state of the memory, i.e. can read before writing, but the decoder (reader) is not.

1.2- THE CONSTRAINTS

1.2.1. WUM

This model has been introduced by Borden [3], whom we quote : "A write-unidirectional memory is a binary storage medium which is constrained, during the updating of the information stored "to either writing 1's in selected bit positions or 0's in selected bit positions and is not permitted to write combinations of 0's and 1's. Such a constraint arises when the mechanism that chooses to write 0's or 1's operates much more slowly than the means of accessing and scanning a word".

1.2.2. WIM

A write-isolated memory is a binary storage medium on which no change of two consecutive positions is allowed when updating. This constraint is dictated by the nowaday technology for writing on some digital disks [21] and is studied in [7].

1.2.3. WRM

Suppose our storage media allows only a limited number, say r, of bit changes when updating. This question, refered to as writing on a reluctant memory, has been considered first by Fellows [9].

1.2.4. WDM

The problem of writing on a memory with defects has been considered by many authors (see [10] and [14]). The model is the following : a set S_0 of positions of the memory are stuck at "0", a set S_1 at "1". The sets S_0, S_1, are known to the encoder only and satisfy $|S_0|+|S_1| = s$.

1.2.5. WEM

This a general model, containing as special cases WOM, WUM, WIM and WRM, where costs are associated to transitions. It is introduced in [1], where the maximal rate achievable with a maximal cost per letter criteria is investigated.

1.3- A GENERAL UPPER BOUND ON CAPACITY

Let us give an upper bound, valid in all four cases. Let V_n be the set of binary n-tuples, $V_n(t)$ the set of possible states of the memory after t utilizations. Consider the following directed graph

$G = (V, E)$ with $V = V_n(t) \cup V_n(t+1)$ and
$E = \{(i, j),$ where $i \in V_n(t), j \in V_n(t+1)$ and $i \to j$ is allowed$\}$.
Now the following is clear :

Proposition 1 $\qquad M \leqslant \underset{i}{Max} \, v(i), \qquad i \in V_n(t), \qquad$ (GB)

where $v(i)$ is the valency of i.

Proof. indeed any state i can be updated to at most $v(i)$ states j. ∎

1.4- TRANSLATION-INVARIANT CONSTRAINTS.
EXACT FORMULAS FOR CAPACITY

Let us denote by $F(x)$ the set of possible states for updating a given x in V_n, with $|F(x)| = v(x)$. We shall say that the constraints are translation-invariant if
$$F(x) = x + F(\underline{0}) : = \{x+y : y \in F(\underline{0})\}.$$
We set $F(\underline{0}) = F$, $|F| = f_n$ and call $F(x)$ the F-set centered at x.

Examples :

WIM. $\qquad F = \{x = (x_1, x_2,... x_n) \in V_n : x_j x_{j+1} = 0$ for $i = 0, 1,..., n-1\}$
WRM. $\qquad F = \{x \in V_n : w(x) \leq r\}$.
$\qquad$ Here F is the Hamming sphere of radius r centered at $\underline{0}$.
WDM. $\qquad F = \{x \in V_n : x_i = 0$ for $i \in S_0 \cup S_1\}$.
That is, F is a cylinder of radius n-s and length s.

We now present a coding strategy based on the notion of good blocks, which is a straightforward extension of the treatment presented in [7] for a WIM and will apply whenever the constraints are translation-invariant.

A subset $B = \{b_1, b_2,... b_m\}$ of Vn is called a good block if :

$$\underset{b \in B}{U} \quad F(b) = V_n . \qquad (1)$$

That is V_n is covered by F-sets centered on the elements of B.

Proposition 2 . If a block B is good, any translate B+t, t ∈ V_n, of B is also good.

Proof. $\underset{b' \in B+t}{U} F(b') = (\underset{b \in B}{U} F(b)) + t = V_n.$ ∎

Proposition 3. If B is good, then :
$$\forall \ x \in V_n \quad \exists \ b \in B \quad \exists \ f \in F : x + f = b.$$

In other words, starting from any state x of the memory, there exists an allowed transition f which transforms x into an element of B (say b).

Proof.   By (1), for all x, there is an i s.t. x is in $F(b_i)$, i.e. $x = b_i + f$ for some f in F. ∎

Proposition 4.   If B_0, B_1,...B_{M-1} are pairwise disjoint good blocks, they yield a WIM-code of size M.

Proof. Put the M messages to be coded in 1-1 correspondence with the blocks. By Proposition 3, whatever the state of the W∗M is, updating will be possible to any message. ∎

We shall now that the upperbound in Proposition 1 is tight. This will a fortiori give the capacity in the more favorable case 1 when writer and reader know the previous state. This result is not difficult to prove in a probabilistic (non constructive) way. We shall rather give here a "semi-constructive" proof, which also helps in obtaining good codes.

In view of Proposition 4, it is intuitive to look for "small" good blocks, so as to be able to pack many of them (e.g. by translation) in V_n. In fact, we shall first use the existence of small good subgroups of V_n (i.e. good blocks which are groups). Then the second step, finding pairwise disjoint good blocks, becomes simple : if G is a good subgroup, $|G| = 2^k$, then there are 2^{n-k} pairwise disjoint good blocks, namely the cosets of G. To that end, we use Theorem 1 of [4], which is established for coverings of V_n by Hamming spheres centered on the elements of a group (group coverings). Its extension to group covering by tiles other than spheres is easy.

Proposition 5. There exists a group covering G of V_n with 2^k sets $F(g_i)$, $g_i \in G$, with

$$k \leq n - \log f_n + \log n + 0(1). \quad ∎$$

This gives :
$M = 2^{n-k} \geq f_n / n.0^{(1)}$, and the following result.

Proposition 6. $R = \lim_{n \to \infty} n^{-1} \log f_n.$ ∎

We now turn to the problem of error correction for the previous described memories. In this paper, we shall only consider the case of WOM's, which was only mentioned in part 1. Let us first recall a few definitions and results.

2.1. INTRODUCTION TO WOM'S

A WOM (write-once memory) is a storage medium consisting of n binary positions, or "wits" with the property that :

- a wit in a "0" state may be updated to the "1" state but is then forever stuck at "1".
- Each wit is initially in the "0" state.

We shall call a {n,m,t}-WOM-code a scheme which allows t successive writings of m arbitrary bits (i.e. one message among 2^m) on a WOM of size n. WOM-codes have been studied from an information-theoretic view point in [23], and constructed using classical coding-theory in [5], [25] (for example, with parameters, {23,11,3}, {2^{m-1},m,$2^{m-2}+2^{m-4}$ +1}). We shall adapt those methods in order to answer a question raised by Rivest and Shamir in [18], namely : how can error-correcting WOM-codes be constructed ?

Let Π denote the set of wits of the WOM; we shall identify Π with the set {1,2,...n}, with n = |Π|. The set of messages we wish to write will be identified with a subset M of a finite abelian group G (in practice G = F^m, the group of binary vectors of length m). We will also need to index the wits of Π by a subset P of G ; more formally :

let σ be a one-to-one mapping of Π onto a subset P of G.

Let ε be the function $\Pi \rightarrow$ {0,1} describing the state of the WOM, i.e. $\varepsilon(\pi) = 0$ when π is unused, and $\varepsilon(\pi) = 1$ when π is used, π ranging over Π.

Reading a message :
The last message c written on the WOM is read by computing
$c(\varepsilon) = \Sigma_{\pi, \varepsilon(\pi)=1} \sigma(\pi)$.

Writing a message :
Given a state ε of the WOM, writing a message c $\in$ M is done by finding a set W $\subset \Pi$ of unused wits such that :
$\Sigma_{\pi\in W} \sigma(\pi) + \Sigma_{\pi, \varepsilon(\pi)=1} \sigma(\pi) = c.$ (2)

In the next section, we shall recall some basic facts from coding theory, and see how they are relevant to writing on WOMs.

2.2. THE TOOLS FROM CLASSICAL CODING THEORY.

2.2.1. Notations. (See [16] for more details).

We denote by supp($\underline{x}$) the support of a binary n-tuple $\underline{x}$, i.e. supp($\underline{x}$) = {i $\in$ {1,2,...n} : x_i=1}. Then |supp($\underline{x}$)| = w($\underline{x}$) is the Hamming weight of $\underline{x}$. A linear code over F={0,1} with length n, dimension k and (minimum) distance d is written C[n,k,d] and its dual $C^{\perp}$[n,r=n-k,∂]. A rxn full rank matrix whose rows generate $C^{\perp}$ is a parity check matrix of C, and we shall write H(C) = [$\underline{h}_1$,$\underline{h}_2$,...$\underline{h}_n$], $\underline{h}_i$ $\in$ F^r. We call ρ(C) the covering radius of C, i.e. the maximal possible distance between vectors in F^n and C. For a survey on covering radius see [24]. Let C be a given [i,k,d]-code. We denote by k(i,d,C) the maximal dimension of a code of length i and distance d containing C. We set k(i,d) = k(i,d,{$\underline{0}$}). A [i,k,d]-code C for which k=k(i,d,C) (resp. k=k(i,d)) will be called maximal (resp. optimal). Clearly the following holds :

$$k \leq k(i,d,C) \leq k(i,d).$$

For i < d, one has k(i,d) = 0. The proof of the following result is in [8].

Proposition 7.
Any code C [n,k,d] has a covering radius ρ satisfying :

$$k(\rho,d) + k \leq k(n,d,C). \qquad \blacksquare \qquad (3)$$

2.2.2. Applications to WOMs.

Suppose the group we use to construct our WOM-code is G =F^r; with every state ε of the WOM , we will associate the set S(ε) $\subset$ P corresponding to the unused wits, i.e. S(ε) = {$\sigma(\pi)$: $\varepsilon(\pi)$ = 0}. So, in the initial state S($\underline{0}$)=P. We will write simply S when no confusion can arise.

Let H(S) be any r x |S| matrix whose columns are the elements of S. (H(S) is defined modulo permutations of its columns). H(S) is the parity-check matrix of a code that we shall denote by C(S) (also defined up to equivalence, i.e. permutation of coordinates ; this will not give any trouble). The following lemmas provide information on those subsets S of P that generate G (and therefore M) and on the cost (in wits) of such generation; their proofs can be found in [B].

Lemma 1. Suppose P generates G. Let ∂ be the minimum distance of the linear code generated by the rows of H(P). Any subset S of P such that |S| > n-∂ generates G. $\blacksquare$

In other words, this last result means that whenever $|S| > n-\partial$, any message of M can be written on the WOM. Furthermore, the following lemma tells us how many wits are needed to write any such message.

<u>Lemma 2.</u> Whenever $S(\varepsilon)$ generates G, any message can be written by using $\rho(C(S))$ or fewer wits. ∎

2.3. A MODEL FOR CORRECTING SINGLE ERRORS

We use the notations of the introduction. When the desired message is $c \in M$ and an error has occured on position π, then the word actually read is : $c \pm \sigma(\pi)$. So the sets M and P should be such that any two couples $(c,p) \in M \times P$ and $(c', p') \in M \times P$ verify :

$$c \pm p \neq c' \pm p'.$$

To achieve this when the size of the WOM is $n = 2^r-1$, we choose a group $G = G_1 \times G_2$, where $G_1 = F^r$ and G_2 is any abelian group.

The set $P \subset G$ should verify the condition :

(P)   the projection on the first coordinate, $pr_1 : P \to G_1$ is a one-to-one mapping between P and $G_1 \backslash \{0\}$.

The set $M \subset G$ of messages should verify :

(M) $M \subset \{0\} \times G_2.$

So when a single error occurs, the message read on the WOM is $c \pm p$ instead of c :

Property (M) ensures that $pr_1(c \pm p) = pr_1(p)$, and property (P) ensures that $pr_1(p)$ uniquely determines p, and hence c.

Next, we want to maximise the number of times the WOM can be reused. To do this, we must find sets P and M verifying (P) and (M) such that any reasonably "large" subset S of P generates M, and furthermore such that only "few" elements of S are required to generate an arbitrary message $c \in M$. This is the object of the next section.

2.4. SINGLE-ERROR-CORRECTING WOM-CODES BUILT WITH BCH CODES

Denote by $C = BCH(2)$ the 2-error-correcting $[2^r-1, 2^r-1-2r, 5]$ BCH code (see[6] chap. 9). If β is a primitive element in $GF(2^r)$, then a parity-check matrix of C is given by

$$H = \begin{pmatrix} 1 & \beta & \beta^2 \dots & \beta^{n-1} \\ 1 & \beta^3 & \beta^6 \dots & \beta^{3(n-1)} \end{pmatrix} = [h_1, h_2, \dots h_n],$$

where every β^i in H must be thought of as a r-tuple (column) of elements in F.

Now take $G_1 = G_2 = F^r$, $P = \{h_i : 1 \leq i \leq n\}$, $M = \{0\} \times F^r$. Then conditions (P) and (M) are fulfilled. The following properties of BCH(2) are well-known to coding theorists :

P1. $\rho(BCH(2)) = 3$.

P2. The dual of BCH(2) is a code with minimum distance ∂ at least $2^{r-1} - 2^{r/2}$ if $r \geq 3$.

Combining P2 and lemma 1, we prove that any S with $|S| \geq 2^{r-1}-2^{r/2}$ generates G.

Now we use proposition 7 to upperbound $\rho(C(S))$.

Proposition 8.

i)   If $|S| > (n/2) + n^{1/2}$, then $\rho(C(S)) \leq 9$

ii)   If, moreover, $|S| > (\sqrt{2}/2)n$, then $\rho(C(S)) \leq 7$. ■

Proposition 8 means that the above scheme yields single-error-correcting WOM-codes, with parameters $\{2^r-1, r, t\}$, where, applying lemma 1 and 2 staightforward averaging :
       $t = n/15.6 + o(n)$.

An estimation of their efficiency (average number of bits written per wit in the worst possible case) is : $\phi \cong r/15.6$.

In [B], a similar study is persued for 3-error-correcting BCH codes. Analogous methods lead to single-error-correcting WOM codes with the following parameters $\{2^r-1, 2r, t\}$, with $t = n/22 + o(n)$ and $\phi \cong r/11$. We conjecture that considering t-error-correcting BCH codes with higher values of t would not increase the efficiency.

REFERENCES

[1] R. AHLSWEDE, Z. ZHANG, *Coding for write-efficient memories*, Indorm. and Control, vol. 83, n°1 (1989) 80-97.

[2] L.BASSALYGO, S. GELFAND and M. PINSKER, *Coding for channels with localized errors*, Oberwolfach Tagungsbericht 21/1989, May 1989.

[3] J.M. BORDEN, *Coding for write-unidirectional memories*. Preprint.

[4] G. COHEN, P. FRANKL, *Good coverings of Hamming spaces with spheres*, Discrete Math. 56 (1985) 125-131.

[5] G. COHEN, P. GODLEWSKI, F. MERKX, *Linear block codes for write-once memories*, IEEE Trans. Inform. Theory, IT-32, N°5 (1986) 697-700.

[6] G. COHEN, G. SIMONYI, *Coding for write-unidirectional memories and conflict resolution*, Discrete Applied Math.24 (1989) 103-114.

[7] G. COHEN, G. ZEMOR, *Write-Isolated Memories*, French-Israeli Conference on combinatorics and algorithms, Nov. 1988, Jerusalem, to appear in Discrete Math.

[8] I. CSISZAR and J. KORNER, *Information Theory*, Academic Press.

[9] M.R. FELLOWS, *Encoding graphs in graphs*, Ph. D. Dissertation, Univ-Calif. San Diego, Computer Science, 1985.

[10] C. HEEGARD and A.A. EL GAMAL, *On the capacity of Computer Memory with Defects*, IEEE Trans. on Inform. Theory, vol. IT-29, n°5, (1983) 731-739.

[11] T. KLOVE, *On Robinson's coding problem*, IEEE Trans. on Inform. Theory, IT-29, n°3 (1983) 450-454.

[12] K.U. KOSCHNICK, *Coding for Write-Unidirectional Memories*, Oberwolfach Tagungsbericht 21/1989, May 1989.

[13] A.V. KUZNETSOV, *Defective channels and defective memories*, Oberwolfach Tagungsbericht 21/1989, May 1989.

[14] A.V. KUZNETSOV and B.S. TSYBAKOV, *Coding in memories with defective cells*, Probl. Peredachi, Inform, vol. 10. n°2 (1974) 52-60.

[15]   L. LOVASZ, On the ratio of optimal integral and fractional covers, Discrete Math. 13 (1975) 383-390.

[16]   F.J. MACWILLIAMS and N.J.A. SLOANE, *The Theory of Error-correcting Codes*, North-Holland, New-York, 1977.

[17]   W. M.C.J. van OVERVELD, *The four cases of WUM-codes over arbitrary alphabets*, submitted to IEEE Trans. on Inform. Theory

[18]   R.L. RIVEST, and A. SHAMIR, *How to reuse a "write-once" memory*, Inform. and Control 55 (1982) 1-19.

[19]   M.R. SCHROEDER, *Number Theory in Science and Communication*, Springer-Verlag Series in Information Sciences, 1984.

[20]   G. SIMONYI, *On Write-Unidirectional Memory Codes*, IEEE Trans. on Inform. Theory, vol. 35, n°3 (1989) 663-669.

[21]   A. VINCK, *personal communication.*

[22]   H.S. WITSENHAUSEN and A.D. WYNER, *On Storage Media with Aftereffects*, Inform. and Control, 56 (1983), 199-211.

[23]   J.K. WOLF, A.D. WYNER, J. ZIV and J. KÖRNER, *Coding for write-once memory*, AT and T Bell Lab.-Tech. J. 63, N°6 (1984) 1089-1112.

[24]   G. D. COHEN, M. G. KARPOVSKY, H. F. MATTSON Jr. and J. R. SHATZ, *Covering radius - survey and recent results*, IEEE Trans. on Inform. Theory 31 (1985) 328-343.

[25]   P. GODLEWSKI, *WOM-codes construits à partir des codes de Hamming*, Discrete Math. 65 (1987) 237-243.

GEOMETRIC PROBLEMS SOLVABLE IN
SINGLE EXPONENTIAL TIME

Joos Heintz[1], Teresa Krick[1], Marie-Françoise Roy[2] and
Pablo Solernó[1]

Abstract. *Let S be a semialgebraic set given by a boolean combination of polynomial inequalities. We present an algorithmical method which solves in single exponential sequential time and polynomial parallel time, the following problems:*

- computation of the dimension of S.

- computation of the number of semialgebraically connected components of S and construction of paths in S connecting points in the same component.

- computation of the distance of S to another semialgebraic set and finding points realizing the distance if they exist.

- computation of the "optical resolution" of S if S is closed (the pelotita and the bolón).

- computation of integer Morse directions of S if S is a regular algebraic hypersurface.

The mentioned time bounds apply also to polynomial inequalities solving. As an application of our method we state an efficient Lojasiewicz inequality and an efficient finiteness theorem.

1. Introduction

The aim of this paper is to give an overview over a series of new algorithmical results in computational semialgebraic geometry which are based on our "efficient" quantifier elimination procedure for real closed fields [HRS 2,3]. Methodical predecessors of this procedure are (in chronological order) [GV 1], [G], [FMG 1], [HRS 0], [S 1], [HRS 1].

We have no space to explain the details of our quantifier elimination algorithm. So we give here only a list of its mathematical and algorithmical ingredients:

(1) some elementary differential geometry, essentially the ideas of Thom and Milnor to bound the number of connected components of semialgebraic sets ([BCR] Th.11.5.2).

(2) some semialgebraic geometry over non archimedean fields which we need for the use of infinitesimals for algorithmical purposes introduced in [GV 1].

(3) some computer algebra for the treatment of polynomial inequalities based on ideas of [BKR] and [CR] and developed in [FGM 1], [RS], [GLRR].

(4) new complexity results in computational algebraic geometry, mainly in the form of an efficient quantifier elimination procedure for algebraically closed fields ([FGM 1], [FGM 2], and [DFGS]). This results are applications of basic tools from commutative algebra, namely

– the Bezout-Inequality ([H], [F])

– the effective (affine) Nullstellensatz ([B], [CGH 1,2], [K 1]; see also [FG], [Te] and the references given there)

Our method resume those of [GV 1] in the points (1) and (2) whereas in the points (3) and (4) new ideas intervene. We think that the tools of [R] may serve as well for the purposes of this paper but we are inhabile to check it.

---

[1] Working Group Noaï Fitchas. Instituto Argentino de Matematica CONICET. Viamonte 1636 (1055) Buenos Aires, ARGENTINA. (mailing address)

[2] I.R.M.A.R. Université de Rennes I, 35042 Rennes Cedex FRANCE.

Our quantifier elimination method for real closed fields can be combined with some basic techniques for the treatment of Nash varieties and Nash functions ([BCR], Ch.8) for the solution of some fundamental algorithmical questions on the topology of semialgebraic sets.

One of our principal results is a single exponential sequential time and polynomial parallel time algorithm which computes the number of semialgebraically connected components of any semialgebraic set. This improves and rectifies previous results of [C 1,2] (see also criticism in [T]). An analogous result in the sequential bit complexity model was simultaneously obtained by D. Yu. Grigor'ev and N.N. Vorobjov (Jr.) [GV 2]. The methods used for solving this computational problem can be extended in different directions. Thus one obtains simple exponential sequential and polynomial parallel time algorithms for the construction of continuous semialgebraic paths between two points lying in the same connected component of a semialgebraic set (Theorem 8) or for the problem of defining the connected components of a semialgebraic set by quantifier free formulas [CGV] (a survey of the known methods and results is given in [GHRSV]).

Although we are not able to reproduce here the rather lengthy proofs for our topological results (see [HRS 4] and [HRS 5]). We sketch here the two of their main tools:

– the construction of sufficiently many Morse directions
– the pelotita and the bolón

Single exponential sequential and polynomial parallel time algorithms for the computation of dimension and distance of semialgebraic sets, for polynomial equations solving, etc. are relatively easy consequences of our general method which yields also an effective Lojasiewicz inequality and an effective finiteness theorem for semialgebraic sets. We arranged our results in an order which permits easy deductions. The aim of this paper consists rather in presentation of results, their motivation and interrelations than in providing complete proofs.

2. Notions and Notations

Throughout this paper we fix a real closed field R and a subring A of R (for example $R := \mathbf{R}$ and $A := \mathbf{Z}$).

Let $X_1, \ldots, X_n$ be indeterminates (variables) over R. To a polynomial $F \in A[X_1, \ldots, X_n]$ we assign its total degree $\deg(F)$ and its number of variables n. For a finite set $\mathcal{F}$ of polynomials of $A[X_1, \ldots, X_n]$ we write $\deg \mathcal{F} := 2 + \sum_{F \in \mathcal{F}} \deg(F)$.

We consider R^n, $n = 0, 1, \ldots$, as a topological space equipped with the euclidean topology. Let $x := (x_1, \ldots, x_n)$ and $y := (y_1, \ldots, y_n)$ be two points of R^n, we write

$$|x - y| := \sqrt{(x_1 - y_1)^2 + \cdots + (x_n - y_n)^2}$$

for their euclidean distance. If r is a positive element of R we write $B(x, r) := \{y \in R^n; |x - y| < r\}$ for the open ball of radius r centered at x.

If $x, y \in R$ (i.e. $n = 1$) we write $[x, y], (x, y), [x, y), (x, y]$ for the closed, open and half open intervals with boundaries x and y.

A *semialgebraic subset* of R^n (over A) is a set definable by a boolean combination of equalities and inequalities involving polynomials from $A[X_1, \ldots, X_n]$.

A semialgebraic set has only finitely many semialgebraically connected components which are all semialgebraic (see [BCR], Def.2.4.2 and Th.2.4.5).

Let $V \subset R^n$, $W \subset R^m$ be semialgebraic sets and $f : V \to W$ a map. We call f *semialgebraic* if its graph is a semialgebraic subset of R^{n+m}.

The image of a semialgebraic set by a semialgebraic function is semialgebraic too. This is called the *Tarski-Seidenberg Principle* (see [BCR], Théorème 2.2.1).

The Tarski-Seidenberg Principle can also be stated in terms of logics. It means that the elementary theory of real closed fields with constants from A admits quantifier elimination ([BCR], Prop.5.2.2).

From this we obtain the following *Transfer Principle* ([BCR], Prop.5.2.3): let R' be a real closed extension of R and let Φ be a formula without free variables (i.e. all the variables are quantified, by "$\exists$" and "$\forall$") in the elementary language $\mathcal{L}$ of real closed fields with constants from A. Then Φ is valid in R' iff Φ is valid in R.

Let $\Phi \in \mathcal{L}$ be a formula in the free variable $X_1, \ldots, X_n$. Φ defines a subset S of R^n. Since the elementary theory of R admits quantifier elimination we see that S is semialgebraic. On the other hand Φ can also be interpreted over R'. Thus Φ defines also a semialgebraic subset of $(R')^n$ which we denote by $S(R')$ (the independence of this notation from the particular defining formula Φ is justified by the Transfer Principle).

We call two formulas in the same free variables $X_1, \ldots, X_n$ *equivalent* if they define the same subset of R^n (this means they are equivalent with respect to the elementary theory of real closed fields).

A formula containing no quantifiers is called *quantifier free*. Thus the semialgebraic sets are those which are definable by quantifier free formulas. A formula is called *prenex* if all their quantifiers occur at the beginning.

Let $\Phi \in \mathcal{L}$ be a formula built up by atomic formulas involving a finite set $\mathcal{F}$ of polynomials of $A[X_1, \ldots, X_n]$. We write $\deg \Phi := \deg \mathcal{F}$. The *length* of Φ is denoted by $|\Phi|$.

An algorithm $\mathcal{N}$ (represented by a suitable family of arithmetical networks; see [Ga]) which for given natural numbers D, n and any input set $\mathcal{F} \subset A[X_1, \ldots, X_n]$ subject to $\deg \mathcal{F} \leq D$ computes an output set $\mathcal{G} \subset A[X_1, \ldots, X_m]$ is called *admissible* if the following conditions are satisfied:

- $\deg \mathcal{G} = D^{n^{o(1)}}$.
- the sequential complexity of $\mathcal{N}$ is $D^{n^{o(1)}}$.
- the parallel complexity of $\mathcal{N}$ is $(n \cdot \log D)^{o(1)}$.

In the case that such an algorithm exists we say that $\mathcal{G}$ is computable from the input $\mathcal{F}$ in *admissible time*.

Under certain circumstances $\mathcal{F}$ and $\mathcal{G}$ may represent the polynomials involved in quantifier free formulas $\Phi, \Psi \in \mathcal{L}$ defining semialgebraic sets $V \subset R^n$ and $W \subset R^m$. If $\mathcal{N}$ is admissible and computes also from the input data Φ the boolean combination of atomic formulas representing Ψ, we say that W is *admissibly computable* from V.

We shall also make use of notions of *critical point*, *Nash function* and *Nash variety*. For precise definitions we refer to [BCR].

3. Algorithmical and mathematical tools. First results

In this section we collect some algorithmical and mathematical tools we need for our geometrical constructions. Without proof, we state first a theorem which expresses a local lifting property for projections of semialgebraic sets. In terms of logics this theorem can also be interpretated as a statement about the local existence of continuous semialgebraic Skolem functions, computable in admissible time.

THEOREM 1. *Let S be a semialgebraic subset of $R^k \times R^n$. It is possible to compute in admissible time the following:*
  – a partition of R^k into semialgebraic sets T_i, $1 \leq i \leq s$.
  – for each $1 \leq i \leq s$ a finite family $(\xi_{ij})_{1 \leq j \leq \ell_i}$ of continuous semialgebraic functions from T_i to R^n such that for each $1 \leq j \leq \ell_i$ the graph of ξ_{ij} belongs to S and such that for each $x \in T_i$ each semialgebraically connected component of $S \cap (\{x\} \times R^n)$ contains at least one point of the graph of some ξ_{ij}. In particular, each connected component of $S \cap (T_i \times R^n)$ contains at least the graph of some ξ_{ij}.   ◊

For the proof of Theorem 1, see [HRS 3], Théorème 6 (weaker versions appear in [G], [HRS 1] and [S 1]).

Theorem 1 has a series of consequences (Corollaries 2, 3, 4):

COROLLARY 2. *Let S be a semialgebraic subset of $R \times R^n$. In admissible time it is possible to compute the following:*
  – a partition of R in semialgebraic intervals T_i, $1 \leq i \leq s$.
  – for each $1 \leq i \leq s$ a finite family $(\xi_{ij})_{1 \leq j \leq \ell_i}$ of continuous semialgebraic functions (curves) from T_i to R^n such that the graph of each ξ_{ij} belongs to S and such that each semialgebraically connected component of $S \cap (\{x\} \times R^n)$ contains at least one point of the graph of some ξ_{ij}. In particular, each connected component of $S \cap (T_i \times R^n)$ contains at least the graph of some ξ_{ij}.

PROOF. Put $k := 1$ in Theorem 1 and observe that any semialgebraic subset of R can be decomposed in admissible time in finitely many intervals (by Thom's Lemma, see [CR]).   ◊

Since the curves ξ_{ij} in Corollary 2 are defined on intervals, their images are semialgebraically connected. This observation will be relevant in later applications of this corollary.

COROLLARY 3. *(Efficient quantifier elimination, [HRS 2,3], [R]). Let Φ be a prenex formula in the language $\mathcal{L}$ of ordered fields with constants in A. Suppose that Φ contains m blocks of quantifiers and n variables. Then it is possible to compute in sequential time $(\deg \Phi)^{n^{0(m)}} |\Phi|^{0(1)}$ and in parallel time $n^{0(m)} (\log \deg \Phi)^{0(1)} + (\log |\Phi|)^{0(1)}$ a quantifier free formula Ψ which is equivalent to Φ.*   ◊

COROLLARY 4. *(Efficient curve selection lemma, [HRS 1], [S 1]). Let S be a semialgebraic subset of R^n and let p be in the topological closure $\bar{S}$ of S. Assume that S and p are definable by quantifier free formulas of $\mathcal{L}$. Then there exists an admissible algorithm which computes positive elements $\delta, \varepsilon \in R$ and continuous semialgebraic curves $\gamma_i : [0, \delta] \to R^n$, $1 \leq i \leq s$, satisfying the following conditions:*
  – $\gamma_i((0, \delta]) \subset S$ and $\gamma_i(0) = p$ for each $1 \leq i \leq s$.
  – each connected component of $B(p, \varepsilon) \cap S$ contains $\gamma_i((0, \delta])$ for at least one i.

PROOF. Apply Theorem 1 to the semialgebraic set S' defined by

$$S' := \{(r,x) \in R \times R^n;\; x \in S,\; |x-p|^2 \le r\}. \qquad \Diamond$$

Let $F \in A[X_1, \ldots, X_n]$ be a polynomial such that the closed semialgebraic set $V := \{F = 0\} := \{x \in R^n;\; F(x) = 0\}$ is bounded, and such that the gradient $\bigtriangledown F := \left(\frac{\partial F}{\partial X_1},\, \frac{\partial F}{\partial X_2}, \ldots, \frac{\partial F}{\partial X_n} \right)$ vanishes nowhere on V. Thus V is a regular hypersurface of R^n and we also say that F is a *regular* polynomial. Let $G \in A[X_1, \ldots, X_n]$ be a linear form (a linear coordinate transformation), and let $g : V \to R$ be the semialgebraic map induced by G on V.

We are going to apply to V and g concepts borrowed from elementary differential geometry: to be more precise, we consider g as a Nash function defined on V (see [BCR], Ch.8).

DEFINITION 6. Let notations be as before. We call $g : V \to R$ a *Morse direction* (or *Morse function*) if all its critical points are nondegenerate (see [M]).

We remark that the set of critical points of a Morse function is finite since it is semialgebraic and nowhere dense ([M]).

As a consequence of Sard's theorem ([GP], [BCR], Th.9.5.2), one obtains that the coefficient vectors of the linear forms of $R[X_1, \ldots, X_n]$ inducing Morse directions on V form a dense subset of R^n (see [BCR], proof of Prop.11.5.1 for details). Later we shall make use of the existence of an admissible algorithm which yields Morse directions of V, for a regular algebraic hypersurface V. Such an algorithm can be obtained combining the fact that being a Morse direction is an elementary property and the mentioned density of Morse directions with efficient quantifier elimination over R (Corollary 3).

We can summarize these observations in the following theorem:

THEOREM 7. *There exists a computable function* $L : \mathbf{N}^2 \to \mathbf{N}$ *with the following properties:*

(i) $L(n,D) = D^{n^{O(1)}}$ *for* n, $D \in \mathbf{N}$

(ii) *for each hypersurface V of R^n given by a regular polynomial $F \in A[X_1, \ldots, X_n]$ with $D := \deg F$, there exists a positive nonsingular matrix $M \in \mathbf{N}^{n \times n}$, with entries bounded by $L(n,D)$, such that for each row $(a_1, \ldots, a_n)$ of M the linear function $a_1 X_1 + a_2 X_2 + \cdots + a_n X_n$ induces a Morse function on V.* $\quad \Diamond$

For details of the proof of Theorem 7 we refer the reader to [S 1] and [S 2].

Corollaries 2, 3, 4 and Theorem 7 represent the key for the following theorem which extends and rectifies the main result of [C 1,2] (see also [T]).

THEOREM 8. *([GV 2], [HRS 4], [HRS 5], [GHRSV]). Let S be a semialgebraic subset of R^n and let x_1, x_2 be points of S. Suppose that S and x_1, x_2 are given by quantifier free formulas of $\mathcal{L}$. There exists an admissible algorithm which decides whether x_1 and x_2 are in the same semialgebraically connected component of S, and if they do so, the algorithm constructs a continuous semialgebraic curve of S connecting x_1 and x_2. In particular the algorithm computes the number of semialgebraically connected components of S.*

For a proof of this theorem see [HRS 4] (in the crucial case of a regular and bounded hypersurface) and [HRS 5] (for the general case). If $\Phi \in \mathcal{L}$ defines the semialgebraic set S, we have:

REMARK. As a consequence of the Thom-Milnor bound on the Betti numbers of semialgebraic sets ([BCR], Théorème 11.5.2) and of the Bezout bound for constructible sets ([H], Theorem 2 and its Corollary 1) one obtains that the number of semialgebraically connected components of S is of order $(\deg \Phi)^{0(n)}$ (see [G], Lemma 1). The main results of [HRS 4], [HRS 5] and [GV 2] can be restated and reproved in a parametrized form. This is done in [CGV]. Thus one obtains an admissible algorithm constructing quantifier free formulas of $\mathcal{L}$ defining the connected components of a given semialgebraic set.

4. Some geometrical constructions computable in admissible time

In this section we consider a series of basic construction and decision problems which are all more or less variants of the following general Theorem 9.

We call a point $x := (x_1, \ldots, x_n) \in R^n$ *real algebraic* if all its coordinates $x_1, \ldots, x_n$ are algebraic over A. We codify real algebraic points by means of Thom's Lemma ([CR]). The algorithmical machinery for dealing with real algebraic points is developed in [FGM 1], [RS] and [GLRR]. The corresponding algorithms are all admissible. In the sequel, when we speak about (real algebraic) points of a semialgebraic set we refer to a point equipped with a data structure, namely a system of inequalities of univariate polynomials codifying the coordinates of the point in the sense of Thom's Lemma.

THEOREM 9. ([S 1], [HRS 1,3]). *Let Φ be a prenex formula in the language $\mathcal{L}$ of ordered fields with constants from A. Suppose that Φ contains m blocks of quantifiers, and n variables from which q are free. Thus Φ defines a (possibly empty) semialgebraic set S contained in R^q.*

There exists an algorithm running in
- *sequential time:* $\quad (\deg \Phi)^{n^{0(m)}} \cdot |\Phi|$
- *parallel time:* $\quad n^{0(m)} \cdot (\log \deg \Phi)^{0(1)} + (\log |\Phi|)^{0(1)}$

and which decides whether Φ is satisfiable in R^q. If this is the case the algorithm computes a set T of real algebraic points of S such that $\#T = (\deg \Phi)^{n^{0(m)}}$ and such that each semialgebraically connected component of S cuts T (in this case we call T a system of representatives for the semialgebraically connected components of S).

In particular, if Ψ is a quantifier free formula the algorithm decides whether the system defined by Ψ is solvable, and, in the affirmative case, computes at least one solution, both in admissible time (apply, for this, Theorem 9 to the formula $\Phi : (\exists X)(\Psi(X))$). $\Diamond$

A complete proof of Theorem 9 is given in [S 1]. A slightly different proof follows easily from the particular quantifier elimination procedure mentioned in Corollary 3 (see [HRS 3] for details). A sequential version of Theorem 9 is contained in [G].

LEMMA 10. *Let S be a semialgebraic subset of R^n defined by a quantifier free formula $\Psi \in \mathcal{L}$. Then the interior $\overset{\circ}{S}$ and the the topological closure $\overline{S}$ of S are definable by quantifier free formulas of $\mathcal{L}$ which are computable from Ψ in admissible time.*

The proof of Lemma 10 is an immediate consequence of the efficient quantifier elimination (Corollary 3) since $\overset{\circ}{S}$ and $\overline{S}$ are definable from Φ by means of two quantifier alternations. ◇

PROPOSITION 11. *Let S_1 and S_2 be two semialgebraic subsets of R^n given by quantifier free formulas Φ_1, $\Phi_2 \in \mathcal{L}$. There exists an admissible algorithm which computes the euclidean distance between S_1 and S_2. This algorithm also computes (if they exist) two real algebraic points $x_1 \in \overline{S_1}$ and $x_2 \in \overline{S_2}$ such that $|x_1 - x_2|$ is the infimum of the distances between points of S_1 and S_2.*

PROOF. By Lemma 10 we may assume without loss of generality that S_1 and S_2 are closed. Apply now Theorem 9 to the semialgebraic set defined by:

$$S' := \{(x_1, x_2) \in S_1 \times S_2; \ |x_1 - x_2|^2 \leq |y_1 - y_2|^2 \quad \text{for all} \quad (y_1, y_2) \in S_1 \times S_2\},$$

which is definable from Φ_1 and Φ_2 using just one quantifier alternation. ◇

The dimension theory of semialgebraic sets we refer to in the next proposition can be found in [BCR], Section 2.8.

PROPOSITION 12. *Let S be a semialgebraic subset of R^n and let $x \in S$ be a real algebraic point. Then $\dim(S)$, the dimension of S, and $\dim_x(S)$, the local dimension of x in S, are computable in admissible time.*

PROOF. The dimension of S is the maximal integer q such that there exists $r \in R$ and an affine A-linear map $M : R^n \to R^q$ with the property that $M(S)$ contains the open ball $B(0, r)$. Thus, for each $0 \leq q \leq n$ we can build up from Φ a formula $\Psi_q \in \mathcal{L}$ which contains no free variables, only two blocks of quantifiers and says that $\dim(S) \leq q$. Applying the decision procedure for the elementary theory of R resulting from Corollary 3 in this situation we conclude that $\dim(S)$ is computable in admissible time.
For the computation of $\dim_x(S)$ in admissible time, one proceeds in an analogous way.
◇

A sequential version of Proposition 12 is given in [G].

We conclude this section with results (the pelotita and the bolón) which characterize the "optical resolution" of closed and bounded semialgebraic sets. These results represent a crucial step in the proof of Theorem 8.

LEMMA 13. *Let S be a semialgebraic subset of R^n and let $f : S \to R$ be a function which can be extended to a Nash function in an open neighborhood of S. Let S' be the semialgebraic set of all points of S which are local extrema with respect to f. Then $E := f(S')$ is a finite subset of R, semialgebraic over A.*

PROOF. E is expressible by a formula of $\mathcal{L}$. Therefore E is semialgebraic over A. Suppose that E is infinite. Then there exist points $a, b \in R$, $a < b$, definable over A, such that $[a, b] \subset E$. We consider the semialgebraic set $T := S' \cap f^{-1}([a, b])$. T is infinite and therefore $\dim(T) > 0$. By [BCR], Proposition 8.1.12 there exists a

disjoint finite decomposition $T = \bigcup_{i=1}^{N} T_i$, where each T_i is a Nash subvariety of R^n,
Nash-diffeomorphic to $(0,1)^{\dim(T_i)}$. Since T is infinite, there exists an index i, say
$i = 1$, such that the restriction map $f|_{T_i}$ is not constant. Let $x_1, x_2 \in T_1$ be such that
$f(x_1) \neq f(x_2)$. Since T_1 is Nash-diffeomorphic to $(0,1)^{\dim(T_1)}$, we consider T_1 as a
Nash variety, and $f|_{T_1}$ as a Nash function from T_1.
By Sard's Theorem ([BCR], Théorème 9.5.2), $f|_{T_1} : T_1 \to R$ has only finitely many
critical values and hence only finitely many local extrema.
On the other hand $f(T_1)$ is semialgebraically connected and contains more than one
point. Thus $f(T_1)$ is an infinite subinterval of $[a, b]$. By the choice of a and b this
implies that $f|_{T_1}$ has infinitely many local extrema. Contradiction. $\Diamond$

We formulate now two propositions each containing a quantitative (new) and a
qualitative (classical) statement. Proposition 14 means qualitatively that the connected
components of a given compact semialgebraic set have positive distance and Proposition
15 says classically that any sufficiently big closed ball intersects any connected component of a given closed semialgebraic set in such a way that this intersection itself is not
empty and connected.

PROPOSITION 14. (The pelotita). *Let S be a closed and bounded semialgebraic subset
of R^n given by a quantifier free formula $\Phi \in \mathcal{L}$, with $D := \deg(\Phi)$. It is possible to
find in admissible time two positive elements $\alpha_1, \alpha_2 \in A$ such that for any $x \in S$ the
set $B\left(x, \frac{\alpha_1}{\alpha_2}\right) \cap S$ is contained in only one semialgebraically connected component of S.
If $A := \mathbf{Z}$ and $R := \mathbf{R}$ one may take $\alpha_1 = 1$ and $\alpha_2 = \ell^{D^{n^{O(1)}}}$ where ℓ is an upper
bound for the absolute value of the coefficients of the polynomials occurring in Φ.*

PROOF. We consider the closed and bounded semialgebraic subset $S \times S$ of R^{2n} and
the continuous semialgebraic map $f : S \times S \to R$ defined by $f(x, y) := |x - y|^2$, where
x and $y \in S$.
Obviously f can be extended to a Nash function on R^{2n}.
The points of $S \times S$ which realize the euclidean distance between semialgebraically
connected components of S are local minima of f. Without loss of generality we may
assume that S has more than one semialgebraically connected component. Thus there
exists by Lemma 13 a positive real algebraic number $\xi \in R$ which realizes the smallest
image value of a positive local minimum of f. From the choice of ξ we have that for
each $0 < \eta < \xi$ and each $x \in S$ the open ball $B(x, \eta)$ cuts at most one semialgebraically
connected component of S.
On the other hand, Theorem 9 (or Corollary 3) implies that ξ is computable in admissible time as a root of an univariate polynomial over A, of degree of order $D^{n^{O(1)}}$.
From the coefficients of this polynomial we get positive numbers $\alpha_1, \alpha_2 \in A$ such that
$|\xi| > \frac{\alpha_1}{\alpha_2}$.
Thus, α_1, α_2 are computable in admissible time and satisfy the condition that for each
$x \in S$ the set $B\left(x, \frac{\alpha_1}{\alpha_2}\right) \cap S$ is contained in only one semialgebraically connected component of A.

The bound for the absolute values of α_1 and α_2 in case of $A := \mathbf{Z}$ and $R := \mathbf{R}$ follows immediately from the proof of Corollary 3 (see [HRS 2,3]). $\diamond$

PROPOSITION 15. (The bolón). *Let S be a closed semialgebraic subset of R^n given by a quantifier free formula $\Phi \in \mathcal{L}$ with $D := \deg(\Phi)$. It is possible to find in admissible time two positive elements $\alpha_1, \alpha_2 \in A$ such that for any $\rho \in R$ with $\rho \geq \frac{\alpha_1}{\alpha_2}$ and any semialgebraically connected component C of S the following holds:*

$\qquad C \cap B(0, \rho)$ is non empty and semialgebraically connected.

If $A := \mathbf{Z}$ and $R := \mathbf{R}$ one may choose as before $\alpha_2 = 1$ and $\alpha_1 = \ell^{D^{n^{O(1)}}}$.

For the lack of space, we will show only that there exists a positive $\rho_0 \in R$ such that for any semialgebraically connected component C of S the set $C \cap B(0, \rho_0)$ is non empty and semialgebraically connected. A complete proof can be found in [HRS 5].

PROOF. We consider the continuous semialgebraic map $f : S \to R$ defined by $f(x) := |x|^2$, where $x \in S$.

By Lemma 13 there exists a positive element $\rho_1 \in R$ which bounds the values of all local minima of f from above.

Consider the decomposition $B(0, \rho_1) \cap S = \mathcal{U}_1 \cup \ldots \cup \mathcal{U}_t$ in semialgebraically connected components. Choose elements $x_1 \in \mathcal{U}_1, \ldots, x_t \in \mathcal{U}_t$. If for $1 \leq i \neq j \leq t$, x_i and x_j are in the same semialgebraically connected component of S, there exists by [BCR], Définition and Proposition 2.5.11, a continuous semialgebraic curve $\gamma_{ij} : [0,1] \to S$ connecting x_i and x_j. Let $\rho_0 > \max\{\rho_1, |\gamma_{ij}(u)|^2; u \in [0,1]; 1 \leq i \neq j \leq t$ with x_i and x_j in the same semialgebraically connected component of $S\}$.

One verifies immediatly that ρ_0 has the desired properties. $\diamond$

5. An effective Łojasiewicz inequality

The efficient quantifier elimination algorithm (Corollary 3) has also implications of rather mathematical nature which we illustrate in this section by an effective version of the Łojasiewicz inequality and an effective finiteness theorem.

Let us first state a quantitative "mathematical" consequence of the quantifier elimination procedure ([HRS 2,3]) (see also [G], [S 1], [HRS 1], [R] for different approachs).

Let Φ be a formula of $\mathcal{L}$. We shall adopt the following notations:

- $\mathcal{F}_\Phi :=$ the set of polynomials over A appearing in Φ
- $d(\Phi) := \max\{\deg(F); F \in \mathcal{F}_\Phi\}$
- $\ell(\Phi) :=$ maximal absolute value of the coefficients of the polynomials $F \in \mathcal{F}_\Phi$.

THEOREM 16. *Let Φ be a prenex formula in n variables and with m quantifier alternations. Let $D := \deg(\Phi)$, $d := d(\Phi)$ and $\ell := \ell(\Phi)$.*

Then, there exists a quantifier free formula $\Psi \in \mathcal{L}$, equivalent to Φ in the elementary theory of R and which satisfies the following conditions:

$$\deg(\Psi) = D^{n^{O(m)}}, \quad d(\Psi) = D^{n^{O(m)}}, \quad \ell(\Psi) = \ell^{D^{n^{O(m)}}}$$

In the particular case of $m = 1$, one can choose Ψ such that:

$$\deg(\Psi) = D^{n^{O(1)}}, \quad d(\Psi) = D^{O(n)}, \quad \ell(\Psi) = \ell^{D^{O(n)}} \quad \diamond.$$

The reader should observe that in the second assertion of Theorem 16, $\deg(\Psi)$, which corresponds to the coarse bound $D^{n^{O(1)}}$, contains information about degree *and* number of the polynomials appearing in Ψ, whereas $d(\Psi)$, to which corresponds the asymptotically optimal bound $D^{0(n)}$, bounds only their degree.

NOTATION. For a semialgebraic function $f : S \to R$, we write
$$\{f = 0\} := \{x \in S;\ f(x) = 0\}.$$

We are ready to state the following version of an effective Lojasiewicz inequality.

THEOREM 17. (Effective Lojasiewicz inequality. General version). *Let S be a closed and bounded semialgebraic set contained in R^q and defined by a prenex formula $\Phi \in \mathcal{L}$. Let $f : S \to R$ and $g : S \to R$ two continuous semialgebraic functions also defined by two prenex formulas $\Phi_1, \Phi_2 \in \mathcal{L}$. Suppose that $\mathcal{F}_{\Phi_1} \cup \mathcal{F}_{\Phi_2} \cup \mathcal{F}_\Phi \subset A[X_1, \ldots, X_n]$ and that each of the formulas Φ_1, Φ_2, Φ contains at most m quantifier alternations. Let $D := \max\{\deg(\Phi), \deg(\Phi_1), \deg(\Phi_2)\}$. Then there exists a universal constant $c_1 \in \mathbb{N}$ (not depending on S, f and g) and a positive element $c_2 \in R$ (depending from them) such that $\{f = 0\} \subset \{g = 0\}$ implies that for all $x \in S$,*
$$|g(x)|^{D^{n^{c_1} \cdot m}} \le c_2 \cdot |f(x)|$$
If $A := \mathbb{Z}$, $R := \mathbb{R}$ and $\ell := \max\{\ell(\Phi), \ell(\Phi_1), \ell(\Phi_2)\}$, then c_2 can be chosen such that $c_2 = \ell^{D^{n^{0(m)}}}$. $\diamond$

THEOREM 18. (Effective Lojasiewicz inequality. Specific version). *Let S be a closed and bounded semialgebraic set contained in R^n and defined by a quantifier free formula $\Phi \in \mathcal{L}$ with $\mathcal{F}_\Phi \subset A[X_1, \ldots, X_n]$. Let $F, G \in A[X_1, \ldots, X_n]$ be polynomials inducing on S continuous semialgebraic functions $f : S \to R$ and $g : S \to R$. Let $D := \max\{\deg(\Phi), \deg(F), \deg(G)\}$. Then there exists a universal constant $c_1 \in \mathbb{N}$ (not depending on S, f and g) and a positive element $c_2 \in R$ (depending from them) such that $\{f = 0\} \subset \{g = 0\}$ implies that for all $x \in S$,*
$$|g(x)|^{D^{c_1 \cdot n}} \le c_2 \cdot |f(x)|.$$

We observe that the constant c_2 can be computed in admissible time from the input $\mathcal{F}_\Phi, F, G$.

If $A := \mathbb{Z}$, $R := \mathbb{R}$ and ℓ is an upper bound for $\ell(\Phi)$ and the absolute values of the coefficients appearing in F and G, then c_2 can be chosen such that $c_2 = \ell^{D^{0(n)}}$.

 $\diamond$

Let us observe that Theorems 17 and 18 differ from classical (qualitative) versions of the Lojasiewicz inequality only by the explicit form of the exponent which bounds the increasement of g. By the way, these exponents are asymptotically optimal as it can easily seen by well known examples (see eg. [K 2] or [S 3]).

The proof of the general version of the effective Lojasiewicz inequality (Theorem 17) is a not too difficult combination of Theorem 16 with the arguments used in [BCR], Section 2.6. Similarly for the specific version of the effective Lojasiewicz inequality whose proof is based on the second part of Theorem 16. For details of the proofs of these theorems we refer to [S 3].

Similar bounds as ours in Theorem 18 have independently been obtained in a case circumscribed to the context of algebraic geometry in [JS] and [K 2].

As a consequence of Theorem 17 and 18 we obtain

THEOREM 19. (Effective finiteness theorem). *Let S be an open semialgebraic set contained in R^n given by a quantifier free formula $\Phi \in \mathcal{L}$.*
There exists an admissible algorithm which finds a decomposition $S = \bigcup_{1 \leq i \leq s} S_i$, given by polynomials $F_{ij} \in A[X_1, \ldots, X_n]$, $1 \leq i \leq s$, $1 \leq j \leq \ell_i$, such that:

- $S_i = \{x \in R^n;\ F_{i1}(x) > 0, \ldots, F_{i\ell_i}(x) > 0\}$ *for each* $1 \leq i \leq s$
- $\displaystyle \sum_{\substack{1 \leq i \leq s \\ 1 \leq j \leq \ell_i}} \deg(F_{ij}) = \deg(\Phi)^{n^{0(1)}}$
- $\max\{\deg(F_{ij});\ 1 \leq i \leq s,\ 1 \leq j \leq \ell_i\} = \deg(\Phi)^{0(n)}$

An analogous statement holds for closed semialgebraic subsets of R^n (compare [BCR], Th.2.7.1). ◊

This theorem means classically that any open semialgebraic set is a finite union of semialgebraic open basic sets.

References

[BKR] Ben-Or M., Kozen D., Reif J.: The complexity of elementary algebra and geometry. J. of Comp. and Syst. Sci. 32 (1986) 251-264

[BCR] Bochnak J., Coste M., Roy M-F.: Géométrie algébrique réelle. Springer-Verlag (1987)

[B] Brownawell W.: Bounds for the degrees in the Nullstellensatz. Ann. Math. Vol. 126 N° 3 (1987) 287-290

[CGH 1] Caniglia L., Galligo A., Heintz J.: Some new effectivity bounds in computational geometry. Proc. AAECC-6 LN Comp. Sci. 357 (1988) 131-151

[CGH 2] Caniglia L., Galligo A., Heintz J.: Borne simple exponentielle pour les degrés dans le théorème des zéros sur un corps de caractéristique quelconque. CR Acad. Sci. Paris, t.307 (1988) 255-258

[C 1] Canny J.: The complexity of motion planning. MIT Thesis 1986, MIT Press (1988)

[C 2] Canny J.: A new algebraic method for robot motion planning and real algebraic geometry. Proc. 28th FOCS (1987) 39-48

[CGV] Canny J., Grigor'ev D.Yu., Vorobjov N.N. (Jr.): Finding connected components of a semialgebraic set in subexponential time. Manuscript Steklov Math. Inst. Leningrad LOMI (1990)

[CR] Coste M., Roy M-F.: Thom's Lemma, the coding of real algebraic numbers and the topology of semialgebraic sets. J. of Symb. Comp. 5 (1988) 121-129

[DFGS] Dickenstein A., Fitchas N., Giusti M., Sessa C.: The membership problem for unmixed polynomial ideals is solvable in single exponential time. To appear in Proc. AAECC 7, Toulouse (1989)

[FGM 1] Fitchas N., Galligo A., Morgenstern J.: Algorithmes rapides en séquentiel et en parallèle pour l'élimination des quantificateurs en géométrie élémentaire. Séminaire Structures Algébriques Ordonnées, Sélection d'exposés 1984-1987 Vol I. Publ. Univ. Paris VII, No. 32, (1990) 29-35

[FGM 2] Fitchas N., Galligo A., Morgenstern J.: Precise sequential and parallel complexity bounds for the quantifier elimination of algebraically closed fields. Journal of Pure and Applied Algebra 67 (1990) 1-14

[FG] Fitchas N., Galligo A.: Nullstellensatz effectif et conjecture de Serre (Théorème de Quillen-Suslin) pour le Calcul Formel. Sém. Structures Alg. Ord. Univ. Paris VII; final version to appear in Math. Nachrichten

[F] Fulton W.: Intersection Theory. Springer Verlag (1984)

[Ga] von zur Gathen J.: Parallel arithmetic computations: a survey. Proc. 13th Conf. MFCS (1986)

[GLRR] González L., Lombardi H., Recio T., Roy M-F.: Sous-résultants et spécialisations de la suite de Sturm. To appear in RAIRO, Inf. Théorique.

[G] Grigor'ev D.Yu.: Complexity of deciding Tarski algebra. J. Symb. Comp. 5 (1988) 65-108

[GHRSV] Grigor'ev D.Yu., Heintz J., Roy M-F., Solernó P., Vorobjov N.N. (Jr.): Comptage des composantes connexes d'un ensemble semi-algébrique en temps simplement exponentiel. To appear in C.R. Acad. Sci. Paris

[GV 1] Grigor'ev D.Yu., Vorobjov N.N. (Jr.): Solving systems of polynomial inequalities in subexponential time. J. Symb. Comp. 5 (1988) 37-64

[GV 2] Grigor'ev D.Yu., Vorobjov N.N. (Jr.): Counting connected components of a semialgebraic set in subexponential time. Manuscript Steklov Math. Inst. Leningrad LOMI (1990)

[GP] Guillemin V., Pollack A.: Differential Topology. Prentice-Hall (1974)

[H] Heintz J.: Definability and fast quantifier elimination over algebraically closed fields. Theor. Comp. Sci. 24 (1983) 239-277

[HRS 0] Heintz J., Roy M-F., Solernó P.: Complexity of semialgebraic sets. Manuscript IAM (1988)

[HRS 1] Heintz J., Roy M-F., Solernó P.: On the complexity of semialgebraic sets (ext. abst.). Proc. IFIP'89, 293-298

[HRS 2] Heintz J., Roy M-F., Solernó P.: Complexité du principe de Tarski-Seidenberg. C.R. Acad. Sci. Paris t.309 (1989) 825-830

[HRS 3] Heintz J., Roy M-F., Solernó P.: Sur la complexité du principe de Tarski-Seidenberg. Bull. de la Soc. Math. de France 118 (1990) 101-126

[HRS 4] Heintz J., Roy M-F., Solernó P.: Single exponential path finding in semialgebraic sets. Preprint (1990)

[HRS 5] Heintz J., Roy M-F., Solernó P.: Construction de chemins dans un ensemble semialgébrique. Manuscript (1990)

[JS] Ji S., Shiffman B.: A global Lojasiewicz inequality for complete intersections in C^n. Preprint (1989)

[K 1] Kollár J.: Sharp effective Nullstellensatz. J. AMS 1 (1988) 963- 975

[K 2] Kollár J.: A Lojasiewicz-type Inequality for Algebraic Varieties. Preprint (1989)

[M] Milnor J.: Morse Theory. Princeton Univ. Press (1963)

[R] Renegar J.: On the computational complexity and geometry of the first order theory of the reals I, II, III Technical Reports 853, 854, 856 (1989)

[S 1] Solernó P.: Complejidad de conjuntos semialgebraicos. Thesis Univ. de Buenos Aires (1989)

[S 2] Solernó P.: Construction de fonctions de Morse pour une hypersurface régulière en temps admissible. Manuscript IAM (1989)

[S 3] Solernó P.: Effective Lojasiewicz inequalities. To appear in AAECC Springer-Verlag (1990)

[Te] Teissier B.: Résultats récents d'algèbre commutative effective. Séminaire Bourbaki 718 (1989)

[T] Trotman D.: On Canny's roadmap algorithm: orienteering in semialgebraic sets. Manuscript Univ. Aix-Marseille (1989)

A DESCRIPTION OF THE [16,7,6] CODES

Juriaan Simonis

Delft University of Technology
Faculty of Mathematics and Informatics
P.O. Box 356, 2600 AJ DELFT, HOLLAND

Abstract: The paper gives a description of the three binary linear [16,7,6] codes by linking these codes to certain semilinear spaces. The basic technique is repeated puncturing and shortening.

I. INTRODUCTION

Let $n(k,d)$ be the smallest length that any binary linear code of dimension k and minimum distance d can have. After Van Tilborg [4] completed the determination of $n(7,d)$ for all values of d, quite a few coding theorists have tried to tackle the first open case for $k = 8$, i.e. whether $n(8,10)$ is equal to 25 or to 26. Recently two papers ([2],[7]) claimed the nonexistence of [25,8,10] codes, but, since the essential part of both papers is a computer search, the question —in my view — is as yet undecided. The main step in Kostova and Manev's paper [2] is the classification, again by computer, of the equivalence classes of [16,7,6] codes.

The present paper proposes to give a complete description of the [16,7,6] codes and their symmetry groups by mathematical means. Apart from that, it has two secondary goals: firstly, to demonstrate the advantage of puncturing and shortening without generator matrix manipulation, and, secondly, to propagate the application of geometric structures like semilinear spaces to existence and classification problems in coding theory.

1.1 Binary linear codes

Let C be a binary linear $[n,k,d]$ code, i.e. a k–dimensional linear subspace of F_2^n of minimum distance d .

If S is the set of the n coordinate positions, then each codeword of C , and, more gene-rally, each vector in the ambient F_2–vector space F_2^n , will be identified with its support in S , i.e. the subset of S consisting of the coordinate positions containing the symbol 1 . In other words, we identify F_2^n with the power set $\mathcal{P}(S)$ of S . Thus we can apply set theoretic notions like union, intersection, inclusion and complement to vectors in F_2^n .

The complement of $X \in F_2^n$ in S will be denoted by $\bar{X}$ .The sum of two vectors in F_2^n is their symmetric difference as subsets of S .

The **weight** of a vector $X \in F_2^n$, i.e. its cardinality as a subset of S , will be denoted by $|X|$.

The **distance** of two vectors $X, Y \in F_2^n$ is given by $d(X,Y) := |X + Y|$, and the distance of $X \in F_2^n$ and a subset $S \subset F_2^n$ by $d(X,S) := \min \{d(X,Y) \mid Y \in S\}$.

So the minimum distance d of the (linear!) code C equals $d(X,C\backslash\{X\})$ for any codeword $X \in C$.

The **covering radius** of the code C is defined by $t(C) := \max \{d(X,C) \mid X \in F_2^n\}$.

Let

$$A_i(C) := \{ X \in C \mid |X| = i \}, \quad i = 0,1,....,n,$$

denote the subsets of C of constant weight. Then the **weight distribution** of C is defined to be the sequence of the $n + 1$ non–negative integers $A_i(C) := |A_i(C)|$.

The **dual** $C^\perp$ of the code C is the orthogonal complement of the linear subspace $C \subset F_2^n$ with respect to the standard inner product on F_2^n . We occasionally use the, redundant, notation $B_i(C) := A_i(C^\perp)$, $B_i(C) := A_i(C^\perp)$.

The weight distributions of C and $C^\perp$ are connected by the celebrated **MacWilliams** identities (cf. [3], p. 127):

$$\sum_{j=0}^{n} K_i(j;n)B_j(C) = 2^{n-k} \cdot A_i(C) \quad (i=0,\dots,n),$$

with

$$K_i(j;n) := \sum_{\alpha=0}^{i} (-1)^\alpha \binom{j}{\alpha}\binom{n-j}{i-\alpha} .$$

II. THE CODES C_T AND C^T

Let C be a binary linear [n,k,d] code, with coordinate position set S, and let $T \subset S$ be any subset, say of cardinality m.

2.1 Definition

i)   The binary linear code

$$C_T := \{ C \cap T \mid C \in C \}$$

of length m is said to be derived from C by **puncturing** (with respect to the complement $\bar{T}$ of T in S).

ii)   The binary linear code

$$c^T := \{ C \in C \mid C \subset T \}$$

of length m is said to be derived from C by **shortening** (with respect to $\bar{T}$).

The following proposition shows that puncturing and shortening are intimately connected:

2.2 Proposition

i)   If we interpret $c^{\bar{T}}$ as a subcode of C, then

$$C_T \cong C/c^{\bar{T}} .$$

ii) The dual codes $(C_T)^\perp$ and $(C^T)^\perp$ (with respect to the standard inner product on $\mathcal{P}(T) \cong F_2{}^m$) satisfy

$$(C_T)^\perp = (C^\perp)^T \quad \text{and} \quad (C^T)^\perp = (C^\perp)_T \,.$$

Proof.

i) The kernel of the surjective linear mapping

$$\gamma\colon C \longrightarrow C_T \,, \quad C \longmapsto C \cap T \,,$$

is equal to $C^{\bar{T}}$.

ii) $X \in (C^\perp)^T \iff X \in C^\perp \wedge X \subset T \iff X \in (C_T)^\perp$.

Substitution of $C^\perp$ for C yields the second equality. □

2.2.1 Corollary. If $\dim(C_T) =: p$ and $\dim(C^T) =: q$, then

$$\dim(C_{\bar{T}}) = k - q \,, \quad \dim((C^\perp)_T) = m - q \,, \quad \dim((C^\perp)_{\bar{T}}) = n - m - k + p \,,$$

$$\dim(C^{\bar{T}}) = k - p \,, \quad \dim((C^\perp)^T) = m - p \,, \quad \dim((C^\perp)^{\bar{T}}) = n - m - k + q \,.$$

2.2.2 Application. Let C be an $[n,k,d]$ code and suppose that no $[n-6,k-4,d]$ code exists. Then any two words in $B_4(C)$ share at most one coordinate position.

Proof. If two words X , $Y \in B_4(C)$ had more than one coordinate position in common, we could choose a 6-set $T \subset S$ such that $X \subset T$ and $Y \subset T$. Hence the dimension of the code $(C^\perp)^T$ would be at least 2, and $C^{\bar{T}}$ would be an $[n-6,b,d]$ code with $b \geq k - 4$.

2.2.2.1 Remark. So, in this case, the elements of S and the words in $B_4(C)$ constitute the points and the lines of a so-called **semilinear space** , i.e. an incidence structure such that any line has at least two points and any two points are on at most one line. Verhoeff's table [6] provides many instances of codes C satisfying the condition of 2.2.2. For instance the $[16,7,6]$ codes, to be discussed in section III, and the $[25,8,10]$ codes whose existence has recently been put into doubt by computer calculations of several authors (cf. [2] and [7]).

A general semilinear space does not have much structure, but , due to the fact that $B_4(C)$ consists of all weight four codewords in a linear code, our semilinear space $(S,B_4(C))$ has additional properties, for instance the **pentagon property**: if four sides of a complete pentagon are present in the geometry, then the fifth side is present as well.

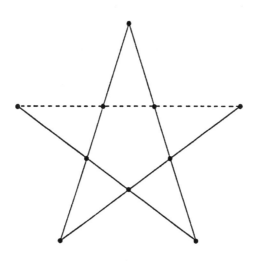

figure 1.

2.3 The minimum distance of C_T

2.3.1 Proposition. For any subset $T \subset S$ we have

$$d(C_T) \geq d - t(C^{\bar{T}}),$$

where $d(C_T)$ is the minimum distance of C_T and $t(C^{\bar{T}})$ is the covering radius of $C^{\bar{T}}$.

Proof. Choose a codeword $X \in C$ such that $|X \cap T| = d(C_T)$. Then

$$d = d(X,C\backslash\{X\}) \leq d(X,C^{\bar{T}}) = |X \cap T| + d(X \cap \bar{T},C^{\bar{T}}) \leq d(C_T) + t(C^{\bar{T}}) \qquad \square$$

2.3.1.1 Example. If $\bar{T} \in A_i(C)$, then C_T is called the **residual code** of C with respect to the codeword $\bar{T}$ (cf. [4]). Since $t(C^{\bar{T}}) \leq \frac{i}{2}$, we have: $d(C_T) \geq d - \frac{i}{2}$.

III. THE [16,7,6] CODES

3.1 The weight distribution

If we claim the nonexistence of a particular code, we implicitly refer to [6].

Let C be a [16,7,6] code. Then

$A_7 = 0$ (If $T \in A_7$, then $C_{\bar{T}}$ would be a [9,6,3] code, and such a code does not exist).

$A_9 = 0$ (If $T \in A_9$, then $C_{\bar{T}}$ would be a [7,6,3] code, because words of weight 2 in $C_{\bar{T}}$ require the existence of words of weight 7 in C. But no [7,6,3] code exists)

$B_1 = 0$ (An element $T \in B_1$ would yield the, nonexisting, [15,7,6] code $C^{\bar{T}}$).

$B_2 = 0$ (An element $T \in B_2$ would yield the, nonexisting, [14,6,6] code $C^{\bar{T}}$).

$B_3 = 0$ (An element $T \in B_3$ would yield the, nonexisting, [13,5,6] code $C^{\bar{T}}$).

We use this information in the MacWilliams equations number 0, 1, 2, 3 and 16:

$$1 + A_6 + A_8 + A_{10} + A_{11} + A_{12} + A_{13} + A_{14} + A_{15} + A_{16} = 128$$

$$16 + 4A_6 - 4A_{10} - 6A_{11} - 8A_{12} - 10A_{13} - 12A_{14} - 14A_{15} - 16A_{16} = 0$$

$$120 - 8A_8 + 10A_{11} + 24A_{12} + 42A_{13} + 64A_{14} + 90A_{15} + 120A_{16} = 0$$

$$560 - 20A_6 + 20A_{10} + 10A_{11} - 24A_{12} - 90A_{13} - 196A_{14} - 350A_{15} - 560A_{16} = 0$$

$$1 + A_6 + A_8 + A_{10} - A_{11} + A_{12} - A_{13} + A_{14} - A_{15} + A_{16} = 128B_{16}$$

The linear combination $10\ eq(0) + 5\ eq(1) + eq(3) - 10\ eq(16)$:

$$-64A_{12} - 120A_{13} - 256A_{14} - 400A_{15} - 640A_{16} = 1280(1 - B_{16})$$

shows us that $B_{16} = 1$, i.e. that C is an even weight code.

For each choice of A_{14} and A_{16}, we have four independent equations in the remaining unknowns A_6, A_8, A_{10} and A_{12}. Since $d(C) \geq 6$ implies that $A_{14} + A_{16} \leq 1$, we are left with just three possible weight distributions for C (and $C^{\perp}$):

case 1) $A_{14} = 0,\ A_{16} = 1,\ B_4 = 20.$

case 2) $A_{14} = 1,\ A_{16} = 0,\ B_4 = 12.$

case 3) $A_{14} = 0,\ A_{16} = 0,\ B_4 = 10.$

3.2 The semilinear spaces B_4

In 2.2.2.1, we have seen, that the 16 elements of S and the B_4 elements of B_4 constitute the points and the lines of a semilinear space.

Let us define

$$\nu_p := |\{\ L \in B_4 \ |\ p \in L\ \}|\ ,\ p \in S,\ \text{and}$$
$$\alpha_i := |\{\ p \in S \ |\ \nu_p = i\ \}|.$$

(So ν_p is the number of lines in the pencil through the point p, and α_i is the number of pencils containing exactly i lines.)

In S, there is no room for more than 5 lines through a given point, so $\nu_p \leq 5$. Moreover, the value $\nu_p = 4$ does not occur. For if the four lines L_1, L_2, L_3 and L_4 pass through the point p, then $S + L_1 + \cdots + L_4$ is a fifth line passing through p.

The usual counting of incidences leads to the following equalities:

$\Sigma\ \alpha_i = 16$ (the number of points),

$\Sigma\ i\ \alpha_i = 4\ B_4$ (the number of flags) and

$\Sigma\ \begin{bmatrix} i \\ 2 \end{bmatrix} \alpha_i = v$: the number of pairs of intersecting lines.

3.3 The description of the [16,7,6] codes

3.3.1 $A_{14} = 0$, $A_{16} = 1$

Since $B_4 = 20$, exactly one line passes through each pair of points. Consequently, the lines are the blocks of the, unique, 2(16,4,1)–design, better known as the affine plane over the field F_4 (cf. [1], p.390). The reader may verify that the lines in F_4^2, indeed, generate a [16,9,4] code with the required weight distribution.

So C is unique, and its symmetry group, isomorphic to the affine collineation group $A\Gamma L(2,4)$, acts transitively on S.

3.3.2 $A_{14} = 1$, $A_{16} = 0$

Let T be the unique word of weight 14.

Since C^T is an even [14,5,6] code with $A_{14}(C^T) = 1$, its weight distribution is fixed. Indeed, the remaining non–zero weights are six and eight, whence

$$A_6(C^T) = A_8(C^T) = \tfrac{1}{2}(2^5 - 1) = 15.$$

In particular, the second MacWilliams equation

$$91 - 5A_6(C^T) - 5A_8(C^T) + 91A_{14}(C^T) = 32B_2(C^T)$$

gives the value $B_2(C^T) = 1$, so exactly one line $L \in B_4(C)$ is not contained in T.

The two points in $L \setminus T$ each have multiplicity $\nu_p = 1$, so $\alpha_1 \geq 2$.

The two points in $T \cap L$ are the only ones for which ν_p could be equal to 5, so $\alpha_5 \leq 2$.

Combining this with the three equations in 3.2 and taking into account that $v \leq \begin{bmatrix} 12 \\ 2 \end{bmatrix}$, the number of all line pairs, we obtain the unique solution

$$\alpha_0 = 0, \ \alpha_1 = 2, \ \alpha_2 = 0, \ \alpha_3 = 12 \ \text{and} \ \alpha_5 = 2.$$

Through each of the two points p, q of T ∩ L passes four lines distinct from L , say L_1, L_2, L_3, L_4 and M_1, M_2, M_3, M_4 respectively. Any pair of lines L_i forms a quadrilateral with exactly one pair of lines M_i , and the sum of these four sides must be one of the three lines that do not intersect L . (Here we use the pentagon property described in 2.2.2.1.) So the semilinear space B_4 has a unique structure (see figure 2).

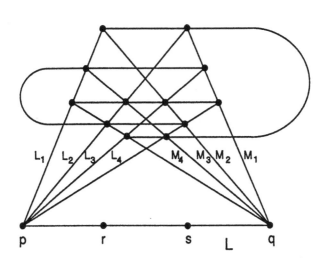

figure 2.

The symmetry group of B_4 has three orbits in S : {p,q} , L \ {p,q} and S \ L .

The dual $\mathcal{D}^\perp$ of the [16,8,4] code $\mathcal{D}$ spanned by B_4 contains one word of weight 14 and six words of weight 12, and these seven words span a six–dimensional subspace $\mathcal{E}$. Three seven–dimensional subspaces of $\mathcal{D}$ contain $\mathcal{E}$. One of these contains words of weight 2 and 4, but the other two are [16,7,6] codes. They are isomorphic under the mapping induced by the interchanging of p and q .

3.3.3 $A_{14} = 0$, $A_{16} = 0$

Let $L \in B_4$ be any line.

For the $[12,7,2]$ code $C_{\bar{L}}$, we have:

$$B_{12}(C_{\bar{L}}) = 1, \quad A_2(C_{\bar{L}}) \leq 1, \quad \text{and} \quad A_{12}(C_{\bar{L}}) = 0.$$

(If $A_{12}(C_{\bar{L}}) = 1$, then all words $\neq L$ in $B_4(C)$ would be contained in $\bar{L}$, i.e. $B_4(C_{\bar{L}}) = 9$. But this is incompatible with the Macwilliams identities for $C_{\bar{L}}$).

These restrictions leave us with just two possible weight distributions for $C_{\bar{T}}$:

i) $A_2(C_{\bar{T}}) = 0$ and $B_4(C_{\bar{T}}) = 1$, or

ii) $\quad A_2(C_{\bar{T}}) = 1$ and $B_4(C_{\bar{T}}) = 3$.

In other words, L intersects six or eight other lines, depending on whether L is contained in a word of $A_6(C)$ or not.

Let α be the number of lines for which the former situation holds.

If $T \in A_6(C)$ and $\dim((C^{\perp})^T) = 1$, then T cannot contain a word of $B_6(C)$, for otherwise $C_{\bar{T}}$ would be a $[10,6,4]$ code.

So α equals the number of words in $A_6(C)$ for which $\dim((C^{\perp})^T) = 1$.

Now we count the number of inclusions $T \subset U$, with $T \in A_6(C)$ and $U \in A_{12}(C)$:

i) If $\dim(\dim((C^{\perp})^T) = 1$ (0), then $\dim(C^{\bar{T}}) = 2$ (1), and $C^{\bar{T}}$ contains three (one) word of weight six. So α words of $A_6(C)$ are contained in three words of $A_{12}(C)$ and $44 - \alpha$ words of $A_6(C)$ are contained in one word of $A_{12}(C)$.

ii) $\quad$ The three-dimensional code C^U contains six words of weight six. So the 10 words of $A_{12}(C)$ each contain six words of $A_6(C)$.

The equation

$$3 \cdot \alpha + 1 \cdot (44 - \alpha) = 6 \cdot 10$$

gives $\alpha = 8$.

Hence the number v of pairs of intersecting lines in $B_4(\mathcal{C})$ is equal to

$$\tfrac{1}{2}(\alpha \cdot 6 + (10 - \alpha) \cdot 8) = 32,$$

and the three equations have only one solution:

$$\alpha_0 = 0 \,,\, \alpha_1 = 0 \,,\, \alpha_2 = 8 \,,\, \alpha_3 = 8 \text{ and } \alpha_5 = 0 \,.$$

The sum of all words in B_4 is a word in B_8 that consists of all points of multiplicity 3. Hence this word must contain the $10 - \alpha = 2$ lines L and M that intersect eight other lines. So L and M are disjoint, and all other lines intersect both L and M. Eventually, we are left with just one possible structure for the semilinear space B_4 (see figure 3).

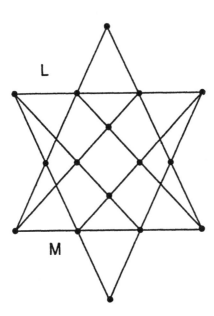

figure 3.

The symmetry group of B_4 has two orbits in S : $L \cup M$ and $S \setminus (L \cup M)$.

The dual code $\mathcal{D}^\perp$ of the [16,8,4] code $\mathcal{D}$ spanned by B_4 has the same weight distribution as $\mathcal{D}$. As a matter of fact, the codes $\mathcal{D}$ and $\mathcal{D}^\perp$ are isomorphic, but $\mathcal{D} \neq \mathcal{D}^\perp$!

Finally, we can retrieve C as the, unique, seven–dimensional subcode of $\mathcal{D}^\perp$ that does not contain any word of weight four.

References

1. Th. Beth, D. Jungnickel and H. Lenz, Design Theory (B.I.–Wissenschaftsverlag, Mannheim, 1985).
2. B.K. Kostova and N.L. Manev, A [25,8,10] code does not exist, preprint.
3. F.J. MacWilliams and N.J.A. Sloane, The Theory of Error–Correcting Codes (North–Holland, New York, 1983).
4. H. van Tilborg, On the uniqueness resp. nonexistence of certain codes meeting the Griesmer bound, Information and Control 44 (1980), 16–35.
5. H. van Tilborg, The smallest length of binary 7–dimensional linear codes with prescribed minimum distance., Discrete Mathematics 33 (1981), 197–207.
6. T. Verhoeff, An updated table of minimum–distance bounds for binary linear codes, IEEE Transactions on Information Theory 33 (1987) 665–680.
7. Ø. Ytrehus and T. Helleseth, There is no binary [25,8,10] code, IEEE Transactions on Information Theory 36 (1990) 695–696.

Periodic Sequences for Absolute Type Shaft Encoders

Kazuhiko YAMAGUCHI*   Hideki IMAI**

*   Department of Computer Science and Information Mathematics
University of Electro-Communications
Chofugaoka, Chofu, Tokyo 182 JAPAN

**   Division of Electrical and Computer Engineering Yokohama National University
Hodogayaku, Yokohama 240 JAPAN

Abstract:

Periodic sequence for single track absolute-type shaft encoder is studied. The absolute-type shaft encoder detects the absolute angle of rotation. Multi-track one having Gray coded sequence are practically in use. This paper searches possibility of single track absolute-type shaft encoders, and obtains good periodic sequences for them.

1. Introduction

A shaft encoder is used to detect the shaft angle of the rotating device. The encoder is called 'absolute-type' if the absolute angle can be determined and is called 'incremental-type' if the angle is determined relatively to the initial angle by accumulating the rotation of the shaft. In general the incremental-type has an advantage with respect to the simplicity of implementation, but the absolute type is required in such cases as the initial angle or the direction of the rotation is not given.

An absolute-type shaft encoder is obtained by applying Gray coding. It has k tracks around the shaft and k reading heads, and the angle of the shaft is divided into 2^k sections. Each section has a unique codeword of k bits, which is determined by Gray coding(Figure 1). This shaft encoder is called 'Multi-track' one.

Another absolute-type shaft encoder is based on a binary periodic sequence having window property and needs only one track around the shaft(Figure 2). We call it single-track absolute-type shaft encoder.

The binary periodic sequence is said to have a window of width k if every bit pattern of length k whose leftmost bit is in a certain period of the sequence is different. We call such bit patterns 'value of window'. A period of the periodic sequence having a window of width k is written on the track of the shaft encoder. (For the sake of the simplicity the phrase 'a period of' will be omitted in the sequel.) Then the shaft angle can be determined absolutely by reading k consecutive bits of the sequence.

For example, a maximum length linear feedback shift register sequences (abbreviated to m-sequence)[1] of period $2^k - 1$ is the periodic sequence having a window of width k. Such a single-track shaft encoder has a practical advantage over the Gray coded encoder, because the encoder can be made more compact.

Several researchers studies the single-track shaft encoder from various view points[2]–[8]. However, single-track shaft encoder has a practical problem which is called 'boundary-error'(Figure 3).

The boundary-error occurs from ambiguous interval of reading heads and/or sections of track, and occurs when each reading head try to read the boundary region of the consecutive two sections. Multi-track one avoids the same problem by using Gray coding. However, this problem is not well studied in single-track cases.

Trivial solution of this problem is to use extra track or tracks for detecting the boundary regions. However the shaft encoder is not single-track one.

Tomlinson[5] proposed a method to determine the angle whether there are boundary errors or not. However this method uses memory cells and cannot determine the angle immediately. We have been studying this problem, and propose an idea to determine the shaft angle with maintaining the operations from the boundary-error [6][8].

This paper studies the realization of single-track shaft encoder based on the idea. New binary sequences for the single-track absolute-type shaft encoders are constructed, and the encoders are boundary-error free.

2. Single track absolute-type shaft encoder

In this chapter, we introduce single-track absolute-type shaft encoder (we abbreviate it STASE). As show in Figure 2, consider STASE having resolution n (=16 counts/turn in the figure.) The resolution n means that n distinct values of angles can be detected by the STASE. The detected values of angle are equal intervals of $2\pi/n$ radian. For this purpose, the track of STASE is divided n 'sections' of equal interval(=$2\pi/n$ radian). A binary periodic sequence of period n is written on the track.

For example, each section is painted 'white' or 'black' that corresponds with the value '0' or '1' of the binary periodic sequence. Using the optical sensors for reading the values, the angle of rotation are determined. Naturally, we can use magnetic or the other sensors instead of optical ones.

In this paper, we use the word 'reading heads' for those sensors and assume them to have ideal characteristic. In this chapter, we neglect the problem of boundary-error. So, we also assume that all sections are painted exactly same intervals and reading heads are placed in the same interval.

Consider to detect the n distinct angle of the shaft with k reading heads. The periodic sequence for a STASE should satisfy the following condition. We call such sequence (n, k)-window sequence.

Definition 1. (n, k)-window sequence

Let $S = a_0, a_1, \cdots, a_{n-1}$ $(a_i = 0, 1)$ be a periodic sequence of period n, and define $W_i = (a_{i+1}, a_{i+2}, \cdots, a_{i+k})^\dagger$ $(i=0,1,\cdots, n-1)$, If the relation $W_i \neq W_j$ is satisfied for any $i, j = 0, 1, \cdots, n-1$ and $i \neq j$. S has window of length k, and is called (n, k)-window sequence. We call k 'window length of S'.

† In the definition of W_i, the term 'moduro n' is omitted in the subscripts of a. Therefore, $a_{i+1 \; moduro \; n}$, $a_{i+2 \; moduro \; n}$, $\cdots$, $a_{i+k \; moduro \; n}$ are denoted as a_{i+1}, a_{i+2}, $\cdots$, a_{i+k}. For the sake of simplicity, we use this notation through this paper.

From this condition, an STASE using (n, k)-window sequence discriminates n distinct values of angle. W_i corresponds with the detected i-th angle of STASE, and this STASE has k-reading heads.

For a given resolution n, the number reading heads, k is constrained by following lemma.

Lemma 1.

Consider (n, k)-window sequence of length n, the window length k must satisfy

$$k \geq log_2 \lceil n \rceil. \tag{1}$$

For a given window length k, the number of distinct window value is less than or equal to 2^k. Therefore, if this number 2^k is smaller than n, there are two or more angles which can not be discriminated. And if window length k is equal to $log_2 \lceil n \rceil$, this is possible minimum number of k, such (n, k)-window sequence is optimal. Because the smallest number of reading heads, that is window length k are required for an STASE from the practical viewpoint. Studies for optimal (n, k)-window sequences gives Theorem 1 [6][8].

Theorem 1.

For any positive integer n, optimal (n, k)-window sequences of period n that satisfies

$$k = log_2 \lceil n \rceil \tag{2}$$

are always exist.

The optimal (n, k)-window sequences is constructed by following.

(1) For $n = 2^k - 1$, the M-sequences of period $2^k - 1$ are the optimal (n, k)-window sequences.

(2) For $n = 2^k$, DeBruijn sequences[1] are the optimal one.

(3) For $2^{(k-1)} < n < 2^k - 1$, the optimal (n,k)-window sequences are constructed by truncating for M-sequences of period $2^k - 1$ [6][8].

3. Boundary error in shaft encoders

An STASE has a problem which we call 'boundary-error'. Theoretically, all sections and reading heads must be exactly placed at the same intervals. However there are ambiguities in the intervals of sections and reading heads of practical STASE. We have also assumed that characteristics for reading heads are identical. There might be ambiguity for characteristics of reading heads.

In these circumstance, the boundary-error may occur when the reading heads are try to read on the boundary between two consecutive sections (Figure 3). Multi-track absolute-type shaft encoder uses Gray coded sequence to avoid the same problem.

This chapter describes the boundary-error of STASE. Let us define a set $\beta_i = \{B_{i(1)}, B_{i(2)}, \cdots \}$. $B_{i(j)}$ is a read vector which is given by reading on boundary region between ith and $i + 1$-th angles (position). And let us denote the value of window W_i at i-th position as

$$W_i = (w_i(0), w_i(1), \cdots, w_i(k-1))$$
$$= (a_i, a_{i+1}, \cdots, a_{i+k-1}).$$

(3)

Then $B_{i(j)}$ is described as

$$B_{i(j)} = (b_{i(j)}(0), b_{i(j)}(1), \cdots, b_{i(j)}(k-1)),$$

(4)

where $b_{i(j)}(h) = w_i(h)$ or $w_{i+1}(h)$,
$$= a_{i+h} \quad \text{or} \quad a_{i+h+1}.$$

All of read vector $B_{i(j)}$ (j=1,2, $\cdots$) can be classified into following four cases.

(1) The angle is correctly detected as the position i or $(i + 1)$, i.e. $B_{i(j)} = W_i$ or W_{i+1}.

(2) The angle is detected as a incorrect position g, i.e. $B_{i(j)} = W_h$, $\forall h$ ($h \neq i$ or $i + 1$).

(3) The angle is detected as a boundary region but it is in two or more sets of boundary regions, i.e. $B_{i(j)} = B_h \ \exists$ g,h such that $B_{g(h)} \in \beta_g$, $g \neq i$

(4) The angle is correctly detected as the position is in the boundary between W_i and W_{i+1}, i.e.

$$B_{i(j)} \in \beta_h \qquad \text{for all } h \neq i, i+1 \tag{5}$$

and

$$B_{i(j)} \neq W_h \quad \text{for } h = 0, 1, \cdots n - 1 \tag{6}$$

The cases (2) and (3) can not be determined the angle correctly, and show the character of boundary-error.

4. Boundary error free STASE

In this chapter, we propose two type of STASE that avoids the boundary-error. We call them Type 1 and Type 2 boundary error free STASE.

4.1 Type 1 Boundary error free STASE

In Chapter 3, we described the characters of boundary-error. The simple solution of boundary error free STASE is realized by using new window sequence which avoid the cases (2) and (3) of boundary-error described in Chapter 3. We call it Type 1 STASE.

Type 1 STASE uses an (n, k)-window sequence and k' $(\geq k)$ reading heads for avoid all boundary-errors of case (2) and (3). We call it (n, k, k')-window sequence. $k'-k$ is the number of extra reading heads for avoid the boundary-error. An (n, k, k')-window sequence has to be an (n, k)-window sequence. And the read values for the STASE using (n, k, k')-window sequence are given as

$$W_i' = (a_i, a_{i+1}, \cdots, a_{i+k'-1})$$

for position i and

$$B_{i(j)}' = (b_{i(j)}'(0), b_{i(j)}'(1), \cdots, b_{i(j)}'(k' - 1))$$

for the boundary between W_i and W_{i+1} which satisfies

$$B_{i(j)} \neq W_h, \ \forall h \ (h \neq i, i+1), \tag{7}$$

$$B_{i(j)} \neq B_{g(h)} \forall h, g \text{ such that } B_{g(h)} \in \beta(g) \ (g \neq i). \tag{8}$$

Since the (n, k, k')-window sequence has to be an (n, k)-window sequence, First we try to construct (n, k, k')-window sequence from optimal (n, k)-window sequence. We evaluate the minimum value of k' for all m-sequence and extended m-sequence (period $n < 128$).

Table 1 shows the results that are the minimum value of k' and the generator polynomial of (n, k, k')-window (M-sequence) sequence. As show in the table, Type 1 STASE using m-sequence needs much more reading heads to avoid boundary-error.

Next, we compute optimal (n, k, k')-window sequence by brute forced computer search for small n(Figure 2). The optimal (n, k, k')-window sequence means the (n, k, k')-window sequence having possible minimum value of k' here. The result shows the number of reading heads of optimal (n, k, k')-window sequence is small. However construction of the optimal sequence for large n is problem.

4.2 Type 2 Boundary error free STASE

In section 4.1, we only discussed the STASE that has the same value between sections' interval and interval of reading heads except those of ambiguities. However, all reading heads of this type STASE are in boundary region at same time. The basic idea of Type 2 STASE is to decide the intervals of the reading heads as only one or no reading head tries to read the boundary region at a time.

Figure 4 shows the Type 2 STASE. In this figure, we assume that the shaft encoder has n sections and k' reading heads. The interval of the section is $l = 2\pi/n$ radian and that of the reading heads is $l(k'-1)/k'$ radian. We assume that the ambiguities of section intervals and reading head intervals are small enough, then this type STASE has only one or no reading head in boundary region.

In this case, the read value of a window becomes the k'-bit vector that contains $(k'-1)$ bits of the sequence and a double read bit of the sequence, i.e.,

$$
\begin{aligned}
W_{i,j}^+ &= (w_i^+(0), w_i^+(1), \cdots, w_i^+(k'-1)) \\
&= (a_i, a_{i+1}, \cdots, a_{i+j-1}, a_{i+j}, a_{i+j}, a_{i+j+1}, \cdots, a_{i+k'-1}),
\end{aligned}
\tag{9}
$$

where i is the position (the value of angle) and j indicates the place of double read bit.

For all i $(=0,1,\cdots,n-1)$,j $(=0,1,\cdots,k'-1)$, if the relation

$$
W_{i,j}^+ \neq W_{h,g}^+
$$

$h \neq i$ $(=0,1,\cdots,n-1)$ and j $(=0,1,\cdots,k'-1)$
are satisfied, we call this sequence $(n, k, k')^+$-sequence. By using the $(n, k, k')^+$-sequence, Type 2 STASE of resolution n with k' reading heads can be constructed.

To compare this type STASE with Type 1 STASE, first the optimum cases of TYPE 2 STASE are evaluated by computer search, which is shown in Table 3. The numbers of reading heads (k') are improved upon Type 1 STASE.

Next we discuss more constructive approaches of the encoder. For the type 2 STASE, we have following lemma.

Lemma 2. Upper bound of optimal k' for Type 2 STASE

For any positive integer n, there exists a Type 2 STASE of resolution n with k' reading heads that satisfies

$$k' \leq 3\lceil log_2 n\rceil. \tag{10}$$

Proof

We show an $(n, k, k')^+$-sequence satisfies Inequality (10). The $(n, k, k')^+$-sequence is optimal (n,k)-window sequence with $k' = 3k$ reading heads. We show the deciding method of the angle for this encoder. From $W_{i,j}^+$, we make three k-bit vectors as follows,

$$
\begin{aligned}
W_{i,j}^+(1) &= (w_i^+(0), \quad w_i^+(1), \quad \cdots, \quad w_i^+(k-1)) \quad (= W_i \text{ if } j > k-1) \\
W_{i,j}^+(2) &= (w_i^+(k), \quad w_i^+(k+1), \quad \cdots, \quad w_i^+(2k-1)) \quad (= W_{(i+k)} \text{ if } j < k \text{ or } j > 2k-1) \\
W_{i,j}^+(3) &= (w_i^+(2k), \quad w_i^+(2k+1), \quad \cdots, \quad w_i^+(3k-1)) \quad (= W_{(i+2k)} \text{ if } j < 2k)
\end{aligned}
\tag{11}
$$

Since we use (n, k)-window sequence, two vectors in $W_{i,j}^+(h)$ indicate the correct position i, $i+k$ or $i+2k$. So we can get the position i by majority logic. Therefore, this sequence is an $(n, k, k')^+$-sequence. The value of k' for optimal $(n, k, k')^+$-sequence is smaller than or equal to $3\lceil log_2 n\rceil$.

□

At last, we show practical construction of Type 2 STASE. We construct $(n, k, k')^+$-window sequence from optimal (n, k)-window sequence as Figure 1, and evaluate minimum value of k' for all m-sequence and extended m-sequence (period $n < 512$). The results shows Type 2 STASE save the reading heads(Table 4).

5. Conclusion

The problem of boundary-error for STASE are discussed. In Chapter 4, two types of boundary-error free STASE are proposed. They are still single-track and absolute-type one.

References

(1) S.W.Golomb: 'Shift Register Sequences', San Francisco,Holden-Day (1967).

(2) E.Petriu: 'Absolute-Type Pseudorandom Shaft Encoder with Any Desired Resolution', Electron. Lett. ,21,5, pp. 215-216 (1985).

(3) G.H.Tomlinson: 'Absolute-Type Shaft Encoder Using Shift Register Sequences', Electron. Lett., 23, 8, pp. 215-216 (1987).

(4) E.M.Petriu: 'New Pseudorandom Encoding Technique for Shaft Encoders with Any Desired Resolution', Electron.Lett.,23,10, pp.215-216(1987).

(5) G. H. Tomlinson, E. Ball: 'Elimination of Errors in Absolute Position Encoders Using M-Sequences', Electron.Lett., 23,25, pp.1372-1374(1987).

(6) E. Nomura, K. Yamaguchi and H. Imai,: 'Absolute encoders based on maximum-length linear feedback shift register sequences', the procedingon the 10th symposium on Information Theory and Its Applications, JA8-4, pp.195-200 (1987) (in Japanese).

(7) E. Nomura, K. Yamaguchi and H. Imai,: 'Error correction for angular detectors using M-sequences', National Convention Record 1988 Institute of Electronics,Information and Communication Engineers, A-196 (1988) (in Japanese).

(8) K. Yamaguchi and H. Imai,: 'A theory of periodic sequence for shaft encoder' 2nd International Colloquiam on Coding Theory, Osaka Japan (1988).

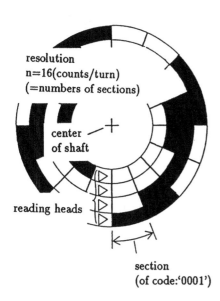

Figure 1.
Multi track absolute-type shaft encoder

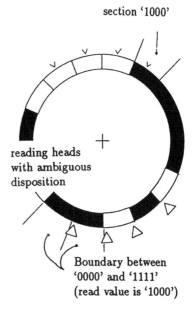

Figure 3.
'Boundary-error' in single track
absolute type shaft encoder

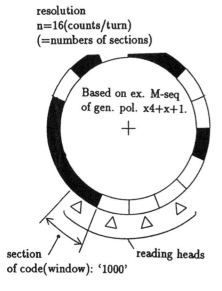

Figure 2.
Single track absolute-type shaft encoder

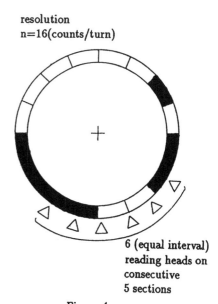

Figure 4.
Type 2 boundary-error free
single track absolute type shaft encoder

Table 1: TYPE1 STASE Using M-sequences

n	k'	g(deg.)
4	4	7(2)
7	7	13(3)
8	8	13(3)
15	14	23(4)
16	16	23(4)
31	24	45(5)
32	27	45(5)
63	31	103(6)
64	31	
127	>31	
128	>31	

n   : resolution of STASE (counts/turn)
k'   : window length (bit)
g   : generator polynomial of M-seq.(in octal form)
(x)   : x is the degree of generator
    23 means $x^4 + x + 1$
seq   : a period of sequence (in octal form)

Table 2: TYPE1 STASE Using Optimal-sequences

n	k'	seq(octal)
3		* not exist
4	2	03
5	3	03
6	3	07
7	4	007
8	4	017
9	5	017
10	5	0037
11	6	0037
12	6	0077
13	7	00067
14	7	00157
15	7	00637
16	7	000637
17	7	001477
18	7	003177
19	8	0003077
20	8	0006177
21	8	0014377
22	9	00014177
23	9	00030337
24	9	00060677
25	9	000147763
26	9	000316077
27	9	000634177
28	9	0001470377
29	9	0003160777

Table 3: TYPE2 STASE Using Otimal-seq.

n	k'	seq(octal)
3	2	1
4	2	03
5	3	03
6	3	07
7	4	007
8	4	017
9	5	017
10	5	0027
11	5	0057
12	5	0137
13	6	00117
14	6	00237
15	6	00477
16	6	001317
17	6	002637
18	6	005477
19	7	0004257
20	7	0010537
21	7	0021277
22	7	00042577

Table 4: TYPE2 STASE Using M-sequences

n	k'	g(deg.)
4	3	7(2)
7	6	13(3)
8	7	13(2)
15	8	23(4)
16	12	23(4)
31	9	45(5)
32	11	67(5)
63	12	103(6)
64	14	103(6)
127	14	203(7)
128	14	203(7)
255	16	435(8)
255	16	703(8)
511	17	1773(9)
512	17	1773(9)

ERROR-CODED ALGORITHMS FOR ON-LINE ARITHMETIC

George Sinaki

Delco Systems Operations, 6767 Hollister Avenue

Goleta, California 93117, USA

1 ABSTRACT

A method is presented for detection and correction of errors in multi-module networks for high-speed numerical computations. The modules execute on-line arithmetic operations in which the results are generated during the digit-serial input of the operands, beginning always with the most significant digit. The error detection/correction method effectively uses low-cost arithmetic codes since the interconnections between on-line modules are of minimum bandwidth. By detecting errors in on-line manner, i.e., as they occur, an effective gracefully degradable organization could be achieved. An uncorrected error at the j-th step leads to restriction of precision in the remaining steps but not to a catastrophic termination of the computation. The code-preserving algorithms are developed and their performance and effectiveness evaluated. Two particular schemes based on residue and AN encoding are discussed.

2 INTRODUCTION

This paper is concerned with the development of a set of error coded basic algorithms for on-line arithmetic. In on-line processing the operands, as well as the results, flow through arithmetic units in a digit-by-digit manner starting with the most significant digit. On-line arithmetic provides a cost-effective approach to achieve higher computational rates by allowing overlap at the digit level between the successive operations [1], [2], [3]. In particular, on-line arithmetic is highly attractive in high-speed multi-module structures for parallel and pipelined computations [4]. Since on-line arithmetic requires relatively long sequences of operations and multiple units in order to achieve cost-effective speed-up over conventional arithmetic, it is important to protect on-line algorithms against hardware failures. If not protected, the hardware failures could quickly contaminate large number of results in progress due to tight coupling of steps at the digit level. By detecting errors, as they occur, an effective, gracefully degradable organization could be achieved. Namely, error at the j-th step would lead to restriction of precision (significance) of the remaining steps but not catastrophic termination.

2.1 Definitions

By on-line algorithms we mean those arithmetic algorithms in which the operands as well as the results flow through the arithmetic unit in a digit-by-digit fashion, most significant digits first. These algorithms are such that, in order to generate the j-th digit of the result, $(j + \delta)$ digits of the corresponding operands are required. δ is called the on-line delay and is preferred to be as small as possible [1], [2].

It is not difficult to see that the use of redundant number representation is mandatory for on-line algorithms. If we were to use a non-redundant number system, then even for simple operations like addition and subtraction there is an on-line delay of $\delta = m$ due to carry propagation (m is the length of the operands). By allowing redundancy in the number representation system it is possible to limit the carry propagation to one digit position, and thus have the selection operate on a truncated value of P_j , called $\hat{P}_j$. This has an important implication that the recursion step time is independent of the precision.

In general, an on-line algorithm is specified recursively in terms of on-line representation of operands, results and some internal values. The recursion is usually of the following form:

$$P_j = f(P_{j-1}, x_{j+\delta}, y_{j+\delta}, z_j)$$

where f is a linear function and P_j is the partial internal result. The output digit is obtained by applying a selection function on the partial result P_j and the current inputs.

$$z_{j+1} = SELECT(\hat{P}_j, x_{j+\delta}, y_{j+\delta})$$

Several of the well known basic algorithms satisfy the on-line property with respect to either the operands or the results. Consider, for example, conventional division which has the on-line property with respect to the quotient digits. Similarly, conventional multiplication has the on-line property with respect to the multiplier. This property has later been extended to the product digits as well.

The totally parallel addition/subtraction [5] can be easily performed in on-line manner with $\delta = 1$. On-line algorithms for multiplication and division were introduced in [1], [2]. An overview of the generalized (with respect to radix) on-line algorithms for addition/subtraction, multiplication and, division has appeared in [3] along with the design of an on-line arithmetic unit. Others have extended on-line algorithms to encompass the on-line square rooting [6], [7] and on-line normalization [8]. Also, systematic methods for derivation of on-line addition/subtraction, multiplication and, division algorithms appears in [9] and for division in [10].

An analysis of the performance of on-line arithmetic in common computational problems such as the evaluation of scalar and vector expressions and recursion systems appears in [4]. Multi-module on-line networks for typical array computations are discussed in [11].

It should be noted that on-line arithmetic is complementary to other approaches that are used to achieve concurrency in execution of algorithms. For example, it can be used in already optimized tree-structured networks. On-line arithmetic is very attractive in reconfigurable networks because of high modularity and simple interconnection. However, there are applications such as non-linear recurrence systems where the on-line approach offers unique improvements in performance [4].

3 ERROR CODED ON-LINE ALGORITHMS

If an error occurs at a certain step of an on-line algorithm and if this error is detected immediately after generation and inhibited from spreading to the next module, then the operation of the following units can be continued although with less precision. Of course, the final results have correspondingly less precision than the original operands. Therefore, the on-line algorithms have capability of "graceful degradation". Of course, if error correction is applied, then such an error would not affect the computation and there would be no loss of precision.

In this paper we present error detection with residue-coded operands. Error detection/correction with AN-coded operands are discussed in [9].

3.1 Residue Encoding

Assume that the operands X and Y are represented by m digits in a radix r redundant number representation system, that is:

$$X = \sum_{i=1}^{m} x_i r^{-i}$$

$$Y = \sum_{i=1}^{m} y_i r^{-i}$$

These two numbers flow through the MAIN unit, shown in Figure 1, digit-by-digit most significant digit first. The MAIN Unit performs the algorithm MAIN-OP on the incoming operands and after certain amount of delay, the result Z appears at the output, again digit-by-digit starting with the

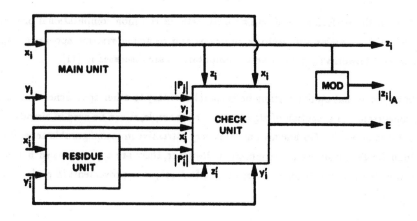

Figure 1: Residue-Coded On-Line Unit

most significant digit, such that:

$$Z = \sum_{i=1}^{m} z_i r^{-i}$$

At the same time the RESIDUE unit receives the residue of the corresponding digits of the MAIN unit. We represent this "residue" operands by X' and Y', such that:

$$X' = \sum_{i=1}^{m} x_i' r'^{-i}$$

$$Y' = \sum_{i=1}^{m} y_i' r'^{-i}$$

The following relation exists between these two sets of operands:

$$x_i' = x_i \bmod A \quad for \ i = 1, 2, ..., m \tag{1}$$

$$y_i' = y_i \bmod A \quad for \ i = 1, 2, ..., m \tag{2}$$

where A is the check modulus. The RESIDUE Unit performs the algorithm RESIDUE-OP. The output of the RESIDUE unit, in a similar manner, is represented by Z' and is:

$$Z' = \sum_{i=1}^{m} z_i' r'^{-i}$$

Notice that the relation $z_i' = z_i \bmod A$ is not necessarily satisfied. After generating z_i and z_i', MAIN and RESIDUE units operate on the next set of inputs. At the same time z_i and z_i', along with some other information, reach the CHECK unit. CHECK unit performs an algorithm called "DETECT-OP". This unit, after performing its algorithm on z_i, z_i' and other received information, decides whether these results agree with each other or not. If the results do not agree then it sets an error flag which inhibits all the operations until the source of error is detected. For example, the current step can be repeated by the MAIN and RESIDUE units and if the error still persists, the operation can be continued with less precision. In what follows we only present in detail the error-coded on-line division algorithm. The corresponding algorithms for addition/subtraction and multiplication are discussed in [9].

Assume that the dividend N and the divisor D are represented by m digits in a radix-r redundant number representation system. Also assume that the residue of each digit with respect to a constant A ($A = 2^a - 1$) is attached to it and is transferred to the on-line DIVIDE unit. Therefore the coded operands are:

$$(N, N') = .(n_1, n_1')(n_2, n_2'), \ldots, (n_m, n_m')$$

$$(D, D') = .(d_1, d_1')(d_2, d_2'), \ldots, (d_m, d_m')$$

$$(Q, [Q]_A) = .(q_1, [q_1]_A)(q_2, [q_2]_A) \cdots, (q_m, [q_m]_A)$$

n_i, d_i and q_i belong to the redundant digit sets, not necessarily the same. In the above equation:

$$[x]_A = x \bmod A$$

The algorithm "MAIN-DIVIDE" which is performed by the MAIN Unit follows:

Algorithm "MAIN DIVIDE"

Step 1 [Initialization]:

$$P_0 = \sum_{i=1}^{\delta} n_i r^{-i}$$

$$D_0 = \sum_{i=1}^{\delta} d_i r^{-i}$$

$$Q_0 = 0$$

Step 2 [Recursion]:
For j=1,2,....,m Do:
Step 2.1 [Selection]:

$$q_j = SELECT(r\hat{P}_{j-1}, \hat{D}_{j-1})$$

$$Q_j = Q_{j-1} + q_j r^{-j}$$

Step 2.2 [Input Digits]:

$$D_j = D_{j-1} + d_{j+\delta}r^{-j-\delta}$$

Step 2.3 [Basic Recursion]:

$$P_j = rP_{j-1} - q_jD_j + n_{j+\delta}r^{-\delta} - Q_{j-1}d_{j+\delta}r^{-\delta} \tag{3}$$

Step 3 [End]

The quotient digit SELECT function can be defined and implemented as in [9]. The algorithm performed by the RESIDUE Unit is similar to the previous one and it is named "RESIDUE-DIVIDE". The radix of the RESIDUE Unit is assumed to be r', preferably as small as possible. The digits n_i' and d_i' are defined as shown in Equations (1) and (2) and belong to the following digit set:

$$n_i', d_i', [q_i]_A \in \{0, 1, 2, ..., (A-1)\} \tag{4}$$

The output of the RESIDUE Unit is the quotient of the residues. It is assumed to be in the following set:

$$q_i' \in \{-\rho', ..., -1, 0, 1, ..., \rho'\} \quad (r'-1) \geq \rho' \geq r'/2$$

The proof of the convergence of the algorithms MAIN and RESIDUE DIVIDE has been given in [9]. In order to establish a detection algorithm, we consider the following equations for the j-th partial remainders in the MAIN and RESIDUE Units:

$$P_j = r^j \sum_{i=1}^{j+\delta} n_i r^{-i} - r^j [\sum_{i=1}^{j} q_i r^{-i}] [\sum_{i=1}^{j+\delta} d_i r^{-i}] \tag{5}$$

and

$$P_j' = r'^j \sum_{i=1}^{j+\delta} n_i' r'^{-i} - r'^j [\sum_{i=1}^{j} q_i' r'^{-i}] [\sum_{i=1}^{j+\delta} d_i' r'^{-i}] \tag{6}$$

Using (5) and (6) an algorithm that can detect an error at each step of the on-line division process is derived. This algorithm is performed on q_i and q_i', generated by the MAIN and RESIDUE Units, and determines whether these quotient digits are correct or not. From (5) and (6) we have:

$$[\sum_{i=1}^{j} q_i r^{-i}] [\sum_{i=1}^{j+\delta} d_i r^{-i}] = \sum_{i=1}^{j+\delta} n_i r^{-i} - r^{-j}P_j$$

and

$$[\sum_{i=1}^{j} q_i' r'^{-i}] [\sum_{i=1}^{j+\delta} d_i' r'^{-i}] = \sum_{i=1}^{j+\delta} n_i' r'^{-i} - r'^{-j}P_j'$$

By dividing these two equations, obtaining the residues of both sides and also assuming:

$$[r]_A = [r']_A = 1 \tag{7}$$

we get the following equation:

$$[[\sum_{i=1}^{j} q_i]_A \star [[\sum_{i=1}^{j+\delta} n_i']_A - [P_j']_A]_A]_A = [[\sum_{i=1}^{j} q_i']_A \star [[\sum_{i=1}^{j+\delta} n_i]_A - [P_j]_A]_A]_A \qquad (8)$$

To simplify (8) the following change of parameters is done:

$$[\sum_{i=1}^{j} q_i]_A = X_j \quad and \quad [\sum_{i=1}^{j} q_i']_A = X_j'$$

$$[\sum_{i=1}^{j+\delta} n_i']_A = Y_j' \quad and \quad [\sum_{i=1}^{j+\delta} n_i]_A = Y_j$$

Therefore (8) becomes:

$$[[X_j]_A \star [[Y_j']_A - [P_j']_A]_A]_A = [[X_j']_A \star [[Y_j]_A - [P_j]_A]_A]_A \qquad (9)$$

In order to detect an error in the division process, equality in (9) is checked by the CHECK Unit using the following algorithm.

Algorithm "DETECT DIVIDE"

Step 1 [Initialization]:

$$X_0 = X_0' = 0$$

$$Y_0 = [\sum_{i=1}^{\delta} n_i]_A$$

$$Y_0' = [\sum_{i=1}^{\delta} n_i']_A$$

Step 2 [Recursion]:
For j=1,2,....,m Do:
Step 2.1 [Input Digits]:

$$X_j = [X_{j-1} + q_j]_A$$

$$X_j' = [X_{j-1}' + q_j']_A$$

$$Y_j = [Y_{j-1} + n_{j+\delta}]_A$$

$$Y_j' = [Y_{j-1}' + n_{j+\delta}']_A$$

Step 2.2 [Check for Error]:

$$Z_j = [X_j \star (Y_j' - [P_j']_A)]_A$$

$$Z_j' = [X_j' \star (Y_j - [P_j]_A)]_A$$

Step 2.3 [Comparison]:

$$If\ (Z_j \neq Z'_j)\ ERROR\ EXIT$$

Step 3 [End]

As mentioned earlier, the radix of the RESIDUE Unit is an important factor in the design of the error-coded units. In the extreme case where $r = r'$, the detection process is merely duplication of the MAIN Unit. On the other hand, there are some lower bounds for r' that should be met. These bounds are calculated next.

Since residue digits (n'_i, d'_i) are assumed to be in radix r' number system we have:

$$n'_i, d'_i \leq r' - 1$$

using (4) we get:

$$A - 1 \leq r' - 1\ \ or\ \ r' \geq A \tag{10}$$

From (7) and (10) we obtain:

$$r' = M'A + 1\ \ for\ \ M' = 1, 2, ...$$

and, assuming that A is a low-cost modulus $(A = 2^a - 1)$, then:

$$r' = M'2^a - M' + 1\ \ for\ \ M' = 1, 2, ...$$

also from (7) we get:

$$r = MA + 1\ \ for\ \ M = 1, 2, ...$$

4 PERFORMANCE EVALUATION

In this section we analyze the effect of imposing error codes on basic on-line arithmetic algorithms. The economic feasibility of arithmetic error codes in a computer system depends on their cost and effectiveness with respect to the set of arithmetic algorithms and their speed requirements. The choice of a specific code from the available alternatives further depends on their relative cost and effectiveness values.

For a given algorithm, word length, and number representation system of the perfect unit the introduction of any code will result in changes that represent the cost of the code. The components of the cost are discussed below in general terms applicable to all arithmetic error codes [13].

1) Word length: The encoding introduces redundant bits in the number representation. A proportional hardware increase takes place in storage arrays, data paths, and processor units. The increase is expressed as a percentage of the perfect design. "Complete duplication" (100 percent increase) is the encoding which serves as the limiting case. In residue encoding, the residue of each digit with respect to A is attached to it and should be carried along with the corresponding digit. Assuming that the operands and the results belong to a redundant number system, the increase in word length is found to be proportional to $(\log_{2\rho} A)$ [9].

2) The Checking Algorithm: This test the code validity of every incoming operand and every result of an instruction. A correcting operation follows when an error-correcting code is used. The cost of the checking algorithm has two interrelated components: the hardware complexity and the time required by checking. The complete duplication case requires only bit by bit comparison; other codes require more hardware and time. Provisions for fault detection in the checking hardware itself are needed and add to the cost.

In the residue scheme, the checking is done by the CHECK unit and consists of comparing the outputs of the RESIDUE and the MAIN units. This operation is performed by the "DETECT DIVIDE" algorithm mentioned before. Therefore, the only extra hardware we require for checking algorithm is the CHECK unit.

Hardware implementation of the CHECK Unit has been considered in [14]. Referring to this reference, we note that the hardware required to implement the CHECK Unit in not complex at all. Also, it proves that the checking procedure does not introduce any delay into the operation of the on-line units.

3) The Arithmetic Algorithms: An encoding usually requires a more complex arithmetic operation than the perfect computer. This cost is expressed by the incremental time and hardware required by new algorithm. The algorithms used by the error coded units are not different from those used by the ordinary units. Therefore, imposing error codes on on-line units does not add any cost of this type. Also, note that we do not require new algorithms for the residue units. The algorithms "RESIDUE OP" are exactly the same as "MAIN OP" algorithms, but they are run on the residue operands.

The gate level design of a residue-coded on-line division unit has been considered in [14]. Figure 2 summarizes the gate ($G_{EC-MAIN}, G_{EC-RES}, G_{CHECK}$) and ROM ($M_{EC-MAIN}, M_{EC-RES}, M_{CHECK}$) requirements of the residue-coded on-line division units with respect to operand length and radix of implementation. Figure 3 represents the percentage (Δ_G) of increase in gate count due to introduction of codes in the on-line division unit. From Figure 2 it can be deduced that the cost of the on-line division unit is minimal when the radix of implementation is equal to 16.

The effectiveness of the proposed scheme in detecting errors is proved in [9].

5 CONCLUSIONS

In this paper we have presented a method for detection and correction of errors in on-line arithmetic algorithms, suitable for multi-module networks. This method is based on low-cost arithmetic error codes and encodes each digit of the operands separately. The encoded operands pass through the arithmetic unit digit-by digit, most significant digit first. The proposed algorithms are such that they preserve the codes, therefore each digit of the result must conform with its code. In this way, and depending on the code used, errors in each single digit of the result can be detected or corrected.

We proposed residue-coded operánds for detection of such errors. In this method residue of every digit of the operands with respect to a constant is attached to it and is sent to the on-line unit. Two separate processors, process the operands and their residues. The result generated can be checked for having the correct residue with respect to the same constant. In this way, we proved that a single error in each digit of the operands and the corresponding results can be detected.

m	r	MAIN UNIT			RESIDUE UNIT			CHECK UNIT			RC-DIVIDE UNIT		RC-DIVIDE UNIT (TOTAL)
		G EC-MAIN	M EC-MAIN	$t^{(\delta_g)}$ EC-MAIN	G EC-RES	M EC-RES	$t^{(\delta_g)}$ EC-RES	G CHECK	M CHECK	$t^{(\delta_g)}$ CHECK	G RC-DIVIDE	M RC-DIVIDE	$T^{(\delta_g)}$ RC-STEP
m=8	4	5544	2656	54	5544	2656	54	7	752	23	11095	6064	54
	16	14200	4608	77	5544	2656	54	7	848	23	19751	8112	77
	64	26952	9088	93	5544	2656	54	7	1232	23	32503	12976	93
	256	43800	23936	116	5544	2656	54	7	2768	23	49351	29360	116
m=16	4	11088	5312	54	11088	5312	54	7	1264	27	22183	11888	54
	16	28400	9216	77	11088	5312	54	7	1360	27	39495	15792	77
	64	53904	18176	93	11088	5312	54	7	1744	27	64999	25232	93
	256	87600	47872	116	11088	5312	54	7	3280	27	98695	56464	116
m=32	4	22176	10624	54	22176	10624	54	7	2288	31	44359	23536	54
	16	56800	18432	77	22176	10624	54	7	2384	31	78983	31440	77
	64	107808	36352	93	22176	10624	54	7	2768	31	127 K	49744	93
	256	171 K	95744	116	22176	10624	54	7	4304	31	192 K	108 K	116
m=64	4	44352	21248	54	44352	21248	54	7	4336	35	88711	46832	54
	16	110 K	36864	77	44352	21248	54	7	4432	35	154 K	62544	77
	64	210 K	72704	93	44352	21248	54	7	4816	35	253 K	98768	93
	256	342 K	187 K	116	44352	21248	54	7	6352	35	385 K	213 K	116

Figure 2: Hardware and Delay Requirements of the Residue-Coded Divide Unit when $r' = 4(\rho' = 3)$ and $A = 3(\alpha = 2)$

| m | r | RC-DIVIDE UNIT | | DIVIDE UNIT | $\Delta_{G(\%)}$ |
		$G_{RC\text{-}DIVIDE}$	$M_{RC\text{-}DIVIDE}$	G_{DIVIDE}	
m=8	4	11095	6064	5544	100.1
	16	19751	8112	14200	39.1
	64	32503	12976	26952	20.6
	256	49351	29360	43800	12.7
m=16	4	22183	11888	11088	100.1
	16	39495	15792	28400	39.1
	64	64999	25232	53904	20.6
	256	98695	56464	87600	12.7
m=32	4	44359	23536	22176	100.0
	16	78983	31440	56800	39.1
	64	127 K	49744	107808	20.6
	256	192 K	108 K	171 K	12.3
m=64	4	88711	46832	44352	100.0
	16	154 K	62544	110 K	40.0
	64	253 K	98768	210 K	20.5
	256	385 K	213 K	342 K	12.6

Figure 3: Comparison of the Gate and Memory Requirements of the RC-Divide and Divide Units

6 REFERENCES

1) Ercegovac, M.D., "A General Method for Evaluation of Functions and Computations in a Digital Computer", Ph.D. Thesis, Report No. 750, Department of Computer Science, University of Illinois, Urbana, Augest 1975.

2) Trivedi, K.S. and Ercegovac, M.D., "On-Line Algorithms for Division and Multiplication" IEEE Trans. on Comput., Vol. C-26, No.7, July 1977.

3) Irwin, M.J., "An Arithmetic Unit for On-Line Computation", Ph.D. dissertation, University of Illinois at Urbana-Champaign, report No. UIUCDCS-R-77-873.

4) Ercegovac, M.D. and Grnarov, A.L., "On The Performance of On-Line Arithmetic", Proc. 1980 International Conference on Parallel Processing, Michigan, 1980.

5) Avizienis, A. "Signed Digit Number Representation for Fast Parallel Arithmetic", IRE Trans. Electron. Comput., Vol. EC-10, pp. 389-400, 1961.

6) Ercegovac, M.D., "An On-Line Square Rooting Algorithm", Proc. 4th IEEE Symp. Comput. Arithmetic, pp. 183-189, October 1978.

7) Oklobdzija, V.G., "An On-Line Higher Radix Square Rooting Algorithm", M.S. Thesis, Computer Science Department, University of California, Los Angeles, June 1978.

8) Grnarov, A.L. and Ercegovac, M.D., "An Algorithm for On-Line Normalization", UCLA Computer Science Department Quarterly, Vol.7, No.3, pp. 81-94, July 1979.

9) Sinaki, G., "Error-Coded Algorithms for On-Line Arithmetic", Technical Report UCLA-ENG-8197, Computer Science Department, University of California, Los Angeles, February 1981.

10) Trivedi, K.S. and Rusnak, J.G., "Higher Radix On-Line Division", Proc. 4th IEEE Symp. Comput. Arithmetic, pp. 164-174, October 1978.

11) Grnarov, A.L. and Ercegovac, M.D., "VLSI-Oriented Iterative Networks for Array Computations", Proc. of International Conference on Circuits and Computers, 1980.

12) Atkins, D.E., "Higher Radix Division Using Estimates of the Divisor and Partial Remainders", IEEE Trans. on Computers, Vol.C-17, No. 10, pp. 925-931, October 1968.

13) Avizienis, A., "Arithmetic Error Codes: Cost and Effectiveness Studies for Application in Digital System Design", IEEE Trans. on Comput., Vol. C-20, No. 11, Nov. 1971, pp. 1322-1331.

14) Sinaki, G. and Ercegovac, M.D., "Design of a Digit-Slice On-Line Arithmetic Unit", in Proceedings of the Fifth Symposium on Computer Arithmetic, Ann Arbor, Michigan, May 1981.

Constructions of Codes Correcting Burst Asymmetric Errors

Yuichi SAITOH   Hideki IMAI

Division of Electrical and Computer Engineering
Faculty of Engineering
Yokohama National University
156 Tokiwadai, Hodogaya-ku, Yokohama, 240 Japan
E-mail: saitoh@imailab.dnj.ynu.ac.jp

A class of q-ary systematic codes correcting burst asymmetric errors is proposed. These codes have approximately $b + \log_q(q-1)k + \log_q \log_q k$ check symbols, where b is the maximal length of correctable burst asymmetric errors and k is the number of information symbols. The codes have less check symbols than ordinary burst-error-correcting codes if $\log_q(q-1)k + \log_q \log_q k < b$. A decoding algorithm for the codes is also presented. Encoding and decoding of the codes are very easy. Further, this paper gives some types of codes obtained by modifications of these codes: codes correcting random errors and burst asymmetric errors, and codes correcting random and burst asymmetric errors. Furthermore, more efficient q-ary burst-asymmetric-error-correcting codes are presented. When $q = 2$, these codes have approximately $b + \log_2 k + \frac{1}{2} \log_2 \log_2 k$ check bits.

1   Introduction

The difference between the probability of mistaking a 1 for a 0 and vice-versa can be large in certain digital systems, such as optical communications and optical disks. Practically, we can assume that only one type of errors, either $1 \to 0$ or $0 \to 1$, can occur in those systems. These errors are called *asymmetric errors*. In this paper, asymmetric errors are assumed to be $1 \to 0$ errors without loss of generality.

These errors can be generalized to the case of q-ary symbols. We consider that i is transmitted and j is received for i and j chosen from the set $\{0, 1, \ldots, q-1\}$, where q is an arbitrary positive integer. An error for which $i < j$ is called *an additive error* and an error for which $i > j$ is called *a subtractive error*. *Asymmetric errors* mean errors which occur in a channel, where additive errors cannot occur in any received word, as shown in Figure 1. Ordinary errors, for which both additive and subtractive

This work was presented in part at the 1990 IEICE National Conference, Tokyo, Japan, March 1990, at the meeting of Technical Group on Information Theory, Yokohama, Japan, March 1990, and in *Electronics Letters*, vol. 26, no. 5, pp. 286–287, March 1990.

errors can occur, are simply called *errors*. Notice that the definition of an asymmetric error in this paper differs from Varshamov's definition [11] if $q > 2$.

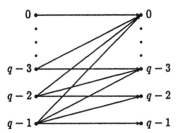

Figure 1: q-ary asymmetric channel

In some semiconductor and magnetic recording memories, a clump of asymmetric errors tends to occur [1], [6]. If the interval between the leftmost error and the rightmost error in a word includes $b - 2$ symbols, the clump of the errors is called *a burst asymmetric error of length b*. In computer systems including such memories, a few ordinary errors may also occur. When we want to make such memories or systems highly reliable, we may use codes correcting ordinary burst errors or codes correcting ordinary random errors and burst errors, but it is more efficient to use codes correcting burst asymmetric errors or codes correcting random errors and burst asymmetric errors.

A code correcting all burst asymmetric errors of length b or less is abbreviated as a b-*BAsEC code*. A b-BAsEC code that can correct t or fewer ordinary errors is also written as a t-*EC/b-BAsEC code*. In this paper, a class of q-ary systematic b-BAsEC codes is proposed. These codes have less check symbols than ordinary burst-error-correcting codes in most cases. A decoding algorithm for the proposed code is presented. This paper also shows constructions of t-EC/b-BAsEC codes. In addition, more efficient b-BAsEC codes and binary codes correcting random and burst asymmetric errors are given.

2 b-BAsEC Codes

Construction 1 Suppose that a codeword of the q-ary systematic b-BAsEC code is the concatenation of $X_1, X_2, \ldots, X_{m-l+1}, X_{m-l+2}, y_1, X_{m-l+3}, y_2, \ldots, X_{m+1}, y_l$, and is denoted by

$$[X_1 X_2 \cdots X_{m-l+1} X_{m-l+2} \, y_1 X_{m-l+3} \, y_2 \cdots X_{m+1} \, y_l], \tag{1}$$

where $m \geq l-2$. The symbols X_i for $i = 1, 2, \ldots, m+1$ denote q-ary b-tuples, i.e., $X_i = [x_1^{(i)} x_2^{(i)} \cdots x_b^{(i)}]$, where $x_j^{(i)}$ for $j = 1, 2, \ldots, b$ are chosen from $\{0, 1, \ldots, q-1\}$. The b-tuples $X_1, \ldots, X_m$ are used as information symbols, and X_{m+1} is defined by

$$X_{m+1} \equiv -\sum_{i=1}^{m} X_i \pmod{q},$$

where the sum is the symbol-by-symbol addition. The symbols y_i for $i = 1, 2, \ldots, l$ are taken from $\{0, 1, \ldots, q-1\}$, and $[y_1 y_2 \cdots y_l]$ is a codeword of a q-ary single-asymmetric-error-correcting code which encodes the integer

$$y = \left[\sum_{i=1}^{m+1} \sum_{j=1}^{b} \{(i-1)b+j\} x_j^{(i)} \right] \bmod p \tag{2}$$

to the q-ary l-tuple, where p is a prime number satisfying $p > (q-1)mb$. $\quad\square$

This code has mb information symbols and $b+l$ check symbols. If we use a generalized Hamming code [8, p. 221] for $[y_1 \cdots y_l]$, $l = \lceil \log_q p \rceil + \mu$ where μ is an integer satisfying $\frac{q^{\mu-1}-1}{q-1} - \mu + 1 < \lceil \log_q p \rceil \le \frac{q^\mu-1}{q-1} - \mu$.

Now assume that the codeword given by (1) is transmitted and $[X_1' X_2' \cdots X_{m-l+1}' \ X_{m-l+2}' \ y_1' \ X_{m-l+3}' \ y_2' \cdots X_{m+1}' \ y_l']$ is received. Then the b-BAsEC code can be decoded as follows:

Decoding Algorithm 1 Step 1: Let the q-ary b-tuple $[s_1 s_2 \ldots s_b] \equiv -\sum_{i=1}^{m+1} X_i' \pmod{q}$. If $s_i = 0$ for all i, then judge that $X_1', \ldots, X_m'$ are the correct information symbols; otherwise go to Step 2.

Step 2: Suppose that y' is the decoded result of $[y_1' \cdots y_l']$ and y'' is the value of the right-hand side of (2) computed from $[X_1' \cdots X_{m+1}']$. Then let $S = (y' - y'') \bmod p$.

Step 3: Calculate

$$
\begin{aligned}
\tau &= s_1 + 2s_2 + 3s_3 + \cdots + bs_b, \\
i &\equiv \frac{S - \tau}{b} \pmod{p}, \\
t &= s_1 + s_2 + \cdots + s_b, \\
u &= \lfloor i/t \rfloor,
\end{aligned}
$$

where $0 \le i \le (q-1)mb$. Find j that satisfies $s_1 + \cdots + s_j = i \bmod t$ and $s_j \ne 0$. Then correct the error by adding $s_{j+1}, \ldots, s_b, s_1, \ldots, s_j$ to the $(bu+j+1)$th symbol,$\ldots$, the $(bu+j+b)$th symbol in $[X_1' \cdots X_{m+1}']$, respectively. $\quad\square$

Theorem 1 *The code from Construction 1 can correct all burst asymmetric errors of length b or less by Decoding Algorithm 1.*

Proof: A burst asymmetric error is simply called *a burst* here. We assume that the length of the burst is b or less.

If a burst occurs in $[X_1' \cdots X_{m+1}']$, then $[s_1 \ldots s_b] \ne [0 \cdots 0]$. Thus if $s_i = 0$ for all i, then $[X_1' \cdots X_{m+1}']$ is considered to be correct.

The number of asymmetric errors in $[y_1 \cdots y_l]$ is at most one because length of a burst is b or less. Therefore y' is always correct.

Consider the case where $[s_1 \ldots s_b] \neq [0 \cdots 0]$. If a burst occurs, its pattern in $[X'_1 \cdots X'_{m+1}]$ appears in $[s_1 \cdots s_b]$. Assume that a burst of pattern $[s_{j+1} \cdots s_b s_1 \cdots s_j]$ has occurred from the $(bu + j + 1)$th to the $(bu + j + b)$th symbols in $[X'_1 \cdots X'_{m+1}]$. Then S satisfies

$$
\begin{aligned}
S = \ & \{(bu + j + 1)s_{j+1} + \cdots (bu + b)s_b \\
& + (bu + 1 + b)s_1 + \cdots + (bu + j + b)s_j\} \bmod p.
\end{aligned}
$$

This equation is also written as

$$
\begin{aligned}
S &= (but + bs_1 + bs_2 + \cdots + bs_j + \tau) \bmod p \\
&= \{b(ut + s_1 + \cdots + s_j) + \tau\} \bmod p.
\end{aligned}
$$

Therefore, letting $i = ut + s_1 + \cdots + s_j$, we have

$$
S = (bi + \tau) \bmod p \text{ and } 0 \leq i \leq (q - 1)mb. \tag{3}
$$

Note that $bi_1 + \tau \not\equiv bi_2 + \tau \pmod{p}$ for i_1 and i_2 such that $0 \leq i_1, i_2 \leq (q - 1)mb < p$ and $i_1 \neq i_2$, i.e., i satisfying (3) is unique. Also u and j satisfying $i = ut + s_1 + \cdots + s_j$ are determined uniquely. Hence, we can find the position of the burst and correct it. $\qquad$ Q.E.D.

It is noted that a b-BAsEC code with k ($< p$) information symbols can be constructed similarly even if $k \bmod b \neq 0$.

3   1-EC/b-BAsEC Codes

We obtain 1-EC/b-BAsEC codes by modifying Construction 1.

Construction 2 Same as Construction 1 for $p > 2(q - 1)(m + 1)b$ except that a q-ary single-error-correcting code is used for $[y_1 \cdots y_l]$. $\qquad \square$

Assume that the transmitted and received words are the same as those in Section 2.

Decoding Algorithm 2 Step 1: Perform Steps 1 and 2 of Decoding Algorithm 1.

Step 2: If the Hamming weight of $[s_1 \cdots s_b]$ is one, then go to Step 3; otherwise perform Step 3 of Decoding Algorithm 1.

Step 3: Calculate

$$
i \equiv \frac{S}{s_j} \pmod{p},
$$

where $s_j \neq 0$ and $-(q - 1)(m + 1)b \leq i \leq (q - 1)(m + 1)b$. If $i > 0$, then correct the error by adding s_j to the i-th symbol in $[X'_1 \cdots X'_{m+1}]$; otherwise correct the error by subtracting s_j from the $(-i)$th symbol. $\qquad \square$

Theorem 2 *The code from Construction 2 can correct a single error and all burst asymmetric errors of length b or less by Decoding Algorithm 2.*

Proof: Similarly to Theorem 1, y' is always correct whether a single error or a burst asymmetric error of length $\leq b$ has occurred. Hereafter a burst asymmetric error of length 2 which consists of a single asymmetric error in $[y_1 \cdots y_l]$ and a single asymmetric error in $[X_1 \cdots X_{m+1}]$ is considered as a single error. By the Hamming weight of $[s_1 \cdots s_b]$, we can correctly distinguish the case where a single error has occurred from the case where a burst error of length > 1 has occurred in $[X_1 \cdots X_{m+1}]$; i.e., the Hamming weight is one if a single error occurs and not one if the burst error occurs. The code can correct all burst asymmetric errors of length b or less from Theorem 1. We now consider the case where a single error s_j has occurred in i'-th symbol of $[X_1' \cdots X_{m+1}']$. If the error is a subtractive error, then $S = i's_j \bmod p$; otherwise $S = -i's_j \bmod p$. Note that $i_1 s_j \not\equiv i_2 s_j \bmod p$ for $i_1 \neq i_2$ and $-(q-1)(m+1)b \leq i_1, i_2 \leq (q-1)(m+1)b$. Therefore i satisfying $S = is_j \bmod p$ is unique. Hence if $i > 0$, then the error is a subtractive error and we should add s_j to i-th symbol in $[X_1' \cdots X_{m+1}']$; otherwise we should subtract s_j from the $(-i)$th symbol. Q.E.D.

4 t-EC/b-BAsEC Codes

By modifying the preceding constructions, we have the following construction of t-EC/b-BAsEC codes.

Construction 3 Let the symbols $X_1, \ldots, X_{m+1}, y$ be the same as those in Construction 1, and let

$$\{[X_1 \cdots X_{m+1} V] \text{ for all } X_1, \ldots, X_m\}$$

be a q-ary *code correcting t or fewer ordinary errors* (t-EC code), where V is a q-ary ν-tuple. Encode V into the integer v and $y + pv$ into the q-ary h-tuple $[z_1 \cdots z_h]$ which is a codeword of a t-EC code. Insert $[z_1 \cdots z_t]$, $[z_{t+1} \cdots z_{2t}]$, $\ldots$, $[z_{\alpha-t+1} \cdots z_\alpha]$ into $[X_1 \cdots X_{m+1}]$ at intervals of b symbols as follows:

$$[X_1 \cdots X_{m-\alpha/t+2}[z_1 \cdots z_t] X_{m-\alpha/t+3}[z_{t+1} \cdots z_{2t}]$$

$$\cdots X_{m+1}[z_{\alpha-t+1} \cdots z_\alpha]] \tag{4}$$

where α is the integer that satisfies $h \leq \alpha < h + t$, $\alpha \bmod t = 0$, and $m \geq \alpha/t - 2$, and if $\alpha > h$, $z_{h+1}, \ldots, z_\alpha$ are ignored. Let the $\{(m+1)b+h\}$-tuple given by (4) be a codeword of the t-EC/b-BAsEC code. □

This code has mb information symbols and $b + h$ check symbols.

Suppose that we transmit the codeword of (4), and receive

$$[X_1' \cdots X_{m-\alpha/t+2}'[z_1' \cdots z_t'] X_{m-\alpha/t+3}'[z_{t+1}' \cdots z_{2t}']$$

$$\cdots X_{m+1}'[z_{\alpha-t+1}' \cdots z_\alpha']] \tag{5}$$

Then the code can be decoded as follows:

Decoding Algorithm 3 Step 1: Suppose that y' and v' are the values of obtained from the decoded result $y' + pv'$ of $[z'_1 \cdots z'_h]$. Let the q-ary b-tuple $[s_1 \cdots s_b] \equiv - \sum_{i=1}^{m+1} X'_i \pmod{q}$. If the Hamming weight of $[s_1 \cdots s_b]$ is greater than t, then go to Step 3; otherwise go to Step 2.

Step 2: Let V' be the decoded result of v'. Judge that the decoded results of $[X'_1 \cdots X'_{m+1} V']$ are the correct information symbols.

Step 3: Let y'' be the value of the right-hand side of (2) computed from $[X'_1 \cdots X'_{m+1}]$. Calculate $S = (y' - y'') \bmod p$ and perform Step 3 of Decoding Algorithm 1. □

Theorem 3 *The code from Construction 3 can correct t or fewer errors and all burst asymmetric errors of length b or less by Decoding Algorithm 3.*

Proof: Similarly to Theorem 1, y' is always correct whether t or fewer errors or a burst asymmetric error of length b or less has occurred. Hereafter a burst asymmetric error of length b or less which consists of t or fewer asymmetric errors in $[y_1 \cdots y_l]$ and t or fewer asymmetric errors in $[X_1 \cdots X_{m+1}]$ is considered as t or fewer errors. By the Hamming weight of $[s_1 \cdots s_b]$, we can correctly distinguish the case where t or fewer errors have occurred from the case where a burst asymmetric error of length $> t$ has occurred in $[X_1 \cdots X_{m+1}]$. If t or fewer errors occur, the code can correct errors because $\{[X_1 \cdots X_{m+1}V]\}$ is a t-EC code. The rest can be proved in a similar manner to the previous theorems. Q.E.D.

5 More Efficient q-ary b-BAsEC Codes

By the following construction, we can construct more efficient q-ary b-BAsEC codes than those from Construction 1.

Construction 4 We use $X_1, \ldots, X_{m+1}$, and y in the same way as we did in Construction 1. The integer y is encoded into a codeword Y of a q-ary code of length l which can detect all unidirectional errors [2], [9, pp.357–360]. Let a codeword of the proposed q-ary b-BAsEC code be $[X_1 \cdots X_{m+1}Y]$. □

Assume that $[X_1 \cdots X_{m+1} Y]$ is transmitted and $[X'_1 \cdots X'_{m+1} Y']$ is received. The code is decoded as follows:

Decoding Algorithm 4 Step 1: Perform Step 1 of Decoding Algorithm 1.

Step 2: If errors have been detected in Y', then judge that $X'_1, \ldots, X'_m$ are the correct information symbols; otherwise perform Step 2 of Decoding Algorithm 1 and go to Step 3.

Step 3: Perform Step 3 of Decoding Algorithm 1. □

Theorem 4 *The code from Construction 4 can correct all burst asymmetric errors of length b or less by Decoding Algorithm 4.*

Proof: Consider the case where a burst asymmetric error of length b or less has occurred. If we detect the error in Y' and obtain $[s_1 \cdots s_b] \neq [0 \cdots 0]$, then the burst error has occurred in both X'_{m+1} and Y' because its length is b or less. Then we know that the error of the pattern $[s_1 \cdots s_b]$ has occurred in X'_{m+1} and we can correct it. The rest can be proved in the same manner as Theorem 1. Q.E.D.

When $q = 2$, if we use a $\lfloor l/2 \rfloor$-out-of-l code for Y, the code has approximately $b + \log_2 k + \frac{1}{2} \log_2 \log_2 k$ check bits because $\log_2 \binom{l}{\lfloor l/2 \rfloor} \approx l - \frac{1}{2} \log_2 l$ [4]. Hence the codes of this section are more efficient than the codes from Construction 1.

6 Binary Codes Correcting Random and Burst Asymmetric Errors

We can also construct codes that can correct t or fewer asymmetric errors and all burst asymmetric errors of length b or less as follows:

Construction 5 Let $X_1, \ldots, X_{m+1}$ be the same as those in Construction 1 for $q = 2$. Let us define y_u by

$$y_u = \left[\sum_{i=1}^{m+1} \sum_{j=1}^{b} \{(i-1)b + j\}^u x_j^{(i)} \right] \bmod p \qquad (6)$$

for $u = 1, 2, \ldots, t$, where p is a prime number satisfying $p > (m+1)b$. Suppose that

$$z = \sum_{u=1}^{t} y_u p^{u-1}$$

is encoded into $[z_1 \cdots z_h]$, which is a codeword of a binary code correcting t or fewer asymmetric errors. Then let a codeword of the proposed code be (4), where α is the integer that satisfies $h \leq \alpha < h + t$, $\alpha \bmod t = 0$, and $m \geq \alpha/t - 2$, and if $\alpha > h$, $z_{h+1}, \ldots, z_\alpha$ are ignored. □

Suppose that the transmitted and received words are given by (4) and (5), respectively. Then the code can be decoded as follows:

Decoding Algorithm 5 Step 1: Perform Step 1 of Decoding Algorithm 1.

Step 2: Suppose that y'_u is the value obtained from the decoded result $\sum_{u=1}^{t} y'_u p^{u-1}$ of $[z'_1 \cdots z'_h]$ for $u = 1, 2, \ldots, t$.

Step 3: If the Hamming weight of $[s_1 \cdots s_b]$ is greater than t, go to Step 5; otherwise go to Step 4.

Step 4: Suppose that y''_u is the value of the right-hand side of (6) computed from $[X'_1 \cdots X'_{m+1}]$ for $u = 1, 2, \ldots, t$. Also assume that $S_u = (y'_u - y''_u) \bmod p$ for all u. Find $i_1, i_2, \ldots, i_t$ which satisfy

$$S_u \equiv i_1^u + i_2^u + \cdots + i_t^u \pmod{p}$$

for all u, and $0 \leq i_1, \ldots, i_t \leq p - 1$. Correct the errors by reversing the $i_j (\neq 0)$-th bit in $[X'_1 \cdots X'_{m+1}]$ for $j = 1, 2, \ldots, t$.

Step 5: Assume that y'' is the value of the right-hand side of (6) for $u = 1$ computed from $[X'_1 \cdots X'_{m+1}]$, and $S = (y'_1 - y'') \bmod p$. Perform Step 3 of Decoding Algorithm 1. □

When $t = 2$, we may perform the following step instead of Step 4 described above.

Step 4: Suppose that y''_1 and y''_2 are the values of the right-hand sides of (6) for $u = 1$ and 2 computed from $[X'_1 \cdots X'_{m+1}]$, respectively. Also assume that $S_1 = (y'_1 - y''_1) \bmod p$ and $S_2 = (y'_2 - y''_2) \bmod p$. Calculate

$$i_1 \equiv \frac{1}{2}(S_1 + \sqrt{2S_2 - S_1{}^2}) \pmod p,$$

$$i_2 \equiv \frac{1}{2}(S_1 - \sqrt{2S_2 - S_1{}^2}) \pmod p,$$

where $0 \le i_1, i_2 \le p - 1$. If $i_2 \ne 0$, correct the errors by reversing the i_1-th bit and the i_2-th in $[X'_1 \cdots X'_{m+1}]$; otherwise correct the error by reversing the i_1-th bit in $[X'_1 \cdots X'_{m+1}]$.

Theorem 5 *The binary code from Construction 5 can correct t or fewer asymmetric errors and all burst asymmetric errors of length b or less by Decoding Algorithm 5.*

Proof: Similarly to Theorem 1, y' is always correct whether t or fewer asymmetric errors or a burst asymmetric error of length b or less has occurred. Henceforth a burst asymmetric error of length b or less which consists of t or fewer asymmetric errors in $[y_1 \cdots y_l]$ and t or fewer asymmetric errors in $[X_1 \cdots X_{m+1}]$ is considered as t or fewer asymmetric errors. Similarly to Theorems 2 and 3, we can distinguish the case where t or fewer asymmetric errors have occurred from the case where burst asymmetric errors have occurred in $[X_1 \cdots X_{m+1}]$. If the Hamming weight of $[s_1 \cdots s_b]$ is greater than t, a burst asymmetric error of length $> t$ has occurred and we should perform Step 5. The proof for this case is the same as Theorem 1.

Consider the case where t or fewer asymmetric errors have occurred. From [11, Theorem 5] and [5, Code 4],

$$\{[X_1 \cdots X_{m+1}] \mid y_u \text{ is a positive integer for } u = 1, 2, \ldots, t\}$$

is a t-asymmetric-error-correcting code. $\{[z_1 \cdots z_h]\}$ is a t-asymmetric-error-correcting code, so that we can obtain correct y_u from $[z'_1 \cdots z'_h]$. Hence we know the correct positions of the errors by $i_1, \ldots, i_t$.

Q.E.D.

7 Asymptotic Optimality of b-BAsEC Codes

The following lower bounds on the number of check symbols in a systematic code that corrects ordinary burst errors is well known [8, p. 110].

Theorem 6 (Reiger bound) *Any q-ary systematic code that corrects all burst errors of length b or less must have at least $2b$ check symbols.*

We can obtain the following lower bound on b-BAsEC codes by a simple generalization of the known bound [8, p. 111].

Theorem 7 *Any q-ary systematic b-BAsEC code with $k \geq b$ information symbols must have at least $b - 1 + \lceil \log_q \{(q-1)(k-b+1)+1\} \rceil$ check symbols.*

From this, we obtain the following theorem on the asymptotic property of the b-BAsEC code from Construction 1.

Theorem 8 *If $\displaystyle \lim_{b \to \infty} \frac{\log_q k}{b} = c$ (constant), then the q-ary b-BAsEC code from Construction 1 and with k information symbols and $n - k$ check symbols is asymptotically optimal in the meaning that*

$$\lim_{b \to \infty} \frac{n-k}{b - 1 + \lceil \log_q \{(q-1)(k-b+1)+1\} \rceil} = 1.$$

Proof: Assume that a generalized Hamming code is used for $[y_1 \cdots y_l]$, and $\gamma = b - 1 + \lceil \log_q \{(q-1)(k-b+1)+1\} \rceil$. By the given condition, we can obtain $(\log_q \log k_q)/b \to 0$ and $\gamma/b \to c+1$ for $b \to \infty$. We also have $n - k = b + l$. We can choose constant numbers δ and ϵ that satisfy

$$l \leq \log_q p + \log_q \log_q p + \delta \leq \log_q k + \log_q \log_q k + \epsilon.$$

Therefore, $(b+l)/b \to c' + 1 \leq c + 1$ for $b \to \infty$. Thus,

$$1 \leq \frac{n-k}{\gamma} = \frac{(n-k)/b}{\gamma/b} \to \frac{c'+1}{c+1} \leq 1;$$

hence $(n-k)/\gamma \to 1$ for $b \to \infty$.   Q.E.D.

Similarly, we can prove that the codes from Construction 4 are asymptotically optimal.

8   Tables of Some Codes

Examples of parameters of the proposed codes with k information symbols and of length n are listed in Tables 1–7. Hamming codes, $\lfloor l/2 \rfloor$-out-of-l codes, BCH codes, and Reed-Solomon codes are used for $[y_1 \cdots y_l]$ and $[z_1 \cdots z_h]$. The upper bounds for the number of check symbols given by Theorem 7 are listed as 'bound' in Tables 1, 2, and 5. According to Theorem 6, in order to correct ordinary burst errors of length b or less, a code must have at least $2b$ check symbols. By the proposed constructions, however, a lot of codes that have less than $2b$ check symbols can be constructed.

The codes in Table 2, for which Y is a codeword of a $\lfloor l/2 \rfloor$-out-of-l code, have approximately $b + \log_2 k + \frac{1}{2} \log_2 \log_2 k$ check bits as mentioned in Section 5. Hence they are more efficient than the codes in Table 1 for which $[y_1 \cdots y_l]$ is a codeword of a Hamming code. However, when k and l are large, encoding and decoding of the $\lfloor l/2 \rfloor$-out-of-l code are very hard. The encoding and decoding methods in References [3] and [4] may make the implementation easier. However, if we use these methods for encoding and decoding of the $\lfloor l/2 \rfloor$-out-of-l code, then the b-BAsEC codes from Construction 4 have approximately $b + \log_2 k + \log_2 \log_2 k$ check bits, which are nearly equal to those in the codes in Table 1. Similarly, even if we use a Berger code [9, pp.359–360] for Y, the codes from Construction 4 are as efficient as those in Table 1. However, the codes from Construction 4 have the advantage that their decoders based on Decoding Algorithm 4 do not need to correct errors in Y and have only to detect them.

9   Conclusion

In this paper, a class of systematic q-ary b-BAsEC codes and its decoding algorithm have been proposed. By the proposed construction, we can obtain b-BAsEC codes which have simple implementation and are more efficient than ordinary burst-error-correcting codes. Simple modifications of b-BAsEC codes have given t-EC/b-BAsEC codes. This paper has also presented constructions of more efficient b-BAsEC codes and binary codes correcting random and burst asymmetric errors. If we modify the proposed construction of b-BAsEC codes, we can obtain two-dimensional-asymmetric-error-correcting codes [10].

Independently of us, S. Park and B. Bose have presented binary b-BAsEC different from our codes very recently [7]. Their codes need approximately $b + \log_2 k + \frac{1}{2}\log_2 \log_2 k$ check bits, where k is the number of information bits. Thus their codes are more efficient than the binary version of our codes from Construction 1, and are approximately as efficient as the binary version of our codes from Construction 4.

References

[1] M. Blaum, Systematic unidirectional burst detecting codes, IEEE Trans. Comput., 37 (1988) 453–457.

[2] B. Bose and T. R. N. Rao, Theory of unidirectional error correcting/detecting codes, IEEE Trans. Comput., C-31 (1982) 521–530.

[3] B. Bose, On unordered codes, in Dig. Papers 17th Int. Symp. Fault-Tolerant Comput. (1987) 102–107.

[4] D. E. Knuth, Efficient balanced codes, IEEE Trans. Inform. Theory, IT-32 (1986) 51–53.

[5] R. J. McEliece, Comment on "A class of codes for asymmetric channels and a problem from the additive theory of numbers", IEEE Trans. Inform. Theory, IT-19 (1973) 137.

[6] B. Parhami and A. Avižienis, Detection of storage errors in mass memories using low-cost arithmetic error codes, IEEE Trans. Comput., C-27 (1978) 302–308.

[7] S. Park and B. Bose, Burst asymmetric/unidirectional error correcting/detecting codes, in Dig. Papers 20th Int. Symp. Fault-Tolerant Comput. (1990).

[8] W. W. Peterson and E. J. Weldon, Jr., Error-Correcting Codes, 2nd. ed. (The MIT Press, 1972).

[9] T. R. N. Rao and E. Fujiwara, Error-Control Coding for Computer Systems (Prentice Hall Int., New Jersey, 1989).

[10] Y. Saitoh, Design of asymmetric or unidirectional error control codes, Master's thesis, Division of Electrical and Computer Eng., Faculty of Eng., Yokohama National University, Yokohama, Japan (1990).

[11] R. R. Varshamov, A class of codes for asymmetric channels and a problem from the additive theory of numbers, IEEE Trans. Inform. Theory, IT-19 (1973) 92–95.

Table 1: Parameters of Binary b-BAsEC Codes from Construction 1

b	p	m	l	k	$n-k$	$(n-k)/b$	bound
16	251	10–15	12	160–240	28	1.75	23
32	509	11–15	13	352–480	45	1.41	40
64	1021	12–15	14	768–960	78	1.22	73
128	2039	13–15	15	1664–1920	143	1.12	138
256	4093	15	17	3840	273	1.07	267

Hamming codes are used for $[y_1 \cdots y_l]$.

Table 2: Parameters of Binary b-BAsEC Codes from Construction 6

b	p	m	l	k	$n-k$	$(n-k)/b$	bound
8	19	2	6	16	14	1.75	11
	31	3	7	24	15	1.88	12
16	67	2–4	8	32–64	24	1.50	20–21
	113	5–7	9	80–112	25	1.56	22
	251	8–15	10	128–240	26	1.63	22–23
32	67	2	8	64	40	1.25	37
	113	3	9	96	41	1.28	38
	251	4–7	10	128–224	42	1.31	38–39
64	251	2–3	10	128–192	74	1.16	70–71

$\lfloor l/2 \rfloor$-out-of-l codes are used for Y.

Table 3: Parameters of Binary t-EC/b-BAsEC Codes

t	b	p	m	h	k	$n-k$	$(n-k)/b$
2	128	4093	22–31	48	2816–3968	176	1.38
3	128	4093	21–31	69	2688–3968	197	1.54
4	128	4093	20–31	88	2560–3968	216	1.69

BCH codes are used for $[z_1 \cdots z_h]$.

70

Table 4: Parameters of Binary Codes Correcting t or Fewer Asymmetric Errors and All Burst Asymmetric Errors of Length b or Less

t	b	p	m	h	k	$n-k$	$(n-k)/b$
2	32	509	12–14	28	384–448	60	1.88
2	64	1021	13–14	30	832–896	94	1.47
3	64	1021	14	48	896	112	1.75
2,3	128	4093	16–30	$18t$	2048–3840	$128+18t$	$1+9t/64$

BCH codes are used for $[z_1 \cdots z_h]$.

Table 5: Parameters of 256-ary b-BAsEC Codes

b	p	m	l	k	$n-k$	$(n-k)/b$	bound
8	65521	2–32	4	16–256	12	1.50	9
16	65521	2–16	4	32–256	20	1.25	17
32	65521	2–8	4	64–256	36	1.13	33
64	65521	2–4	4	128–256	68	1.06	65
128	65521	2	4	256	132	1.03	129

Reed-Solomon codes are used for $[y_1 \cdots y_l]$.

Table 6: Parameters of 256-ary 1-EC/b-BAsEC Codes

b	p	m	l	k	$n-k$	$(n-k)/b$
8	65521	2–15	4	16–120	12	1.50
16	65521	2–7	4	32–112	20	1.25
32	65521	2–3	4	64–96	36	1.13

Reed-Solomon codes are used for $[y_1 \cdots y_l]$.

Table 7: Parameters of 256-ary t-EC/b-BAsEC Codes

t	b	p	m	h	k	$n-k$	$(n-k)/b$
$2 \le t \le 7$	16	65521	4–15	$2+4t$	64–240	$18+4t$	$1.13+t/4$
$2 \le t \le 15$	32	65521	4–7	$2+4t$	128–224	$34+4t$	$1.06+t/8$

Reed-Solomon codes are used for $[z_1 \cdots z_h]$.

A Construction Method for m-ary Unidirectional Error Control Codes

— Application for m-ary *Berger* Check to Asymmetric Error Control Codes —

Tae Nam AHN∗   Kohichi SAKANIWA†   T.R.N. RAO‡

∗ Computer Department
Korea Army Headquarters
Seoul, KOREA

†Department of Electrical & Electronic Engineering
Tokyo Institute of Technology
O-okayama, Meguro-ku, Tokyo 152, JAPAN

‡The Center for Advanced Computer Studies.
University of Southwestern Louisiana
Lafayette, Louisiana 70504-4330, U.S.A.

Abstract : Unidirectional error control codes have been studied extensively for application to VLSI memories. Also, studies on multiple-valued logic and memories received considerable attention since by employing multi-valued (m-ary) symbol representation, one can drastically reduce the interconnection area in VLSI chips as compared to the *binary* representation which requires more than 70% chip area for interconnection. In this context, m-ary unidirectional error control codes will be of great importance in improving reliability of such multiple-valued logic and memory systems.

In this paper, we propose a new construction method for m-ary t-unidirectional error correcting and *all*-unidirectional error discriminating (t-UEC & *all*-UEDis)codes. We first show that a t-UEC & *all*-UEDis code can be characterized as a t-asymmetric error correcting (t-AEC) code with an added capability for *discriminating* errors between *positive* errors and *negative* errors. Noting this fact, we propose a simple construction method for t-UEC & *all*-UEDis codes starting from m-ary t-AEC codes.

1 Introduction

It is well established that error correcting/detecting codes improve the reliability of various kind of communication systems and computing systems, and thus a number of error correcting/detecting codes have been developed. Up until now, *binary* error correcting/detecting codes have been of major concern to most researchers since binary transmission and binary logic have dominated the fields of digital communication systems and computing systems[1].

In recent years, however, multi-level communication systems, such as 256QAM, have been developed and used in practical applications for information compaction and/or bandwidth reduction[2]. It is also noted that studies on multiple-valued logic and memories have received considerable attention since by employing multi-valued symbol representation one can drastically reduce the interconnection area in VLSI chips as compared to the *binary* representation which requires more than 70% chip area for interconnections[3][4]. In this context, it is considered that researches on multi-valued error correcting/detecting codes are of great importance in improving reliability of such multiple-valued communication and computing systems.

In this paper, we focus our attention mainly on the multiple-valued logic and memory systems in which *unidirectional* error control codes are thought to have major role in improving its reliability. It should be noted, however, that very few studies have been reported on the m-ary *unidirectional* error control codes while some are on the m-ary *asymmetric* error correcting codes[5][6][7][8]. The m-ary unidirectional error detecting code proposed by *Bose* and *Pradhan*[9] for applications to multi-valued logic and byte error control is a specific example.

In this context, this paper proposes a new construction method for m-ary t-unidirectional error correcting and *all*-unidirectional error discriminating (t-UEC & *all*-UEDis) codes. We first show that a t-UEC & *all*-UEDis code can be characterized as a t-asymmetric error correcting (t-AEC) code with an added capability for *discriminating* errors between *positive* type and *negative* type errors. Noting this fact, we propose a simple construction method for t-UEC & *all*-UEDis codes as follows :

1. We append an m-ary *Berger* check to a given m-ary t-AEC code C_A to obtain the capability of *discriminating* errors between positive error set and negative error set.

2. To reduce the number of m-ary *Berger* check digits, we employ a suitable partitioning of the base code C_A according to the weight of each codeword.

The required number of *Berger* check digits for the proposed construction method is given by $\lceil \log_m N_P \rceil$, where N_P is the number of partitions on C_A, and attains its minimum (which is zero) when C_A is an m-ary *constant weight* code. These codes can be systematic if we employ a systematic asymmetric error control code C_A as the base code and therefore are suitable for application for m-ary memories.

2 m-ary Error Correcting/Discriminating Codes

As in the *binary* channel, three major types of errors are of great importance, that is, *symmetric, unidirectional* and *asymmetric* errors. *Symmetric* error means that both *positive* and *negative* symbol errors can occur simultaneously in a single word. On the other hand, *unidirectional* error, which is important in logic and memory systems, means that either *positive* or *negative* symbol errors, but not both types of errors will occur in a single word. *Asymmetric* error, which is important in both memory systems and optical communication systems, means that only *positive* (or *negative*) symbol errors can occur in any word, but the type is known *a priori*.

In addition to the type of errors, in the m-ary channel there is variety in weights by which the error weight is counted. This is an important point that differs from the *binary* channel in which only the *Hamming* weight is meaningful. Though the *Hamming* and *Lee* weight remain to be important for m-ary codes, it must be noted that there are multi-level channels that do not match with those weights, as has been correctly pointed out by *Berlekamp*[10]. To fill up this shortage of weights in the design of m-ary error control codes, an investigation on weights and distances, other than *Hamming* and *Lee* distances, has been done by the authors[11]. Thus according to the combination of error types and weights, by which the magnitude of error is counted, we can define number of error sets and corresponding error control codes, which shall provide a wide variety of choice to a system designer in designing memory and logic while balancing considerations of reliability vs. complexity.

In this section, we briefly review the theory of m-ary error correcting/*discriminating* code over the *integer ring*[11][12]. In the channel model we treat in this paper, an error symbol is added to the sending symbol by usual addition over the *integer ring*. More precisely, the sending alphabet is given by $Z_m = \{0, 1, \ldots, m-1\}$ and the error alphabet is an arbitrary subset of $Z = \{0, \pm 1, \pm 2, \ldots\}$. A code C is defined as n-tuples with the elements in Z_m, i.e.,

$$C = \{c = (c_0, c_1, \ldots, c_{n-1}) \mid c_i \in Z_m\} \ (\subset Z_m^n).$$

It is noted that codes constructed based on this channel model can also be adopted to channels in which the sending and the received alphabets are the same and exhibit the same error control capabilities.

A most general way to define error sets for block codes may be as follows.

[**Definition 1**] (Error Correcting/Discriminating Set)

(1) Error *discriminating* sets E_1, E_2 are defined as arbitrary subsets of Z^n satisfying

$$\{0\} \subseteq E_i \subset Z^n, \quad i = 1, 2.$$

(2) Error correcting set E_T is defined as

$$E_T \triangleq E_1 \cap E_2.$$ ¶

Then by using these error sets, error correcting/discriminating code is defined as follows.

[Definition 2] (E_T-Error-Correcting/(E_1, E_2)-Error-Discriminating Code)

A code **C** is called E_T-*error-correcting*/(E_1, E_2)-*error-discriminating* (E_T-EC/(E_1, E_2)-EDis) code, if **C** can correct any error belonging to E_T and simultaneously *discriminate (distinguish)* any errors between E_1 and E_2. ¶

Fig. 1 shows the conceptual relationship among the error sets. The classes of error-correcting, error-detecting, error-correcting and detecting codes come out as special cases out of the general class of E_T-EC/(E_1, E_2)-EDis codes. For example, if $E_1 = E_T \subseteq E_2 = E_D$, then the code **C** is E_T-EC/E_D-ED code. If $E_1 = \{0\}$, then **C** is E_2 error detecting code and so on.

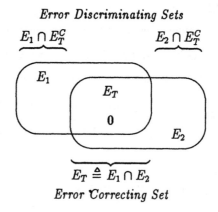

$$E_T \triangleq E_1 \cap E_2$$
Error Correcting Set

Fig. 1   Relationship among E_1, E_2 and E_T

[Definition 3] (Error Difference Set $\Delta(E_1, E_2)$)

For a pair of error discriminating sets (E_1, E_2) given in Definition 1, the error difference set $\Delta(E_1, E_2)$ is defined by

$$\Delta(E_1, E_2) \triangleq \{\pm(u - v)|u \in E_1, v \in E_2\}.$$ ¶

Then, we have following theorem which states the necessary and sufficient condition for an m-ary code over the integer ring to be E_T-EC/(E_1, E_2)-EDis[11][12].

[**Theorem 1**] (A Necessary and Sufficient Condition for E_T-EC/(E_1, E_2)-EDis)

A code C is an E_T-error-correcting/(E_1, E_2)-error-discriminating code *if and only if*

$$x - y \notin \Delta(E_1, E_2) \quad \text{for } \forall x, y \in C, \ x \neq y. \qquad \P$$

By letting

$$E_1 = E_+ \triangleq Z_+^n, \quad E_2 = E_- \triangleq -Z_+^n, \quad \text{where} \quad Z_+ \triangleq \{0, 1, 2, \ldots\}$$

we immediately get the following condition for (E_+, E_-)-error-discriminating $((E_+, E_-)$-EDis), which is also called *all-unidirectional error discriminating* (*all*-UEDis) codes hereafter.

[**Corollary 1**] (A Necessary and Sufficient Condition for (E_+, E_-)-EDis)

A code C is an *all-unidirectional error discriminating* code *if and only if*

$$x \nleq y \text{ and } x \ngeq y \quad \text{for } \forall x, y \in C, \ x \neq y \qquad (1)$$

where

$$x \leq y \overset{\text{def}}{\Longleftrightarrow} x_i \leq y_i \text{ for } \forall i. \qquad \P$$

It is interesting to note that the condition (1) is also a necessary and sufficient condition for *all-unidirectional error detecting* (*all*-UED) codes. Thus, *all*-UED codes are therefore *all*-UEDis codes.

3   Code Construction Method

Based on the discussion given in the previous section, we give here a simple construction method for $t(\geq 0)$-*unidirectional* error correcting and *all-unidirectional* error *discriminating* (t-UEC & *all*-UEDis) codes. As observed in the previous section, if a t-*asymmetric* error correcting (t-AEC) code C has (E_+, E_-)-error-*discriminating* capability, the code C becomes a t-UEC & *all*-UEDis code.

It is shown below that (E_+, E_-)-error-*discriminating* capability can be added to any m-ary code C^1 by appending m-ary *Berger* check to it[2].

---

[1] C can be whole Z_m^k. See also Remark-2 at the end of this section.

[2] m-ary *Berger* checks were discussed in Bose and Pradhan[9].

Thus our discussion in this section starts by assuming that we have a t-AEC code $\mathbf{C}_A$. It is noted that any asymmetric error control code $\mathbf{C}_A$ can be employed for our construction method while an example of $\mathbf{C}_A$ is given in the following section.

We first describe our code construction algorithm in which a partition on $\mathbf{C}_A$ is introduced to reduce the number of digits to be appended and then show the decoding algorithm.

[Code Construction Method] :

Step 1 : Start with an asymmetric error control code $\mathbf{C}_A$ by any previously known method.

Step 2 : Let

$$Q(u) \triangleq \sum_{i=0}^{k-1} u_i, \quad \text{for} \quad u = (u_0, u_1, \ldots, u_{k-1})$$

and induce a partition $\{P_0, P_1, \ldots, P_{N_P-1}\}$ on $\mathbf{C}_A$ such that

i) u and v $(\in \mathbf{C}_A)$ are in the same partition *if and only if* $Q(u) = Q(v)$,

ii) $Q(u) < Q(v)$ for $u \in P_i$, $v \in P_j$ $(i < j)$, and

iii) $P_i \neq \emptyset$ for $0 \leq i \leq N_P - 1$.

Step 3 : Let

$$S_D \triangleq \{Q(u) \mid u \in \mathbf{C}_A\} \quad \text{and} \quad S_R \triangleq \{0, 1, \ldots, N_P - 1\}$$

and define a *monotonically increasing* mapping $\nu(\cdot)$ $(S_D \longmapsto S_R)$ by

$$\nu(Q(u)) = i, \quad \text{for } u \in P_i.$$

By using this mapping, define the *modified m-ary Berger* check[3] for $u \in \mathbf{C}_A$ by

$$a = B(u) \triangleq \phi\{\nu(Q(u))\}, \quad \text{where} \quad \phi(i) \triangleq \;\ll N_P - 1 - i \gg_m \tag{2}$$

and $\ll z \gg_m$ denotes the m-ary representation of $z \in Z$ in radix m. It is noted that the number of digits required to represent $a = B(u)$ is $\lceil \log_m N_P \rceil$, i.e.,

$$B(u) \in Z_m^\beta \quad \text{where} \quad \beta \triangleq \lceil \log_m N_P \rceil. \tag{3}$$

---

[3] The m-ary *Berger* check for $u = (u_0, u_1, \ldots, u_{k-1})$ is conventionally defined as[9]
$$B(u) \triangleq \;\ll \sum_{i=0}^{k-1}(m - 1 - u_i) \gg_m$$
and in general requires $\lceil \log_m (m-1)k + 1 \rceil$ digits for its representation.

Step 4 : Define the code C_U by

$$C_U \triangleq \{x = (u, a) \mid a = B(u), u \in C_A\}. \tag{4} \quad \P$$

Before giving the actual decoding algorithm for the code C_U, we shall show that the code C_U actually satisfies the condition of Corollary 1 given in the previous section. In what follows, we let

$$B(\cdot) \triangleq \phi(\nu(\cdot)) \ : \ S_D \longmapsto \{a = B(u) \mid u \in C_A\}$$

for simplicity. Then $B^{-1}(\cdot)$ is given by

$$B^{-1}(a) = \nu^{-1}\{N_P - 1 - D(a)\}, \quad \text{where} \quad D(a) \triangleq \sum_{i=0}^{\beta-1} a_i m^i \tag{5}$$

and is well defined for a such that $D(a) \in S_R \ (= \{0, 1, \ldots, N_P - 1\})$. It may be also convenient to extend the *domain* of $B^{-1}(\cdot)$ to whole Z_m^β by defining

$$B^{-1}(a) \triangleq N_P - 1 - D(a), \quad \text{for} \quad N_P \leq D(a). \tag{6}$$

Then we have following lemma.

[**Lemma 1**]

$$u \leq v \ (u, v \in Z^k) \ \implies \ \left. \begin{array}{l} Q(u) \leq Q(v) \\ \text{and} \quad Q(u) < Q(v) \ \text{if} \ u \neq v \end{array} \right\} \tag{7}$$

$$a \leq b \ (a, b \in Z_m^\beta) \ \implies \ \left. \begin{array}{l} B^{-1}(a) \geq B^{-1}(b) \\ \text{and} \quad B^{-1}(a) > B^{-1}(b) \ \text{if} \ a \neq b \end{array} \right\} \tag{8}$$

Proof : Eq.(7) is trivial from the definition of $Q(z)$. To prove Eq.(8), assume that $D(a), D(b) \in S_R$ and let $a = B(Q(u))$ and $b = B(Q(v))$. Then

$$D(a) = N_P - 1 - \nu(Q(u)) \quad \text{and} \quad D(b) = N_P - 1 - \nu(Q(v))$$

and we immediately have

$$a \leq b \ \implies \ D(a) \leq D(b) \implies \nu(Q(u)) \geq \nu(Q(v))$$
$$\implies B^{-1}(a) = Q(u) \geq Q(v) = B^{-1}(b),$$

since $\nu(\cdot)$ is monotonically increasing.

For the other cases such as (i) $D(a) \notin S_R, D(b) \in S_R$, (ii) $D(a), D(b) \notin S_R$, it is easy to see that similar proofs remain valid if we note Eq.(5). (Q.E.D.)

From this lemma, we can easily establish that the code $\mathbf{C}_U$ actually satisfies the condition of Corollary 1.

[**Lemma 2**] The code $\mathbf{C}_U$ given above is an *all-unidirectional error discriminating* code, i.e., it satisfies the condition shown in Eq.(1).

Proof : By contradiction. Suppose that

$$x \leq y \text{ or } x \geq y \quad \text{for } \exists x, y \in \mathbf{C}_U, \ x \neq y.$$

If $x = (u, a) \leq y = (v, b)$, i.e., $u \leq v$ and $a \leq b$, then

$$Q(u) \leq Q(v) \quad \text{and} \quad Q(u) = B^{-1}(a) \geq B^{-1}(b) = Q(v)$$

by Eqs.(7) and (8) of Lemma 1 and we must have $Q(u) = Q(v)$. But this is impossible since we have either $u \neq v$ or $a \neq b$. Similar proof holds for $x \geq y$. ¶

Now, let us consider the decoding algorithm for the code $\mathbf{C}_U$. Let $x = (u, a)$ $(\in \mathbf{C}_U)$ denote the transmitted codeword and $x' = (u', a')$ the received word where $a = B(Q(u))$ and $x' = x + e = (u + e_1, a + e_2)$. Then we have following three cases.

Case (i) *No errors*, i.e., $e = (e_1, e_2) = (0, 0)$. We clearly have

$$Q(u')(= Q(u)) = B^{-1}(a')(= B^{-1}(a)). \tag{9}$$

Case (ii) *Positive errors*. We have two possible cases :

(ii-1) $e_1 \geq 0$ $(e_1 \neq 0)$ and $e_2 = 0$: In this case, $a' = a$ and we have from Eq.(7) that

$$B^{-1}(a') = B^{-1}(a) = Q(u) < Q(u' = u + e_1),$$

and therefore

$$Q(u') > B^{-1}(a'). \tag{10}$$

(ii-2) $e_1 \geq 0$ and $e_2 \geq 0$ $(e_2 \neq 0)$: In this case, further two cases are possible.
(ii-2-a) $a' = (a'_0, a'_1, \ldots, a'_{\beta-1}) \notin Z_m^\beta$, that is

$$a_i' \geq m \quad \text{for} \quad \exists i. \tag{11}$$

(ii-2-b) $a' \in Z_m^\beta$. For this case, $B^{-1}(a')$ is well defined and we easily have from Eqs.(7) and (8) that

$$Q(u' = u + e_1) \geq Q(u) = B^{-1}(a) > B^{-1}(a' = a + e_2)$$

and therefore

$$Q(u') > B^{-1}(a'). \tag{10}$$

Case (iii) *Negative errors.* We again have two possible cases that exactly correspond to those of Case(ii) and we have the following results :

(iii-1) $e_1 \leq 0$ $(e_1 \neq 0)$ and $e_2 = 0$:

$$Q(u') < B^{-1}(a'). \tag{12}$$

(iii-2) $e_1 \leq 0$ and $e_2 \leq 0$ $(e_2 \neq 0)$: Either

$$a_i' < 0 \quad \text{for} \quad \exists i. \tag{13}$$

or

$$Q(u') < B^{-1}(a'). \tag{12}$$

Note here that all possible cases are exhausted in the above consideration and the corresponding conditions Eq.(9), Eq.(10), Eq.(11), Eq.(12) and Eq.(13) are *all distinct.* Thereby we can determine the error direction by using these distinct conditions which are summarized in Step 1 of the decoding algorithm given below.

[Decoding algorithm for C_U] : Let $x = (u, a)$ be the transmitted codeword and $x' = (u', a')$ be the received word.

Step 1 : Determine the error direction (Discriminate unidirectional errors) :

(i)   No unidirectional errors *if* : $a' \in Z_m^\beta$ and $Q(u') = B^{-1}(a')$.

(ii)   Positive errors *if* :

P-1. $a_i' \geq m$ for $0 \leq \exists i \leq \beta - 1$, or

P-2. $a' \in Z_m^\beta$ and $Q(u') > B^{-1}(a')$.

(iii)   Negative errors *if* :

N-1. $a_i' < 0$ for $0 \leq \exists i \leq \beta - 1$, or

N-2. $a' \in Z_m^\beta$ and $Q(u') < B^{-1}(a')$.

Step 2 : Correct the error e_1 by using the decoding algorithm for the base t-asymmetric error correcting code $\mathbf{C}_A$, knowing the error direction of e_1. ¶

Remark-1 : P-1 and N-1 in the above decoding algorithm will not occur in the channel in which the sending and received alphabets are the same. ¶

Remark-2 : As is easily seen, if we let $\mathbf{C}_A = Z_m^k$, the resulting code $\mathbf{C}_U$, though it has no error correcting capability any more, is a unidirectional error *discriminating* code with code length $n = k + \lceil \log_m (m-1)k + 1 \rceil$. This is exactly the same code as given by Bose and Pradhan[9] although they treated it only as a unidirectional error *detecting* code. ¶

4 Illustrative Example

In this section, we shall give a simple example to illustrate a special feature of the proposed code construction method.

Suppose that we are required to construct a *double* unidirectional error correcting and *all* unidirectional error discriminating (2-UEC & *all*-UEDis) code of alphabet size 5 with respect to the weight[7]

$$W(z) = \max\{H_S(z), H_S(-z)\}, \quad \text{where} \quad H_S(z) \triangleq \sum_{z_i > 0} z_i. \tag{14}$$

Then we can employ $(4, 2)$ Reed-Solomon (RS) code over $GF(5)$ as the base asymmetric error correcting code $\mathbf{C}_A$ (Table 1 shows all codewords). It is known that this code is capable of correcting up to *double asymmetric* error with respect to the weight $W(\cdot)$ given in Eq.(14) [7].

Following the construction algorithm given in the previous section, we easily get the desired 2-UEC & *all*-UEDis $(6, 2)$ code $\mathbf{C}_U$ as in Table 1. Next we shall demonstrate how this code $\mathbf{C}_U$ is decoded. Let the transmitted codeword be

$$x = (u, a) = (2, 1, 3, 4; 0, 2)$$

and the received word

Case-1 : $x' = (u', a') = (3, 2, 3, 4; 1, 2)$, i.e., $e = (e_1, e_2) = (1, 1, 0, 0; 1, 0)$
Case-2 : $x' = (u', a') = (2, 1, 3, 2; 0, 1)$, i.e., $e = (e_1, e_2) = (0, 0, 0, -2; 0, -1)$

Table 1. Unidirectional Error Control Code C_U Obtained from the Base Code C_A

$u \in C_A$	$Q(u)$	$i = \nu(Q(u))$	$B(u) = \phi(i)$	$x = (u, B(u)) \in C_U$
$(0,0,0,0)$	0	0	$(1,3)$	$(0,0,0,0;1,3)$
$(1,1,1,1)$	4	1	$(1,2)$	$(1,1,1,1;1,2)$
$(0,2,3,1)$	6	2	$(1,1)$	$(0,2,3,1;1,1)$
$(1,0,2,3)$	6	2	$(1,1)$	$(1,0,2,3;1,1)$
$(3,1,0,2)$	6	2	$(1,1)$	$(3,1,0,2;1,1)$
$(2,3,1,0)$	6	2	$(1,1)$	$(2,3,1,0;1,1)$
$(0,4,1,2)$	7	3	$(1,0)$	$(0,4,1,2;1,0)$
$(2,0,4,1)$	7	3	$(1,0)$	$(2,0,4,1;1,0)$
$(1,2,0,4)$	7	3	$(1,0)$	$(1,2,0,4;1,0)$
$(4,1,2,0)$	7	3	$(1,0)$	$(4,1,2,0;1,0)$
$(0,1,4,3)$	8	4	$(0,4)$	$(0,1,4,3;0,4)$
$(3,0,1,4)$	8	4	$(0,4)$	$(3,0,1,4;0,4)$
$(4,3,0,1)$	8	4	$(0,4)$	$(4,3,0,1;0,4)$
$(1,4,3,0)$	8	4	$(0,4)$	$(1,4,3,0;0,4)$
$(2,2,2,2)$	8	4	$(0,4)$	$(2,2,2,2;0,4)$
$(0,3,2,4)$	9	5	$(0,3)$	$(0,3,2,4;0,3)$
$(4,0,3,2)$	9	5	$(0,3)$	$(4,0,3,2;0,3)$
$(2,4,0,3)$	9	5	$(0,3)$	$(2,4,0,3;0,3)$
$(3,2,4,0)$	9	5	$(0,3)$	$(3,2,4,0;0,3)$
$(1,3,4,2)$	10	6	$(0,2)$	$(1,3,4,2;0,2)$
$(2,1,3,4)$	10	6	$(0,2)$	$(2,1,3,4;0,2)$
$(4,2,1,3)$	10	6	$(0,2)$	$(4,2,1,3;0,2)$
$(3,4,2,1)$	10	6	$(0,2)$	$(3,4,2,1;0,2)$
$(3,3,3,3)$	12	7	$(0,1)$	$(3,3,3,3;0,1)$
$(4,4,4,4)$	16	8	$(0,0)$	$(4,4,4,4;0,0)$

Then by Step 1 of the decoding algorithm given in the previous section, we get $a' \in Z_m^2$ for both Case-1 & 2 and

$$Q(u') = 12 > 4 = B^{-1}(a'), \quad \text{for Case-1,}$$
$$Q(u') = 8 < 12 = B^{-1}(a'), \quad \text{for Case-2.}$$

Thus it is found that there are *positive* errors for Case-1 and *negative* errors for Case-2.

The error correction is done in Step 2 of the decoding algorithm and $e_1 = (1,1,0,0)$ for Case-1 and $e_1 = (0,0,0,-2)$ for Case-2 will be corrected by using the *double asymmetric error correcting capability* of the base code C_A(i.e., $(4,2)$ RS code over $GF(5)$). The error correcting procedure of the base code C_A uses *Newton's Identity* and is rather simple[7], but is omitted here(see also [6] and [10]).

We shall conclude this section with a comment. If we try to guarantee a comparable error control capability by using Reed-Solomon code, for instance, we may adopt $(6,2)$ RS code over $GF(7)$ which is capable of correcting *double symmetric* error with respect to *Hamming* weight. It should be noted, however, that as is easily seen, $(6,2)$ RS code over $GF(7)$ has only five codewords, namely,

$$(0,0,0,0,0,0), \ (1,1,1,1,1,1), \ (2,2,2,2,2,2), \ (3,3,3,3,3,3) \ \text{and} \ (4,4,4,4,4,4)$$

that consist of elements all less than 5.[4]

5 Conclusion

Focusing the attention on the multiple-valued logic and memory systems, we have proposed a simple construction method for m-ary t-*unidirectional* error correcting and *all unidirectional* error *discriminating* codes starting from t-*asymmetric* error correcting codes. In view of that very few construction methods for m-ary unidirectional error control codes exist prior to this, the proposed method is considered to be a useful and also practical for multiple-valued memory systems.

The rate of the obtained codes depends on both the weight distribution and the rate of the employed base asymmetric error control code. The total number of check digits for the obtained code $\mathbf{C}_U$ is given by

$$Check(\mathbf{C}_A) + \lceil \log_m N_P \rceil$$

where $Check(\mathbf{C}_A)$ is the number of check digits for the base asymmetric error control code $\mathbf{C}_A$ and N_P is the number of partitions which is determined by the weight distribution of $\mathbf{C}_A$ and attains its minimum (of zero) when $\mathbf{C}_A$ is a constant weight code. Thus it remains for a future study to find m-ary asymmetric error control codes with a smaller weight distribution.

Acknowledgments: The authors are grateful to Professor S. Tsujii, Dr. T. Uyematsu and Dr. H. Jinushi of Tokyo Institute of Technology for many useful suggestions.

---

[4] We have $(24,16)$ BCH code over $GF(5)$ with the minimum *Hamming* distance 5 [10]. But it leaves nothing when this is shortened to a code length of $n = 6$.

References

[1] T.R.N. Rao and E. Fujiwara : *Error-Control Coding for Computer Systems*, Prentice Hall, 1989

[2] T. Murase : "256QAM 400Mb/s Microwave Radio System," *Japan Telecommunications Review*, vol.**30**, No.2, pp.23–30, April 1988

[3] J. T. Butler : "Multiple-Valued Logic," *Computer*, vol.21-4, pp.13–15, April 1988

[4] K. C. Smith : "Multiple-Valued Logic : A Tutorial and Appreciation," *Computer*, vol.**21**-4, pp.17–27, April 1988

[5] R.R. Varshamov : "A Class of Codes for Asymmetric Channels and Problem from Additive Theory of Numbers," *IEEE Trans. on Information Theory*, vol.**IT-19**, pp.92–95, January 1973.

[6] R.J. McEliece : "A Comment on A Class of Codes for Asymmetric Channels and a Problem from the Additive Theory of Numbers," *IEEE Trans. on Information Theory*, vol.**IT-3**, p.137, January 1973.

[7] P.H. Delsarte and P.H. Piret : "Spectral Enumerators for Certain Additive-Error-Correcting Codes over Integer Alphabets," *Information and Control*, vol.48, pp.193-210, March 1981.

[8] K. Sakaniwa, T.N. Ahn and T.R.N. Rao : "A Construction Method for m-ary Error Correcting and Discriminating Codes," *Proc. The 11th Symposium on Information Theory and Its Applications (SITA '88)*, pp.137–142, Dec. 1988 (in Japanese)

[9] B. Bose and D.K. Pradhan : "Optimal Unidirectional Error Detecting/Correcting Codes," *IEEE Trans. on Computers*, vol.**C-31**, no.6, pp.564–568, June 1982

[10] E.R. Berlekamp : *Algebraic Coding Theory*, McGraw-Hill, NY, 1968

[11] K. Sakaniwa, T.N. Ahn and T.R.N. Rao : "A New Theory of m-ary Error Correcting/Discriminating Codes (Part-I, II, III and IV)," *IEICE (the Institute of Electronics, Information and Communication Engineers of Japan) Technical Report*, **CS88-64**,

Nov. 1988, **CS88-65**, Nov. 1988, **CS88-98**, Jan. 1989, **CS89-53**, July 1989 (in Japanese)

[12] L.P. To and K. Sakaniwa : "Theory and Construction of m-ary Error Discriminating Codes," *Proc. The 11th Symposium on Information Theory and Its Applications (SITA '89)*, pp.121–125, Dec. 1989 (in Japanese)

FEASIBLE CALCULATION OF THE GENERATOR FOR COMBINED LFSR SEQUENCES

Lu Peizhong   Song Guowen

Reserch   Institute   of
Telecommunication   Techniques

231 Guyuan Road,Shanghai
Postcode 200434
People's Republic of China

Abstract

We have a finite number of linear feedback shift registers (LFSR) with known generating polynomials over a commutative ring R. $S_R(f(x))$ denotes the R module of all homogeneous LFSR sequences in R generated by $f(x)$.

The purpose of our paper is to determine the generating polynomial of the recurrence sequences obtained by multiplying the outputs of these LFSRs. When R is a finite field, we present a new explicit and computationally feasible method for determining the polynomial $h(x)$, without factoring the polynomials $f_i(x)$, such that

$$S_R(h(x)) = S_R(f_1(x))...S_R(f_n(x))$$

To this end we apply tensor products of matrices. We find that the polynomial $h(x)$ is just the minimal polynomial of the tensor product of these companion matrices of $f_i(x)$.

1. Introduction

A common type of running key generator employed in stream cipher systems consists of n linear feedback shift registers whose output sequences are combined in a nonlinear function F (in Fig.1) to produce the key stream. Considerable interest has been paid to the problem of analysing the sequences that result from nonlinear combinations, in particular, products of LSFR sequences. R.A. Rueppel and O.J.Staffelbach [3] have disscussed the conditions which guarantee the products of linear recurring sequences to attain maximum linear complexity. Zierler and Mills[1], Lidl[2] found the LFSR which could produce the same sequences generated by any linear combination of products of all possible output sequences of not necessarily distinct LFSRs. They obtained the theoretical generating polynomial of the product sequence of n linear recurring sequences which have no multiple roots in their generating polynomials. But their papers did not give an easy way for the explicit determination of the polynomial of the product sequence. They must factorize the polynomials over finite fields, and then build the desired polynomial in the extension field. So, it is difficult to obtain the desired polynomial in a computer if we only use their results. In this paper, we use the concepts of tensor products in matrix algebra to discuss the problem. We obtain some results on general rings. Then we turn our attentions to finite fields. The main result in this paper is that the product sequence Y in Fig.1 is generated by the minimal polynomial of the matrix $T=T_1 \odot T_2 \odot \cdots \odot T_n$, where T_i is the companion matrix of the sequence x_i. Thus we can compute the polynomial by using a computer.

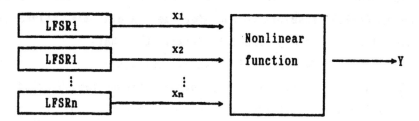

Fig 1. Common type of running key generator

2. Tensor products and general results

Let R be a commutative ring with a unit. Let S denote the R module consisting of all sequences $S=(s_0,s_1,\cdots\cdots,)$, $s_i \in R$, where addition and multiplication are performed componentwise. Let $f(x)$ be a monic polynomial, $f(x)=x^n+c_{n-1}x^{n-1}+\cdots+c_0$. $S_R(f(x))=\{s \in S \mid \sum_{i=0}^{n} c_i s_{k-i}=0, k \geqslant n\}$ $S_R(f(x))$ is clearly a free R-module with rank n. $S_R(f(x))S_R(g(x))$ is defined to the R-module spanned by all the products st , with s in $S_R(f(x))$,t in $S_R(g(x))$.

Let L be left shift map on S, L: $(s_0,s_1,s_2,\ldots) \rightarrow (s_1,s_2,s_3,\ldots\ldots)$. For any monic polynomial $f(x)$ in R[x], L can be considered as an endomorphism of $S_R(f(x))$.

The following theorem describes the set $S_R(f(x))$.

Theorem 1. Let R be a finite commutative ring. E is a subset of S, then $E=S_R(f(x))$ for some polynomial $f(x)$ in $R[x]$, if and only if E is a finitely generated free R module, and $L(E) \subseteq E$. ■

Corollary 1. Let R be a finite field, then for any tow monic polynomials $f(x), g(x)$ in $R[x]$, there exists a monic polynomial $h(x)$ in $R[x]$, such that $S_R(h(x))=S_R(f(x))S_R(g(x))$. ■

Theorem 2. Let R be a commutative ring, $E=S_R(f(x))S_R(g(x))$, $f(x)$, $g(x)$ be monic polynomials. Let T_f and T_g be the companion matrices of $f(x)$ and $g(x)$ respectively, which are rational canonical blocks. $T_f \odot T_g$ is tensor product (also called Kronecker product) of $T_f, T_g, h(x)$ is the characteristic polynomial of $T_f \odot T_g$, then E is a submodule of $S_R(h(x))$.

■

3. The case R=GF(q)

In this section, we repeatedly use the properties of tensor products of matrices over fields. We list some properties in the following:

(a) If A, B, C, D are matrices over a field, then $(A \odot B)(C \odot D)=AB \odot CD$.

(b) $(A \odot B) \odot C=A \odot (B \odot C)$.

(c) If A, B are similar to C, D respectively, denote $A \sim C$, $B \sim D$, then $A \odot B \sim C \odot D$.

(d) For any matrices A, B, $A \odot B \sim B \odot A$.

Theorem 3. Let $f(x)$ be a monic polynomial without multiple roots, and $f(0) \neq 0$, $g(x)$ be a monic nonconstant polynomial, $n=\deg(f(x))$, $m=\deg(g(x))$. Let $r_1, r_2, \cdots, r_n$ be the roots of $f(x)$ in its splitting field. Then the minimal polynomial $h(x)$ of the matrix $T_f \odot T_g$ is the least common multiple of the polynomials, $r_i^m g(x r_i^{-1}), i=1, \cdots, n$. It is just the polynomial $h(x)$, such that $S_R(h(x))=S_R(f(x))S_R(g(x))$.

Proof: T_f has n distinct characteristic values $r_1, r_2, \cdots, r_n$ in the splitting fields of $f(x)$, then over the extension field,

$$T_f \quad \sim \quad \begin{bmatrix} r_1 & 0 & \cdots & 0 \\ 0 & r_2 & \cdots & 0 \\ \multicolumn{4}{c}{\cdots\cdots\cdots} \\ 0 & 0 & \cdots & r_n \end{bmatrix}$$

$$
T_f \odot T_g \sim
\begin{bmatrix}
r_1 T_g & 0 & \cdots & 0 \\
0 & r_2 T_g & \cdots & 0 \\
& \cdots\cdots\cdots & & \\
0 & 0 & \cdots & r_n T_g
\end{bmatrix}
$$

Let $d_i(x)$ be the minimal polynomial of $r_i T_g$, $m_i = \deg(d_i(x))$, then $d_i(r_i T_g) = 0$, and $m_i \leqslant m$. Let $b_i(x) = r_i^{-m} d_i(r_i x)$, then $b_i(T_g) = 0$. Because T_g is a rational canonical block, its minimal polynomial is $g(x)$. Thus $m \leqslant m_i$. Thus $m = m_i$ and $r_i^{-m} d_i(r_i x) = g(x)$. Thus $d_i(x) = r^{-m} g(x r^{-1})$. Thus the minimal polynomial of $T_f \odot T_g$ is the common least multiple. Then $S_R(f(x)) S_R(g(x))$ is the subset of $S_R(h(x))$. To complete the proof, we use induction on the degree n of $f(x)$. If n=1, then $f(x) = x-r$, $r \neq 0$, $r \in GF(q)$. Thus $h(x) = rg(r^{-1} x)$, $\dim S(h(x)) = m$. Since $s = (1, r, r^2, \cdots) \in S(f(x))$. Let $b = (0, \cdots, 0, 1, *, \cdots) \in S_R(g(x))$. Then $sb, L(sb), \cdots, L^{m-1}(sb)$ are linearly independent over R. Therefore $\dim(S_R(f(x)) S_R(g(x)) \geqslant m$. This shows that $S_R(f(x)) S_R(g(x)) = S_R(h(x))$. Let $f(x) = (x-r_1)(x-r_2)\cdots(x-r_n)$, for $r_i \in K$, where K is an extension field of $GF(q)$. Then

$$
S_R(f(x)) S_R(g(x)) = (S_R(f(x)/(x-r_1)) S_R(g(x)) + S_R((x-r_1)) S_R(g(x))
$$
$$
= S_R((h(x))
$$

by induction and lemma 8.68 in [2]. ∎

Applying this theorem, we can obtain the desired polynomial $h(x)$ without factoring $f(x)$ and $g(x)$. We have an algorithm to obtain $h(x)$. The algorithm is performed when we diagonalize the matrix $xI - T_f \odot T_g$ by elementary operations of the matrix's rows and columns.

For nonconstant polynomials $f_1(x), \cdots, f_n(x)$ over R, we define $f_1(x) \vee \cdots \vee f_n(x)$ to be the monic polynomial whose roots are the distinct elements of the form $\alpha_1 \cdots \alpha_n$, where each α_i is a root of $f_i(x)$ in the splitting field of $f_1(x) \cdots f_n(x)$ over R. The following results can be considered as corollaries of Theorem 3.

Corollary 3. If $f_i(x)$ is a monic polynomial without multiple roots for $i = 1, \cdots, n$, then we have $S_R(f_1(x)) \cdots S_R(f_n(x)) = S_R(f_1(x) \vee \cdots \vee f_n(x))$.

Proof: We have only to prove the result in the case of n=2. Let $f_1(x) = (x-r_1) \cdots (x-r_n)$, $f_2(x) = (x-d_1) \cdots (x-d_m)$. Then $r^m f_1(x r_i^{-1}) = (x-r_i d_1) \cdots (x-r_i d_m)$. Therefore, $h(x) = [r_1^m f_2(x r_1^{-1}), \ldots, r_n^m f_2(x r_n^{-1})]$ ∎

Corollary 4. If $f(x) \in R[x]$ is a nonconstant monic polynomial without

multiple roots, and $f(0)\neq 0$, k is a positive integer, then $S_R(f(x)^k)=$
$S_R((x-1)^k)S_R(f(x))$.

Proof: $[r_1^k(xr_1^{-1}-1)^k,\ldots,r_n^k(xr_n^{-1}-1)^k]=[(x-r_1)^k,\ldots,(x-r_n)^k]=f(x)^k$ ∎

In general case, let $f(x)=p_1^{a_1}(x)\cdots p_k^{a_k}(x)$ $g(x)=q_1^{b_1}(x))\cdots q_t^{b_t}(x)$, $p_i(x)$, $q_i(x)$ are irreducible polynomials over $R=GF(q)$. Let $h_{ij}(x) \in R[x]$ such that $S_R(h_{ij}(x))=S_R(p_i^{a_i}(x))S_R(q_j^{b_j}(x))$. Then

$$S_R(f(x))S_R(g(x))= \sum_{i,j} S_R(p_i^{a_i}(x))S_R(q_j^{b_j}(x))$$
$$=S_R([h_{11}(x),\cdots,h_{ij}(x),\cdots,h_{kt}(x)]) \tag{1}$$

Zierler and Mills [1] have proved the following results.

Lemma 1. Let a,b be positive integers, p be the characteristic of $GF(q)$ $a-1 = \sum_v j_v p^v$, $b1 = \sum_v k_v p^v$, $0\leqslant j_v<p$, $0\leqslant k_v<p$. Let r be the smallest nonnegative integer such that $j_v+k_v< p$, for all $v\geqslant r$. Let $a\vee b= p^r + \sum_{i\geqslant r} (j_i+k_i)p^i$. Then $S_R((x-1)^a)S_R((x-1)^b) =S_R((x-1)^{a\vee b})$ ∎

By Lemma1 and Corrollary 4 ,

$$S_R(p^a(x))S_R(q^b(x))=S_R((p(x) \vee q(x))^{a\vee b}). \tag{2}$$

Thus, $$S_R(f(x))S_R(g(x)) =S_R([\cdots,(p_i(x) \vee q_j(x))^{a\vee b} ,\cdots]) \tag{3}$$

So far, the determination of the polynomial in (3) involves the factorization of $f(x)$ and $g(x)$ and the calculation of the roots in their extension fields. It is not easily. Now, we want to overcome this. We first give an interesting lemma in number theory which describes the number $a\vee b$.

Lemma 2. $a\vee b$ is the smallest nonnegative integer n, such that, for any i, if $n+1\leqslant i+b\leqslant a+b-1$, then $\binom{n}{i}$ = 0 (mod p)

Proof: We prove the result in 4 steps.
(i) For any nonnegative integer n,k, let $n=\sum_{i=0}^{l} n_i p^i$ and $k=\sum_{i=0}^{l} k_i p^i$,for

$0\leqslant n_i<p, 0\leqslant k_i<p$. Then

$$\binom{n}{k} \equiv \prod_{i=0}^{l}\binom{n_i}{k_i} \pmod{p} .$$

In order to show this we use the fact that $(1+x)^p\equiv 1+x^p \pmod{p}$. If $0\leqslant r< p$ then $(1+x)^{ap+r} \equiv (1+x^p)^a(1+x)^r \pmod{p}$. Comparing coefficients of x^{bp+s} (where $0\leqslant s < p$) on both sides yields

$$\binom{ap+r}{bp+s} \equiv \binom{a}{b} \binom{r}{s} \pmod{p}$$

The result (i) now follows by induction.

(ii) $n \leqslant a+b-1$

If m satisfies the property that, for any i, if $m+1 \leqslant i+b < a+b-1$ involves

$\binom{m}{i} \equiv 0 \pmod{p}$, then any positive integer, larger than m, will satisfies

the same property. In particular, when $m=a+b-1$, m satisfies the property.
Now all the constraint conditions on i have disappeared. Thus $n \leqslant a+b-1$

(iii) $n > a \vee b-1$

Let $i=a-1$, so $a \vee b-1 \leqslant i+b-1 \leqslant a+b-2$, and $a \vee b-1= (p-1)+(p-1)p+\cdots(p-1)p^{r-1}$
$+ \sum_{v \geqslant r} (j_v+k_v)p^v$, $i=a-1=\sum_v j_v p^v$. Then

$$\binom{a \vee b-1}{i} \equiv \binom{p-1}{j_0} \cdots \binom{p-1}{j_{r-1}} \prod_{v>r} \binom{j_v+k_v}{j_v} \pmod{p}$$

Because of $j_v+k_v < p$, for all $v > r$, and if $j_v+k_v=0$ then $j_v=0$. Thus

$$\binom{a \vee b-1}{i} \neq 0 \pmod{p}$$

(iv) $n \leqslant a \vee b$

When $r=0$, $a \vee b=1+\sum_{v \geqslant r} (j_v+k_v)p^v = a+b-1$. So $n \leqslant a \vee b$.

When $r>0$, we let $a \vee b= \sum_{v \geqslant 0} d_v p^v$, $0 \leqslant d_v < p$. So, if $v<r$, then $d_v =0$. For

all i, if $a \vee b \leqslant i+b-1 \leqslant a+b-2$, we can let $i= \sum_{v \geqslant r} j_v p^v + \sum_{v=0} i_v p^v$ where,

for some $u<r$, we have $i_u \neq 0$.   Therefore

$$\binom{a \vee b}{i} \equiv \prod_{v=0}^{r-1} \binom{d_v}{i_v} \prod_{v \geqslant r} \binom{d_v}{j_v} \pmod{p}$$

Because of

$$\binom{d_u}{i_u} = \binom{0}{i_u} \equiv 0 \pmod{p}, \text{ we get } \binom{a \vee b}{i} \equiv 0 \pmod{p} .$$

Then , $n \leqslant a \vee b$.

From (iii) ,(iv), we get $n=a \vee b$ ∎

Let us use J_a to represent the canonical rational block matrix of
the polynomial $(x-1)^a$.

Lemma 3. The minimal polynomial of the matrix $J_a \odot J_b$ is $(x-1)^{a \vee b}$.
Proof: J_a can be transformed into Jordan canonical form

$$J_a \quad \sim \quad \begin{bmatrix} 1 & 1 & 0 & \cdots & 0 \\ 0 & 1 & 1 & \cdots & 0 \\ & \cdots & & \cdots & \\ 0 & 0 & 0 & \cdots & 1 \end{bmatrix}$$

By property (c), we know

$$J_a \odot J_b \quad \sim \quad \begin{bmatrix} J_b & J_b & 0 & \cdots & 0 \\ 0 & J_b & J_b & \cdots & 0 \\ & \cdots & & \cdots & \\ 0 & 0 & 0 & \cdots J_b \\ 0 & 0 & 0 & \cdots J_b \end{bmatrix}$$

It is clear that the characteristic polynomial of $J_a \odot J_b$ is $(x-1)^c$, for $c \leqslant ab$. Let $J = J_b$, and

$$A = \begin{bmatrix} J-I & J & 0 & \cdots & 0 \\ 0 & J-I & J & \cdots & 0 \\ & & \cdot & & \\ & & & \cdot & \\ & & & & \cdot \\ 0 & 0 & 0 & \cdots & J-I \end{bmatrix}$$

We get

$$A^2 = \begin{bmatrix} (J-I)^2 & 2(J-I)J & J^2 & \cdots \\ 0 & (J-I)^2 & 2(J-I)J & \cdots \\ \diagdown & \diagdown & \diagdown & \cdots \\ & & & \end{bmatrix}$$

In general, A^n has $a \times a$ block matrices. Every block matrix in A^n is a $b \times b$ submatrix. In its first block line, the i-th block submatrix is denoted A_i

$$A_i = \binom{n}{i} (J-I)^{n-i} J^i, \quad 0 \leqslant i < a.$$

Therefore, $A^n = 0$ if and only if $A_i = 0$, for $0 \leqslant i < a$. Because of $(J-I)^b = 0$ and the invertible of J, we get that $A^n = 0$ if and only if

$$\binom{n}{i} (J-I)^{n-i} = 0, \quad \text{for} \quad n-b+1 \leqslant i < a.$$

The minimal polynomial of J is $(x-1)^b$, so, $(J-I)^m \neq 0$, for $m < b$. Thus $A^n = 0$ if and only if

$$\binom{n}{i} = 0 \pmod{p}, \quad \text{for} \quad n+1 \leqslant i+b \leqslant a+b-1.$$

By Lemma 2, we get $n = a \vee b$ ∎

Theorem 4. Let T_p, T_q be companion matrices of $P^a(x)$, $Q^b(x)$ respectively, then the minimal polynomial of $T_p \odot T_q$ is $(P(x) \vee Q(x))^{a \vee b}$.

Proof: Let P,Q be companion matrices of $P(x), Q(x)$ respectively. From Theorem 3, the minimal polynomials of $J_a \odot P$, $J_b \odot Q$ are $P^a(x), Q^b(x)$ respectively. Let $n = \deg P(x)$, $m = \deg(q(x))$, then $J_a \odot P$ is a $na \times na$ matrix. Thus the minimal polynomial is just the charcteristic polynomial. Therefore, $J_a \odot P \sim T_p$. Similarly, $J_b \odot Q \sim T_q$. We have the following relations

$$
\begin{aligned}
T_p \odot T_q &\sim (J_a \odot P) \odot (J_b \odot Q) \\
&= ((J_a \odot P) \odot J_b) \odot Q \\
&\sim (J_a \odot (J_b \odot P)) \odot Q \\
&= ((J_a \odot J_b) \odot P) \odot Q \\
&\sim (J_a \odot J_b) \odot (P \odot Q)
\end{aligned}
$$

We normalize the matrices $J_a \odot J_b$ and $P \odot Q$ in rational canonical form. By Lemma 3 and Theorem 3, we can let,

$$
J_a \odot J_b \sim
\begin{bmatrix}
J_{s_1} & 0 & \ldots & 0 \\
0 & J_{s_2} & \ldots & 0 \\
& \ldots\ldots\ldots & & \\
0 & 0 & \ldots & J_{s_n}
\end{bmatrix}
$$

where $s_1 \leqslant s_2 \leqslant \cdots \leqslant s_k$, and $s_k = a \vee b$.

$$
P \odot Q \sim
\begin{bmatrix}
N_1 & 0 & \ldots & 0 \\
0 & N_2 & \ldots & 0 \\
& \ldots\ldots\ldots & & \\
0 & 0 & \ldots & N_t
\end{bmatrix}
$$

where N_i is rational canonical block.

The minimal polynomial of N_i is $m_i(x)$ which is a divisor of $P(x) \vee Q(x)$. $m_t(x) = P(x) \vee Q(x)$. Thus,

$$T_p \odot T_q \quad \sim \quad \begin{bmatrix} J_{s_1} \odot N_1 & & & 0 \\ & \cdots\cdots & & \\ & & J_{s_i} \odot N_j & \\ & \cdots\cdots & & \\ 0 & & & J_{s_k} \odot N_t \end{bmatrix}$$

Comparing the elementary divisors of the matrix, and by the nonmultiple roots property of $P(x) \vee Q(x)$ we know, that the minimal polynomial of $T_p \odot T_q$ is the minimal polynomial of $J_{s_k} \odot N_t$, which is $(P(x) \vee (Q(x))^{a \vee b}$ by Lemma 3 . ∎

Now we get the main result of the paper.

Theorem 5. Let $f(x) = P^a_1(x) \cdots P^a_k(x), g(x) = Q^b_1(x) \cdots Q^b_t(x)$ be the monic polynomials in $F[x], f(0)g(0) \neq 0$, T_f, T_g be companion matrices of $f(x)$, $g(x)$ respectively. Let $h(x)$ be the minimal polynomial of the tensor product $T_f \odot T_g$, then $S_R(h(x)) = S_R(f(x))S_R(g(x))$

Proof:

$$T_f \quad \sim \quad \begin{bmatrix} P^a_1 & 0 & \cdots & 0 \\ 0 & P^a_2 & \cdots & 0 \\ & \cdots\cdots\cdots & & \\ 0 & 0 & \cdots & P^a_k \end{bmatrix}$$

$$T_g \quad \sim \quad \begin{bmatrix} Q^a_1 & 0 & \cdots & 0 \\ 0 & Q^a_2 & \cdots & 0 \\ & \cdots\cdots\cdots & & \\ 0 & 0 & \cdots & Q^a_t \end{bmatrix}$$

where P^a_i , Q^b_i are the companion rational matrices of the polynomials $P^a_i(x)$ and $Q^b_i(x)$ respectively. Then

$$T_f \odot T_g \quad \sim \quad \begin{bmatrix} \ddots & & 0 \\ & P^a_i \odot Q^b_j & \\ 0 & & \ddots \end{bmatrix}$$

and thus the minimal polynomial $h(x)$ is the least common multiple of $(P_i(x) \vee Q_j(x))^{a_i \vee b_j}$. Therefore, we have proved the thoerem. ∎

Reference

[1] N. Zierler and W. H. Mills, Products of Linear Recurring Sequences, Journal of Algebra,vol 27,147-157 (1973)

[2] R. Lidl, Finite Fields,Addison Wesley ,1983

[3] R. A. Rueppel and O. J. Staffelbach,   Products of Linear Recurring Sequences with Maximum Complexity.IEEE Trans.IT,vol33,pp124-131,1987.

[4] S. M. Jennings, Multiplexed Sequences:Some Properties of the minimun polynomial. Crypoto'83

(5) N. Jacobson, Basic Algebra II , W. H. Freeman and Company, 1980.

SUBSTITUTION OF CHARACTERS IN q-ARY m-SEQUENCES

István Vajda

Technical University of Budapest, H-1111 Budapest, Stoczek u.2.

Tibor Nemetz

Math. Inst. of the HAS, H-1053 Budapest, Realtanoda u. 13-15.

Abstract: This paper discusses a simple way of increasing the linear complexity of maximal length q-ary sequences. This is attained by using character substitution tables. The achievable maximum of increase is determined and it is shown that a portion of about $1/e$ of all substitution tables share this maximum. The mean value and the variance of the linear complexity is derived for the sequence's output by randomly chosen substitutions. The special case of permutations as substitutions are investigated, as well. At the end of the paper we propose an extension of the notion of linear complexity.

1. INTRODUCTION

The maximal length sequences (m-sequences) are typical modules in complex sequence generators used for producing pseudorandom sequences with increased algorithmic complexity [3],[4],[5],[7]. These more complex sequences are usually applied in cryptography, and as such, they are the targets of algorithmic attacks aiming at reproducing the generated sequences from their smallest observed segments. The attacker is assumed to know the rule of generation , except some of its parameters. The design of sequences is mainly based on the largeness of linear complexity. In this paper we concentrate on this issue, keeping in mind that the known generation methods are usually producing periodic sequences over GF(q).

Let us recall that the linear complexity of a sequence over GF(q) is

the degree of minimal degree linear recursion with coefficients from GF(q)
by which the sequence can be reproduced [3]. Formally, the linear
complexity of the sequence $\{b_i\}$ is LC, if there is a sequence
$$c_0, c_1, \ldots, c_{LC} \ , \ c_i \in GF(q), \ i=1,2,\ldots,q, \ c_0 c_{LC} \neq 0,$$
and it has the shortest length among all such sequences satisfying $c_0=1$,

$$b_{j+LC} = -\sum_{i=1}^{LC} c_i b_{j+LC-i} \quad ,j \geq 0 \ . \tag{1}$$

Perhaps the simplest way to turn an m-sequence into a more complex
pseudo-random one is to apply the same substitution T to all characters of
the m-sequence. This paper is devoted to the analysis of such generators.
If the m-sequence consists of q-ary characters, then the substitutions
considered are arbitrary (not necessarily one-to-one) mappings

$$T : GF(q) \Rightarrow GF(q).$$

There are q^q such mappings and only q^2 of them are affine (which do
not increase linear complexity). The number q^q-q^2 of nonlinear
substitutions is large enough even if the characters are of half-bytes
size: $16^{16}-16^2 > 10^{19}$. Many of them may increase linear complexity. We
investigate them through randomly chosen tables. This way the linear
complexity LC also becomes a random variable LC. We derive formulas for the
statistical mean and the variance of this variable. The special case, when
only permutations are selected, is also considered.

Stronger generators can be expected when using polyalphabetic
substitution i.e. a set of substitution tables from which a new table is
selected for each substitution, according to an auxiliary selector
sequence.

In the final section of the paper the definition of linear complexity
is slightly extended, and some consequences are examined.

2. POLYNOMIAL REPRESENTATION OF THE SUBSTITUTION TABLE

An arbitrary mapping $T: GF(q) \Rightarrow GF(q)$ can be described by a polynomial
$t(x)$ over GF(q) having degree at most $q-1$. We will demonstrate this by

using the interpolation polynomials

$$
e_i(x) = \begin{cases} -x \cdot (x^{q-2} + \alpha_i x^{q-3} + \ldots + \alpha_i^{q-3} x + \alpha_i^{q-2}) & , \text{if } \alpha_i \neq 0 \\ -(x^{q-1} - 1) & , \text{if } \alpha_i = 0 \end{cases} , \tag{2}
$$

defined for all $\alpha_i \in GF(q)$. The identity $(x - \alpha_i) e_i(x) = -x(x^{q-1} - 1)$ holds obviously for all α_i. Furthermore simple substitution yields $e_i(\alpha_i) = -(q-1)$ for $\alpha_i \neq 0$. This sums up to

$$
e_i(\gamma) = \begin{cases} 1, & \text{if } \gamma = \alpha_i \\ 0, & \text{otherwise} \end{cases} . \tag{3}
$$

Let us use the notation $T(\alpha_i) = \beta_i$, where $\alpha_1, \alpha_2, \ldots, \alpha_q$ are the different elements of $GF(q)$. Then the polynomial representation of substitution T is the following:

$$
t(x) = \sum_{i=1}^{q} \beta_i e_i(x) . \tag{4}
$$

Here $t(x)$ is a polynomial over $GF(q)$ with degree at most $q-1$, i.e.
$$
t(x) = t_0 + t_1 x + \ldots + t_{q-1} x^{q-1}, \quad t_i \in GF(q).
$$

It can be checked easily that there exists only one polynomial over $GF(q)$ of degree $\leq q-1$, drawn through the points $(\alpha_1, \beta_1), (\alpha_2, \beta_2), \ldots, (\alpha_q, \beta_q)$, and therefore this polynomial is $t(x)$ constructed above. Consequently one q-ary polynomial $t(x)$ of degree $\leq q-1$ can be assigned uniquely to the each of mappings T. Thus randomly selecting one polynomial from the set of all polynomials over $GF(q)$ with degree $\leq q-1$ corresponds to a random selection of one substitution table for the set of q-ary characters. We conclude this section by noting that a random choice of such a polynomial is equivalent to choosing its coefficients independently with uniform distribution.

3. SIMPLE SUBSTITUTION

Any sequence $\{b_i\}$ over $GF(q)$ with period $N = q^r - 1$ can be written in the power-of-α representation

$$b_i = \sum_{d \in D} h_d \alpha^{di} \quad , \text{ for all } i \; , \tag{5}$$

where $D \subseteq \{1, 2, \ldots, q-1\}$ and α is a primitive element of GF(q). In principle, representation (5) can be obtained by applying Discrete Fourier Transformation (DFT) to the sequence $\{b_i\}$. The coefficients $\{h_d\}_{d \in D}$ are the nonzero spectrum components. It is known that the linear complexity of the sequence $\{b_i\}$ is equal to the number of nonzero spectrum components [6], i.e.

$$LC = |D| . \tag{6}$$

Let the sequence $\{a_i\}$ be a maximum length sequence of binary words, i.e. a q-ary m-sequence with period Q-1, where $q = 2^m$ and $Q = q^r$. Then the linear complexity of the sequence $\{t(a_i)\}_{i=1}^{\infty}$ generated by the power function $t(x) = x^k$, $1 \le k < q-1$ is given by $r^{w(k)}$, where $w(k)$ is the binary weight of integer k, furthermore the sets of nonzero spectral components for sequences corresponding to mappings $t(x) = x^k$, $1 \le k \le q-1$, are disjoint [5],[6]. From here it follows for the linear complexity LC_1 in this case

$$LC_1 = \sum_{k=0}^{q-1} \kappa_{\{t_k \neq 0\}} \cdot r^{w(k)} \quad , \tag{7}$$

where κ is the usual indicator function. Particularly it follows from (7) that, when $t_0 \cdot t_1 \cdots t_{q-1} \neq 0$ and $r > 1$ that the linear complexity is equal to

$$\sum_{k=0}^{2^m-1} r^{w(k)} = 1 + \binom{m}{1} r + \binom{m}{2} r^2 + \ldots + \binom{m}{m-1} r^{m-1} + r^m = (r+1)^m, \tag{8}$$

see [8].

Obviously $LC_1 \le (1+r)^m$ always holds, and this maximum complexity is achieved by any polynomial with no zero coefficients. Their number is $(q-1)^q$, i.e. they constitute

$$(1 - \frac{1}{q})^q \approx \frac{1}{e} \tag{9}$$

portion of all random mapping.

Formula (7) permits to calculate the expected value and variance of the linear complexity in a simple way, using the concluding remark of the previous section:

$$E(LC_1) = \sum_{k=0}^{q-1} E(\kappa_{\{t_k \neq 0\}}) \cdot r^{w(k)} = (1+r)^m E(\kappa_{\{t_o \neq 0\}}) \qquad (10)$$

$$Var(LC_1) = \sum_{k=0}^{q-1} D^2(\kappa_{\{t_k \neq 0\}}) \cdot [r^{w(k)}]^2 = (1+r^2)^m D^2(\kappa_{\{t_o \neq 0\}}), \qquad (11)$$

where $E(\cdot)$, resp. $D^2(\cdot)$ denote expectation resp. variance of the indicator variables. This gives rise to

Theorem 1.: In the case of simple substitution with randomly chosen substitution table, the statistical mean and the variance of the LC_1 can be given by

$$E(LC_1) = (1 - \frac{1}{q})(r+1)^m , \qquad (12)$$

and

$$Var(LC_1) = \frac{q-1}{q^2}(r^2+1)^m , \qquad (13)$$

resp., if $r \geq 2$.

From formulas (8),(9) and (10) it follows

Corollary 1. of Th.1.: In the case of simple šubstitution with randomly chosen substitution table

$$P(LC_1 > E(LC_1)) \approx 1/e. \qquad (14)$$

Formulas (12) and (13) can be used to give upperbound on the tail of probability distribution function of r.v. LC_1 applying the Chebyshev's Inequality. It says that

$$P \left(\left| \frac{LC_1 - E(LC_1)}{E(LC_1)} \right| > \varepsilon \right) \leq \frac{Var(LC_1^2)}{E(LC_1)^2 \varepsilon^2} \qquad (15)$$

For m-sequences with byte characters and using byte substitution table, (i.e. $q = 2^8$), the following numerical results can be obtained from (15)

$r \cdot m$	Prob. $/\varepsilon=0.1$	Prob. $/\varepsilon=0.5$	$E(LC_1)$
16	$3.5 \cdot 10^{-3}$	$1.4 \cdot 10^{-4}$	6535
32	$1.8 \cdot 10^{-2}$	$7.1 \cdot 10^{-4}$	$3.9 \cdot 10^{5}$
64	$6.7 \cdot 10^{-2}$	$2.7 \cdot 10^{-3}$	$4.3 \cdot 10^{7}$

For example if $r=4$ and $m=8$ the probability of the event that the linear complexity of the generated sequence using randomly chosen byte substitution table is smaller than $5^8/2 \cong 195322$ is less than $7.1 \cdot 10^{-4}$.

The case of $r=1$ needs special attention. Then the elements of the basic sequence $1, \gamma, \gamma^2, \ldots, \gamma^{Q-1}$ are the different nonzero elements of the field $GF(Q)$, and the result $LC=Q-1$ can be obtained.

In the binary case ($q=2$) the mean and the variance of the LC_1 are $(M+1)/2$ and $(M^2+1)/4$, respectively, where $M=r \cdot m$.

For the other extreme case when $q=Q$, we get $E(LC_1)=(2^M-1)^2/2^M \approx 2^M-2$. From this there follows an interesting consequence of Theorem 1:

Corollary 2. of Th.1.: The average of LC_1 for Q-ary random sequences of length $Q-1$ is $\approx Q-2$, where $Q=2^M$.

4. POLYALPHABETIC SUBSTITUTION

Let us apply k different substitution transformations $T_1, T_2, \ldots, T_k$ periodically to substitute the consecutive elements of a q-ary m-sequence with period k, i.e. the elements of the subsequence $a^{(j)} = a_j, a_{j+k}, a_{j+2k}, \ldots$ are mapped by T_j, $j=1,2,\ldots,k$. Here we suppose, that g.c.d.$(k,Q-1)=1$. This way a Simple Polyalphabetic Substitution scheme (SPS) emerges.

It is obvious to question how the linear complexity (LC_2) changes as the value of parameter k increases. A partial solution can be obtained by investigating the case of choosing the tables randomly by uniform, independent distribution. In this case the expected value of the linear complexity LC_2 can be bounded from above. Applying similar arguments leading to results (12) and (13) we get

$$E(LC_2) \le k \cdot (1 - \frac{1}{q^k})(r + 1)^m \quad . \tag{16}$$

The polyalphabetic substitution rule with k substitution tables can be refined (General Polyalphabetic Substitution, GPS). Let's form a string of mappings $T'_1, T'_2, \ldots, T'_s$ over $\{T_1, T_2, \ldots, T_k\}$ with all T_i appearing at least once (i.e. $s \ge k$), where g.c.d.$\{s, Q-1\} = 1$ and s is the smallest period and apply this new sequence periodically. It is interesting to note that such a modification may result in an increase of the average LC by a multiple of s at most.

One possible realization of such polyalphabetic substitution could be built from a linear feedback shift register (LFSR1) producing the basic sequence and an other LFSR (LFSR2) generating the selector sequence. In cryptographic applications we can suppose that the attacker knows the structure of generator except the actual phase of the basic sequence and the selector sequence, i.e. the initial content of LFSR1 and LFSR2 at the starting time. As for the set of substitution tables it is a "must" to keep them secret.

If we choose mappings from the set of permutations from the set of all possible substitutions, the relative frequency of the characters will correspond to the uniform distribution.

In the following section we give a lower bound on the mean value of LC (LC_3) for the case of randomly chosen permutations.

5. PERMUTATIONS

Using permutations of the field elements as a mapping, all $\beta_1, \beta_2, \ldots, \beta_q$ in formula (4) are different elements of GF(q). From formulas (3) and (4) it is easy to see that if q>2 then

$$t_j = (\underline{\beta}, \underline{\rho}^{(j)}) = \sum_{i=1}^{q} \beta_i \rho_i^{(j)} \tag{17}$$

for some $\rho_i^{(j)}$, where $\rho_i^{(j)} \in GF(q)$, $j=0, 1, \ldots, q-1$. Because the coefficients of x^{q-1} in each polynom $e_i(x)$ are equal to 1, i.e. $\rho_i^{(q-1)} = 1$ for all i, therefore $t_{q-1}=0$. We are interested in the probability $P(t_j \neq 0)$, when $\underline{\beta}$ is a random permutation of the elements of $GF(q)$ and $j<q-1$. Here we establish a lower bound.

__Theorem 2:__ $P(t_j \neq 0) \geq 1/2$ for all $j<q-1$.

__Proof:__ The proof uses the fact, that to each β, with $(\underline{\beta},\underline{\rho}^{(j)})=0$, a permutation $\underline{\beta}'$ can be constructed for which $(\underline{\beta}',\underline{\rho}^{(j)}) \neq 0$. Furthermore, all different β -s result in different $\underline{\beta}'$.

To this end, let i_1 and i_2 be the indices of two different nonzero elements of $\underline{\rho}^{(j)}$, $1 \leq i_1, i_2 \leq q$. From (3) it is easy to see that such pair always exists for $0 \leq j < q-1$. We use this two indexes for our construction. Let β an arbitrary permutation with $(\underline{\beta},\underline{\rho}^{(j)})=0$. Then let $\underline{\beta}'$ be the vector obtained from $\underline{\beta}$ by interchanging the two elements at indexes i_1 and i_2. The contribution of the terms with indexes i_1 and i_2 to the value of t_j will then be changed by $(\beta_{i_1} - \beta_{i_2})(\rho_{i_1}^{(j)} - \rho_{i_2}^{(j)})$, while all other terms remain unchanged. Because all elements of $\underline{\beta}$ are different, therefore $(\underline{\beta}',\underline{\rho}^{(j)}) \neq 0$. Obviously all the $q!$ possible permutations will be examined this way, one time each. The above rule automatically ensures that all $\underline{\beta}'$ vectors will be different.□

We guess that $P(t_j=0) \propto 1/q$ provided q is large enough.

From here, since $t_{q-1}=0$, one can derive the following:

__Corollary to Theorem 2:__ Using the above notations the mean value of the linear complexity for randomly chosen permutations is

$$E(LC_3) \geq \frac{1}{2}\left[(r+1)^m - r^m\right]. \qquad (18)$$

6. ON THE DEFINITION OF LINEAR COMPLEXITY

Let $\{b_i\}$ be a binary sequence from which a q-ary sequence $\{B_i\}$ can be generated by taking the consecutive disjunct m-bit segments from the binary sequence, that is $\{B_i\}$ is a sequence over V_2^m, where V_2^m stands for the set of binary m-tuples. In this section we slightly generalize the definition

(1) of linear complexity.

Let the generalized linear complexity of the sequence B be defined by the degree of the shortest linear recursion with coefficients from the set of binary matrices with size mxm, $m \geq 1$. Our proposition is motivated by the fact that in this case the linearity of the operations is preserved and at the same time, the complexity of the operations is essentially given by the shortest period, due to the possible parallel processing.

Definition: The generalized linear complexity is GLC if there exists a sequence $C=C_0, C_1, \ldots, C_{GLC}$ of binary matrices of size mxm with shortest length from those , for which C_0 is invertible and

$$B_i = -C_0^{-1}(C_1 B_{i-1} + C_2 B_{i-2} + \ldots + C_{GLC} B_{i-GLC}) ,$$ (19)

$i \geq GLC$.

The elements of $GF(2^m)$ can be represented by binary m-tuples using a basis of $GF(2^m)$, i.e. by the elements of V_2^m. This way the definition (1) of the linear complexity can be written into the form of (23) with the restriction that the coefficients of the recursion can be taken only from a subset of invertible matrices. Thus, considering the sequences over $GF(2^m)$ as sequences over V_2^m it follows at once that

GLC$\leq$ LC . (20)

Of course, GLC=LC for binary cases. It is known that the average of linear complexity over all binary sequences, $\{b_i\}$ with a given length H is closely H/2, see [7]. From the Corollary 2 of Th.1. it follows that the same quantity for generalized linear complexity of 2^m-ary sequences, $\{B_i\}$ of length H' with H'=H/m=2^m-1 is equal to $\approx$H'-2, i.e. close to H in bits. This means that the size of the segment to be observed is half in bits in the case when the linear reproduction of the binary sequence is attempted. It is interesting to examine this observation in the case of GLC, using natural notations GLC_b and GLC_B for complexities of sequences b and B , resp. Our last result states an inequality between them.

Theorem 3.: It holds for arbitrary binary sequence b and corresponding sequence B that

$$GLC_B \leq \lceil GLC_b/m \rceil +1.$$ (21)

The proof of this theorem is constructive utilizing the basic definition of the LFSR and the underlying recursion. It is omitted due to space limitations.

REFERENCES

[1] Zierler, N., "Linear recurring sequences", J. SIAM, Vol. 7, 1959.

[2] Golomb, S. W., Shift-Register Sequences. San Francisco: Holden Day, 1967.

[3] Massey, J. L., "Shift-Register Synthesis and BCH Decoding", IEEE Trans. Inform. Theory, Vol. IT-15, Jan. 1969. 4.

[4] Key, E. L., "An analysis of the structure and complexity of nonlinear binary sequence generators", IEEE Trans. Inform. Theory, IT-22, No. 6., Nov. 1976.

[5] L. Brynielson, "On the Linear Complexity of Combined Shift Register Sequences", Advances in Cryptology - Eurocrypt '85, Lecture Notes in Computer Science, No. 219, Springer, 1985.

[6] Simon, M, K., Omura, J. K., Scholtz, R. A., Levitt, B. K., Spread Spectrum Communications, Vol 1., Ch. 5., Computer Science Press, 1985.

[7] Rueppel, R. A., New Approaches to Stream Ciphers. Communication and Control Series of Engineering, Springer Verlag, 1986.

[8] T. Siegenthaler, Methoden für den Entwurf von Stream Cipher-Systemen, Doctoral Thesis, ETH Zürich, 1986

Pseudo-Polyphase Orthogonal Sequence Sets with Good Cross-Correlation Property

Naoki Suehiro

Faculty of Engineering, TokyoEngineering University

Katakura, Hachioji, Tokyo 192, Japan

ABSTRACT

This paper proposes a class of pseudo-polyphase orthogonal sequence sets with good cross-correlation property. Each set, composed of N pseudo-polyphase orthogonal sequences, is introduced from a maximum length sequence (m-sequence) by the inverse DFT, where N is the period of sequences.

A periodic sequence is called an orthogonal sequence, when the autocorrelation function is 0 in every term except for period-multiple-shift terms. It is known that a polyphase periodic sequence is transformed into an orthogonal sequence by the DFT or by the inverse DFT. There are N way for transforming a shifted m-sequence by the inverse DFT matrix, because an m-sequence is a periodic sequence of period N. So, we obtain N pseudo-polyphase orthogonal sequences by transforming the shifted m-sequences with the inverse DFT.

The absolute values of $(N-1)$ terms in any obtained sequence are the same value $\sqrt{\frac{N+1}{N}}$. The absolute value of remained one term in the sequence is $\sqrt{\frac{1}{N}}$. So, the obtained sequences can be called a pseud-polyphase orthogonal sequence.

The absolute values of $(N-1)$ terms in any crosscorrelation function between two different sequences in a set are the same value $\frac{\sqrt{N+1}}{N}$. The absolute value of the remained one term is $\frac{1}{N}$. So, these sequences have good crosscorrelation property.

1 Introduction

Binary sequences with good correlation properties have been studied for asynchronous SSMA systems[3][5]. However, considering the development of hardware equipments, we should study non-binary sequence. Of course, non-binary sequences can be with better correlation properties than binary sequences. When the absolute value of all elements in a sequence are the same, the sequence is called a polyphase sequence.

A periodic sequence is called an orthogonal sequence, when the auto- correlation function is 0 in every term except for period-multiple-shift terms. A set of FZC sequences[1][2][4] is composed of $(N-1)$ polyphase orthogonal sequences of period N with very good cross-correlation property, when N is prime. Suehiro and Hatori also proposed a class of polyphase orthogonal sequences named "modulatable orthogonal sequences." A set of the modulated sequences is composed of $(N-1)$ polyphase orthogonal sequences of period N^2 with very good cross-correlation property, when N is prime[7]. A filter for eliminating co-channel interferences can be designed for each receivers in the asynchronous SSMA system using the modulated polyphase sequences[8]. Suehiro also proposed a class of real-valued orthogonal sequences with a filter for each receivers for eliminating co-channel interferences in the asynchronous SSMA system[9].

In this paper, when the absolute value of almost all elements in a sequence are the same and the absolute value of the rest elements are smaller, the sequence is called a pseudo-polyphase sequence. The purpose of this paper is to propose a new class of pseudo-polyphase orthogonal sequence sets with good cross-correlation property for asynchronous SSMA systems. A set of pseudo-polyphase orthogonal sequences is composed of N sequences of period N. The absolute value peak of the cross-correlation function between any two sequences in the set is $\frac{\sqrt{N+1}}{N}$.

A set of FZC sequences of period prime N is composed of $(N-1)$ polyphase orthogonal sequence with ideal cross-correlation. When N is not prime, the number of sequences, in a FZC sequence set with good cross-correlation property, is very small. On the other hand, a set of the proposed sequences of period $N = p^n - 1$ is composed of N pseudo-polyphase orthogonal sequences with almost ideal cross-correlation, where p is a prime number. For the proposed sequence set, N need not prime. The number of sequence in the proposed set is slightly larger than the set of FZC sequences of prime period, and far larger than the set of the modulatable polyphase orthogonal sequences.

2 Genaral Method for Making Orthogonal Sequences

In this chapter, a general method for making orthogonal sequences[6] is discussed.
Let F_N be an N dimensional DFT (Discrete Fourier Transform) matrix, as

$$F_N = [f_N(i,j)] ,$$

where i and j are the row number and the column number,

$$f_N(i,j) = \frac{1}{\sqrt{N}} \exp\left(-\frac{2\pi\sqrt{-1}}{N} ij\right) ,$$

$$0 \leq i \leq N-1, \quad 0 \leq j \leq N-1 .$$

Let A and B are vectors, which represent periodic sequences of period N, as

$$A = \begin{pmatrix} a_1 \\ a_2 \\ . \\ . \\ . \\ a_N \end{pmatrix} , \quad B = \begin{pmatrix} b_1 \\ b_2 \\ . \\ . \\ . \\ b_N \end{pmatrix} .$$

Let $c_1, c_2, \ldots, c_N, d_1, d_2, \ldots$ and d_N be defined as

$$F_N A = \begin{pmatrix} c_1 \\ c_2 \\ . \\ . \\ c_N \end{pmatrix}, \quad F_N B = \begin{pmatrix} d_1 \\ d_2 \\ . \\ . \\ d_N \end{pmatrix}.$$

Then, the cross-correlation function between two periodic sequences represented by A and B is represented by a vector as

$$\frac{1}{\sqrt{N}} F_N^{-1} \begin{pmatrix} c_1 \overline{d_1} \\ c_2 \overline{d_2} \\ . \\ . \\ c_N \overline{d_N} \end{pmatrix} \tag{1}$$

where $\overline{d_1}$ is the complex conjugate for d_1.

When $A = B$, the auto-correlation function is obtained. The obtained auto-correlation function is represented by a vector

$$\frac{1}{\sqrt{N}} F_N^{-1} \begin{pmatrix} |c_1|^2 \\ |c_2|^2 \\ . \\ . \\ |c_N|^2 \end{pmatrix}.$$

So, when $|c_i| = 1$ for $^\vee c_i$, the auto-correlation function is represented by the vector

$$\begin{pmatrix} 1 \\ 0 \\ . \\ . \\ 0 \end{pmatrix}.$$

A periodic sequence is called an orthogonal sequence, when the auto- correlation function is 0 for every term except for period-multiple-shift terms. We can obtain an orthogonal sequence represented by the vector A by transforming any polyphase periodic sequence with the inverse DFT matrix as

$$\begin{pmatrix} a_1 \\ a_2 \\ . \\ . \\ a_N \end{pmatrix} = F_N^{-1} \begin{pmatrix} c_1 \\ c_2 \\ . \\ . \\ c_N \end{pmatrix}, \quad \text{where } |c_i| = 1 \text{ for } ^\vee c_i. \tag{2}$$

3 Pseudo-Polyphase Orthogonal Sequence Set

In this chapter, a class of pseudo-polyphase orthogonal sequence sets is proposed. Each set, composed of N pseudo-polyphase orthogonal sequences, is introduced from a maximum length sequence (m-sequence) by the DFT, where N is the period of sequences.

There are two versions of m-sequence. one is on the finite field, and the other is on the complex number field. For example, (1, 1, 0) is an m-sequence of period 3 on the finite field. On the other hand, $((-1)^1, (-1)^1, (-1)^0) = (-1, -1, 1)$ is an m-sequence of period 3 on the complex-number field. In this paper, "an m-sequence" means an m-sequence on the complex-number field.

There are N ways for transforming a shifted m-sequence of period N by an N-dimensional inverse DFT matrix. For example,

$$F_3^{-1}\begin{pmatrix} -1 \\ -1 \\ 1 \end{pmatrix}, \quad F_3^{-1}\begin{pmatrix} 1 \\ -1 \\ -1 \end{pmatrix} \quad \text{and} \quad F_3^{-1}\begin{pmatrix} -1 \\ 1 \\ -1 \end{pmatrix}.$$

Because an m-sequence is a polyphase sequence, we can obtain N orthogonal sequences of period N, by transforming a shifted m-sequence with an N-dimensional inverse DFT matrix. Now, we should discuss about the absolute value of each element in each obtained orthogonal sequence.

Let X be an m-sequence of period N as

$$X = \begin{pmatrix} x_1 \\ \cdot \\ \cdot \\ \cdot \\ x_N \end{pmatrix}.$$

Let

$$Y = \begin{pmatrix} y_1 \\ \cdot \\ \cdot \\ \cdot \\ y_N \end{pmatrix}$$

be defined as $Y = F_N^{-1}X$. Then Y is an orthogonal sequence.

Let $Y@\overline{Z}$ be defined as

$$Y@\overline{Z} = \begin{pmatrix} y_1\overline{z_1} \\ \cdot \\ \cdot \\ \cdot \\ y_N\overline{z_N} \end{pmatrix},$$

then

$$Y@\overline{Y} = \begin{pmatrix} |y_1|^2 \\ \cdot \\ \cdot \\ \cdot \\ |y_N|^2 \end{pmatrix}. \tag{3}$$

On the other hand, it is well known that the auto-correlation function for an m-sequence of period N is

$$(1, -\frac{1}{N}, \ldots, -\frac{1}{N}).$$

For example, the auto-correlation function for

$$(W_3^1, W_3^1, W_3^0, W_3^1, W_3^2, W_3^2, W_3^0, W_3^2)$$

is

$$(1, -\frac{1}{8}, \ldots, -\frac{1}{8}).$$

So, we obtain the following formula by using (1)

$$\frac{1}{\sqrt{N}} F_N^{-1}(Y \mathbin{@} \overline{Y}) = \frac{1}{N} \begin{pmatrix} N \\ -1 \\ \cdot \\ \cdot \\ -1 \end{pmatrix}$$

$$= \frac{1}{N} \begin{pmatrix} -1 \\ \cdot \\ \cdot \\ -1 \end{pmatrix} + \frac{1}{N} \begin{pmatrix} N+1 \\ 0 \\ \cdot \\ \cdot \\ 0 \end{pmatrix}.$$

Accordingly,

$$Y \mathbin{@} \overline{Y} = \frac{1}{\sqrt{N}} F_N \begin{pmatrix} -1 \\ \cdot \\ \cdot \\ -1 \end{pmatrix} + \frac{1}{\sqrt{N}} F_N \begin{pmatrix} N+1 \\ 0 \\ \cdot \\ \cdot \\ 0 \end{pmatrix}$$

$$= \frac{1}{N} \begin{pmatrix} -N \\ 0 \\ \cdot \\ \cdot \\ 0 \end{pmatrix} + \frac{1}{N} \begin{pmatrix} N+1 \\ \cdot \\ \cdot \\ N+1 \end{pmatrix}$$

$$= \frac{1}{N} \begin{pmatrix} 1 \\ N+1 \\ \cdot \\ \cdot \\ N+1 \end{pmatrix}.$$

So, we obtain the following formula from the formula (3).

$$
\begin{pmatrix} |y_1| \\ |y_2| \\ . \\ . \\ . \\ |y_N| \end{pmatrix} = \begin{pmatrix} \sqrt{\frac{1}{N}} \\ \sqrt{\frac{N+1}{N}} \\ . \\ . \\ . \\ \sqrt{\frac{N+1}{N}} \end{pmatrix} . \tag{4}
$$

The absolute values of $(N-1)$ terms in any obtained orthogonal sequence are the same value $\sqrt{\frac{N+1}{N}}$. The absolute value of the remained one term in the sequence is $\sqrt{\frac{1}{N}}$.

4 Cross-Correlation Property

In this chapter, the cross-correlation property for the pseudo-polyphase orthogonal sequences, proposed in previous chapter, is discussed.

Let

$$
X = \begin{pmatrix} x_0 \\ x_1 \\ . \\ . \\ . \\ x_{N-1} \end{pmatrix}
$$

be an m-sequence of period N, and let X_i be defined as

$$
X_i = \begin{pmatrix} x_{0+i} \\ x_{1+i} \\ . \\ . \\ . \\ x_{N-1+i} \end{pmatrix} .
$$

Of course, X is a periodic sequence of period N, so, $x_{N+k} = x_k$.

It is well known that, when $i \neq j$,

$$
X_i \textcircled{a} \overline{X_j} = \begin{pmatrix} x_{0+i}\overline{x_{0+j}} \\ x_{1+i}\overline{x_{1+j}} \\ . \\ . \\ . \\ x_{N-1+i}\overline{x_{N-1+j}} \end{pmatrix}
$$

is also an m-sequence. So, because of the formula (1), the cross-correlation function, between any two pseudo-polyphase orthogonal sequences in a set, is obtained by transforming an m-sequence with the inverse DFT matrix. The absolute values of $(N-1)$ terms in any crosscorrelation function between two different sequences in a set are the same value $\frac{\sqrt{N+1}}{N}$. The absolute value of the remained one term is $\frac{1}{N}$.

References

FOR PAPERS IN JOURNALS:

[1] D.C. Chu, Polyphase codes with good periodic correlation properties, IEEE Trans. on Information Theory, IT-18 (1972) 531-532.

[2] R.L. Frank, Comments on 'Polyphase codes with good correlation properties', IEEE Trans. on Information Theory, IT-19 (1973) 244.

[3] R. Gold, Optimal binary sequences for spread spectrum multiplexing, IEEE Trans. on Information Theory, IT-13 (1967) 619-621.

[4] D.V. Sarwate, Bounds on crosscorrelation and autocorrelation of sequences, IEEE Trans. on Information Theory, IT-25 (1979) 720-724.

[5] D.V. Sarwate and W.B. Pursley, Cross-correlation properties of pseudo-random and related sequences, Proc. IEEE 68 (1980) 593-619.

[6] N. Suehiro and M. Hatori, Polyphase periodic sequences without crosscorrelation and their application to asynchronous SSMA systems, Trans. of IECE Japan, 68-A (1985)1087-1093, (in Japanese).

[7] N. Suehiro and M. Hatori, Modulatable orthogonal sequences and application to SSMA systems, IEEE Trans. on Information Theory, IT-34 (1988) 93-100.

[8] N. Suehiro, Elimination of co-channel interferences in an asynchronous SSMA system using modulatable orthogonal sequences, Proc. 4th Joint Swedish-Soviet International Workshop on Information Theory (1989) 180-184.

[9] N. Suehiro, Modulatable real-valued orthogonal sequences and their application to an asynchronous SSMA system with a method for eliminating co-channel interferences, Proc. of Bilkent Intern. Conf. on New Trends in Comm. Cont. and Signal Processing (1990) 546-552.

Real-Valued Bent Function and Its Application to the Design of Balanced Quadriphase Sequences with Optimal Correlation Properties

Shinya MATSUFUJI *   Kyoki IMAMURA **

* Faculty of Science and Engineering, Saga University, Saga 840 Japan
** Faculty of Computer Science and Systems Engineering,
   Kyushu Institute of Technology, Iizuka, Fukuoka 820 Japan

Abstract   A real-valued bent function is newly defined and shown to be useful to design polyphase sequences with optimal correlation properties. As an application of practical importance, balanced quadriphase sequences with optimal correlation properties are designed and shown to be the sum of two binary {1,-1} bent sequences.

1. Introduction

Many binary pseudorandom sequences with low periodic correlation properties are known, such as Gold sequences[1], Kasami sequences[2], and bent sequences[3,4] and are used in spread spectrum multiple access communications. The binary bent sequences constructed by using binary bent functions have the period 2^n-1, n being a multiple of 4, the family size of $\sqrt{2^n}$ and the optimal periodic correlation properties in the sense of Welch's lower bound[5]: The maximum absolute value of the out-of-phase periodic auto-correlation and of the periodic cross-correlation for a set of sequences with period N is asymptotically lower bounded by $\sqrt{N}$, when the number of sequences in the set (i.e., family size) becomes large.

The bent function was generalized to the q-ary case, q being an integer, and the bent sequence to the p-ary case, p being a prime[6,7].

In this paper we firstly introduce the real-valued bent function of which the domain is the same as the previous p-ary bent function but the range is the set of real numbers instead of integers modulo p. Secondly possible correlation values of the sequences constructed by using the real-valued bent functions are determined and shown to be optimal. Thirdly a family of balanced quadriphase sequences of optimal correlation properties is constructed by using the real-valued bent functions. Although quadriphase sequences are practically important, balanced quadriphase sequences with optimal correlation properties have not been known [8][9].

2. Real-Valued Bent Function

Let p be a prime, V_p^m the set of m-tuples over the integers modulo p and $\omega = \exp(j2\pi/p)$, $j = \sqrt{-1}$.

A function $g(\underline{y})$ from V_p^m to V_p^1, where $\underline{y} \in V_p^m$, is called bent[6,7] if its Fourier transform

$$G(\underline{\lambda}) = p^{-m/2} \sum_{\underline{y} \in V_p^m} \omega^{g(\underline{y}) - \underline{\lambda}\underline{y}^T} \tag{1}$$

has unit magnitude for all $\underline{\lambda} \in V_p^m$, where superscript T denotes the transposition. The Fourier transform is useful in the analysis and the design of pseudorandom sequences and we have

$$\omega^{g(\underline{y})} = p^{-m/2} \sum_{\underline{y} \in V_p^m} G(\underline{\lambda}) \omega^{\underline{\lambda}\underline{y}^T} \tag{2}$$

In this paper we introduce the real-valued bent function defined as

Definition 1 A real-valued function $g(\underline{y})$ defined on V_p^m, where $\underline{y} \in V_p^m$, is called a real-valued bent function if $G(\underline{\lambda})$ defined by (1) has unit magnitude for all $\underline{\lambda} \in V_p^m$.

We will call $G(\underline{\lambda})$ as the Fourier transform of $g(\underline{y})$ and the formula (2) also holds.
A general class of real-valued bent functions is given by the following theorem.

Theorem 1 The function $g_{\underline{A}}(\underline{y})$ defined on V_p^m, where m is even and $\underline{A} \in V_p^m$ a constant vector used as a parameter to identify the function, is a real-valued bent function if

$$g_{\underline{A}}(\underline{y}) = \sum_{i=1}^{m/2} [y_i y_{i+m/2} + A_i y_i + A_{i+m/2} y_{i+m/2}]$$

$$+ \sum_{i=2}^{m/2} h_i (y_{1+m/2}, \ldots, y_{i-1+m/2}) y_i$$

$$+ \sum_{i=1}^{m/2} h_{i+m/2} (y_{1+m/2}, \ldots, y_{i+m/2}), \tag{3}$$

where $\underline{y} = (y_1, y_2, \ldots, y_m)$, $\underline{A} = (A_1, A_2, \ldots, A_m)$ and $h_i(y_{1+m/2}, \ldots, y_{i-1+m/2})$ is a mapping from V_p^{i-1} to V_p^1 and $h_{i+m/2}(y_{1+m/2}, \ldots, y_{i+m/2})$ a mapping from V_p^i to R, the set of real numbers.

Proof

Let $m = 2k$. Substitution of (3) into (2) gives

$$G_A(\underline{\lambda}) = p^{-k}[\sum_{y_{k+1} \in V_p^1} \omega^{h_{k+1}(y_{k+1}) + y_{k+1}(A_{k+1} - \lambda_{k+1})} \sum_{y_{k+1} \in V_p^1} \omega^{y_1(y_{k+1} + A_1 - \lambda_1)}]$$

$$\prod_{i=2}^{k}[\sum_{y_{k+i} \in V_p^1} \omega^{h_{k+i}(y_{k+i}, \ldots, y_{k+i}) + y_{k+i}(A_{k+i} - \lambda_{k+i})} \sum_{y_i \in V_p^1} \omega^{y_i(y_{k+i} + A_i + h_i(y_{k+i}, \ldots, y_{k+i-1}) - \lambda_i)}]$$

$$= \prod_{i=1}^{k} \omega^{h_{k+i}(y_{k+1}^0, \ldots, y_{k+i}^0) + y_{k+i}^0(A_{k+i} - \lambda_{k+i})}$$

$$= \omega^u, \tag{4}$$

where we used the notation $G_A(\underline{\lambda})$ for $G(\underline{\lambda})$ and $\underline{\lambda} = (\lambda_1, \lambda_2, \ldots, \lambda_{2k})$, $y_{k+1}^0 = \lambda_1 - A_1$, $y_{k+i}^0 = \lambda_i - A_i - h_i(y_{k+1}^0, \ldots, y_{k+i-1}^0)$ for $2 \le i \le k$ and $0 \le u < p$.

Q.E.D.

Let us notice that our real-valued bent function $g_A(\underline{y})$ in (3) can be written as

$$g_A(y) = \hat{g}_A(y) + h(y_{k+1}, \ldots, y_{2k}), \tag{5}$$

where

$$\hat{g}_A(\underline{y}) = \sum_{i=1}^{k}(y_i y_{k+i} + A_i y_i + A_{k+i} y_{k+i}) + \sum_{i=2}^{k} y_i h_i(y_{k+1}, \ldots, y_{k+i-1}) \tag{6}$$

is a p-ary bent function and $h(y_{k+1}, \ldots, y_{2k})$ a real-valued function from V_p^k to R. We can show that (3) and (5) are equivalent to each other, since $h(y_{k+1}, \ldots, y_{2k})$ can be decomposed as the last term on the left side of (3).

3. Real-Valued Bent Sequences

Let $\{g_A(\underline{y}), A \in V_p^{n/2}\}$ be a set of $p^{n/2}$ real-valued bent functions defined by (5) with $n = 2m = 4k$. Let α be a primitive element of $GF(p^n)$.

We can construct a family $\{a_A(i), A \in V_p^{n/2}\}$ of $p^{n/2}$ real-valued sequences of period $p^n - 1$ by

$$a_A(i) = f_A(\alpha^i), \tag{7}$$

where

$$f_A(x) = g_A[L(x)] + \text{tr}(\sigma x), \tag{8}$$
$$\sigma \in GF(p^n) \setminus GF(p^{n/2}) \tag{9}$$
$$x \in GF(p^n) \tag{10}$$

and tr( ) is a trace from $GF(p^n)$ to $GF(p)$, $g_A[L(x)]$ a real-valued bent function defined by (5) with

$$L(x) = (\text{tr}(\beta_1 x), \ldots, \text{tr}(\beta_{n/2} x)) \tag{11}$$

and $\{\beta_1, \ldots, \beta_{n/2}\}$ a basis of $GF(p^{n/2})$. Since the definition of $\{a_A(i)\}$ is the same as the p-ary bent sequence except that real-valued bent function is used instead of p-ary bent function, we will call the above $\{a_A(i)\}$ as a family of real-valued bent sequences in this paper.

We can compute the values of the periodic correlation functions for $\{a_A(i)\}$ defined by

$$R_{AB}(\tau) = \sum_{x \in V_p^n} \omega^{f_A(x) - f_B(\alpha^\tau x)} - 1 \tag{12}$$

$(0 \le \tau \le p^{n/2})$, where we assumed

$$f_A(0) = 0 \tag{13}$$

for all $A \in V_p^{n/2}$.

We can prove the following theorem.

Theorem 2   The values of $R_{A B}(\tau)$ are of the form $\omega^u p^{n/2} - 1$, where u is a real number satisfying $0 \le u < p$, except that $R_{AA}(0) = p^n - 1$, $R_{AA}(\tau) = -1$ for $\tau = (p^{n/2} + 1)j$, $1 \le j \le p^{n/2} - 2$ and $R_{A B}(\tau) = -1$ for $\tau = (p^{n/2} + 1)j$, $0 \le j \le p^{n/2} - 2$.

Note: This theorem shows that the family of real-valued bent sequences satisfies the Welch's lower bound and has the optimal correlation properties.

Proof

The periodic correlation function $R_{A B}(\tau)$ can be written as

$$R_{AB}(\tau) = \sum_{\mu \in GF(p^n)} F_A(\mu) F_B^*(\alpha^\tau \mu) - 1, \tag{14}$$

where $F_A(\mu)$ is the trace transform of $f_A(x)$ defined by

$$F_A(\mu) = p^{-n/2} \sum_{x \in GF(p^n)} \omega^{f_A(x) - \mathrm{tr}(\mu x)}$$

(15)

and the inverse formula holds

$$\omega^{f_A(x)} = p^{-n/2} \sum_{\mu \in GF(p^n)} F_A(\mu) \omega^{\mathrm{tr}(\mu x)}$$

(16)

and the superscrtipt * denotes the complex conjugate.

From (8) we have

$$F_{\underline{A}}(\mu) = p^{-n/2} \sum_{x \in GF(p^n)} \omega^{g_{\underline{A}}[L(x)]} \omega^{\mathrm{tr}[(\sigma - \mu)x]}$$

(17)

Substitution of

$$\omega^{g_{\underline{A}}[L(x)]} = p^{-n/4} \sum_{\underline{\lambda} \in V_p^{n/2}} G_{\underline{A}}(\underline{\lambda}) \omega^{\underline{\lambda} L(x)^T}$$

$$= p^{-n/4} \sum_{\underline{\lambda} \in V_p^{n/2}} G_{\underline{A}}(\underline{\lambda}) \omega^{\mathrm{tr}[\phi(\underline{\lambda})x]}$$

(18)

into (17) gives

$$F_{\underline{A}}(\mu) = p^{-3n/4} \sum_{\underline{\lambda} \in V_p^{n/2}} G_{\underline{A}}(\underline{\lambda}) \sum_{\mu \in GF(p^n)} \omega^{\mathrm{tr}[\{\sigma + \phi(\underline{\lambda}) - \mu\}x]}$$

$$= \begin{cases} p^{n/4} G_{\underline{A}}(\underline{\lambda}) & \text{if } \mu = \sigma + \phi(\underline{\lambda}) \\ 0 & \text{otherwise,} \end{cases}$$

(19)

where $G_A(\underline{\lambda})$ is the Fourier transform of $g_A[L(x)]$ and $\phi(\underline{\lambda})$ is a one-to-one mapping from $V_p^{n/2}$ to $GF(p^{n/2})$ defined by $\phi(\underline{\lambda}) = \lambda_1 \beta_1 + \ldots + \lambda_{n/2} \beta_{n/2}$ for $\underline{\lambda} = (\lambda_1, \ldots, \lambda_{n/2})$. Similliary we have

$$F_{\underline{B}}^*(\alpha^\tau \mu) = \begin{cases} p^{n/4} G_{\underline{B}}^*(\underline{\lambda}) & \text{if } \mu = \alpha^\tau[\sigma + \phi(\underline{\lambda})] \\ 0 & \text{otherwise.} \end{cases}$$

(20)

As shown in the Appendix 1, we have the following lemma.

Lemma 1 Let $\tau \neq 0$, and $\underline{\lambda}_A, \underline{\lambda}_B \in V_p^{n/2}$.

$$\sigma + \phi(\underline{\lambda}_A) = \alpha^\tau[\sigma + \phi(\underline{\lambda}_B)]$$

(21)

has no solution (λ_A, λ_B) for $\tau = (p^{n/2}+1)j$, $1 \le j \le p^{n/2}-2$ and one unique solution (λ_A, λ_B) otherwise.

From (14), (19), (20) and **Lemma 1**, we have

$$\mathbf{R}_{AB}(\tau) = \begin{cases} -1 & \text{if } \tau = (p^{n/2}+1)j, \ 1 \le j \le p^{n/2}-2 \\ p^{n/2} G_A(\lambda_A) G_B(\lambda_B) - 1 & \text{otherwise } (\tau \ne 0). \end{cases} \tag{22}$$

where (λ_A, λ_B) is the unique solution of (21). From (14), (15), we have

$$\mathbf{R}_{AB}(0) = p^{n/2} \sum_{\underline{y} \in V_p^{n/2}} \omega_{n/2}^{g_A(\underline{y}) - g_B(\underline{y})} - 1$$

$$= p^{n/2} \sum_{\underline{y} \in V_p^{n/2}} \omega_{n/2}^{(\underline{A} - \underline{B})\underline{y}^T} - 1$$

$$= \begin{cases} p^n - 1 & \text{if } \underline{A} = \underline{B} \\ -1 & \text{if } \underline{A} \ne \underline{B}. \end{cases} \tag{23}$$

Q.E.D.

Corollary 1   We have

$$\sum_{\tau=0}^{p^n-2} \mathbf{R}_{AB}(\tau) = 1. \tag{24}$$

Proof

Since $F_A(0) = 0$ from (19), we have

$$F_A(0) = \sum_{x \in \mathrm{GF}(p^n)} \omega^{f_A(x) - f_B(\alpha^\tau x)} = 0. \tag{25}$$

Using (25), we obtain

$$\sum_{\tau=0}^{p^n-2} \mathbf{R}_{AB}(\tau) = \sum_{\tau=0}^{p^n-2} \sum_{x \in \mathrm{GF}(p^n) \setminus \{0\}} \omega^{f_A(x) - f_B(\alpha^\tau x)}$$

$$= \sum_{x \in \mathrm{GF}(p^n) \setminus \{0\}} \omega^{f_A(x)} + \sum_{y \in \mathrm{GF}(p^n) \setminus \{0\}} \omega^{-f_B(y)}$$

$$= (-1)(-1) = 1.$$

Q.E.D.

4. Balanced Quadriphase Sequences with Optimal Correlation Properties

In this section we will successfully apply real-valued bent sequences to design a set of quadriphase sequences taking values from the alphabets $\{+1, -1, +j, -j\}$. The implementation of quadriphase sequence can be decomposed into two biphase sequences.

Since each of the 4 alphabets $\{+1, -1, +j, -j\}$ can be realized by $(-1)^a j^b = (-1)^{a+b/2}$, the real-valued bent function having the range $\{0, 1/2, 1, 3/2\}$ will be used to construct quadriphase sequences in the following manner. Let $p = 2$, $n = 0$ (mod4), and α be a primitive element of $GF(2^n)$. Let us write real-valued sequence $f_A(x)$ in (7) as

$$f_A(X) = \hat{f}_A(X) + h(X), \tag{26}$$

where $\hat{f}_A(x)$ is a binary bent sequence of period 2^n-1 and $h(x)$ is a mapping from $V_2^{n/4} = \{tr(\beta_{1+n/4}\, x), \ldots, tr(\beta_{n/2}\, x)\}$ to $\{0, 1/2\}$. We define a set of $2^{n/2}$ quadriphase sequences of period 2^n-1, $\{c_A(i),\ A \in V_2^{n/2}\}$ by

$$C_A(i) = (-1)^{f_A(\alpha^i)} \tag{27}$$

The set of quadriphase sequences has optimal correlation properties from Theorem 2 and the out-of-phase auto-correlation and cross-correlation takethe following 5 values, -1, $\pm 2^{n/2}-1$ and $\pm j2^{n/2}-1$.

We will find the condition that the sequence of (27) is balanced, i.e. , in one period each of the symbol -1, $\pm j$ appears 2^{n-2} times and the symbol $+1$ one time less. Let us define

$$
\begin{aligned}
e_{00} &= \#\{X \mid \hat{f}_A(X) = 0,\ h(X) = 0\} \\
e_{01} &= \#\{X \mid \hat{f}_A(X) = 0,\ h(X) = 1/2\} \\
e_{10} &= \#\{X \mid \hat{f}_A(X) = 1,\ h(X) = 0\} \\
e_{11} &= \#\{X \mid \hat{f}_A(X) = 1,\ h(X) = 1/2\} .
\end{aligned}
\tag{28}
$$

In one period the symbols $+1, +j, -1$ and $-j$ appear $e_{00}-1$, e_{01}, e_{10}, e_{11} times, respectively we can show that

$$e_{00} = e_{01},\quad e_{10} = e_{11}, \tag{29}$$

since from (25), we have $0 = e_{00}+je_{01}-e_{10}-je_{11} = (e_{00}-e_{10})+j(e_{01}-e_{11})$. Therefore the sequence (27) is balanced if and only if

$$\#\{X \mid h(X) = 0\} = \#\{X \mid h(X) = 1/2\}. \tag{30}$$

Above discussion gives the following theorem.

Theorem 3 The set of $2^{n/2}$ quadriphase sequences of period $2^n\text{-}1$ defined by (27) and (26) has optimal correlation properties and is balanced if and only if (30) holds.

The following sufficient condition for (30) is obvious.

Corrollary 2 If $h(x) = (1/2)\text{tr}(\beta_{i+n/4}x)$ $(1 \leq i \leq n/4)$, then the quadriphase sequence (27), (26) is balanced.

The quadriphase sequence (27), (26) can be decomposed into two biphase sequences and we have

$$C_A(i) = k(-1)^{\widehat{f}_A(\alpha^i)+2h(\alpha^i)} + jk(-1)^{\widehat{f}_A(\alpha^i)} \tag{31}$$

where $k = (1-j)/2$. Notice that $\{\widehat{f}_A(\alpha^i)+2h(\alpha^i)\}$ and $\{\widehat{f}_A(\alpha^i)\}$ are binary bent sequences.

The design method in this section can be extended to the $p.q$-phase sequences if we assume that, $f_A(\ )$ is a p-ary bent function and the range of $h(\ )$ in (7) is $\{0, 1/q, \ldots, (q-1)/q\}$.

5. Conclusion

The p-ary bent function and p-ary bent sequence are generalized to real-valued bent function and real-valued bent sequence. A new class of real-valued bent function is given. The set of real-valued bent sequences is shown to have the optimal correlation properties.

The real-valued bent sequence is applied to design a set of balanced quadriphase sequences with optimal correlation properties.

References

[1] R.Gold,"Optimum binary sequences for Spread spectrum multiplexing",IEEE, Transaction on Information Theory, Vol.IT-13,pp619-621, October 1974

[2] T.Kasami,"Weight Distribution Formula for Some Class of Cyclic Codes", Coordinated Science Laboratory,Illinois Urbana Technical Report R-285, April 1966

[3] J.D.Olsen, R.A.Scholz, L.R.Welch, "Bent Function Sequences",IEEE E, Transaction on Information Theory,Vol.IT-28,pp.858-868, Nov.,1982

[4] Simon, Omura, Scholtz, Levit,"Spread Spectrum Communications", Vol.1, Chap.5, Computer Science Press,1985

[5] L.R.Welch,,"Lower Bounds on the Maximum Correlation of Signals", IEEE, E, Transaction on Information Theory, Vol.IT-20,pp397-399, May 1974

[6] P.V.Kumar,"On Bent Sequences and Generalized Bent Functions",Ph.D. Dissertation,University of Southern California,1983

[7] P.V.Kumar,R.A.Scholtz,L.R.Welch, "Generalized Bent Functions and Their Properties", Journal of Combinational Theory, Vol.40, pp.90-107, Sept. 1985

[8] S.M.Krone,D.V.Sarwate,"Quadriphase Sequences for Spread-Spectrum Multiple-Access Communication", IEEE Transaction on Information. Theory, Vol.IT-30, pp520-529, May 1984

[9] Patrick Sole, "A Quaternary Cyclec Code and a Family of Quadriphase Sequences with Low Correlation Properties", Lecture note in Computer Science Coding Theory and Applications, Vol.388, pp.193-201, Nov. 1988.

Appendix 1   Proof of Lemma 1:

Since $\{1, \sigma\}$ is a basis for $\mathbf{GF}(p^n)$ over $\mathbf{GF}(p^{n/2})$, we can write

$$\alpha^{\tau} = b_0 + b_1\sigma$$
$$\sigma^2 = c_0 + c_1\sigma \tag{A-1}$$

where $b_0, b_1, c_0, c_1 \in \mathbf{GF}(p^{n/2})$.   Substitution of (A-1) into (21) gives

$$\sigma + \phi(\lambda_A) = [b_0\phi(\lambda_B) + b_1c_0] + [b_0 + b_1c_1 + b_1\phi(\lambda_B)]\sigma. \tag{A-2}$$

If $b_1 \neq 0$, i.e., $\alpha^{\tau} \notin \mathbf{GF}(p^{n/2})$, then (A-2) has a unique solution for $\phi(\lambda_A)$, $\phi(\lambda_B) \in \mathbf{GF}(p^{n/2})$. If $b_1 = 0$, and $\tau \neq 0$, i.e., $\alpha^{\tau} \in \mathbf{GF}(p^{n/2}) \setminus \{0,1\}$, then (A-2) has no solution for $\phi(\lambda_A)$, $\phi(\lambda_B)$.

Q.E.D.

Coded Modulation with Generalized Multiple Concatenation of Block Codes

G. Schnabl   M. Bossert

AEG Mobile Communication GmbH
Wilhelm-Runge-Str.11, D-7900 Ulm, F.R.Germany

Abstract

Block coded modulation schemes can be obtained by generalized concatenation of the modulation and block codes of length N. The modulation is considered as a code in the euclidean space $I\!R^2$. One obtains a concatenated code in the euclidean space $I\!R^{2N}$ using binary block codes. In this paper the obtained code is taken as an inner code, which is concatenated with block codes once again. The advantage of this multiple concatenation compared to single concatenation is the use of very short (easy to decode) outer codes such as repetition codes and even weight codes.

The construction of multiple concatenated codes is presented. A decoding method is given which uses soft decision decoding for the outer codes. As examples twice concatenated codes are designed for QPSK and 8-PSK modulation.

Computer simulations have been performed in order to compare multiple concatenated codes to single concatenated codes of equal length and rate. The multiple concatenated codes investigated here are found to be better than the single concatenated codes as far as the symbol error probability as well as the decoding effort is concerned.

1 Introduction

In [2] [6] [8] [9] [10] block coded modulation schemes were constructed by partitioning the modulation signal set into subsets and applying block codes to the enumerations (labels) of the subsets. This can also be looked at as a generalized concatenation of codes [13]. The modulation which is a code in the euclidean space serves as the inner code. It is partitioned into subsets and it is then concatenated with block codes. Good performance can in this case only be obtained with rather long outer block codes requiring an important decoding effort. The partition of modulation schemes with small signal sets is relatively simple. The concatenation of modulation with short block codes creates a multidimensional signal space which can be partitioned in a more sophisticated way. A second concatenation with short block codes results in coded modulation schemes which require little decoding effort. The purpose of this paper is to investigate the properties of these *generalized multiple concatenated* block codes in the euclidean space. In the sequel, we show that multiple concatenated codes are easy to decode and that there are multiple concatenated codes which have better error correction properties than conventional single concatenated block coded modulation schemes.

2 Code Construction

2.1 Single Concatenation of Codes

Generalized concatenation (GC) of codes as described in [13] and in [7], Ch.18 §8.2, is used here. An inner code $B^{(1)}$ is partitioned into subsets $B_j^{(i)}$, where j enumerates these subsets. This enumeration is protected with outer codes.

In case of coded modulation the inner code $B^{(1)}(I\!\!R^2; M = 2^{k_1}, \delta_b^{(1)})$ is the modulation scheme which is a code in the euclidean space $I\!\!R^2$. It has M codewords and $\delta_b^{(1)}$ is its minimum squared euclidean distance. $B^{(1)}$ is partitioned into disjunct subsets $B_j^{(2)}(I\!\!R^2; 2^{k_2}, \delta_b^{(2)})$ with $\delta_b^{(2)} > \delta_b^{(1)}$ and each subset is partitioned into subsets again, etc.. The subsets are enumerated by $a^{(1)}, a^{(2)}, \ldots$. In this way, finally, an s-th order partition of $B^{(1)}$ is obtained:

$$
\begin{aligned}
B^{(1)} &= \bigcup_{a^{(1)}} B_{a^{(1)}}^{(2)}(I\!\!R^2; 2^{k_2}, \delta_b^{(2)}) , & a^{(1)} &\in GF(2)^{k_1-k_2}, \\
B_{a^{(1)}}^{(2)} &= \bigcup_{a^{(2)}} B_{a^{(1)}a^{(2)}}^{(3)}(I\!\!R^2; 2^{k_3}, \delta_b^{(3)}) , & a^{(2)} &\in GF(2)^{k_2-k_3}, \\
&\;\;\vdots & &\;\;\vdots \\
B_{a^{(1)}\ldots a^{(S-2)}}^{(S-1)} &= \bigcup_{a^{(2)}} B_{a^{(1)}\ldots a^{(S-1)}}^{(S)}(I\!\!R^2; 2^{k_S}, \delta_b^{(S)}) , & a^{(S-1)} &\in GF(2)^{k_{S-1}-k_S}, \\
B_{a^{(1)}\ldots a^{(S-1)}}^{(S)} &= \bigcup_{a^{(S)}} \{b_{a^{(1)}a^{(2)}\ldots a^{(S)}}\} , & a^{(S)} &\in GF(2)^{k_S}.
\end{aligned}
\tag{1}
$$

Every codeword $b \in B^{(1)}$ is fully determined by its enumeration:

$$
(a^{(1)}, a^{(2)}, \ldots, a^{(S)}) \iff b \in B^{(1)} .
$$

As an example figure 1 shows the enumerations of QPSK and 8-PSK, which are equi-

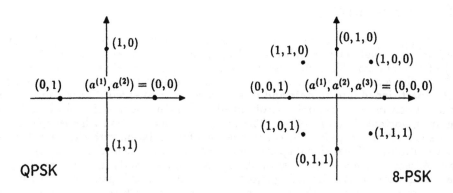

Figure 1: Enumeration of QPSK and 8-PSK

valent to the subset partitioning in [5] [11]. Binary codes can be partitioned in the same way using the Hamming distance instead of the squared euclidean distance. Only signal sets with $M = 2^k$ are considered here; $M = q^k$ can be treated in the same way.

Let us apply s outer codes $A^{(i)}(2^{k_i-k_{i+1}}; n_a, M_a^{(i)}, d_a^{(i)})$ of length n_a over $GF(2)^{k_i-k_{i+1}}$ to the enumerations $a^{(i)}, i = 1, \ldots, s$. $d_a^{(i)}$ denotes the minimum distance of $A^{(i)}$. One obtains a matrix:

$$\begin{pmatrix} a_1^{(1)} & a_1^{(2)} & \cdots & a_1^{(S)} \\ a_2^{(1)} & a_2^{(2)} & \cdots & a_2^{(S)} \\ \vdots & \vdots & \ddots & \vdots \\ a_{n_a}^{(1)} & a_{n_a}^{(2)} & \cdots & a_{n_a}^{(S)} \end{pmatrix} \Longleftrightarrow \begin{pmatrix} b_1 \\ b_2 \\ \vdots \\ b_{n_a} \end{pmatrix} = c \in C$$

where the i-th column of the matrix is a codeword of $A^{(i)}$. Every row determines one codeword $b_j \in B^{(1)}, j = 1, \ldots, n_a$. Every codeword c of the new generalized concatenated code C contains n_a codewords of the inner code $B^{(1)}$. We have constructed a GC-code

$$C\left(\mathbb{R}^{2 \cdot n_a}; \prod_{i=1}^{S} M_a^{(i)}, \delta_c \geq \min_{i=1,\ldots,S}\{d_a^{(i)} \cdot \delta_b^{(i)}\}\right), \qquad (2)$$

see [13] and [7], Ch.18 §8.2, for the derivation of its minimum distance δ_c.

As an example we take QPSK as the inner code $B^{(1)}(\mathbb{R}^2; 2^2, \delta_b^{(1)} = 2)$ which is partitioned according to figure 1 into two subsets $B_0^{(2)}(\mathbb{R}^2; 2^1, \delta_b^{(2)} = 4) = \{(1,0),(-1,0)\}$ and $B_1^{(2)} = \{(0,1),(0,-1)\}$ which are enumerated by $a^{(1)} \in GF(2)$. The codewords in each subset are enumerated by $a^{(2)} \in GF(2)$. We have a second order partition.
Using binary block codes of length 4 as outer codes,

$A^{(1)}(2; 4, 2^3, 2)$,   even weight code,
$A^{(2)}(2; 4, 2^4, 1) = \mathbb{F}^4$,   uncoded,

we obtain the GC-code $C(\mathbb{R}^8; 2^7, \delta_c = 4)$.
Thus, each codeword consists of four QPSK signals.

2.2   Multiple Concatenation of Codes

Further concatenation of the code from eq.(2) with block codes leads to a multiple concatenated code. Let us denote the concatenated code from eq.(2) by $C^{(1)}$ and take it as a new inner code. Before concatenating $C^{(1)}$ with new outer codes we have to partition $C^{(1)}$ into subsets. This can be done in the following way. Subsets of $C^{(1)}$ are obtained by concatenation of the old inner code $B^{(1)}$ from eq.(1) with subsets of the old outer codes $A^{(i)}$. Thus, we take the same partition of the inner code $B^{(i)}$ as for the construction of $C^{(1)}$. We partition every old outer code $A^{(i)}$ into disjunct subsets $A_{j_i}^{(i)}$ which are enumerated by $j_i = 0, \ldots, m_i - 1$:

$$A^{(i)}(q; n_a, M_a^{(i)}, d_a^{(i)}) = \bigcup_{j_i=0}^{m_i-1} A_{j_i}^{(i)}(q; n_a, M_{a j_i}^{(i)}, d_{a j_i}^{(i)}), \quad i = 1, \ldots, s. \qquad (3)$$

Now, if we construct new concatenated codes using $A_{j_i}^{(i)}$ as outer codes instead of $A^{(i)}$ we obtain $\prod_{i=1}^{S} m_i$ different GC codes $C_j^{(2)}$ which are enumerated by j:

$$C_j^{(2)}\left(\mathbb{R}^{2 \cdot n_a}; M_c^{(2)} = \prod_{i=1}^{S} M_{a j i}^{(i)}, \delta_c^{(2)} \geq \min_{i=1,\ldots,S}\{d_{a j i}^{(i)} \cdot \delta_b^{(i)}\}\right), \quad j = 1, \ldots, \prod_{i=1}^{S} m_i.$$

We have $C_j^{(2)} \subset C^{(1)}$, since $A_{j_i}^{(i)} \subset A^{(i)}$. Thus, we have partitioned $C^{(1)}$ into subcodes $C_j^{(2)}$. Since the subsets $C_j^{(2)}$ shall have a higher minimum distance than $C^{(1)}$, $\delta_c^{(2)} > \delta_c^{(1)}$, the partitions of the $A^{(i)}$ should be chosen in such a way that
$\min_{i=1,\ldots,S}\{d_{a_{j_i}}^{(i)} \cdot \delta_b^{(i)}\} > \min_{i=1,\ldots,S}\{d_a^{(i)} \cdot \delta_b^{(i)}\}$. At the end of this section we show in an example how to partition the old outer codes $A^{(i)}$ into subcodes.
We have $\prod_{i=1}^{S} m_i$ different subsets. Instead of j the subcodes $C_{g^{(1)}}^{(2)}$ can be enumerated by $g^{(1)}$:

$$ C^{(1)} = \bigcup_{g^{(1)}} C_{g^{(1)}}^{(2)} , \quad g^{(1)} \in GF(2)^\mu , $$

for the special case of $\prod_{i=1}^{S} m_i = 2^\mu$. The enumeration $g^{(1)}$ of the subcodes can be done in different ways.

Higher order partitions of $A^{(i)}$ result in a higher order, say, t-th order partition of $C^{(1)}$ into subsets $C_{g^{(1)}\ldots g^{(i-1)}}^{(i)}(I\!R^{2 \cdot n_a}; M_c^{(i)}, \delta_c^{(i)}), i = 2,\ldots,t$. The t-th order partition of $C^{(1)}$ can now be concatenated with t new outer codes $G^{(i)}(2^{\mu_i}; n_g, M_g^{(i)}, d_g^{(i)})$. Form an $(n_g \times t)$ matrix of enumerations:

$$ \begin{pmatrix} g_1^{(1)} & g_1^{(2)} & \cdots & g_1^{(t)} \\ g_2^{(1)} & g_2^{(2)} & \cdots & g_2^{(t)} \\ \vdots & \vdots & \ddots & \vdots \\ g_{n_g}^{(1)} & g_{n_g}^{(2)} & \cdots & g_{n_g}^{(t)} \end{pmatrix} \Longleftrightarrow \begin{pmatrix} c_1 \\ c_2 \\ \vdots \\ c_{n_g} \end{pmatrix} = h \in H $$

where the i-th column of the matrix is a codeword of $G^{(i)}$. Every row determines one codeword $c_j \in C^{(1)}, j = 1,\ldots,n_g$. Every codeword h of the new concatenated code H contains n_g codewords of the inner code $C^{(1)}$. A twice concatenated code is obtained:

$$ H\left(I\!R^{2 \cdot n_a \cdot n_g}; M_h = \prod_{i=1}^{t} M_g^{(i)}, \delta_h \geq \min_{i=1,\ldots,t}\{d_g^{(i)} \cdot \delta_c^{(i)}\}\right) . \tag{4} $$

As an example we take the concatenated code $C(I\!R^8; 2^7, \delta_c = 4)$ from the example at the end of section 2.1. The partition is done by the partition of its outer codes. The first old outer code $A^{(1)}(2; 4, 2^3, 2)$ is partitioned into the repetition code $A_0^{(1)}(2; 4, 2^1, 4)$ and its three cosets $A_{j_1}^{(1)}$, $j_1 = 1,\ldots,3$. The second old outer code $A^{(2)}(2; 4, 2^4, 1)$ is partitioned into the even weight code $A_0^{(2)}(2; 4, 2^3, 2)$ and its coset, the odd weight code $A_1^{(2)}$. Concatenating $B^{(1)}$ with the outer codes $A_{j_i}^{(i)}$, $i = 1, 2$, one obtains $C_{g^{(1)}}^{(2)}(I\!R^8; 2^4, \delta_c^{(2)} = 8) \subset C^{(1)}(I\!R^8; 2^7, \delta_c^{(1)} = 4)$. Since we use $m_1 = 4$ subsets of $A^{(1)}$ and $m_2 = 2$ subsets of $A^{(2)}$ we obtain $m_1 \cdot m_2 = 8$ subsets $C_{g^{(1)}}^{(2)} \subset C^{(1)}$. Thus, they can be enumerated by $g^{(1)} \in GF(2)^3$. The enumeration $g^{(1)}$ of the subcodes can for example be obtained as a direct sum of the binary equivalents j_i^* of j_i as $g^{(1)} = (j_1^*|j_2^*) \in GF(2)^3$.
In a second step one fixed codeword $a_{j_1,l_1}^{(1)} \in A_{j_1}^{(1)}, l_1 = 0, 1$, is chosen and each $A_{j_2}^{(2)}$ is partitioned into four cosets $A_{j_2,l_2}^{(2)}, l_2 = 0,\ldots,3$, of the repetition code. Concatenating $B^{(1)}$ with $a_{j_1,l_1}^{(1)}, l_1 = 0, 1$, and $A_{j_2,l_2}^{(2)}, l_2 = 0,\ldots,3$, results in eight subsets $C_{g^{(1)}g^{(2)}}^{(3)}(I\!R^8; 2, 16) \subset C_{g^{(1)}}^{(2)}$. These subsets are enumerated by $g^{(2)} \in GF(2)^3$, e.g. by

the direct sum $g^{(2)} = (l_1^* | l_2^*)$. Finally, the two codewords in $C_{g^{(1)}g^{(2)}}^{(3)}$ are enumerated by $g^{(3)} \in GF(2)$.

One has obtained a third order partition of $C^{(1)}$:

$$c_{g^{(1)}g^{(2)}g^{(3)}} \in C_{g^{(1)}g^{(2)}}^{(3)}(I\!\!R^8; 2, 16) \subset C_{g^{(1)}}^{(2)}(I\!\!R^8; 2^4, 8) \subset C^{(1)}(I\!\!R^8; 2^7, 4) \ .$$

The third order partition of $C^{(1)}$ is concatenated with the following new outer codes:

$G_1^{(1)}(2^3; 8, 2^3, d_g^{(1)} = 8)$,   repetition code,

$G_1^{(2)}(2^3; 8, 2^{3 \cdot 7}, d_g^{(2)} = 2)$,   even weight code,

$G_1^{(3)}(2; 8, 2^8, d_g^{(3)} = 1) = I\!\!F^8$,   uncoded.

We obtain the twice concatenated code

$$H_1\left(I\!\!R^{64}; 2^{32}, \delta_{h1} \geq 16\right) \ . \tag{5}$$

Each codeword consists of 32 QPSK signals.

Choosing the outer codes:

$G_2^{(1)}(2^3; 16, 2^{3 \cdot 5}, d_g^{(1)} = 8)$,   3 extended BCH(16,5)-codes,

$G_2^{(2)}(2^3; 16, 2^{3 \cdot 15}, d_g^{(2)} = 2)$,   even weight code,

$G_2^{(3)}(2; 16, 2^{16}, d_g^{(3)} = 1) = I\!\!F^8$,   uncoded,

we obtain the twice concatenated code

$$H_2\left(I\!\!R^{128}; 2^{76}, \delta_{h2} \geq 16\right) \ . \tag{6}$$

Each codeword consists of 64 QPSK signals. (Note, that there would be a better code than $G_2^{(1)}$ but for the sake of easy decoding we have chosen the 3 extended BCH(16,5)-codes.)

Multidimensional signal spaces have recently been used for trellis coded modulation [12] [4]. Treating multidimensional signal spaces as concatenated codes makes it easier to partition them and to detect received multidimensional signals even for high dimensions! Taking a modulation as the inner code and choosing outer codes $A^{(i)} = I\!\!F^n$, a partition of the $2n$-dimensional signal space is obtained by simply partitioning the binary block codes $A^{(i)}$ of length n. The detection of the $2n$-dimensional signals is best done using decoding methods for generalized concatenated codes or by look-up tables for small n.

3   Decoding Procedure

In [13] a multi-stage decoding procedure for generalized concatenated codes in the euclidean space has been given. It uses erasure decoding of the outer codes. In [1] the decoding procedure was modified introducing soft decision decoding of the outer codes. This multi-stage decoding method is used here.

Let a codeword $c \in C$ according to eq. (2) be transmitted, $c = (b_1, b_2, \ldots, b_{n_a})$. Let $r = (r_1, r_2, \ldots, r_{n_a})$ be the received word (distorted by errors). The decoding procedure has s steps determining $(a_1^{(i)}, a_2^{(i)}, \ldots, a_{n_a}^{(i)})$ in the i-th step. dist_E denotes the squared euclidean distance.

Soft Decision Decoding

i-th step:
$a^{(1)}, \ldots, a^{(i-1)}$ have been determined in previous steps.

1. inner decoding: decode r_j, $j = 1, \ldots, n_a$, according to $B_{a^{(1)} \ldots a^{(i-1)}}^{(i)}$ as $\hat{b}_j$.
   Determine the second nearest codeword $\hat{b}_j^{(2)}$ and the distance difference
   $\Delta_j = \mathrm{dist}_E(\hat{b}_j^{(2)}, r_j) - \mathrm{dist}_E(\hat{b}_j, r_j)$.
2. obtain the enumeration $(a_j^{(1)}, \ldots, a_j^{(i-1)}, \hat{a}_j^{(i)}, \ldots, \hat{a}_j^{(S)})$ of $\hat{b}_j$, $j = 1, \ldots, n_a$.
3. outer decoding: soft decision decoding of $(\hat{a}_1^{(i)}, \hat{a}_2^{(i)}, \ldots, \hat{a}_{n_a}^{(i)})$ according to $A^{(i)}$
   as $(a_1^{(i)}, a_2^{(i)}, \ldots, a_{n_a}^{(i)})$ using $(\Delta_1, \Delta_2, \ldots, \Delta_{n_a})$ as soft decision information.

In this decoding procedure the determination of $\hat{b}_j^{(2)}$ has to be done for all $j = 1, \ldots, n_a$ in each step since $\hat{b}_j^{(2)}$ may differ from step to step even when $\hat{b}_j$ stays the same. Δ_j is a better reliability criterion than $\delta_j = \mathrm{dist}_E(\hat{b}_j, r_j)$ as shown in [1] [3].

In the case of a multiple concatenated code the inner code (which is a concatenated code itself) can be decoded using the method described above or the inner code can be maximum-likelihood-(ML-)decoded using a look-up table. In the case of the twice concatenated codes simulated in the next section the inner (concatenated) code was ML-decoded whereas the second concatenation was decoded using the above multi-stage decoding scheme.
Multiple concatenated codes are usually constructed using short outer codes. These outer codes can simply be ML-decoded by comparison of the detected word to all codewords.

4 Simulation Results

Computer simulations of twice concatenated codes have been performed in order to evaluate their error correction properties. An AWGN-channel was assumed for the simulations. The codes have been decoded according to section 3. The twice concatenated codes are compared to single concatenated codes of similar length and rate. QPSK modulation as well as 8-PSK modulation have been considered. The probability of QPSK and 8-PSK symbol errors has been determined for the various coded modulation schemes. The code $H_1(I\!R^{64}; 2^{32}, \delta_{h1} \geq 16)$ from eq.(5) has been simulated. A comparable single concatenated code is constructed by concatenating QPSK with the outer BCH-codes, $A^{(1)} = \mathrm{BCH}(2; 31, 2^{10}, 12)$ and $A^{(2)} = \mathrm{BCH}(2; 31, 2^{21}, 5)$. The code $C_1(I\!R^{62}; 2^{31}, \delta_{c1} \geq 20)$ is obtained [1]. C_1 was simulated in [1] using a soft-decision algorithm to decode the outer BCH-codes.
H_1 is decoded using a look-up table for the inner code $C^{(i)}$ and ML-decoding for the outer repetition and even weight codes $G_1^{(i)}$. Figure 2 shows the symbol error probability for C_1, H_1 and for uncoded modulation (BPSK). Both codes have a rate of 0.5. H_1 is found to be better than C_1. In addition H_1 requires less decoding effort than C_1. A code $H_1^*(I\!R^{56}; 2^{28}, \delta_{h*} \geq 16)$, which is similar to H_1 but uses outer codes $G^{(i)}$ of length 7, yields approximately equal results as H_1.

As a second example the code $H_2(I\!R^{128}; 2^{76}, \delta_{h2} \geq 16)$ from eq.(6) was used. The three extended $\mathrm{BCH}(2; 16, 2^5, 8)$ codes which form the outer code $G_2^{(1)}$ were ML-decoded independently from each other. A comparable single concatenated code is construc-

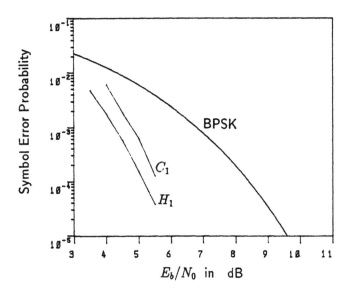

Figure 2: Symbol error probability of coded QPSK

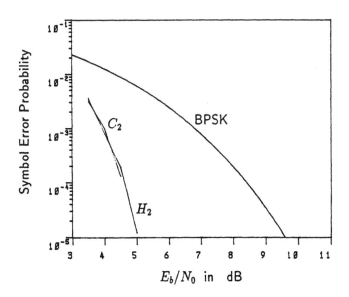

Figure 3: Symbol error probability of coded QPSK

ted by concatenating QPSK with the outer BCH-codes, $A^{(1)}$ =BCH(2; 63, 2^{16}, 23) and $A^{(2)}$ =BCH(2; 63, 2^{45}, 7). The code $C_2(I\!\!R^{126}; 2^{61}, \delta_{c2} \geq 28)$ is obtained [1]. C_2 was simulated in [1] .The results are plotted in figure 3. Both codes have a rate of approximately 0.5. Nearly identical symbol error probabilities are obtained for both codes whilst the decoding effort is much lower in case of the twice concatenated code H_2. In case of C_2 two BCH-codes of length 63 have to be soft-decision decoded whereas in case of H_2 the three BCH(2; 16, 2^5, 8) codes as well as the repetition and even weight code can be ML-decoded.

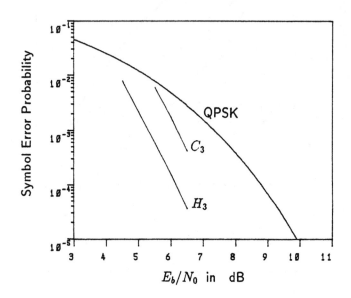

Figure 4: Symbol error probability of coded 8-PSK

A final example is given for 8-PSK modulation. Let us denote the 8-PSK signal set by $B^{(1)}(I\!\!R^2; 2^3, \delta_b^{(1)} = 0.586)$. As indicated in figure 1 and in [11] a third order partition of $B^{(1)}$ can be obtained as:

$$b_{a^{(1)}a^{(2)}a^{(3)}} \in B^{(3)}_{a^{(1)}a^{(2)}}(I\!\!R^2; 2, 4) \subset B^{(2)}_{a^{(1)}}(I\!\!R^2; 2^2, 2) \subset B^{(1)}(I\!\!R^2; 2^3, 0.586).$$

with the enumerations $a^{(1)}, a^{(2)}, a^{(3)} \in GF(2)$. Using the following codes as outer codes:
$A^{(1)}(2; 4, 2^1, 4)$,   repetition code,
$A^{(2)}(2; 4, 2^3, 2)$,   even weight code,
$A^{(3)}(2; 4, 2^4, 1) = I\!\!F^4$,   uncoded,
we obtain the concatenated code $C^{(1)}(I\!\!R^8; 2^8, \delta_c = 2.34)$. $C^{(1)}$ can be partitioned according to section 2.2 by using subsets of $A^{(i)}$ as outer codes.

$$c_{g^{(1)}g^{(2)}g^{(3)}} \in C^{(3)}_{g^{(1)}g^{(2)}}(I\!\!R^8; 2^4, 8) \subset C^{(2)}_{g^{(1)}}(I\!\!R^8; 2^7, 4) \subset C^{(1)}(I\!\!R^8; 2^8, 2.34)$$

We obtain an enumeration $g^{(1)} \in GF(2)$; $g^{(2)} \in GF(2)^3$; $g^{(3)} \in GF(2)^4$. The code $C^{(1)}$ is concatenated with the following new outer codes,
$G_3^{(1)}(2; 8, 2^4, d_g^{(1)} = 4)$,   extended Hamming code,
$G_3^{(2)}(2^3; 8, 2^{3 \cdot 7}, d_g^{(2)} = 2)$,   even weight code,

$G_3^{(3)}(2^4; 8, 2^{4 \cdot 8}, d_g^{(3)} = 1),$ uncoded,

which are applied to the enumerations $g(i)$. One obtains the twice concatenated code

$$H_3 \left(I\!\!R^{64}; 2^{57}, \delta_{h3} \geq 8 \right) .$$

Its outer codes, $G_3^{(i)}$, are ML-decoded.

A comparable single concatenated code $C_3(I\!\!R^{62}; 2^{57}, \delta_{c3} = 4)$ was constructed in [1] by concatenating 8-PSK with the outer codes $A^{(1)} = \text{BCH}(2; 31, 2^5, 16)$ and $A^{(2)} = \text{BCH}(2; 31, 2^{21}, 5)$ and $A^{(3)}(2; 31, 2^{31}, 1)$ (uncoded). C_3 was simulated in [1]. In figure 4 the symbol error probability is shown for both codes, C_3 and H_3. Both codes have a rate of approximately 0.6. H_3 is found to be much better than C_3.

5 Conclusions

The construction of generalized multiple concatenated codes in the euclidean space has been described. Twice concatenated codes were designed for QPSK and 8-PSK modulation. Computer simulations have been performed in order to determine their error correction properties. The symbol error probability of the twice concatenated codes is found to be better than that of single concatenated codes of about equal length and rate. At the same time the twice concatenated codes require much less decoding effort. In contrast to the single concatenated codes they use only short block codes which can be ML-decoded.

References

[1] Bossert, M., *Concatenation of Block Codes*, Deutsche Forschungsgemeinschaft Report (1988).

[2] Calderbank, A.R., *Multi-Level Codes and Multi-Stage Decoding*, IEEE Trans. Comm. 37 (1989) 222–229.

[3] Forney Jr., G.D., *Generalized Minimum Distance Decoding*, IEEE Trans. Inf. Theory 12 (1966) 125–131.

[4] Forney Jr., G.D., Gallager, R.G., Lang, G.R., Longstaff, F.M. Qureshi, S.U., *Efficient Modulation for Band–Limited Channels*, IEEE J. Sel. Areas in Comm. 2 (1984) 632-647.

[5] Imai, H., Hirakawa, S., *A New Multilevel Coding Method Using Error-Correcting Codes*, IEEE Trans. Inf. Theory 23 (1977) 371–377.

[6] Kasami, T., Takata, T., Fujiwara, T., Lin, S., *A Concatenated Coded Modulation Scheme for Error Control*, IEEE Trans. Comm. 38 (1990) 752–763.

[7] MacWilliams, F.J., Sloane, N.J.A., *The Theory of Error-Correcting Codes* (North Holland, New York, 1977).

[8] Pottie, G.J., Taylor, D.P., *Multi-Level Channel Codes Based on Partitioning*, IEEE Trans. Inf. Theory 35 (1989) 87–98.

[9] Sayegh, S.I., *A Class of Optimum Block Codes in Signal Space*, IEEE Trans. Comm. 30 (1986) 1043–1045.

[10] Takata, T., Ujita, S., Fujiwara, T., Kasami, T., Lin, S., *Linear Structure and Error Performance Analysis of Block Coded PSK Modulation Codes*, Trans. of IEICE of Japan J73-A (1990).

[11] Ungerboeck, G., *Channel Coding with Multilevel/Phase Signals*, IEEE Trans. Inf. Theory 28 (1982) 55–66.

[12] Wei, L.-F., *Trellis-Coded Modulation with Multidimensional Constellations*, IEEE Trans. Inf. Theory 33 (1987) 483–501.

[13] Zinoviev, V.A., Zyablov, V.V., Portnoy, S.L., *Concatenated Methods for Construction and Decoding of Codes in Euclidean Space*, Preprint, USSR Academy of Science, Institute for Problems of Information Transmission (1987).

TRELLIS CODED MODULATION BASED ON TIME-VARYING MAPPING AND ENCODERS FOR UTILIZING A CHANNEL INTERSYMBOL INTERFERENCE

Ryuji Kohno, Choonsik Yim, and Hideki Imai

Division of Electrical and Computer Engineering
Faculty of Engineering
Yokohama National University
156 Tokiwadai, Hodogaya-ku, Yokohama JAPAN 240

ABSTRACT   This paper proposes and investigates a trellis coded modulation using time-varying mapping and encoders which is very useful to increase minimum Euclidean free distance(d_{free}) and obtain large coding gain by intentionally utilizing intersymbol interference(ISI) in a channel. If M-ary modulation signals are transmitted in a channel with ISI, the received signals will be mapped into more than M signal points due to ISI,i.e. a channel memory. In the presence of ISI, optimum allocation of distorted signal points due to ISI is derived so as to maximize Euclidean distance in the signal points at every instant considering the trellis diagram. In order to maintain the allocation in the state transition, a method to utilize time-varying mapping and encoders is described. In particular, the coded 4PSK using the time-varying mapping and encoders with a code rate $\frac{1}{2}$ in the presence of ISI, $1 + \frac{1}{2}D$ can achieve much larger d_{free} and lower BER than the Ungerboeck's coded 4PSK using the time-invariant mapping and encoder in a channel without ISI, providing that their decoding complexity is equivalent. The improvement can increase as a constraint length of the encoder.

1. INTRODUCTION

In communication systems, we have studied such techniques and theory that originated in the interplay between coding theory and signal processing, which have different origins and routes of development because of their different algebraic fields. Combined techniques between coding theory and signal processing can improve the system performance, such as combination of coding & modulation (coded modulation), equalizing & decoding, and so on [1][2][3].

Coded modulation is a useful technique to improve BER performance without spreading signal frequency bandwidth. Trellis coded modulation (TCM), that is combination of a trellis code on a finite field and multilevel/phase modulation[4], has been applied to a voice-band data modem with over 14.4kb/s and considered for mobile communication. In general, bit error rate (BER) performance of coded modulation scheme is determined by the minimum free Euclidean distance, d_{free} that is derived by a trellis diagram in decoding and a code rate and a constraint length of encoder,i.e. memory in an encoder. In a channel with intersymbol interference (ISI), d_{free} will be able to be increased if we utilize ISI,i.e. memory in a channel as well as that in an encoder. Most of practical channels have a memory such as a telephone subscriber, a mobile radio communication, and a magnetic recording channels, where ISI due to a channel memory is used to be eliminated by an equalizer and so on.

The purpose of this study is to utilize ISI due to a channel memory combining with an encoder memory in order to increase d_{free} of a coded modulation. We assume that an adaptive equalizer is used for the purpose of obtaining a desired ISI for the coded modulation. In a conventional approach to design a coded modulation, a fixed pair of an encoder and mapping has been selected among available encoders and mapping without considering ISI. In this paper, before selecting an encoder and mapping we first derive such a good allocation of received signal points as to maximize Euclidean distance of each a pair of branches of paths converging to the same state in a trellis diagram at every instant. If such Euclidean distance of a pair of branches at each instant is maximized considering distortion of received signal points due to ISI, d_{free} will be able to be maximum.

There is not always such a fixed pair of an encoder and mapping that the received signal points can be satisfied with the good allocation. In order to realize such a good allocation of received signal points by combination of encoder, mapping and ISI, we propose a trellis coded modulation system using time-varying mapping and encoder. Some proper pairs of encoders and mapping are switched periodically in the system.

In the following sections, first a model of a coded modulation system in a channel with ISI is described. In section 3, a trellis coded modulation system using time-varying mapping and encoder is proposed. We mention the good allocation of distorted signal points as to maximize d_{free}. In section 4, we evaluate the proposed system in comparison with a conventional system. d_{free} and theoretical BER corresponding to various ISI models, contraint lengths and code rates are derived. Moreover, BER performance based on computer simulation is shown. Finally, it is confirmed that the proposed system in a channel with ISI can achieve much larger d_{free} and lower BER than the Ungerboeck's TCM system in a channel without ISI, providing that their decoding complexity is equivalent.

2. A TRELLIS CODED MODULATION SYSTEM IN THE PRESENCE OF INTERSYMBOL INTERFERENCE

Fig.1 shows a model of a trellis coded modulation system in a channel with ISI. Data are encoded by a trellis encoder and the output codewords are mapped into modulated signal points in order to increase effective Euclidean distance between signal points, that is a trellis coded modulation (TCM). In general, TCM can improve BER performance without spreading signal frequency bandwidth. Transmitted signals are distorted in a channel by ISI and additive white Gaussian noise (AWGN). Then, received signals are decoded by Viterbi algorithm. Even if a pulse signal having no interference between adjacent sampling points is transmitted, a received signal will be distorted as shown in Fig.1.1 (a), where T is a sampling interval. We call it ISI and express ISI of (a) by a shift register model shown in (b). In general, ISI can be written by the polynomial,

134

$$H(D) = 1 + \sum_{i=1}^{N} a_i D^i, \tag{1}$$

where D is a delay operator. By using the same D, a convolutional encoder with a code rate of $R = \frac{1}{2}$ and a contraint length of $CL = 3$ shown in Fig.1.2 is written by

$$\begin{cases} g_1(D) = 1 + D + D^2, \\ g_2(D) = 1 + D^2. \end{cases} \tag{2}$$

The relationship between transmitted and received signal points is illustrated by Fig.1.3 in the case of 4PSK and ISI of $H(D) = 1 + \frac{1}{2}D$. The received signal points have no noise, but are distorted due to the ISI. In this case, the received signal points result in those of 16QAM. Since total memory of an encoder and a channel is 3 in the case of $CL = 3$ and $H(D) = 1 + aD$, received signals are decoded by using trellis diagram with 8 states shown in Fig.1.4, where S_{ij} means that i and j are a present and a previous states of the encoder. In this decoding, we can consider a pair of a present and a previous states of the encoder as a state. The decoding complexity of the above-mentioned case is two times as much as that in the case of a channel without ISI. Provided that their decoding complexity is equivalent, the TCM systems in a channel with ISI and without ISI should be compared under the condition of the same total memory of an encoder and a channel. The TCM system using an encoder of $R = \frac{1}{2}$, $CL = 4, 5$ and 4PSK in a channel without ISI has maximum $d^2_{free} = 12.0, 14.0$, respectively, while the TCM system using an encoder of $R = \frac{1}{2}$, $CL = 3, 4$ and 4PSK in a channel with ISI, $H(D) = 1 + \frac{1}{2}D$ achieves maximum $d^2_{free} = 12.5, 15.0$, respectively.

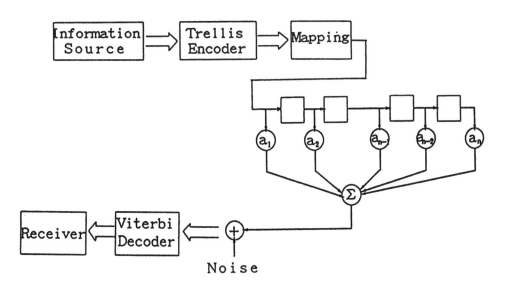

Fig.1 A coded modulation system in a channel with ISI

135

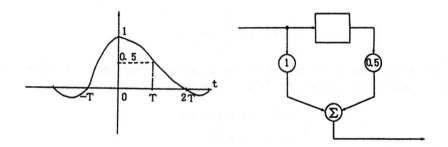

Fig.1.1 An example of intersymbol interference, ISI

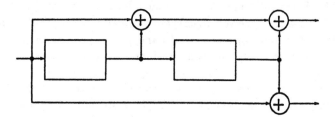

Fig.1.2 A convolutional encoder with R=1/2 and CL=3

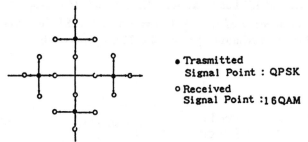

● Trasmitted
  Signal Point : QPSK
○ Received
  Signal Point :16QAM

Fig.1.3 Transmitted and received signal points in the case of 4PSK and ISI=1+0.5D

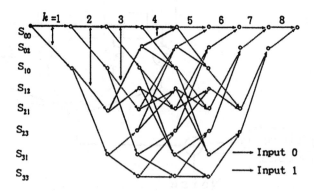

Fig.1.4 Trellis diagram for decoding a received signal from an encoder with CL=3
through a channel with ISI=1+aD

3. TRELLIS CODED MODULATION SYSTEM USING TIME-VARYING MAPPING AND ENCODER

If we use a TCM system in a channel with ISI not considering proper utilization of memory in an encoder and a channel, d_{free} will not be increased so long. In this section, we first consider such optimum allocation of distorted signal points by ISI as to maximize d_{free}. In order to realize the allocation, we propose utilization of time-varying mapping and encoder. For the sake of simplicity, we explain the proposed idea by using the case of a code rate $R = \frac{1}{2}$, 4PSK, and ISI $H(D) = 1 + \frac{1}{2}D$. However, it can be easily extended to a general case.

3.1 Maximization of Minimum Free Euclidean Distance

Fig.2 shows a model of the proposed TCM system using time-varying mapping and encoder for a channel with ISI. In a conventional system, an equalizer is employed to eliminate or compensate for ISI. In the proposed system, however, an adaptive equalizer is used in order to obtain a desired ISI for combination of memory of an encoder and a channel.

Let a constraint length of encoder be CL. Since the number of states of the encoder is 2^{CL-1} and the number of memory in the channel is $N = 1$, the total number of states is 2^{CL}. In Fig.1.4, $CL = 3$ and there are $2^3 = 8$ states. A pair of paths first converge to the same state at instant $CL + 1$ in a trellis diagram. Each a branch corresponds to a signal point. If distorted signal points due to ISI are allocated so as to maximize Euclidean distance between each a pair of branches of the converged paths at every instant, d_{free} will be maximum.

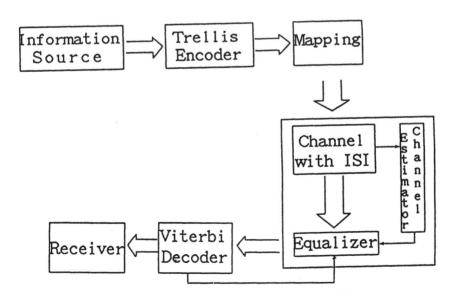

Fig.2 Trellis Coded Modulation Scheme Using Time-varying Mapping and Encoder for a Channel with ISI

Therefore, we consider such optimum allocation of distorted signal points due to ISI. Fig.3 (i) and (ii) show groups of the distorted signal points having the same output at the present and the previous instant, respectively. The optimum allocation of distorted signal points due to ISI can be determined as shown in Fig.4 by considering these groups and the trellis diagram in Fig.1.4. In Fig.4, a pair of connected signal points by an arrow correspond to a pair of branches of the two paths converging to the same state in Fig.1.4. At instant $k = 1$, the pair of signal points in Fig.4 achieves maximum distance between a pair of branches in Fig.1.4 corresponding to the signal points, because the pair of signal points has longest distance in Fig.3(ii). At instant $k = 2$, two pairs of signal points, which correspond to two pairs of branches converging to the same state in Fig.1.4, are selected so as to have longest distance as shown in Fig.4 considering Fig.3(ii). At instant $k \geq 3$, optimum pairs of signal points are determined in the same way.

If $CL \geq 4$, the number of paths will be 2^k at instant k of $4 \geq k \geq CL$. Eight paths converging to the same state at instant $CL + 3$ can be considered as a group and eight signal points corresponding to eight branches of the paths can be allocated to those of (iii) in Fig.4. If some combination of encoder and mapping realizes such optimum allocation of distorted signal points due to ISI, d_{free}^2 will be written by

$$
d_{free}^2 = 2^2 + \sqrt{5}^2 + \sqrt{5}^2 (CL - 2) + 1^2
$$
$$
= 5(CL - 1) + 10. \tag{3}
$$

However, a fixed pair of an encoder and mapping cannot always realize such allocation.

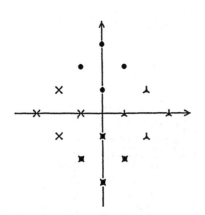

(i)Groups of Signal Points having the Same Output at the Present instance

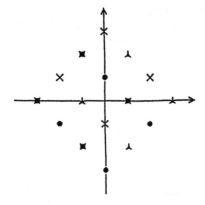

(ii)Groups of Signal Points having the Same Output at the Previous instance

Fig.3 Classification of groups of signal points

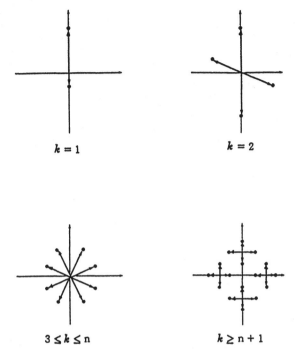

$k = 1$ $k = 2$

$3 \leq k \leq n$ $k \geq n + 1$

Fig.4 Combination of received signal points as to maximize the distance at each instant

3.2 Time-varying Mapping and Encoder for Utilizing ISI

In order to realize the above-mentioned optimum allocation of signal points, we propose to use some encoders and mapping which are switched periodically, i.e. time-varying encoder and mapping. We can derive the following conditions concerning encoders and mapping by considering transition of the allocation in Fig.4.

(1) Instance $1 \leq k \leq 3$

In order to realize the allocation in Fig.4 (i)(ii)(iii), we should use the encoder with the generator polynomial written as

$$\begin{cases} g_1(D) = 1 + D + D^2 + \cdots \\ g_2(D) = 1 + D^2 + \cdots \end{cases} \quad (4)$$

Mapping should be Gray mapping at instant 1, set partitioning at instant 2, and rotated Gray mapping by $\frac{\pi}{2}$ at instant 3.

(2) Instance $4 \leq k \leq (CL - 1)$ if $CL \geq 5$

In order to realize (iii), we should use the encoder with the polynomial written as

$$\begin{cases} g_1(D) = 1 + D^{k-3} + D^{k-2} + D^{k-1} + \cdots \\ g_2(D) = 1 + D^{k-3} + D^{k-1} + \cdots \end{cases} \quad (5)$$

Mapping should be Gray mapping.

(3) Instance $k = CL$

In order to realize (iii), we should use the encoder with the polynomial written as

$$
\begin{cases}
g_1(D) = 1 + D^{CL-2} + D^{CL-1} + \cdots \\
g_2(D) = 1 + D^{CL-1} + \cdots
\end{cases}
\tag{6}
$$

Mapping should be rotated Gray mapping by $\frac{\pi}{2}$.

(4) Instance $(CL + 1) \leq k$

There is no restriction in an encoder and mapping in order to realize (iv).

In the case of $ISI = 1 + \frac{1}{2}D$ and a code rate $R = \frac{1}{2}$, the time-varying mapping and encoding can realize such an optimum allocation as to increase d_{free} as long as possible. In more general case, the time-varying mapping and encoding will realize such allocation as to have longer d_{free} than a fixed pair of mapping and an encoder.

3.3 Combination of the Proposed TCM and Equalization

An adaptive equalizer can provide the desired ISI and channel measurement information to the proposed TCM, while the decoder can provide the error-corrected data and reliability of the decoded data to the equalizer. If decoded data and their reliability are utilized to update filter coefficients of the equalizer, stable adaptation of equalizer will be realized because misadjustment of the filter coefficients can be reduced. We have presented some schemes of combining an equalizer and a decoder that can improve the whole performance by interactively exchanging information between a decoder and an equalizer [2][3]. The schemes are available to the proposed TCM as well.

In a channel with undesired ISI, without ISI, and time-varying ISI such as a mobile communication channel, an adaptive equalizer can provide the desired ISI in a receiver. Moreover, an equalizer can provide the desired ISI in a transmitter so as to be used for the proposed TCM. Such system should be compared with a conventional one having the same average transmission power and decoding complexity.

4. SYSTEM EVALUATION

In this section, d_{free} and BER performance are derived to evaluate the proposed system in comparison with a conventional TCM system.

4.1 Minimum Free Distance Corresponding to Constraint Length and ISI Length

In the previous section, we explained the method to utilize ISI by using the case of modulation: 4PSK, a code rate: $R = \frac{1}{2}$, a constraint length: CL, and ISI model: $H(D) = 1 + \frac{1}{2}D$. Table 1 shows d_{free}^2 of the proposed and a conventional TCM systems for the ISI model

$$
H(D) = 1 + \sum_{i=0}^{n} a^{2^i} D^{2^i},
\tag{7}
$$

where n and a are defined as ISI length and ISI coefficient.

Table 1. d_{free}^2 of the proposed TCM

CL	NO ISI	a	n=0	n=1	n=2
3	10.0	0.2	13.08	14.09	12.21
		0.5	15.0	16.75	17.62
		0.8	19.08	26.33	27.45
4	12.0	0.2	17.44	15.12	13.25
		0.5	20.0	18.0	19.62
		0.8	26.24	32.89	31.01
5	14.0	0.2	21.8	16.16	14.41
		0.5	25.0	19.25	21.62
		0.8	32.8	39.45	34.61
6	16.0	0.2	26.16	17.19	15.57
		0.5	30.0	20.5	23.62
		0.8	38.36	46.0	38.17

Although general ISI model is written by (1), it is clear that ISI model written by (7) is more useful to increase d_{free}. In Table 1, no ISI means the TCM system in a channel without ISI, i.e. the conventional system. Provided that decoding complexity is equivalent, the proposed and the conventional systems having the same total memory in an encoder and a channel should be compared. For instance, the proposed systems in the case of $CL = 3$ and $n = 0$ have the same total memory as that of the conventional one in the case of $CL = 4$, while the proposed systems have larger d_{free} than that of the conventional one. d_{free} increases as ISI coefficient a. Even if ISI length n increases, d_{free} will not always increase. Consequently, we confirm that the proposed system has much larger d_{free} than the conventional one which is not utilizing ISI in a channel.

In the proposed scheme, ISI has a role of "natural amplifier". We have interest in performance of coded modulation schemes when they have equal average power of transmitted signals and decoding complexity. Even if transmitted 16QAM signals are received through ISI channel in a conventional TCM scheme, ISI will not be utilized but equalized to be detected as 16QAM signals. The decoding complexity is equal to the proposed TCM but the average power of transmitted signals is higher than that of the proposed TCM. Therefore, we did not compare the proposed 4PSK-TCM with a conventional 16QAM-TCM.

4.2 BER Performance Corresponding to Constraint Length, ISI Coefficient and ISI length

Although d_{free} increases as ISI coefficient a in Table 1, BER will not always decrease in general because BER depends on distribution of paths in a trellis diagram. BER of the conventional system in a channel without ISI, $P_{e,1}$ is satisfied with

$$P_e \leq \sum_{d=d_{free}}^{\infty} Q(\frac{d}{\sqrt{2}\sigma}) \equiv \sum_{d=d_{free}}^{\infty} P_{e,1}(d), \tag{8}$$

where σ is a standard deviation of AWGN and $Q()$ is a Gaussian error function.

On the other hand, BER of the proposed system in a channel with ISI, P_e is satisfied with

$$P_e \leq \sum_{d=d_{free}}^{\infty} \left[2\sqrt{\frac{1}{2}Q(\frac{a^n E_s}{\sigma}) + \frac{1}{n}\sum_{l=0}^{n-1}\frac{1}{2^{n-l}}\sum_{i=0}^{n-l}Q(\frac{X_i E_s}{\sigma})} \right]^d \equiv \sum_{d=d_{free}}^{\infty} P_{e,2}(d), \quad (9)$$

where E_s is a signal energy, i.e. $E_s = 1$, σ^2 is noise variance,

$$X_i = \sum_{k=0}^{n-l} c_{i,k} a^{k+l}, \quad (10)$$

and for $i = 1, 2, \cdots, 2^n$,

$$c_{i,k} = \begin{cases} 1, & k = 1 \\ (-1)^{\lfloor \frac{i-1}{4} \rfloor}, & k = 2, 3, \cdots, n+1 \end{cases} \quad (11)$$

For the sake of simplicity, $P_{e,1}(d_{free})$ and $P_{e,2}(d_{free})$ are plotted as approximate upper bounds of (8) and (9) in Figs.5 and 6. Fig.5 shows BER vs. S/N ratio performance of the proposed system using ISI of $H(D) = 1 + \frac{1}{2}D$ and the conventional one without ISI. The proposed system improves BER over the conventional one.

Fig.6 shows BER vs. ISI coefficient a performance of the proposed system using ISI of $H(D) = 1 + \sum_{i=0}^{n} a^{2^i} D^{2^i}$ corresponding to a constraint length CL, where S/N ratio = $5.0dB$. From this figure, we note that optimum ISI coefficient and ISI length are $a = \frac{1}{2}$ and $n = 0$; optimum ISI model is $H(D) = 1 + \frac{1}{2}D$ in the case of a code rate $R = \frac{1}{2}$ and 4PSK, providing that the total memory of encoder and channel or decoding complexity is equivalent.

4.3 Computer Simulation

Since BER depends on distribution of paths in a trellis diagram as well as d_{free}, computer simulation is necessary to obtain exact BER performance. In simulation, BER is calculated as the number of errors within 100,000 data bits. A code rate R is $\frac{1}{2}$ and codewords are mapped into signal points of 4PSK. Fig.7 shows BER vs. S/N Ratio of the proposed, the conventional systems in a channel with ISI of $H(D) = 1 + \frac{1}{2}D$ and the conventional one in a channel without ISI. It is noted that the proposed system using time-varying mapping and encoder can achieve S/N improvement of about $1.8dB$ at BER=10^{-3} over the conventional systems. Therefore, time-varying mapping and encoder are important to improve BER performance.

5. CONCLUSION

We discussed utilizing memory in a channel,i.e. ISI as well as that in an encoder in order to increase minimum free distance, d_{free} of coded modulation. First, we approached to find such good allocation of distorted signal points due to ISI as to maximize

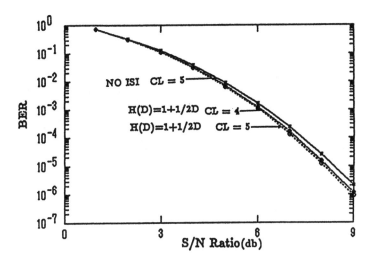

Fig.5 BER v.s. S/N Ratio of the proposed and the conventional systems: ISI $H(D) = 1 + 1/2D$

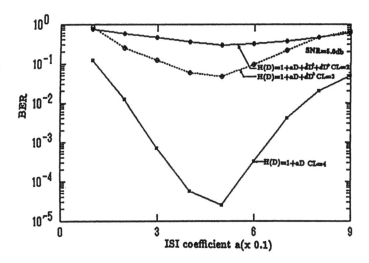

Fig.6 BER v.s. ISI coefficient a corresponding to constraint length CL: ISI $H(D) = 1 + \sum_{i=0}^{n} a^{2^i} D^{2^i}$

Euclidean distance of the distorted signal points at every instant, which correspond to branches of the paths converging to the same state in a trellis diagram. Subsequently, in order to realize the allocation, time-varying mapping and encoder that are periodically switched combination of mapping and encoders were proposed. The condition switching mapping and encoders was described. d_{free} and BER were calculated in some combination of codes and ISI models. We investigated the desired ISI model that the proposed scheme can perform large d_{free} and low BER.

In particular, it was noted that the coded 4PSK using the time-varying mapping and encoder with a code rate $\frac{1}{2}$ in the presence of ISI $1 + \frac{1}{2}D$ can achieve much larger d_{free} and lower BER than the Ungerboeck's coded 4PSK using the time-invariant mapping and encoder in a channel without ISI, providing that their decoding complexity is equivalent. The improvement can increase as a constraint length of the encoder. Therefore, this paper clarified that the time-varying mapping and encoders is important to utilize total memory of an encoder and a channel for decoding or correcting errors.

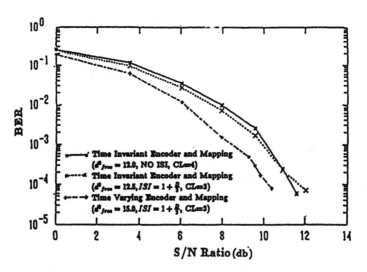

Fig.7 BER v.s. S/N Ratio of the proposed and the conventional systems: ISI $H(D) = 1 + 1/2D$ and $CL = 3, 4$

REFERENCE

[1] Imai.H and Hirakawa.S.,"A new multilevel coding method using error-correcting codes," IEEE Trans. Inform. Theory, vol.IT-23, May 1975.

[2] Kohno.R, Imai.H, and Hatori.M,"Design of an automatic equalizer including a decoder of error-correcting code," IEEE Trans. Commun. vol.COM-33, pp.1142-1146, 1985.

[3] Kohno.R, Imai.H, and Pasupathy.S,"An automatic equalizer including a Viterbi decoder for a trellis coded modulation system," Proc. IEEE International Conference on Acoustics, Speech and Signal Processing, pp.1368-1371, 1989.

[4] Ungerboeck.G.,"Channel coding with multilevel/phase signals," IEEE Trans. Inform. Theory, vol.IT-28,1,pp.55-67, Jan. 1982

USE OF THE ALGEBRAIC CODING THEORY IN NUCLEAR ELECTRONICS

N.M. Nikityuk
Joint Institute for Nuclear Research, Dubna USSR6
Moscow, Head Post Office, P.O. Box 79

Introduction

The method of coding previously suggested in paper [1] used for the creation of fast tracking processors [2,3], parallel encoders with wide functional possibilities and economic majority coincidence units for large number of inputs n > 30 [3-5]. These devices are algebraic in structure and have a number of advantages with the usual method of construcing special-purpose processors (SP) for fast event selection registered in multichannel charged particle detectors (MCPD). These advantages are due to the of finite field algebra and, in particular, Galois field algebra [3-5], where multiplication, division and power raising are executed simpler than in the field with a position number system.. New results of studies of the development and use of the syndrome codind method in nuclear electronics are described.

I. New type of time-to-digital converters

Along with parallel methods of data registration and data processing in MCPD, sequential methods are widely used which differ in economic and simple registration electronics. Among the are counters, shift registers and circular counters. Shift registers are used in systems of data

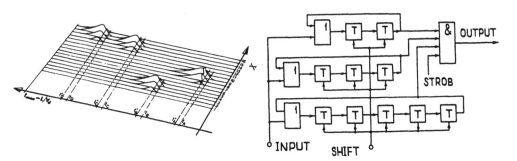

Fig.1. Development jf signals in drift chamber when multi-track events are registered

Fig.2. Principal scheme of the counters in the resudial systems.
T - triggers, 1 - OR,   - AND

registration from drift chambers [6]. However, shift registers containing some hundred bits are required to register several time intervals and such systems are not effective for a large number of channel registrations. There are systems where RAMs are used as shift

registers [7], but the time resolution T_R for such devices is large. As shown below, coders and decoders used in technical error-correcting codes [8] and in signature analysers [9] can be applied for the creation of effectivelly time-to-digital converters of the 2start-stop2 type. It

should be noted that timing measurements been allways an important in particle physics experiments. These measurements allow multitrack events to be registered. As an example, the development of signals in drift chamber in time is shown in fig.1. The tracks of charged particles can be restored if we know three coordinates of events5 X, Y and t_s - the propagation time of signals realative to start pulse at the ends of the detector [10].

1. *Circular counters in the resudual system.* If a shift register having n bits is divided into L unequal parts such that p_1, p_2,p_3 ..p_1 are the numbers of bits in each part respectively, then an error-correcting counter can be constructed in residual system. The numbers p_1 are mutually simple ones. The compression coefficient of such a counters equals the product of p_1 divided on by their sum. Fig. 2 gives the scheme of such a counter in the residual system with $p_1 = 2$, $p_2 = 3$ and $p_3 = 5$ (M = 30 and $k_c = 30310$). If one more on module $p_4 = 7$ is added, $K_c = 210317$ = 12 [11]. The operation of the counter is explained in table 1, where the states of the triggers over one period of M = 30 shift pulses are presented. The signal of event is supposed to coincide with the first synchropulse. As it follows from table 1 the signals at the inputs of the AND element coincide after the 30-th synchropulse. In other words the scheme in fig. 2 executes the function of digital delay.

Table 1

State of the schift registes

Triggers Time	1	2	3	4	5	6	7	8	9	10
1	1	0	1	0	0	1	0	0	0	0
2	0	1	0	1	0	0	1	0	0	0
3	1	0	0	0	1	0	0	1	0	0
4	0	1	1	0	0	0	0	0	1	0
5	1	0	0	1	0	0	0	0	0	1
6	0	1	0	0	1	1	0	0	0	0

Morever, table 1 can be considered semiltaneously as a coding matrix H^T, where rows are changed by columns. Such matrix have interesting properties. Two coding matrix $H^T_{30,10}$ and $H^T_{217,17}$ are shown in fig 3. As according to the syndrome coding method, information bits equal to zero, we use the theorem [12] to evaluate the properties of the coding matrix5 a linear (n,k)- code with parity-check matrix $H^T_{n,\gamma} = [h_0,h_1,...$ $.h_{n-1}$ where h_1 - are vector-columns, 1 =

7	1 0	1 0 0	0 1 0 0 0
8	0 1	0 1 0	0 0 1 0 0
9	1 0	0 0 1	0 0 0 1 0
10	0 1	1 0 0	0 0 0 0 1
11	1 0	0 1 0	1 0 0 0 0
12	0 1	0 0 1	0 1 0 0 0
....
....
....
28	0 1	1 0 0	0 0 1 0 0
29	1 0	0 1 0	0 0 0 1 0
30	0 1	0 0 1	0 0 0 0 1

0, 1,2 ..n-1, with dimensions (n-k)x1 and γ = n-k parity check bits, has maximum coding distance d when any d-1 columns of of the matrix $H_{n,\gamma}^T$ are linear independed. Therefore for synonymous registration two time intervals with the help of the above scheme it is necessary to have four mutually linearly independent columns in the matrix $H_{30,10}^T$ as d = 5 for t = 2. This theorem can be also interpreted as follows. For synonymous registration two independent events (t = 2), it is nessasary and enough the modulo 2 sum or best the bullean sum of any two columns of the coding matrix $H_{n,\gamma}^T$ should be different. The calculations show that this condition is fullfield except some sums. For example, the 2-nd, 3-d, and 17-th and 18-th columns of the matrix $H^T{}_{30,10}$ are added according to the sum rules, we obtain a similar result

2-nd column → 0101001000   17-th column → 1001001000
3-d column → 1000100100   18-th column → 0100100100
———————   ———————
1101001100   1101101100

Since all columns of the matrix $H^T{}_{30,10}$ are different and the number of parity bits is equal to 10, then this matrix represents a superimposed correcting code with 4 < d < 5 . If it an extended matrix is constructed by adding one more module $p_4 = 7$, then we get the coding matrix $H^T{}_{210,17}$,

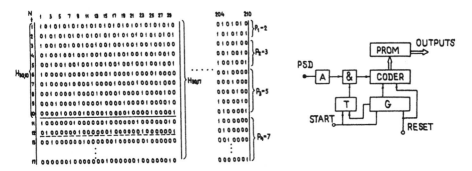

Fig.3. *Coding matrix for super-*
imposed codes

Fig.4. *Block-diagram of the time-to-*
digital converters. PSD - position-
sensitive detector, T - trigger, A-
amplifier, G - bundle generator

in which the sum of two arbitrary columns is different. It is should be noted that adding one module increases the coding distance by one. The shortened matrix $H^T_{30,12}$ is designated as shaded line in fig 3. The calculations is show that one can get the coding matrix with d = 5 if the 12-th row of the matrix $H^T_{30,17}$ is changed for the 11-th row. Then we get an optimum coding matrix with K_c = 30/11 and d = 5. From simple reasonings a simple relation can be obtained between number of moduli of the residual system ϕ and coding distance d = 2ϕ - 2, or ϕ = [(d + 2)]/2, where brackets denote the nearest integer. So, ϕ = 4 or 5 for d = 5.

Let us consider the coding of two time intervals. For simplicity let p_1 be 3 and p_2 be 4. Assume that events are registered at the 2-th and 4-th clock times. Then we have he table 2. After 3x4 = 12 clocks the process of coding comes to an end and the code 0100011 which carries information on events at the 2-nd and 5-th clocks appears in the circular counters. This fact can be verified in the following manner. If the inputs of the counters are closed after the cycle of measurement, which equal 12 clocks of the bundle generator, and then the counters are shifted, the pulses in high bits coincide after the 2-nd and 4-th clock times. These events are marked as * and #, respectively.

Table 2

Coding of two events

Triggers	1	2	3	4	5	6	7
Times							
1	0	0	0	0	0	0	0
*2	1	0	0	0	0	0	0
3	0	1	0	0	1	0	0
4	0	0	1	0	0	1	0
#5	1	0	0	1	0	0	1
6	0	1	0	1	1	0	0
7	0	0	1	0	1	1	0
8	1	0	0	0	0	1	1
9	0	1	0	1	0	0	1
10	0	0	1	1	1	0	0
11	1	0	0	0	1	1	0
12	0	1	0	0	0	1	1
1	0	0	1	1	0	0	1
2	1	0	0	1	1	0	0
3	0	1	0	0	1	1	0
4	0	0	1	0	0	1	1

3. *Time-to digital converters constructed on the base of the theory of binary BCH-codes.* Using the property of BCH-codes correcting t mistakes, devices can be constructed for coding t intervals during the cycle of measurement. For simplicity we consider the Galois field $GF(2^4)$ generated by the irreducable polynomial $X^4 + X + 1$ assuming that t = 2. Then the generated polynomial $g(X)$ = $m_1(X)m_3(X)$ = $(X^4 + X^3 + X^2 + X + 1)(X^4 + X + 1)$ = $X^8 + X^7 + X^6 + X^4 + 1$. The equation of carrying from the high bit to the low-order bits takes the form $X^8 = X^7 + X^6 + X^4 + 1$. Using this equation, we can determine the states of a 8-bit encoder at 15 clock times as m = 4 and $2^4 - 1$ = 15. The states of the 8-bit coder are given in table 3. It is supposed that the events are registered at the 2-nd and 12-th clock times. Then we get table 3. All triggers are reset in the initial state. The START signal initiates the bundle generator G and the trigger T opens

Table 3.
Coding of two events

Triggers	0 1 2 3 4 5 6 7
Time	
1	0 0 0 0 0 0 0 0
*2	1 0 0 0 0 0 0 0
3	0 1 0 0 0 0 0 0
4	0 0 1 0 0 0 0 0
5	0 0 0 1 0 0 0 0
6	0 0 0 0 1 0 0 0
7	0 0 0 0 0 1 0 0
8	0 0 0 0 0 0 1 0
9	0 0 0 0 0 0 0 1
10	1 0 0 0 1 0 1 1
11	1 1 0 0 1 1 1 0
#12	1 1 1 0 0 1 1 1
13	1 1 1 1 0 0 0 0
14	0 1 1 1 1 1 0 0
15	0 0 1 1 1 1 1 0

an AND element.
At the 2-nd clock time after amplifying the signal is registered in the 0-th bit of the coder shifted up to the 9-th clock time. At the 10-th clock time the signal is carried from the high to the low-order bit, and therefore the code 10001011 is formed. In order to determine the states of the triggers of the coder after the 11-th clock time, it is nessecary to shift the code 10001011 ⟶ 01000101 and to the right and to add result with the code 100010111.

$$
\begin{array}{r}
01000101 \\
+\ \underline{10001011} \\
11001110 \qquad \text{mod 2}
\end{array}
$$

and so on. After the 12-th clock time the second event is accidentally registered, and coding occurs up to 15-th clock times as is done in the theory of binary BCH codes. It should be noted that the duration of a synchropulses are selected in order to provide a minimum loss of useful events. According to the theory of the BCH-codes, the word 00111110 carries information on the time of the two events registered in the position-sensitive detector. This word can be readout in a serial or a parallel code and decoded by PROM. The efficiency of such a coding system increases with number M. So, K_c = 1024/30 for M = 1024 and t = 3. Besides, for complete coding it is unnessecary to use stop-signals because synchropulses are produced by the bundle generator. A principal scheme of the coder for our example is given in fig. 5.

    II. Economic parallel coding of weak electric and light signals.
    Another significat feature of the coding matrix is the possibility to construct a parallel encoders with useful properties. As shown [1,3-5], if one position-sensitive detector, for example, set for each column and a registration channel for each row the position and the number of ones in the coding matrix determine the structure of a parallel encoder (without memory) for t signals, where t > 1. Besides if a superimposed codes is describet by coding matrix, amplifiers-mixers can be used to calculate a syndrome code instead of modulo-2 adders. If light signals are coded photomultipliers can be used. For example it is necessary to creat a scintillation hodoscope with M = 30 scintillators and the multiplicity t < 2. What minimum number of photomultipliers are needed. Let us consider the constructed matrix $H_{30,11}$.
Fig.6 shows schemes for calculation of the 1-st, 10-th and 11-th synd-

rome bits. Optical fibers can be used for coding. The method of

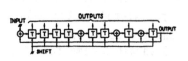

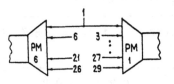

Fig.5. Principal scheme of time-to -digital converter for two events

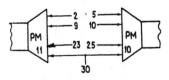

Fig.6. Schemes for calculating the 1-st, 10-th and 11-bits of the syndrome

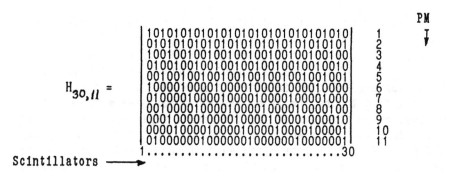

coding only for t = 1 signal is given in [14]. The coding matrix can be created as a mask, where black squares are scintillaters and white squares are usual glass (fig 7).

III. A solution of the problem of a „ghost" in pixel detectors
    The use of the method of syndrome coding allows one to solve effectively the problem of a registration ghost in pixel detectors or in hodoscope calorimeters as it shown in fig. 8, where the number of a pixels n = 64. The use of priority encoders for coordinate registration even for t = 2 does not lead to the solution the problem if data are read out on registers X and Y. There are two methods for solving this problem suggested by the author. The first method allows iteration codes to be used [3]. If a 2-dimensional iteration code with coding distance $d = d_1 d_2 > 5$ is used the coordinate X_1, Y_1 and X_2, Y_2 or X_1, Y_2 and X_2, Y_1 can be synonumously decoded by decoding a syndrome with the help of

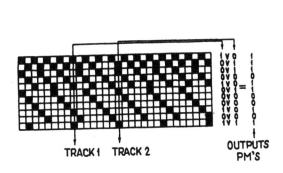

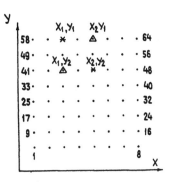

Fig.7. Coding mask-detector for t = 2 and M = 30

Fig.8. To the question of the identification of ghost by the method of syndrome coding

PROM or PLA. But is known that iteration codes are difficult for decoding. Using the second method, n pixels are umbered as degree the Galois (GF2^m) field elements. If we take the binary BCH-code with parameters d = 5, m = 6 and n = 2^6 - 1 = 63, then the coordinate of two events can be calculated fast and simply. In detail see [3-5].

IV. Use of the theory of RS-codes

There are many experiments where both coordinate and their images should registered. For example, the events with two clusters are given. As known, in the coding theory a burst of errors β in length is determined by the error vector. Thus, the burst of errors 6 in length can look as follows:

```
        0011111100000      001000010000      001100110000
           Burst 1            Burst 2           Burst 3
```

Similar configurations of bursts (clusters) take place in MCPD when several neighbouring position-sensitive detectors are fired from one charged particle. The author has suggested to use the theory of RS-codes for registering such events. If the lengh of a cluster is assumed to be β < m, the compression coefficient K_c = $(2^m - 1)/2t$. Let the MDCP have n = 60 channel registration, which are divided into 15 group with 4 bits in each group. For t = 2 and m = 4 the coding matrix (parity check matrix) takes the form of [11, 14]. In the matrix (I) α^0 - α^{14} the elements of the Galois field GF(2^4) which generated by irreducable polynomial X^4+ X + 1.

* 0	α^0	α^0	α^0	α^0
1	α^1	α^2	α^3	α^4
* 2	α^2	α^4	α^6	α^8
3	α^3	α^6	α^9	α^{12}
4	α^4	α^8	α^{12}	α^1
5	α^5	α^{10}	α^0	α^5
6	α^6	α^{12}	α^3	α^9
7 $H_=^T$	α^7	α^{14}	α^6	α^{13}
8	α^8	α^1	α^9	α^2
9	α^9	α^3	α^{12}	α^6
10	α^{10}	α^5	α^0	α^{10}
11	α^{12}	α^9	α^6	α^3
12	α^{12}	α^9	α^6	α^3
13	α^{13}	α^{11}	α^9	α^7
14	α^{14}	α^{13}	α^{12}	α^{11}

There are 15 nonzero elements5 α^0 = 1000, α^1 = 0100 (root of the polinomial), α^2 = 0010, α^3 = 0001, α^4 = 1 + α = 1100, α^5 = 0101, α^6 = 0011, α^7 = 1101, α^8 = 1010, α^9 = 0101, α^{10} = 1110, α^{11} = 0111, α^{12} = 1111, α^{13} = 1011 and α^{14} = 1001.

(1) As assumed the in algebraic coding of RS-codes the nonzero component of coordinate vector of fired position-sensitive detectors e(X) is given by a pair of elements Y_1 and X_1 which are the image and coordinate of clusters, respectively. For t events we have

$$S_j = \sum_{i=1}^{t} Y_1 X_1^j, \quad 1 \ll j \ll 2t. \quad (2)$$

For t < 5 the table method [3,17-18] can be used for the solution of eq. (2) and for t > 5 the well known Peterson algorithm [11] or Chien algorithm [18]. It should be noted that along with the images and coordinates of clusters the number of events registered in MDCP should be ferst calculated. The algorithm of a majority coincidence circuit is based on the property of the L_t matrix. The matrix

$$L_t = \begin{vmatrix} S_1 & S_2 & S_3 \dots S_t \\ S_2 & S_3 & S_4 \dots S_{t+1} \\ \dots\dots\dots\dots\dots\dots\dots\dots \\ S_t & S_{t+1} & S_{t+2} \dots S_{2t+1} \end{vmatrix} \quad (3)$$

is singular if S_j is made from t different nonzero pairs X_1, Y_1 and matrix (3) is nonsingular if S_j is made from smaller than t nonzero pairs (X_1, Y_1) [8]. In other words, the properties of the determinant L_t are such that if, e.g., t = 1, then $detL_1$ = 0 , but all other determinants are zero. But if t = 2, then $detL_1$ = 0, $detL_2$ = 0, but $detL_3$ = 0 and so on. For t = 3 we have

$$L_3 = \begin{vmatrix} S_1 & S_2 & S_3 \\ S_2 & S_3 & S_4 \\ S_3 & S_4 & S_5 \end{vmatrix} \qquad \begin{aligned} detL_1 &= S_1 \\ detL_2 &= S_1 S_3 + S_2^2 \\ detL_3 &= S_1 S_3 S_5 + S_1 S_4^2 + S_2^2 S_5 + S_3^3 \end{aligned} \quad (4)$$

An efficient and fast method is given to calculate determinant in the

GF(2^m) using the PROM in [3] .

Continue to consider our example. Let t be 2, $X_1 = \alpha^0$ and $X_2 = \alpha^2$, $Y_1 = \alpha^7$ = 1101 and $Y_2 = \alpha^{11}$ = 0111. Let as examine three case.

a). Cluster α^{11} = 0111 registered at position $X_1 = \alpha^2$. From (I) we have $S_1 = \alpha^{11}\alpha^2 = \alpha^{13}$, $S_2 = \alpha^{11}\alpha^4 = \alpha^{15} = a^0$, $S_3 = \alpha^{11}\alpha^6 = \alpha^2$, $S_4 = \alpha^{11}\alpha^8 = \alpha^4$ In addition from (4) we have the following relations: $detL_1 = \alpha^{13} = 0$, $detL_2 = \alpha^{13}\alpha^2 + \alpha^0 = $, $detL_3 = \alpha^{13}\alpha^2 \alpha^6 + \alpha^{13}\alpha^8 + \alpha^0\alpha^6 + \alpha^6 = 0$,

b) Assume that $X_1 = \alpha^0$, $Y_1 = \alpha^7$ and $X_2 = \alpha^2$, $Y_2 = \alpha^{11}$, i.e. t = 2. Then we get $S_1 = \alpha^7\alpha^0 + \alpha^{11}\alpha^4 = \alpha^5$, $S_2 = \alpha^7\alpha^0 + \alpha^{11}\alpha^4 = \alpha^9$, $S_3 = \alpha^7\alpha^0 + \alpha^{11}\alpha^6 = \alpha^{12}$ and $S_4 = \alpha^7\alpha^0 + \alpha^{11}\alpha^8 = \alpha^3$.

To find X_1 and X_2, it is necessary to solve the quadratic equation

$$X + \sigma_1 X + \sigma_2 = 0. \tag{5}$$

According to the Peterson algorithm, we have

$$S_j\sigma_t + S_{j+1}\sigma_{t-1} + \cdots + \ldots S_{j+t} = 0.$$

In the specific case when t = 2

$$\sigma_1 = \frac{S_3 S_2 + S_1 S_2 S_4}{S_2^3 + S_1 S_2 S_3}, \qquad \sigma_2 = \frac{S_2 S_4 + S_3^2}{S_2^2 + S_1 S_3}. \tag{6}$$

From eq. (2) we obtain a relation to S_j, X_1 and Y_1

$$S_1 = X_1 Y_1 + X_2 Y_2 \text{ and } S_3 = X_1^3 Y_1 + X_2^3 Y_2 \tag{7}.$$

From eq. (6) we have $\sigma_1 = \alpha^8$ and $\sigma_2 = \alpha^2$. One can verify that for these values of σ_1,σ_2 and $X_1 = \alpha^0$, $X_2 = \alpha^2$ eq. (4) is equal to zero. From eq. (6) S_1, S_3, X_1 and X_2 we can calculate Y_1 and Y_2. A the block-diagram of the special-purpose processor is given in fig. 9. The groups of inputs are numbered by 0 - 14 - degrees of the Galois GF(2^4) field elements8 16-30 are amplifiers-shapers, 31-90 the schemes of multiplication of t he input symbols considered as elements of the GF(2^4) by constants α^0 - α^{14}. The syndrome S_1 - S_4 is calculated with the help of the parity checkers (91). The PROMs (92) and (93) are used to calculate X_1,X_2 and Y_1,Y_2. If ECL-microcircuits are used of SP then the time of multiplication and parity checking do not exeed 6 ns. Besides, the author

[19-21] has shown that in galois field GF(2^m) such operation as multiplication of several elements and power raising can be executed fast.

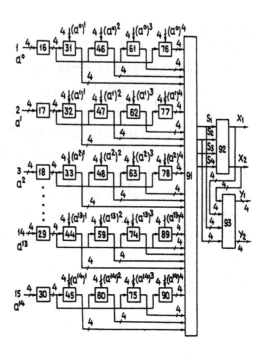

*Fig.9. Blocr-diagram of the special-purpose
processors*

Conclusion

Several aspects of using the syndrome coding method in nuclear
alectronics are described. It is shown that using the algebraic coding
theory practically new devices can be created having a number of valuable
properties5 time-to-digital converters which can code several intervals
simultaneously and possess high resolution, digital delays, parallel
encoders for coding ligt signals t > 1 and special-purpose processors for
the registration of a cluster coordinates and images. The coding matrix
suggested by the author for the superimposed code is close to an optimum
and has higher parameters than the known [13].

References

1. Nikityuk N.M., Radzhabov R.S., Shafranov M.D. A new method of information registration from multiwire proportional chambers. Nucl.
   Instr. and Meth., 1978, v.155, p.
2. Gustafsson L., Hagberg E. A fast trigger processor, for a scattering
   experiment implemented in FASTBUS. Nucl. Instr. and Meth., 1988,

v. A265 No. 3, p. 521.

3. Nikityuk N.M. The method of syndrome coding and its application for data compression and processing in high energy physics experiments.

4. Nikityuk N.M. A special purpose processor having algebraic structure for fast event calculation. JINR, No. P10-87-254, Dubna, 1987.

5. Nikityuk N.M. Some questions of using the algebraic coding theory for construction of special-purpose processors in high energy physics spectrometers. JINR, No. E10-89-362, Dubna, 1989. Submitted to the International Conference AAECC 7, Toulouse Cedex, France.

6. Etkin A., Kramer A. The Brookhaven National Laboratory's multiparticle spectrometer drift chamber system, IEEE Trans. on Nucl. Sci. 1980, v. NS-27, No. 1, p. 139.

7. Pernika M. Nucl. Instr. and Meth., 1978, v. 156, No. 1, p. 311.

8. Peterson W., Weldon E. Error correcting codes,

9. Frohwerk R.A. Hewlett Packard Journal, 1977 v.28, No.9, p. 2.

10. Giorgi M., Gasparini E., Centro S. et al. PDC5 A wire chamber read-out on 6-bit fast ADC. Nucl. Instr. and Meth., 1984, v. 223, No. 1, p. 113.

11. Nikityuk N.M. Syndrome coding method for sequential systems of data processing in high energy physics spectrometers. JINR, No. P-10-88-742, Dubna, 1988.

12. Chetagurov Y.A., Rudnev Yu. P. Increasing the safety of digital devices by the abundance coding method. 2Energiya2, Moscow, 1974, p. 32.

13. Kautz W.H., Singleton. Nonrandom binary superimposed codes. IEEE Trans. on Inf. Theory, 1964, v. IT-10, No. 4, p. 363.

14. Niimura N., Yamada K., Kobota T. Position sensitive neutron detectors. Nucl. Instr. and Meth., 1983, v. 211,No. 1, p. 203.

15. Gordon J.A. Very simple method to find the minimum polynomial of the arbitrary nonzero element of a finite field. Electr. Letters, 1976, v. 12, No. 25, p. 663.

16. Nikityuk N.M. Theoretical foundations of creation of special-purpose processors for processing events with clusters. JINR, No. P10-88-854, Dubna, 1988.

17. Chien R.T., Cunningham B.D., Oldham I.B. Hybrid method for finding roots of a polynomial with application to BCH-coding. IEEE Trans. on Inf. Theory, 1969, v. IT-15, No. 2, p. 329.

18. Okano H., Imai H. Construction method of high-speed decoders using ROMs for Bose-Chaudhuri-Hocquenghem and Reed-Solomon codes. IEEE Trans. on Computers, 1987, vol. C-36, No. 10, p. 1165.

19. Gaidamaka R.I., Nikiryuk N.M. Application of analytical transformations and calculation on computers for synthesis of switching functions and solution of the problem of divising universal dynamically programmed logic modules. JINR, No. E10-88-53. Submitted to the Int. Conf. ISSAC-88 and AECC-6, Roma, July, 1988.

20. Nikityuk N.M. Combined operations in Galois field $GF(2^m)$ and their application. JINR, No. P11-87-54, Dubna, 1987.

21. Nikityuk N.M. Fast algorithms for execution of multiplication over Galois field $GF(2^m)$. JINR, No. P11-88-852.

Some Ideas about Fault-Tolerant Chinese Remaindering

John Abbott
School of Mathematical Sciences
Bath University
Bath BA2 7AY
email: jaa@maths.bath.ac.uk

Abstract

We present some algorithms for performing Chinese Remaindering allowing for the fact that one or more residues may be erroneous — we suppose also that an *a priori* upper bound on the number of erroneous residues is known. A specific application would be for residue number codes (as distinct from quadratic residue codes). We generalise the method of Ramachandran, and present two general algorithms for this problem, along with two special cases one of which uses the minimal number Chinese Remainderings (for two errors), and the other uses 12 compared with the lower bound of 10 Chinese Remainderings. These algorithms are best suited to the case where errors are unlikely; we compare with a method based on continued fractions. Our methods use only the standard Chinese remaindering operation and equality testing of the reconstructed values.

Introduction and Notation

This paper will concentrate on the problem over the integers and for simplicity the presentation will be in those terms, however, the result applies in any effective Euclidean domain with a compatible measure (i.e. there is a uniquely determined "smallest remainder"). We investigate fault-tolerant Chinese Remainder Algorithms where a large integer value is to be determined from its residues modulo several primes (it suffices to use mutually coprime moduli) some of which may be corrupted. Apparently, the first paper to consider this problem is [Stone63], however the algorithm implicit in this paper was very inefficient. Subsequently, more efficient algorithms have been described, for example [Mandelbaum72] and [B&M74]. The paper [Ramachandran83] presents an efficient and elegant method, as well as giving a brief summary of some other methods. We present a restricted generalization of this elegant method to multiple errors, along with some alternatives.

Let N denote the non-negative integer which we are endeavouring to find, and $m_1 < \cdots < m_{n+r}$ be the moduli where $M := m_1 m_2 \cdots m_n$ is assumed to be larger than N (thus the product of any n moduli exceeds N) and r is the number of "excess" residues whose presence will allow us to correct errors. We shall write $r_1, \ldots, r_{n+r}$ for the corresponding residues, and e for an upper bound on the number of residues which may be in error; this must be known in advance! Note that for any value of e we obtain $r > 2e$ [Stone63] otherwise we cannot guarantee to recover. As usual, to allow recovery of a positive or negative integer, N, we need $2|N| < M$.

Continued Fraction Method

We describe here a method based on continued fraction expansion — there is some similarity with the "Euclidean Algorithm" method for Reed-Solomon codes (e.g. [Mandelbaum82]). This method works only over the integers, and is particularly well suited to the case where r is small, and errors occur frequently. This is in direct contrast to our generalizations of Ramachandran's algorithm which are best suited to the case when errors are rare or r is large. Let e_i be the error in residue r_i, so at most e of the e_i are non-zero. To apply this algorithm we certainly need $2e < r$ (the proper condition will become apparent) though if the moduli are all of similar magnitude then $2e < r$ will be sufficient. Reconstruct N^* from all the $n + r$ residues, this means that

$$\frac{N^*}{M_{all}} = \frac{N}{M_{all}} + \sum \frac{e_j^*}{m_j}$$

where $M_{all} = m_1 m_2 \cdots m_{n+r}$, and e_j^* is non-zero iff e_j is non-zero. Notice that the correct term is at most $1/M_{ext}$ where $M_{ext} = m_{n+1} m_{n+2} \cdots m_{n+r}$ whereas the error term is at least $1/m_{n+r}$, and indeed the error term is a close appoximation to N^*/M_{all} with small denominator. By theorem 184 of [H&W79] the error term will appear as a continued fraction convergent provided $2M_{err}^2 < M_{ext}$: this is the proper condition for applicability of this method, and is more restrictive than just $2e < r$. Now we compute the continued-fraction expansion of N^*/M, until we obtain the convergent with greatest denominator not exceeding $\sqrt{M_{ext}}$. This will be the error term whose denominator tells us which residues are wrong: the denominator is just the product of the moduli of the faulty residues. We can then subtract the error term to obtain N/M_{all}, and thus N. This algorithm does not appear to have been given explicitly before.

The mainstay for our results is the following simple and useful lemma:

Lemma

Let $S_1, S_2, \ldots, S_k$ each be subsets of n of the residues (and moduli) whose union contains at least n correct ones. The integers reconstructed from all the S_i are the same (and strictly less than M) if and only if $S_1 \cup S_2 \cup \cdots S_k$ contains no incorrect residues.

Proof

If $\bigcup S_i$ contains only true residues then clearly the reconstructed integers must all be N because the product of any n moduli we care to choose exceeds N. For the converse, suppose $N' := reconstruct(S_1) = reconstruct(S_2) = \cdots = reconstruct(S_k)$ and that $N' < M$, then each residue in $\bigcup S_i$ is just the residue of N'. Now, the n correct residues contained in the union will construct N. Thus $N' \equiv N \mod \prod m_i$ where the product is of the n correct moduli. So $\prod m_i$ divides $N' - N$, but the latter has magnitude less than $M \leq \prod m_i$. Hence $N' = N$.

The Case $e = 1$, Ramachandran's method.

We recall the case when $e = 1$ — note that r must be at least 2, otherwise there is no solution. The solution given in [Ramachandran83] is simple and elegant. We make a sequence of reconstructions from sets of n residues until we obtain two values the same (necessarily from successive reconstructions). We show how to pick the indices of the residues used in each reconstruction. Arbitrarily start with $\{0, \ldots, n-1\}$. Subsequent sets may be derived by adding (mod $n + r$) alternately $\lfloor r/2 \rfloor$ and $\lceil r/2 \rceil$ to each set of indices. This produces a succession of length $\lceil \frac{2(n+r)}{r} \rceil$. For example, in the case $n = 8$ and $r = 5$, and indexing the residues from 0 to 12, our succession could begin:

$$
\begin{array}{l}
\{ \quad 0 \ \ 1 \ \ 2 \ \ 3 \ \ 4 \ \ 5 \ \ 6 \ \ 7 \qquad\qquad\qquad\qquad \} \\
\{ \qquad\quad\ 2 \ \ 3 \ \ 4 \ \ 5 \ \ 6 \ \ 7 \ \ 8 \ \ 9 \qquad\qquad\quad \} \ \text{ by adding } \lfloor 5/2 \rfloor \\
\{ \qquad\qquad\qquad\ 5 \ \ 6 \ \ 7 \ \ 8 \ \ 9 \ \ 10 \ \ 11 \ \ 12 \ \} \ \text{ by adding } \lceil 5/2 \rceil \\
\{ \ 0 \ \ 1 \qquad\qquad\qquad\quad 7 \ \ 8 \ \ 9 \ \ 10 \ \ 11 \ \ 12 \ \} \qquad \text{and so on.}
\end{array}
$$

Any two of these n-tuples contain at least $n + 1$ different members, of which at most one can be faulty. Thus the lemma above applies, and we can say that if any two reconstructed integers are the same (and less than M) then their value is N. Also each index is omitted exactly twice (from adjacent sets), so we are guaranteed two correct reconstructions.

Lower Bound for $e = 1$

A trivial lower bound for the length of the succession can be deduced from the facts that each member of the succession excludes r of the moduli, and that each of the $n + r$ residues must be absent from two or more n-tuples (since we must guarantee to exclude the faulty residue from two or more reconstructions). Hence the lower bound is $\lceil \frac{2(n+r)}{r} \rceil$ sets in the succession which is precisely what Ramachandran's method achieves.

General Case: $e \geq 1$

Before examining the problem in detail, we make a few general observations. We saw in the case $e = 1$ that two n-tuples were required to ensure that n distinct correct residues could be found in their union. Now we can write down $B(n, e) := \sum_{i=1}^{e} \binom{e}{i} \binom{n-1}{n-i} \geq 2^e - 1$ distinct n-tuples whose union contains precisely $n - 1$ correct residues: select $n - 1$ of the correct residues. We can pick i incorrect residues in $\binom{e}{i}$ ways, and then fill the remaining $n - i$ places in an n-tuple in $\binom{n-1}{n-i}$ ways selecting from our set of $n - 1$ correct residues; this makes a total of $\binom{e}{i} \binom{n-1}{n-i}$ n-tuples with exactly i incorrect residues, and i can be any number between 1 and e.

However, if we can pick the n-tuples in our succession judiciously so that any two n-tuples employed during our search contain at least $n + e$ distinct residues then, by the lemma, it will suffice to have two agreeing reconstructions below the bound M. In this general case, the trivial lower bound becomes $2 \binom{n+r}{e} / \binom{r}{e}$.

Two general methods

We shall explain two ways to decompose the "multiple error" situation into several "single error" subproblems. However, we cannot use Ramachandran's method directly because we wish to be able to correct single errors and detect multiple errors. An algorithm to achieve this is:

Algorithm 1: single error correction, multiple error detection

Input: $n + r$ residues with at most e faulty, must have $r > e$.
Output: Either an indication of failure (if more than one residue is faulty) or the value of the integer represented.
(1) Let $k = \lceil (n + r)/(r - e) \rceil$.
(2) Partition the $n + r$ residues into k sets of $r - e$ residues such that each residue appears in at least one set.
(3) For each set S in the partition do:
   Perform $\lceil (n + e)/n \rceil$ reconstructions ensuring that each residue outside S is used at least once;
   if these are all the same (and less than M) then return with success.
(4) Return with failure — more than 1 residue was faulty.

Correctness follows from the lemma. Certainly if one residue is incorrect then the algorithm will succeed in at most $\lceil (n + r)/(r - e) \rceil \lceil (n + e)/n \rceil$ reconstructions, because one of the sets S must contain the faulty residue. If two or more residues are faulty then the algorithm may fail since the sets of $n + e$ residues used in the reconstructions may all contain an erroneous residue. Note that the algorithm will never claim success and return a false value since any set of $n + e$ residues must always include n true residues, and the lemma says that the reconstructions will differ (or be too large) if a wrong residue is present. If no residue is wrong the algorithm will return with success after $\lceil (n + e)/n \rceil$ reconstructions from n residues. We comment that if $e < n$ then we will make two reconstructions of size n; however, there are $n - e$ residues common to both reconstructions and we can share this calculation between the two reconstructions.

Direct Decomposition (pigeonhole principle).

Let $e' = \lceil (e + 1)/2 \rceil$, the least integer greater than $e/2$. We may apply a direct decomposition if $n + r \geq e'(n + e + 1)$. We split the problem into e' subproblems with at least $n + e + 1$ residues in each subproblem, and by the pigeonhole principle some subproblem has at most one erroneous residue. We apply algorithm 1 to each subproblem in turn until one succeeds; and we are guaranteed that some subproblem must lead to success because the algorithm above always succeeds on any subproblem with at most one faulty residue, while subproblems with many faulty residues may fail. We may assume that $e < n$ since otherwise $r \geq n(e + 1)$ when $e + 2$ disjoint n-tuples will suffice to recover the answer. In the worst case this direct decomposition uses $e' \lceil \frac{n+r}{n+r-e'(n+e)} \rceil \lceil \frac{n+e}{n} \rceil$ reconstructions. For fixed e we see that this estimate is greatest when $n + r$ is least, i.e. when $n + r = e'(n + e + 1)$. In this case $e'(n + e + 1)\lceil \frac{n+e}{n} \rceil$ reconstructions could be necessary. If there is no faulty residue the first call to algorithm 1 succeeds after 2 reconstructions of size n. This method is not very efficient when errors are present: e.g. if $n = 20$ and $e = 3$ then $r \geq 48$; and if $r = 48$ then the lower bound is 6 reconstructions but this decomposition could need 96.

Recursive ("cascade") Decomposition

The other "cascade" decomposition is the natural recursive solution — this can be applied only when $r > 2e$. Suppose we have a means of solving the general "k error" problem (for $k = 1, \ldots, e - 1$) including detecting when e residues are at fault. Then we can solve the "e error" problem by removing some residues (hoping this will remove at least one faulty residue) and trying to solve the "$k = e - 1$ error" problem on the remaining residues. More formally, partition the $n+r$ residues into $\lceil (n+r)/s \rceil$ sets of size s (to be determined) ensuring each residue appears in at least one partition. Try solving the "$e-1$ error" problem on the complements of these sets until one of the subproblems succeeds, when its answer is returned. Certainly one of the subproblems must succeed since either there are no faulty residues and the first subproblem succeeds or there is a faulty residue which appears in one of the partition sets whose complement contains at most $e - 1$ faulty residues and the "$e - 1$ error" subproblem on that complement will succeed.

We can pick s to minimise the worst case computing time. For example when $e = 2$, algorithm 1 will perform about $(n + r - s)/(r' - s)$ loops when it fails, and algorithm 1 will be called about $(n+r)/s$ times. In this case we find the minimum worst time when $s = n + r - \sqrt{(n+r)(n+e)}$ which is about $r'/2$ when r' is small. Similarly, at level k we find $s \approx r'/k$ when r' is small. This gives the rather disappointing overall complexity of $e^e (\frac{n+e}{r'})^e (\frac{n+e}{n})$ — comparing with the lower bound which is at most $2(\frac{n+e}{r'})^e$ we see an extra factor of about e^e.

Two Special Cases

We shall discuss two highly efficient but rather specialised methods for solving this problem. The first method needs about 16% more reconstructions than optimal, and under suitable conditions the second method achieves the lower bound.

(1) Special Case $n = r$ (both even) and $e = 2$

Suppose $n = r$ are both even, $2m$ say. We shall write $r_0, \ldots, r_{4m-1}$ for the residues (and moduli). Let A denote $\{r_0, \ldots, r_{m-1}\}$ and B, C and D the three remaining quarters. Then we find that the 6 distinct n-tuples in $X := \{A \cup B, C \cup D, A \cup C, B \cup D, A \cup D, B \cup C\}$ contain each pair (r_i, r_j) at least once, and by symmetry exclude each pair at least once. Similarly if α, β, γ and δ denote those residues with subscripts being 0,1,2 and 3 modulo 4 respectively, then again the six n-tuples in $Y := \{\alpha \cup \beta, \ldots, \gamma \cup \delta\}$ contain each pair of residues at least once. We observe that this is slightly suboptimal as the lower bound requires only five n-tuples.

Clearly the union of any two n-tuples from X contains $3m = n + m$ residues, as does the union of any two from Y. The union of an n-tuple from X with one from Y must also contain $3m = n + m$ residues because the intersection of such n-tuples contains m residues. So provided $m \geq 2$ we can apply the lemma to show that these twelve n-tuples solve the problem — the lower bound for this case is ten.

We regret that we have been unable to find a generalisation of this method to higher values of e. In an implementation we would share the cost of the Chinese Remaindering of the 16 intersections.

(2) Prime Spaces (hyperplane method)

• Hyperplane method when $e = 2$

Now we look at a special case for $e = 2$, and will generalise to larger e later. Here we restrict to $r = p$ for some prime p, and $n = (p-1)r$. This gives us a total of p^2 moduli. Thus we can lay out the moduli in a square of side p which may be identified with the plane F_p^2 where F_p denotes the field of p elements. We shall describe how to pick a set of r-tuples (corresponding to those moduli which will be left out) such that each pair of moduli occurs in exactly one r-tuple. Note that two distinct moduli correspond to two distinct points in the plane; and through these points there is exactly one line, which necessarily passes through a total of $p = r$ points (i.e. the line corresponds to an r-tuple of moduli).

Consider the set of all lines passing through $(0,0)$. Apart from $(0,0)$ itself, every point is contained in precisely one line; and each line contains $p-1$ points other than $(0,0)$. Since the plane contains $p^2 - 1$ points other than $(0,0)$, there are $p+1$ distinct lines. The lines can be characterised by picking any point on them (other than $(0,0)$ of course): for example, $(1,0)$, and $(x,1)$ for $x \in F_p$ are possible representatives. Also each line has p different translates which can be reached by adding some fixed multiple of the vector $(1,0)$ to each point on the line — add multiples of $(0,1)$ to the first line listed (passing through $(1,0)$), as a line translated along itself does not change.

The $p^2 + p$ lines enumerated above give us a collection of r-tuples of moduli where each pair clearly occurs in exactly one r-tuple. Our n-tuples will be the complements of these lines: let's call them C_1 up to C_{p^2+p}, and the corresponding lines H_1 through H_{p^2+p}. To employ the lemma above we need to know how big $C_i \cup C_j$ is when $i \neq j$. There are only two cases: when the lines H_i and H_j intersect, and when they don't. If they do intersect then only a single point is in the intersection, and hence $C_i \cup C_j$ contains $n + r - 1$ points; otherwise $C_i \cup C_j$ contains all the $n + r$ points. We know that $r > e = 2$, so the lemma tells us that if any two of the C_i give the same reconstructions (with value less than M) then they must both be correct. Unfortunately, the erroneous pair of residues appears in just one line H_j, say. This means that all C_i with $i \neq j$ contain either or both faulty residues — but to make the lemma work we'll need two correct reconstructions!

We must find a second set of n-tuples which eliminate each pair exactly once. We shall do this by permuting the positions of some of the moduli in the plane. We shall assume that $r > 9$. Consider the four lines passing through the point pairs $\{(0,0),(0,1)\}$, $\{(0,0),(1,0)\}$, $\{(2,0),(2,1)\}$ and $\{(0,2),(1,2)\}$. These form a square, and any line in the plane must intersect these lines in at least two distinct places, and at most four distinct places. We perform a cyclical shift along these four lines; for example to effect the shift along the line through $(0,0)$ and $(0,1)$, the residue in $(0,0)$ moves to $(0,1)$, that in $(0,1)$ moves to $(0,2)$ and so on, until the residue in $(0,p-1)$ moves to $(0,0)$. It is important that the shifts do not take one intersection point to another, so that there is no possibility of a point being returned to its original line after all shifts have been completed.

Let us investigate the n-tuples derived, as above, from this new arrangement of residues in the plane. We shall denote these n-tuples by C_1' through C_{p^2+p}'. As before, each 2-tuple of residues is excluded from precisely one C_j', and $C_i' \cup C_j'$ contains at least $n + r - 1$ residues whenever $i \neq j$. We need also that $C_i \cup C_j'$ be large enough. By construction, there is a C_k which satisfies $n + 2 \leq |C_k \cup C'_j| \leq n + 4$, and if $i \neq k$ then $|C_i \cup C_k| \geq n + 8$; hence $|C_i \cup C'_j| \geq n + 4$. In other words, any two different sets S_1 and S_2 chosen from the C_is or the C'_is have union containing $n + 2$ or more elements. Now using the lemma we can say that two agreeing reconstructions (with value less than M) must necessarily be correct, and that there must be two such, provided no more than two residues were faulty.

From the point of view of implementation, we observe that there is an obvious pairing of the C_i with the C'_j such that most of the elements are the same (i.e. $|C_i \cup C'_j| \leq n + 4$), so there is the possibility of sharing some of the Chinese Remaindering computation. Notice that the lower bound on the number of reconstructions for this problem is $2\binom{p^2}{2}/\binom{p}{2}$ which simplifies to $2p(p + 1)$, being precisely the number this method uses.

If our assumption does not hold (i.e. the prime p is small) then we must use a different argument — or determine explicit methods for the finitely many cases excluded from the above. A less efficient variant of the method just described, is to perform the cyclical shifts along the "axes", those lines through the origin generated by a vector all except one of whose components are 0. In this case we may change only a single residue in some lines, and so the lemma tells us that we need 2 such shifts in addition to the original, giving a total of $3p(p + 1)$ reconstructions.

• **Hyperplane method when $e > 2$**

This scheme generalises to $e > 2$ but becomes suboptimal because, for example, three points from an e-tuple may be collinear. Where before we were counting lines in F_p^2, we shall now count hyperplanes in F_p^e. However, note that a hyperplane is determined by its normal (a line), and which of p possible translates it is. The conditions on n and r become: for some prime p, $r = p^{e-1}$ and $n = r(p - 1)$. In F_p^e we see there are $(p^e - 1)/(p - 1)$ lines through the origin, and so there are $p(p^e - 1)/(p - 1)$ hyperplanes altogether.

As before, the hyperplanes correspond to the residues that will be left out; their complements will be our n-tuples. Let us call the hyperplanes $H_1, H_2, \ldots$, and their complements $C_1, C_2, \ldots$. Then each e-tuple of residues will occur in one or more of the H_i. Any two distinct hyperplanes intersect in at most an $e - 2$ dimensional subspace which contains p^{e-2} points, thus for all $i \neq j$ we have $|C_i \cup C_j| \geq p^e - p^{e-2} = n + (p - 1)p^{e-2}$.

As in the case when $e = 2$ we want to find a permutation of some of the residues so that the hyperplanes corresponding to the new arrangement all differ in at least e residues from those corresponding to the original arrangement. If we can find a set S of $2e - 1$ vectors in F_p^e such that any subset of e vectors is linearly independent, then there is an elegant construction: any hyperplane, H, can be parallel to at most $e - 1$ linearly independent vectors, hence there are certainly e vectors in S which are not parallel to H. So if we can place $2e - 1$ lines in F_p^e with one line being parallel to each vector in S then any hyperplane must intersect at least e of these lines, and at most $2e - 1$. For example, in F_p^3 each line has p^2 translates, and if the space already contains k lines then at most kp of the translates will intersect with any of the lines already there. In general then, we need $p^{e-2} \geq 2e - 1$ to able to place $2e - 1$ lines in the space in a first-fit fashion. Once the lines have been placed we perform cyclic shifts as the case when $e = 2$.

To apply the lemma we must ensure than any pair of n-tuples contains $n + e$ or more different residues. This is automatically true if the two n-tuples are from the same arrangement, and we wish to ensure it if they are not. A sufficient condition is that $|C_i \cup C_j| \geq n + 3e - 1$ for any $C_i \neq C_j$ derived from the same arrangement. Suppose C_1 and C_2 come from different arrangements, then by construction there is a C_3 from the same arrangement as C_2 and which satisfies $|C_1 \cup C_3| \leq n + 2e - 1$. The condition above says that $|C_2 \cup C_3| \geq n + 3e - 1$ hence $|C_1 \cup C_2| \geq n + e$.

We have shown that if $(p-1)p^{e-2} \geq 3e - 1$ then the cyclic permutation along the $2e-1$ lines will give us a solution requiring only $2(p^e - 1)/(p-1) = 2p^e + 2p^{e-1} + \Theta(p^{e-2})$ reconstructions, which is only slightly greater than the optimal $2\binom{p^e}{e}/\binom{p^{e-1}}{e} = 2p^e + e(e-1)(p-1) + \Theta(p^{-e+2})$. The inequality relating p and e is true for $e = 3$ and $p \geq 5$, and for $e \geq 4$ and $p \geq 3$ — for $p = 2$ we need $e \geq 7$.

If $2e - 1$ linearly independent vectors cannot be found, or they cannot be placed in F_p^e without intersection then we cannot get so close to the lower bound. There is an obvious range of compromises depending on how many linearly independent vectors can be found (certainly $e + 1$ since the axes and main diagonal will suffice) and placed in F_p^e. We can always resort to performing e cyclic shifts along the axes, giving a total of $(e + 1)(p^e - 1)/(p - 1)$ reconstructions which is definitely suboptimal by at least a factor of $(e + 1)/2$ — note that any hyperplane intersects at least one of the axes.

Some failing cases for the hyperplane method

Unfortunately, it is not always possible to find a set of $2e - 1$ linearly independent vectors in F_p^e: for example if $p = 2$ and $e = 3$ then any set of vectors such that each subset of three is a basis can have no more than four members being, in essence, $(1, 0, 0)$, $(0, 1, 0)$, $(0, 0, 1)$ and $(1, 1, 1)$. It is easy to see that for $p = 2$ and any e there can be no more than $e + 1$ such vectors. The same is true for $p = 3$: without loss of generality we may pick the "axes" as e of the vectors, then any other vector must have all coordinates non-zero, and can be scaled to have first coordinate equal to 1. Suppose we could have two such extra vectors $\underline{x} := (1, x_2, \ldots, x_e)$ and $\underline{y} := (1, y_2, \ldots, y_e)$, then if for any i we have $y_i = x_i$ we get find a linear dependence between $\underline{x}$, $\underline{y}$ and the $e - 2$ axes which are zero in positions 1 and i. Hence for all i we have $y_i \neq x_i$ and both x_i and y_i are non-zero, so necessarily $x_i = -y_i$, giving a linear relation between $\underline{x}$, $\underline{y}$ and the axis in direction 1.

Similarly if $e > p + 1$ then there are only $e + 1$ such vectors. Without loss of generality we may assume the axis vectors are included. Suppose there were two other vectors $\underline{x}$ and $\underline{y}$ (necessarily with all coordinates non-zero) that could be added to the set. Then the plane $\{\lambda\underline{x} + \mu\underline{y} : \lambda, \mu \in F_p\}$ contains p^2 points. Also if we consider the values in some fixed coordinate position we see that each field element $0, 1, \ldots, p - 1$ occurs exactly p times. Thus there is a total of ep zeroes including the e zeroes in $\underline{0}$. The remaining $e(p - 1)$ zeroes are distributed amongst the other $p^2 - 1$ vectors in the plane. The pigeon-hole principle implies that some vector has two or more zeroes. Now there is clearly a linear relation between $\underline{x}$, $\underline{y}$ and $e - 2$ axis vectors.

We note, in contrast, that it is possible to find infinitely many vectors in Z^n (for $n > 1$) such that any subset of n is linearly independent. One such set is $\{(x^0, \ldots, x^{n-1}) : x \in Z\}$; the linear independence of any subset of n vectors from this set follows from the non-singularity of Vandermonde matrices.

Acknowledgments

Thanks to James Davenport and Geoff Smith for their helpful and insightful comments, and also to the referees.

References

[B&M73] Barsi F & Maestrini P "Error Correcting Properties of Redundant Residue Number Systems", IEEE TR Comp C-22, pp 307–315. Mar 1973

[B&M74] Barsi F & Maestrini P "Error Detection and Correction by Product Codes in Residue Number Systems", IEEE Tr Comp C-23, pp 915–923 Sep 1974

[H&W79] Hardy G H & Wright E M, "An Introduction to the Theory of Numbers", Clarendon Press, Oxford 1979

[Mandelbaum72] Mandelbaum D, "Error Correction in Residue Arithmetic", IEEE Tr Comp vol C-21, pp 538–545. June 1972

[Mandelbaum82] Mandelbaum D, "Decoding of Erasure and Errors for Certain RS Codes by Decreased Redundancy", IEEE Trans Inf Thy IT-28 pp 330-336

[Ramachandran83] Ramachandran V, "Single residue error correction in residue number systems", IEEE Tr Comp vol C-32, pp 504–507. May 1983

[Stone63] Stone J J, "Multiple Burst Error Correction with the Chinese Remainder Theorem", J. SIAM vol 11(1) pp 74–81, March 1963

[Y&L73] Yau S & Liu Y "Error Correction in Redundant Residue Number Systems", IEEE Tr Comp C-22, pp 5–11. Jan 1973

On a Categorial Isomorphism between a class of Completely Regular Codes and a class of Distance Regular Graphs

Josep Rifà-Coma *

Abstract

In [5] we established some relations between Distance Regular Graphs and Completely-Regular Codes in order to show the non-existence of a class of Distance-Regular Graph. In [6] we introduced the concept of Propelinear Code which is the algebraic structure associated to an e-Latticed, Distance-Regular Graph.

In this paper we present a functorial isomorphism between two categories, the category of e-Latticed, Distance Regular Graphs and the category of Completely Regular Binary Propelinear Codes. Starting from this categorial isomorphism it is easy to translate some properties from one category to the other, specially the non-existence properties of certain types of graphs (see 5.1).

We also show in this paper that the linear structure is the only possible propelinear structure in Golay Codes G_{23} and G_{24}. From this fact we deduce the uniqueness of certain types of graphs given by its parameters (see 5.2 and cf. [2] Chap.11).

This paper was originally conceived to try to answer a question asked by Prof. Brouwer (see theorem 11).

The categorial treatment owes itself to talks that I have had with Prof. O. Moreno.

*e-mail: iinf5@ccuab1.uab.es, Dpt. d'Informàtica, Universitat Autònoma de Barcelona. 08193-Catalonia (Spain)

1 Introduction

1.1 Completely-Regular Codes

Let F^n be an n-dimensional vector space over the Galois field $GF(q)$ of $q = p^r$ elements, p being a prime number.

The *weight* , $W_H(V)$, of a vector $V \in F^n$ is the number of its non-zero coordinates. The *Hamming distance* between two vectors $V, S \in F^n$ is $d(V, S) = W_H(V - S)$.

A q-ary *block-code* of length n, is a subset C of F^n. We call the elements of C code words. When C is a vector subspace of F^n it will be called *linear code*.

The *minimum distance* d of a code C is the minimum value of $d(V, S)$, where $V, S \in C$, and $V \neq S$.

The *error correcting* capability e of a code C is $\lfloor \frac{d-1}{2} \rfloor$. In this case C is an e-error correcting code.

We define $B(V, p) = Card\{c \in C | d(V, c) = p\}$, the number of code words at distance p from $V \in F^n$, (see [3]). Let $\rho(V)$ be the distance from $V \in F^n$ to the code C, that is to say: $\rho(V) = min\{p | 0 \leq p \leq n, B(V, p) \neq 0\}$. The *covering radius* ρ of a code C is the maximum value of $\rho(V)$, for all $V \in F^n$

An e-error-correcting code $C \subset F^n$ is called *completely regular* if $\forall V \in F^n, \forall p, 0 \leq p \leq n$, the value of $B(V, p)$ depends only on $\rho(V)$, that is to say, $\forall V \in F^n$ such that $\rho(V) = u$, its value is a constant number $B(V, p) = b_{up}$. We refer to the numbers b_{up} as the *outer distribution numbers* of a completely regular code C, (see [3]).

1.2 Distance-Regular Graphs

A *graph* $\Gamma(X, A)$ consists of a vertex set X and *edge* set A, where every edge is a subset of cardinality two of X. Let Γ be a connected undirected graph on X. A *path* of length r from x to y, $(x, y \in X)$, is a sequence of vertices $x_0 = x, x_1, x_2, \ldots, x_{r-1}, x_r = y$ such that each (x_i, x_{i+1}) is an edge of Γ. The *distance* from x to y is the minimum length of paths from x to y, and is denoted by $d(x, y)$. The *diameter* of Γ is the maximum distance between any two vertices and denoted by ρ. Let $\Gamma(X, A)$ and $\Gamma'(Y, B)$ be two graphs. A map $g : X \longrightarrow Y$ is a *graph-morphism* if and only if given two adjacent elements $a, b \in X$, that is to say, $(a, b) \in A$, then $(g(a), g(b)) \in B$. A graph is called a *Distance-Regular Graph* if for every integers $i, j, k \in \{0, 1, \ldots, \rho\}$, the number of $z \in X$ such that $d(x, z) = i$ and $d(z, y) = j$ is a constant number, denoted by p_{kij}, whenever $d(x, y) = k$, and $p_{kij} = p_{kji}$. The constant number p_{011} is called the *valency* of the graph. If Γ is a distance-regular graph the so called intersection matrix $B_1 = (p_{k1j})$ is a tridiagonal matrix with non-zero off-diagonal entries:

$$B_1 = \begin{pmatrix} a_0 & c_1 & \cdots & 0 & 0 \\ b_0 & a_1 & \cdots & 0 & 0 \\ 0 & b_1 & \cdots & 0 & 0 \\ \vdots & \vdots & \ddots & \vdots & \vdots \\ 0 & 0 & \cdots & c_{\rho-1} & 0 \\ 0 & 0 & \cdots & a_{\rho-1} & c_\rho \\ 0 & 0 & \cdots & b_{\rho-1} & a_\rho \end{pmatrix}$$

Or shorter, $B_1 = \begin{pmatrix} * & c_1 & \cdots & c_{\rho-1} & c_\rho \\ a_0 & a_1 & \cdots & a_{\rho-1} & a_\rho \\ b_0 & b_1 & \cdots & b_{\rho-1} & * \end{pmatrix}$ where

$$\begin{cases} a_i = p_{i1i}, & i = 0, 1, \ldots, \rho \\ b_i = p_{i1(i+1)}, & i = 0, 1, \ldots, \rho - 1 \\ c_{i+1} = p_{(i+1)1i}, & i = 0, 1, \ldots, \rho - 1 \end{cases}$$

1.3   e-Latticed Distance-Regular Graphs

Let F be the field $Z/2$. We can consider a fixed basis $e_1, e_2, \ldots, e_n$, in F^n, and we can represent each element of F^n making use of its coordinates in the basis $\{e_i\}$. We will use the graph structure F^n, where the vertices are all the elements in F^n, and two vertices are adjacent if and only if they are at unit distance apart (distance means Hamming's distance).

This graph structure is called *n-cube*.

Definition 1 *A distance-regular graph is e-latticed, of valency n, if the parameters p_{i1j} of its intersection matrix B_1 coincide with those of n-cube for $i + j \le 2 \cdot e - 1$.*

That is to say, a distance-regular graph will be e-latticed if its parameters satisfy:

$$\begin{cases} p_{i1j} = n - i & \text{for } 0 \leq i = j - 1 \leq e - 1 \\ p_{i1i} = 0 & \\ p_{i1j} = i & \text{for } 0 \leq i = j + 1 \leq e \end{cases}$$

Remark: The n-cube F^n is an n-latticed distance regular graph.

1.4 Hamming's Coverings

Definition 2 *Let Γ, Γ' be two connected undirected graphs on X. A map $\theta : \Gamma \longrightarrow \Gamma'$ is called an e-local isomorphism if $\forall\, i = 0, 1, \ldots, e - 1$ and $\forall\, u \in \Gamma$, θ induces a graph isomorphism:*

$$\theta : B_i(u) \longrightarrow B_i(\theta(u))$$

where $B_i(u) = \Gamma_0(u) \bigcup \Gamma_1(u) \bigcup \ldots \bigcup \Gamma_i(u)$, and $\Gamma_j(u) = \{x \in \Gamma | d(x, u) = j\}$

Remark: If $\theta : \Gamma \longrightarrow \Gamma'$ is an e-local isomorphism, then its restriction $\theta : \Gamma_i(u) \longrightarrow \Gamma'_i(\theta(u))$, for every $i = 0, 1, \ldots, e - 1$, is a graph isomorphism.

Definition 3 *An e-covering of a connected, undirected graph Γ is a surjective, map from the n-cube F^n to Γ which is also an e-local isomorphism.*

Definition 4 *Let θ be an e-covering, $\theta : F^n \longrightarrow \Gamma$, and let $\alpha \in \Gamma$ be the image of $0 \in F^n$, $\theta(0) = \alpha$. Say $C = \theta^{-1}(\alpha)$. C will be called the associated code of e-covering θ.*

Remark:

- In [5] we proved that starting from a distance regular e-latticed graph Γ, $(e \geq 3)$, of valency n, $\forall\, \alpha \in \Gamma$, we can construct an e-covering, $e \geq 3$, $\theta_\alpha : F^n \longrightarrow \Gamma$, such that the associated code $C = \theta_\alpha^{-1}(\alpha)$ becomes a completely regular code.

- We also showed in [5] that $\forall \alpha \in \Gamma$, the covering θ_α constructed in [5] is the unique covering such that $\theta_\alpha(e_i) = \alpha_i$, where $\{e_i\}$ is a prefixed basis in F^n, and α_i are the n vertices of Γ at a distance one apart from α.

1.5 Propelinear Codes

Definition 5 *Let C be a subset of F^n, $(F = \mathcal{Z}/2)$, $0 \in C$. We will call propelinear code to code C such that:*

$\forall \, v \in C$, *there exists an isometry* $\pi_v : F^n \longrightarrow F^n$ *such that;*

1. $v + \pi_v(s) \in C$ *if and only if* $s \in C$.

2. $\forall \, w \in C$, $\pi_v \circ \pi_w = \pi_m$, *where* $m = v + \pi_v(w) \in C$.

We will call $PC(n,e)$-code any propelinear code $C \subset F^n$, where $d = min\{d(a, v + \pi_v(a)) | \forall \, a \in F^n, \forall \, v \in C, a \neq v + \pi_v(a)\}$, and $e = \lfloor \frac{d-1}{2} \rfloor$.

Remark:

- Every linear code is a propelinear code if we take $(\forall \, v \in C)$, $\pi_v = Id$.

- The following code $C \subset F^4$ is not a linear, but a propelinear code: Let $C = \{0, v, w, s\} \subset F^4$, where $0 = (0,0,0,0)$, $v = (1,1,0,0)$, $w = (0,1,1,0)$, $s = (0,1,0,1)$. For every vector in C, the isometries are given by: $\pi_0 = Id.; \pi_v = \pi_{12} \circ \pi_{34}; \pi_w = \pi_{23} \circ \pi_{14}; \pi_s = \pi_{13} \circ \pi_{24}$, where each π_{ij} means the transposition that changes the coordinates i and j.

Definition 6 *Let C be a $PC(n,e)$-code. $\forall \, x \in C, s \in F^n$, we define $x \star s = x + \pi_x(s)$.*

Proposition 7 *With the $\star$ operator C has a group structure, non-necessarily commutative. Also the operator $\star$ makes F^n as a C-set (that is to say C operates on F^n, on the left, and $(x \star y) \star s = x \star (y \star s)$, $\forall \, x, y \in C, s \in F^n$)*

Proof: It can be seen in [6] ∎

2 The Category $\mathcal{G}$ of e-Latticed Distance Regular Graphs

We can consider the set of connected, without loops, undirected, e-latticed, distance-regular graphs, $e \geq 3$, of valency n, and we suppose that we have fixed a concrete vertex α in each of them. We also suppose that we have enumerated with α_1, α_2, ..., α_n the n vertices that are at a distance one apart from α. We can write $\Gamma(e, \alpha, \alpha_i)$ to describe a graph of this kind. *This set of graphs is the set of objects of the category $\mathcal{G}$.*

For two graphs $\Gamma(e, \alpha, \alpha_i)$, $\Gamma'(e', \beta, \beta_i) \in \mathcal{G}$, *the set* $\mathcal{M}(\Gamma, Gamma')$ *of morphisms consists of all the 2-local isomorphisms* $f : \Gamma \longrightarrow \Gamma'$, *such that* $f(\alpha) = \beta$.

3 The Category $\mathcal{C}$ of Propelinear Completely Regular Codes

We can consider the set of all propelinear (with a fixed propelinear structure), e-error correcting binary completely regular codes, $e \geq 3$, of length n, and we suppose that we have fixed a concrete basis $e_1, e_2, \ldots, e_n$ in F^n formed by the vectors of weight one. We can write $X(e, e_i)$ to describe a code of this kind.

Definition 8 *A propelinear morphism* $f : X \longrightarrow Y$ *between two propelinear codes of length* n *is an isometry* $f : F^n \longrightarrow F^n$, *such that* $f(X)$ *is in* Y, *and* $\forall\ x \in X, y \in F^n$, *is* $f(x \star y) = f(x) * f(y)$.

($\star$ and $*$ are, respectively, the propelinear structure in X and Y)

The set of codes $X(e, e_i)$ are the objects of category $\mathcal{C}$. For two codes X, Y, the set $\mathcal{M}(X, Y)$ of morphisms consists of all the propelinear morphisms $f : X \longrightarrow Y$.

3.1 the $\Omega(C)$-graph associated to a Propelinear Code

Let C be a propelinear $PC(n, e)$-code, starting from the group $C_\star$ (C is not a subgroup of F^n, because the operator $\star$ is not defined in F^n), we can construct a quotient set $\Omega(C)$ formed by the right cosets $C \star a$, where $a \in F^n$.

We can consider the set $\Omega(C)$ as a vertex set and we define the adjacency relation between vertices in the following way:

Two vertices $A, A' \in \Omega(C)$ are adjacent if and only if there exists representatives in the classes A and A' which are at unit Hamming's distance apart. The vertices $C \star e_i$ are the n vertices in $\Omega(C)$ at a distance one apart from C.

We can consider in $\Omega(C)$ a vertex underlined. This is the vertex associated with the class C in $\Omega(C)$. We can also enumerate in $\Omega(C)$ the n vertices at a distance one apart from C as $\alpha_i = C \star e_i$. Thus $\Omega(C)(e, C, \alpha_i) \in \mathcal{G}$.

Definition 9 *We will call $\Omega(C)$ the coset-graph associated to propelinear code C.*

Remark:

- We showed in [6] that if C is an e-error correcting, propelinear code, ($e \geq 3$), then $\Omega(C)$ is a distance-regular graph $\Omega(C) \in \mathcal{G}$ if and only if C is a completely regular code C. The diameter ρ of the graph coincides with the covering radius of the code.

- Also we proved in [6], $\forall \beta \in \Gamma$, $D = \theta_\alpha^{-1}(\beta)$ is a completely regular code with the same outer distribution numbers as C. In the case that Γ is $\Omega(C)$ then $\theta_\alpha^{-1}(\beta)$ coincides with some class $C \star s$ in C.

The next proposition improves this last remark.

Proposition 10 *Let C be a completely regular $PC(n,e)$-code, $e \geq 3$. Then every class $C \star s$ in $\Omega(C)$ is the translation of a propelinear $PC(n,e)$-code D, which has the same outer distribution numbers as C. This propelinear code is the associated code to the graph $\Omega(C)$ when we take as a fixed vertex the vertex $C \star s$.*

Proof: Let $s \in F^n$ be a fixed element and let $C \star s$ be its class in $\Omega(C)$. Let D be the code defined by $D = (C \star s) + s = \{c + \pi_c(s) | \forall c \in C\}$. For every $v \in D, v = (c \star s) + s$, where $c \in C$, we will take $\pi_v = \pi_c$, and we will define the $*$ operator in D: $\forall v \in D, x \in F^n \Rightarrow v * x = v + \pi_v(x) = c + \pi_c(s) + s + \pi_c(x)$.
Firstly we are going to show that D is a propelinear code.

- We will prove that for all $v \in D$, $v + \pi_v(m) \in D$ if and only if $m \in D$.

  If $m \in D$, then $m = d + \pi_d(s) + s$, for some $d \in C$. Thus we will have
  $v + \pi_v(m) = v + \pi_v(d + \pi_d(s) + s) = c + \pi_c(s) + s + \pi_c(d) + \pi_c \circ \pi_d(s) + \pi_c(s) = c + \pi_c(d) + \pi_c \circ \pi_d(s) + s = g + \pi_g(s) + s$, where $g = c + \pi_c(d) \in C$.

So $v + \pi_v(m) \in D$.

If $v + \pi_v(m) \in D$ we want to see that $m \in D$.

$v + \pi_v(m) \in D \Rightarrow v + \pi_v(m) = d + \pi_d(s) + s$ for some element $d \in C$. Moreover $v = c + \pi_c(s) + s$, where $c \in C$.

Thus we can write, $d + \pi_d(s) + s = v + \pi_v(m) = c + \pi_c(s + m) + s$. Hence

$$d + \pi_d(s) = c + \pi_c(s + m) \tag{1}$$

We know that C_* is a group (see Prop. 7) and given $d \in C$ there exists $g \in C$ which is the inverse element of c. We have $c \star g = 0$, that is to say, $c + \pi_c(g) = 0$, and $\pi_c(g) = c$, or also $\pi_g(c) = g$.

From Eq.(1) we obtain: $\pi_g(c + \pi_c(s + m)) = \pi_g(d + \pi_d(s)) \Rightarrow g + s + m = \pi_g(d) + \pi_g \circ \pi_d(s)$. So $m = g + \pi_g(d) + \pi_g \circ \pi_d(s) + s$. But $g + \pi_g(d) \in C$, hence $m \in D$.

- It remains to prove that for all $m, n \in D$ is $\pi_m \circ \pi_n = \pi_p$, where $p = m + \pi_m(n) \in D$.

If $m \in D$ then $m = c + \pi_c(s) + s$, where $c \in C$, and $\pi_m = \pi_c$.

If $n \in D$ then $n = d + \pi_d(s) + s$, where $d \in C$, and $\pi_n = \pi_d$.

We have $\pi_m \circ \pi_n = \pi_c \circ \pi_d = \pi_{c \star d} = \pi_h$, where $h = c + \pi_c(d) \in C$.

$p = m + \pi_m(n) = c + \pi_c(s) + s + \pi_c(d + \pi_d(s) + s) = c + s + \pi_c(d) + \pi_{c \star d}(s) = h + \pi_h(s) + s$.

The conclusion is $p \in D$, and $\pi_p = \pi_h = \pi_m \circ \pi_n$.

Now D is a propelinear code and its outer distribution numbers are the same as those of $C \star s$, and coincide with the outer distribution numbers of C (see remark 3.1). From D we can construct the associated graph $\Omega(D)$.

Every class in $\Omega(D)$ is of type: $D * t = \{m + \pi_m(t) | \ \forall \ m \in D\} = \{c + \pi_c(s) + s + \pi_c(t) | \forall c \in C\} = \{c + \pi_c(s + t) + s | \forall c \in C\} = (C \star (s + t)) + s$.

So we can conclude that every class in $\Omega(D)$ is the translation of a class in $\Omega(C)$. ∎

Theorem 11 *Let G_{23} be the binary Golay code $(23, 12, 7)$ and let G_{24} be the extended binary Golay code $(24, 12, 8)$.*

The linear structure is the only propelinear structure in G_{23} or G_{24}.

Proof: Let C be G_{23} or G_{24}.

Assume there is another propelinear structure $\star$ in C, apart from the linear one. The aim of this proof is to show that for every $v \in C$ is $\pi_v = Id$. But the vectors of weight 7 or 8, in C generate C, hence it is sufficient to show $\pi_v = Id$ for every $v \in C$ with weight 7 or 8.

For every $v \in C$, π_v is an isometry on F^n, but $\forall v, s \in C$, $v \star s \in C$. Thus $\pi_v(s) \in C$ and π_v is an isometry on C.

Let $e \in \{e_i\}$, where $\{e_i\}$ is the prefixed basis in F^n.

It follows from proposition 10 that $D = (C \star e) + e$ is a propelinear code with the same outer distribution numbers as C.

It is well-known (see [4]) that any code with the same parameters as the binary Golay code is isomorphic to it, so D is a linear code.

Take $v, w \in C$ to obtain $v + \pi_v(e) + e$ and $w + \pi_w(e) + e \in D$, and also $v + w + \pi_{v+w}(e) + e \in D$. Thus $\pi_v(e) + \pi_w(e) + \pi_{v+w}(e) + e \in D$, hence

$$\pi_v(e) + \pi_w(e) + \pi_{v+w}(e) + e = 0 \tag{2}$$

since the minimum distance in C is $d \geq 7$.

Say $\pi_v(e) = e' \neq e$, and apply Eq.(2) to obtain for all $w \in C$, $\pi_w(e) = e$, or $\pi_w(e) = e'$. Also we obtain from Eq.(2) that $\forall v, w \in C$ is $\pi_{v+w} = \pi_v \circ \pi_w$. The particular case $v = w$ shows that $\pi_v \circ \pi_v = Id$.

We conclude that for every $e \in \{e_i\}$ there exists $e' \in \{e_i\}$, such that for every $v \in C$, and irrespective of v, $\pi_v(e) = e$ or $\pi_v(e) = e'$, and in this case $\pi_v(e') = e$.

For all $v \in C$ let $Sup(v)$ be the subset of $\{e_i\}$, $Sup(v) = \{e_i \mid weight(v + e_i) = weight(v) - 1\}$.

Let $e' \neq e$,

- $\forall v \in C$, if $e \in Sup(v)$, and $e' \in Sup(v)$ then $\pi_v(e) = e$ and $\pi_v(e') = e'$.

  Indeed, if $\pi_v(e) = e'$ then the weight $W_H(v + \pi_v(e) + e)$ would be equal to $W_H(v) - 2$ and this is not possible in the Golay code.

- $\forall v \in C$, if $e \notin Sup(v)$, and $e' \notin Sup(v)$ then $\pi_v(e) = e$ and $\pi_v(e') = e'$.

  The same argument as above.

- $\forall v \in C$, if $e \in Sup(v)$, and $e' \notin Sup(v)$ then $\pi_v(e) = e'$ and $\pi_v(e') = e$.

Indeed, suppose $\pi_v(e) = e$, $\pi_v(e') = e'$ and take $w \in C$ such that $\pi_w(e) = e'$.

- – If $e \in Sup(w)$ then $e' \notin Sup(w)$. Thus if $h = v + w$, we will have $e \notin Sup(h)$ and $e' \notin Sup(h)$, hence $\pi_h(e) = e$. But $\pi_h(e) = \pi_{v+w}(e) = \pi_v \circ \pi_w(e) = \pi_v(e') = e'$. A contradiction.

- – If $e \notin Sup(w)$ then $e' \in Sup(w)$. Thus if $h = v + w$, we will have $e' \notin Sup(h)$ and $e \in Sup(h)$, hence $\pi_h(e) = e$. But $\pi_h(e) = \pi_{v+w}(e) = \pi_v \circ \pi_w(e) = \pi_v(e') = e'$. A contradiction.

- $\forall v \in C$, if $e \notin Sup(v)$, and $e' \in Sup(v)$ then $\pi_v(e) = e'$ and $\pi_v(e') = e$.

The same argument as above.

We continue with the proof of theorem. Let $v \in C$ be a vector of weight 7 or 8. After the previous assertions we have that π_v is an isometry fixing all the coordinates $\{e_i\}$, except that in $Sup(w)$, where $w = v + \pi_v(v)$. Say $w = e_1 + e_2 + \ldots + e_k + e_1' + \ldots + e_k'$, where $\pi_v(e_i) = e_i'$, and $weight(w) = 0, 8, 12, 16$.

- If $weight(w) = 8$, let $s \in C$ be a vector of weight 7 or 8 (see [4]) such that $e_1, e_2 \in Sup(s)$, and $e_i \notin Sup(s)$ (for all $e_i \neq e_1, e_2$). Then $s + \pi_v(s) \in C$, and $weight(s + \pi_v(s)) = 4$. A contradiction.

- The other cases $weight(w) = 12, 16$ give a contradiction to.

Hence $\pi_v = Id$, and we conclude the proof. ∎

4 An isomorphism between the categories $\mathcal{C}$ and $\mathcal{G}$

4.1 A Functor between the Category $\mathcal{C}$ and $\mathcal{G}$

Lemma 12 *Let $C(e, e_i)$, $D(e', e_i')$ be two objects in $\mathcal{C}$ category. Let $f : C \longrightarrow D$ be a propelinear morphism. We can define a map $f' : \Omega(C)(e, C, \alpha_i) \longrightarrow \Omega(D)(e', D, \beta_i)$, which is a $\mathcal{G}$-morphism.*

Proof: We define $f'(C \star a) = D * f(a).$ ∎

Taking into account this lemma we can deduce the following proposition,

Proposition 13 *There exists a functor $\mathcal{F}$ between the categories C and $\mathcal{G}$.*

Let $C(e, e_i)$, $D(e', e_i')$ be two objects of the category , and let $f : C \longrightarrow D$ be a propelinear morphism.

We can define $\mathcal{F} : C \longrightarrow \mathcal{G}$ such that $\mathcal{F}(C) = \Omega(C)$ and for each C-morphism $f : C \longrightarrow D$ a $\mathcal{G}$-morphism $\mathcal{F}(f) = f'$, where $f' : \Omega(C)(e, C, \alpha_i) \longrightarrow \Omega(D)(e', D, \beta_i)$ is defined like in lemma 12.

4.2 A Functor between the Categories $\mathcal{G}$ and C

We want to see that there exists a functor $\mathcal{S}$ between the category $\mathcal{G}$ and the category C. If $\Gamma(e, \alpha, \alpha_i), \Gamma'(e', \beta, \beta_i) \in \mathcal{G}$, there exists an unique e-covering and an unique e'-covering, respectively, (see [5]), $\theta : F^n \longrightarrow \Gamma$, and $\theta' : F^n \longrightarrow \Gamma'$ such that $\theta^{-1}(\alpha) = C$, and $\theta'^{-1}(\beta) = D$ are propelinear codes, and such that $\theta(e_i) = \alpha_i$, $\theta'(e_i) = \beta_i$.

Define $\mathcal{S}(\Gamma) = C$, and $\mathcal{S}(\Gamma') = D$.

Let $\Gamma(e, \alpha, \alpha_i)$, $\Gamma'(e', \beta, \beta_i) \in \mathcal{G}$, and let f be a $\mathcal{G}$-morphism, $f : \Gamma \longrightarrow \Gamma'$, for which $f(\alpha) = \beta$.

We will prove that there exists a functor $\mathcal{S}$ that associates Γ to a code $C = \mathcal{S}(\Gamma)$, and Γ' to a code $D = \mathcal{S}(\Gamma')$, and such that $\mathcal{S}(f) : C \longrightarrow D$ is a propelinear morphism.

Lemma 14 *Take $\Gamma \in \mathcal{G}$, and let $\theta : F^n \longrightarrow \Gamma$ and e-covering ($e \geq 3$). Then for each $v \in C = \theta^{-1}(\alpha)$, there exists a single isometry π_v such that $\forall s \in F^n, \theta(s) = \theta(v \star s)$.*

Proof: Is direct, but it can be seen in [6]. ∎

Proposition 15 *There exists a functor $\mathcal{S}$ between the categories $\mathcal{G}$ and C.*

Proof: With the above notation, given the $\mathcal{G}$-morphism $f : \Gamma(e, \alpha, \alpha_i) \longrightarrow \Gamma'(e', \beta, \beta_i)$, such that $f(\alpha) = f(\beta)$ and $f(\alpha_i) = z_i$, where z_i are the n vertices in Γ' at a distance one apart from β, (that is to say, z_i are the same as β_i but, possibly in another order).

- Construct $f \circ \theta$.

  $f \circ \theta$ is a covering $F^n \longrightarrow \Gamma'$, and is the only one that $f \circ \theta(0) = \beta$ and $f \circ \theta_i(e_i) = z_i$.

  Let g_i be the n vectors of weight one in F^n such that $\theta'(g_i) = z_i$.

  Consider the isometry $\mu : F^n \longrightarrow F^n$ which consists in the permutation which changes e_i to g_i.

- Define $S(f) = \mu$.

  μ is an isometry on F^n and $\mu(C) = D$.

$$
\begin{array}{ccc}
F^n & \xrightarrow{\theta} & \Gamma \\
\downarrow \mu & & \downarrow f \\
F^n & \xrightarrow{\theta'} & \Gamma'
\end{array}
$$

$\theta' \circ \mu$ and $f \circ \theta$ are two coverings with the same values on 0 and e_i. Thus $\theta' \circ \mu = f \circ \theta$, hence $\forall \, v \in C, \forall s \in F^n$

$$
\begin{aligned}
\theta' \circ \mu(v \star s) & = \theta' \circ \mu(v + \pi_v(s)) = \theta'(\mu(v) + \mu \circ \pi_v(s)) \\
& = \theta'(\mu(v) + \mu \circ \pi_v \circ \mu^{-1}(\mu(s))) \qquad (3) \\
f \circ \theta(v \star s) & = (\text{see lemma 14}) = f(\theta(s)) = \theta' \circ \mu(s) = \theta'(\mu(s)) \quad (4)
\end{aligned}
$$

Now, by lemma 14 we know that starting with $\mu(v) \in D$, there exists a single isometry $\pi_{\mu(v)}$ such that for each $\mu(s) \in F^n$ is $\theta'(\mu(s)) = \theta'(\mu(v) + \pi_{\mu(v)}\mu(s))$.

From Eq.(3) and Eq.(4) we conclude that $\mu \circ \pi_v \circ \mu_{-1} = \pi_{\mu(v)}$.

Hence $\mu(v \star s) = \mu(v + \pi_v(s)) = \mu(v) + \mu \circ \pi_v(s) = \mu(v) + \pi_{\mu(v)} \circ \mu(s) = \mu(v) \star \mu(s)$.

Thus $S(f) = \mu$ is C-morphism.

- Moreover, it is easy to prove that $\forall \Gamma \in \mathcal{G}$, is $S(Id_\Gamma) = Id_{S(\Gamma)}$, and $\forall f \in \mathcal{M}(\Gamma, \Gamma'); g \in \mathcal{M}(\Gamma', \Gamma'')$ then $S(g \circ f) = S(g) \circ S(f)$.

Now the proposition is proved. ∎

4.3 Isomorphism between the Categories $\mathcal{C}$ and $\mathcal{G}$

It can be easily proved that the functors previously described are inverses, that is to say if $\mathcal{F} : \mathcal{C} \longrightarrow \mathcal{G}$ and $\mathcal{S} : \mathcal{G} \longrightarrow \mathcal{C}$ are these functors, then $\mathcal{F} \circ \mathcal{S} = Id_{\mathcal{G}}$ and $\mathcal{S} \circ \mathcal{F} = Id_{\mathcal{C}}$.

5 Applications

5.1 On the non-existence of certain types of graphs

5.1.1 Graphs related with perfect codes.

Let Γ be an e-latticed distance regular graph, $e \geq 3$, with intersection matrix:

$$
B_1 = \begin{pmatrix}
* & 1 & 2 & \ldots & e \\
0 & 0 & 0 & \ldots & n-e \\
n & n-1 & n-2 & \ldots & *
\end{pmatrix}
$$

In this case Γ is an e-latticed graph with diameter $\rho = e$. So the associated code C is a completely regular binary e-error correcting code with covering radius e. This is the case in which the code C is a perfect code, and so, taking into account the well-known results on perfect codes, we can deduce the non-existence of this type of graph for $e \geq 4$, unless $n = 2 \cdot e + 1$ and $|C| = 2$ (in this case C is the repetition code [see 5.2.1.2]) or $n = e$ and $|C| = 1$ (in this case C is the trivial zero code [see 5.2.1.1]).

5.1.2 Graphs related with uniformly packed codes.

Let Γ be an e-latticed distance regular graph, $e \geq 3$, with intersection matrix $(c \neq e + 1)$:

$$
B_1 = \begin{pmatrix}
* & 1 & 2 & \ldots & e & c \\
0 & 0 & 0 & \ldots & 0 & n-c \\
n & n-1 & n-2 & \ldots & n-e & *
\end{pmatrix}
$$

In this case Γ is an e-latticed graph with diameter $\rho = e + 1$. So the associated code C is a completely regular binary (e)-error correcting code with covering radius ρ. This is the case in which the code C is an uniformly packed code, and so, taking into account the well-known results on uniformly packed codes (see [7]), we can deduce the non-existence of this type of graph for $e \geq 4$, unless $n = 2 \cdot \rho$ and $|3| = 2$ (in this case C is the repetition code and the given graph is the folded n-cube [see 5.2.2.1).

5.2 Uniqueness of certain graphs given by its parameters

5.2.1 Graphs related with perfect codes.

Let Γ be an e-latticed distance regular graph, $e \geq 3$, with intersection matrix:

$$B_1 = \begin{pmatrix} * & 1 & 2 & \ldots & e \\ 0 & 0 & 0 & \ldots & n-e \\ n & n-1 & n-2 & \ldots & * \end{pmatrix}$$

The only possible cases in which such graph exists were quoted in 5.1.1

1. **Case $n = e$, and $|C| = 1$.** The code C is the trivial zero code and there exists a single propelinear structure on code C, with these parameters. Its functorialy associated graph has the following intersection matrix:

$$B_1 = \begin{pmatrix} * & 1 & 2 & \ldots & n \\ 0 & 0 & 0 & \ldots & 0 \\ n & n-1 & n-2 & \ldots & * \end{pmatrix}$$

This graph is the n-cube and is uniquely determined by its intersection matrix.

2. **Case $n = 2 \cdot e + 1$ and $|C| = 2$.** The code C is the repetition code formed by two codewords, the codeword $(0, 0, ..., 0)$ and the codeword $(1, 1, ..., 1)$. There exists a single propelinear structure on code C.

The functorialy associated graph has the following intersection matrix:

$$B_1 = \begin{pmatrix} * & 1 & 2 & \ldots & e-1 & e \\ 0 & 0 & 0 & \ldots & 0 & e+1 \\ n & n-1 & n-2 & \ldots & e+2 & * \end{pmatrix}$$

This graph is the folded n-cube and it is uniquely determined by its parameters.

3. **Case $e = 3$.** No perfect code exists for $e \geq 4$. In the case $e = 3$ we know the existence of the binary Golay code G_{23}. The functorially associated graph Γ has the following intersection matrix:

$$B_1 = \begin{pmatrix} * & 1 & 2 & 3 \\ 0 & 0 & 0 & 20 \\ 23 & 22 & 21 & * \end{pmatrix}$$

The Golay code G_{23} has a single propelinear structure, which is the linear structure (see theorem 11). Making use of the categorial isomorphism between C and G and the well-known result that the binary Golay 3-error correcting code G_{23} is uniquely determined by its parameters (see [4]), we can deduce that this graph, Γ, is *uniquely determined* by its intersection matrix, (cf. [2], chapter 11).

5.2.2 Graphs related with uniformly packed codes.

Let Γ be an e-latticed distance regular graph, $e \geq 3$, with intersection matrix $(c \neq e + 1)$:

$$B_1 = \begin{pmatrix} * & 1 & 2 & \cdots & e & c \\ 0 & 0 & 0 & \cdots & 0 & n-c \\ n & n-1 & n-2 & \cdots & n-e & * \end{pmatrix}$$

The only possible cases in which such a graph exists were quoted in 5.1.2

1. **Case $n = 2 \cdot (e + 1)$ and $|C| = 2$.** The code C is the repetition code formed by two codewords, the codeword $(0, 0, ..., 0)$ and the codeword $(1, 1, ..., 1)$. There exists only one propelinear structure on code C. The functorially associated graph has the following intersection matrix:

$$B_1 = \begin{pmatrix} * & 1 & 2 & \cdots & e & n \\ 0 & 0 & 0 & \cdots & 0 & 0 \\ n & n-1 & n-2 & \cdots & n-e & * \end{pmatrix}$$

This graph is the folded n-cube and it is uniquely determined by its parameters.

2. **Case** $e = 3$. No uniformly packed code exists for $e \geq 4$, (see [7]). In the case $e = 3$ we know the existence of the extended binary Golay code G_{24}. The functorially associated graph has the following intersection matrix:

$$
B_1 = \begin{pmatrix} * & 1 & 2 & 3 & 24 \\ 0 & 0 & 0 & 0 & 0 \\ 24 & 23 & 22 & 21 & * \end{pmatrix}
$$

The extended Golay code G_{24} has a single propelinear structure, which is the linear structure (see theorem 11). Making use of the categorial isomorphism between $\mathcal{C}$ and $\mathcal{G}$ and the well-known result that the extended binary Golay 3-error correcting code G_{24} is uniquely determined by its parameters (see [4]), we can deduce that this graph, Γ, is *uniquely determined* by its intersection matrix, (cf. [2], chapter 11).

References

[1] E.Bannai, T.Ito, Algebraic Combinatorics I. (The Benjamin-Cummings Publishing Co.,Inc., California, 1984).

[2] A.Brouwer, A.M.Cohen, A.Neumaier, Distance Regular Graphs. (Springer-Verlag, 1989).

[3] P.Delsarte, An algebraic approach to the association schemes of coding theory, Philips Research Reps., Suppl. 10, (1973).

[4] F.J.Macwilliams, N.J.A.Sloane, The Theory of Error-Correcting Codes, (North-Holland Publishing Company, 1977).

[5] J.Rifà , L.Huguet, Classification of a Class of Distance-Regular Graphs via Completely Regular Codes. Discrete Applied Mathematics 26 (1990) 289-300.

[6] J.Rifà, Distance Regular Graphs and Propelinear Codes. Springer-Verlag, LNCS 357, (1989) 341-355.

[7] H.C.J. Van Tilborg, Uniformly Packed Codes, Doctoral Dissertation, Eindhoven University, (1976).

SINGLE EXPONENTIAL PATH FINDING IN SEMIALGEBRAIC SETS
PART I: THE CASE OF A REGULAR BOUNDED HYPERSURFACE

Joos Heintz[2]   Marie-Francoise Roy[1]   Pablo Solernó[2]

Abstract. *Let V be a bounded semialgebraic hypersurface defined by a regular polynomial equation and let x_1, x_2 be two points of V. Assume that x_1, x_2 are given by a boolean combination of polynomial inequalities. We describe an algorithm which decides in single exponential sequential time and polynomial parallel time whether x_1 and x_2 are contained in the same semialgebraically connected component of V. If they do, the algorithm constructs a continuous semialgebraic path of V connecting x_1 and x_2. By the way the algorithm constructs a roadmap of V. In particular we obtain that the number of semialgebraically connected components of V is computable within the mentioned time bounds.*

1. Introduction

Let R be a real closed field containing a subring A. Let $X_1, \ldots, X_n$ be indeterminates over A.

We consider the following path finding problem for semialgebraic sets: let V be a semialgebraic subset of R^n containing two points x_1 and x_2. Suppose that V and x_1, x_2 are described by boolean combinations of inequalities involving polynomials $F_1, \ldots, F_s \in A[X_1, \ldots, X_n]$. Let $D := \deg(F_1) + \cdots + \deg(F_s)$ be the sum of the total degrees of $F_1, \ldots, F_s$.

Our aim is to design (by means of an arithmetical network over A; see [Ga], [FGM]) an algorithm which decides whether x_1 and x_2 lie in the same semialgebraically connected component of V, and, if they do, constructs a continuous semialgebraic path connecting x_1 and x_2. By such an algorithm the exact number of semialgebraically connected components can easily be determined. Moreover if V is closed and bounded a roadmap construction of V is required.

By cylindrical algebraic decomposition it is possible to solve algorithmically this path finding problem for semialgebraic sets. Unfortunately this technique has its shortcomings since it produces quite elevated worst case complexity results: sequential time bounds of order $D^{n^{O(n)}}$ and parallel time bounds of order $n^{O(n)}(\log D)^{O(1)}$ (see [FGM] for details). This method has been used in [SS] for the solution of path finding problems in robotics.

However it is possible to obtain more realistic (we shall say "admissible") complexity bounds for the path finding problem: the sequential time complexity (i.e. the size of the corresponding arithmetical network) is of order $D^{n^{O(1)}}$ and the parallel time complexity (the depth of the network) is of order $n^{O(1)}(\log D)^{O(1)}$.

---

[1] IRMAR. Université de Rennes I. 35042 Rennes Cedex. FRANCE.
[2] Working Group Noaï Fitchas. Instituto Argentino de Matemática CONICET. Viamonte 1636 (1055) Buenos Aires, ARGENTINA. / Facultad de Ciencias Exactas y Naturales. Dpto. de Matemática (Univ. de Buenos Aires), Ciudad Universitaria (1428) Buenos Aires, ARGENTINA.

In this paper we present the core of this complexity result considering the case where V is a bounded regular hypersurface of R^n given by a polynomial $F \in A[X_1, \ldots, X_n]$ with non vanishing gradient on V (Theorem 12). The methods developed in [GV 1], [HRS 2,3] allow to transfer this algorithmical result for bounded regular hypersurfaces to arbitrary semialgebraic sets (see Theorem 13 below and for proofs [HRS 4]).

The paper uses ideas of elementary differential geometry and combines them with techniques from computational algebraic geometry. This method has been introduced in [GV 1], [G] and was extended in [S 1], [HRS 1,2,3]. More specifically our tools are the ones used in the efficient quantifier elimination procedure for real closed fields ([HRS 2,3]).

Our algorithms are based on elementary techniques of computational linear algebra and hence efficiently parallelizable. In this point they differ substantially from the algorithmical methods of [GV 1] and [G], which use efficient polynomial factorization. At this moment this has no efficient parallel counterpart.

A first attempt to obtain efficiently parallelizable single exponential time bounds for the path finding problem and roadmap constructions in semialgebraic sets has been made in [C 2] and [C 3]. However in this work the semialgebraically input sets are subject to considerable geometric and computational restrictions: they are in general position or already equipped with a Whitney Stratification (see also the comments on the proofs of [C 2] and [C 3] in [T]).

Let us also point out that in the bit-model a single exponential sequential time algorithm for the path finding problem with a somewhat more precise complexity bound of $D^{O(n^{20})}$ has been independently obtained by D. Yu Grigor'ev and N.N. Vorobjov in [GV 2].

It is not too difficult to derive from our methods a "parametrized" version of our main Theorem 12. Taking into account the results of [HRS 4] (or [GV 2]) one obtains thus an admissible algorithm which solves the problem of defining the semialgebraically connected components of a given semialgebraic set. Details have been done in [CGV] for the sequential time complexity in the bit-model.

Surveys of these results are given in [HKRS] and [GHRSV].

2. Notions and Notations

Throughout this paper we fix a real closed field R and a subring A of R (for example $R := \mathbf{R}$ and $A := \mathbf{Z}$).

Let $X_1, \ldots, X_n$ be indeterminates (variables) over R. To a polynomial $F \in A[X_1, \ldots, X_n]$ we assign its total degree $\deg(F)$ and its number of variables n. For a finite set $\mathcal{F}$ of polynomials of $A[X_1, \ldots, X_n]$ we write $\deg \mathcal{F} := 2 + \sum_{F \in \mathcal{F}} \deg(F)$.

We call an element $z \in R$ *real algebraic* if it is algebraic over A.

We consider R^n, $n = 0, 1, \ldots$, as a topological space equipped with the euclidean topology. Let $x := (x_1, \ldots, x_n)$ and $y := (y_1, \ldots, y_n)$ be two points of R^n, we write

$$|x - y| := \sqrt{(x_1 - y_1)^2 + \cdots + (x_n - y_n)^2}$$

for their euclidean distance. If r is a positive element of R we write $B(x, r) := \{y \in R^n; |x - y| < r\}$ for the open ball of radius r centered at x.

If $x, y \in R$ (i.e. $n = 1$) we write $[x, y], (x, y), [x, y), (x, y]$ for the closed, open and half open intervals with boundaries x and y.

A *semialgebraic subset* of R^n (over A) is a set definable by a boolean combination of equalities and inequalities involving polynomials from $A[X_1, \ldots, X_n]$.

A semialgebraic set has only finitely many semialgebraically connected components which are all semialgebraic (see [BCR], Def. 2.4.2 and Th. 2.4.4).

Let $V \subset R^n$, $W \subset R^m$ be semialgebraic sets and $f : V \to W$ a map. We call f *semialgebraic* if its graph is a semialgebraic subset of R^{n+m}.

The image of a semialgebraic set by a semialgebraic function is semialgebraic too. This fact is called the *Tarski-Seidenberg Principle* (see [BCR], Théorème 2.2.1).

The Tarski-Seidenberg Principle can also be stated in terms of logics. It means that the elementary theory of real closed fields with constants from A admits quantifier elimination ([BCR], Prop. 5.2.2).

From this we obtain the following *Transfer Principle* ([BCR], Prop. 5.2.3): let R' be a real closed extension of R and let Φ be a formula without free variables (i.e. all the variables are quantified) in the elementary language $\mathcal{L}$ of real closed fields with constants from A. Then Φ is valid in R' iff Φ is valid in R.

Let $\Phi \in \mathcal{L}$ be a formula in the free variables $X_1, \ldots, X_n$. Φ defines a subset S of R^n. Since the elementary theory of R admits quantifier elimination we see that S is semialgebraic. On the other hand Φ can also be interpreted over R'. Thus Φ defines also a semialgebraic subset of R'^n which we denote by $S(R')$ (the independence of this notation from the particular defining formula Φ is justified by the Transfer Principle).

We call two formulas in the same free variables $X_1, \ldots, X_n$ *equivalent* if they define the same subset of R^n (this means they are equivalent with respect to the elementary theory of real closed fields).

A formula containing no quantifiers is called *quantifier free*. Thus the semialgebraic sets are those which are definable by quantifier free formulas. A formula is called *prenex* if all its quantifiers occur at the beginning.

Let $\Phi \in \mathcal{L}$ be a formula built up by atomic formulas involving a finite set $\mathcal{F}$ of polynomials of $A[X_1, \ldots, X_n]$. We write $\deg \Phi := \deg \mathcal{F}$. The *length* of Φ is denoted by $|\Phi|$.

An algorithm $\mathcal{N}$ (represented by a suitable family of arithmetical networks over A, see [Ga], [FGM]) which for given natural numbers D, n and any input set $\mathcal{F} \subset A[X_1, \ldots, X_n]$ subject to $\deg \mathcal{F} \leq D$ computes an output set $\mathcal{G} \subset A[x_1, \ldots, X_m]$, is called *admissible* if the following conditions are satisfied:

  - $\deg \mathcal{G} = D^{n^{O(1)}}$.
  - the sequential complexity of $\mathcal{N}$ is $D^{n^{O(1)}}$.
  - the parallel complexity of $\mathcal{N}$ is $(n \cdot \log D)^{O(1)}$.

In the case that such an algorithm exists we say that $\mathcal{G}$ is computable from the input $\mathcal{F}$ in *admissible time*.

Under certain circumstances $\mathcal{F}$ and $\mathcal{G}$ may represent the polynomials involved in quantifier free formulas $\Phi, \Psi \in \mathcal{L}$ defining semialgebraic sets $V \subset R^n$ and $W \subset R^m$. If $\mathcal{N}$ is admissible and computes also from the input data Φ the boolean combination of atomic formulas representing Ψ, we say that W is *admissible computable* from V.

Note that the data structures corresponding to polynomials and to finite sets of polynomials are considered as coefficients vectors written in dense form.

We shall also make use of the notions of *critical point, Nash function* and *Nash variety*. For precise definitions, we refer to [BCR].

3. Algorithmical and mathematical tools

In this section we collect some algorithmical and mathematical tools we need for our later roadmap construction. We state first a theorem which expresses a local property for projections of semialgebraic sets. In terms of logics this theorem can also be interpretated as a statement about the local existence of continuous semialgebraic Skolem functions, computable in admissible time.

THEOREM 1 ([HRS 3]). *Let S be a semialgebraic subset of $R^k \times R^n$. It is possible to compute in admissible time the following items:*

- *a partition of R^k into semialgebraic sets T_i, $1 \leq i \leq s$.*
- *for each $1 \leq i \leq s$ a finite family $(\xi_{ij})_{1 \leq j \leq \ell_i}$ of continuous semialgebraic functions from T_i to R^n such that for each $1 \leq j \leq \ell_i$ the graph of ξ_{ij} belongs to S and such that for each $x \in T_i$ each semialgebraically connected component of $S \cap (\{x\} \times R^n)$ contains at least one point of the graph of some ξ_{ij}. In particular, each connected component of $S \cap (T_i \times R^n)$ contains at least the graph of some ξ_{ij}.* $\Diamond$

For a proof of Theorem 1, see [HRS 3], Théorème 7. A weaker version of this theorem appears in [G], [HRS 1] and [S 1].

Theorem 1 has a series of consequences (Corollaries 2, 3, 4):

COROLLARY 2. *Let S be a semialgebraic subset of $R \times R^n$. In admissible time it is possible to compute the following items:*

- *a partition of R in semialgebraic intervals T_i, $1 \leq i \leq s$.*
- *for each $1 \leq i \leq s$ a finite family $(\xi_{ij})_{1 \leq j \leq \ell_i}$ of continuous semialgebraic functions (curves) from T_i to R^n such that the graph of each ξ_{ij} belongs to S and such that for each semialgebraically connected component of $S \cap (\{x\} \times R^n)$ contains at least one point of the graph of some ξ_{ij}. In particular, each connected component of $S \cap (T_i \times R^n)$ contains at least the graph of some ξ_{ij}.*

Proof. Put $k := 1$ in Theorem 1 and observe that any semialgebraic subset of R can be decomposed in admissible time in finitely many intervals. $\Diamond$

Since the curves ξ_{ij} in Corollary 2 are defined on intervals, their images are semialgebraically connected. This observation will be relevant in later applications of this corollary.

COROLLARY 3 (efficient quantifier elimination, [HRS 2,3], [R]). *Let Φ be a prenex formula in the language $\mathcal{L}$ of ordered fields with constants in A. Suppose that Φ contains m blocks of quantifiers and n variables. Then it is possible to compute in sequential time $(\deg \Phi)^{n^{O(m)}} |\Phi|^{O(1)}$ and in parallel time $n^{O(m)} (\log \deg \Phi)^{O(1)} + (\log |\Phi|)^{O(1)}$ a quantifier free formula Ψ which is equivalent to Φ.* $\Diamond$

COROLLARY 4 (efficient curve selection lemma, [HRS 1], [S 1]). *Let S be a semialgebraic subset of R^n and let p be in the topological closure $\bar{S}$ of S. Assume that S and p are definable by quantifier free formulas of $\mathcal{L}$. Then there exists an admissible algorithm*

which computes positive elements $\delta, \varepsilon \in R$ and continuous semialgebraic curves γ_i : $[0, \delta] \to R^n$, $1 \leq i \leq s$, satisfying the following conditions:
- $\gamma_i((0, \delta]) \subset S$ and $\gamma_i(0) = p$ for each $1 \leq i \leq s$.
- each semialgebraically connected component of $B(p, \varepsilon) \cap S$ contains $\gamma_i((0, \delta])$ for at least one i.

Proof. Apply Theorem 1 to the semialgebraic set S' defined by
$$S' := \{(r, x) \in R \times R^n; \; x \in S, \; |x - p|^2 \leq r\}. \quad \Diamond$$

Let K be the fraction field of A. In the next lemma we consider an arithmetical network over K.

LEMMA 5. Let $\mathcal{J}$ be a zero dimensional ideal of $K[X_1, \ldots, X_n]$ given by a finite set $\mathcal{F}$ of generators. Denote by $Z(\mathcal{J})$ the set of zeros of $\mathcal{J}$ contained in R^n. Let U be a new indeterminate. In admissible time it is possible to compute polynomials $P, P_1, \ldots, P_n \in K[U]$ such that for each element $x \in Z(\mathcal{J})$ there exists a root u of P contained in R satisfying $x = (P_1(u), \ldots, P_n(u))$. $\quad \Diamond$

For a proof of Lemma 5 see [HRS 3], Proposition 7, or [C 1].

Let $F \in A[X_1, \ldots, X_n]$ be a polynomial such that the closed semialgebraic set $V := \{F = 0\} := \{x \in R^n; \; F(x) = 0\}$ is bounded, and such that the gradient $\nabla F := \left(\dfrac{\partial F}{\partial X_1}, \ldots, \dfrac{\partial F}{\partial X_n} \right)$ vanishes nowhere on V. Thus V is a regular bounded hypersurface of R^n.

Let $G \in A[X_1, \ldots, X_n]$ be a linear form and let $g : V \to R$ be the semialgebraic map induced by G on V.

We are going to apply to V and g concepts borrowed from elementary differential geometry. To be more precise, we consider V as a Nash subvariety of R^n and g as a Nash function defined on V (see [BCR], Chapitre 8); therefore the reader interested in more generality may replace in the next statements (Observation 6, Lemma 7, Propositions 8 and 9) the semialgebraic set V by any bounded and closed Nash subvariety of R^n and g by any Nash function mapping V into R.

The local inversibility theorem for Nash functions ([BCR], Proposition 2.9.5) implies the following basic fact:

OBSERVATION 6. Let x be a point of V and let $y_1, \ldots, y_{n-1}$ be a system of local coordinates for x. Suppose that x is not a critical point of g. Then there exists $1 \leq i \leq n - 1$ such that $g, y_1, \ldots, y_{i-1}, y_{i+1}, \ldots, y_{n-1}$ is a system of local coordinates for x.

LEMMA 7. Let t be an element of R such that t is not a critical value of g and such that the fibre $g^{-1}(t)$ is non empty. Then there exists a positive element $\varepsilon \in R$ such that for the semialgebraically connected components $C_1, \ldots, C_s$ of $g^{-1}((t - \varepsilon, t + \varepsilon))$ and for each $t' \in (t - \varepsilon, t + \varepsilon)$ the sets $C_1 \cap g^{-1}(t'), \ldots, C_s \cap g^{-1}(t')$ are non empty and semialgebraically connected.

Proof. Since the semialgebraically connected components of semialgebraic sets definable by polynomials of previously bounded degree are themselves uniformly semialgebraically definable we may argue by the Transfer Principle. Thus we may assume without loss of generality that $R := \mathbf{R}$. Thus V is compact.

By assumption no point of the fibre $g^{-1}(t)$ is critical. By Observation 6 we may therefore choose for each $x \in g^{-1}(t)$ an open neighborhood U_x in V, a positive element $\varepsilon_x \in \mathbf{R}$ and local coordinates of the form $g|_{U_x}, y_2, \ldots, y_{n-1}$ which map U_x homeomorphically onto the connected chart $(t - \varepsilon_x, t + \varepsilon_x) \times (-\varepsilon_x, \varepsilon_x)^{n-2}$ (here $g|_{U_x}$ denotes the restriction of g to U_x). From the particular form of this chart we infer that U_x and $U_x \cap g^{-1}(t')$ are connected for any $t' \in (t - \varepsilon_x, t + \varepsilon_x)$. Since $g^{-1}(t)$ is compact there exist finitely many points $x_1, \ldots, x_p \in g^{-1}(t)$ such that $U_{x_1}, \ldots, U_{x_p}$ cover $g^{-1}(t)$. Again by compactness we see that we may choose a positive $\varepsilon < \min\{\varepsilon_{x_j} : 1 \le j \le p\}$ such that for each $t' \in (t - \varepsilon, t + \varepsilon)$ the fibre $g^{-1}(t')$ is contained in $U_{x_1} \cup \ldots \cup U_{x_p}$ (this can be seen by the following indirect argument: suppose that there exists a sequence $(t_i)_{i \in \mathbf{N}}$ of real numbers converging to t such that for each $i \in \mathbf{N}$ there is a point $y_i \in (V \cap g^{-1}(t_i)) \backslash (U_{x_1} \cup \ldots \cup U_{x_p})$; since V is compact we may suppose that the sequence $(y_i)_{i \in \mathbf{N}}$ converges to a point $y \in (V \cap g^{-1}(t)) \backslash (U_{x_1} \cup \ldots \cup U_{x_p})$. Contradiction).

For $1 \le j \le p$ let $U_j := U_{x_j} \cap g^{-1}((t - \varepsilon, t + \varepsilon))$.

Choosing ε small enough and taking into account the "open ball" form of the charts $U_1, \ldots, U_p$ one can achieve the following: for each $1 \le j, j' \le p$ and for each $t' \in (t - \varepsilon, t + \varepsilon)$ the sets U_j are connected, $U_j \cap g^{-1}(t')$ is non empty and connected and its topological closure in $g^{-1}(t')$ coincides which $\overline{U_j} \cap g^{-1}(t')$.

For each $t' \in (t - \varepsilon, t + \varepsilon)$ we consider the following undirected graph $G_{t'}$: its vertex set is $\{1, \ldots, p\}$ and two vertices $j, j' \in \{1, \ldots, p\}$ form a "primitive edge" of $G_{t'}$ if $U_j \cap U_{j'} \cap g^{-1}(t') \ne \phi$. The edges of the graph $G_{t'}$ are defined by taking the transitive closure of the relation of $\{1, \ldots, p\}$ induced by the primitive edges. The connected components of the graph $G_{t'}$ describe the (topological) connected components of $g^{-1}(t')$ as unions of the connected sets $U_1 \cap g^{-1}(t'), \ldots, U_p \cap g^{-1}(t')$. We claim that $G_t = G_{t'}$ for all $t' \in (t - \varepsilon, t + \varepsilon)$ if $\varepsilon > 0$ is chosen small enough. Thus G_t describes the connected components of $g^{-1}((t - \varepsilon, t + \varepsilon))$ as unions of the connected sets U_j, $1 \le j \le p$. This implies Lemma 7.

Now we are going to prove our claim. First we show that G_t is a subgraph of $G_{t'}$ for all $t' \in (t - \varepsilon, t + \varepsilon)$ with $\varepsilon > 0$ sufficiently small. For this purpose let $t' \in (t - \varepsilon, t + \varepsilon)$ and let $j, j' \in \{1, \ldots, p\}$ be two vertices of G_t forming a primitive edge. Thus $U_j \cap U_{j'} \cap g^{-1}(t) \ne \phi$. Since $U_j \cap U_{j'}$ is a non empty open subset of the chart U_j its image by g contains an open interval of the form $(t - \varepsilon, t + \varepsilon)$. Therefore $U_j \cap U_{j'} \cap g^{-1}(t') \ne \phi$ for $t' \in (t - \varepsilon, t + \varepsilon)$. This means that j and j' form a primitive edge of $G_{t'}$ and our assertion follows now easily.

Let us now show the converse, namely that $G_{t'}$ is a subgraph of G_t for all $t' \in (t - \varepsilon, t + \varepsilon)$ where ε is chosen sufficiently small. Suppose that this is not the case. Then there exist vertices $j, j' \in \{1, \ldots, p\}$ and a sequence $(t_i)_{i \in \mathbf{N}}$ of real values converging to t such that $U_j \cap U_{j'} \cap g^{-1}(t_i) \ne \phi$ for all $i \in \mathbf{N}$ and such that $U_j \cap U_{j'} \cap g^{-1}(t) = \phi$. Thus we may choose without loss of generality a sequence $(y_i)_{i \in \mathbf{N}}$ of points $y_i \in U_j \cap U_{j'} \cap g^{-1}(t_i)$ converging to a point $y \in \overline{U_j} \cap \overline{U_{j'}} \cap g^{-1}(t)$. Let $k \in \{1, \ldots, p\}$ such that $y \in U_k \cap g^{-1}(t)$. Since $U_k \cap g^{-1}t$ is a neighborhood of y in $g^{-1}(t)$ and since the set $\overline{U_j} \cap g^{-1}(t)$, which contains y, is the closure of $U_j \cap g^{-1}(t)$ in $g^{-1}(t)$, we conclude that $U_j \cap U_k \cap g^{-1}(t) \ne \phi$.

By the same argument one infers that $U_{j'} \cap U_k \cap g^{-1}(t) \neq \phi$. Thus $\{j, k\}$ and $\{j', k\}$ are primitive edges of G_t which implies that $\{j, j'\}$ is an edge of G_t. Contradiction.

PROPOSITION 8. *Let* $t_1 < t_2$ *be two elements of* R. *Suppose that the interval* (t_1, t_2) *doesn't contain any critical value of* g. *Let* $C_1, \ldots, C_s$ *be the semialgebraically connected components of* $g^{-1}((t_1, t_2))$. *Then for each* $t \in (t_1, t_2)$, *the sets* $C_1 \cap g^{-1}(t), \ldots, C_s \cap g^{-1}(t)$ *are the semialgebraically connected components of* $g^{-1}(t)$. *In particular we obtain that the number of semialgebraically components remain constant when* t *ranges over* (t_1, t_2).

Proof. By the Transfer Principle we may assume $R := \mathbf{R}$.
In a first step of the proof we consider arbitrary real numbers t_1', t_2' with $t_1 < t_1' < t_2' < t_2$. Let $C_1', \ldots, C_{s'}'$, be the connected components of the compact semialgebraic set $g^{-1}([t_1', t_2'])$. We claim that for any $t \in [t_1', t_2']$ the sets $C_1' \cap g^{-1}(t), \ldots, C_{s'}' \cap g^{-1}(t)$ are the connected components of $g^{-1}(t)$. To see this we observe that $[t_1', t_2']$ can be covered by finitely many open intervals with the properties stated in Lemma 7. These intervals can be arranged in a chain of succesively overlaping members. Looking at the common values contained in any two of these overlaping intervals we infer our claim just by glueing connected sets together.
Proposition 8 follows now easily from our claim by considering any ascending chain of closed intervals $[t_1', t_2'], [t_1'', t_2''], \ldots$ converging to (t_1, t_2). ◇

PROPOSITION 9. *Let notations and assumptions be as in Proposition 8. Then for each* $1 \leq j \neq j' \leq s$ *and* $k = 1, 2$, *the following is true:*
(i) $\overline{C_j} \cap g^{-1}(t_k)$ *is semialgebraically connected*
(ii) *each point of* $\overline{C_j} \cap \overline{C_{j'}} \cap g^{-1}(t_k)$ *is a critical point of* g.

Proof. By the Transfer Principle, one may again assume that $R = \mathbf{R}$.
We show first (i): Suppose on the contrary that there exists a connected component of $g^{-1}((t_1, t_2))$, say C_1, such that $\overline{C_1} \cap g^{-1}(t_1)$ is disconnected. Then there exist open subsets $\mathcal{O}, \mathcal{U}$ of $\mathbf{R}^n$ such that $\mathcal{O} \cap \mathcal{U} = \phi$, $\overline{C_1} \cap g^{-1}(t_1) \cap \mathcal{O} \neq \phi$, $\overline{C_1} \cap g^{-1}(t_1) \cap \mathcal{U} \neq \phi$ and $\overline{C_1} \cap g^{-1}(t_1) \subset \mathcal{O} \cup \mathcal{U}$. By compactness one sees then that there exists $t_1 < t < t_2$ such that $C_1 \cap g^{-1}(t) \cap \mathcal{O} \neq \phi$, $C_1 \cap g^{-1}(t) \cap \mathcal{U} \neq \phi$ and $C_1 \cap g^{-1}(t) \subset \mathcal{O} \cup \mathcal{U}$. This contradicts the conclusion of Proposition 8.
Now we show (ii): Suppose on the contrary that there exist two different connected components of $g^{-1}((t_1, t_2))$, say C_1 and C_2, such that $g^{-1}(t_1) \cap \overline{C_1} \cap \overline{C_2}$ contains a point x which is not critical for g. By Observation 6, we may chose an open neighborhood $\mathcal{U}$ of x in V, a real number $0 < \varepsilon < t_2 - t_1$ and local coordinates of the form $g|_{\mathcal{U}}, y_2, \ldots, y_{n-1}$ which maps $\mathcal{U}$ homeomorphically onto $(t_1 - \varepsilon, t_1 + \varepsilon) \times (-\varepsilon, \varepsilon)^{n-2}$. Thus $\mathcal{U} \cap g^{-1}((t_1, t_2))$ is mapped onto $(t_1, t_1 + \varepsilon) \times (-\varepsilon, \varepsilon)^{n-2}$ which is connected. Since x is in the closure of C_1 and C_2 we see that $C_1 \cap \mathcal{U} \neq \phi$ and $C_2 \cap \mathcal{U} \neq \phi$. Choose points $x_1 \in C_1 \cap \mathcal{U}$ and $x_2 \in C_2 \cap \mathcal{U}$. The images of x_1 and x_2 in $(t_1, t_1 + \varepsilon) \times (-\varepsilon, \varepsilon)^{n-2}$ can be connected by a continuous path. Thus x_1 and x_2 can be connected by a continuous path in $\mathcal{U} \cap g^{-1}((t_1, t_2))$. This implies $C_1 = C_2$. Contradiction. ◇

We shall need the following concept of M-directions (M-functions) which share some basic properties with Morse functions.

DEFINITION. Let notations be as before. We call $g : V \to R$ an *M-direction* (or *M-function*) of V if the set of critical points of g is nowhere dense.

Note that the set of critical points of an M-function is finite, since this set is semialgebraic and nowhere dense.

OBSERVATION 10. As a consequence of Sard's theorem ([BCR], Théorème 9.5.2), one obtains that the coefficient vectors of the linear forms of $R[X_1, \ldots, X_n]$ inducing M-directions on V form a dense subset of R^n (see [BCR], proof of Proposition 11.5.1 for details). Later we shall make use of the existence of an admissible algorithm which yields M-directions of V with *integer* coordinates.

Such an algorithm can be obtained combining the fact that being an M-direction is an elementary property and the mentioned density of M-directions with efficient quantifier elimination over R (Corollary 3).

A detailed account of this argument can be found in [S 2]. Compare also [HRS 3] Théorème 3, where a description of an admissible algorithm for finding M-directions which doesn't make use of Sard's theorem and efficient quantifier elimination is given (in fact, this algorithm constructs M-directions with coordinates contained in a real quadratic field extension of $\mathbf{Q}$).

4. The roadmap algorithm

During this section let $V := \{F = 0\}$ be a regular bounded hypersurface of R^n defined by a polynomial $F \in A[X_1, \ldots, X_n]$ of degree $d := \deg(F)$, whose gradient $\nabla F := \left(\dfrac{\partial F}{\partial X_1}, \ldots, \dfrac{\partial F}{\partial X_n} \right)$ vanishes nowhere on V (we say in this case that F is a *regular* polynomial). The variables $X_1, \ldots, X_n$ induce on V continuous semialgebraic coordinate functions $\pi_1, \ldots, \pi_n$.

DEFINITION. Let $G \in A[X_1, \ldots, X_n]$ be a linear form and let $g : V \to R$ be the semialgebraic map induced by G on V. A piecewise parametrized continuous semialgebraic curve σ of V is called a *roadmap of V with respect to g* if the following conditions are satisfied:

(i) for each $t \in R$, any semialgebraically connected component of $g^{-1}(t)$ cuts σ.

(ii) any two points of σ which lie on the same semialgebraically connected component of V lie on the same semialgebraically connected component of σ.

This definition of roadmap is closely related to the regular and bounded hypersurface V and the semialgebraic map g.

As a matter of fact any semialgebraic set of dimension one can be represented in admissible time as an union of continuously parametrized semialgebraic curves. This can easily be deduced from Corollary 2 (see [HRS 4] for details).

By Observation 10 we are able to construct (in admissible time) a $\mathbf{Z}$-linear combination of the variables $X_1, \ldots, X_n$ which induces on V an M-direction. Thus we suppose from now on that the coordinate function $\pi_n : V \to R$ is an M-direction of V. For the sake of notational simplicity we write $\pi := \pi_n$.

In this section we are going to describe an algorithm $\mathcal{N}(A, n)$ (depending on the parameters A and n) which constructs a roadmap of V with respect to the M-direction

π and which, for any two points x_1 and $x_2 \in V$, given by quantifier free formulas Φ_1 and Φ_2 of the elementary language $\mathcal{L}$, decides whether x_1 and x_2 lie in the same semialgebraically connected component of V. If x_1 and x_2 do so, the algorithm finds a continuous semialgebraic curve of V joining them. We shall represent the algorithm $\mathcal{N}(A, n)$ by a family of arithmetical networks over A depending on the parameter $D := \deg(\Phi_1) + \deg(\Phi_2) + d$ when n is fixed. We shall apply $\mathcal{N}(A, n)$ recursively in n changing at each call the base ring A. In order to obtain an admissible overall complexity in terms of arithmetical operations and sign evaluations in the original base ring, we shall use only constructions which are uniform on A.

In the next section we will show that the algorithm $\mathcal{N}(A, n)$ runs in admissible time in the inputs: n, d, for the construction of the roadmap, and n, $\deg(\Phi_1) + \deg(\Phi_2) + d$ for the decision and the construction of the curve.

During this section we suppose that for any subring A' of R and any bounded hypersurface of R^{n-1} defined by a regular polynomial $F' \in A'[X_1, \ldots, X_{n-1}]$ such an algorithm $\mathcal{N}(A', n-1)$ is already given.

We subdivide the description of our algorithm in several steps.

<u>Step 1</u>: The set $\mathcal{B} := \{x \in V; x \text{ is a critical point of } \pi\}$ is definable by the following quantifier free formula Φ:

$$F = 0 \wedge \frac{\partial F}{\partial X_1} = 0 \wedge \ldots \wedge \frac{\partial F}{\partial X_{n-1}} = 0$$

Since π is an M-direction, the set $\mathcal{B}$ is finite. Applying efficient quantifier elimination (Corollary 3) to the formulas $(\exists X_2) \ldots (\exists X_n)\Phi, \ldots, (\exists X_1) \ldots (\exists X_{n-1})\Phi$, we compute (in admissible time) univariate polynomials in $X_1, \ldots, X_n$ which represent the coordinates of the elements of $\mathcal{B}$ as their roots, suitably codified by Thom's Lemma (see [CR]). In this sense we are able to compute $\mathcal{B}$ "explicitly" and to handle its elements. In particular we are able to evaluate the sign of a given polynomial at any point of $\mathcal{B}$ (see [HRS 3] or [RS] for details). Observe also that the coordinates of the points of $\mathcal{B}$ are algebraic over A and zeros of polynomials of degree $d^{n^{O(1)}}$. In such a way we obtain the critical values of π explicitly.

<u>Step 2</u>: Using Corollary 2 we compute (in admissible time) the following:
- a partition of $\pi(V)$ in intervals T_i, $1 \le i \le s$, definable over A.
- for each $1 \le i \le s$ a finite family of continuous semialgebraic curves $\xi_{ij} : T_i \to V$, $1 \le j \le \ell_i$, such that for any $t \in T_i$ each semialgebraically connected component of $\pi^{-1}(t)$ contains at least one point of the form $\xi_{ij}(t)$, where $1 \le j \le \ell_i$, and such that $\pi \circ \xi_{ij}(t) = t$ for all $1 \le j \le \ell_i$ and all $t \in T_i$. Since V is a closed and bounded semialgebraic subset of R^n, each ξ_{ij} has a unique continuous semialgebraic extension to the boundary of T_i (by Corollary 3, this extension can be computed in admissible time). Thus we may assume that all intervals T_i are closed and have disjoint interiors. For $t \in \pi(V)$ let $V_t^+ := \{x \in V; \pi(x) > t\}$ and $V_t^- := \{x \in V, \pi(x) < t\}$. Given a point $x \in V$ corresponding to a critical value $t := \pi(x)$ such that $x \in \mathcal{B}$ or x is an endpoint of a curve ξ_{ij}, we construct (in admissible time) two families of continuous semialgebraic curves $(\xi_k^{(x)})_{1 \le k \le m_x}$ and $(\tilde{\xi}_k^{(x)})_{1 \le k \le \tilde{m}_x}$ as follows: assume first that x is contained in

the topological closure of V_t^+. Using the effective curve selection lemma (Corollary 4) we find (in admissible time) a closed interval T_x^+ of the form $T_x^+ = [t, t^+]$ with $t < t^+$, a positive element $\varepsilon_x^+ \in R$ and continuous semialgebraic curves $\xi_k^{(x)} : T_x^+ \to V$, $1 \le k \le m_x$, satisfying the following conditions:

- $\xi_k^{(x)}((t, t^+]) \subset V_t^+$, $\xi_k^{(x)}(t) = x$ and $\pi \circ \xi_k^{(x)}(t') = t'$ for each $t' \in T_x^+$ and each $1 \le k \le m_x$.
- each semialgebraically connected component of $B(x, \varepsilon_x^+) \cap V_t^+$ contains at least one $\xi_k^{(x)}((t, t^+])$.

If x is not contained in the closure of V_t^+ we put $m_x := 0$.

Assume now that x is contained in the closure of V_t^-. By the effective curve selection lemma we find (in admissible time) a closed interval T_x^- of the form $T_x^- = [t^-, t]$ with $t^- < t$, a positive element $\varepsilon_x^- \in R$ and continuous semialgebraic curves $\tilde{\xi}_k^{(x)} : T_x^- \to V$, $1 \le \tilde{k} \le \tilde{m}_x$, satisfying the analogous conditions as before.

If x is not contained in the closure of V_t^- we put $\tilde{m}_x = 0$.

Let $\mathcal{K}$ be the set of all curves of the form ξ_{ij}, where $1 \le i \le s$, $1 \le j \le \ell_i$, or of the form $\xi_k^{(x)}, \tilde{\xi}_k^{(x)}$, where $x \in \mathcal{B}$, $1 \le k \le m_x$, $1 \le \tilde{k} \le \tilde{m}_x$.

Without loss of generality we may suppose that the curves contained in $\mathcal{K}$ have domains whose interior is not empty.

$\mathcal{K}$ is computable in admissible time and so is also the set of all "endpoints" of the curves contained in $\mathcal{K}$, namely

$$C := \left\{ \begin{array}{c} x \in V; \text{ there exists } \eta \in \mathcal{K} \text{ with domain } [t_1, t_2] \text{such that} \\ x = \eta(t_1) \text{ or } x = \eta(t_2) \end{array} \right\}$$

Summarizing all this we have constructed in admissible time the following items
- a finite set $\mathcal{K}$ of continuous semialgebraic curves of V having domains with non empty interiors.
- a finite set C consisting of all endpoints of the curves contained in $\mathcal{K}$.

Note in particular that the domains of the curves of $\mathcal{K}$ cover $\pi(V)$ and that C contains all the critical points of π. Observe also that the coordinates of the points of C are algebraic over A and that we have $\#C \cap \mathcal{K} = d^{n^{O(1)}}$ (# denotes cardinality).

Subdividing the domains of the curves contained in $\mathcal{K}$ and increasing eventually C we may suppose that no curve of $\mathcal{K}$ ranges over an interval whose interior contains an element of $\pi(C)$. This implies that two curves of $\mathcal{K}$ defined on intervals whose interiors overlap, have the same domain. Moreover we may assume that at most one boundary point of the domain of any curve of $\mathcal{K}$ is a critical value.

Step 3: We are going to describe a roadmap σ of V with respect to π. For this purpose let us consider an arbitrary element t of $\pi(C)$ which is not a critical value of π. Note that under this circumstance $\pi^{-1}(t)$ is a bounded hypersurface of R^{n-1} defined by the regular polynomial $F(X_1, \ldots, X_{n-1}, t) \in A[t][X_1, \ldots, X_{n-1}]$. Thus we call $\pi^{-1}(t)$ a regular fibre. For each two points x, $x' \in C \cap \pi^{-1}(t)$ we decide whether x and x' are in the same semialgebraically connected component of $\pi^{-1}(t)$ by means of the algorithm $\mathcal{N}(A[x, x'], n - 1)$. If they do we find a continuous semialgebraic curve of $\pi^{-1}(t)$ connecting x and x'.

We define σ as the join of these connecting curves (for t ranging over all non critical values contained in $\pi(C)$) with the curves of $\mathcal{K}$ constructed in Step 2. Thus σ is a piecewise parametrized continuous semialgebraic curve of V.

LEMMA 11. σ is a roadmap of V with respect to π.

Proof. From Step 2 in the construction of σ it is clear that for each $t \in R$ any semialgebraically connected component of $\pi^{-1}(t)$ cuts σ. Let x_1, x_2 be two different points of σ lying in the same semialgebraically connected component of V. We are going to show that x_1 and x_2 lie in the same connected component of σ. From Steps 2 and 3 in the construction of σ we see that we may suppose without loss of generality that $x_1, x_2 \in C$. Thus we can choose curves $\eta_1, \eta_2 \in \mathcal{K}$ such that x_1 is an endpoint of η_1 and x_2 is an endpoint of η_2.

Suppose first that η_1 and η_2 are defined on the same interval $[t, t']$ and that x_1 and x_2 are in the same semialgebraically connected component of $\pi^{-1}([t, t'])$. From Step 2 in the construction of σ we deduce that (t, t') contains no element of $\pi(C)$ and in particular no critical value of π.

Suppose first that t and t' are not critical values of π. By hypothesis $\eta_1([t, t'])$ and $\eta_2([t, t'])$ are contained in the same connected component C of $\pi^{-1}([t, t'])$. From Proposition 8 we infer that $C \cap \pi^{-1}(t)$ is a semialgebraically connected component of $\pi^{-1}(t)$ containing the points $\eta_1(t)$, $\eta_2(t) \in C$. Therefore $\eta_1(t)$ and $\eta_2(t)$ are connected by a continuous semialgebraic path of σ (following Step 3 in the construction of our roadmap). This implies that $\eta_1([t, t'])$, $\eta_2([t, t'])$ and hence also x_1, x_2 are contained in the same semialgebraically connected component of σ.

Assume now that t is a critical value of π. Then, by Step 2 in the construction of the roadmap σ, t' is not a critical value. First suppose that $\eta_1((t, t'])$ and $\eta_2((t, t'])$ are contained in the same semialgebraically connected component C of $\pi^{-1}((t, t'])$. Since $(t, t']$ contains no critical value of π we deduce from Proposition 8 that $C \cap \pi^{-1}(t')$ is semialgebraically connected in the regular fibre $\pi^{-1}(t')$. Furthermore $C \cap \pi^{-1}(t')$ contains the points $\eta_1(t')$, $\eta_2(t')$ which are elements of C. Thus, as before, we see that $\eta_1(t')$ and $\eta_2(t')$ are connected by a continuous semialgebraic arc of σ. This implies that x_1 and x_2 are contained in the same semialgebraically connected component of σ. Now suppose that $\eta_1((t, t'])$ and $\eta_2((t, t'])$ are contained in distinct semialgebraically connected components C and C' of $\pi^{-1}((t, t'])$. Since $(t, t']$ contains no critical point and π is an M-function, we infer from Proposition 9 that the semialgebraically connected components of $\pi^{-1}([t, t'])$ are unions of the topological closures of the semialgebraically connected components of $\pi^{-1}((t, t'])$ and eventually some critical points contained in $\pi^{-1}(t)$. Since by hypothesis $\eta_1([t, t'])$ and $\eta_2([t, t'])$ are semialgebraically connected in $\pi^{-1}([t, t'])$ there exist distinct semialgebraically connected components $C_1, \ldots, C_s$ of $\pi^{-1}((t, t'])$ with $C_1 = C$, $C_s = C'$ such that $\overline{C}_1 \cap \overline{C}_2 \neq \phi, \ldots, \overline{C}_{s-1} \cap \overline{C}_s \neq \phi$. By Proposition 9, the sets $\overline{C}_1 \cap \overline{C}_2, \ldots, \overline{C}_{s-1} \cap \overline{C}_s$ consist only of critical points contained in $\pi^{-1}(t)$.

Thus by Step 2 in the construction of σ there exist continuous semialgebraic curves $\theta_1, \theta_2, \theta_2', \ldots, \theta_{s-1}, \theta_{s-1}', \theta_s \in \mathcal{K}$ defined on $[t, t']$ such that the following holds:

- $\theta_1([t, t'])$ is contained in $\overline{C}_1$,

  $\theta_2([t, t'])$ and $\theta_2'([t, t'])$ are contained in $\overline{C}_2$,

$\vdots$

$\theta_{s-1}([t,t'])$ and $\theta'_{s-1}([t,t'])$ are contained in $\overline{C}_{s-1}$,

$\theta_s([t,t'])$ is contained in $\overline{C}_s$, and

$-$ $\theta_1(t) = \theta_2(t)$, $\theta'_2(t) = \theta_3(t), \ldots, \theta'_{s-2}(t) = \theta_{s-1}(t)$,

$\theta'_{s-1}(t) = \theta_s(t)$.

On the other hand $\eta_1(t')$ and $\theta_1(t')$ are elements of C lying in the same semialgebraically connected component $C_1 \cap \pi^{-1}(t')$ of the regular fibre $\pi^{-1}(t')$. Thus, by Step 3 of the construction of σ, the points $\eta_1(t')$ and $\theta_1(t')$ are connected by a continuous semialgebraic subarc of σ. This implies that $\eta_1(t')$ and $\theta_1(t')$ are in the same semialgebraically connected component of σ. Similarly one shows that $\theta_2(t')$ and $\theta'_2(t'), \ldots, \theta_{s-1}(t')$ and $\theta'_{s-1}(t')$, $\theta_s(t')$ and $\eta_2(t')$ lie pairwise in the same semialgebraically connected component of σ.

From this we conclude that one and the same connected component of σ contains all the endpoints of $\eta_1, \theta_1, \theta_2, \theta'_2, \theta_3, \ldots, \theta'_{s-2}, \theta_{s-1}, \theta'_{s-1}, \theta_s, \eta_2$. The same holds for the images of these curves. Thus x_1 and x_2 are in the same semialgebraically connected component of σ.

Now let $x_1 \neq x_2$ be arbitrary points of C lying in the same semialgebraically connected component of V. By [BCR], Définition et Proposition 2.5.11, there exists a continuous semialgebraic curve $\gamma : [0,1] \to V$ such that $\gamma(0) = x_1$ and $\gamma(1) = x_2$.

Let r be the number of times that $\pi \circ \gamma$ "crosses" values of $\pi(C)$, i.e. let r be the number of semialgebraically connected components of the set $\{z \in [0,1]; \pi \circ \gamma(z) \in \pi(C)\}$.

The case $r = 1$ is already treated.

Suppose that $r > 1$ and that the assertion is true for distinct points $x_1, x_2 \in C$ which can be connected by a continuous semialgebraic curve which crosses less than r times values of $\pi(C)$. Choose $z \in [0,1]$ such that $t := \pi \circ \gamma(z)$ is an element of $\pi(C)$ different from $\pi(x_1) = \pi \circ \gamma(0)$ and $\pi(x_2) = \pi \circ \gamma(1)$. By Step 2 of the construction of σ there exists $x \in C$ lying in the same semialgebraically connected component of $\pi^{-1}(t)$ as $\gamma(z)$. Thus x_1, x and x_2 lie in the same semialgebraically connected component of V and we can connect x_1 with x and x with x_2 by two continuous semialgebraic curves γ' and γ'' such that $\pi \circ \gamma'$ and $\pi \circ \gamma''$ cross less than r times values of $\pi(C)$. From our induction hypothesis we conclude now that x_1, x and x_2 lie on the same semialgebraically connected component of σ. This ends the proof of Lemma 11. $\diamond$

To finish the description of our algorithm $\mathcal{N}(A, n)$ it suffices to explain how a point $x \in V$, given by a quantifier free formula of $\mathcal{L}$, can be connected with the roadmap σ. From this the reader infers easily an algorithm which for two points x_1, x_2 of V, given by quantifier free formulas of $\mathcal{L}$, decides whether x_1 and x_2 lie on the same semialgebraically connected component of V, and, if they do so, finds a continuous semialgebraic curve connecting x_1 and x_2. Let x be a point of V defined by a quantifier free formula Φ of $\mathcal{L}$. By Corollary 3 we compute x explicitly from Φ in admissible time. If $t := \pi(x)$ is a critical value of π we find by the effective curve selection lemma (Corollary 4) in admissible time a continuous semialgebraic curve $\theta : T \to V$, defined on a closed interval T, such that the following holds:

$-$ the boundary points of T are t and some non critical value t'.

$-$ $\theta(t) = x$.

Replacing the point x by $\theta(t')$ we may assume without loss of generality that t is not a critical value of π. Thus $\pi^{-1}(t)$ is a bounded hypersurface of R^{n-1} defined by the regular polynomial $F(X_1, \ldots, X_{n-1}, t) \in A[t][X_1, \ldots, X_{n-1}]$. For each curve $\eta \in \mathcal{K}$ whose domain contains t we compute $\eta(t)$. Thus we obtain a finite set $\mathcal{D}$, explicitly computable in admissible time, which for each semialgebraically connected component of $\pi^{-1}(t)$ contains at least one point (see Step 2 in the construction of the roadmap σ). For each point $x' \in \mathcal{D}$ we decide by means of the algorithm $\mathcal{N}(A[x, x'], n-1)$ whether x' and x lie on the same semialgebraically connected component of $\pi^{-1}(t)$. Finally we find a point $x' \in \mathcal{D}$ and a continuous semialgebraic curve τ in $\pi^{-1}(t)$ connecting x and x'. Since x' is a point of σ the curve τ connects x with the roadmap σ.
This ends the description of the roadmap algorithm $\mathcal{N}(A, n)$.

Finally we observe that the algorithm $\mathcal{N}(A, n)$ can also be used to compute the number of semialgebraically connected components of V exactly. For this purpose we note that the set $\mathcal{C}$ computed in Step 2 of our construction of the roadmap σ cuts each semialgebraically connected component of V in at least one point. By means of the algorithm $\mathcal{N}(A, n)$ we can decide whether two points of $\mathcal{C}$ lie in the same semialgebraically connected component of V. Thus we can find a system of points of $\mathcal{C}$ representing each semialgebraically connected component of V just once. The cardinality of this system is exactly the number of semialgebraically connected components of V.

5. Complexity of the roadmap algorithm

It is clear that iterating the procedure of the last section we obtain a recursive construction of an algorithm denoted by $\mathcal{N}(A, n)$ which accepts as input a regular polynomial $F \in [X_1, \ldots, X_n]$ defining a bounded hypersurface V of R^n and points $x_1, x_2 \in V$ given by quantifier free formulas $\Phi_1, \Phi_2 \in \mathcal{L}$. The algorithm $\mathcal{N}(A, n)$ computes for this input a roadmap σ and decides whether x_1 and x_2 are in the same semialgebraically connected component of V, and, if they do, constructs a continuous semialgebraic curve joining x_1 and x_2. The algorithm $\mathcal{N}(A, n)$ is realizable by an arithmetical network over A whose size and depth depends on n and $D := \deg F + \deg \Phi_1 + \deg \Phi_2$.

We fix the inputs V, x_1, x_2 (given by F, Φ_1, Φ_2) and the parameters n, D for the moment.

Reviewing the constructions of the last sections we observe that the sequential and parallel complexities of $\mathcal{N}(A, n)$ (the size and the depth of the arithmetical network realizing the algorithm for fixed n and D) doesn't really depend on the ground ring A but only on the parameters n and D. Thus, let $S(n, D)$ be the sequential and $P(n, D)$ the (simultaneous) parallel complexity of the algorithm $\mathcal{N}(A, n)$ applied to an input with parameters of size n and D.

In the last section, we constructed in admissible time the following items:
– a variable transformation corresponding to an M-direction π of V.
– certain continuous semialgebraic curves of V (e.g. the curves of $\mathcal{K}$).
– certain points of V, definable over $\mathcal{L}$, with algebraic coordinates.

All the constructions of this "preprocessing" are realizable by arithmetical networks over A of size $D^{n^{O(1)}}$ and depth $n^{O(1)}(\log D)^{O(1)}$. Then we used $D^{n^{O(1)}}$ calls of algorithms of the form $\mathcal{N}(A[x, x'], n-1)$, where x and x' were certain points of V definable over $\mathcal{L}$ (e.g. from $\mathcal{C}$) with $t := \pi(x) = \pi(x')$ and t being not a critical value of π.

Such an algorithm $\mathcal{N}(A[x, x'], n-1)$ is realizable by an arithmetical network over $A[x, x']$. Its inputs are the regular polynomial $F(X_1, \ldots, X_{n-1}, t)$ and the points $x, x' \in A[x, x']^n$. This input has parameters of size $n-1$ and $\deg(F)+2n$, which are bounded by $n-1$ and D. Thus the arithmetical network *over* $A[x, x']$ which realizes this algorithm has sequential complexity $S(n-1, D)$ and parallel complexity $P(n-1, D)$.

We have to transform this algorithm into an arithmetical network *over* A.

Let K be the fraction field of A. Since x and x' are explicitly given they are zeros of a 0-dimensional ideal of $K[X_1, \ldots, X_n]$ with known generators whose number and degrees are of order $D^{n^{O(1)}}$. Using Lemma 5 we construct in sequential time $D^{n^{O(1)}}$ and parallel time $n^{O(1)}(\log D)^{O(1)}$ an algebraic element (over A) u of R, suitably codified by Thom's Lemma as a zero of an effectively given univariate polynomial $P \in A[U]$ with $\deg(P) = D^{n^{O(1)}}$, such that $K[x, x'] \subset K[u]$.

We interprete now our algorithm as an arithmetical network over $K[u]$. From [HRS 3], Proposition 3, we deduce that arithmetical operations and sign determinations in $K[u]$ can be performed by an arithmetical network over A of size $D^{n^{O(1)}}$ and depth $n^{O(1)}(\log D)^{O(1)}$. This implies that the algorithm $\mathcal{N}(A[x, x'], n-1)$ in question can be executed by an arithmetical network *over* A in sequential time $D^{n^{O(1)}} \cdot S(n-1, D)$.

Let us first estimate the sequential complexity $S(n, D)$ of the algorithm $\mathcal{N}(A, n)$. We have to consider the "preprocessing", which can be realized in sequential time $D^{n^{O(1)}}$, and $D^{n^{O(1)}}$ calls of algorithms of type $\mathcal{N}(A[x, x'], n-1)$, which can be executed by arithmetical networks over A of size $D^{n^{O(1)}} S(n-1, D)$.

Thus we obtain the following recursion formula:

$$S(n, D) = D^{n^{O(1)}} S(n-1, D).$$

From this formula one infers easily that $S(n, D) = D^{n^{O(1)}}$.

Let us now estimate the parallel complexity $P(n, D)$ of the algorithm $\mathcal{N}(A, n)$.

The "preprocessing" can be executed in parallel time $n^{O(1)}(\log D)^{O(1)}$ and the $D^{n^{O(1)}}$ calls of the algorithms $\mathcal{N}(A[x, x'], n-1)$ can be made simultaneously. We are going to estimate the parallel complexity of such a call. Let $x, x' \in V$ and u be as before, and let $\mathcal{A}$ be the arithmetical network over A which realizes the algorithm $\mathcal{N}(A[x, x'], n-1)$. First observe that $\mathcal{A}$ simulates arithmetical operations and sign tests in $K[u]$ which come from the original version of the algorithm $\mathcal{N}(A[x, x'], n-1)$ as arithmetical network over $K[u]$. Here the elements of $K[u]$ are interpretated as polynomials of degree $D^{n^{O(1)}}$ in u and represented by their coefficient vectors which are suitably codified over A. Thus the element u, algebraic over A, may be replaced by an indeterminate U. We may now read our arithmetical network as follows: $\mathcal{A}$ computes first certain polynomials $G_1, \ldots, G_s \in K[U]$ with $\deg(G_1)+\cdots+\deg(G_s) = D^{n^{O(1)}}$, evaluates then the signs of $G_1(u), \ldots, G_s(u)$ and finally simulates the original version of the algorithm $\mathcal{N}(A[x, x'], n-1)$, given as an arithmetical network over $K[u]$. Taking into account that the degrees of $G_1, \ldots, G_s$ are bounded by $D^{n^{O(1)}}$ and that the sequential complexity of $\mathcal{A}$ is of order $D^{n^{O(1)}}$, we can realize the computation of the coefficients of $G_1,\ldots,G_s$ by an arithmetical network of size $D^{n^{O(1)}}$ and depth $n^{O(1)}(\log D)^{O(1)}$ using interpolation in $D^{n^{O(1)}}$ points of $\mathbf{Z}$ (see [VSBR]). The evaluation of the signs of $G_1(u),\ldots,G_s(u)$, which can be done simultaneously, costs $n^{O(1)}(\log D)^{O(1)}$ parallel steps. Using these data one simulates the original version of the algorithm $\mathcal{N}(A[x,x'],n-1)$ given as an arithmetical network over $K[u]$ by an arithmetical network over A of depth $P(n-1,D)$. If we add to this "preprocessing" and to the $D^{n^{O(1)}}$ calls of the algorithms the complexity of decide if two points of C can be joined in σ (which can be reduced to the computations of the connected components of a graph, represented here by the points of C), we obtain the following recursion formula:

$$P(n,D) = n^{O(1)}(\log D)^{O(1)} + P(n-1,D).$$

From this we infer the parallel complexity bound: $P(n,D) = n^{O(1)}(\log D)^{O(1)}$.

We summarize these complexity results in the following

THEOREM 12. *Let V be a bounded hypersurface of R^n given by a regular polynomial $F \in A[X_1,\ldots,X_n]$ and let x_1, x_2 be points of V definable by two quantifier free formulas Φ_1, Φ_2. Let $g : V \to R$ be an M-direction of V given by a linear form $G \in A[X_1,\ldots,X_n]$. There exists an admissible algorithm which constructs, from the input F, Φ_1, Φ_2, G, a roadmap of V with respect to g and which decides whether x_1 and x_2 lie in the same semialgebraically connected component of V. If this is the case, the algorithm constructs a continuous semialgebraic curve contained in V connecting x_1 and x_2. In particular, the number of semialgebraically connected components of V is computable in admissible time.*

$\Diamond$

6. Conclusion

Let $V \subset R^n$ be a semialgebraic set defined by a quantifier free formula $\Phi \in \mathcal{L}$. Let $g : V \to R$ be a direction of V induced by a linear form $G \in A[X_1,\ldots,X_n]$. If V is closed and bounded it makes sense to define the notion of roadmap of V with respect to g in the same way as we did for bounded regular hypersurfaces of R^n. If V doesn't satisfy these conditions we are not going to speak about roadmaps. However, we can obtain the following result for the case of an arbitrary semialgebraic set:

THEOREM 13. *([HRS 4]. See also [GV 2] for an analogous result in the sequential bit-complexity model). Let $V \subset R^n$ be a semialgebraic set defined by a quantifier free formula Φ. Let x_1, x_2 be two points of V defined by quantifier free formulas $\Phi_1, \Phi_2 \in \mathcal{L}$. Then there exists an admissible algorithm which decides whether x_1 and x_2 are in the same semialgebraically connected component of V. If they do, the algorithm computes a continuous semialgebraic curve contained in V connecting x_1 and x_2. In particular the algorithm computes the exact number of semialgebraically connected components of V.*

If V is closed and bounded and if there is given a direction $g : V \to R$ induced by a linear form $G \in A[X_1,\ldots,X_n]$, the algorithm is able to compute a roadmap of V with respect to g.

195

The proof of this theorem (see [HRS 4]) is based on Theorem 12 and combines efficient computations with infinitesimals over R developped in [GV 1] and [HRS 3].

195

Acknowledgments: Part of this work was realized during a stay of the authors at the University of Bonn, invited by Prof. M. Karpinski.

References

[BCR] Bochnak J., Coste M., Roy M.-F.: Géométrie algébrique réelle. Springer Verlag (1987).

[C 1] Canny J: Some algebraic and geometric computations in PSPACE. ACM Symposium on the theory of computation, 460-467 (1988).

[C 2] Canny J.: The complexity of motion planning. M.I.T. Thesis 1986, M.I.T. Press (1988).

[C 3] Canny J.: A new algebraic method for robot motion planning and real geometry. Proc. 28th IEEE Symp. on Found. of Comp. Science (FOCS), 39-48 (1987).

[CGV] Canny J., Grigor'ev D., Vorobjov N.: Defining connected components of a semialgebraic set in subexponential time. Manuscript Steklov Mathematical Inst., Leningrad, LOMI (1990).

[CR] Coste M., Roy M.-F.: Thom's lemma, the coding of real algebraic numbers and the topology of semialgebraic sets. J. Symbolic Computation 5, 121-129 (1988).

[FGM] Fitchas N., Galligo A., Morgenstern J.: Algorithmes rapides en séquentiel et en parallèle pour l'élimination des quantificateurs en géométrie élémentaire. Sém. Structures Algébriques Ordonnées, Sélection d'exposés 1984-1987 Vol I. Publ. Univ. Paris VII, No. 32, 29-35 (1990).

[Ga] von zur Gathen J.: Parallel arithmetic computations; a survey. Proc. 13th Conf. MFCS (1986).

[G] Grigor'ev D.: Complexity of deciding Tarski algebra. J. Symbolic Computation 5, 65-108 (1988).

[GV 1] Grigor'ev D., Vorobjov N.: Solving systems of polynomial inequalities in subexponential time. J. Symbolic Computation 5, 37-64 (1988).

[GV 2] Grigor'ev D., Vorobjov N.: Counting connected components of a semialgebraic set in subexponential time. Manuscript Steklov Mathematica Inst., Leningrad, LOMI (1990).

[GHRSV] Grigor'ev D., Heintz J., Roy M.-F., Solernó P., Vorobjov N.: Comptage des composantes connexes d'un ensemble semi-algébrique en temps simplement exponentiel. To appear in C.R. Acad. Sci. Paris.

[HRS 1] Heintz J., Roy M.-F., Solernó P.: On the complexity of semialgebraic sets. Proc. IFIP'89, 293-298 (1989).

[HRS 2] Heintz J., Roy M.-F., Solernó P.: Complexité du principe de Tarski-Seidenberg. Comptes Rendus de l'Acad. de Sciences Paris, 309, 825-830 (1989).

[HRS 3] Heintz J., Roy M.-F., Solernó P.: Sur la complexité du principe de Tarski-Seidenberg. Bulletin de la Soc. Math. de France 118, 101-126 (1990).

[HRS 4] Heintz J., Roy M.-F., Solernó P.: Construction de chemins dans un ensemble semi-algébrique (II). Preprint (1990).

[HKRS] Heintz J., Krick T., Roy M.-F., Solernó P.: Geometric problems solvable in single exponential time. Preprint (1990).

[R] Renegar J.: On the computational complexity and geometry of the first order theory of the reals. Technical Report 856, Cornell Univ. (1989).

[RS] Roy M.-F., Szpirglas A.: Sign determination on 0-dimensional sets. To appear in Proc. MEGA'90, Castiglioncello (1990).

[SS] Schwartz J., Sharir M.: On the piano movers' problem: II General Techniques for calculating topological properties of real algebraic manifolds. Adv. Appl. Math. 298-351 (1983).

[S 1] Solernó P.: Complejidad de conjuntos semialgebraicos. Thesis Univ. de Buenos Aires (1989).

[S 2] Solernó P.: Construction de fonctions de Morse pour une hypersurface régulière en temps admissible. Preprint (1990).

[T] Trotman D.: On Canny's roadmap algorithm: orienteering in semialgebraic sets. Manuscript Univ. Aix-Marseille (1989).

[VSBR] Valiant L., Skyum S., Berkowitz S., Rackoff C.: Fast parallel computation of polynomials using few processors. SIAM J. Comp. 12, 641-644 (1983).

On the complexity of algebraic power series

M.E.Alonso[*]   T.Mora[†]   M.Raimondo[†]

1   Introduction

A classical computational model do deal with "computable" power series consists in giving an algorithm to compute their coefficients and to consider suitable truncations in order to perform the required operations. In the case of algebraic power series, i.e. when the series $f(X)$ is given by a polynomial $G(X_1, , X_n, T)$ s.t.$G(X_1, , X_n, f(X)) = 0$, these coefficients can be computed for instance using [K-T]. Since there is in general more than one series vanishing at the origin and satisfying the above identity, one must compute enough terms of the Taylor expansion of f, which, at least, permit to distinguish it from the other roots of G. It is clear that computational problems arise naturally in case the series f should be used for further calculations, e.g. to determine the solution h of a polynomial depending on the X variables and on f.

In order to avoid these problems, in [AMR], we introduced a purely symbolic model of computation, based on the notion of Locally Smooth Systems (LSS), and we showed that these systems have good computational properties: standard bases and normal forms can be calculated in the ring of algebraic power series and it is possible to give effective versions of classical theorems like the Weierstrass Preparation Theorem and the Noether Normalization Lemma.

The aim of this paper is to look for suitable measures for the complexity of algebraic power series. In [R1], [R2] R.Ramanakoraisina defines the complexity of an analytic function f satisfying a polynomial equation to be the degree $c(f)$ of its minimal polynomial; and he shows that this definition satisfies all the required properties of complexities. Since we are interested in the local properties of f (at the origin) we consider here a notion which takes care of

[*]Departamento de Algebra, Facultad de Ciencias Matemáticas, Universidad Complutense, Madrid, SPAIN - Partially supported by CICYT-PB860062 (Spain) and by Accion Integrada Espana-Italia

[†]Dipartimento di Matematica, Universitá di Genova, ITALY - Partially supported by MPI Funds(40% and 60%) , by C.N.R. (progetto strategico 'Matematica Computazionale') and by Azione Integrata Italia-Spagna

both the degree and the multiplicity defined as the minimum order $e(f)$ of defining polynomials.

We introduce then a notion of complexity for algebraic series represented in model of computation of LSS's. To do this we introduce suitable costs for Locally Smooth Systems by means of the length $\lambda(f)$(i.e. the number of extra variables) and the degree $\delta(f)$ (i.e. the product of the degrees of the involved polynomials). We define then the complexity of f : $\xi(f) = (\lambda(f), \delta(f))$ as the minimum of the costs of all Locally Smooth Systems representing f .

The main result of this note shows how these costs can be estimated in terms of degree and multiplicity : $\lambda(f) \leq e(f)$ and $\delta(f) \leq c(f)^{\lambda(f)} \leq c(f)^{e(f)}$. Conversely, we have that $o(f) \leq c(f) \leq \delta(f)$, where $o(f)$ is the order of f at the origin.

Then we introduce the complexity ξ' as the minimum cost of Standard Locally Smooth Systems defining f, and we find bounds for ξ' in terms of ξ and of the maximum degree of involved polynomials. Using this we show that the cost of representing the Weierstrass form of a distinguished polynomial of order b can be estimated by $(b(r+1), D^{b(r+1)})$ where r is the lenght of the involved Locally Smooth Sytem and with D depending on the degrees of the data. Finally, using the above estimates, we find a test to check whether an algebraic function is indeed a rational function.

2 Notation and preliminaries

Let K be a subfield of the field of complex numbers, let $X = (X_1, ..., X_n)$ a set of variables and $K[[X]]_{alg}$ the algebraic closure of $K[X]$ in $K[[X]]$, which is the set of algebraic formal power series, and which is also the henselization of the ring of polynomials with respect to the maximal ideal corresponding to the origin.

Let us recall from [AMR] the notion of Locally Smooth Systems. We say that a system of polynomials $\mathbf{F} = (F_1, ..., F_r)$ is a LSS if the F_i's are polynomials in $K[X_1, ..., X_n, Y_1, ..., Y_r]$ vanishing at the origin and s.t. the Jacobian of the F_i's with respect to the Y_j's at the origin is a lower triangular non singular matrix, i.e. we can write:

$$F_i(X, Y_1, ..., Y_r) = \sum_{j=1}^{r} c_{ij} Y_j + H_i(X, Y_1, ..., Y_r)$$

with $H_i \in (X, (Y_1, ..., Y_r)^2)$ and (c_{ij}) a non-singular lower triangular r by r matrix. Under this assumption, by the Implicit Function Theorem, there are unique algebraic series $f_1, ..., f_r \in K[[X]]_{alg}$ s.t. $f_j(0) = 0 \ \forall j$, and $F_i(X, f_1, ..., f_r) = 0 \ \forall i$. We will also say that $\mathbf{F} = (F_1, ..., F_r)$ is a LSS for the f_i's (or defining the f_i's; or that the f_i's are given by the LSS $\mathbf{F}$ etc.).

The key point of our approach in [AMR] was to look for results in $K[[X]]_{alg}$ by working with suitable, and computable, extensions of $K[X]$. Namely, given a LSS **F** , we then consider the rings $K[X_1,,X_n,f_1,....,f_r] := K[X,\mathbf{F}]$ and $K[X_1,,X_n,f_1,....,f_r]_{loc} := K[X,\mathbf{F}]_{loc}$ viewed as a subring of $K[[X]]_{alg}$ and the evaluation map

$$\sigma_F : K[X_1,,X_n,Y_1,....,Y_r]_{loc} \rightarrow K[[X]]$$

defined by $\sigma_F(Y_i) = f_i$. We have:

$$K[X,\mathbf{F}]_{loc} := K[X,f_1,....,f_r]_{loc} \simeq \frac{K[X,Y_1,....,Y_r]_{loc}}{(F_1,...,F_r)}$$

(Where, for any K-algebra A, with $K[Z] \subset A \subset K[[Z]]$, we denote: $A_{loc} = \{ \frac{a}{1+b} , a, b \in A , b(0) = 0 \}$).

The classical approach to compute with algebraic series (cf [K-T]) consists in representing them as solutions of polynomial equations, i.e. a series $f(X) \in K[[X]]_{alg}$ is given by a polynomial $G(X_1,,X_n,T)$ s.t. $G(X,f(X)) = 0$, and, since there is in general (also in case G is irreducible) more than one series vanishing at the origin and satisfying G, by an algorithm which computes the Taylor expansion of f up to order d, $\forall$ d, or, at least, enough terms in order to distinguish f from the other roots of G. In this paper we introduce suitable measures of the complexities for algebraic power series in both representations and we will compare them. We first recall from [AMR] how the two computational models are compatible (cf [AMR] Appendix and Proposition 2.3).

Theorem 1 *(a)(Artin-Mazur) Let $f \in K[[X]]_{alg}$, $G \in K[X,T]$ such that $G(X,f(X)) = 0$ and assume that an algorithm to compute the Taylor expansion of f up to order d, $\forall d$, is given . Then it is possible to compute a locally smooth system $(F_1,...,F_r)$ defining algebraic series $f_1,...,f_r$, with $f_1 = f$.*

(b) Conversely, let $\mathbf{F} = (F_1,...,F_r)$ be a LSS in $K[X,Y_1,...,Y_r]$ defining the series $f_1,...,f_r \in K[[X]]_{alg}$, and let $d_i = deg(F_i)$ and $d = \prod_{i=1}^{r} d_i$. Then: given $h(X) = H(X,f_1(X),..,f_r(X)) \in K[X,\mathbf{F}]_{loc}$ represented by $H \in K[X,Y_1,...,Y_r]_{loc}$ with $H = \frac{H_0}{1+H_1}$ and $deg(H_0)$ and $deg(H_1)$ bounded by m, there exist a polynomial $Q \in K[X,T]$ with $deg(Q) \leq (m+1)d$ s.t. $Q(X,h(X)) = 0$. (Note: $h = \sigma_F(H) \in K[X,\mathbf{F}]_{loc}$).

Remark 1 *In our computational model all the ring operations with series turn out almost automatically, since our computational tool are the "rings" $[X,\mathbf{F}]_{loc}$. Suppose that f and g are given by distinct LSS's (respectively $\mathbf{F}$ and $\mathbf{G}$ then it is enough to merge them. More precisely, let $f = \sigma_F(F) \in [X,\mathbf{F}]_{loc}$ with $F \in K[X,Y_1,...,Y_r]_{loc}$ and $g = \sigma_G(G) \in [X,\mathbf{G}]_{loc}$ with $G \in K[X,Y_1,...,Y_s]_{loc}$ then any rational function $h = H(f,g) \in K[X]_{alg}$ with $H = \frac{H_0}{1+H_1}$, $H_0(0) = H_1(0) = 0$ can be represented by the LSS $\mathbf{H} := (\mathbf{F}, \mathbf{G}) = (F_1,...,F_r,F_{r+1},...,F_{r+s}) \subset K[X,Y_1,...,Y_r,Y_{r+1},...,Y_{r+s}]$ with $F_{r+i} = G_i(Y_{r+1},...,Y_{r+s})$ via the evaluation $h = \sigma_H(H(F,G)) \in [X,\mathbf{F}]_{loc}$.*

Of course, in many cases it will be not necessary to add so many extra variables: we will propose a test for this in the last section.

3 The complexity of algebraic power series

We recall a notion of complexity which has been recently introduced for Nash functions by R. Ramanakoraisina (cf [R1] and [R2]). Let $f \in K[[X]]_a lg$; we define the complexity $c(f)$ of f as

$$c(f) = min\{degP, \ where \ P \in K[X,T] \ and \ P(X, f(X)) = 0\}$$

In [R1],[R2] it is shown that: $c(f + g) \leq c(f)c(g)$, $c(fg) \leq 2c(f)c(g)$, $c(f^2) \leq 2c(f)$ and that $c(\frac{\partial f}{\partial X_i}) \leq c(f)^2$

Let us remark that, if P is irreducible, then $c(f) = deg(P)$. Take in fact the Zariski closure in C^{n+1} of the analytic germ $(X, f(X))$, then $W = \{(x,t) \in C^{n+1} : P(x,t) = 0\}$. If Q is s.t. $Q(X, f(X)) = 0$ then Q vanishes on W, hence $degQ \geq degP$.

Let $o(g)$ denote the order of vanishing of a function g at the origin of the coordinates, defined as the lowest degree for which, in the Taylor expansion of g, there is a non-zero coefficient. We introduce the *representative multiplicity* of $f \in K[[X]]_{alg}$ $(e(f))$ as

$$e(f) = min\{o(P); \ where P \in K[X,T] and P(X, f(X)) = 0\}.$$

Recall that, for a local ring R, $e(R)$ denotes its multiplicity (cf [Z-S] Ch.VIII 10). Let $P(X,T)$ an irreducible polynomial defining f, then $e(f) = e(\frac{K[X,T]_{loc}}{P(X,T)})$.

Let $\mathbf{F} = (F_1, ..., F_r)$ be a LSS and let $d_i = deg(F_i)$ and $d = \prod_{i=1}^{r} d_i$ we will say that d is the degree of $\mathbf{F}$, $deg(\mathbf{F}) = d$ and that r, i.e. the number of new variables, is its lenght, $l(\mathbf{F}) = r$. Moreover we introduce the cost of $\mathbf{F}$ as: $cost(\mathbf{F}) = (l(\mathbf{F}), deg(\mathbf{F}))$, and we order costs lexicographically:

$cost(\mathbf{F}) < cost(\mathbf{G}) \Leftrightarrow l(\mathbf{F}) < l(\mathbf{G})$ or $l(\mathbf{F}) = l(\mathbf{G})$ and $deg(\mathbf{F}) < deg(\mathbf{G})$.

Moreover we will denote by $V_F \subset C^{n+r}$ the algebraic variety defined by the ideal $(F_1, ..., F_r)$. By definition V_F is non-singular at the origin of the coordinates. Nevertheless V_F can be reducible, therefore we will consider also its irreducible component W_F through the origin. We also remark that, in general, W_F may be not definable by r polynomial equations; i.e. it may be not a complete intersection.

Example 1 *Let*

- $F_1 = 3Y_1 - 4X + Y_2^2 - 2XY_2 + 2Y_1^2 - 2XY_1 - X_2 - X_2Y_1$
- $F_2 = Y_2 - 2Y_1 + X + XY_2 - Y_1^2$

Then (F_1, F_2) is not a prime ideal in $K[X, Y_1, Y_2]$ and the associated prime at the origin is

$$\wp = (F_1, F_2, X^3 + 3X^2 - Y_1Y_2 + 3X - Y1 - Y2)$$

On the other hand we have:

$c(f_1) = 4$ *defined by:* $X^4 + 4X^3 - T^3 + 6X^2 - 3T^2 + 4X - 3T.$
$c(f_2) = 5$ *defined by:* $X^5 + 5X^4 + 10X^3 - 3T^2 + 5X - 3T.$
Moreover we have that if $W = V(\wp)$ *then* $deg W = 5$, *but for every defining LSS* G , $deg(G) \geq 6$.

We are going now to introduce complexities of algebraic power series in the LSS model; it turns out convenient to introduce this notion for a *set* of such functions.

We say that f is defined *via* the LSS **H** if $f \in \{h_1, ..., h_r\}$, where the $h_i's$ are the algebraic power series defined by **H**. Similarly, we say that $\{f_1, ..., f_s\}$ are defined via **H** if $\{f_1, ..., f_s\} \subset \{h_1, ..., h_r\}$ as a set.

Definition 1 *We call cost* ξ *of* $\{f_1, ..., f_s\}$ *the minimum of cost(***H***) where* $f_1, ..., f_s$ *are defined via* **H** . *We write:*

$$\xi(f_1, ..., f_s) = (\lambda(f_1, ..., f_s), \delta(f_1, ..., f_s)) = (length(\boldsymbol{H}*), deg(\boldsymbol{H}*)) = cost(\boldsymbol{H}*)$$

where **H**∗ *reaches the minimum. Moreover, we set :* $d(f_1, ..., f_s) = deg(Z)$ *where* Z *is the Zariski closure of the germ* $(X, f_1(X), ..., f_s(X))$ *in* $\boldsymbol{C}^{n+s}$.

Remark 2 *Notice that our definition does not strictly satisfy the notion of complexity given by Benedetti and Risler (cf[BR]), while it is straightforward to verify the following formulas, (the proofs come out easily by the merging procedure described in Remark 1):*

1. $\xi(f, g) \leq (\lambda(f) + \lambda(g), \delta(f)\delta(g))$
2. $\xi(f + g) \leq (\lambda(f) + \lambda(g) + 1, \delta(f)\delta(g))$
3. $\xi(fg) \leq (\lambda(f) + \lambda(g) + 1, 2\delta(f)\delta(g))$
4. $\xi(f^2) \leq (\lambda(f) + 1, 2\delta(f))$
5. $\xi(\frac{\partial f}{\partial X_i})) \leq (2\lambda(f), \delta(f)^2)$

Proposition 1 *1) If* **F** *and* **G** *define the same algebraic functions* $g_1, ..., g_r$ *then* $W_F = W_G$
    2) $d(f_1, ..., f_s) \leq deg(W_F)$, *where* $f_1, ..., f_s$ *are defined via* **F**
    3) $d(f_1, ..., f_s) \leq \delta(f_1, ..., f_s)$

Proof. 1) It is clear, since W_F and W_G are irreducible algebraic varieties with the same germ at the origin of the coordinates.

2) Let U (resp. Z) be the Zariski closure of $(X, g_1(X), ..., g_r(X))$ in $\boldsymbol{C}^{n+r}$ (resp of $(X, f_1(X), ..., f_s(X))$ in $\boldsymbol{C}^{n+s}$. Then $f_1, ..., f_s \subset g_1, ..., g_r$ as sets, and w.l.o.g. assume that $f_1 = g_{r-s+1}, ..., f_s = g_r$. We consider the projection π : $\boldsymbol{C}^{n+r} \longrightarrow \boldsymbol{C}^{n+s}$ to the first n and the last s factors $(x, y_1, ..., y_{r-s}, y_{r-s+1}, ..., y_r) \longmapsto (x, y_{r-s+1}, ..., y_r)$. Then $\pi(U) \subset Z$ and:

$$C^{n+r} \supset U \subset V_F$$
$$\downarrow \qquad \downarrow$$
$$C^{n+s} \supset Z$$
$$\downarrow \qquad \downarrow$$
$$C^n \qquad C^n$$

By 1) we obtain that $U = W_F$, moreover Z is the Zariski closure of $\pi(U)$ and then: $deg(U) = deg(Z)deg(\pi)$ and hence $deg(W_F) \geq deg(Z) = d(f_1,...,f_s)$.
3) is clear.

4   Estimating complexities

The following theorem shows us that the measures of complexity for algebraic power series we have introduced in the two models are compatible, i.e. we can estimate $\xi(f)$ in term of $c(f)$ and $e(f)$.

We will furthermore assume that the base field K is algebraically closed.

Theorem 2 *Let* $f \in K[[X]]_{alg}$, *then:*
   A) $\lambda(f) \leq e(f)$
   B) $\delta(f) \leq c(f)^{\lambda(f)} \leq c(f)^{e(f)}$

Proof.A) For a local ring R let $e(R)$ denote its multiplicity.

Let $P(X,T)$ an irreducible polynomial defining f, such that $e(f) = e(\frac{K[X,T]_{loc}}{P(X,T)})$.

If $e := e(f) = 1$ we have finished. By induction suppose that there exist k variables $Y_1, ..., Y_k$ and an ideal I_k such that $\frac{K[X,Y_1,...,Y_k]}{I_k}$ is an integral extension of $\frac{K[X,T]}{P(X,T)}$ and such that $e_k := e(\frac{K[X,Y_1,...,Y_k]_{loc}}{I_k}) \leq e + 1 - k$. Let $R_k := \frac{K[X,Y_1,...,Y_k]_{loc}}{I_k}$. If $e_k > 1$, since we know by hypothesis that there is an analytic smooth branch through the origin, the local ring R_k is not unibranche (cf. the proof of Artin-Mazur as in [AMR]) so there exists an integer function h in the normalization of it such that h assumes two different values at two distinct branches of it . Let us consider $R' := \frac{K[X,Y_1,...,Y_k]_{loc}}{I_k}[h] = \frac{K[X,Y_1,...,Y_k]_{loc}[Y_{k+1}]}{I_{k+1}}$ and let $R_{k+1} := \frac{K[X,Y_1,...,Y_k,Y_{k+1}]_{loc}}{I_{k+1}}$. By the projection formula ([Z-S]) we have $e(R_k) = e(R_{k+1}) + \sum[R'/\wp_i : K]e(\wp_i)$ and hence $e(R_k) > e(R_{k+1})$. The procedure then halts when $e_k = 1$, then $k = \lambda(f)$ and $1 \leq e-k+1 = e-\lambda(f)+1$.

   B) Let Z the Zariski closure of $(X, f(X))$ as above and let W_H be the irreducible component through the origin of V_H where **H** is a LSS defining f and such that W_H is dominated by the normalization Z' of Z. Then there exists a dominating morphism $\psi : W_H \longrightarrow K^n$ and the degree of W_H is given by $[K(W_H) : K(X_1, ..., X_n)] = [K(W_H) : K(Z)][K(Z) : K(X_1, ..., X_n)] = [K(Z) : K(X_1, ..., X_n)] = c(f)$. Hence $deg(W_H) = c(f)$. Suppose now that **H** is minimal, as constructed in A), and that $r = \lambda$. By Heintz' results on definability (cf. [H] Prop.3), there exist $n + r + 1 = n + \lambda(f) + 1$ polynomials $g'_j s$ of degree bounded by $c(f)$ such that $W_H = g_1 = ... = g_{n+\lambda(f)+1} = 0$ as a set . By means of a generalization of Heintz proof (it is enough to choose projections which distinguish not only points but also the tangents at the origin), we obtain that the $g_j's$ can be chosen in order that the variety $g_1 = ... = g_{n+\lambda(f)+1} = 0$ is non singular at the origin of coordinates, then W_H is actually its irreducible component through the origin. Computing the Jacobian determinant of the $g_i's$ w.r.t. the $Y_j's$ variables we know that it has rank $\lambda(f)$ and therefore we can choose the corresponding polynomials $g_1', ..., g_{\lambda(f)}'$ which give a LSS H' defining the $f_i's$ and with $deg(H') \leq c(f)^{\lambda(f)} \leq c(f)^{e(f)}$.

On the other side we know that the complexity of f can be estimated by the degree of any LSS defining f (cf Theorem 1 b)). We will give an improvement of this result and will also show how to compute $c(f)$ in term of the LSS, for this we need the following lemma.

Lemma 1 *Let $G(X,T), F(X,T) \in K[X,T]$ be polynomials of degree d and m respectively s.t. F is a factor of G and let $h(X) \in K[[X]]_{alg}$ be s.t. $G(X, h(X)) = 0$. Then if the Taylor expansion of $F(X, h(X))$ vanishes up to order dm we have that $F(X, h(X)) = 0$.*

Proof. In fact take an irreducible factor G_1 of G with $G_1(X, h(X)) = 0$, if $\{F = 0\}$ and $\{G_1 = 0\}$ do not have a common component there exists a set of linear forms H_j through the origin such that $\{F = 0\} \cap \{G_1 = 0\} \cap \{H_1 = ... = H_{n-1=0}\}$ is a finite set of points, whose multiplicity at the origin is greater than dm, in contradiction with Bézout theorem.

Proposition 2 *Let F a LSS defining $f_1, ..., f_r$. Then we have:*
   a) $c(f_i) \leq deg(F) \; \forall \, i$
   b) $d(f_1, ..., f_r) \leq \prod_i^r c(f_i)$
   And there is an algorithm to compute $c(f_i) \; \forall \, i$.

Proof. Let $V_F = V(F_1, ..., F_r) \subset K^{n+r}$, $W_i = Zar.cl.\{(X, f_i(X)), X \in K^n\} \subset K^{n+1}$, $W = Zar.cl\{(X, f_1(X), ..., f_r(X)), X \in K^n\} \subset K^{n+r}$, and $\pi_i : K^{n+r} \longrightarrow K^{n+1}$ the projection $(X, Y_1, ..., Y_r) \mapsto (X, Y_i)$. Then

$$
\begin{array}{ccc}
W & \subset & V_F & \subset & K^{n+r} \\
& & \downarrow & & \downarrow \\
W_i & \subset & \pi_i(V_F) & \subset & K^{n+1}
\end{array}
$$

So, $c(f_i) \leq deg(\pi_i(V_F) \leq deg(V_F)deg(F)$, by Bézout inequality.
   As for b) let $V_i = W_i \times K^{r-1} \subset K_{n+r}$, then we observe that $deg(V_i) = c(f_i)$ and that W is an irreducible component of $W' = V_1 \cap ... \cap V_r$, and apply Bézout inequality. Moreover: $\pi_i(V_F)$ is defined by the ideal $J = (F_1, ..., F_r) \cap K[X, Y_i]$ which can be calculated by an elimination Groebner basis computation. Now, J is a principal ideal, say $J = (Q)K[X, Y_i]$. Therefore we only have to compute the irreducible factor Q' of Q such that $Q'(X, f_i(X)) = 0$. This can be done using Lemma 4.

Corollary 1 *Let* **F** *a LSS defining* $f_1, ..., f_r$, $h = \sigma_F(H)$ *with* $H \in K[X,Y]$, $d = deg(F)$ *and* $m = deg(H)$, *then* :
   i) $o(h) \le c(h) \le md$
   ii) $o(f_i) \le c(f_i) \le d \forall i$ *provided* $f_i \neq 0$.

Remark 3 *In the Example above, we have:* $e(f_1) = e(f_2) = 1$ $c(f_1) = 4$; $\xi(f_1) = (1,4)$; $c(f_2) = 5$; $\xi(f_2) = (1,5)$; $d(f_1, f_2) = 5$; $\xi(f_1, f_2) = (2,6)$; $\delta(f_1, f_2) = 6$.
   Moreover,let us consider: $f(X) = X\sqrt{1-X}$; $g(X) = 1 - \sqrt{1-X}$. *Then* $c(f) = 3$, $e(f) = 2$ *and* $\xi(f) = \xi(f,g) = (2,4)$.

5   Standard Locally Smooth Systems

In [AMR] an important role in order to perform computations in the ring $K[[X]]_{alg}$ has been played by the notion of standard locally smooth system (SLSS) We recall (in a simplified version) this definition: **G** is a standard locally smooth system (SLSS) if:
   1) $\mathbf{G} = (G_1, ..., G_r)$ is a LSS for the functions $f_1, ..., f_r$
   2) $f_i \neq 0 \forall i$
   3) $G_i = Y_i(1 + Q_i) - R_i$ with Q_i , $R_i \in (X,Y)$,
$R_i \in K[X, Y_1, ..., Y_{i-1}, Y_{i+1}, ..., Y_r]$ and $in(R_i) = in(f_i) \in K[X]$.
   The introduced notion turns out to be quite important both for its own sake, since it directly gives explicit information on the $f_i's$, and because it is a computational tool for standard bases, Weierstrass Preparation Algorithm and elimination algorithms (cf [AMR]). In this section we look for bounds for the costs of SLSS's.
   To do this, we introduce the complexity ξ' of $\{f_1, ..., f_s\}$ as the minimum of the cost(**H**) where the $f_i's$ are defined via a SLSS **H**, and we write:

$$\xi'(f_1, ..., f_s) = (\lambda'(f_1, ..., f_s), \delta'(f_1, ..., f_s)) = (length(H'), deg(H')),$$

where H' reaches the minimum. We further assume that $f_i \neq 0 \forall$ i and that $f_1, ..., f_s$ are linearly independent. We also introduce the natural numbers
   $\Delta = Sup\{d_i, d_i = deg(F_i)\}$
   $\Omega = \prod_{i=1}^{s} o(f_i)$.

Lemma 2 *Let* $\mathbf{F} = (F_1, ..., F_r)$ *be a LSS defining* $f_1, ..., f_r$, *where* $f_i \neq 0$ *for every* i, *and let* $\mathbf{G} = (G_1, ..., G_r)$ *be a SLSS obtained by* **F** , *then*
   $o(f_i) \le deg(G_i) \le \Delta + o(f_i) - 1$

Proof. G_i is obtained by the corresponding F_i, as follows: while there is a term t, $deg(t) \le o(f_i)$ depending on some Y_j, we substitute an occurence of Y_j by $Y_j - F_j$. Each such substitution can at worst introduce terms of degree $o(f_i) + \Delta - 1$.

Proposition 3 *With the above notation, we have:*

a) $\lambda' = \lambda$

b) $Sup\{\Omega, \delta(f_1, ..., f_s)\} \le \delta'(f_1, ..., f_s) \le Inf\{2^\lambda \Delta^\lambda, \Delta^\lambda \delta(f_1, ..., f_s)\}$.

Proof. Let us take **F** minimal and **G** obtained from **F** as in the above Lemma. Let $d_i = degG_i$. Then $d_i \le o(f_i) + \Delta - 1 \le d_i + \Delta - 1$ and $\prod_{h=1}^\lambda (d_h + \Delta - 1) \le Inf\{2^\lambda \Delta^\lambda, \Delta^\lambda \delta(f_1, ..., f_s)\}$.

6 Weierstrass Preparation Theorem

In [AMR] §5 we gave constructive versions of Weierstrass Preparation and Division theorems, we are going now to bound the complexity of these constructions.

Let $X' = (X_1, ..., X_{n-1})$ so $X = (X', X_n)$ and let g be an algebraic series distinguished in X_n , say $g(0, X_n) = X_n^b + $ *higher degree terms*, then there exist a unit v and series $h_i \in K[[X']]_{alg}$ such that $g = v(X_n^b + \sum_{i<b} h(X')X_n^i)$. Let us represent data and output in our computational model.

DATA: a LSS $\mathbf{F_0}$ defining series $f_{0,i}$ with $cost(F_0) = (r, d_0)$, $deg(F_{0,i}) = d_{0,i}$ and $D_0 = sup\{d_{0,i}\}$; $G_0 \in K[X, Y_1, ..., Y_r]$ such that $\sigma_{F_0}(G) = g$, $deg(G_0) = m$ and $o(g(0, X_n)) = b$.

OUTPUT: a LSS $\mathbf{H} :=$
$(H_{1,0}, ..., H_{r,0}, H_0, H_{1,1}, ..., H_{r,1}, H_1, ..., H_{1,b-1}, ..., H_{r,b-1}, H_{b-1})$ in $K[X', U_{1,0}, ..., U_{r,0}, U_0, U_{1,1}, ..., U_{r,1}, U_1, ..., U_{1,b-1}, ..., U_{r,b-1}, U_{b-1}]$ defining algebraic series $h_{1,0}, ..., h_{r,0}, h_0, h_{1,1}, ..., h_{r,1}, h_1, ..., h_{1,b-1}, ..., h_{r,b-1}, h_{b-1} \in K[[X']]_{alg}$ and such that $h_i = \sigma_H(U_i)$ for all i's.

Theorem 3 *Let* $D = sup\{b(m+1), b(2\Delta_0 + 1)\}$,
then $cost(H) \le (b(r+1), D^{b(r+1)})$.

Proof. Let us recall from [AMR] the lines of the algorithm.

(1) Construct a new LSS $\mathbf{F_2}$ defining series $f_{2,i}$ which differ from the $f_{0,i}$'s only by polynomials of degrees $\le b$ and such that $T(f_{0,i}) > X_n^b$ and a new $\mathbf{G_2}$ such that $\sigma_{F_2}(G) = g$.

(2) Let **F** the SLSS costructed by $\mathbf{F_2}$ and G such that $\sigma_F(G) = g$.

(3) Let $U = (U_{1,0}, ..., U_{r,0}, U_0, U_{1,1}, ..., U_{r,1}, U_1, ..., U_{1,b-1}, ..., U_{r,b-1}, U_{b-1})$ a new set of variables and let

- $P := X_n^b - \sum_{j=0}^{b-1} U_j X_n^j$

- $P_i := Y_i - \sum_{j=0}^{b-1} U_{ij} X_n^j \; \forall i = 1, ..., r$.

(4) Apply Buchberger reduction (with respect to a suitable term ordering) to $G, F_1, ..., F_r$, we obtain polynomials $H_0, ..., H_{d-1}, H_{1,0}, ..., H_{1,b-1}, ..., H_{r,0}, ..., H_{r,b-1} \in (X_1, ..., X_{n-1}, U)K[X_1, ..., X_{n-1}, U] = K[X', U]$ s.t.:

- $G - \sum_{j=0}^{b-1} U_j X_n^j \in (P, P_1, ..., P_r)$

- $F_i - \sum_{j=0}^{b-1} U_{ij} X_n^j \in (P, P_1, ..., P_r) \ \forall \ i$

Let us now examinate the costs of these constructions:

(1) $length(F_2) \leq r$; $deg(F_{2,i}) \leq bd_{0,i}$; $\Delta_2 = sup\{deg(F_{2,i})\} \leq b\Delta_0$ and $deg(G_2) \leq bm$.

(2) $length(F) \leq r$; $deg(F_i) \leq 2b\Delta_0$; $G = G_2$ and $\Delta := sup\{deg(F_i), deg(G)\} \leq sup\{bm, 2b\Delta_0\}$

(4) $length(H) \leq b(r+1)$, $deg(H_i)$ and $deg(H_{ij})$ are bounded by $\Delta + b$ and therefore $deg H) \leq (\Delta + b)^{b(r+1)} \leq D^{b(r+1)}$.

We conclude this section remarking that, by the above result we obtain a single exponential complexity also for the Weierstrass Division Theorem, however the main drawback consists in the large number of extra variables needed to represent the Weierstrass polynomials: by an initial form computation we can cancel those which are zero, in the next section we propose a test to check if they are in fact rational functions.

7 Applications

We apply now the above results, and precisely the Proposition 2 and the Corollary 1, in order to obtain some informations on the rationality of power series. This is clearly of great interest since, as we have remarked in the previous section, the cost grows exponentially on the number of involved functions.

Let f be a power series in X, $f = \sum_{a \in IN^n} f_a X^a$, with $f_a \in K$; then we write $f_{(i)} := \sum_{|a|=i} f_a X^a$, so that $f = \sum_{i=0}^{\infty} f(i)$.

Proposition 4 *Let $h \in K[[X]]_{alg}$ such that $c(h) \leq t$,
then: $h \in K[X] \Longleftrightarrow h_{(j)} = 0 \ \forall j : t < j \leq t^2$.*

Proof. The same proof of Prop. 2.7 of [AMR] works.

Corollary 2 *It is possible to check whether $h \in K[X]_{alg}$ is a rational function.*

Proof. Let $Q \in K[X, T]$ be an irreducible polynomial such that $Q(X, h(X)) = 0$. Let $s = deg_T(Q) \leq c(h) = deg(Q) \leq t$ and write

$$Q(X, T) = a_0(X)T^s + a_1(X)T^{s-1} + ... + a_{s-1}(X)T + a_s(X) \in K[X, T]$$

Let us further introduce the following polynomial $Q* \in K[X, T]$:
$Q*(X, T) = a_0^s Q(X, \frac{T}{a_0}) = T^s + a_1 T^{s-1} + a_0 a_2 T^{s-2} + ... + a_0^{s-2} a_{s-1} T + a_0^{s-1} a_s = T^s + \sum_{i=1}^{s} a_0^{i-1} a_i T^{s-i}$ Then we obtain that $deg(Q*) \leq s(t - s + 1) \leq \frac{(t+1)^2}{4}$
.Let us consider $k = ha_0 \in K[[X]]_{alg}$. Now, if $h(X) = \frac{l(X)}{g(X)} \in K[X]_{loc}$, it is easy to see that g is a factor of a_0 and therefore we obtain that $k(X) = \frac{a0(X)}{g(X)} f(X) \in K[X]$. It is straightforward to see that k is a root of $Q*$. Then

$c(k) \leq \frac{(t+1)^2}{4}$. Conversely, suppose that k is a polynomial root of Q_*, then $h = \frac{k}{a_0} \in K(X) \cap K[[X]]_{alg} = K[X]_{loc}$. In order to apply Proposition 4 , we need to check whether $k(j) = 0$ for $\frac{(t+1)^2}{4} \leq j \leq \frac{(t+1)^4}{16}$, this can be done using suitable linear systems (with the coefficients of a_0 as unknowns), once we know the Taylor expansion of h up to degree $\frac{(t+1)^4}{16}$ and we know that $deg(a_0) \leq t - s \leq t$. More precisely we write a_0 as a polynomial of degree $t - 1$ (or if is possible we use a better estimate) with unknown coefficients U_a with $a = (a_1, ..., a_n)$. We then consider starting from $k(b) := \frac{\partial^\rho k}{\partial X^\rho}$ where $b = |b| = [\frac{(t+1)^2}{4}]$ an enough number of equations (at least $\geq max(\binom{t+n-1}{n}, b^2 - b))$ involving the unknowns U and the Taylor coefficients of h up to degree d^2 . The problem is then reduced to compute whether this system has a non-zero solution, i.e. to compute the vanishing of a suitable determinant

Remark 4 *If we dispose of an efficient factorization modulus, we can test whether an algebraic function is polynomial (resp. rational) in the following way.*

1. *Given f produce a polynomial $Q(X, T)$ over which it vanishes.*

2. *Factorize Q to check whether it has a polynomial factor which is linear in T.*
   Then check whether it is of the form: $T - f(x)$.

3. *If not, construct Q_* and factorize it, etc.as in 2).*

References

[AMR] M.E.Alonso, T.Mora, M.Raimondo.*A computational model for algebraic power series.* J. Pure and Appl. Algebra, to appear.

[B-R] R.Benedetti, J.J.Risler.*Real Algebraic and Semialgebraic sets.* Hermann,Paris 1990.

[H] J.Heintz. *Definability and fast quantifier elimination in algebraically closed fields.* Theoretical Computer Science 24 (1983).

[K-T] H.T.Kung, J.F.Traub.*All Algebraic Functions can Be Computed Fast.* J. ACM 25 (1978).

[R1] R.Ramanakoraisina. *Complexité des fonctions de Nash.* Comm. Algebra 17 (1989).

[R2] R.Ramanakoraisina.*Bézout Theorem for Nash functions.* Preprint U.E.R. Math.Univ. Rennes (1989).

[Z-S] O.Zariski, P.Samuel.*Commutative Algebra Vol II.* Van Nostrand 1960.

LOCAL DECOMPOSITION ALGORITHMS

Maria Emilia Alonso - Univ. Complutense Madrid
Teo Mora - Univ. Genova
Mario Raimondo - Univ. Genova

INTRODUCTION

For an ideal $I \subset P := k[X_1,...,X_n]$, many algorithms using Gröbner techniques are a direct consequence of the definition itself of Gröbner basis: among them we can list algorithms for computing syzygies, dimension, minimal bases, free resolutions, Hilbert function and other algebro-geometric invariants of an ideal; the explicit knowledge of syzygies allows moreover to compute ideal intersection and quotients.

Other algorithms require a specific property of Gröbner bases w.r.t. some special orderings (elimination orderings, i.e. orderings s.t. if $m \in <X_1,...,X_d>$ and $n < m$ then $n \in <X_1,...,X_d>$): by computing a Gröbner basis of I w.r.t. such an ordering, one gets Gröbner bases for $I_d := I \cap k[X_1,...,X_d]$ and, if $I_d = (0)$, for $I k(X_1,...,X_d)[X_{d+1},...,X_n]$; thus, one obtains effective elimination and, as a consequence, algorithms for the computation of radicals, primary decomposition, equidimensional decomposition; primality, radicality, equidimensionality tests; solutions of systems of equations, i.e. description of the variety of zeroes (e.g. by giving the primary decomposition of the ideal, and, for each primary J, describing the associate irreducible component by giving the generic zero of Rad(J) in an explicit field extension of k); elimination provides also tests for dimension, ideal intersection and quotients.

Such decomposition algorithms, together with other algorithms based on elimination, have been introduced in [GTZ]; while the algorithms in [GTZ] have some weaknesses, which we have remarked in the paper, no alternative to them is known at this moment: an alternative has been proposed by [BGS], but we don't have sufficient information on it, so we could not use it. In contrast much research has been recently devoted to radical computation: [KR] and [KL] use schemes similar to the one of [GTZ], [EH] applies cohomological techniques and the Jacobian ideal, [GH] applies ideal quotients. We propose here a new algorithm for radical computation; it is a variant of the algorithm in [GH] which doesn't require generic position.

Instead of considering "global" varieties, i.e. varieties defined in an affine or projective space, one can study "local" varieties, by considering appropriate local rings:

      a) the localization of the polynomial ring at the origin: description of a variety near a point

      b) the ring of (convergent - algebraic) formal power series: analytical branches of a variety near a point

      c) the localization at a prime ideal of a coordinate ring: description of a variety "near a subvariety"

Standard bases in local rings are a direct generalization of Gröbner bases in the sense that the definition and the basic properties generalize the corresponding definition and properties of Gröbner bases. The tangent cone algorithm which works in $k[\underline{X},\underline{Y}]_{1+(\underline{X})}$ is a generalization-modification of Buchberger algorithm which is based on these properties and which by suitable representation techniques can be used to compute standard bases in all the three cases listed above [MPT].

As a consequence, on a "representable" local ring many problems can be solved in essentially the same way as on polynomial rings; this allows the computation of algebro-geometric invariants (syzygies, dimension, free resolutions, Hilbert function, regularity, system of parameters) and of ideal intersections and quotients [MOR].

The elimination properties of Gröbner bases are however not directly generalizable. The main reason for that is a practical one: standard basis computation requires orderings s.t. $X_i < 1$ for each i and then one has, say, $XY < Y < X < 1$, so that elimination orderings cannot be used for standard basis computations.

The basic case of effective "local elimination" is the following:

given a basis F of an ideal I in $B := k[\underline{X},\underline{Y}]_{1+(\underline{X},\underline{Y})} = k[\underline{X}]_{1+(\underline{X})}[\underline{Y}]_{1+(\underline{Y})}$, to compute a basis of $J := I \cap A$ where $A := k[\underline{X}]_{1+(\underline{X})}$, and, if $J = 0$, of $I k(\underline{X})[\underline{Y}]_{1+(\underline{Y})}$.

A recent generalization of the tangent cone algorithm [MPT] allows to compute standard bases in $A[\underline{Y}]$ w.r.t. to an ordering s.t. $X_i < 1 < Y_j$ $\forall i,j$ and if $m \in <\underline{X}>$ and $n < m$ then $n \in <\underline{X}>$. Since w.l.o.g. $F \subset A[\underline{Y}]$, if L is the ideal in $A[\underline{Y}]$ generated by F, then $L \cap A = I \cap A = J$; moreover, if $J = (0)$, $L k(\underline{X})[\underline{Y}]_{1+(\underline{Y})} = I k(\underline{X})[\underline{Y}]_{1+(\underline{Y})}$. So by the generalized tangent cone algorithm, the above local elimination problem is solvable.

As a direct consequence one gets elimination algorithms for localizations at the origin and (by essentially using the Weierstrass and Noether algorithms in [AMR]) for rings of algebraic series w.r.t. generic coordinates.

As a consequence, the whole host of elimination-based algorithms is at least in principle available in the local case too. Most of the paper is devoted to show how the general scheme for decomposition algorithms in polynomial rings introduced in [GTZ] can be generalized to the local case.

This scheme allows equidimensional decomposition and equidimensionality tests, and reduces the problems of primary decomposition, radical computation, radicality test, primality and primariety tests to (resp.) factorization, squarefree algorithm, squarefree test, irreducibility test for univariate polynomials over algebraic extensions of a suitable field, which is a trascendental extension of k in the case of polynomial rings,

localizations at the origin and at prime ideals, and which unfortunately is the fraction field of an algebraic power series ring, in the corresponding case.
So we get:
all decompositions algorithms in cases a) and c),
equidimensional decomposition, equidimensionality test, radical computation, radicality test in case b)
while, by lack of a factorization algorithm for polynomials over an algebraic power series ring, we can only reduce, in case b), primary decomposition, primality and primariety tests to univariate polynomial factorization.

1 DECOMPOSITION ALGORITHMS

Let A be a Noetherian ring, $I \subset A$ be an ideal; let $I = \mathfrak{q}_1 \cap \ldots \cap \mathfrak{q}_s$ be an irredundant primary decomposition; $\forall i$, let $\mathfrak{p}_i$ the prime associated to $\mathfrak{q}_i$.

We recall that each $\mathfrak{q}_i$ is called a primary component of I, each $\mathfrak{p}_i$ is called an associated prime of I; $\mathfrak{p}_i$ and $\mathfrak{q}_i$ are called <u>embedded</u> if $\exists j$ s.t. $\mathfrak{p}_j \subset \mathfrak{p}_i$, are called <u>isolated</u> otherwise; the isolated primary components and all the associated primes are uniquely determined by I, while the embedded primary components are not; a primary component $\mathfrak{q}_i$ s.t. $\dim(\mathfrak{q}_i) = \dim(I)$ is isolated, while the converse is false; the radical of I is the intersection of the isolated associated primes of I; I is <u>unmixed</u> (or: equidimensional) if $\dim(\mathfrak{p}_i) = \dim(I)$ for each associated prime $\mathfrak{p}_i$ of I.

We apply the common label of "decomposition algorithms" to algorithms solving the following problems:
a) **primary decomposition**: given $I \subset A$, return a list $[\mathfrak{q}_1,...,\mathfrak{q}_s]$ s.t. $\mathfrak{q}_i$ is primary $\forall i$ and $I = \mathfrak{q}_1 \cap \ldots \cap \mathfrak{q}_s$ is an irredundant primary decomposition of I
b) **prime decomposition**: given $I \subset A$, return the list $[\mathfrak{p}_1,...,\mathfrak{p}_s]$ of the associated primes of I (actually we don't know any algorithm to solve this problem without computing the primary decomposition: we have listed it for completeness)
c) **primality and primariety tests**: given $I \subset A$, decide whether I is prime (primary)
d) **equidimensional decomposition**: given $I \subset A$, return a list $[\mathfrak{u}_0,...,\mathfrak{u}_d]$ s.t.
   i) $\forall i$, either $\mathfrak{u}_i$ is unmixed and $\dim(\mathfrak{u}_i) = i$ or $\mathfrak{u}_i = (1)$
   ii) $I = \mathfrak{u}_0 \cap \ldots \cap \mathfrak{u}_d$
   iii) $\forall i$, if $\mathfrak{u}_i \neq (1)$, then $\mathfrak{u}_i$ doesn't contain $\mathfrak{u}_{i+1} \cap \ldots \cap \mathfrak{u}_d$
e) **top-dimensional component**: given $I \subset A$, compute Top(I), the intersection of those primary components $\mathfrak{q}_i$ of I s.t. $\dim(\mathfrak{q}_i) = \dim(I)$
f) **equidimensionality test**: given $I \subset A$, decide whether I is unmixed, i.e. $I = Top(I)$
g) **radical computation**: given $I \subset A$, compute Rad(I)
h) **equidimensional radical decomposition**: given $I \subset A$, compute the equidim. decomposition of Rad(I).
i) **radicality test**: given $I \subset A$, decide whether I is radical.

2 THE GTZ DECOMPOSITION SCHEME FOR POLYNOMIAL IDEALS

In [GTZ] a general scheme has been proposed for decomposition algorithms on ideals in $k[X_1,...,X_n]$; before describing it, we need to introduce some notation and a preliminary result.

In dependence of a fixed subset of variables $\{X_1,...,X_d\}$, by $-^e$ and $-^c$ we denote ideal extension and contraction between $k[X_1,...,X_n]$ and $k(X_1,...,X_d)[X_{d+1},...,X_n]$.

If $\{X_1,...,X_d\}$ is a set of independent variables for I, then
1) I^{ec} is the intersection of the primary components of I for which $\{X_1,...,X_d\}$ is a set of independent variables.
2) if $I^e = \mathfrak{u}_1 \cap \ldots \cap \mathfrak{u}_r$ is the primary decomposition of I^e, then $I^{ec} = \mathfrak{q}_1 \cap \ldots \cap \mathfrak{q}_r$ is the primary decomposition of I^{ec}, where $\mathfrak{q}_i = (\mathfrak{u}_i)^c$ $\forall i$
3) moreover if $\mathfrak{m}_i$ is the prime associated to $\mathfrak{u}_i$ and $\mathfrak{p}_i$ the one associated to $\mathfrak{q}_i$ then $\mathfrak{m}_i = \mathfrak{p}_i{}^e$, $\mathfrak{p}_i = \mathfrak{m}_i{}^c$
4) I^{ec} is prime (primary, radical) iff I^e is such
5) $Rad(I^{ec}) = (Rad(I^e))^c$
If moreover I is s.t. $\dim(I) = d$ and $\{X_1,...,X_d\}$ is a maximal set of independent variables for I, then:
1) I^e is a zero-dimensional ideal
2) I^{ec} is the intersection of those isolated primary components $\mathfrak{q}$ of I s.t. $\dim(\mathfrak{q}) = d$ and $\{X_1,...,X_d\}$ is a maximal set of independent variables for $\mathfrak{q}$.
We can now describe a general version of the GTZ scheme for decomposition algorithms:

   choose a set of independent variables, which, after permutation, we can assume to be $X_1,...,X_\delta$
   compute $s \in k[X_1,...,X_\delta]$ s.t. $I^{ec} = I : s^* =: \cup_j I : s^j$
   if I^e is zero-dimensional **then**

compute the decomposition of the zero-dimensional ideal I^e by means of a special algorithm
else
     compute the decomposition of I^e, by recursive application of the algorithm
compute the decomposition of I^{ec} by contraction of the components of I^e
if $I^{ec} \neq I$, **then**
     compute $t \in N$ s.t. $I = I^{ec} \cap (I, s^t)$
     compute the decomposition of (I, s^t), by recursive application of the algorithm
     compute the decomposition of I by joining the two decompositions of I^{ec} and (I, s^t).

There are several possible schemes to choose the independent variables, which strongly influence the structure of the computation tree; [GTZ] proposes the following two choices, where $\{X_1, \ldots, X_d\}$ is a maximal set of independent variables:
i) choose $\{X_1, \ldots, X_d\}$ so that I^e is zero-dimensional
ii) choose successively $X_1, X_2, \ldots, X_d$ obtaining $s_i \in k[X_1, \ldots, X_i]$, $t_i \in N$ s.t.

$$I = I \, k(X_1, \ldots, X_d)[X_{d+1}, \ldots, X_n] \cap (I, s_1^{t_1}) \cap \ldots \cap (I, s_d^{t_d})$$

The different constructions required by the algorithm can be performed by suitable Gröbner basis computations:
- inspection of the leading terms of a G-basis of I (w.r.t. any term-ordering) allows to find a maximal set of independent variables for I
- if $<$ is an elimination ordering on $<X_1, \ldots, X_n>$ over $<X_1, \ldots, X_d>$ (by this we mean that $m \in <X_1, \ldots, X_d>$ and $n < m$ implies $n \in <X_1, \ldots, X_d>$) and G is a Gröbner basis of I w.r.t. $<$, then $G \subset k[X_1, \ldots, X_d][X_{d+1}, \ldots, X_n]$ is a Gröbner basis of I^e w.r.t. the restriction of $<$ to $<X_{d+1}, \ldots, X_n>$;
- moreover if $s \in k[X_1, \ldots, X_d]$ is any multiple of all the squarefree parts of the leading coefficients of the elements of G, then $I^{ec} = I : s^*$
- if G is a G-basis of $L \subset k(X_1, \ldots, X_d)[X_{d+1}, \ldots, X_n]$ and, $\forall g \in G$, $prim(g) \in k[X_1, \ldots, X_d][X_{d+1}, \ldots, X_n]$ is the primitive polynomial associated to g, let $G_0 := \{prim(g) : g \in G\}$; then $L^c = (G_0) : h^*$ where h is any multiple of all the squarefree parts of the leading coefficients of the elements in G_0
- if s is s.t. $I : s^* = I^{ec}$, let G be a basis of I^{ec} and let t be the minimal power of s s.t. $s^t g \in I \, \forall g \in G$ (which can be easily tested if a G-basis of I is known); then $I = I^{ec} \cap (I, s^t)$
- to test if $I = I^{ec}$, if a G-basis of both ideals for the same ordering is known, one has to check equality of the associated monomial ideals; if a G-basis of I is known, one has to check whether $g \in I \, \forall g$ in any basis of I^{ec}.

The primality, radical, primary tests in the 0-dimensional case are obtained by performing a generic change of coordinates so to have the ideal in a suitable generic position (the first coordinates of each zero must be different; we will refer to this notion as GTZ-position) and then by checking if the Gröbner basis of the ideal satisfies the corresponding structure theorems; this requires irreducibility tests (for primality and primariety) or squarefree tests (for radicality) for polynomials in $k(X_1, \ldots, X_d)[Y]$. Radical computation requires computing the squarefree part of polynomials in $k(X_1, \ldots, X_d)[Y]$, while primary decomposition requires factorization.

Before discussing the different instantiations of this scheme, let us point out its main weakness: let $I = \mathfrak{q}_1 \cap \ldots \cap \mathfrak{q}_s$ be an irredundant primary decomposition and $\mathfrak{p}_i$ the prime associated to $\mathfrak{q}_i$. Assume moreover that $\mathfrak{q}_i \cap k[X_1, \ldots, X_d] = (0)$ iff $i \leq r$, so that $I^{ec} = \mathfrak{q}_1 \cap \ldots \cap \mathfrak{q}_r$ is an irredundant primary decomposition. One has $(I, s^t) = (\mathfrak{q}_1, s^t) \cap \ldots \cap (\mathfrak{q}_s, s^t)$, where the ideals are not necessarily primary. Because of the choice of s, one has $(\mathfrak{q}_i, s^t) = \mathfrak{q}_i$ if and only if $i > r$, while, for $i \leq r$, $\mathfrak{q}_i$ is strictly contained in $(\mathfrak{q}_i, s^t)$; therefore for $i < r$, the primary components of $(\mathfrak{q}_i, s^t)$, while irredundant in the decomposition of (I, s^t), are necessarily redundant in the primary decomposition of I.
Let us see now the different instantiations of this scheme:
a) primary decomposition
if $[\mathfrak{n}_1, \ldots, \mathfrak{n}_r]$ is the primary decomposition of I^e, and $\mathfrak{q}_i = (\mathfrak{n}_i)^c$, then $[\mathfrak{q}_1, \ldots, \mathfrak{q}_r]$ is a primary decomposition of I^{ec}; if moreover $[\mathfrak{q}_{r+1}, \ldots, \mathfrak{q}_s]$ is a primary decomposition of (I, s^t), then $[\mathfrak{q}_1, \ldots, \mathfrak{q}_s]$ is a (not necessarily irredundant) primary decomposition of I.
c) primality and primariety tests
I is prime (primary) iff $I = I^{ec}$ and I^e is prime (primary)
d) equidimensional decomposition
For this we choose $\{X_1, \ldots, X_d\}$ so that I^e is zero-dimensional, and therefore I^{ec} is unmixed with $\dim(I^{ec}) = d$.
If $[\mathfrak{u}_0, \ldots, \mathfrak{u}_\delta]$, $\delta \leq d$, is the equidimensional decomposition of (I, s^t), then the equidimensional decomposition of I is $[\mathfrak{v}_0, \ldots, \mathfrak{v}_d]$, where:

for $i < \delta$: $\vartheta_i := (1)$ if $I^{ec} \subseteq \mathfrak{u}_i$, $\vartheta_i := \mathfrak{u}_i$ otherwise

for $\delta < i < d$: $\vartheta_i := (1)$

if $\delta < d$: $\vartheta_\delta := (1)$ if $I^{ec} \subseteq \mathfrak{u}_\delta$, $\vartheta_\delta := \mathfrak{u}_\delta$ otherwise; $\vartheta_d := I^{ec}$

if $\delta = d$: $\vartheta_d := \mathfrak{u}_d \cap I^{ec}$

e) top-dimensional component

It can be done by the following variant of the GTZ-scheme:

> $J := I, L := (1)$
> **While** $\dim(J) = \dim(I)$ **do**
>> **choose** a maximal set of independent variables $\{X_{i_1},...,X_{i_d}\}$
>> **compute** $s \in k[X_{i_1},...,X_{i_d}]$ s.t. $J^{ec} = J : s^*$
>> $L := L \cap J^{ec}$
>> **compute** $t \in N$ s.t. $J = J^{ec} \cap (J,s^t)$
>> $J := (J,s^t)$
> $Top(I) := L$

f) equidimensionality test

It is performed by means of the algorithm above, followed by a test whether $I = Top(I)$

g) radical computation

We don't know of any algorithm for radical computation which doesn't pass through the computation of the equidimensional radical decomposition, while we heard that recently Vasconcelos obtained such an algorithm (actually, a closed formula for the radical).

h) equidimensional radical decomposition

Since $Rad(I) = Rad(I^{ec}) \cap Rad(I,s^t)$, $Rad(I^{ec}) = Rad(I^e)^c$, $Rad(I,s^t) = Rad(I,s)$, one can apply the scheme above, avoiding the useless computation of t. The equidimensional decomposition of $Rad(I)$ is obtained by merging $Rad(I^{ec})$ with an equidimensional decomposition of $Rad(I,s)$, according to the scheme above; $Rad(I^{ec})$ is obtained by contracting $Rad(I^e)$.

i) radicality test

We don't know any test based on the GTZ-scheme which doesn't require at least the computation of $Rad(I,s)$ and checking whether $I = I^{ec} \cap Rad(I,s) = Rad(I)$. The best we know runs as follows: choose $\{X_1,...,X_d\}$ to be a <u>maximal</u> set of independent variables, so that I^e is 0-dimensional. Then:

> if I^e is not radical, I is not radical
> if I^e is radical and $I = I^{ec}$, then I is radical
> if I^e is radical and $I \neq I^{ec}$, then one has to compute $Rad(I,s)$ and check whether $I = I^{ec} \cap Rad(I,s)$, since

the latter ideal is equal to $Rad(I)$.

3 DECOMPOSITION SCHEMES FOR IDEALS IN GENERIC POSITION

Let I be s.t. $\dim(I) = d$; if I is in Noether position (which means that $X_{d+1},...,X_n$ are integral mod. I), then $\{X_1,...,X_d\}$ is a maximal set of independent variables and I^{ec} is the top dimensional component of I. So if we apply the GTZ scheme to an ideal in Noether position, choosing $\{X_1,...,X_d\}$ as independent variables, we have $I = I^{ec} \cap (I,s^t)$ and either $(I,s^t) = (1)$ or $\dim(I,s^t) < d$. Some of the decomposition algorithms are therefore simplified:

d) equidimensional decomposition

If $[\mathfrak{u}_0,...,\mathfrak{u}_\delta]$, $\delta < d$, is the equidimensional decomposition of (I,s^t), then the equidimensional decomposition of I is $[\vartheta_0,...,\vartheta_d]$, where:

> for $i \leq \delta$: $\vartheta_i := (1)$ if $I^{ec} \subseteq \mathfrak{u}_i$, $\vartheta_i := \mathfrak{u}_i$ otherwise
> for $\delta < i < d$: $\vartheta_i := (1)$
> $\vartheta_d := I^{ec}$

e) top-dimensional component

$Top(I)$ is equal to I^{ec}

f) equidimensionality test

I is equidimensional iff $I = I^{ec}$.

We recall, that, given I, there is a Zariski open set U of $k^{n \times n}$ s.t. for each $c = (c_{ij} : i,j = 1...n) \in U$, denoting $L_c : k[X_1,...,X_n] \to k[X_1,...,X_n]$ the change of coordinates s.t. $L_c(X_i) = \Sigma c_{ij} X_j$, then $L_c(I)$ is in Noether position. Therefore if we perform a random change of coordinates on I (and then again on $(I,s^t)...$), probabilistically each ideal we deal with is in Noether position, and at each recursion the dimension drops at least one.

The equidimensional decomposition algorithm is correct under the weaker assumption (which can be tested) that either $(I,s^t) = (1)$ or $\dim(I,s^t) < d$; on the other side the equidimensionality test and the top-dimensional component algorithm would give a wrong answer if the random change of coordinates is not sufficently generic.

For primary decomposition, a scheme which repeatedly performs a random change of coordinates will have at most d recursive calls against $\sum_{i=0}^{d} \binom{n}{i}$ calls of the GTZ scheme. However such an approach repeatedly destroys sparsity of the data.

h) equidimensional radical decomposition

Giusti and Heintz [GH] have recently proposed a radical equidimensional decomposition algorithm for polynomial ideals, with "admissible" complexity, which is based on a stronger notion of genericity.

Let I be s.t. $\dim(I) = d$ and assume I is in Noether position; let $I = \mathfrak{q}_1 \cap \ldots \cap \mathfrak{q}_s$ be an irredundant primary decomposition; $\forall i$, let $\mathfrak{p}_i$ the prime associated to $\mathfrak{q}_i$.

Consider $I_{d+1} := I \cap k[X_1,\ldots,X_{d+1}]$ and let $\mathfrak{u}_i := \mathfrak{q}_i \cap k[X_1,\ldots,X_{d+1}]$, $\mathfrak{v}_i := \mathfrak{p}_i \cap k[X_1,\ldots,X_{d+1}] = \mathrm{Rad}(\mathfrak{u}_i)$, so that $I_{d+1} = \mathfrak{u}_1 \cap \ldots \cap \mathfrak{u}_s$ is a (perhaps redundant) primary decomposition.

We say I is in GH position [GH] if $\mathfrak{v}_j \subset \mathfrak{v}_i$ implies $\mathfrak{p}_j \subset \mathfrak{p}_i$. In geometrical terms, this means that the projections of the irreducible components of the variety defined by I are not containing each other.

In [GH] it is proved that, given I, there is a Zariski open set U s.t. for each $c = (c_{ij} : i,j=1\ldots n) \in U$, denoting $L_c : k[X_1,\ldots,X_n] \to k[X_1,\ldots,X_n]$ the change of coordinates s.t. $L_c(X_i) = \Sigma\, c_{ij}\, X_j$, then $L_c(I)$ is in Noether and in GH position.

If I is in GH position, let us assume that $\mathfrak{q}_i \cap k[X_1,\ldots,X_d] = (0)$ if and only if $i \leq r$ and $\exists\, j \leq r : \mathfrak{p}_j \subset \mathfrak{p}_i$ if and only if $r < i \leq u$.

Then:

$I^{ec} = \mathfrak{q}_1 \cap \ldots \cap \mathfrak{q}_r$ is an irredundant primary decomposition

$\mathrm{Rad}(I) = \mathrm{Rad}(I^{ec}) \cap \mathfrak{p}_{u+1} \cap \ldots \cap \mathfrak{p}_s$

Moreover, denoting $J := \mathrm{Rad}(I^{ec})$ and $L := I : J^* = \cup_m I : J^m$, one has $\dim(L) < \dim(I)$, $\mathrm{Rad}(I) = J \cap \mathrm{Rad}(L)$, giving a radical equidimensional decomposition algorithm for an ideal in GH-position.

4 RADICAL COMPUTATION BY IDEAL QUOTIENTS

We have already pointed out that the GTZ-scheme introduces spurious components in the decomposition of I, i.e. irredundant components in the decomposition of (I,s^t), which are redundant in the one of I. For radical computation, one would moreover like to discard embedded components without having to compute them explicitly. We propose here a different scheme for radical equidimensional decomposition, which has the advantage over the GTZ-scheme, both of avoiding the introduction of "spurious" components and of removing the components embedded in $\mathrm{Rad}(I^{ec})$. It is essentially a variant of the algorithm in [G-H] which is no more of "admissible" complexity, but doesn't require linear changes of coordinates and so it doesn't destroy sparsity of the data. It is based on the following well-known result:

Let $\mathfrak{q}$ be a primary ideal, $\mathfrak{p}$ be its associated prime, $g \in A, I \subset A$; then:

if $g \in \mathfrak{p}$, then $\mathfrak{q} : g^* = (1)$; if $g \notin \mathfrak{p}$, then $\mathfrak{q} : g^* = \mathfrak{q}$.

if $I \subset \mathfrak{p}$, then $\mathfrak{q} : I^* = (1)$; if I is not contained in $\mathfrak{p}$, then $\mathfrak{q} : I^* = \mathfrak{q}$.

if $I \subset \mathfrak{q}$, then $\mathfrak{q} : I = (1)$; if I is not contained in $\mathfrak{q}$, then $\mathfrak{q} : I^*$ is $\mathfrak{p}$-primary.

Proposition 1 Let I be s.t. $\dim(I) = d$ and let $\{X_1,\ldots,X_d\}$ be a maximal set of independent variables for I.

Let $Y := \Sigma_{i>d}\, c_i\, X_i$, $c_i \in k$, $c_{d+1} \neq 0$, and let $g \in k[X_1,\ldots,X_d][Y]$ be a primitive polynomial s.t. $(g) = \mathrm{Rad}(I^e) \cap k(X_1,\ldots,X_d)[Y]$. Let $J := \mathrm{Rad}(I^{ec}) \cap \mathrm{Rad}(I : g^*)$, $L := I : J^*$.

Then the isolated primary components of I are the union of the components of I^{ec}, of the isolated components of $I : g^*$ and of the isolated components of L, and in particular, $\mathrm{Rad}(I) = \mathrm{Rad}(I^{ec}) \cap \mathrm{Rad}(I : g^*) \cap \mathrm{Rad}(L)$. Moreover $I \subset k[X_1,\ldots X_d,Y,X_{d+2},\ldots,X_n]$ is in GH-position if and only if $L = (1)$.

Proof: Let $I = \mathfrak{q}_1 \cap \ldots \cap \mathfrak{q}_s$ be an irredundant primary decomposition and $\mathfrak{p}_i$ the prime associated to $\mathfrak{q}_i$.
Let us index the components so that:

$\mathfrak{q}_1,\ldots,\mathfrak{q}_r$ are those components $\mathfrak{q}_i$ s.t. $\mathfrak{q}_i \cap k[X_1,\ldots,X_d] = (0)$

$\mathfrak{q}_{r+1},\ldots,\mathfrak{q}_u$ are those components $\mathfrak{q}_i$ s.t. $\exists\, j \leq r$ s.t. $\mathfrak{p}_j \subset \mathfrak{p}_i$

so that

$I^e = \mathfrak{q}_1{}^e \cap ... \cap \mathfrak{q}_r{}^e$, $I^{ec} = \mathfrak{q}_1 \cap ... \cap \mathfrak{q}_r$ are irredundant primary decompositions,

$\text{Rad}(I) = \text{Rad}(I^{ec}) \cap \mathfrak{p}_{u+1} \cap ... \cap \mathfrak{p}_s$.

Since $g \in I^e \cap k[X_1,...,X_n]$, $g \in \mathfrak{q}_i{}^{ec} = \mathfrak{q}_i \subset \mathfrak{p}_i \ \forall \ i \leq r$; moreover since for $r < i \leq u$, $\exists \ j \leq r$ s.t. $\mathfrak{p}_j \subset \mathfrak{p}_i$, we can conclude also that $g \in \mathfrak{p}_i \ \forall \ i \leq u$ and therefore $(\mathfrak{q}_i : g^*) = (1)$ for each $i \leq u$, while for $i > u$, we have either $(\mathfrak{q}_i : g^*) = \mathfrak{q}_i$, or $(\mathfrak{q}_i : g^*) = (1)$.

Therefore $I : g^* = (\mathfrak{q}_{u+1} : g^*) \cap ... \cap (\mathfrak{q}_s : g^*)$. Let us now reindex the components $\mathfrak{q}_i$, $u < i \leq s$, so that $\mathfrak{q}_i$ is an isolated component of $I : g^*$ for $u < i \leq w$, an embedded component of $I : g^*$ for $w < i \leq t$. So:

$I^{ec} \cap (I : g^*) = \mathfrak{q}_1 \cap ... \cap \mathfrak{q}_r \cap \mathfrak{q}_{u+1} \cap ... \cap \mathfrak{q}_t$, $J = \mathfrak{p}_1 \cap ... \cap \mathfrak{p}_r \cap \mathfrak{p}_{u+1} \cap ... \cap \mathfrak{p}_w$

and $J \subset \mathfrak{p}_i$ if and only if $i \leq t$, so that $(\mathfrak{q}_i : J^*) = \mathfrak{q}_i$ for $i > t$, $(\mathfrak{q}_i : J^*) = (1)$ for $i \leq t$.

Therefore $L = \mathfrak{q}_{t+1} \cap ... \cap \mathfrak{q}_s$.

This proves the first statement. To prove the second one we need to show that $(\mathfrak{q}_i : g^*) = \mathfrak{q}_i \ \forall i > u$ if and only if I is in GH-position.

Denote then $\mathfrak{u}_i := \mathfrak{q}_i \cap k[X_1,...,X_d,Y]$, $\mathfrak{v}_i := \mathfrak{p}_i \cap k[X_1,...,X_d,Y] = \text{Rad}(\mathfrak{u}_i)$. Then $\mathfrak{v}_i{}^e = (1)$ for $i > r$, while $\mathfrak{v}_i$ is principal for $i \leq r$, so it has an irreducible generator $g_i \in k[X_1,...,X_d,Y]$; then the irreducible factors of g are exactly the g_i's, $\text{SQFR}(g) = g_1 ... g_r$.

For $i > u$, one has $(\mathfrak{q}_i : g^*) = (1)$ if and only if $g \in \mathfrak{p}_i \cap k[X_1,...,X_d,Y] = \mathfrak{v}_i$; then, since $\mathfrak{v}_i$ is prime, at least an irreducible component of g is in $\mathfrak{v}_i$, so$(\mathfrak{q}_i : g^*) = (1)$ if and only if $\exists \ j \leq r$ s.t. $\mathfrak{v}_j \subset \mathfrak{v}_i$, if and only if I is not in GH-position.

We obtain therefore the following algorithm for radical equidimensional decomposition:

$I' := I$, List $:= []$
Repeat
    $L := I'$
    While $L \neq (1)$ **do**
        choose a maximal set of independent variables for L, say, w.l.o.g., $X_1,...,X_d$
        List $:=$ **Append**$(\text{Rad}(L^{ec}),\text{List})$
        choose random $Y := \Sigma_{i>d} \ c_i \ X_i$
        compute $g \in k[X_1,...,X_d][Y]$ a primitive polynomial s.t. $(g) = L^e \cap k(X_1,...,X_d)[Y]$
        $L := L : g^*$
    let J be the intersection of the ideals in List
    $I' := I' : J^*$
until $I' := (1)$

We remark that actually the computation of g is free: in fact, the computation of $\text{Rad}(L^e)$ can be performed by the algorithm in [GM]: this returns a random $Y = \Sigma_{i>d} \ c_i \ X_i$, and a lexicographical Gröbner basis G (satisfying the Shape Lemma [GM]) of $\text{Rad}(L^e) \subset k(X_1,...,X_d)[Y,X_{d+2},...,X_n]$; g is then the primitive polynomial associated to the single element in $G \cap k(X_1,...,X_d)[Y]$. We remark also that this is the only point where a random change of a single coordinate is required, and that the following computations are again performed in the original coordinate frame.

The reason why we lose the "admissible" complexity of [GH] is in the quotient computations, for which we use "non-admissible" Gröbner basis computations

Since with this scheme embedded components are consistently removed and no spurious component is introduced, the inclusion tests required by the GTZ-scheme are no more required: for radical computation, one has just to intersect all the ideals in List, while for equidimensional radical decomposition, one has just to intersect, for each $\delta \leq d$, the δ-dimensional ideals in List.

Also the radicality test is simplified:

    I is radical if and only if each 0-dimensional ideal L^e computed by the algorithm is such and $I = \text{Rad}(I)$.

So, unlike in the GTZ-scheme where (I,s) could be non-radical while I were such, here non-radicality is detected early, provided that some isolated component of I is not prime.

Finally by substituting the computation of $\text{Rad}(I^e)$ with the computations of its primary decomposition (which too gives a lex G-basis of I^e in $k(X_1,...,X_d)[Y,X_{d+2},...,X_n]$, so the computation of g is still free), we can compute all isolated primary components of I.

5 DECOMPOSITION IN LOCALIZATIONS AT THE ORIGIN

5.1 Generalities

First of all we recall the following: let $P = k[X_1,...,X_n]$, let $<$ be a tangent cone ordering [MPT] on $<X_1,...,X_n>$; generalizing Gröbner basis theory, we can define for $f \in P$, $T(f)$ to be the maximal (w.r.t. $<$) term, $M(f)$ the maximal monomial, in the expansion of f. Then we define $Loc(P) := Loc(P,<) := \{f/1+g : f, g \in P, T(g) < 1\} \subset k(X_1,...,X_n)$ and we can extend $T(-)$, $M(-)$ to elements of $Loc(P)$ by $T(q) = T(f)$, $M(q) = M(f)$ for $q = f/1+g$.

If $I \subset Loc(P)$ is an ideal, $h \in Loc(P)$, we say:

$\{h_1,...,h_t\} \subset I$ is a standard basis of I if $\{M(h_1),...,M(h_t)\}$ generate $M(I) = (M(h) : h \in I) \subset P$

q is a normal form of h w.r.t. I iff $q - h \in I$ and either $q = 0$ or $M(q) \notin M(I)$

The tangent cone algorithm allows to compute standard bases of ideals and normal forms of elements in $Loc(P)$ (for details cf. [MPT]).

Let now $P = k[Z_1,...,Z_m]$, $Q = k[Z_1,...,Z_m,Y_1,...,Y_s]$, $R = k(Z)[Y]$. By an elimination ordering on $<Z,Y>$ over $<Z>$ we intend a term ordering $<$ on $<Z,Y>$ s.t.

1) the restriction $<_0$ to $<Z>$ is a tangent cone ordering

2) if $m \in <Z>$ and $n < m$ then $n \in <Z>$, which implies, in particular, $1 < Y_i\ \forall i$.

An elimination ordering is a tangent cone ordering, so that it is possible to compute standard bases w.r.t. $<$ for ideals in $Loc(Q)$ which, by 2), is equal to $Loc(P)[Y]$. Also, if G is a standard basis of $I \subset Loc(Q)$ w.r.t. $<$:

1) $G \cap Loc(P)$ is a standard basis w.r.t. $<_0$ of $I \cap Loc(P)$

2) if $I \cap Loc(P) = (0)$, then G is a Gröbner basis of I R, w.r.t. the restriction of $<$ to $<Y>$.

As a consequence we have the following local elimination algorithm:
given a tangent cone ordering $<$ on $<Z,Y>$ and $F \subset Q$ generating $I \subset Loc(Q,<)$, denoting $<_Z$, $<_Y$ the restrictions of $<$ to $<Z>$ and $<Y>$, respectively:

choose an elimination ordering $<_e$ on $<Z,Y>$ over $<Z>$ whose restriction to $<Z>$ coincides with $<_Z$

compute a standard basis G for the ideal generated by F in $Loc(Q,<_e) = Loc(P,<_Z)[Y]$.

Then:

$G \cap Loc(P,<_Z)$ is a standard basis for $I \cap Loc(P,<_Z)$

if $I \cap Loc(P,<_Z) = (0)$, then G is a Gröbner basis of (F) $k(Z)[Y]$ and a basis of I $Loc(k(Z)[Y],<_Y)$.

Also, for $I, J \subset Loc(P)$, $f \in Loc(P)$ we can compute $I \cap J$, $I : f^*$ either with the GTZ-algorithm or through syzygy computation.

If $X_i < 1\ \forall i$, to compute $I : f^*$ for $I \subset Loc(P)$, we can also apply an algorithm proposed by [BGS] for homogeneous polynomial ideals, which computes both $I : f^*$ and t s.t. $I : f^* = I : f^t$. One computes a standard basis $G = (g_1,...,g_r)$ of $(I, f - T) \subset k[X_1,...,X_n,T]$ w.r.t. a tangent cone ordering s.t. $T < m\ \forall m \in <X_1,...,X_n>$; if one writes $g_i = T^{t_i} h_i$, where h_i is not divisible by T, then:

$t := \max\{t_i\}$, $I : f^* = (\psi(h_1),...,\psi(h_r))$, where $\psi(-)$ is the morphism which evaluates T to f.

However, unlike the homogeneous case, $(\psi(h_1),...,\psi(h_r))$ is not necessarily a standard basis of $I : f^*$.

5.2 Localizations at the origin

We are interested now in decomposition algorithms in $P_0 = k[X_1,...,X_n]_0$, the localization of $P = k[X_1,...,X_n]$ at the origin, i.e. the local ring of those rational functions which are defined at the origin, whose maximal ideal is $m := (X_1,...,X_n)$. We recall that, if $<$ is a tangent cone ordering s.t. $X_i < 1$ for each i, then P_0 coincides with $Loc(P)$.

Let us begin discussing the relationship among the primary decomposition of ideals in P and in P_0; let us denote $-^E$ and $-^C$ ideal extension and contraction between $P = k[X_1,...,X_n]$ and $P_0 = k[X_1,...,X_n]_0$. It is well-known that:

- if $I \subset k[X_1,...,X_n]_0$ there is at least one $J \subset k[X_1,...,X_n]$ s.t. $I = J^E$
- J^E is a proper ideal if and only if $J \subset (X_1,...,X_n)$
- if J is prime (primary, radical) then J^E is such
- $Rad(J^E) = (Rad(J))^E$
- if $J = \mathfrak{q}_1 \cap ... \cap \mathfrak{q}_s$ is an irredundant primary decomposition with $\mathfrak{q}_i \subset (X_1,...,X_n)$ if and only if $i \leq t$, then $J^E = \mathfrak{q}_1^E \cap ... \cap \mathfrak{q}_t^E$ is a primary decomposition
- each primary component of J is contained in $(X_1,...,X_n)$ if and only if $J = I^C$.

There is therefore an obvious approach to decomposition algorithms for ideals I in P_0, which consists in computing, say, the primary decomposition of an ideal $J \subset P$ s.t. $J^E = I$ and discarding the components not passing through the origin. If $J = I^C$, there are no such components; however we don't know of any algorithm which allows to compute I^C without at least a partial decomposition; moreover the behaviour of the tangent cone algorithm w.r.t. extraneous components is not well known; apparently, if F is the input basis and G the output one, the components not vanishing at the origin in (F) and (G) (considered as polynomial ideals) are totally unrelated: the algorithm can as well remove extraneous components as introduce new ones. Because of this, we don't know how to obtain a decomposition algorithm which, while following the [GTZ] scheme, is able to discard automatically the extraneous components.

Our approach to decomposition algorithms in localizations at the origin is therefore essentially the obvious procedure above, improved by some elementary tricks in order to detect extraneous components as early as possible and to avoid to introduce new ones. Our main result will be a technique based on ideal quotients to remove extraneous components of maximal dimension.

In what follows we assume to have an ideal $I \subset P_0$ given through a basis $\{g_1,...,g_s\} \subset P$ and we denote by J the ideal in P generated by $\{g_1,...,g_s\}$; remark that by a standard basis computation we can compute $d = \dim(I)$.

5.3 Removing extraneous components of maximal dimension

Let $\{X_1,...,X_\delta\}$ $(\delta \geq d)$ be a maximal set of independent variables for J; let $Y := \Sigma_{i=\delta+1..n} c_i X_i$, $c_i \in k$ and let $g \in k[X_1,...,X_\delta][Y]$ be the primitive polynomial s.t. $(g) = J^e \cap k(X_1,...,X_\delta)[Y]$. Recall that $-^e$ and $-^c$ denote ideal extension and contraction between $k[X_1,...,X_n]$ and $k(X_1,...,X_d)[X_{d+1},...,X_n]$, $-^E$ and $-^C$ denote ideal extension and contraction between $k[X_1,...,X_n]$ and $k[X_1,...,X_n]_0$.

Let $g = g_0\, g_1$ where g_0 is the product of the factors of g vanishing at the origin, g_1 the product of those not vanishing at the origin.

Let $J = \mathfrak{q}_{11} \cap ... \cap \mathfrak{q}_{1s_1} \cap \mathfrak{q}_{21} \cap ... \cap \mathfrak{q}_{2s_2} \cap \mathfrak{q}_{31} \cap ... \cap \mathfrak{q}_{3s_3} \cap \mathfrak{q}_{41} \cap ... \cap \mathfrak{q}_{4s_4}$ be an irredundant primary decomposition, where:

$\mathfrak{q}_{1j} \cap k[X_1,...,X_\delta] = (0)$, $\mathfrak{q}_{1j} \subset (X_1,...,X_n)$

$\mathfrak{q}_{2j} \cap k[X_1,...,X_\delta] \neq (0)$, $\mathfrak{q}_{2j} \subset (X_1,...,X_n)$

$\mathfrak{q}_{3j} \cap k[X_1,...,X_\delta] = (0)$, $\mathfrak{q}_{3j}$ not contained in $(X_1,...,X_n)$

$\mathfrak{q}_{4j} \cap k[X_1,...,X_\delta] \neq (0)$, $\mathfrak{q}_{4j}$ not contained in $(X_1,...,X_n)$

and let $\mathfrak{p}_{ij} = \mathrm{Rad}(\mathfrak{q}_{ij})$.
Then

$$I = J^E = \mathfrak{q}_{11}{}^E \cap ... \cap \mathfrak{q}_{1s_1}{}^E \cap \mathfrak{q}_{21}{}^E \cap ... \cap \mathfrak{q}_{2s_2}$$

$$J^e = \mathfrak{q}_{11}{}^e \cap ... \cap \mathfrak{q}_{1s_1}{}^e \cap \mathfrak{q}_{31}{}^e \cap ... \cap \mathfrak{q}_{3s_3}{}^e.$$

Also if $\delta > d$, $s_1 = 0$ i.e. there are no primary components $\mathfrak{q}$ s.t. $\mathfrak{q} \cap k[X_1,...,X_\delta] = (0)$, $\mathfrak{q} \subset (X_1,...,X_n)$.

Proposition 2 Denote $J_1 := J : g_1{}^*$. Then:

1) $J_1 = \mathfrak{q}_{11} \cap...\cap \mathfrak{q}_{1s_1} \cap \mathfrak{q}_{21} \cap...\cap \mathfrak{q}_{2s_2} \cap (\mathfrak{q}_{31} : g_1{}^*) \cap...\cap (\mathfrak{q}_{3s_3} : g_1{}^*) \cap (\mathfrak{q}_{41} : g_1{}^*) \cap...\cap (\mathfrak{q}_{4s_4} : g_1{}^*)$

2) There is a Zariski non-void open set U of $k^{n-\delta}$ s.t.

i) if $(c_{\delta+1},...,c_n) \in U$, then:

$J_1 = \mathfrak{q}_{11} \cap ... \cap \mathfrak{q}_{1s_1} \cap \mathfrak{q}_{21} \cap ... \cap \mathfrak{q}_{2s_2} \cap (\mathfrak{q}_{41} : g_1{}^*) \cap ... \cap (\mathfrak{q}_{4s_4} : g_1{}^*)$

so that $J_1{}^E = \mathfrak{q}_{11}{}^E \cap ... \cap \mathfrak{q}_{1s_1}{}^E \cap \mathfrak{q}_{21}{}^E \cap ... \cap \mathfrak{q}_{2s_2}{}^E = I$ and each component of J_1, for which $\{X_1,...,X_\delta\}$ is a maximal set of independent variables, passes through the origin

ii) if $(c_{\delta+1},...,c_n) \notin U$, then $\exists i$ s.t. $\mathfrak{q}_{3i}$ is among the primary components of J_1

iii) if $\delta > d$, either $g_0 = 1$ or $(c_{\delta+1},...,c_n) \notin U$.

<u>Proof</u>: Both $\mathfrak{q}_{1j} \cap k[X_1,...,X_\delta,Y]$ and $\mathfrak{q}_{3j} \cap k[X_1,...,X_\delta,Y]$ are hypersurfaces and so are generated by a single primitive polynomial in $k[X_1,...,X_\delta][Y]$, say, resp., h_{1j} and h_{3j}; the h_{ij}'s divide g.
Clearly $g_1 \notin \mathfrak{p}_{1j}$, $g_1 \notin \mathfrak{p}_{2j}$, so $\mathfrak{q}_{1j} : g_1{}^* = \mathfrak{q}_{1j}$, $\mathfrak{q}_{2j} : g_1{}^* = \mathfrak{q}_{2j}$. As a consequence:
$J_1 = \mathfrak{q}_{11} \cap...\cap \mathfrak{q}_{1s_1} \cap \mathfrak{q}_{21} \cap...\cap \mathfrak{q}_{2s_2} \cap (\mathfrak{q}_{31} : g_1{}^*) \cap...\cap (\mathfrak{q}_{3s_3} : g_1{}^*) \cap (\mathfrak{q}_{41} : g_1{}^*) \cap...\cap (\mathfrak{q}_{4s_4} : g_1{}^*)$
The first statement is therefore proved.
Moreover $(\mathfrak{q}_{3j} : g_1{}^*) = \mathfrak{q}_{3j}$ if and only if h_{3j} vanishes for $X_i = 0$, $Y = 0$. Since $\mathfrak{q}_{3i}$ has only finitely many zeroes $(0,...,0,x_{\delta+1},...,x_n)$ whose first δ coordinates are zeroes, there is then a non-void Zariski open set U of $k^{n-\delta}$ s.t. if $(c_{\delta+1},...,c_n) \in U$, h_{3i} doesn't vanish for $X_i = 0$, $Y = 0$. Therefore, for $(c_{\delta+1},...,c_n) \in U$, h_{3i} divides g_1 so that $g_1 \in \mathfrak{p}_{3i}$ $\forall i$ and $(\mathfrak{q}_{3j} : g_1{}^*) = (1)$ proving the second statement.

So we get the following algorithm to (probabilistically) remove extraneous components of maximal dimension:

> **While** $\dim(I) < \dim(J)$ **do**
>> **choose** a maximal set of independent variables for J, say, w.l.o.g., $X_1,...,X_\delta$
>> **compute** a Gröbner basis of J^e
>> **Repeat**
>>> **choose random** $Y := \Sigma_{i>d} c_i X_i$
>>> **compute** $g \in k[X_1,...,X_d][Y]$ a primitive polynomial s.t. $(g) = J^e \cap k(X_1,...,X_d)[Y]$ -
>>> *this is done by an obvious variant of the FGLM conversion algorithm [FGLM], cf. [GM]*
>> **until** $g(0) \neq 0$
>> $J := J : g^*$
> **Repeat**
>> **choose** a maximal set of independent variables for J, say, w.l.o.g., $X_1,...,X_d$
>> **compute** a Gröbner basis of J^e
>> **choose random** $Y := \Sigma_{i>d} c_i X_i$
>> **compute** $g \in k[X_1,...,X_d][Y]$ a primitive polynomial s.t. $(g) = J^e \cap k(X_1,...,X_d)[Y]$
>> **factor** $g = g_0 g_1$
>> $J := J : g_1^*$
> **until** $g_0 \neq 1$

5.4   Reduction to the zero-dimensional case

We use the same notation and assumptions of the paragraph above and we assume that $g_0 \neq 1$ and that $\delta = d$, so that $J_1^e = \mathfrak{q}_{11}^e \cap ... \cap \mathfrak{q}_{1s_1}^e \cap (\mathfrak{q}_{31} : g_1^*)^e \cap ... \cap (\mathfrak{q}_{3s_3} : g_1^*)^e$ and, if Y is sufficiently generic, $\mathfrak{q}_{3i} : g_1^* = (1)$. J_1^e is then a zero-dimensional ideal in $k(X_1,...,X_d)[X_{d+1},...,X_n]$ and so to it we can apply the GTZ zero-dimensional decomposition algorithms.

If we are interested in the radical computation and/or test we can then apply the scheme of §4, using only ideal quotient computations.

If we are interested in the other decomposition problems or we like to follow the GTZ-scheme also for radicals, we can assume J_1 is given through a Gröbner basis G for an elimination ordering over $<X_1,...,X_d>$. So, applying the first version of the GTZ-scheme we obtain $s \in k[X_1,...,X_d]$, $t \in N$ s.t. $J_1 = J_1^{ec} \cap (J_1, s^t)$.

First of all remark that if $s(0) \neq 0$, then $(J_1, s^t)^E = (1)$ so $I = (J_1^{ec})^E$ and there is no need to iterate further. Otherwise, let $u, v \in k[X_1,...,X_d]$ be s.t. $s = uv$, $u(0) \neq 0$, $v(0) = 0$. Then $(J_1, s^t)^E = (J_1, v^t)^E$, so we can apply the iteration to (J_1, v^t), avoiding the introduction of those extraneous components which are related to the factors of u.

If Y is not sufficiently generic, then it is possible that some components of J^{ec} are not passing through the origin; if L is a component of J^e s.t. L^c is not passing through the origin, then this can be detected during the computation of L^c, since then in any basis of L^c there is f s.t. $f(0) \neq 0$.

6   DECOMPOSITION IN ALGEBRAIC POWER SERIES RINGS

6.1 Generalities

Let $k[[X_1,...,X_n]]_{alg}$ denote the subring of $k[[X_1,...,X_n]]$ consisting of those formal power series which are algebraic, i.e. which satisfy a polynomial equation.

Let $g_1,...,g_s \in k[[X_1,...,X_n]]_{alg}$ and let us consider the ring $B = \{f\ g^{-1} : f, g \in k[\underline{X}, g_1,..., g_s], g$ invertible$\}$. Remark that if $h_1,...,h_s \in B$ and I, J denote respectively the ideals generated by $\{h_1,...,h_s\}$ in $k[[X_1,...,X_n]]_{alg}$ and in B, then $I = J\ k[[X_1,...,X_n]]_{alg}$, $J = I \cap B$, so for many purposes it is sufficient to compute in the ring B. For such rings a computational model has been introduced in [AMR]:

Theorem 1 There are polynomials $F_1,..., F_r \in k[X_1,...,X_n, Y_1,....,Y_r] =: P$ and a tangent cone ordering $<_\sigma$ on $<X_1,...,X_n, Y_1,.....,Y_r>$, whose restriction $<$ to $<X_1,...,X_n>$ is anticompatible with the usual degree s.t.
> 1) there are <u>unique</u> $f_1,..., f_r \in k[[X_1,...,X_n]]_{alg}$ s.t. $f_j \neq 0$ and $f_j(0) = 0$ $\forall j$, and $F_i(\underline{X},f_1,...,f_r) = 0$ $\forall i$.
> 2) $F_i = Y_i (1+Q_i) - R_i$ with $Q_i, R_i \in (\underline{X},\underline{Y})$, $R_i \in k[\underline{X},Y_1,...,Y_{i-1},Y_{i+1}..., Y_r]$ and $M_\sigma(R_i) = M(f_i)$
> 3) $\{F_1,...,F_r\}$ is a standard basis for the ideal it generates in Loc(P) w.r.t. $<_\sigma$ and $M_\sigma(F_i) = Y_i$

4) $g_j \in k[X_1,...,X_n, f_1,...,f_r]$ and can be explicitly represented in it

5) $k[\underline{X},\underline{F}]_{loc} := \{f\, g^{-1} : f, g \in k[\underline{X}, f_1,..., f_r], g \text{ invertible}\}$ satisfies $B \subset k[\underline{X},\underline{F}]_{loc} \subset k[[\underline{X}]]_{alg}$.
$F = (F_1,...,F_r)$ is called a *standard locally smooth system* (SLSS) over < defining $f_1,..., f_r$.

The evaluation map $\sigma_F : Loc(P) \to k[\underline{X},\underline{F}]_{loc}$ defined by $\sigma_F(Y_i) = f_i$ satisfies $Ker(\sigma_F) = (F_1,...,F_r)\, Loc(P)$, so that $k[\underline{X},\underline{F}]_{loc} \approx Loc(P)/(F_1,...,F_r)$.
Finally, if each g_i is given by its minimal polynomial and its Taylor expansion, F can be explicitly computed.

Because of the result above, in our approach to decomposition algorithms in $k[[X_1,...,X_n]]_{alg}$, we will assume to compute within $k[\underline{X},\underline{F}]_{loc}$, for some SLSS F. Within this computational model, an elimination algorithm is available according to the following:

Proposition 3 Let $Q := P[Z_1,...,Z_s]$; let $<_e$ be an elimination ordering in $<\underline{X},\underline{Y},\underline{Z}>$ over $<\underline{X},\underline{Y}>$ s.t. its restriction to $<\underline{X},\underline{Y}>$ is $<_\sigma$. Denote σ both the evaluation map $\sigma_F : Loc(P) \to k[\underline{X},\underline{F}]_{loc}$ and its polynomial extension $\sigma : Loc(Q) = Loc(P)[Z_1,...,Z_s] \to k[\underline{X},\underline{F}]_{loc}[Z_1,...,Z_s]$.
Let $I \subset k[\underline{X},\underline{F}]_{loc}[Z_1,...,Z_s]$, $\{B_1,...,B_t\} \subset Q$ be s.t. $I = (\sigma(B_1),...,\sigma(B_t))$, and $J := (B_1,...,B_t,F_1,...,F_r)\, Loc(Q)$.
Let $G := (A_1,...,A_u,F_1,...,F_r)$ be a standard basis of J w.r.t. $<_e$ (which can be computed by the tangent cone algorithm).
Then $G' := G \cap Loc(P)$ is a standard basis of $J \cap Loc(P)$ and $\sigma(G')$ is a basis of $I \cap k[\underline{X},\underline{F}]_{loc}$ and therefore of $I\, k[[\underline{X}]]_{alg}[\underline{Z}] \cap k[[\underline{X}]]_{alg}$.
Moreover if $G \cap Loc(P) = (F_1,...,F_r)$, then $I \cap k[\underline{X},\underline{F}]_{loc} = (0)$, $I\, k[[\underline{X}]]_{alg}[\underline{Z}] \cap k[[\underline{X}]]_{alg} = (0)$ and $\sigma(G)$ is a Gröbner basis of $I\, K[\underline{Z}]$, where K is the fraction field of $k[[\underline{X}]]_{alg}$.

As a consequence of the above result, for $I, J \subset k[[\underline{X}]]_{alg}$, $f \in k[[\underline{X}]]_{alg}$ we can compute $I \cap J$, $I : f^*$ with the GTZ-algorithm.
In order to apply the decomposition schemes to ideals we need to reduce to 0-dimensional polynomial ideals over a field; this is possible in this computational model, by the following:

Theorem 2 Given $G_1,...,G_s \in Loc(P)$ and denoting $g_i := \sigma_F(G_i)$, $I = (g_1,...,g_s) \subset k[[\underline{X}]]_{alg}$, then there is an algorithm which computes

a) a linear change of coordinates $L : k[[X]]_{alg} \dashrightarrow k[[X]]_{alg}$

b) $d := dim(I)$

c) a SLSS $H := (H_1,...,H_u) \subset k[X_1,...,X_d,\underline{U}] =: Q$ defining algebraic series in $k[[X_1,...,X_d]]_{alg}$

d) $A_1,..., A_t \in Q[X_{d+1},...,X_n]$

s.t. denoting, with a slight abuse of notation, $\sigma_H : Loc(Q)[X_{d+1},...,X_n] \to k[X_1,...,X_d,H]_{loc}[X_{d+1},...,X_n]$ the extension of the evaluation morphism σ_H, $a_j := \sigma_H(A_j)$, $J := (a_1,..., a_t)\, k[[X_1,...,X_d]]_{alg}[X_{d+1},...,X_n]$ one has:

1) $L(I)$ is in Noether position (i.e. $X_{d+1},...,X_n$ are integral mod. $L(I)$)

2) $J = L(I) \cap k[[X_1,...,X_d]]_{alg}[X_{d+1},...,X_n]$

3) $(A_1,..., A_t, H_1,...,H_u)$ is a standard basis in $Loc(Q)[X_{d+1},...,X_n]$ for an elimination ordering over $Loc(Q)$ whose restriction to $Loc(Q)$ is $<_\sigma$

Corollary 1 With notation of Theorem 2:
   i) I is prime (primary, radical, equidimensional) iff J is such
   ii) $Rad(L(I)) = Rad(J)\, k[[X_1,...,X_n]]_{alg}$

For details on the above results, we refer the reader to [AMR].
Since our decomposition algorithms for algebraic power series rings require Noether position, and so a change of coordinates and loss of sparsity, we take the most advantage of genericity, by using the stronger notion introduced in [GH].
So, let I be a d-dimensional ideal in Noether position, let
$J := I \cap k[[X_1,...,X_d]]_{alg}[X_{d+1},...,X_n]$,
$J = \mathfrak{q}_1 \cap ... \cap \mathfrak{q}_s$ be an irredundant primary decomposition and $\forall i$, let $\mathfrak{p}_i$ the prime associated to $\mathfrak{q}_i$.
$J_{d+1} := J \cap k[[X_1,...,X_d]]_{alg}[X_{d+1}]$
$\mathfrak{u}_i := \mathfrak{q}_i \cap k[[X_1,...,X_d]]_{alg}[X_{d+1}]$, $\mathfrak{v}_i := \mathfrak{p}_i \cap k[[X_1,...,X_d]]_{alg}[X_{d+1}] = Rad(\mathfrak{u}_i)$,

so that $J_{d+1} = \mathfrak{u}_1 \cap \ldots \cap \mathfrak{u}_s$ is a (perhaps redundant) primary decomposition.

We say I is in GH position if $\mathfrak{v}_j \subset \mathfrak{v}_i$ implies $\mathfrak{p}_j \subset \mathfrak{p}_i$

The same argument as in [GH] for the polynomial case, shows that, given I, there is a Zariski open set U of $k^{n \times n}$ s.t. for each $c = (c_{ij} : i,j = 1\ldots n) \in U$, denoting $L_c : k[[X_1,\ldots,X_n]]_{alg} \to k[[X_1,\ldots,X_n]]_{alg}$ the change of coordinates s.t. $L_c(X_i) = \Sigma c_{ij} X_j$, then $L_c(I)$ is in Noether and in GH position.

6.2 Radical computation

Let I be an ideal in $k[X_1,\ldots,X_d,H]_{loc}[X_{d+1},\ldots,X_n]$ s.t. dim(I) = d and $I \cap k[X_1,\ldots,X_d,H]_{loc} = (0)$. It is easy to remark that Prop. 1 holds for I, if we substitute $k[X_1,\ldots,X_n]$ with $k[[X_1,\ldots,X_d]]_{alg}[X_{d+1},\ldots,X_n]$ and $k(X_1,\ldots,X_d)[X_{d+1},\ldots,X_n]$ with $K[X_{d+1},\ldots,X_n]$, where K is the fraction field of $k[[X_1,\ldots,X_d]]_{alg}$ and -^e and -^c denote ideal extension and contraction between $k[[X_1,\ldots,X_d]]_{alg}[X_{d+1},\ldots,X_n]$ and $K[X_{d+1},\ldots,X_n]$. Therefore the decomposition scheme of §4 can be applied to this situation too, if each step can be effectively performed.

Let us therefore remark that:

- if we are given $I \subset k[\underline{X},F]_{loc}$, by the algorithm and with notation of Theorem 2, we can compute d = dim(I), a change of coordinates L, an SLSS $H \subset k[X_1,\ldots,X_d,\underline{U}] =: Q$ and $\{A_1,\ldots,A_t\} \subset Q[X_{d+1},\ldots,X_n]$ satisfying the conditions of the theorem, so that in particular L(I) and J are in Noether position.

- J^e is a 0-dimensional ideal in $K[X_{d+1},\ldots,X_n]$ and with an appropriate choice of the elimination ordering in Loc(Q) during the application of the above algorithm, $\{A_1,\ldots,A_t\} \subset Q[X_{d+1}]$ consists of a single element $H \in Q[X_{d+1}]$ s.t. $h = \sigma(H) \in k[X_1,\ldots,X_d,H]_{loc}[X_{d+1}]$ is the primitive generator of $J^e \cap K[X_{d+1}]$.

- if L is sufficiently generic, then J^e is in GTZ position; then $Rad(J^e) = J^e + (SQFR(h))$. This can be tested by checking whether $J^e + (SQFR(h))$ satisfies the structure theorem for a 0-dimensional radical ideal in GTZ-position.

- if this is the case, then one can choose $Y := X_{d+1}$ and $g := SQFR(h)$; otherwise Y and g will be freely obtained during the computation of $Rad(J^e)$ by the algorithm in [G-M].

- $Rad(J^{ec}) = (Rad(J^e))^c$ can be obtained, exactly as in the polynomial case, by quotient with an element in $k[X_1,\ldots,X_d,H]_{loc}$ which can be read from a Gröbner basis of $Rad(J^e)$

- the computations to be performed in $K[X_{d+1},\ldots,X_n]$ are Gröbner basis computations and computation of the squarefree associates of univariate polynomials. Both are available in $K[X_{d+1},\ldots,X_n]$ since the coefficients are in the computable ring $k[X_1,\ldots,X_d,H]_{loc} \subset K$

- if L is sufficiently generic, then L(I) is in GH-position, and $Rad(J) = Rad(J^{ec}) \cap Rad(J : g^*)$. The final division test is therefore expected to return the ideal (1)

6.3 Equidimensional decomposition and test

Since L(I) and J are in Noether position and I is equidimensional if and only if J is such, we can apply the results of §3 to obtain an equidimensional test, a top-dimensional component and an equidimensional decomposition algorithm.

6.4 Other decomposition algorithms

If we assume to have decomposition algorithms for zero-dimensional ideals in $K[X_{d+1},\ldots,X_n]$ where K is the field of fractions of $k[[X_1,\ldots,X_d]]_{alg}$, then the GTZ decomposition scheme can be applied to the case of algebraic power series, since after a suitable change of coordinates, which puts the ideal in Noether position, all computations required by the GTZ scheme can be performed in this case too.

However the zero-dimensional algorithms for primality, primarieties tests and primary decompositions require the ability of factorize univariate polynomials over K: we don't know of any factorization algorithm over K.

We remark however that Lazard [LAZ] has proposed an algorithm for solving zero-dimensional systems of equations over a computable field (with no factorization requirements) by applying Duval techniques. A Duval approach can be applied in $K[X_{d+1},\ldots,X_n]$, since it requires just ring operations and our coefficients are in $k[X_1,\ldots,X_d,H]_{loc}$, and its interest for concrete applications should be investigated.

7 DECOMPOSITION IN LOCALIZATIONS AT PRIME IDEALS

7.1 Generalities

Let $Q := k[X_1,\ldots,X_m]$; let $H \subset \mathfrak{m} \subset Q$ be two ideals, with $\mathfrak{m}$ prime. Let us denote $A := Q_{\mathfrak{m}} / \mathfrak{m} Q_{\mathfrak{m}}$, $\pi : Q_{\mathfrak{m}} \to A$ the canonical projection.

First of all for our purposes we can assume H = (0). In fact let $I \subset A$ be an ideal and let $L := \pi^{-1}(I)$ (remark that if I is given by $f_1,\ldots,f_t \in Q$ s.t. $\{\pi(f_1),\ldots,\pi(f_t)\}$ generate I, then $L = (f_1,\ldots,f_t) + H Q_{\mathfrak{m}}$). Then:

I is prime (radical, primary, equidimensional) if and only if L is such

if $L = L_1 \cap \dots L_t$ is a primary decomposition, then $I = \pi(L_1) \cap \dots \pi(L_t)$ is a primary decomposition

$\pi(\mathrm{rad}(L)) = \mathrm{rad}(I)$

if $L = \mathfrak{u}_0 \cap \dots \cap \mathfrak{u}_d$ is an equidimensional decomposition, then $I = \pi(\mathfrak{u}_0) \cap \dots \cap \pi(\mathfrak{u}_d)$ is such.

By standard techniques [MOR] we can moreover assume that $\mathfrak{m}$ is maximal; for our purposes however we will be forced to assume moreover that $\mathfrak{m}$ is in GTZ-position; this allows an improvement in the representation proposed in [MOR], which will be crucial for our applications.

We can assume therefore to deal with the local ring $A := Q_{\mathfrak{m}}$, with maximal ideal $m := \mathfrak{m}Q_{\mathfrak{m}}$, with $\mathfrak{m} = (p(X_1), X_2 - q_2(X_1), \dots, X_m - q_m(X_1))$ where p is irreducible in $k[X_1]$ and $\deg(q_i) < \deg(p)$ for each i. Let then $P := k[Z, Y_1, \dots, Y_m]$ and let $<$ be a tangent cone ordering s.t. $t < 1$ iff $t \in (Y_1, \dots, Y_m)$. Define $\phi : P \to Q$ by $\phi(Z) = X_1, \phi(Y_1) = p(X_1), \phi(Y_i) = X_i - q_i(X_1)$; ϕ induces a surjective morphism $\mathrm{Loc}(P) \to Q_{\mathfrak{m}}$, which we will still denote by ϕ, so that:

$\mathrm{Ker}(\phi) = (p(Z) - Y_1)$

$\phi(Y_1, \dots, Y_m) = (p(X_1), X_2 - q_2(X_1), \dots, X_m - q_m(X_1)) = \mathfrak{m}Q_{\mathfrak{m}}$.

Let us consider the univariate polynomial ring $S := k(Y_1)[Z]$; in it the polynomial $p(Z) - Y_1$ is clearly irreducible so that $\mathfrak{k} := S/(p(Z) - Y_1)$ is a field. We have an inclusion from P into $k(Y_1)[Z, Y_2, \dots, Y_m] = S[Y_2, \dots, Y_m]$ and by quotienting both rings for the respective ideals generated by $p(Z) - Y_1$ we get an inclusion from $P/\mathrm{Ker}(\phi)$ into $S[Y_2, \dots, Y_m]/(p(Z) - Y_1)$. Since the first ring is isomorphic to Q, while the second ring is isomorphic to $B := \mathfrak{k}[Y_2, \dots, Y_m]$, if we denote z a root of $p(Z) - Y_1$ in $\mathfrak{k}$, we get an immersion $\lambda : Q \to B$ which is defined by $\lambda(X_1) = z, \lambda(X_i) = Y_i + q(z)$. Denoting by $\psi : S[Y_2, \dots, Y_m] \to \mathfrak{k}[Y_2, \dots, Y_m]$ the canonical projection, we have then $\psi(a) = \lambda(\phi(a)) \; \forall a \in P$.

A relation between primary decomposition in Q (and so in A) and in B is given by the following proposition, where we will denote by $-^E$ and $-^C$ ideal contraction and extension between Q and A.

Proposition 4 Let $J \subset Q$ be an ideal s.t. for each primary component $\mathfrak{q}$ of J^E $\dim(\mathfrak{q}) > 0$ and $\mathfrak{q} \cap k[X_1] = 0$. Then $(J B \cap Q)^E = J^E$.

If $J B = \mathfrak{u}_1 \cap \dots \cap \mathfrak{u}_t$ is a primary decomposition in B and $\mathfrak{q}_j := \mathfrak{u}_j \cap Q$, then $J^E = \mathfrak{q}_1^E \cap \dots \cap \mathfrak{q}_t^E$ is a primary decomposition in A

<u>Proof</u> : Let $g \in Q$ be s.t. $\lambda(g) \in J B$, let $h \in P$ be s.t. $\phi(p) = g$. Since $\lambda(g) \in J B$, then $h \in \psi^{-1}(J B) \subset k(Y_1)[Z, Y_2, \dots, Y_n]$; so $\exists t, \exists u \in k[Y_1]$, with $u(0) \neq 0$, s.t. $Y_1^t u h \in \phi^{-1}(J)$, and $\phi(u) p^t g \in J, p^t g \in J^E$ (since $\phi(u)$ is invertible). Since no power of p^t is in any primary component of J^E, $g \in J^E$; hence $J B \cap Q \subset J^E$; since $J \subset J B \cap Q$, we obtain $(J B \cap Q)^E = J^E$

The second statement is then obvious.

Proposition 5 Let I be an ideal of A, let $I = \mathfrak{q}_1 \cap \dots \cap \mathfrak{q}_s$ be an irredundant primary decomposition; let us index the primary components so that $\mathfrak{q}_i \cap K[X_1] = (0)$ if and only if $1 \leq i \leq r$. and let $I_1 := I : p^*$. Then:

    i) $I_1 = \mathfrak{q}_1 \cap \dots \cap \mathfrak{q}_r$

    ii) If $I \neq I_1$ the following conditions are equivalent:

        a) $\forall j \leq s$, if $\dim(\mathfrak{q}_j) > 0$, then $\mathfrak{q}_j \cap k[X_1] = 0$

        b) $\dim(I : I_1) = 0$

        c) $r = s - 1$, $\mathfrak{q}_s$ is m-primary

<u>Proof</u>: If $i \leq r$, then $\mathfrak{q}_i \cap K[X_1] = (0)$, so $p \notin \mathrm{Rad}(\mathfrak{q}_i)$ and $(\mathfrak{q}_i : p^*) = \mathfrak{q}_i$. Since A is local, $\mathfrak{q}_i \subset m \; \forall i$; therefore if $j > r$, $(0) \neq \mathfrak{q}_j \cap k[X_1] \subset \mathfrak{m} \cap k[X_1] = (p(X_1))$, so that $\mathfrak{q}_j \cap k[X_1] = (p(X_1)^t q(X_1))$ for some $q \in K[X_1]$ not divisible by p; then $q \notin \mathfrak{m}$ and is invertible in A, so, since $p^t q \in \mathfrak{q}_j$, one has $p^t \in \mathfrak{q}_j$, $(\mathfrak{q}_j : p^*) = (1)$.

As a consequence $I_1 = I : p^* = (\mathfrak{q}_1 : p^*) \cap \dots \cap (\mathfrak{q}_s : p^*) = \mathfrak{q}_1 \cap \dots \cap \mathfrak{q}_r$, proving i).

Since $I_1 \subset \mathfrak{q}_i$, $i = 1 \dots r$, and $I_1 \not\subset \mathfrak{q}_j$, $j > r$ (otherwise $\mathfrak{q}_j$ would be redundant), one has $(\mathfrak{q}_i : I_1) = (1)$ for $i \leq r$, $(\mathfrak{q}_i : I_1)$ is $\mathrm{Rad}(\mathfrak{q}_i)$-primary for $i > r$; and $\mathrm{Rad}(I : I_1) = \mathrm{Rad}(\mathfrak{q}_{r+1}) \cap \dots \cap \mathrm{Rad}(\mathfrak{q}_s)$, so that $\dim(I : I_1)$ is the maximum of the dimensions of those $\mathfrak{q}_j$'s s.t. $\mathfrak{q}_j \cap k[X_1] \neq 0$. This shows the equivalence of a) and b) and the implication c) $\Rightarrow$ b). The implication a) $\Rightarrow$ c) is obvious, since A is local, so a 0-dimensional ideal is m-primary, and there is at most one m-primary component in an irredundant decomposition of I.

To obtain decomposition algorithms in A, we would make use of Props. 4 and 5; to do so, we need to assume the equivalent conditions of Prop. 5. Let us remark that, if $I \cap Q$ is in Noether position, then a) holds; so a) is a genericity condition; since we already assume a (weaker) genericity condition, namely that $\mathfrak{m}$ is in GTZ-position, there is no cost in assuming this stronger condition. The algorithm we will adopt is as follows:

first we compute I_1, then, if $I \neq I_1$, we check if b) holds (by computing $I : I_1$); if this is the case we obtain a decomposition of I_1; if $I \neq I_1$, we finally compute the m-primary component of I. Let us now discuss how to effectively perform each of these steps.

7.2 Removing the zero-dimensional component and checking genericity

First of all, if $g(X_1,...,X_m) \in Q$ and $\xi(g) \in P$ is a normal form of $g(Z, Y_2 - q_2(Z),...,Y_m - q_m(Z))$ w.r.t. $(p(Z) - Y_1)$, then $\phi(\xi(g)) = g$.

So let $I \subset A$ be generated by $\{g_1,...,g_t\} \subset Q$, let $h_i := \xi(g_i) \in P$ and $I := (h_1,..., h_r, p(Z) - Y_1) \subset Loc(P)$, so that $I = \phi^{-1}(I)$. It is clear that $I_1 := (I : Y_1^*). = \phi^{-1}(I_1)$, so that also $I : I_1 = \phi^{-1}(I : I_1)$. Therefore I_1 can be obtained by computing I_1, and $\dim(I : I_1) = \dim(I : I_1)$ can be obtained by computing $I : I_1$.

7.3 Decomposition of ideals with no m-primary component

Since $I_1 = I : p^*$ satisfies the assumptions of Prop. 4, to compute its decomposition, we have to compute the decomposition of an ideal in the polynomial ring B, provided we are able also to compute $L \cap Q$ for an ideal $L \subset B$. This second problem is solved by means of the following lemma, which reduces it to some evaluations by ϕ and to compute ideal contractions from $k(Y_1)[Z,Y_2,...,Y_n]$ to $k[Z,Y_1,...,Y_n]$.

Lemma 2 Let $L = (f_1,...,f_t) \subset B$; let $L_1 := \psi^{-1}(L) = (f_1,...,f_t,p-Y_1) \subset S[Y_2,...,Y_n]$; let $L_2 := L_1 \cap P$; let $L_3 := \phi(L_2) \subset Q$. Then $L_3 = L \cap Q$.

Proof: let $f \in L_3$, let $g = \xi(f) \in P$; then $g \in L_2$ (since $p-Y_1 \in L_2$ and so $L_2 = \phi^{-1}(L_3)$) and therefore $g \in L_1$, $\lambda(f) = \psi(g) \in L$. Conversely let $g \in Q$ be s.t. $\lambda(g) \in L$; let $g' := \xi(g) \in P$; then $\psi(g') = \lambda(\phi(g')) = \lambda(\phi(\xi(g))) = \lambda(g)$ so that $g' \in L_1 \cap P = L_2$ and $g = \phi(g') \in L_3$.

About decomposition in Q, we face the same situation as in § 5, namely recognizing "extraneous" components and avoiding their introduction. This was done, in § 5, by recognizing elements g s.t. $g(0) \neq 0$, since, for such a g and for $I \subset P_0$, we have $I : g^* = I$ and $(I,g^t) = (1)$. So here we must recognize the elements of B which play the same role.

First of all remark, that if we are given $f \in B = \bar{P}[Y_2,...,Y_n]$ we obtain easily an element $g \in P$ s.t.
  i) g is in normal form w.r.t. $p(Z) - Y_1$ (i.e. its degree in Z is less than $\deg(p)$)
  ii) g is not divisible by Y_1
  iii) $\psi(g)$ is associated to f
In fact $\bar{P}[Y_2,...,Y_n] = k(Y_1)[z,Y_2,...,Y_n]$, so there are primitive polynomials $g_1(Z,Y_1,...,Y_n)$ s.t. $g_1(z,Y_1,...,Y_n)$ is associated to f; we choose one which is in normal form w.r.t. $p(Z) - Y_1$ and divide out Y_1.

The following lemma shows that the elements $f \in B$ we are looking for are those s.t. $\phi(g)(0) \neq 0$, with g defined as above.

Lemma 3 Let $\mathfrak{a} \subset B$ be an ideal, let $\mathfrak{b} := \mathfrak{a} : \psi(g)) = \mathfrak{a} : f^*$, $\mathfrak{c} := (\mathfrak{a},\psi(g)^t) = (\mathfrak{a}, f^t)$. Then if $\phi(g)(0) \neq 0$:
  1) $(\mathfrak{a} \cap Q)^E = (\mathfrak{b} \cap Q)^E$
  2) $(\mathfrak{c} \cap Q)^E = (1)$
Proof: In fact denote $\mathfrak{a}_1 := \psi^{-1}(\mathfrak{a})$, $\mathfrak{a}_2 := \mathfrak{a}_1 \cap P$, $\mathfrak{a}_3 := \phi(\mathfrak{a}_2) = \mathfrak{a} \cap Q$ and similarly for $\mathfrak{b}$ and $\mathfrak{c}$.
Then $\mathfrak{b}_1 = \mathfrak{a}_1 : g^*$, $\mathfrak{b}_2 = \mathfrak{a}_2 : g^*$, $\mathfrak{b}_3 = \mathfrak{a}_3 : \phi(g)^*$; $\mathfrak{c}_1 = (\mathfrak{a}_1,g^t)$, $\mathfrak{c}_2 = (\mathfrak{a}_2,g^t)$, $\mathfrak{c}_3 = (\mathfrak{a}_3,\phi(g)^t)$.
However $\phi(g)$ is invertible in A, whence the thesis.

7.4 Computation of the m-primary component

Let I, I_1 be as in Prop. 5 and assume $I \neq I_1$ and the equivalent conditions of Prop. 5.ii) are satisfied. In this case I has an m-primary component $\mathfrak{q}_s$, $I = I_1 \cap \mathfrak{q}_s$. Let us again denote $I := \phi^{-1}(I)$, $I_1 := \phi^{-1}(I_1)$.
Let $\wp := (Y_1,...,Y_m) \subset Loc(P)$. Define a degree δ on P by $\delta(Z) = 0$, $\delta(Y_i) = 1$ if $Y_i < 1$. Let $<$ be a tangent cone ordering on P s.t. $Z > 1$ and $\delta(t_1) < \delta(t_2)$ implies $t_1 > t_2$. Remark that $f \in \wp^t$ if and only if $\delta(M(f)) \geq t$.

Proposition 6 Let F be a standard basis of I_1 w.r.t. $<$ and let s be s.t. $\delta(M(f)) \leq s \ \forall f \in F$. Then:
  1) $m^r \cap I_1 \subset I$ if and only if $\wp^r \cap I_1 \subset I$
  2) there is $t \in N$ s.t. $\wp^t I_1 \subset I$
  3) $\wp^{t+s} \cap I_1 \subset I$
  4) let $L_0 := I + m^{t+s}$. Then L_0 is m-primary and $I = L_0 \cap I_1$

<u>Proof:</u>   1) $\wp^r \cap I_1 \subset \phi^{-1}(m^r) \cap I_1 = \phi^{-1}(m^r \cap I_1) \subset \phi^{-1}(I) = I$.
$m^r \cap I_1 = \phi(\wp^r) \cap I_1 = \phi(\wp^r \cap I_1) \subset \phi(I) = I$.

   2) Since $\mathfrak{q}_s$ is m-primary, there is t s.t. $m^t \subset \mathfrak{q}_s$ and then $m^t \cap I_1 \subset I$, so $\wp^t I_1 \subset \wp^t \cap I_1 \subset I$.

   3) if $f \in \wp^{t+s} \cap I_1$ and $(1+q) f = \Sigma g_i f_i$ is a standard representation, then $\delta(M(g_i)) + \delta(M(f_i)) \geq \delta(M(f))$, so $\delta(M(g_i)) \geq (t + s) - s = t$, so $g_i \in \wp^t$, $f \in \wp^t I_1 \subset I$.

   4) L_0 is obviously m-primary. Since $\wp^{t+s} \cap I_1 \subset I$, then $m^{t+s} \cap I_1 \subset I$.
Hence $I = I + (m^{t+s} \cap I_1) = (I + m^{t+s}) \cap I_1 = L_0 \cap I_1$.

Therefore, to compute the m-primary component of I, we need only to compute t s.t. $\wp^t I_1 \subset I$. This can be done by finding, $\forall i$, t_i s.t. $Y_i^{t_i} f \in I$ for each $f \in F$ and then taking $t := \Sigma t_i$.

To solve the latter problem, one finds t s.t. $Y_i^t M(f) \in M(I)$; then one computes a normal form g of $Y_i^t f$ w.r.t. a standard basis of I; if g is different from 0, then by recursive application of the same procedure one finds u s.t. $Y_i^u g \in I$ and then $Y_i^{u+t} f \in I$.

REFERENCES

[AMR]   M.E.Alonso, T. Mora, M. Raimondo, *A computational model for algebraic power series*, J. Pure Appl. Algebra, to appear; a preliminary version appeared as *Computing with algebraic series*, Proc. ISSAC 89, ACM (1989)

[BGS]   D. Bayer, A. Galligo, M. Stillman, conferences at COCOA I (1986), Luminy (1988) and elsewhere

[EH]   D. Eisenbud, C. Huneke, *A Jacobian method for finding the radical of an ideal*, Preprint (1989)

[FGLM]   J.C. Faugére, P. Gianni, D. Lazard, T. Mora, *Efficient change of ordering for Gröbner bases of zero dimensional ideals*, submitted to J. Symb. Comp.

[GM]   P.Gianni, T. Mora, *Algebraic solution of systems of polynomial equations using Gröbner bases*, AAECC5, L. Notes Comp. Sci. 356 (1989)

[GTZ]   P. Gianni, B. Trager, G. Zacharias, *Gröbner bases and primary decomposition of polynomial ideals*, J. Symb. Comp. 6 (1988), 149-167

[GH]   M. Giusti, J. Heintz, *Un algorithme - disons   rapide" - pour la décomposition d'une variété algébrique en composantes irréductibles et équidimensionnelles*, Proc. MEGA '90, Birkhauser, to appear

[KR]   A. Kandri Rody, *Radical of ideals in polynomial rings*

[KL]   T. Krick, A. Logar, *Membership problem, representation problem and the computation of the radical for one-dimensional ideals*, Proc. MEGA '90, Birkhauser, to appear

[LAZ]   D. Lazard, *Solving zero-dimensional algebraic systems*, Preprint (1989)

[MPT]   T.Mora, G.Pfister, C.Traverso, *An introduction to the tangent cone algorithm*, Issues in non-linear geometry and robotics, JAI Press, to appear

[MOR]   T. Mora, *La queste del saint $Gr_a(A_L)$*, Proc. AAECC 7, Disc. Appl. Math., to appear

An Asymptotically Fast Probabilistic Algorithm for Computing Polynomial GCD's over an Algebraic Number Field

Lars Langemyr*
Wilhelm-Schickard-Institut für Informatik
Universität Tübingen, Auf dem Sand 13
W–7400 Tübingen, Germany

Abstract

We give a probabilistic algorithm for computing the greatest common divisor (GCD) of two polynomials over an algebraic number field. We can compute the GCD using $O(l \log^5(l))$ expected binary operations where l is the size of the GCD given by standard estimations. Since we require time $\Omega(l)$ just to write down the GCD, the algorithm is close to optimal.

1 Introduction

In Langemyr and McCallum [9] we gave a deterministic algorithm for computing the greatest common divisor (GCD) of two polynomials over an algebraic number field. The algorithm extended the ideas by Brown [2] and Collins for the case of integral polynomial GCD's to polynomials over algebraic number fields. The algorithm worked by computing in modular images given by distinct rational primes, and then applying the Chinese remainder theorem. One major feature of [9] was that we showed how to avoid the costly factorizations of the minimal polynomial modulo the prime numbers as required by Weinberger and Rothschild [17] in their factorization algorithm for polynomials over algebraic number fields. The deterministic computing time bound [9, Theorem 5.1] was essentially $O((nml)^2)$, where m is the degree of the algebraic extension and n is the maximal degree of the two input polynomials, and l is the size of the GCD given by standard estimations. In [9, Page 444] we also indicated that a realistic computing time bound for the algorithm would be essentially $O(l^2)$ given some assumptions. In the present paper this argument is made more precise: By using probabilistic techniques and a test division we overcome the problem that we need more modular images to verify that the degrees of the modular GCD's are correct than to restore the coefficients in the GCD (provided that the degrees are correct). By using asymptotically fast underlying arithmetic we further improve the computing time bound from $O(l^2)$ to $O(l \log^5(l))$.

*Supported by STU, ESPRIT BRA 3012 CompuLog and Fakultetsnämnden KTH. *Present address:* Numerical Analysis and Computing Science, The Royal Institute of Technology, S–100 44 Stockholm, Sweden. The material was improved and extracted from the author's PhD thesis [8] while affiliated with Tübingen. *E-mail:* larsl@nada.kth.se

Since we require time $\Omega(l)$ just to write down the GCD, the algorithm is close to optimal. Similar asymptotically fast results have been established for arithmetic complexity by Moenck [1] and for integral univariate polynomial GCD's by Schönhage [14]. We will see that the concept of dynamic evaluation (Duval [6]) is essential for obtaining such a tight computing time bound for our problem. We also change computational model from the computer word model in [9] to the bitwise computation model [1].

Our probabilistic computing time bound specializes to the probabilistic time bound for integer polynomial GCD for the heuristic algorithm given by Schönhage [14]. Smedley [16] extended the results of Schönhage to polynomials over algebraic number fields, he obtained (by reasoning similarly to [9, Page 444]) the computing time bound $O(l^2)$. The present paper shows that one can obtain computing time bounds at least as good as for the heuristic algorithm by using the older Brown-Collins approach, and by the remarks above we are not likely to do essentially better than this unless somebody shows that the a priori bound is not tight. Then however the algorithm can be improved accordingly, by using the new a priori bound, and thus remain close to optimal.

Practical experiments with a similar algorithm (without asymptotically fast subalgorithms) are reported in [9, Section 6] and [8, Chapter 10]. It turns out that for dense and large inputs the modular algorithm is always very much faster than its best classical rival. A theoretical comparison with a classical algorithm (the Subresultant Polynomial Remainder Sequence) is reported in [8, Chapter 6]. The analysis is further improved in a recent manuscript by the author [10], where we obtain the computing time bound $O(n^3 m^3 l^2)$ binary operations, using the above notation and classical subalgorithms only.

2   Notation and Introductory Results

Let $r(y)$ be a monic irreducible polynomial over $\mathbf{Z}$, the rational integers. Let $m = \delta r$, meaning the polynomial degree of r is m. Let $E = |r|$, where $|r|$ is the maximum absolute value of the coefficients of $r(y)$. Let α be a root of $r(y)$, and adjoin α to $\mathbf{Q}$, the rational numbers. We obtain the algebraic number field $\mathbf{Q}(\alpha)$. Since $r(y)$ is monic, $\mathbf{Z}[\alpha]$ is closed under multiplication, and consequently forms an integral domain. Any element in $\mathbf{Q}(\alpha)$ can be represented using a rational number and an element in $\mathbf{Z}[\alpha]$. We represent elements in $\mathbf{Z}[\alpha]$ by using the canonical isomorphism $\mathbf{Z}[\alpha] \simeq R = \mathbf{Z}[y]/(r(y))$. We will use $\log(\cdot)$ for the base 2 logarithm and $\ln(\cdot)$ for the natural logarithm. We always use the base 2 logarithm in O-expressions when the constant factor does not matter.

Let $F(x), G(x) \in R[x]$, of degree at most n, with leading coefficients $\bar{a}$ and $\bar{b}$ respectively. Now consider only $\bar{a} \in R$. We represent $\bar{a}$ as a polynomial in y of degree less than m. Using the Extended Euclidean Algorithm [7, Exercise 3, Section 4.6.1] on $\bar{a}(y)$ and $r(y)$ for integral polynomials we can obtain $a'(y) \in R$ such that that $a'(y)\bar{a}(y) = a = \mathrm{res}_y(a', \bar{a})$, where a is a rational integer. We can similarly obtain b from b' and $\bar{b}$. Now let $f(x)$ and $g(x)$ be the associates of $F(x)$ and $G(x)$ obtained by multiplying with a' and b' respectively. $f(x)$ and $g(x)$ have leading rational integer coefficients a and b respectively. By adapting a result from [17] it was shown in [9, Lemma 2.1] that there exists an associate $h(x) \in R[x]$ of the monic GCD of $f(x)$ and $g(x)$ with leading integral coefficient $c = \gcd(a, b)\Delta$, where Δ is the discriminant of $r(y)$. We have the following.

Proposition 1 *Let $F(x)$ and $G(x)$ be polynomials over R, and let $\bar{a}$ and $\bar{b}$ be their* leading coefficients. *Then there exists an associate* $h(x) \in R[x]$ *of the monic GCD of* $F(x)$ *and* $G(x)$ *over* $\mathbf{Q}(\alpha)$ *with leading coefficient* $\gcd(a, b)\Delta$, *where* $a = \mathrm{res}_y(\bar{a}, r)$ *and* $b = \mathrm{res}_y(\bar{b}, r)$.

In the following we will therefore consider only $f(x)$ and $g(x)$ and then in the final section insert the size of the coefficients in $f(x)$ and $g(x)$ in terms of $F(x)$ and $G(x)$ according to the above, thereby obtaining a time bound which is general.

3 Size Bounds

This section essentially quotes results from [9]. Let $f(x)$ and $g(x)$ in $R[x]$ be represented by integral polynomials in x and y of degree less than m in y and of degree at most n in x. For $q \in R$, by $|q|$ we denote a bound on the absolute value on the integers in the representation of q. We extend this notation to polynomials $f(x)$ over R by letting $|f|$ be the maximum $|f_i|$, over the coefficients f_i of f. By [9, Proposition 4.1] we then have:

Proposition 2 *Let* $h(x)$ *be the associate of the GCD characterized in Proposition 1. We then have* $|h| \leq K_1$,

$$K_1 = \sqrt{(m+1)^{3m-3}E^{m-3}m^{3m+3}} \cdot 2^n(n+1)D(1+E)^{m-1},$$

where $D = \max(|f|, |g|)$ *and the other parameters* n, m *and* E *have been defined above in terms of the input. Hence* $\log(K_1)$ *is* $O(k_1)$, *where*

$$k_1 = n + \log(D) + m\log(mE).$$

4 Homomorphism

Our algorithm works by using the Weinberger-Rothschild homomorphism [17], which we describe below. Consider an odd rational prime p, which does not divide $ab\Delta$. We will consider the ring $\tilde{R} = R/(p)$, which is not necessarily an integral domain. It can be written as the sum of fields, however. Let $\tilde{r}(y) = r(y) \bmod p$. Let $\tilde{R} = \mathbf{Z}[y]/(p, \tilde{r}(y))$ and let

$$\phi: R \to \tilde{R}$$

denote the canonical "mod p" homomorphism. (Denote the natural extension of ϕ to $R[x]$ by ϕ also.) Since $p \nmid \Delta$ we conclude that $\tilde{r}(y)$ is square-free. Let $\tilde{r}_1(y), \ldots, \tilde{r}_t(y)$ be the distinct, monic irreducible factors of $\tilde{r}(y)$ over the finite field with p elements $\mathbf{Z}_p$, i.e., $r(y) \equiv \prod_{j=1}^{t} \tilde{r}_j(y) \pmod{p}$. We have the "Chinese remainder" isomorphism

$$\psi: \tilde{R} \to \tilde{R}_1 \oplus \cdots \oplus \tilde{R}_t, \tag{1}$$

where the ψ_i are the component functions of ψ and $\tilde{R}_j$ is the finite field $\mathbf{Z}[y]/(p, \tilde{r}_j(y))$ often denoted $\mathrm{GF}(p^{\delta \tilde{r}_j})$. Denote the natural extension of ψ_j to $R[x]$ by ψ_j also.

5 Lucky Primes

We now characterize the primes for which the images of $f(x)$ and $g(x)$ in $\tilde{R}_i$ have a GCD which is congruent to $h(x)$ modulo p and $\tilde{r}_i$. Let $\tilde{f}_i(x)$ and $\tilde{g}_i(x)$ denote the result of applying $\psi_i \phi$ to $f(x)$ and $g(x)$ respectively. We say that p is lucky if the degree of the GCD's of $\tilde{f}_i$ and $\tilde{g}_i$, for $i = 1, \ldots, t$, over the finite fields $\tilde{R}_i$ are all equal to the degree of $h(x)$. From [9] we have:

Proposition 3 *Let p be an odd rational prime such that p does not divide $ab\Delta$. Let k be the degree of $h(x)$. Then $\delta \, gcd(\tilde{f}_i, \tilde{g}_i) > k$ for some i if and only if p divides the non-zero integer $L_4 = \mathrm{res}(r(y), \mathrm{psc}_k(f, g))$.*

psc_k means the k-th principal subresultant coefficient. For a definition see Loos [11]. For our present purposes its enough to know that it is a submatrix of the Sylvester Matrix of $f(x)$ and $g(x)$ over R. This enables us to obtain a bound on L_4. A proof of the result can be found in [9].

Proposition 4 *Consider $L_4 = \mathrm{res}_y(r(y), \mathrm{psc}_i(F(x), G(x)))$. We have $|L_4| \leq K_4$*

$$K_4 = (n+1)^{2nm} D^{2nm} m^{2nm} (1 + E)^{2(m-1)nm}.$$

The length of the integer K_4 is at most $O(k_4)$ bits, where

$$k_4 = nm \log(nDE^m).$$

Further by [9, Lemma 2.2], if p is lucky, then the monic GCD's of $\tilde{f}_i(x)$ and $\tilde{g}_i(x)$ over $\tilde{R}_i$ for $i = 1, \ldots, s$, times $\psi_i \phi(c)$ are congruent to $h(x)$ modulo p and $\tilde{r}_i(y)$.

6 Random Primes

Up to now we have mostly quoted results from [9]. We now start extending the results. Our bounds $O(k_4)$ on the length of the product of unlucky primes and $O(k_1)$ on the length of the coefficients in the GCD tell us that establishing that all primes chosen are lucky by computing modular GCD's and checking their degrees requires far more primes than restoring the coefficients of the GCD with the Chinese remainder algorithm (provided that all primes used are lucky). We overcome this problem by choosing random primes, and then, at the end of the algorithm performing a division test. By Proposition 3 a prime is lucky iff it does not divide $ab\Delta$ or $L_4 = \mathrm{res}(r(y), \mathrm{psc}_k(f(x), g(x)))$. We devise a technique for choosing primes which are lucky with probability $> 1/2$. Let $Q = ab\Delta K_4 \geq ab\Delta L_4$. We take the smallest σ such that the product of the primes less than σ exceeds Q. Then the number of primes below σ, $\pi(\sigma)$, is the maximum number of distinct primes which can possibly divide $ab\Delta L_4$. By Rosser and Schoenfeld [13, Theorem 10, $d = 101$] we can obtain this by letting $\sigma = 1.2 \ln(Q)$, for $\sigma \geq 101$. Suppose we had $2\pi(\sigma)$ primes. Then a random random prime would divide Q with probability $< 1/2$. We try accomplish this: By [13, Corollary 3] we have that

$$\pi(2\sigma) - \pi(\sigma) > \frac{3\sigma}{5 \ln(\sigma)}, \qquad \text{for } \sigma \geq 20\tfrac{1}{2}$$

by [13, Corollary 2]. By letting $\sigma = x$, $\sigma = 2x$ and $\sigma = 4x$, summing the three inequalities, we obtain

$$\pi(8\sigma) - \pi(\sigma) > \frac{9\sigma}{5\ln(\sigma)} > \frac{9}{5}\frac{1}{1.26}\pi(\sigma) > \pi(\sigma), \qquad \text{for } \sigma \geq 113.6, \qquad (2)$$

since $x/\ln(x)$ is monotone increasing, and for the middle inequality by applying [13, Corollary 1]. Thus we have that $\pi(8\sigma) > 2\pi(\sigma)$, for $\sigma > 113.6$. It thus suffices to choose random primes less than $9.6\ln(Q)$ ($9.6 = 1.2 \cdot 8$) in order to obtain probability $< 1/2$ that a random prime divides Q, provided that $\sigma > 113.6$. Denote this set of consecutive smallest primes by S. We note that $\ln(Q) = O(k_4)$.

We now want to choose random primes from S such that they form a product greater than $2K_1$, where K_1 is a bound on the length of the integer coefficients in the GCD. The probability $P(i)$ of choosing i lucky primes out of s, $i = 0, \ldots, s$ is

$$P(i) = \frac{\binom{\pi(\sigma)}{i}\binom{\pi(\sigma)}{s-i}}{\binom{2\pi(\sigma)}{s}}.$$

The primes are said to be hypergeometrically distributed. The average of this distribution is $s/2$. Thus if we choose twice as many primes as we actually need to restore the coefficients of $h(x)$ we obtain probability $> 1/2$ that at least half of the primes chosen do not divide Q. We accomplish this in two steps. The first step is to apply [13, Theorem 10, $d = 101$]: It suffices to take primes below $1.2\ln(2K_1)$ to obtain a product exceeding $2K_1$. The second step is to realize that for obtaining twice as many primes as the smallest ones forming a product exceeding $2K_1$, it suffices to take the primes below $8 \cdot 1.2\ln(2K_1)$ by (2). By [13, Corollary 1] there may be $1.26(8 \cdot 1.2\ln(2K_1))/\ln(8 \cdot 1.2\ln(2K_1))$ primes below $8 \cdot 1.2\ln(2K_1)$. Expanding the expressions we see that we choose $O(k_1)$ primes forming a $O(k_1)$ bit product by choosing random primes of length $O(\log(k_4)) = O(\log(k_1))$ bits. We have to have $\sigma > 113.6$ but this has no implications for the asymptotic analysis of the algorithm.

7 Dynamic Evaluation

We now introduce some machinery needed to avoid having to explicitly factor $\tilde{r}(y)$ over $\mathbf{Z}_p$. We use the concept of Dynamic evaluation [6]. The basic idea is the following: By (1) computations in $\tilde{R}$ and in $\oplus_{i=1}^{t}\tilde{R}_i$ are isomorphic. We must not divide by zero in any of the fields $\tilde{R}_i$. Trying to invert $a \in \tilde{R}$ the condition $\exists i\,(\psi_i a = 0)$ can be detected by computing $r'(y) \leftarrow \gcd(\tilde{r}(y), a(y))$. The condition is satisfied iff $r'(y)$ is non trivial. We say that a is a unit in $\tilde{R}$ iff $r'(y)$ is constant.

Consider computing the remainder of the polynomials $f_1(x)$ and $f_2(x)$ over $\tilde{R}$. Assume that the leading coefficients a_1 and a_2 are units in $\tilde{R}$. We can then apply a polynomial division algorithm [1, Theorem 8.7] since a_2 is a unit in $\tilde{R}$. By Cantor and Kaltofen [3] we can multiply polynomials of degree n over $\tilde{R}$ in $O(n\,\mathcal{L}(n))$ operations in $\tilde{R}$.[1] Similarly we can multiply elements in $\tilde{R}$ using $O(m\,\mathcal{L}(m))$ operations in $\mathbf{Z}_p$. Applying [1, Theorem 8.7] and the computing time bounds above we obtain:

---

[1]In order to shorten the formulae we use the notation $\mathcal{L}(x) = \log x \log\log x$ and $\mathcal{L}_2(x) = \log^2 x \log\log x$.

Lemma 5 *Let $f_1(x)$ and $f_2(x)$ be polynomials over $\tilde{R}$, with leading coefficients a_1 and a_2 both units in $\tilde{R}$. Then we can compute a remainder $f_3(x)$ such that $\psi_i f_3(x)$ is equal to the remainder of $\psi_i f_1(x)$ and $\psi_i f_2(x)$ over $\tilde{R}_i$ for each $i = 1, \ldots, t$ in $O(n\,\mathcal{L}(n)m\,\mathcal{L}(m))$ operations in $\mathbf{Z}_p$.*

If f_3 is going to be used again for computing further remainders (as in a Polynomial Remainder Sequence [11] (PRS)) we cannot do this unless its leading coefficient is a unit in $\tilde{R}$. In order to establish this we may have to split $\tilde{R}$ into several domains $\tilde{R} = \oplus_{i=1}^{t'} \tilde{R}_i'$, such that $\tilde{R}_i' = \mathbf{Z}_p[y]/(\tilde{r}_i'(y))$, where $\tilde{r} = \prod_{i=1}^{t'} \tilde{r}_i'$. We would obtain a homomorphism ψ' similar to (1). Then we could have that the leading coefficient of $\psi_i' f_3$ is a unit for each $i = 1, \ldots, t'$. This can be accomplished by using the following procedure: Write $f_3(x) = \sum_{i=0}^{n_3} f_i^* x^i$. Set $j \leftarrow 0$; $r'(y) \leftarrow \tilde{r}(y)$. For each $i \leftarrow n, n-1, \ldots$, until $r'(y) = 1$ or $i = 0$ compute $\hat{r}(y) \leftarrow \gcd(f_i^*, r'(y))$, if $\hat{r}(y)$ is non trivial then let $j \leftarrow j+1$; $\tilde{r}_j(y) \leftarrow r(y)/\hat{r}(y)$ and $r'(y) \leftarrow \hat{r}(y)$. Finally if $i = 0$ and $r'(y) \neq 1$ then set $j \leftarrow j+1$; $\tilde{r}_j \leftarrow r'(y)$. This requires no more than $O(nm\,\mathcal{L}_2(m))$ operations in $\mathbf{Z}_p$, by [1, Theorem 8.19]. Then compute $f_3(x) \bmod \tilde{r}_j'(y)$, for each j obtained. This can also be done in $O(nm\,\mathcal{L}_2(m))$ operations in $\mathbf{Z}_p$, by [1, Theorem 8.10].

Lemma 6 *Given a polynomial $f(x)$ over $\tilde{R}$ of degree n we can determine a factorization $\tilde{r}_i'(y), i = 1, \ldots, t'$ of $\tilde{r}(y)$, where $\delta\,r = m$, such that the leading coefficient of $f(x) \bmod \tilde{r}_i'$ is a unit in $\tilde{R}_i'$ or $f(x) \bmod \tilde{r}_i' = 0$ for $i = 1, 2, \ldots, t'$ in $O(nm\,\mathcal{L}_2{}^2(m))$ operations in $\mathbf{Z}_p$.*

We can compute a PRS of $\tilde{f}(x)$ and $\tilde{g}(x)$ over $\tilde{R}$ using Lemma 6 above to perform the remainder computation. The next remainder computation in the PRS is again achieved by applying the Lemma, but with $\tilde{R}$ replaced by each $\tilde{R}_i'$ obtained in the previous step. The strategy used in [9] was that we proved that the leading coefficient f_3 is a nonunit in some step in a modular remainder sequence of $\tilde{f}(x)$ and $\tilde{g}(x)$ only for a finite number of primes. The strategy of Lemma 6 works for any lucky prime however. We can also extend the fast GCD algorithm of [1, Figure 8.7] Consider Step 6 where the remainder computation takes place. We can replace the remainder computation with the procedure of Proposition 6. If a split of $\tilde{R}$ occurs at this point the computations splits in one direction for each $\tilde{R}_i'$ obtained. As pointed out by Duval [6] since the cost of the arithmetic in $\tilde{R}$ is more than linearly dependent on m the cost of performing different computations in $\oplus_{i=1}^{t'} \tilde{R}_i'$ is less than the cost of computing in $\tilde{R}_i$ had the split not occurred. We obtain the following proposition:

Proposition 7 *We can compute the GCD of polynomials over $\tilde{R}$ of degree n in time bounded by the time to perform $O(n\,\mathcal{L}_2(n))$ operations in $\tilde{R}$. We obtain the same output as if we had computed in $\tilde{R}_i$ for $i = 1, \ldots, t$.*

What computing time could be obtained if the polynomial $r(y)$ was factored over $\mathbf{Z}_p$ for enough primes p? The best deterministic algorithm known to us was given by Shoup [15]. Given a polynomial $\tilde{r}(y) \in \mathbf{Z}_p(y)$ of degree m we can determine its irreducible factors using

$$O(m^2\,\mathcal{L}(m)\,\mathcal{L}_2(m) + p^{1/2}\log(p)m^2\,\mathcal{L}(m)(\log(p) + \log(m)))$$

deterministic operations in $\mathbf{Z}_p$. The product of the primes would have to exceed the bound K_1 on the integer coefficients in $h(x)$, so that they can be restored using the

Chinese remainder theorem. From section 6 the primes may have length $O(\log(k_4))$ bits. Thus using the above deterministic factoring algorithm we obtain a computing time bound of

$$O(m^2 \log^2(m) k_1 k_4^{1/2} \log^2(k_1)\epsilon)$$

binary deterministic operations. (ϵ stands for a polynomial expression in $\log\log(k_1)$). Since this grows faster than $l = nmk_1$ we cannot expect to use the Shoup deterministic factoring algorithm for obtaining computing time $O(nmk_1 \log^5(k_1))$ for computing the GCD.

For the probabilistic case we use the results of Rabin [12]. The expected number of steps in Z_p for determining the irreducible factors of $r(y) \in Z_p[y]$ of degree m is $O(m^3 \mathcal{L}(m)^2 \log(p_i))$. Again we see that factoring of all modular polynomials may require

$$O(m^3 \log^2(m) \log^2(k_1) k_1 \epsilon)$$

binary operations. Since this grows faster than nmk_1 (in m) we cannot expect to use the Rabin probabilistic factoring algorithm either for obtaining computing time $O(nmk_1 \log^5(k_1))$. We conclude that it is unlikely that we can factor the modular polynomials quickly enough to obtain computing time $O(nmk_1 \log^5(k_1))$ for our GCD problem. Thus with present factoring algorithms for finite fields it seems we have to use the dynamic evaluation strategy for obtaining the result stated in our abstract.

8 Probabilistic Algorithm

We now have enough tools for giving the algorithm. The algorithm first chooses two times the number of random primes needed to restore the coefficients of the GCD. Each prime has probability $> 1/2$ of being lucky. This means that with probability $> 1/2$ at least the number of primes needed to restore the coefficients of GCD are lucky. By Proposition 3 the degree of the modular GCD's for which this is not the case is greater than k. Thus the modular GCD's with non-minimal degree can be discarded. If the product of primes for the modular GCD's with minimal degree is sufficient for restoring the GCD we restore, otherwise we fail. By proposition 3 we might still only have chosen unlucky primes so the computed GCD may not be the correct one. We therefore have to use trial division. The trial division is easily accomplished by also lifting the cofactors, and then performing modular trial multiplication for a few more primes. We have to introduce a new index in the notation of section 4. We have a sequence of primes $\{p_i\}_{i=1}^s$, for which $\bar{R}_i = R/(p_i)$. Further new notation is defined in the algorithm description.

Algorithm 8 *Algebraic Integer Polynomial GCD. Probabilistic Modular Algorithm.*

- *Inputs:* $r(y)$, a monic irreducible polynomial over $\mathbf{Z}$, $f(x)$ and $g(x)$, non-zero polynomials over R, with leading rational integer coefficients a and b respectively.

- *Output:* With probability $> 1/2$ the associate $h(x)$ of the GCD of $f(x)$ and $g(x)$ over the quotient field F of R which has leading coefficient $c = \gcd(a,b)\Delta$, where $\Delta = \mathrm{discr}(r)$. The coefficients of $h(x)$ then lie in R, by Proposition 1. With probability $< 1/2$ failure.

(1) [Compute bounds.] Compute $\Delta = \mathrm{discr}(r)$. Compute (as described in Proposition 2) a bound K_1 for $|h|$, and and a bound K_4 (as described in Proposition 4) for the integer L_4.

(2) [Compute prime list.] Compute $2\log(2K_1)$ distinct random rational primes $\{p_i\}_{i=1}^{s}$, in the interval $2 < p_i < 9.6\ln(ab\Delta K_4)$. This can be implemented by computing all primes less than $9.6\ln(ab\Delta K_4)$, with the Sieve of Erathostenes, and then using random bits to select from this list of primes. (We will see in the computing time analysis that the fact that we compute many more primes than we need does not affect the computing time bound of the algorithm.) Initialize an array $\{t_i\}_{i=1}^{s}$ to $t_i \leftarrow 0$.

(3) [Check primes.] For $i \leftarrow 1, 2, \ldots, s$ set $t_i \leftarrow 1$ if $p_i \mid \Delta$.

(4) [Compute modular images.] For $i \leftarrow 1, 2, \ldots, s$ unless $t_i = 1$ compute $\bar{f}_i(x) \leftarrow f(x) \bmod p_i$ and $\bar{g}_i(x) \leftarrow g(x) \bmod p_i$ and $\bar{r}_i(y) \leftarrow r(y) \bmod p_i$. If $\delta \bar{f}_i \neq \delta f$ or $\delta \bar{g}_i \neq \delta g$ then set $t_i \leftarrow 1$.

(5) [Compute modular GCD.] Apply Proposition 7, to $\bar{f}_i(x)$ and $\bar{g}_i(x)$ over $\bar{R}_i$ for each $i = 1, 2, \ldots, s$. Obtain monic $\hat{h}'_{ij}(x) \in R'_{ij}[x]$, where $\bar{R}_{ij} = \mathbf{Z}_{p_i}[y]/(r'_{ij}(y))$, and $\bar{R}_i \simeq \bar{R}_{i1} \times \cdots \times \bar{R}_{it'_i}$. If the degrees of $\hat{h}'_{ij}(x)$ are different for $j = 1, 2, \ldots, t'_i$, then set $t_i \leftarrow 1$ and skip the rest of this step for i. Then obtain $h_i^*(x)$ by solving the system of congruences $h_i^*(x) \equiv \hat{h}'_{ij}(x) \pmod{r'_{ij}(y)}$, for $i = 1, 2, \ldots, t'_i$. Also obtain the monic cofactors $\bar{f}_i(x) \leftarrow \bar{f}_i(x)/h_i^*(x)$ and $\bar{g}_i(x) \leftarrow \bar{g}_i(x)/h_i^*(x)$.

(6) [Lucky check.] Let k^* be the minimum of δh_i^* for all $i \leftarrow 1, 2, \ldots, s$ except when $t_i = 1$. Choose a subset of quadruples $\{(p'_i, h'_i, \bar{f}'_i, \bar{g}'_i)\}_{i=1}^{l}$ satisfying $\pi = \prod_{i=1}^{s} p'_i > 2K_1$ and $\delta h'_i = k^*$ from $\{(p_i, h_i^*, \bar{f}_i, \bar{g}_i)\}_{i=1}^{s}$. If there are not enough primes p'_i, then stop and return failure.

(7) [Apply the Chinese remainder algorithm.] For $1 \leq i \leq l$ let $c_i = c \bmod p'_i$, $a_i = a \bmod p'_i$, $b_i = b \bmod p'_i$, and $\Delta_i = \Delta \bmod p'_i$, solve the system of congruences.

$$h'(x) \equiv c_1 \ h'_1(x) \pmod{p'_1}$$
$$\vdots$$
$$h'(x) \equiv c_s \ h'_l(x) \pmod{p'_l}$$
$$\bar{f}'(x) \equiv a_1\Delta_1 \ \bar{f}'_1(x) \pmod{p'_1}$$
$$\vdots$$
$$\bar{f}'(x) \equiv a_l\Delta_l \ \bar{f}'_l(x) \pmod{p'_l}$$
$$\bar{g}'(x) \equiv b_1\Delta_1 \ \bar{g}'_1(x) \pmod{p'_1}$$
$$\vdots$$
$$\bar{g}'(x) \equiv b_l\Delta_l \ \bar{g}'_l(x) \pmod{p'_l}$$

for elements $h'(x)$, $\bar{f}'(x)$, and $\bar{g}'(x) \in \mathbf{Z}_\pi[y, x]$ of degree k^* using the Chinese remainder algorithm as in Step 6 of the Algorithm in [9].

(8) [Perform Division Test.] Select primes not dividing $ab\Delta$ forming a product of at least $K_3 = nmK_1(1 + E)^{m-1}$ from previously not used primes from the interval $2 < p_i < 9.6\ln(ab\Delta K_4)$, and verify that the products $h'(x)\bar{f}'(x)$ and $h'(x)\bar{g}'(x)$ are equal to $c\Delta f(x)$ modulo these primes. If this is not the case, return failure. Otherwise set $h(x) \leftarrow h'(x)$, and return. $\square$

We prove a lemma on the behavior of the algorithm.

Lemma 9 *Algorithm 8 fails with probability less than 1/2. It returns the correct GCD $h(x)$ with probability greater than 1/2.*

Proof. We note that by section 6 we have probability $> 1/2$ that at most $s/2$ of the primes divide $ab\Delta L_4$. Failure to obtain this number of primes results in detection when computing modular images of $f(x)$, $g(x)$ and Δ, or can be detected directly in step 5 or 6 by using Proposition 3. Finally if all primes used turn out to divide L_4 then this will be detected by the multiplication check. By [8, Lemma 6.1] K_3K_4 is a bound on the integers in the products of $h'(x)\bar{f}'(x)$ and $h'(x)\bar{g}'(x)$. The product has been checked for primes forming a product of K_4 when computing modular GCD's. $\square$

8.1 Time Analysis

We use the bitwise computation model [1] for our analysis. Throughout this section ϵ denotes different polynomials in $\log\log(nm\log(DE))$. We remember that $k_1 = n + \log(D) + m\log(mE)$, and $k_4 = nm\log(nDE^m)$. Consider Step 1. We can compute the discriminant by adapting [5] with fast arithmetic. A detailed analysis can be found in [8, Corollary 4.12]. An a priori bound for the length of the discriminant is $O(m\log(mE))$ bits ([5, Page 520]). Thus the computing time bound for the discriminant is

$$O(m^2\log(mE)\log^2(\log(E))\epsilon).$$

The cost of computing the bounds in Step 1 can be ignored. For Step 2 we have

$$\sum_{p \text{ prime} \wedge p<n} \frac{1}{p} \le \log\log(n) + B,$$

by [13, Theorem 5], where B is constant. Using this result in an analysis of the Sieve of Erathostenes results in the cost $O(k_4\log(k_4)\epsilon)$ binary operations for computing the primes less than $9.6\ln(ab\Delta K_4)$. The cost of computing $ab\Delta \bmod p_i$ for $i = 1, 2, \ldots, s$ is $O(k_1\log^2(k_1)\epsilon)$ binary operations, by [1, Theorem 8.9] in Step 3. Similarly for Step 4 we obtain $O(nmk_1\log^2(k_1)\epsilon)$ binary operations. In step 5 we apply Proposition 7. The total time for the dynamic evaluation of the remainder sequence in Step 5 is

$$\sum_{i=1}^{s} O(n\log^2(n)m\log^2(m)\log(p_i)\epsilon) =$$

$$O(nm\log^2(n)\log^2(m)k_1\epsilon).$$

Using a preconditioned version of [1, Theorem 8.13] we lift the result back into $\bar{R}$ if the degrees of all branches are equal. This cost is at most $O(nm\log^2(m)\epsilon)$ operations in $\mathbf{Z}_{p_i}$.

Summing up we obtain

$$\sum_{i=1}^{s} O(nm\log^2(m)\log(p_i)\log^2(\log(p_i))\epsilon) = O(nm\log^2(m)k_1\epsilon).$$

The cost of Step 6 can be ignored. For step 7 we use [1, Theorem 8.21], obtaining $O(nmk_4\log^2(k_4)\epsilon)$ binary operations. The final multiplication test can be done in $O(nmk_1\log^2(k_1)\epsilon)$ binary operations [8, Proposition 9.2]. We have proved the following theorem.

Theorem 10 *Algorithm 8 computes the GCD of $f(x)$ and $g(x)$ with probability $> 1/2$, otherwise it returns failure. It stops in*

$$O(nmk_1\log^2(n)\log^2(k_1)\epsilon)$$

binary operations.

We thus see that despite computing the primes forming a $O(k_4)$ bit product, this cost does not dominate the cost of the algorithm. We have the following simple corollary.

Corollary 11 *By applying Algorithm 8 until failure is not returned we can compute the GCD of $f(x)$ and $g(x)$ in*

$$O(nmk_1\log^2(n)\log^2(k_1)\epsilon)$$

expected binary operations.

We thus obtain a computing time bound of $O(l\log^5(l))$ expected binary steps, where $l = nmk_1$ is the bound for the length of the GCD. We see that we can compute (in expected time) the GCD almost as quickly as we can write it down. Our algorithm is therefore close to optimal. For integer polynomial GCD's we obtain the following corollary:

Corollary 12 *Let $f(x)$ and $g(x)$ be polynomials over $\mathbf{Z}$, satisfying $D = \max(|f|,|g|)$. By applying Algorithm 8 until failure is not returned we can compute the GCD of $f(x)$ and $g(x)$ in*

$$O(n(n+\log(D))\log^2(n)\log^2(n\log(D))\epsilon)$$

expected binary operations.

A recent result due to Schönhage [14] implies a similar bound for the Heuristic algorithm for integer polynomial GCD's [4].

9   General Case

We finally finish the thread started in Proposition 1. We consider general inputs $F(x)$ and $G(x)$ over R. Our aim is to compute an associate of the GCD with integer leading coefficient so that the monic associate of the GCD over $\mathbf{Q}(\alpha)$ can be easily obtained. We convert $F(x)$ and $G(x)$ to $f(x)$ and $g(x)$ with rational integer leading coefficients according to Proposition 1. We then apply Algorithm 8 to $f(x)$ and $g(x)$. We need a lemma on the length of integers in $f(x)$ and $g(x)$.

Lemma 13 *Let $F(x)$ and $G(x)$ be polynomials over R with $\hat{D} = \max(|F|, |G|)$. Then the integers in the associates $f(x)$ and $g(x)$ according to Proposition 1 have coefficients of length $O(k_2)$. We then have*

$$\max(|f|, |g|) = m(m\hat{D})^{2m}(1 + mE)^{2(m-1)}\hat{D}(1 + E)^{m-1}$$

We can compute $f(x)$ and $g(x)$ in $O(nmk_2 \log^2(k_2))$ binary operations, where $k_2 = n + m\log(m\hat{D}E)$.

This result is proved in [8]. We can now apply Algorithm 8 to $f(x)$ and $g(x)$. We obtain the following corollary to Theorem 10.

Corollary 14 *Let $F(x)$ and $G(x)$ be polynomials over R with degree at most n with $\hat{D} = \max(|F|, |G|)$. We can compute the associate $h(x)$ (Proposition 1) of the GCD of $F(x)$ and $G(x)$ in*

$$O(nmk_2 \log^2(n) \log^2(k_2)\epsilon)$$

binary operations by first applying Lemma 13, and then Algorithm 8.

Again we note that if we accept $O(k_2)$ as a bound on the integers in the GCD we can compute the GCD of $F(x)$ and $G(x)$ in time $O(d\log^5(d))$ binary operations, where d is the number of bits in the GCD.

Scott McCallum worked with me on the problem [9]. Thanks to Johan Håstad for providing essential advice. Joachim von zur Gathen and James Davenport pointed to the Dynamic evaluation strategy. George Collins suggested a modular multiplication test.

References

[1] A. Aho, J. E. Hopcroft, and J. D. Ullman. *The Design and Analysis of Computer Algorithms.* Addison-Wesley, Reading, Mass., 1974.

[2] W. S. Brown. On Euclid's algorithm and the computation of polynomial greatest common divisors. *Journal of the ACM*, 18(4):478–504, October 1971.

[3] D. G. Cantor and E. Kaltofen. Fast multiplication over arbitrary rings. 1986. Manuscript.

[4] B. W. Char, K. O. Geddes, and G. H. Gonnet. Gcdheu: heuristic polynomials gcd algorithm based on integer gcd computation. In *Proc. EUROSAM '84*, pages 285–296, Springer-Verlag, 1984. Lecture Notes in Computer Science 174.

[5] G. E. Collins. The calculation of multivariate polynomial resultants. *Journal of the ACM*, 18:515–532, 1971.

[6] D. Duval. *Diverse questions relatives au CALCUL FORMEL AVEC DES NOMBRES ALGÉBRIQUES.* PhD thesis, L'université scientifique, technologique, et médicale de Grenoble, Grenoble, April 1987.

[7] D. E. Knuth. *The Art of Computer Programming II: Seminumerical Algorithms.* Addison-Wesley, Reading, Mass., 1981.

[8] L. Langemyr. *Computing the GCD of two Polynomials Over an Algebraic Number Field*. PhD thesis, NADA, Royal Institute of Technology, Stockholm, 1988.

[9] L. Langemyr and S. McCallum. The computation of polynomial greatest common divisors over an algebraic number field. *J. Symbolic Comp.*, 8:429–448, 1989.

[10] L. Langemyr. An Analysis of the Subresultant Algorithm over an Algebraic Number Field. 1990. Manuscript.

[11] R. G. K. Loos. Generalized polynomial remainder sequences. In B. Buchberger, G. E. Collins, and R. G. K. Loos, editors, *Computer Algebra, Symbolic and Algebraic Computation*, pages 115–137, Springer-Verlag, Wien-New York, 1982.

[12] Michael O. Rabin. Probabilistic algorithms for finite fields. *SIAM Journal on Computing*, 9(2):273–280, May 1980.

[13] J. B. Rosser and L. Schoenfeld. Approximate formulas for some functions of prime numbers. *Illinois J. Math.*, 6:64–94, 1962.

[14] A. Schönhage. Probabilistic computation of integer polynomial GCDs. *J. of Algorithms*, 9:365–371, 1988.

[15] V. Shoup. *On the Deterministic Complexity of Factoring Polynomials over Finite Fields*. Technical Report 782, Computer Science Department, University of Wisconsin-Madison, July 1988.

[16] T. J. Smedley. A new modular algorithm for computation of algebraic number polynomial Gcds. In *Proc. ISSAC '89*, pages 91–94, ACM, July 1989.

[17] P. J. Weinberger and L. P. Rothschild. Factoring polynomials over algebraic number fields. *ACM Transactions on Mathematical Software*, 2(4):335–350, December 1976.

SOME ALGEBRA WITH FORMAL MATRICES.

Bernard MOURRAIN

Centre de Mathématiques

Ecole Polytechnique

F 91128 PALAISEAU Cedex (France)

mourrain@polytechnique.fr

Abstract

The aim of this paper is to give an explicit description of identities satisfied by matrices ($n \times n$ over a field $\mathbf{k}$ of characteristic 0) in order to be able to compute with formal matrices ("forgetting" their representations with coefficients). We introduce a universal free algebra where all formal manipulations are made. Using classical properties of an ideal of identities in an algebra with trace, we reduce our problem to the study of identities among multilinear traces. These are closely linked with the action of the algebra $\mathbf{k}[S_m]$ of the symmetric group on the mth tensor product of $E = \mathbf{k}^n$. Proving a theorem about the kernel of this action and its effective version, we can decompose all identities of matrices in an explicit way as linear combinations, substitutions, product or traces of the well-known Cayley-Hamilton identity. This leads to an algorithm for reducing to a canonical form modulo the ideal of identities of matrices in the free algebra.

1 The matrices and their identities.

In this article, we present a new method, which deals with matrices as variables and reduces a non-commutative polynomial in these variables with products of traces as coefficients, to a canonical form modulo the ideal of identities of matrices. A paper of C. Procesi [7] handles this problem in the case of 2×2 matrices. Here, we want to give an effective version of a structure theorem for this ideal of identities ([6, 8]) . The links with a theorem in the algebra of the symmetric group will be used, after one has reduced the problem to the case of identities among multilinear traces. But first, let us define the working environment.

1.1 Notations.

• **Algebra with trace.**

Let $\mathbf{k}$ be a field and $\mathbf{A}$ an associative unitary $\mathbf{k}$-algebra. We assume futher more, that this algebra has a "trace" Tr. By this, we mean a $\mathbf{k}$-linear application from $\mathbf{A}$ to $\mathbf{k}$ such that for all elements A, $B \in A$ we have $Tr(AB) = Tr(BA)$ (invariant under circular permutations of products). We note I_A the unit of the algebra.

We want to discribe the identities that could exist between elements of A and their trace to be able to compute modulo the ideal they generate. For this purpose, we introduce the following spaces.

● **Universal Representation.**

We are going to build a formal k-algebra which will be associative, unitary and with a linear application called "trace" so that it looks like an algebra of matrices.

Given a set of letters $\mathbf{X} = \{X_1, X_2, \ldots\}$ (we suppose it infinite for convenience), we look at the set $\mathbf{X}^*$ of words made with $\mathbf{X}$ (free monöid with base $\mathbf{X}$). The empty word is noted I and $\check{\mathbf{X}}^* = \mathbf{X}^* \backslash \{I\}$.

On this monöid $\mathbf{X}^*$, we can take the following equivalence relation $\Re$:
Let m_1, m_2 be two elements of $\mathbf{X}^*$. We say that $m_1 \Re m_2$ iff the words m_1 and m_2 have the same length and there exists a circular permutation of the letters which transform m_1 in m_2.
We note the quotient of $\mathbf{X}^*$ by this relation $T(\mathbf{X}^*) = \mathbf{X}^*/\Re$. The class of an element $m = X_{i_1} \ldots X_{i_k}$ of $\mathbf{X}^*$ is noted $T(X_{i_1} \ldots X_{i_k})$.
Especially, I is alone in its class, $T(I) = \{I\}$.
We note $\mathbf{T_X} = k[T(X^*)]$, the free associative commutative k-algebra generated by all the elements of $T(\mathbf{X}^*)$ ie the commutative polynomials in variables $T(X_{i_1} \ldots X_{i_k}) \in T(\mathbf{X}^*)$ ("polynomials in trace").

Now consider the non-commutative, associative, unitary, free k-algebra generated by $\mathbf{X} = \{X_1, X_2, \ldots\}$ noted $k\langle \mathbf{X} \rangle$ (that is the set of functions with finite support from $\mathbf{X}^*$ to k). (The element I is the unit of this algebra). We take $\mathbf{T_X}\langle \mathbf{X} \rangle = \mathbf{T_X} \otimes_k k\langle \mathbf{X} \rangle$ (scalars are just extended to $\mathbf{T_X}$).
An element of $\mathbf{T_X}\langle \mathbf{X} \rangle$ is of the form:

$$\sum_{(i_1,\ldots,i_s)} t_{i_1,\ldots,i_s} X_{i_1} \ldots X_{i_s} + t_0 I \text{ with } t_{i_1,\ldots,i_s}, t_0 \in \mathbf{T_X}.$$

Playing with words, we consider T as a projection from $\mathbf{X}^*$ to $\mathbf{T_X}$. It extends canonically to the $\mathbf{T_X}$-module $\mathbf{T_X}\langle \mathbf{X} \rangle$ as a linear application in the following way:

$$T : \mathbf{T_X}\langle \mathbf{X} \rangle \rightarrow \mathbf{T_X}$$
$$\sum_{(i_1,\ldots,i_s)} t_{i_1,\ldots,i_s} . X_{i_1}, \ldots, X_{i_s} + t_0 I \mapsto \sum_{(i_1,\ldots,i_s)} t_{i_1,\ldots,i_s} T(X_{i_1} \ldots X_{i_s}) + t_0 T(I).$$

This application is called "universal trace" which is, as the usual trace of matrices, invariant under circular permutation in products.

Definitions: For all word m of $\mathbf{X}^*$, we define the partial degree in X_i (deg_{X_i}), as the number of times X_i occurs in m. The total degree is the length of the word. We take $deg_{X_i}(I) = deg(I) = 0$. The same definitions applies to elements in $T(\mathbf{X}^*) = \mathbf{X}^*/\Re$ (it is independant of the order).

The partial (resp total) degree of $t.m \in \mathbf{T_X}\langle X \rangle$ is the sum of the partial (resp total) degree of t and m (it is also the maximal power of λ that appears when we replace a variable (resp all variables) X_i by λX_i). More generally, the degree of an element of $\mathbf{T_X}\langle X \rangle$ is the maximal degree of the terms it contains.

• **The identities.**

In order to study the identities satisfied by elements of $\mathbf{A}$ we are going to substitute the formal arguments $X_1, X_2, \ldots$ by arbitrary elements $\underline{A} = \{A_1, A_2 \ldots\}$ of $\mathbf{A}$.

This substitution $\gamma_{\underline{A}}$ is defined from X^* to $\mathbf{A}$ by $\gamma(X_{i_1} \ldots X_{i_s}) = A_{i_1} \ldots A_{i_s}$. Especially, we have $\gamma_{\underline{A}}(I) = I_\mathbf{A}$ the unit element of $\mathbf{A}$.

Of course, taking the trace Tr of $\mathbf{A}$, the application $Tr \circ \gamma_{\underline{A}}$ is invariant under circular permutations in X^*, so that it factors through $T(X^*)$. We note $\gamma_{T,\underline{A}}$ this application from $T(X^*)$ to k. By construction, we have $\gamma_{T,\underline{A}} \circ T = Tr \circ \gamma_{\underline{A}}$.

We take for $\gamma_{X,\underline{A}}$ the natural extension $\gamma_{T,\underline{A}} \otimes \gamma_{\underline{A}}$ of this substitution on $\mathbf{T_X}\langle X \rangle$ given by:

$$\gamma_{X,\underline{A}} : \mathbf{T_X}\langle X \rangle \rightarrow \mathbf{A}$$
$$\sum_{(i_1,\ldots,i_s)} t_{i_1,\ldots,i_s} X_{i_1} \ldots X_{i_s} + t_0 I \mapsto \sum_{(i_1,\ldots,i_s)} t_{i_1,\ldots,i_s}(\underline{A}) A_{i_1} \ldots A_{i_s} + t_0(\underline{A}) I_\mathbf{A}.$$

By $t(\underline{A})$, we mean $\gamma_{T,\underline{A}}(t)$.

We say that $P \in \mathbf{T_X}\langle X \rangle$ (resp T_X) is an identity of $\mathbf{A}$, if for any family $\underline{A}$ of elements of $\mathbf{A}$, $\gamma_{X,\underline{A}}(P)$ (resp $\gamma_{T,\underline{A}}(P)$) vanishes. We note it: $P \equiv 0[\mathbf{A}]$. The set of all identities in variables X of $\mathbf{A}$ is $I_{X,\mathbf{A}}$.

• **The case of matrices.**

In the case where $\mathbf{A} = \mathbf{M}_n(k)$, this can be done in one step. Consider generic matrices $U_1, U_2 \ldots$ where

$$U_1 = [u^1_{i,j}]_{1 \leq i,j \leq n}, U_2 = [u^2_{i,j}]_{1 \leq i,j \leq n} \ldots$$

($u^k_{i,j}$ being independant variables on k) and define the substitution γ from X^* to $\mathbf{M}_n(k[u^l_{i,j}])$ as $\gamma(X_{i_1} \ldots X_{i_s}) = U_{i_1} \ldots U_{i_s}$. This allows us to build (as before) the substitutions γ_T from T_X to $k[u^l_{i,j}]$ and γ_X from $\mathbf{T_X}\langle X \rangle$ to $\mathbf{M}_n(k[u^l_{i,j}])$.

We have $\gamma(I) = I_n$ (the identity matrix of $\mathbf{M}_n(k)$).

$$\gamma_X : \mathbf{T_X}\langle X \rangle \rightarrow \mathbf{M}_n(k[u^l_{i,j}])$$
$$\sum_{(i_1,\ldots,i_s)} t_{i_1,\ldots,i_s} X_{i_1}, \ldots, X_{i_s} + t_0 I \mapsto \sum_{(i_1\ldots i_s)} t_{i_1,\ldots,i_s}(U) U_{i_1} \ldots U_{i_s} + t_0(U) I_n.$$

To simplify notations, the substitution of an element $P(X_1, \ldots, X_s)$ is noted $P(U_1, \ldots, U_s)$.

The kernels of the two substitutions γ_X, γ_T, noted $I_{X,n}$ and $\Gamma_{X,n}$, are characterised by the following properties:

$$P(X_1, \ldots, X_s) \in \mathbf{I}_{\mathbf{X},n} \ (\text{resp } \Gamma_{\mathbf{X},n})$$
$$\Leftrightarrow$$
$$\gamma_{\mathbf{X}}(P) = 0 \ (\text{resp } \gamma_T(P) = 0)$$
$$\Leftrightarrow$$
$$P([u_{i,j}^1], \ldots, [u_{i,j}^s]) = 0$$
$$\Leftrightarrow$$
$$\forall A_1, \ldots, A_s \in \mathbf{M}_n(\mathbf{k}), \ P(A_1, \ldots, A_s) = 0$$
$$\Leftrightarrow$$

P is an identity in variables $X_1, X_2, \ldots, X_s$ of $\mathbf{M}_n(\mathbf{k})$

notation: $P \equiv 0[\mathbf{M}_n(\mathbf{k})]$.

We call $\mathbf{T}_U\langle \mathbf{U} \rangle$ the image of $\mathbf{T}_{\mathbf{X}}\langle \mathbf{X} \rangle$ by $\gamma_{\mathbf{X}}$ and $\mathbf{T}_U$ those of $\mathbf{T}_X$.

So we can sum up the situation by the following commutative diagram:

$$
\begin{array}{ccccccccc}
0 & \longrightarrow & \mathbf{I}_{\mathbf{X},n} & \longrightarrow & \mathbf{T}_{\mathbf{X}}\langle \mathbf{X} \rangle & \xrightarrow{\gamma_{\mathbf{X}}} & \mathbf{T}_U\langle \mathbf{U} \rangle & \longrightarrow & 0 \\
 & & \downarrow T & & \downarrow T & & \downarrow T_r & & \\
0 & \longrightarrow & \Gamma_{\mathbf{X},n} & \longrightarrow & \mathbf{T}_X & \xrightarrow{\gamma_T} & \mathbf{T}_U & \longrightarrow & 0
\end{array}
$$

1.2 Basic properties.

In this part, we present basic properties of the identities of an algebra A. We assume that $characteristic(\mathbf{k}) = 0$. Remember that A is an associative non-commutative unitary algebra with a trace Tr. We note $\mathbf{I}_{\mathbf{X},A}$ the ideal of identities : $\mathbf{I}_{\mathbf{X},A} = \{P \in \mathbf{T}_{\mathbf{X}}\langle \mathbf{X} \rangle \ ; \ P \equiv 0[A]\}$ and we assume that it is not the whole space $\mathbf{T}_{\mathbf{X}}\langle \mathbf{X} \rangle$ (A $\neq 0$).

• $\mathbf{I}_{\mathbf{X},A}$ (resp $\Gamma_{\mathbf{X},A}$) is stable by substitutions.

Let $P(X_1, \ldots, X_m)$ belong to $\mathbf{T}_{\mathbf{X}}\langle \mathbf{X} \rangle$ (resp $\mathbf{T}_{\mathbf{X}}$). Then $P(X_1, \ldots, X_m) \equiv 0[A]$ implies that for all $Q_1, \ldots, Q_m$ in $\mathbf{T}_{\mathbf{X}}\langle \mathbf{X} \rangle$, $P(Q_1, \ldots, Q_m) \equiv 0[A]$.
(By $P(Q_1, \ldots, Q_m)$, we mean substitutions in $\mathbf{T}_{\mathbf{X}}\langle \mathbf{X} \rangle$ of variables $X_1, \ldots, X_m$ by elements $Q_1, \ldots, Q_m \in \mathbf{T}_{\mathbf{X}}\langle \mathbf{X} \rangle$).

• $\mathbf{I}_{\mathbf{X},A}$ (resp $\Gamma_{\mathbf{X},A}$) is homegeneous in each variable.

Let $P(X_1, \ldots, X_m)$ belong to $\mathbf{I}_{\mathbf{X},A}$ (resp $\Gamma_{\mathbf{X},A}$) of degree d_1 in X_1 and t be a scalar in k, then

$$P(t \ X_1, X_2, \ldots, X_m) = P_0(X_1, \ldots, X_m) + t \ P_1(X_1, \ldots, X_m) + \ldots + t^{d_1} \ P_{d_1}(X_1, X_2, \ldots, X_m).$$

Taking enougth distinct values of t in k and inversing the resulting Van der Monde system, we see that each P_j is in $\mathbf{I}_{\mathbf{X},A}$ (resp $\Gamma_{\mathbf{X},A}$). This can be repeat with each variable $X_2, \ldots, X_m$.

• $I_{X,A}$ (resp $\Gamma_{X,A}$) is stable by polarization.

Let $P(X_1, \ldots, X_m)$ belong to $I_{X,A}$ (resp $\Gamma_{X,A}$) homogeneous and of degree d_1 in X_1 with $d_1 \neq 0$. Introducing new parameters $t_1, \ldots, t_{d_1}$ and new variables $X_1^1, \ldots, X_{d_1}^1$, we define the complete "polarization" of P in X_1 as the coefficient of the product $t_1 \ldots t_{d_1}$ in the development in powers of (t_i) of $P(t_1 X_1^1 + \ldots + t_{d_1} X_{d_1}^1, X_2, \ldots, X_m)$.

This polynomial, denoted by $P^{(d_1)}$, is homogeneous of degree 1 in each variables $X_1^1, \ldots, X_{d_1}^1$. In other words, it is multilinear in such variables. As it is the homogeneous part of $P(t_1 X_1^1 + \ldots + t_{d_1} X_{d_1}^1, X_2, \ldots, X_m)$ of degree 1 in $(X_i^1)_{1 \leq i \leq d_1}$, it is also an identity.

Conversely, $P(X_1, X_2, \ldots, X_m)$ can be obtained by substitution of $P^{(d_1)}(X_1^1, \ldots, X_{d_1}^1, X_2, \ldots, X_m)$: As $P^{(d_1)}(X_1, \ldots, X_1, X_2, \ldots, X_m)$ is the coefficient of $t_1 \ldots t_{d_1}$ in

$$P((t_1 + \ldots + t_{d_1}) X_1, X_2, \ldots, X_m) = (t_1 + \ldots + t_{d_1})^{d_1} P(X_1, X_2, \ldots, X_m)$$

we have

$$P(X_1, X_2, \ldots, X_m) = \frac{1}{d_1!} . P^{(d_1)}(X_1, \ldots, X_1, X_2, \ldots, X_m)$$

This can be repeat for the other variables in oder to obtain a multilinear polynomial in new variables X_j^i, which can give back, by substitutions by the first variables X_i, the polynamial P up to a scalar $((d_1!) \ldots (d_m!))$. WE call this global substitutions, the inverse substitution of the polarization of P. These different steps give us the following property:

Proposition 1.1 — $I_{X,A}$ *(resp $\Gamma_{X,A}$) is linearly generated by the multilinear elements it contains and their substitutions.*

• **Relations between $I_{X,A}$ and $\Gamma_{X,A}$.**

Suppose that the trace $(A, B) \mapsto Tr(AB)$ defines a non-degenerated form (this the case for matrices $n \times n$ for all $n \in \mathbf{N}$). Then we have the following equivalence:

$$P(X_1, \ldots, X_m) \in I_{X,A} \Leftrightarrow T(P(X_1, \ldots, X_m) . X_{m+1}) \in \Gamma_{X,A}.$$

Moreover, the polynomial $P \in \mathbf{T_X}\langle \mathbf{X} \rangle$ is uniquely determined when we know the element $T(P(X_1, \ldots, X_m) . X_{m+1}) \in \mathbf{T_X}$.

See [4], [5] and [6] for a generall discussion of these items.

In conclusion, these steps show that the knowledge of all mulitlinear identities of $\mathbf{T_X}$ allow us to reconstruct the other, by identifications of the products by new variable in traces, subtitutions or linear combinations.

1.3 An example, the Cayley-Hamilton identity.

We recall here, perhaps the oldest identity known about matrices :

Theorem 1.2 — *There exists a unique sequence* $\Theta_1, \Theta_2, \ldots$ *of elements of* T_X *with respective degree* $1, 2 \ldots$ *such that for all* n

$$X^n + \Theta_1(X)X^{n-1} \ldots + \Theta_n(X)I = 0$$

is an identity of $M_n(k)$.

For any matrix $A \in M_n(k)$, the polynomial $t^n + \Theta_1(A)t^{n-1} \ldots + \Theta_n(A)Id_n$ is the characteristic polynomial $det(tId_n - A)$. Here is the expression of the first identitites :

$$
\begin{aligned}
G_1(X) &= X - T(X)\,I \\
G_2(X) &= X^2 - T(X)X + \tfrac{1}{2}[T(X)^2 - T(X^2)]\,I \\
G_3(X) &= X^3 - T(X)X + \tfrac{1}{2}[T(X)^2 - T(X^2)]X - \tfrac{1}{3}[T(X^3) - \tfrac{3}{2}T(X^2)T(X) + \tfrac{1}{2}T(X)^3]I
\end{aligned}
$$

We can construct them by induction using the formula

$$G_n(X) = XG_{n-1}(X) - \frac{1}{n}T(XG_{n-1}(X))\,I.$$

For more information, see Fadeev's method in [2].

1.4 Multilinear elements of T_X and permutations.

We are now working with the algebra $A = M_n(k)$ of matrices $n \times n$ over k. We note $I_{X,n}$ (resp $\Gamma_{X,n}$) the identities of $M_n(k)$ in $T_X\langle X\rangle$ (resp in T_X). As we already have seen, the multilinear elements of $\Gamma_{X,n}$ allow us to discribe all the identities of matrices. So let P be a non-null multilinear polynomial in T_X of degree m.

$$P(X_1, \ldots, X_m) = \sum_{\alpha \in A} \lambda_\alpha . T_\alpha$$

with $\lambda_\alpha \in k$ and $T_\alpha = T(X_{i_1} \ldots X_{i_k}) \ldots T(X_{i_u} \ldots X_{i_m})$.

Here, a variable X_{i_l} of P appears one and only one time in each term T_α. All these multilinear polynomials with variables in $\{X_1, \ldots, X_m\}$ form the set T_X^m.

A usual way to write down T_α is to associate it to the permutation in S_m :

$$\sigma = (i_1, \ldots, i_k) \ldots (i_u, \ldots, i_m)$$

(also invariant under circular permutations of products between parentheses) so that we will write it

$$T_\sigma = T(X_{i_1} \ldots X_{i_k}) \ldots T(X_{i_u} \ldots X_{i_m})$$

This defines the following isomorphism of k-vector spaces between $k[S_m]$ and T_X^m:

$$
\begin{aligned}
\Omega : \quad k[S_m] &\longrightarrow T_X^m \\
\sum_{\sigma \in S_m} \lambda_\sigma . \sigma &\longmapsto P = \sum_{\sigma \in S_m} \lambda_\sigma . T_\sigma
\end{aligned}
$$

Remember the isomorphism of k-vector spaces $M_n(k) \sim E \otimes E$ where E be the k-vector space k^n. A matrix is sum of matrices of rank 1, $u \otimes v$ with $u, v \in E$. For such elements $u \otimes v, u' \otimes v'$, we have $(u \otimes v)(u' \otimes v') = \langle v, u' \rangle u \otimes v'$ and $Tr(u \otimes v) = \langle u, v \rangle$ ($\langle u, v \rangle$ is the canonical scalar product of u and v on E).

So for all permutations, $\sigma = (i_1, \ldots, i_k) \ldots (i_u, \ldots, i_m)$ in S_m,

$$T_\sigma(u_1 \otimes v_1, \ldots, u_m \otimes v_m)$$
$$= T(u_{i_1} \otimes v_{i_1} \ldots u_{i_k} \otimes v_{i_k}) \ldots T(u_{i_u} \otimes v_{i_u} \ldots u_{i_m} \otimes v_{i_m})$$
$$= \langle u_{i_2}, v_{i_1} \rangle \ldots \langle u_{i_1}, v_{i_k} \rangle \ldots \langle u_{i_{u+1}}, v_{i_u} \rangle \ldots \langle u_{i_u}, v_{i_m} \rangle$$
$$= \langle u_1, v_{\sigma^{-1}(1)} \rangle \ldots \langle u_m, v_{\sigma^{-1}(m)} \rangle$$

We remark that any multilinear element of T_X is an identity iff it vanishes on the generating set of matrices of rank 1, $u \otimes v$ with $u, v \in E$. So we have the following equivalences:

$$\sum_{\sigma \in S_m} \lambda_\sigma T_\sigma \in \Gamma_{X,n}$$
$$\Leftrightarrow$$
$$\sum_{\sigma \in S_m} \lambda_\sigma T_\sigma(u_1 \otimes v_1, \ldots, u_m \otimes v_m) = 0, \ \forall u_i, v_i \in E$$
$$\Leftrightarrow$$
$$\sum_{\sigma \in S_m} \lambda_\sigma \langle u_1, v_{\sigma^{-1}(1)} \rangle \ldots \langle u_m, v_{\sigma^{-1}(m)} \rangle = 0, \ \forall u_i, v_i \in E$$
$$\Leftrightarrow$$
$$\sum_{\sigma \in S_m} \lambda_\sigma v_{\sigma^{-1}(1)} \otimes \ldots \otimes v_{\sigma^{-1}(m)} = 0, \ \forall v_i \in E$$

This means that the element $\sum_{\sigma \in S_m} \lambda_\sigma \sigma$ of the algebra of the symmetric group $k[S_m]$ is an identity for its action on $T^m(E) = E \otimes \ldots \otimes E$ as we will see now.

2 The algebra of the symmetric group.

2.1 Action of $k[S_m]$ on the tensor product $T^m(E)$.

Let $k[S_m]$ be the algebra of the symmetric group that is $\oplus_{\sigma \in S_m} k.\sigma$ with the product induced by the group product of S_m. Let E $(= k^n)$ be a k-vector space of dimension n. We note $T^m(E) = E \otimes_k \ldots \otimes_k E$, the mth tensor product of E.

Let $(e_1, \ldots, e_n)$ be a basis of E. Then a basis of $T^m(E)$ is made with all distincts elements of the form $e_{i_1} \otimes \ldots \otimes e_{i_m}$. Let $F_k = \{j; i_j = k\}$. We also write this tensor product

$$e_{i_1} \otimes \ldots \otimes e_{i_m} = e_{F_1 \sqcup \ldots \sqcup F_n}.$$

where $F_1 \sqcup \ldots \sqcup F_n$ is the partition of $(1, \ldots, m)$ (with perhaps empty sets) such that e_i appears at places in the tensor product corresponding to the indices in F_i.

We now consider the action of $k[S_m]$ on $T^m(E)$ as following :

Let $\sigma \in S_m$ and $\underline{u} = u_1 \otimes \ldots \otimes u_m \in T^m(E)$ (with $u_i \in E$) then the action of σ on $\underline{u}$ is defined by

$$\sigma.u_1 \otimes \ldots \otimes u_m = u_{\sigma^{-1}(1)} \otimes \ldots \otimes u_{\sigma^{-1}(m)}$$

(the vector u_i is at the place $\sigma(i)$ in the tensor product). This extends naturally to linear combinations and gives us an action of $k[S_m]$ on $T^m(E)$.

If $m > n$, some elements of $k[S_m]$ have the special property that they vanish on $T^m(E)$. For example,

$$\xi_{n+1} = \sum_{\sigma \in S_{n+1}} \epsilon(\sigma)\sigma.$$

where $\epsilon(\sigma)$ is the signature of σ and S_{n+1} the subset of permutations of S_m that only change $(1,\ldots,n+1)$.

We call them identities of $k[S_m]$ for its action on $T^m(E)$. They make a k-vector space, denoted by I_n^m.

In the case where $m \leq n$, there is no identity except 0, because the permutations of $e_1 \otimes \ldots \otimes e_m$ are distinct elements of a base of $T^m(E)$. Up to now, we work implicitly with $m > n$.

We remark that for any permutations $\alpha, \beta \in S_m$, $\alpha\xi_{n+1}\beta$ is still an identity (factors in ξ_{n+1} are just permuted).

Let $\Xi_{n+1}^m = \sum_{\alpha,\beta \in S_m} k \ \alpha\xi_{n+1}\beta$, be the left and right ideal generated by ξ_{n+1}. We have $\Xi_{n+1}^m \subset I_n^m$.

2.2 Combinatorial tools.

We give here, some tools which will help us to manipulate elements of $k[S_m]$.

First, we order S_m with the reverse lexicographic order : We say $\sigma < \sigma'(\sigma, \sigma' \in S_m)$ iff there exists $l \in \{0,\ldots,m-1\}$ such that

$$\sigma(m) = \sigma'(m),\ldots,\sigma(m-l+1) = \sigma'(m-l+1) \text{ and } \sigma(m-l) < \sigma'(m-l).$$

Let $s = \sum_{\sigma \in S_m} \lambda_\sigma \sigma$ be a non-zero element of $k[S_m]$. The support of $s \neq 0$ is

$$supp(s) = \{\sigma \in S_m; \lambda_\sigma \neq 0\}$$

The leading term of s is the maximal element of $supp(s)$ (for $<$). It is noted $l(s)$.

We say that $s \in k[S_m]$ can be reduced by another element s' if $l(s) = l(s')$ and for some $\mu \in k^*$ we have $l(s - \mu s') < l(s)$.

We associate to an element $\sigma \in S_m$, the sequence of number $(\sigma(1),\ldots,\sigma(m))$. More generally, a sequence of integers will be a finit suite $\nu = (\nu_1,\ldots,\nu_m)$ and a subsequence of s, a subfamily $(\nu_{i_1},\ldots,\nu_{i_n})$ of ν.

We define the width of an element $\sigma \in S_m$ as the maximal length of all increasing subsequences of $(\sigma(1),\ldots,\sigma(m))$. If $width(\sigma) = n$, we can find a sequence $(i_1,\ldots,i_n)$ with $i_1 < i_2 < \ldots < i_n$ and $\sigma(i_1) < \sigma(i_2) < \ldots < \sigma(i_n)$.

In the following, we will also look at decreasing subsequences of $(\sigma(1),\ldots,\sigma(m))$ $(\sigma(i_1) > \sigma(i_2) > \ldots > \sigma(i_r))$. with $i_1 < i_2 < \ldots < i_r$.

• **Partitions in decreasing subsequences.**

The minimal number of elements of a partition of $(\sigma(1),\ldots,\sigma(m))$ in decreasing subsequences is related with the width of σ. This number is, of course, greater or equal to $width(\sigma)$ because two indices of an increasing subsequences cannot belong to the same decreasing subsequence.

Lemma 2.1 — *The minimal number of elements of a partition of $(1,\ldots,m)$ in decreasing sequences of $\sigma \in S_m$ is the maximal length $width(\sigma)$ of an increasing subsequences.*

An example: consider the sequence $(7,3,5,4,1,2,6)$. An increasing subsequences of maximal length is for instance $(3,4,6)$ and we can build three decreasing subsequences with all indices : $(7,6)$, $(5,4,1)$, $(3,2)$.

Proof: We show it for any sequences $\nu = (\nu_1,\ldots,\nu_k)$ of distinct integers taking the same definitions as for permutations.

Let $(\nu_{i_1},\ldots,\nu_{i_n})$ be an increasing sequence of length $width(\nu) = n$. Suppose, moreover that ν_{i_n} is minimal between all last terms of increasing subsequences of length n. All these last terms, written in decreasing order, are $\nu_{i_n^1} > \ldots > \nu_{i_n^1} > \nu_{i_n^0} = \nu_{i_n}$.

In fact, they appear in this order in the sequence ν. Otherwise, we can find amoung them a subsequence $(\nu_{i_n}, \nu_{i'_n})$ of ν such that $\nu_{i_n} < \nu_{i'_n}$. As ν_{i_n} is the last term of an increasing subsequence $(\nu_{i_1},\ldots,\nu_{i_n})$, the subsequence $(\nu_{i_1},\ldots,\nu_{i_n},\nu_{i'_n})$ is also increasing, which is a contradiction $(n = width(\nu))$.

So, we remove the decreasing subsequences $(\nu_{i_n^1},\ldots,\nu_{i_n^1},\nu_{i_n^0})$ from ν. The remaining sequence ν' is then of width $n-1$ (the last term of increasing families of length n are not there). We end the proof by induction on ν'. So the minimal number of decreasing subsequences is lower than $width(\nu)$. As it must be greater than $width(\nu)$, equality holds.

A way to compute fastly an increasing sequence of maximal length is given in [3].

• **Permutations of decreasing sequences.**

A first remark can be done concerning permutations of decreasing sequences.

Let $\sigma \in S_m$. Let $F_1 = (i_1^1,\ldots,i_{k_1}^1),\ldots,F_l = (i_1^l,\ldots,i_{k_l}^l)$ be disjoint decreasing subsequences of $(1,\ldots,m)$ such that for all $j \in \{1,\ldots,l\}$,

$$\sigma(i_1^j) > \ldots > \sigma(i_{k_j}^j).$$

Let $\alpha \in S_m \backslash \{id\}$ be a permutation, that keeps globally fixed each (F_j) and that fixes all other indices $(\in (1,\ldots,m)\backslash \cup_j F_j)$.

We are going to show that $\sigma\alpha > \sigma$:
Take a family $F \in (F_1,\ldots,F_l)$ that is really changed by α. We write it $F = (i_1,\ldots,i_k)$. Suppose that

$$\alpha(i_k) = i_k,\ldots,\alpha(i_{l+1}) = i_{l+1} \text{ but } \alpha(i_l) \neq i_l$$

Then $\alpha(i_l) \in \{i_1,\ldots,i_{l-1}\}$ and $\sigma\alpha(i_l) \in \{\sigma(i_1),\ldots,\sigma(i_{l-1})\}$. As F is a decreasing sequence, this implies that

$$\sigma\alpha(i_k) = \sigma(i_k),\ldots,\sigma\alpha(i_{l+1}) = \sigma(i_{l+1}) \text{ and } \sigma\alpha(i_l) > \sigma(i_l)$$

Putting all families together, we get $\sigma\alpha > \sigma$.

2.3 Reduction of identities.

The aim of the following lines is to show that the set of identities is in fact $\Xi_{n+1}^m = \sum_{\alpha,\beta \in S_m} k.\alpha \xi_{n+1}$ and that we are able to reduce an element of $k[S_m]$ to a "canonical form" modulo this right and left ideal.

Lemma 2.2 — *Let σ belong to S_m. If $width(\sigma) \geq n+1$, then we can find α, $\beta \in S_m$ such that σ is the the leading term of $\alpha \xi_{n+1} \beta$.*

Let $\sigma \in S_m$ with $width(\sigma) \geq n+1$. We can find a subsequence $i_1 < \ldots < i_{n+1}$ of $(1,\ldots,m)$ such that $j_1 = \sigma(i_1) < \ldots < j_{n+1} = \sigma(i_{n+1})$.

Let take $\alpha, \beta \in S_m$ such that $\alpha(l) = j_l, \beta^{-1}(l) = i_l$ for $1 \leq l \leq n+1$ and find the other indices in order to havee $\alpha\beta = \sigma$. For $1 \leq l \leq n+1$ and $\nu \in S_{n+1}$, we have $\alpha\nu\beta(i_l) = j_{\nu(l)}$.

To compare $\alpha\nu\beta$ with $\alpha\beta$ in a reverse lexicographic order, we only need to look at subsequences that differs between them: This can only happen on the subsequences $(i_1,\ldots,i_{n+1})$. But $(j_1,\ldots,j_{n+1}) > (j_{\nu(1)},\ldots,j_{\nu(n+1)})$ ($>$ is the reverse lexicographic order) for all permutations ν in $S_{n+1} - \{id\}$ because $j_1 < \ldots < j_{n+1}$. So $\alpha\nu\beta < \alpha\beta = \sigma$ when $\nu \neq id$ and $\sigma = \alpha\beta$ is the leading term of $\sum_{\nu \in S_{n+1}} \epsilon(\sigma)\alpha\sigma\beta$.

Remark: We could have taken instead of ξ_{n+1} any element $\zeta = \lambda_{id} id + \ldots$ of $k[S_{n+1}]$ with $\lambda_{id} \neq 0$!

This yields that as soon as a term σ of an element $s = \sum_\sigma \lambda_\sigma \sigma$ is of width greater than $n+1$, it can be reduced by an element $\alpha\xi_{n+1}\beta$ of X_{n+1}^m: $s = \lambda\alpha\xi_{n+1}\beta + s'$ with $l(s') < l(s)$.

By reducing recursively all terms (starting from the leading one) by elements $\alpha\xi_{n+1}\beta$ while it is possible, we decompose s as the sum of an element of Ξ_{n+1}^m and an element of $\sum_{width(\sigma)\leq n} k\,\sigma$: $k[S_m] = \Xi_{n+1}^m + \sum_{width(\sigma)\leq n} k\,\sigma$.

Lemma 2.3 — *Let $s \neq 0$ be an identity $\in I_n^m$, then the width n_0 of its leading term is greater than $n+1$.*

If not, then $s = \lambda_0\sigma_0 + \ldots$ with $\lambda_0 \neq 0$ and σ_0 the leading term of s of width lower than $n+1$. Using lemma 2.1, we can find a partition of $(1,\ldots,m)$ in n_0 subsequences (with $n_0 \leq n$:

$$F_1 = (i_1^1,\ldots,i_{k_1}^1),\ldots,F_{n_0} = (i_1^{n_0},\ldots,i_{k_{n_0}}^{n_0})$$

such that $i_1^j < \ldots < i_{k_j}^j$ and $\sigma_0(i_1^j) > \ldots > \sigma_0(i_{k_j}^j)$ for $1 \leq j \leq n_0$.

Let $e_1,\ldots,e_{n_0}$ be independant vectors of E ($n_0 \leq n$). Take

$$\underline{e} = e_{F_1 \sqcup F_2 \sqcup \ldots \sqcup F_{n_0}} \in T^m(E)$$

(In places corresponding to indices in F_j, we take e_j). We have

$$\sigma_0(\underline{e}) = e_{\sigma_0(F_1) \sqcup \sigma_0(F_2) \sqcup \ldots \sqcup \sigma_0(F_{n_0})}$$

If $s(\underline{e}) = 0$, then (as $\{e_{i_1} \otimes \ldots \otimes e_{i_m}\}$ is a base of $T^m(E)$) we can find $\sigma \neq \sigma_0 \in supp(s)$ such that $\sigma(\underline{e}) = \sigma(\underline{e})$. Then taking $\alpha = \sigma_0^{-1}\sigma (\neq id)$, we must have $\alpha(F_1) = F_1,\ldots,\alpha(F_{n_0}) = F_{n_0}$ and according to the remark on permutations of decreasing subsequences page 9, $\sigma = \sigma_0\alpha > \sigma_0$ whichs contradicts the fact that σ_0 is the leading term of s.

Theorem 2.4 — *Let s belongs to* $k[S_m]$. *The following properties are equivalent :*

- s *is an identity of* $T^m(E)$.

- s *belongs to the space* Ξ^m_{n+1} *"generated" by* ξ_{n+1}.

- *the remainder of* s *in a complete reduction by elements* $\alpha\xi_{n+1}\beta$ $(\alpha, \beta \in S_m)$ *is zero.*

In other words, $I^m_n = \Xi^m_{n+1}$ *and* $k[S_m] = I^m_n \oplus \sum_{width(\sigma)\leq n} k \ \sigma$.

Proof:
An identity $s \in I^m_n$ can be decompose as $s = s' + r$ with $s' \in \Xi^m_{n+1}$ and r the remainder of s in its reduction by elements of Ξ^m_{n+1} which is in $\sum_{width(\sigma)\leq n} k \ \sigma$. The leading term of r is of width lower than $n + 1$. According to lemma (2.3), $r = s - s' \in I^m_n$ which is an identity must be null.
So $I^m_n = \Xi^m_{n+1}$ and $I^m_n \cap \sum_{width(\sigma)\leq n} k \ \sigma = \{0\}$ (lemma (2.3)).
Using another representation of permutations as Young Diagrams (see [Knu]), one can "estimate" the dimension of $k[S_m]/I^m_n$, or equivalently the number of permutations of width $\leq n$. It is the sum for all diagrams of S_m with less than n columns of the "hooks lengths".

3 Cayley-Hamilton, the keystone.

Let
$$F_n(X_1,\ldots,X_{n+1}) = \sum_{\sigma \in S_{n+1}} \epsilon(\sigma) \ T_\sigma.$$

This polynomial F_n defines a unique polynomial G_n such that

$$F_n(X_1,\ldots,X_{n+1}) = T(G_n(X_1,\ldots,X_n).X_{n+1})$$

This polynomial is closely linked with the Cayley-Hamilton identity and is use in the following

Theorem 3.1 (Procesi-Razmyslov) — *The ideal of identities* $\Gamma_{X,n}$ *(resp* $I_{X,n}$*) is the ideal of* T_X *(resp* $T_X\langle X\rangle$*) stable by substitutions (resp and trace) generated by* F_n *(resp* G_n*).*

This so-called Razmyslov-Procesi theorem is proved in the first case using Youngs Diagrams (see [8]) and in the latter with the "mysterious" Capelli identity (see [6] and [11]). Here, we want to give an effective version of this theorem based on the reduction in $k[S_m]$, defined in section 2.

Theorem 3.2 — *Let* $P(X_1,\ldots,X_m)$ *belong to* $\Gamma_{X,n}$ *(resp* $I_{X,n}$*). We can construct monomials* M^i_l, N^i_l *(resp and* A^j, B^j, M'^j_l, N'^j_l*) in* $X_1,\ldots,X_m$ *and scalars* λ_i *(resp and* μ_i*) $\in k^*$ such that*

$$P(X_1,\ldots,X_m) = \sum_i \lambda_i F_n(M^i_1,\ldots,M^i_{n+1})T(N^i_1)T(N^i_2)\ldots$$

$$(resp \quad P(X_1,\ldots,X_m) = \sum_i \lambda_i N^i_0.F_n(M^i_1,\ldots,M^i_{n+1})T(N^i_1)T(N^i_2)\ldots$$
$$+ \sum_j \mu_j A^j.G_n(M'^j_1,\ldots,M'^j_n).B^jT(N'^j_1)T(N'^j_2)\ldots)$$

3.1 The identities F_n and G_n.

The polynomial F_n which corresponds to the element ξ_{n+1} of $k[S_m]$, throught the isomorphism Ω page 6 is an identity according to the first sections. In fact, it is the multilinear identity of lowest degree $(n+1)$ in T_X. So $G_n(X_1 \ldots X_n) \in T_X\langle X\rangle$ is the only multilinear identity of degree n (up to a scalar) in $X_1, \ldots, X_n$. Then, the complete polarization of the Cayley-Hamilton identity define in 1.3 is a multiple of G_n. Looking to one of its monomials (for instance $X_1 X_2 \ldots X_n$), we easily check that the two polynomials are equal.

3.2 The substitutions of F_n.

We want to show that all polynomials $\sum_{\sigma \in S_{n+1}} \epsilon(\sigma) T_{\alpha\sigma\beta}$ are identities of $M_n(k)$ that are build from F_n.

Proposition 3.3 *For all permutations α, β in S_m, we can find monomials M_i, N_j such that*

$$\sum_{\sigma \in S_{n+1}} \epsilon(\sigma) T_{\alpha\sigma\beta} = F_n(M_1, \ldots M_{n+1}) T(N_1) T(N_2) \ldots$$

Proof: Let $\alpha, \beta \in S_m$. For convenience, we define

$$i_1 = \beta^{-1}(1), \quad \ldots, \quad i_{n+1} = \beta(n+1)$$
$$j_1 = \alpha(1), \quad \ldots, \quad j_{n+1} = \alpha(n+1)$$

so that for all σ in S_{n+1}, for all l in $\{1, \ldots, n+1\}$, we have $\alpha\sigma\beta(i_l) = j_{\sigma(l)}$. If k belongs to $\{1, \ldots, m\} - \{i_1, \ldots, i_{n+1}\}$, then $\alpha\sigma\beta(k) = \alpha\beta(k)$.
Take a cycle $a = (a_1, \ldots, a_p)$ of $\nu = \alpha\beta$.

1. Either it does not contain an element of $\{j_1, \ldots, j_{n+1}\}$. Then it is also a cycle of $\alpha\sigma\beta$ because its indices are not changed by σ.

   Let $N_1 = X_{a_1} \ldots X_{a_p}$. The term $T(N_1)$ is a factor of all $T_{\alpha\sigma\beta}$ ($\sigma \in S_{n+1}$).

2. Or $a = (a_1, \ldots, a_p)$ is of the form

$$(j_{l_1}, a_1^{l_1}, \ldots, a_{k_{l_1}}^{l_1}, j_{l_2}, a_1^{l_2}, \ldots, a_{k_{l_2}}^{l_2}, \ldots, j_{l_d}, a_1^{l_d}, \ldots, a_{k_{l_d}}^{l_d})$$

   with $l_s \in \{1, \ldots, n+1\}$ and $a_q^p \in \{1, \ldots, m\} - \{j_1, \ldots, j_{n+1}\}$.
   Let $\sigma_0^a = (l_1, \ldots, l_d) \in S_{n+1}$ and $M_{l_q} = X_{j_{l_1}} X_{a_1^{l_1}} \ldots X_{a_{k_{l_1}}^{l_1}}$ for $1 \leq q \leq d$ so that we have $T(X_{a_1} \ldots X_{a_p}) = T(M_{l_1} \ldots M_{l_d})$.

We define σ_0 as the product of σ_0^a for all cycles a of the form (2): $\sigma_0 = (l_1, \ldots, l_d) \ldots$ If we consider σ_0 as a permutation of the monomial M_i, the product of all terms $T(X_{a_1} \ldots X_{a_p})$ with $(a_1, \ldots, a_p$ of type (2) can also be written $T_{\sigma_0}(M_1, \ldots, M_{n+1})$. Looking to all cycles of $\nu = \alpha\beta$, we have

$$T_{\alpha\beta} = T_{\sigma_0}(M_1, \ldots, M_{n+1}) T(N_1) T(N_2) \ldots$$

where the monomials M_i (resp N_j) are defined as in (2) (resp (1)).

For all σ in S_{n+1}, for $1 \leq l \leq n+1$, $\alpha\sigma\beta(i_l) = j_{\sigma(l)}$ and $\alpha\sigma\beta(k) = \alpha\beta(k)$ if $k \in \{1, \ldots, m\} - \{i_1, \ldots, i_{n+1}\}$.

$$\alpha\sigma\beta = (j_{l_1}, \overbrace{a_1^{l_1}, \ldots, a_{k_{l_1}}^{l_1}}^{M_{l_1}}, j_{\sigma(l_2)}, \overbrace{a_1^{\sigma(l_2)}, \ldots, a_{k_{l_2}}^{\sigma(l_2)}}^{M_{\sigma(l_2)}}, \ldots) \ldots$$

and

$$T_{\alpha\sigma\beta} = T_{\sigma\sigma_0}(M_1, \ldots, M_{n+1})T(N_1)T(N_2) \ldots$$

Finally,

$$\sum_{\sigma \in S_{n+1}} \epsilon(\sigma)T_{\alpha\sigma\beta} = \sum_{\sigma \in S_{n+1}} \epsilon(\sigma)(T_{\sigma\sigma_0}(M_1, \ldots, M_{n+1}))T(N_1)T(N_2) \ldots$$
$$= \pm F_n(M_1, \ldots, M_{n+1})T(N_1)T(N_2) \ldots$$

This proves proposition (3.3).

3.3 A generator of the identities.

According to section 2, the reduction of a multilinear element $\sum_{\sigma \in S_m} \lambda_\sigma \sigma$ of $\Gamma_{X,n}$ give us one of its decomposition in Ξ_{n+1}^m as

$$\sum_{\sigma \in S_m} \lambda_\sigma \sigma = \sum_i \mu_i \left(\sum_{\sigma \in S_{n+1}} \epsilon(\sigma)\alpha^i \sigma \beta^i \right).$$

According to the last section, we can construct monomials M_j^i, N_j^i such that

$$\sum_{\sigma \in S_m} \lambda_\sigma \sigma = \sum_i \mu_i F_n(M_1^i, \ldots, M_{n+1}^i)T(N_1^i)T(N_2^i) \ldots$$

An element P of $\Gamma_{X,n}$ is a linear combination of substitutions of multilinear elements (the inverse substitutions of the polarization of P). So that P has also a decomposition of the form

$$\sum_i \mu_i F_n(M_1^i, \ldots, M_{n+1}^i)T(N_1^i)T(N_2^i) \ldots$$

where the monomials also noted M_j^i, N_j^i, are known explicitly from those defined in the previous case.

Let $P(X_1, \ldots, X_m)$ belong to $I_{X,n}$. As we just have seen it, $T(P(X_1, \ldots, X_m).X_{m+1}) \in \Gamma_{X,n}$ can be decomposed as

$$\begin{aligned}T(P(X_1 \ldots X_m).X_{m+1}) =\ & \sum_i \mu_i F_n(M_1^i, \ldots, M_{n+1}^i)T(N_1^i.X_{m+1})T(N_2^i) \ldots \\ & + \sum_j \mu_j T(G_n(M_1^{\prime j}, \ldots, M_n^{\prime j})\, A^j X_{m+1} B^j)T(N_1^{\prime j})T(N_2^{\prime j}) \ldots\end{aligned}$$

As the trace is invariant under circular permutations of products, the polynomial P is uniquely determined as

$$P = \sum_i \mu_i N_1^i.F_n(M_1^i, \ldots, M_{n+1}^i)T(N_2^i) \ldots + \sum_j \mu_j B^j G_n(M_1^{\prime j}, \ldots, M_n^{\prime j})A^j.T(N_1^{\prime j})T(N_2^{\prime j}) \ldots$$

where all monomials $M_l^i, N_l^i, A^j, B^j, M_l^{\prime j}, N_l^{\prime j}$ can be constructed explicitly. This ends the demonstration of theorems (3.1) and (3.2).

4 An algorithm.

We have now all the tools to describe a method of reduction in $\mathbf{T_X}\langle X\rangle/\mathbf{I_{X,n}}$. The field k is a field of characteristic 0 where we are able to compute (for instance the rationnal field $\mathbf{Q}$). Let us fixe the dimension n of the space of matrices.

Input : A polynomial $P(X_1,\ldots,X_p)$ of $\mathbf{T_X}\langle X\rangle$.

Output : Another polynomial $P'(X_1,\ldots,X_p)$ which is a "reduced" form of P :
$P - P'$ is an identity of $M_n(\mathbf{k})$ and $P' = 0$ iff P is an identity.

1. **Homogeneous decomposition.** Select all homogeneous components of P of degree $\geq n+1$: These are coefficients of the different powers of $(t_i)_{1\leq i\leq p}$ with total degree less than $n+1$ in $P(t_1\,X_1,\ldots,t_p\,X_p)$.

2. **Polarization.** For every homogeneous component (that we still note $P(X_1,\ldots,X_p)$), polarize it. This is obtained by succesively taking the coefficients of $t_i^1\ldots t_i^{d_i}$ in $P^{(d_1)\ldots(d_{i-1})}(\ldots,t_i^1\,X_i^1\ldots t_i^{d_i}X_i^{d_i},\ldots)$. When it's finished, rename the variables $X_1^1,X_2^1,\ldots,X_1^2,\ldots$ that really appear above as $Y_1,Y_2,\ldots,Y_m$. Put in a corner, the real name of these variables : $Y_1 = X_1^1, Y_2 = X_2^1,\ldots$

3. **Reduction.** The polarized polynomial is still called $P(Y_1,\ldots,Y_m)$.
   Let $Q(Y_1,\ldots,Y_m,Y_{m+1}) = T(P(Y_1,\ldots,Y_m)Y_{m+1})$. We have to reduced Q by all substitutions of $F_n(Y_1,\ldots,Y_{n+1})$. The polynomial Q is a sum of terms $\lambda.T_\sigma$ with $\sigma \in S_m$ and $\lambda \in \mathbf{k}\backslash\{0\}$. Such a term can be reduced if $width(\sigma) \geq n+1$. Until there is no more monomials of this kind, do the following things :

   - Take a monomial $\lambda.T_\sigma$ such that $width(\sigma) \geq n + 1$. We can find an increasing subsequence $(i_1,\ldots,i_{n+1})$ of $(1,\ldots,m)$ such that $j_1 = \sigma(i_1) < \ldots < j_{n+1} = \sigma(i_{n+1})$. Let α,β be the permutations such that $\beta^{-1}(1) = i_1,\ldots,\beta^{-1}(n+1) = i_{n+1}$ and $\alpha(1) = j_1,\ldots,\alpha(n+1) = j_{n+1}$ and $\alpha\beta = \sigma$
     We have to substract to P the element of $\mathbf{T_X}$: $\lambda(\sum_{\nu\in S_{n+1}} \epsilon(\nu)\,T_{\alpha\nu\beta})$.

   - Decompose $\alpha\beta$ in a product of cycles.

   - For each cycle where some j_l appears, and which is of the form
     $$(j_{l_1},a_1^{l_1},\ldots,a_{k_{l_1}}^{l_1},j_{l_2},a_1^{l_2},\ldots,a_{k_{l_2}}^{l_2},\ldots,j_{l_d},a_1^{l_d},\ldots,a_{k_{l_d}}^{l_d})$$
     with $a_q^p \in \{1,\ldots,m\}\backslash\{j_1,\ldots,j_{n+1}\}$, define $M_{l_1} = X_{j_{l_1}}X_{a_1^{l_1}}\ldots X_{a_{k_{l_1}}^{l_1}}$,
     $M_{l_2} = X_{j_{l_2}}X_{a_2^{l_2}}\ldots X_{a_{k_{l_2}}^{l_2}},\ldots$ When all indices $(j_l)_{1\leq l\leq n+1}$ have been treated, we have
     $$M_1 = X_{j_1}X_{a_1^1}\ldots X_{a_{k_1}^1},\ldots,M_{n+1} = X_{j_{n+1}}X_{a_1^{n+1}}\ldots X_{a_{k_{n+1}}^{n+1}}.$$

   - For the remaining cycles, where no indices $j_1,\ldots j_{n+1}$ appears, take the product of the corresponding element in $\mathbf{T_X}$. We note it T_ρ.

- Go on with $Q - \lambda\, F_n(M_1,\ldots,M_{n+1}).T_\rho$ instead of Q.

4. When this is finished, we have a polynomial Q' in $\mathbf{T}_X$. Take the unique element of $\mathbf{T}_X\langle X\rangle$ such that $Q'(Y_1,\ldots,Y_m,Y_{m+1}) = T(P'(Y_1,\ldots,Y_m)Y_{m+1})$.

5. Substitute all variables Y_i by the initial variables X_l^j and all variables X_l^j introduced in polarization by the corresponding variables X_j. This gives us the request polynomial $P'(X_1,\ldots,X_p) \in \mathbf{T}_X\langle X\rangle$.

The step 3 gives a faithful correspondance (using the isomorphism Ω) between reduction in $k[S_m]$ by elements $\alpha\xi_{n+1}\beta$ and reduction in $\mathbf{T}_X^{m+1}$ by the substitutions of F_n. As asserted in section 2, this loop stops with a polynomial whose terms $T_{\sigma'}$ are such that $width(\sigma') \leq n$.

As we are in characteristic 0, it is not a problem to transform a polynomial in a multilinear one and to do the converse. This allow us to use reduction in the multilinear case but we can expect a direct reduction working with sequences of non-distinct numbers instead of permutations.

References

[1]   S. A. Amitsur, *The T-ideals of the free ring*, J. Lond. Math. Soc. 30 (1955) 470-475.

[2]   F. R. Gantmacher *Theory of matrices*, (Dunod 1966).

[3]   D. E. Knuth, *The art of Computer Programming*, Vol. 3 (Addison-Wesley 1968).

[4]   E. Formanek, *Polynomial Identities of matrices*, Agebraists'hommage. Papers in ring theory and related topics. Contemp. Math. (1981) 41-79.

[5]   J. Pierce, *Associative Algebra*, (Springer-verlag 1986).

[6]   C. Procesi, *The invariant theory of $n \times n$ matrices*, Adv. in Math. 19 (1976) 306-381.

[7]   C. Procesi, *Computing with 2×2 matrices*, J. of Alg. 87 (1984) 342-359.

[8]   Y. P. Razmyslov, *Trace identities of full matrix algebras over a field of characteristic zero*, Translation : Math. USSR Izv. 8 (1974) 727-760.

[9]   A. Regev, *Young Tableaux and P.I. Algebra*, Astérisque 87/88 (1981) 335-352.

[10]   L. H. Rowen, *Polynomials Identities in Ring*, (Academic press. New-york 1980).

[11]   H. Weyl, *The classical groups*, (Princeton University Press, Princeton N. J. 1946).

Implicitization of Rational Parametric Curves and Surfaces

Michael Kalkbrener

Research Institute for Symbolic Computation (RISC)

Johannes Kepler University Linz, Austria

Abstract

In this paper we use Gröbner bases for the implicitization of rational parametric curves and surfaces in 3D-space. We prove that the implicit form of a curve or surface given by the rational parametrization

$$x_1 := \frac{p_1}{q_1} \qquad x_2 := \frac{p_2}{q_2} \qquad x_3 := \frac{p_3}{q_3},$$

where the p's and q's are univariate polynomials in y_1 or bivariate polynomials in y_1, y_2 over a field K, can always be found by computing

$$GB(\{q_1 \cdot x_1 - p_1, \ q_2 \cdot x_2 - p_2, \ q_3 \cdot x_3 - p_3\}) \cap K[x_1, x_2, x_3],$$

where GB is the Gröbner basis with respect to the lexical ordering with $x_1 \prec x_2 \prec x_3 \prec y_1 \prec y_2$, if for every $i, j \in \{1,2,3\}$ with $i \neq j$ the polynomials p_i, q_i, p_j, q_j have no common zeros. This result leads immediately to an implicitization algorithm for arbitrary rational parametric curves.

Furthermore, we present an algorithm for the implicitization of arbitrary rational parametric surfaces and prove its termination and correctness.

1 Introduction

The automatic conversion of parametrically defined varieties into their implicit form is of fundamental importance in geometric modeling. The reason for this is that implicit and parametric representations are appropriate for different classes of problems. For instance, it is universally recognized that the parametric representation is best suited for generating points along a variety, whereas the implicit representation is most convenient for determining whether a given point lies on a specific variety. It is also well-known that the problem of intersecting two varieties is greatly simplified if one variety can be expressed implicitly and the other parametrically.

For some time the implicitization problem has been deemed unsolvable in the CAD literature ([4] or [11]). In 1984 the problem has been solved for rational parametric curves in 2D and rational parametric surfaces in 3D by using resultants (see [10]). Resolvents have been applied to find the implicit representation of rational parametric cubic curves in 3D ([5]). Recently, algorithms based on resultants have been developed for solving the implicitization problem for rational parametric surfaces ([3] and [9]). Arnon and Sederberg used Gröbner bases for the implicitization of polynomial parametric varieties of dimension $n - 1$ in n-dimensional space ([1]). In 1987 Buchberger generalized their method to the case of polynomial parametric varieties of arbitrary dimension ([2]). Recently, we applied Gröbner bases to the most general problem, the implicitization of rational parametric varieties of arbitrary dimension in arbitrary dimensional space ([8]). Many of the implicitization methods are highlighted in [7].

In this paper we use Gröbner bases for the implicitization of rational parametric curves and surfaces in 3D-space. In contrast to the algorithms for the implicitization of m-dimensional varieties in n-dimensional space presented in [8] the algorithms in this paper work without introducing new variables. Therefore they solve the implicitization problem in 3D-space much faster than the general algorithms. (A comparision of the computing times of our implementations in Maple can be found in [8]).

In this paper we prove that the implicit form of a curve or surface given by the rational parametrization

$$x_1 := \frac{p_1}{q_1} \qquad x_2 := \frac{p_2}{q_2} \qquad x_3 := \frac{p_3}{q_3},$$

where the p's and q's are univariate polynomials in y_1 or bivariate polynomials in y_1, y_2 over a field K, can always be found by computing

$$GB(\{q_1 \cdot x_1 - p_1, \ q_2 \cdot x_2 - p_2, \ q_3 \cdot x_3 - p_3\}) \cap K[x_1, x_2, x_3],$$

where GB is the Gröbner basis with respect to the lexical ordering with $x_1 \prec x_2 \prec x_3 \prec y_1 \prec y_2$, if for every $i, j \in \{1, 2, 3\}$ with $i \neq j$ the polynomials p_i, q_i, p_j, q_j have no common zeros. Since we can always assume that p_i and q_i are relatively prime ($i = 1, 2, 3$), the above condition is always satisfied, if the p's and q's are univariate. Therefore, the above result leads immediately to an implicitization algorithm for arbitrary rational parametric curves.

Furthermore, we present an algorithm for the implicitization of arbitrary rational parametric surfaces and prove its termination and correctness.

In section 2 we state the problems we are concerned with. In section 3 a few theorems are proved which are necessary for showing the correctness of the algorithms, which we present in section 4.

2 Problems

Throughout the paper let K be a field and $\bar{K}$ the algebraic closure of K.

Let J be an ideal and $g_1, \ldots, g_m$ polynomials in $K[x_1, \ldots, x_n]$. $V(J)$ denotes the *variety* of J, i.e. the set

$$\{a \in \bar{K}^n \mid f(a) = 0 \text{ for every } f \in J\}.$$

Instead of $V(Ideal(\{g_1, \ldots, g_m\}))$ we will often write $V(\{g_1, \ldots, g_m\})$.

Let L be a field with $K \subseteq L$. Then $(a_1, \ldots, a_n) \in L^n$ is a *generic point* of J if for every $f \in K[x_1, \ldots, x_n]$:

$$f \in J \quad \text{iff} \quad f(a_1, \ldots, a_n) = 0.$$

It is well-know that an ideal is prime if and only if it has a generic point with coordinates in a universal domain (see for instance [12]).

In this paper we want to solve the following two problems:

Implicitization Problem for Rational Parametric Curves:

given: rational parametrization of a curve

$$x_1 = \frac{p_1}{q_1} \qquad x_2 = \frac{p_2}{q_2} \qquad x_3 = \frac{p_3}{q_3},$$

where $p_1, p_2, p_3 \in K[y_1]$, $q_1, q_2, q_3 \in K[y_1] - \{0\}$ and p_i and q_i are relatively prime ($i = 1, 2, 3$).

find: implicit representation of this curve, i.e. polynomials $g_1, \ldots, g_m$ in $K[x_1, x_2, x_3]$ such that

$$V(\{g_1, \ldots, g_m\}) = V(P'),$$

where P' is the prime ideal in $K[x_1, x_2, x_3]$ with

$$\left(\frac{p_1}{q_1}, \frac{p_2}{q_2}, \frac{p_3}{q_3}\right) \in K(y_1)^3$$

as generic point.

Implicitization Problem for Rational Parametric Surfaces:

given: rational parametrization

$$x_1 = \frac{p_1}{q_1} \qquad x_2 = \frac{p_2}{q_2} \qquad x_3 = \frac{p_3}{q_3},$$

where $p_1, p_2, p_3 \in K[y_1, y_2]$, $q_1, q_2, q_3 \in K[y_1, y_2] - \{0\}$ and p_i and q_i are relatively prime ($i = 1, 2, 3$).

decide: whether the parametric object is a surface, i.e. whether the transcendence degree of

$$K(\frac{p_1}{q_1}, \frac{p_2}{q_2}, \frac{p_3}{q_3})$$

(over K) is 2. In this case

find: implicit representation of this surface, i.e. a polynomial g in $K[x_1, x_2, x_3]$ such that

$$V(\{g\}) = V(P'),$$

where P' is the prime ideal in $K[x_1, x_2, x_3]$ with

$$(\frac{p_1}{q_1}, \frac{p_2}{q_2}, \frac{p_3}{q_3}) \in K(y_1, y_2)^3$$

as generic point.

Example 1 *For the rational parametrization*

$$x_1 = \frac{2y_2}{1 + y_1^2 + y_2^2} \qquad x_2 = \frac{2y_1 y_2}{1 + y_1^2 + y_2^2} \qquad x_3 = \frac{y_2^2 - y_1^2 - 1}{1 + y_1^2 + y_2^2}$$

the implicit representation

$$x_1^2 + x_2^2 + x_3^2 - 1$$

of the unit sphere is a solution of the above problem. □

3   Theorems

Throughout the paper let $p_1, p_2, p_3 \in K[y_1, y_2]$ and $q_1, q_2, q_3 \in K[y_1, y_2] - \{0\}$ such that p_i and q_i are relatively prime $(i = 1, 2, 3)$. Let

$$f_1 := q_1 \cdot x_1 - p_1, \qquad f_2 := q_2 \cdot x_2 - p_2, \qquad f_3 := q_3 \cdot x_3 - p_3,$$

$$I := Ideal(\{f_1, f_2, f_3\}) \text{ in } K[x_1, x_2, x_3, y_1, y_2]$$

and let $Q_1, \ldots, Q_r$ be primary ideals in $K[x_1, x_2, x_3, y_1, y_2]$ such that $Q_1 \cap \ldots \cap Q_r$ is a reduced primary decomposition of I. Furthermore, P denotes the prime ideal in the polynomial ring $K[x_1, x_2, x_3, y_1, y_2]$ which has

$$(\frac{p_1}{q_1}, \frac{p_2}{q_2}, \frac{p_3}{q_3}, y_1, y_2) \in K(y_1, y_2)^5$$

as generic point.

Theorem 1 *There exists an $i \in \{1, \ldots, r\}$ with*

$$Q_i = P$$

and for every $j \in \{1, \ldots, r\} - \{i\}$:

$$Q_j \cap K[y_1, y_2] \neq \{0\}.$$

Proof: In this proof we use the following notation:

For a given ideal F in $K[x_1, x_2, x_3, y_1, y_2]$ the ideal in $K(y_1, y_2)[x_1, x_2, x_3]$ generated by F is denoted by F^*.

Obviously, I^* is a zero-dimensional prime ideal. By [6] p.92, there exists exactly one element i of $\{1, \ldots, r\}$ with

$$Q_i \cap K[y_1, y_2] = \{0\}.$$

Furthermore, $I^* = Q_i^*$. By [6] p.47, P^* is a zero-dimensional prime ideal. As $I \subseteq P$,

$$P^* = I^* = Q_i^*.$$

Using [6] p.92 again,

$$Q_i = P. \quad \square$$

For the rest of the paper let us assume that

$$Q_1 = P$$

and that $Q_2, \ldots, Q_r$ are ordered in such a way that there exists a $v \in \{1, \ldots, r\}$ such that

$$Q_1, \ldots, Q_v \text{ are isolated primary components and}$$

$$Q_{v+1}, \ldots, Q_r \text{ are embedded primary components.}$$

Obviously,

$$V(I) = V(P_1) \cup \ldots \cup V(P_v), \tag{1}$$

where P_i is the radical of Q_i for $i = 1, \ldots, r$.

By Krull's Primidealkettensatz (see for instance [6] p.179),

$$dim(P_j) \geq 2 \quad (j = 1, \ldots, v), \tag{2}$$

where $dim(P_j)$ denotes the dimension of P_j.

Definition: Let $(b_1, b_2) \in \bar{K}^2$. We denote the number of elements in the set

$$\{i \in \{1, 2, 3\} \mid p_i(b_1, b_2) = q_i(b_1, b_2) = 0\}$$

by $zero(b_1, b_2)$.

Example 2 *We consider again the parametrization*

$$x_1 = \frac{2y_2}{1 + y_1^2 + y_2^2} \qquad x_2 = \frac{2y_1 y_2}{1 + y_1^2 + y_2^2} \qquad x_3 = \frac{y_2^2 - y_1^2 - 1}{1 + y_1^2 + y_2^2}$$

of the unit sphere. Then for $(0,0)$, $(i,0) \in \bar{Q}^2$, where Q denotes the field of rational numbers:

$$zero(0,0) = 0 \quad and \quad zero(i,0) = 3. \quad \square$$

Theorem 2 *Let $j \in \{2, \ldots, v\}$ and $(a_1, a_2, a_3, b_1, b_2)$ the generic point of the prime ideal P_j in $K[x_1, x_2, x_3, y_1, y_2]$. Then*

$$b_1, b_2 \in \bar{K} \text{ and } dim(P_j) \leq zero(b_1, b_2).$$

Proof: First of all, we know from Theorem 1 that the transcendence degree of $K(b_1, b_2)$ is smaller than 2.

Let us assume that the transcendence degree of $K(b_1, b_2)$ is 1.

Let $i \in \{1, 2, 3\}$. From the fact that p_i, q_i are relatively prime it follows that (b_1, b_2) is no common zero of p_i and q_i. As f_i is an element of P_j, a_i is algebraically dependent on $\{b_1, b_2\}$. Thus, $dim(P_j) = 1$. This is a contradiction to (2).

Therefore,

$$b_1, b_2 \in \bar{K}.$$

If (b_1, b_2) is no common zero of p_i and q_i then a_i is algebraically dependent on $\{b_1, b_2\}$. Thus, the transcendence degree of $K(a_1, a_2, a_3, b_1, b_2)$ is less equal $zero(b_1, b_2)$. Therefore,

$$dim(P_j) \leq zero(b_1, b_2). \quad \Box$$

Theorem 3

$$V(I) \neq V(P)$$

implies

that there exists a $(b_1, b_2) \in \bar{K}^2$ with $zero(b_1, b_2) \geq 2$.

Proof: If $V(I) \neq V(P)$ then we obtain from (1) that v is greater equal 2. Let $(a_1, a_2, a_3, b_1, b_2)$ be the generic point of P_2. By Theorem 2 and (2),

$$(b_1, b_2) \in \bar{K}^2 \text{ and } zero(b_1, b_2) \geq 2. \quad \Box$$

4 Algorithms

If for every $(i, j) \in \{(1, 2), (1, 3), (2, 3)\}$

$$p_i, q_i, p_j, q_j \text{ have no common zeros}$$

then, by Theorem 3,

$$V(I \cap K[x_1, x_2, x_3]) = V(P \cap K[x_1, x_2, x_3]).$$

In this case it follows from the elimination property of Gröbner bases that we can obtain the implicit form of the curve or the surface given by

$$x_1 = \frac{p_1}{q_1} \qquad x_2 = \frac{p_2}{q_2} \qquad x_3 = \frac{p_3}{q_3}$$

by computing

$$\{g_1, \ldots, g_m\} := GB(\{q_1 \cdot x_1 - p_1, \; q_2 \cdot x_2 - p_2, \; q_3 \cdot x_3 - p_3\}) \cap K[x_1, x_2, x_3],$$

where GB has to be computed using the lexical ordering determined by $x_1 \prec x_2 \prec x_3 \prec y_1 \prec y_2$.

In particular, if a polynomial parametric surface or a rational parametric curve is given we obtain from Theorem 3:

Corollary 1
a) (Parametrization by polynomial functions:)
If $q_1 = q_2 = q_3 = 1$ then $V(I) = V(P)$.
b) (Rational parametrization of curves:)
If $p_1, p_2, p_3, q_1, q_2, q_3 \in K[y_1]$ then $V(I) = V(P)$.

Hence, the simple algorithm described above solves the implicitization problem for rational parametric curves.

It is an easy consequence of Theorem 2 that

$$I \cap K[x_1, x_2, x_3] = \{0\}$$

iff

there exists a $(b_1, b_2) \in \bar{K}^2$ with $zero(b_1, b_2) = 3$.

Therefore, if there exists such a (b_1, b_2) then every technique from elimination theory must fail in finding an implicit representation.

Example 3 *The implicit equation of the unit sphere cannot be found by computing*

$$GB(\{q_1 \cdot x_1 - p_1, \; q_2 \cdot x_2 - p_2, \; q_3 \cdot x_3 - p_3\}) \cap K[x_1, x_2, x_3],$$

where the p's and q's are defined as in Example 1 or 2:
Since there exists a $(b_1, b_2) \in \bar{K}^2$ with $zero(b_1, b_2) = 3$ (see Example 2),

$$Ideal(\{q_1 \cdot x_1 - p_1, \; q_2 \cdot x_2 - p_2, \; q_3 \cdot x_3 - p_3\}) \cap K[x_1, x_2, x_3] = \{0\}$$

and therefore

$$GB(\{q_1 \cdot x_1 - p_1, \; q_2 \cdot x_2 - p_2, \; q_3 \cdot x_3 - p_3\}) \cap K[x_1, x_2, x_3] = \emptyset. \quad \square$$

The same problem is addressed in [3] and [9]. In these papers parametrizations of that kind are called parametrizations with base points. Resultant techniques are used to compute implicit representations.

In this paper we use Gröbner bases for solving the implicitization problem for rational parametric surfaces.

Definition: Let h, g be polynomials in $K[x_1, x_2, x_3, y_1]$ such that g has no non-trivial factor in $K[y_1]$ and there exists a polynomial p in $K[y_1]$ with $h = g \cdot p$. Then

$$h_{/y_1} := g.$$

implicit_surface (in: $p_1, p_2, p_3, q_1, q_2, q_3$; out: g)

input: $p_1, p_2, p_3 \in K[y_1, y_2]$, $q_1, q_2, q_3 \in K[y_1, y_2] - \{0\}$ and

$$p_i \text{ and } q_i \text{ are relatively prime} \quad (i = 1, 2, 3).$$

output: $g \in K[x_1, x_2, x_3]$ such that if the transcendence degree of

$$K(\frac{p_1}{q_1}, \frac{p_2}{q_2}, \frac{p_3}{q_3})$$

is 2 then

$$g \notin K \text{ and } V(\{g\}) = V(P'),$$

where P' is the prime ideal in $K[x_1, x_2, x_3]$ with the generic point

$$(\frac{p_1}{q_1}, \frac{p_2}{q_2}, \frac{p_3}{q_3}),$$

and

$$g = 1$$

otherwise.

for every $(i, j) \in \{(1, 2), (1, 3), (2, 3)\}$ **do**
$\quad G_{(i,j)} := GB(\{f_i, f_j\}) \cap K[x_1, x_2, x_3, y_1]$, where $f_k := q_k \cdot x_k - p_k$ $\quad (k = 1, 2, 3)$
$\quad F_{(i,j)} := \{h_{/y_1} \mid h \in G_{(i,j)}\}$
$G := GB(F_{(1,2)} \cup F_{(1,3)} \cup F_{(2,3)} \cup \{f_1, f_2, f_3\}) \cap K[x_1, x_2, x_3]$
$g := gcd(G)$

where GB has to be computed using the lexical ordering determined by $x_1 \prec x_2 \prec x_3 \prec y_1 \prec y_2$.

Example 4 *Again we consider the unit sphere given by*

$$x_1 = \frac{2y_2}{1 + y_1^2 + y_2^2} \qquad x_2 = \frac{2y_1 y_2}{1 + y_1^2 + y_2^2} \qquad x_3 = \frac{y_2^2 - y_1^2 - 1}{1 + y_1^2 + y_2^2}.$$

Using implicit_surface *we obtain*

$G_{(1,2)} := \{x_2 + y_1^2 x_2 - x_1 y_1 - y_1^3 x_1\},$

$F_{(1,2)} := \{-x_2 + x_1 y_1\},$

$G_{(1,3)} := \{x_1^2 + 2x_1^2 y_1^2 - y_1^2 - 1 + y_1^4 x_1^2 + x_3^2 + y_1^2 x_3^2\},$

$F_{(1,3)} := \{x_1^2 y_1^2 + x_1^2 - 1 + x_3^2\},$

$G_{(2,3)} := \{-x_2^2 - 2y_1^2 x_2^2 + y_1^4 + y_1^2 - y_1^4 x_2^2 - y_1^2 x_3^2 - y_1^4 x_3^2\},$

$F_{(2,3)} := \{y_1^2 x_2^2 + x_2^2 - y_1^2 + y_1^2 x_3^2\},$

$G := \{x_1^2 + x_2^2 + x_3^2 - 1\},$

$g := x_1^2 + x_2^2 + x_3^2 - 1,$ *the implicit representation of the unit sphere.* ◻

As termination of the algorithm is obvious it remains to prove its correctness.

Proof of correctness:

Let

$$\bar{I} := Ideal(F_{(1,2)} \cup F_{(1,3)} \cup F_{(2,3)} \cup \{f_1, f_2, f_3\}),$$

$\bar{P}$ a prime ideal in $K[x_1, x_2, x_3, y_1, y_2]$ with $\bar{I} \subseteq \bar{P}$ and $P \neq \bar{P}$ and let $(a_1, a_2, a_3, b_1, b_2)$ be the generic point of $\bar{P}$.

Assumption: $dim(\bar{P}) > 1$.

Then,

$$P \not\subseteq \bar{P}.$$

As $I \subseteq \bar{I} \subseteq \bar{P}$ there exists an $i \in \{2, \ldots, v\}$ with $P_i \subseteq \bar{P}$. By Theorem 2,

$$b_1, b_2 \in \bar{K}.$$

As $dim(\bar{P}) > 1$ there exist $j, k \in \{1, 2, 3\}$ such that $j \neq k$ and $\{a_j, a_k\}$ is algebraically independent over K. Since p_j and q_j are relatively prime and $gcd(f_j, f_k)$ divides p_j and q_j,

$$gcd(f_j, f_k) = 1.$$

Thus, $Ideal(\{f_j, f_k\}) \cap K[x_1, x_2, x_3, y_1] \neq \{0\}$ and therefore there exists a non-zero polynomial $f(x_j, x_k, y_1) \in F_{(j,k)}$. By definition of $F_{(j,k)}$,

$$f(x_j, x_k, b_1) \neq 0.$$

This is a contradiction to the fact that $\{a_j, a_k\}$ is algebraically independent over K.

Thus, P is the only prime ideal that is a superideal of $\bar{I}$ and has a dimension greater than 1. Hence, $\bar{I}$ can be written in the form

$$P \cap R,$$

where R is an ideal in $K[x_1, x_2, x_3, y_1, y_2]$ with $dim(R) < 2$. Therefore,

$$\bar{I} \cap K[x_1, x_2, x_3] = (P \cap K[x_1, x_2, x_3]) \cap (R \cap K[x_1, x_2, x_3]) \text{ and } dim(R \cap K[x_1, x_2, x_3]) < 2. \tag{3}$$

It follows from the elimination property of Gröbner bases that

$$G \text{ is a basis of } \bar{I} \cap K[x_1, x_2, x_3].$$

Case:
$$\text{the transcendence degree of } K(\frac{p_1}{q_1}, \frac{p_2}{q_2}, \frac{p_3}{q_3}) \text{ is 2.}$$

In this case $P \cap K[x_1, x_2, x_3]$ is a prime ideal of dimension 2. Thus, there exists an $h \in K[x_1, x_2, x_3] - K$ with $Ideal(\{h\}) = P \cap K[x_1, x_2, x_3]$. As $G \subseteq P \cap K[x_1, x_2, x_3]$,

$$h \text{ divides } gcd(G).$$

Let $p \in K[x_1, x_2, x_3]$ such that $gcd(G) = h \cdot p$. Obviously, p divides every polynomial in $R \cap K[x_1, x_2, x_3]$. As the dimension of $R \cap K[x_1, x_2, x_3]$ is less than 2, p is a non-zero constant. Thus,

$$V(\{gcd(G)\}) = V(\{h\}) = V(P \cap K[x_1, x_2, x_3]) = V(P').$$

Case:
$$\text{the transcendence degree of } K(\frac{p_1}{q_1}, \frac{p_2}{q_2}, \frac{p_3}{q_3}) \text{ is less than 2.}$$

In this case $dim(P \cap K[x_1, x_2, x_3])$ is less than 2 and therefore, by (3), $dim(\bar{I} \cap K[x_1, x_2, x_3])$ is less than 2. Hence,

$$gcd(G) = 1. \quad \square$$

Some of the Gröbner bases computations in implicit_surface can be replaced by other elimination methods, for instance by computations of Sylvester resultants:
We can replace

$$G_{(i,j)} := GB(\{f_i, f_j\}) \cap K[x_1, x_2, x_3, y_1]$$

by

$$G_{(i,j)} := \{resultant(f_i, f_j)\}, \text{ where } f_i \text{ and } f_j \text{ are considered as polynomials in } y_2.$$

Since resultants seem to have a better run-time behaviour, this could lead to a speed-up of the algorithm.

References

[1] D.S. Arnon and T.W. Sederberg, Implicit Equation for a Parametric Surface by Gröbner Basis, in: V.E. Golden, ed., *Proceedings of the 1984 MACSYMA User's Conference*, General Electric, Schenectady, New York (1984) 431-436.

[2] B. Buchberger, Applications of Gröbner Bases in Non-Linear Computational Geometry, in: *Proc. Workshop on Scientific Software*, IMA, Minneapolis (1987).

[3] E.W. Chionh, Base Points, resultants, and the implicit representation of rational surfaces, Ph.D. thesis, Department of Computer Science, University of Waterloo, Canada (1990).

[4] I.D. Faux and M.A. Pratt, *Computational Geometry for Design and Manufacture* (Ellis Horwood, Chichester, 1981).

[5] R.N. Goldman, The method of resolvents: A technique for the implicitization, inversion, and intersection of non-planar, parametric, rational cubic curves, *Computer Aided Geometric Design 2* (1985) 237-255.

[6] W. Gröbner, *Algebraic Geometry II* (Bibliographisches Institut Mannheim, 1970).

[7] C.M. Hoffmann, *Geometric and Solid Modeling: An Introduction* (Morgan Kaufmann Publishers Inc., 1989).

[8] M. Kalkbrener, Implicitization by Using Gröbner Bases, Technical Report RISC-Series 90-27, Univ. of Linz, Austria (1990).

[9] D. Manocha and J.F. Canny, Implicitizing Rational Parametric Surfaces, Technical Report UCB/CSD 90/592, Computer Science Division, University of California, Berkeley, California, U.S.A. (1990).

[10] T.W. Sederberg, D.C. Anderson, and R.N. Goldman, Implicit Representation of Parametric Curves and Surfaces, *Computer Vision, Graphics, and Image Processing 28* (1984) 72-84.

[11] H.G. Timmer, A Solution to the Surface Intersection Problem, Technical Report NAS2-9590, McDonnell-Douglas Corp. (1977).

[12] B.L. van der Waerden, *Algebra II* (Springer, Berlin Heidelberg New York, 1967).

An inequality about irreducible factors of integer polynomials (II)

Philippe Glesser and Maurice Mignotte

Abstract

We give a new upper bound for the height of an irreducible factor of an integer polynomial. This paper also contains a new bound for the general case of polynomials with complex coefficients.

These bounds are very useful in algorithms to factorize polynomials with integer (or algebraic) coefficients.

Suppose that F is an integer polynomial and that P is factor of F. It is well-known (see [M 1]) that the maximum $H(P)$ of the moduli of the coefficients of P is bounded above by

(*) $\qquad H(P) < 2^d (D+1) H(F)$, where $d = \deg(P)$ and $D = \deg(F)$.

In [M 2] it is proved that the term 2^d cannot be much improved without any further hypothesis. But, in [M 3] , it is showed that this term can be improved when both P is irreducible and the measure of F is not too big. Here, using an inequality of Ph. Glesser, we improve the estimates of [M 3]. Our simplest result is (the measure is defined in § 1) :

THEOREM . – Let P be an irreducible integer polynomial of degree d which divides an integer polynomial F. Then

$$H(P) \leq (e/2)^{\sqrt{d}} (d + 2\sqrt{d} + 1)^{0.5 + \sqrt{d}} M(F)^{1 + \sqrt{d}} .$$

The estimate of this theorem is better than (*) when the measure of F is not too big . The inequality $M(F) < (D+1) H(F)$, shows that this is true when the coefficients of F are not too big.

1. Introduction

1. Let $F = \sum_{i=0}^{n} a_i X^i$, $a_n \neq 0$, be a polynomial with complex coefficients, we use the following classical notations :

$$H(F) = \max_{0 \leq i \leq n} |a_i| , \text{ the height of } F , \qquad L(F) = \sum_{i=0}^{n} |a_i| ,$$

$$\| F \| = \left(\sum_{i=0}^{n} |a_i|^2 \right)^{1/2} \qquad \left(= \left(\frac{1}{2\pi} \int_0^{2\pi} |F(e^{i\theta})|^2 d\theta \right)^{1/2} \right),$$

$$|F| = \max \{ |F(z)| ; |z| = 1 \} .$$

A very important notion was introduced by K. Mahler : the measure of F , which is defined by $M(F) = |a_n| \prod_{j=1}^{n} \max \{ 1 , |z_j| \}$, where $z_1 , \ldots , z_n$ are the complex roots of F . All these sizes, except the measure, have the same order of magnitude up to a factor bounded by the degree :

$$H(F) \leq \| F \| \leq |F| \leq (n+1)H(F) .$$

2. If P is a polynomial over the complex numbers, the relations between the coefficients of P and its roots lead at once to the inequality

(1) $\quad L(P) \leq 2^d M(P)$, where $d = \deg(P)$.

In some sense, this relation is the best possible even if P has integer coefficients : the equality holds for $P = (X+1)^d$.

But what happens when P is irreducible ? The example of the polynomial

$$E(X) = a (X+1)^d + 2 ,$$

where a is a large odd integer, shows that (1) cannot be much improved when $M(P)$ is big, even if P is irreducible. (See [M 3] .)

But what happens if both P is irreducible and its measure not too big ? The following result, proved in [M 3], shows that (1) is not the best possible in that case.

THEOREM A . – Let F be a non zero integer polynomial and P an irreducible factor of F, then

$$\| P \| \le e^{\sqrt{d}} (d + 2\sqrt{d} + 2)^{1+\sqrt{d}} M(F)^{1+\sqrt{d}} , \text{ where } d = \deg P.$$

In the present paper, we refine the estimates of [M 3] .

3. The estimates like (1) have applications to the following question : given an integer polynomial F, find a number B such that the coefficients of any irreducible factor P of F have absolute values bounded above by B . This problem is important in modern algorithms for the factorization of polynomials : for the Berlekamp - Zassenhaus algorithm (see Knuth, v. 2), in the famous Lenstra - Lenstra - Lovasz algorithm (see [L. L. L.]), and also for factoring polynomials with algebraic coefficients. The case of irreducible polynomials is the only one used in these algorithms and our result decreases their cost in an obvious way.

Suppose that F is an integer polynomial and that P is some integer polynomial which divides F , then the following inequality are well - known (see [M 1])

$$(2) \quad L(P) \le 2^d M(P) \le 2^d M(F) \le 2^d \| F \| , \text{ where } d = \deg(P).$$

Notice that theorem A is better than (2) for $M(F) \le e^{\sqrt{d}/2}$, when d is large enough.

4. The fact that (2) is almost the best possible for general integer polynomials has been proved in [M 2] . For any $d > 0$, there exists an integer polynomial F, divisible by $(X + 1)^d$, of height equal to 1 with $\deg(F) \ll d^2 \text{Log } d$. This shows that the constant 2 in inequality (2) cannot be replaced by $2 - \varepsilon$, for any fixed $\varepsilon > 0$, even when $M(F)$ is small.

5 . In practice, to apply theorem A (or the estimate (2)) we can use Landau's inequality, $M(F) \le \| F \|$, or - better - compute directly some upper bound for $M(F)$. See [C.M.P.] .

2. A new inequality about factors of polynomials

The following result of Philippe Glesser will play a key rôle in this section.

Theorem 1. – Let $P \in C[X]$ be a polynomial of degree p whose roots belong to the unit circle. Then, for $\rho > 1$ and any real θ,

$$|P(\rho e^{i\theta})| \geq \left(\frac{\rho+1}{2}\right)^p |P(e^{i\theta})|.$$

> Put $z_0 = e^{i\theta}$. One verifies easily that $\dfrac{|\rho z_0 - z|}{|z_0 - z|} \geq \dfrac{\rho+1}{2}$ for $|z| \leq 1$. Applying this inequality for each root of P leads to the result. <

THEOREM 2. – Let P and Q be polynomials with complex coefficients of respective degrees p and q, where $p > q$. Then

$$\|P\| M(Q) \leq 2^{-q} \frac{(p+q)^{p+q}}{p^p q^q} \|PQ\|.$$

> Put $F = PQ$. Let $z_1, \ldots, z_k$ be the roots of F which lie outside of the unit circle and let B be the product of the Blaschke factors relative to these roots (recall that the Blaschke factor relative to a point α is $B(\alpha, z) = (\bar{\alpha}z - 1)/(z - \alpha)$). Consider the polynomial

$$F^*(z) = F(z)B(z) \quad .$$

With obvious notations, put $F^* = P^* Q^*$. By theorem 1,

$$|P^*(e^{i\theta})| \leq \left(\frac{2}{\rho+1}\right)^p |P^*(\rho e^{i\theta})|.$$

But

$$|P^*(\rho e^{i\theta})| \leq \frac{|P^* Q^*(\rho e^{i\theta})|}{\min\{|Q^*(\rho e^{i\theta})|\}} \quad .$$

Notice that the leading coefficient of Q^* is equal to $M(Q)$ and that any point of the circle $|z| = \rho$ is at a distance at least $\rho - 1$ of any root of Q^* (indeed all the roots of Q^* lie in the unit disk), so that $\min\{|Q^*(z)|; |z| = \rho\} \geq M(Q)(\rho-1)^q$. This implies

$$\|P\|^2 \leq \left(\frac{2}{\rho+1}\right)^{2p} \frac{1}{M(Q)^2(\rho-1)^{2q}} \frac{1}{2\pi} \int_0^{2\pi} |F^*(\rho e^{i\theta})|^2 d\theta$$

$$\leq \left(\frac{2}{\rho+1}\right)^{2p} \frac{\rho^{2(p+q)}}{M(Q)^2(\rho-1)^{2q}} \|F^*\|^2.$$

Take $\rho = (p+q)/(p-q)$. The theorem follows from the relation $\| F * \| = \| F \|$. $<$

3. Construction of some multiple of P with low height

THEOREM 3. – Let P be a polynomial with integer coefficients, irreducible, of degree $d \geq 2$. Let N be an integer, $N \geq d$. Then there exists a non zero polynomial G, with integer coefficients, divisible by P, of degree at most N and which satisfies

$$H(G) \leq \left((N+1)^{d/2} M^N \right)^{\frac{1}{N+1-d}} ,$$

where M is the measure of P.

$>$ Apply the sharpening of Siegel's lemma obtained by Bombieri and Vaaler, [B. V.].

$<$

4. A relation between the height and the measure of an irreducible integer polynomial

Let P be an irreducible integer polynomial of degree d. Let N be an integer $> d$ and consider the polynomial G constructed in theorem 3. We get the following result.

THEOREM 4. – Let P be an irreducible integer polynomial, of degree d and measure $\leq M$. Then, for any integer $N \geq d$, we have

$$\| P \| \leq C_N . \left((N+1)^{d/2} M^N \right)^{\frac{1}{N+1-d}} , \text{ where } C_N = \frac{N^N (N+1)^{1/2}}{d^d (2(N-d))^{N-d}}$$

and

$$H(P) \leq K_N \left((N+1)^{d/2} M^N \right)^{\frac{1}{N+1-d}}, \text{ where } K_N = \binom{N-[(d+1)/2]}{[d/2]} (N+1)^{1/2}.$$

$>$ The upper bound of $\| P \|$ comes from the estimate of § 2, whereas that of $H(P)$ is a consequence of the main result of [G]. $<$

Remark . – We give bounds for $\| P \|$ and $H(P)$ since the first quantity appears in the L.L.L. factorization algorithm whereas the second is used in Berlekamp's.

Taking some suitable values of N we get the following corollaries.

COROLLARY 1 . – Let P be an irreducible integer polynomial of degree d and measure $\leq M$. Then

$$\| P \| \leq (e/2)^{\sqrt{d}} (d + 2\sqrt{d} + 1)^{0.5 + \sqrt{d}} M^{1 + \sqrt{d}}$$

and

$$H(P) \leq \frac{7/6}{(2\pi(\sqrt{d} - 1))^{1/2}} (e/2)^{\sqrt{d}} (d + 3\sqrt{d} + 2)^{0.5 + \sqrt{d}} M^{1 + \sqrt{d}} .$$

> We choose $N = d + [\sqrt{d}]$. The details of computation are not difficult. We omit them. <

We wrote down corollary 1 because of its relative simplicity. This corollary improves theorem A above essentially by a factor $2^{-\sqrt{d}}$. The following result is sharper than corollary 1 in the range $M \geq e^{\sqrt{d}/2}$, and sharper than (1) for $M \leq \exp(d/(5 \log 2 d))$.

COROLLARY 2 . – Let P be an irreducible integer polynomial of degree d and measure at most M . Then

$$\| P \| \leq \sqrt{(2d)} M . \exp \left\{ 2 \left(d (1 + \log(1 + \sqrt{d}/2)) \log((2d)^{1/2} M) \right)^{1/2} \right\} .$$

> For $M \leq e$, corollary 2 is implied by corollary 1 . Whereas in the range $M \geq e^d/\sqrt{d}$ it is implied by (1) . Moreover, (1) implies corollary 2 for $d \leq 4$. Thus we suppose that

$$e \leq M \leq e^d/\sqrt{d} \quad \text{and} \quad d \geq 5 .$$

In theorem 4 , put $N = d + x$, then we get

$$\| P \| \leq (d + x + 1)^{1/2} M \, e^x (1 + \frac{d}{2x})^x ((d + x + 1)^{1/2} M)^{d/(x+1)} .$$

Choose now

$$x = \left(\frac{d \log((2d)^{1/2} M)}{1 + \log(1 + \frac{\sqrt{d}}{2})} \right)^{1/2} .$$

Then $[\sqrt{d}] \leq x \leq d - 1$, and the conclusion follows easily. <

References

[B. V.] E. Bombieri , J. D. Vaaler . – On Siegel's lemma , Invent. Math. , vol. 73, 1983, p. 539 - 560 .

[C.M.P.] L. Cerlienco , M. Mignotte , F. Piras . – Computing the measure of a polynomial, J. Symb. Comp. , vol. 4, n° 1, 1987 , p. 21 - 34.

[G] Ph. Glesser . – Majoration de la norme des facteurs d'un polynôme ; *Annales de la Faculté des Sci. Toulouse* , to appear.

[L.L.L.] A. K. Lenstra , H. W. Lenstra Jr, L. Lovàsz . – Factoring polynomials with rational integer coefficients, Math. Ann. , v. 261 , 1982 , p. 515 - 531 .

[M1] M. Mignotte . – An inequality about factors of polynomials , Math. of Comp., v. 28, 1974 , p. 1153 - 1157 .

[M2] M. Mignotte . – An inequality about irreducible factors of integer polynomials , J. Number Th. , v. 30, n° 2, 1988, pp. 156-166 .

[M3] M. Mignotte . – Sur la répartition des racines des polynômes , Journées de Théorie des Nombres, Caen, septembre 1980 .

THE SYMPLECTIC TRILINEAR MAPPINGS; AN ALGORITHMIC APPROACH OF THE CLASSIFICATION; CASE OF THE FIELD GF(3)

L. BENETEAU, INSA, Mathématique,31077 Toulouse Cedex, FRANCE

Abstract: *Let V and W be vector spaces over a commutative field K with n = dimension(V). The set AT(n,m,K) of the skew-symmetric trilinear mappings from V^3 to W whose images has rank m is provided with natural actions of both linear groups GL(V) and GL(W). We consider the problem of counting orbits. For m =1,2,3, the number a(n,m) of orbits of elements of AT(n,m,GF(3)) is shown to coincide with the total number STH(n,m) of pairwise non-isomorphic Hall systems whose rank and 3-order are respectively n+1 and n+m. For m >3, STH(n,m) is at least 3 + a(n,m). A computational approach was used to obtain a set of representatives of AT(5,2,GF(3)), and correspondingly an exhaustive list of the order 3^7 Hall systems: there are 13 such systems. Two algorithms were used: one proceeding to random changes of basis for partial determination of orbits, and another one computing invariants in order to show that some elements are not related.*

I- INTRODUCTION: Let K be a commutative field of characteristic >2 and let V(n,K) be the n-dimensional vector space over K. A trilinear mapping t from V^3 to the vector space W= V(m,K) is said to be *"symplectic"*, or *"alternate"*, if t(x,x,y) = 0 = t(x,y,y) holds. Then t(x,y,z) is "skew-symmtric" in its arguments x,y,z since: t(y,x,z) = -t(x,y,z) = t(x,z,y), and: t(z,y,x) = -t(y,z,x) = t(y,x,z) = -t(x,y,z). Denote by **AT(n,m,K)** the set of the symplectic trilinear mappings t whose image has rank m. For any couple of elements g and h from the linear groups GL(V) and GL(W), the mapping: x,y,z---------> ht(g(x),g(y),g(z)) is element from AT(n,m,K). This defines an action from the direct product of GL(V) and GL(W), onto AT(n,m,K). Our aim is to determine the orbits in some special cases, **or equivalently to classify the surjective linear mappings from the exterior product $PE^3(n,K)$ = $\bigwedge^3 V(n,K)$ onto V(m,K).** Now any two such mappings have the same kernel if and only if one may obtain one of them by composing the other one with some suitable linear automorphism of V(m,K). Hence classifying the elements from AT(m,m,K) can be dealt with by **determining the orbits within the family of the m-codimensional subspaces of $PE^3(n,K)$ under the natural action of the linear group of V(n,K).**

Concerning **symplectic trilinear** *forms* (namely in the case of AT(n,1,K)), partial results are known; let us mention an **explicit classification for n = 6, due to Revoy** (1977) who has also studied the case n = 7). **When K is algebraically closed, or if K is the field of real numbers, the orbits are known and in finite number when the dimension is at most 8. As soon as n > 5 the size of the orbits in AT(n,1,K) depends on the field.** If K is infinite and if n >8, there is an infinity of orbits. Other works can also be found about the trilinear forms, devoted to their isometry group structure, particularly those of Aschbacher on the Lie groups of type E_6.

In section III are presented results about the classification of AT(n,m,K) (*without restriction on m*). We shall be mainly concerned with the case where K is the three-element field GF(3). The classification of symplectic trilinears mappings has then a special meaning in the theory of designs, and this was our initial motivation for the study of AT(n,m,K). For |K|=3 we show that one may translate any classification of AT(n,m,K) in a result concerning some **"Hall Triple Systems** " (or: **"Hall systems").** These systems are the **2-(3^s,3,1) designs** in which any subspace generated by three non collinear points is isomorphic to the affine plane AG(2,3). **Such a system E has an order of the form Card(E) = 3^s, where s is to be refered to as "the 3-order"** of E. **There is an** injective correspondence from the family of the classes of AT(n,m,GF(3)) into the family of Hall systems of rank n+1 and of 3-order n+m (defined up to isomorphism). For m ≤ 3, this correspondence is one-to-one. It turns out that there are 13 Hall systems of ordre 3^7, and the six new ones are given explicit descriptions.

Just as the affine spaces are related to the vector spaces, the Hall systems are related to the GF(3)-loops. The related theory, that validates our transfer theorem, is not to be developed here, neither is the translation of the problems and the results in terms of FISCHER groups (see the bibliography). Our purpose here is to state several classification theorems in AT(n,m,K) at their natural level of generality, and to derive a complete catalogue of the Hall systems of order at most 3^7, thus completing previous partial results (see[1][2][3][4][5]). The crucial stage was the determination of the classes of AT(5,2,GF(3)). This was done by an alternate use of two algorithms. The first one performs changes of basis at random, *searching for a relation* between an element t of AT(5,2,GF(3)) and a given family $\underline{R}$ of elements from AT(5,2,GF(3)). The other algorithm, on the countrary, shows by *computing invariants* that elements are not linked.

II KINDSHIP BETWEEN HALL SYSTEMS AND THE SYMPLECTIC TRILINEAR MAPPINGS OVER THE THREE-ELEMENT FIELD GF(3):

The **Hall systems** are the "**linear spaces**" **(E,L)** (sets E of "points" endowed with a family L of subsets called *"lines"* or *"blocks"*) such that the subspaces generated by three points are isomorphic to an affine space AG(r,3) with $r \leq 2$. **If an element u is chosen as an "origin" of such a system, for any scalar k from the field GF(3) and for any two elements x,y from E one defines relatively to u an external product k,x ----> k.x and an addition x,y ----> x+y** characterised by the facts that, *for any two distint elements x,y, the set {x ,y, 2(x+y)} is a line*, and that *the three following equalities hold: 0.x = u and 1.x = x = x + u*. These two laws define "a **GF(3)-loop**". The axioms of vector spaces over GF(3) are satisfied, except possibly the associativity of the addition (that holds only in the special case of the affine spaces over GF(3)). As an instance of a **non affine Hall system**, take F as the set-product W x V where W = V(2,GF(3)) and V = V(5,GF(3)), with canonical basis { u,v } and { e_1, e_2, e_3, e_4, e_5 } respectively; for each vector x from V denote by x_i its components with respect to the basis: { e_1, e_2, e_3, e_4, e_5 }. Consider the binary law X,Y--->X.Y that sends any couple X = (a,x) and Y = (b,y) from F to the element X.Y = (d-a-b,-x-y) where d is defined by: d = $(x_1y_2-x_2y_1)(y_3-x_3)u$ $+(x_2y_3-x_3y_2)(y_4-x_4)v$.   The subsets of the form {X,X.Y,Y} for X and Y distincts, turn F into a Hall System.

Consider again some general Hall system (E,L). Since the associativity of the additive law does not hold in general, one calls the **"associator"** of x,y,z the difference: T(x,y,z) = ((x+y)+z) - (x+(y+z)). This mapping T satisfies: T(x,x,y) = T(x,y,y) = O, (where the zero vector is 0 = u);and: T(ax,y,z) = aT(x,y,z) = T(x,ay,z) = T(x,y,az), for each a of GF(3). *Nevertheless T cannot be considered as a symplectic trilinear mapping* . First because we have: T(x+y,a,b) = T(x,a,b) + T(y,a,b) + δ, where δ = (T(T(x,a,b),x,y) + T(T(y,a,b),y,x)), *that does not vanish necessarily*; and overall *the law (+) cannot be viewed as a sum in a vector space, since it is not associative* !...In order to make up for these inconveniences *one decides to restrict oneself to the case where T(T(x,a,b),x,y) = O holds;* this identity characterises the so-called **Hall systems "of class 2"** (see [1],[5]). Then δ = O identically. Let D = D(E) be the subloop generated by all the T(x,y,z). If c,d,e are in D one has: T(x,y,e) = O, and T(x+c,y+d,z+e) = T(x,y,z). Hence (x+y)+e =x+(y+e) and (x+c)+(y+d)=(x+y)+(c+d). So the addition in D is associative, and **in the quotient A(E) = E/D (set of the classes x' = x+D) one may define natural laws** (x'+y'= (x+y)' and kx'=(kx)') **that turn A(E) into a GF(3) vector space** (any associator is killed by passing to the factorset E/D). Since T(x,y,z) depends only on the classes x',y',z', one has the:

Theorem of factorization: If (E,L) is a Hall system of class 2 whose rank is $n+1$, its associator-mapping T can be unically factorized in a symplectic trilinear mapping t from $A(E)$ to $D(E)$ obeying $t(x',y',z')= T(x,y,z)$ for all x,y,z in E; besides:

(i) – $A(E)$ is a GF(3) vector space of dimension n;

(ii) – $D(E)$ is of finite dimension m, with $m \leq n(n-1)(n-2)/6$;

(iii) – the trilinear mapping t has an image of rank m;

(iv) – E is finite of cardinal $3^{(n+m)}$ (its 3-order is $(n+m)$).

Proof: If the e_i's for $i = 1,2,...n$, make up a minimal generating system of $(E,+)$ then from [2] the e_i's make up a basis of $A(E)$. Moreover $D(E)$ is then generated by the $T(e_i,e_j,e_k)$ for: $1 \leq i < j < k \leq n$, and $|E| = |A(E)||D(E)|$.

By identifying $A(E)$ to $V(n,GF(3))$ and $D(E)$ to $V(m,GF(3))$, t becomes some element of $AT(n,m,GF(3))$. **Any symplectic trilinear mapping of $AT(n,m,GF(3))$ arises in this way, and t determines (E,L) up to isomorphism.** So one may state the:

Transfer theorem: For $n \geq 3$ and $m \geq 1$ there is a one-to-one canonical correspondence between: (1) –the isomorphy classes of the Hall systems of classe 2, of rank $n+1$ and of 3-order $n+m$; and (2) –the orbits of symplectic trilinear mappings from $AT(n,m,GF(3))$.

Proof: One has an exact sequence of GF(3)-loops of the form:

$$0 \text{ --------> } N \text{ -----> } L_n \text{ --------> } E \text{ ------> } 0$$

where L_n is the free object on n generators in the category of the GF(3)-loops of class at most 2, and N is an arbitrary subloop of $D(L_n)$ (N is a normal subloop of L_n, since the class is 2). Till the end of the proof let us set $K = GF(3)$. By factorizing through the "abelian factor" $E/D(E) \equiv V(n,GF(3))$ and then through $PE^3(n,K) = \bigwedge^3 V(n,K)$ one obtains:

$$
\begin{array}{ccccccc}
E^3 & \longrightarrow & A(E)^3 & \longrightarrow & V(n,GF(3))^3 & \longrightarrow & PE^3(n,GF(3)) \\
\downarrow{\scriptstyle Id} & & \downarrow{\scriptstyle t} & & \downarrow{\scriptstyle t'} & & \downarrow{\scriptstyle Id} \\
E^3 & \xrightarrow{\;T\;} & D(E) & \longrightarrow & V(m,GF(3)) & \xrightarrow{\;\underline{t}'\ (linear)\;} & PE^3(n,GF(3))
\end{array}
$$

If one replaces E by L_n then the corresponding linear mapping of departure $PE^3(n,K) = \bigwedge^3 V(n,K)$ is a isomorphism of rank $(n(n-1)(n-2)/3)$; indeed the subloop $D(L_n)$ is known to admit for basis the $n(n-1)(n-2)/6$ associators $t(e_i,e_j,e_k) = T(e_i,e_j,e_k)$ for $1 \leq i < j < k \leq n$ if the e_i's for $i = 1,2,...n$, are generators of L_n. *Hence one may identify the subloop $D(L_n)$ with $PE^3(n,K) = \bigwedge^3 V(n,K)$.* The kernel of $\underline{t'}$ is then identified with the subloop N of $D(L_n)$. So t or t' determines E up to isomorphism. Any two vector subspaces N and M of $PE^3(n,K)$ correspond to isomorphic Hall systems iff they are in correspondence by some loop-automorphism f of L_n. Now these automorphisms coincide with the extensions of the linear automorphisms of $V(n,K)$, which completes the proof.

Let $a(n,m)$ be the number of orbits of $AT(n,m,GF(3))$.

Theorem of the number of classes: $a(n,m)$ **is also the maximum number of pairwise non isomorphic Hall systems of classe 2, whose rank and 3-order are n+1 and n+m respectively. If m = 1,2,3 the number of isomorphy classes of all the Hall systems of rank n+1 and 3-order n+m is exactly $a(n,m)$. But it is larger than 3 + $a(n,m)$ if m $\geq$ 4.**

Proof: If $n \geq 3$ and $3 \geq m \geq 1$ any Hall system of rank n+1 and of 3-order n+m with $3 \geq m \geq 1$, is of class 2 (see [5]), therefore there is then a one-to-one correspondence between: (1)-the isomorphy classes of *all* the Hall systems of rank n+1 and of 3-order n+m; and (2)-the orbits of symplectic trilinear mappings of $AT(n,m,GF(3))$.. But **if m $\geq$ 4 there is at least 3 Hall systems that are not of classe 2** (constructed by direct product from systems of order 3^8 described in: [1],[3] [5]).

IV - THE BEGINNING OF A CLASSIFICATION: Consider a symplectic trilinear mapping t of $AT(n,m,K)$. For any vector subspace P of V, its **"orthogonal"**: $P^\perp$ is here the set of the x's from V verifying $t(x,u,v) = 0$ for any couple u,v of elements of P. If $P^\perp \cap P$ is not reduced to O, let us say that P is **singular.** If $P^\perp$ contains P , let us say that P is **totally isotropic.**

Theorem 1: For any commutative field K, AT(5,1,K) consists of exactly two classes. Moreover the symplectic trilinear mappings t of AT(n,1,K) such that V admits a totally isotropic hyperplane make up exactly k classes , where k is the largest integer such that $k \leq (n-1)/2$. If n is odd then there is only one of these classes for which V is non singular; otherwise there is none.

<u>Proof:</u> When t is a non vanishing form, there exists a non singular 3-dimensional vector subspace P; V is then direct sum of $P^\perp$ and P; this settles our classification for m=1 and $n \leq 5$. The two classes of AT(5,1,K) have for representatives t and t' defined by the vanishing of the 40 scalars $t(e_i, e_j, e_k)$ and $t'(e_i, e_j, e_k)$ for $1 \leq i < j < k \leq 5$, except the three following scalars: $t(e_1, e_2, e_3) = 1 = t(e_3, e_4, e_5) = t'(e_1, e_2, e_3)$. If t is a symplectic trilinear form such that V admits a totally isotropic hyperplane M, and if z is an element from the complement of M in V, then the orbit of t depends only on the symplectic *bilinear* form g:(x,y------> t(x,y,z)) of M. Now this form, defined up to multiplication by a non-vanishing coefficient, does not depend on the choice of z. And there are k possible values for the dimension of the "radical" of g.

Corollary: The Hall systems whose rank and 3-order are both n+1, and that admit an affine hyperplane make up exactly k isomorphy classes, where k is the integer part of (n-1)/2. If n is odd then there is only one of these classes whose Hall system is undecomposable in direct product; otherwise there is none.

Theorem 2: One has: $a(3,1) = 1 = a(4,m)$ for m = 1,2,3,4. and $a(5,1) = 2$. Moreover $a(3,m) = 0$ for $m \geq 2$, and $a(4,m) = 0$ for $m \geq 5$. Moreover, for these values of the couple (n, m), a(n,m) can also be viewed as the number of orbits of AT(n,m,K) for an arbitrary commutative field K.

Proof: If the e_i's, for i = 1,2,3,4, make up a basis of the K-vector space V, for any symplectic trilinear mapping t of departure V^3, the rank m of the image of t is at most 4, and *one may choose the e_i's so that there are exactly m non vanishing vectors among the four images* $t(e_i, e_j, e_k)$ for $1 \leq i < j < k \leq 4$, so that AT(4,m,K) comprises just one orbit.

Theorem 3: For any commutative field K of characteristic different from 2, the elements t of AT(6,1,K) for which one of the hyperplanes is singular make up exactly 4 classes. If K is quadratically closed, there are no other classes. If K = GF(3), there is exactly one additional class, so that: $a(6,1) = 5$.

This can be derived from a previous work [7] that describes completely by duality the classes of AT(6,1,K)

Theorem 4 (REVOY,1977): For any commutative field K the non vanishing vectors of $\wedge^3$ V(6,K) make up q classes exactly for the action of the linear group, with q= $(3 + [K^* : K^{*2}])$ <u>if the characteristic of K is > 2</u>; otherwise:

$q = (2 + [K^* : K^{*2}] +[K : p(K)])$, where p is the additive homomorhism defined by x -----> p(x) $= x^2 + x$.

Theorem 5: One has: $a(5,2) = 6$ and $a(5,3) \geq 17$ (see [6]).

The methods that were used in the proof of the foregoing evaluations are presented in the next section and in [6]. The transfer theorem allows us to derive *evaluations of the cardinal numbers of some isomorphy classes of Hall systems,* that are illustrated by the following table, where the *"dimension"* is by convention: (rank(E) - 1).

3-order=	4	5	6	7	8
dimension 3:	1=a(3,1)	O	O	O	O
4:	1 Affine	1=a(4,1)	1=a(4,2)	1=a(4,3)	1 (+ 3)
5:	0	1 Affine	2= a(5,1)	6=a(5,2)	a(5,3)≥17
6:	0	0	1 Affine	5=a(6,1)	???
7:	0	0	0	1 Affine	???
8:	0	0	0	0	1 Affine
9:	0	0	0	0	0

The "diagonal of affine spaces" corresponds to the case where the
3-order coïncides with the dimension . Under this diagonal, there are no
possible spaces : the 3-order is never strictly larger than the dimension
(sed [1]). Above , from the theorem of cardinal number of classes, the
number STH(n,m) of isomorphy classes of Hall systems of dimension n and of
3-order n+m equals a(n,m) only if m < 4. On the contrary a(4,4) = 1
holds, while the total number of isomorphy classes of Hall systems of rank 5
and of order 3^8 is 4 (see [2][5]).

IV -AN ALGORITHMIC APPROACH OF THE CLASSIFICATIONE IN AT(5,2,GF(3)):
Pick up two basis { e_1, e_2, e_3, e_4, e_5 } and { u,v } of V = V(5,GF(3)) and
W = V(2,GF(3)). Consider S = AT(5,2,GF(3)) as the set of symplectic
trilinears mappings t of V^3 onto W. Set: $t(e_i,e_j,e_k)$ = e_{ijk}. Each element t of
S may be represented by the vector from W^{10} whose components are the
e_{ijk}'s for i < j < k, or equivalently by the 20-vector of $(GF(3))^{20}$ obtained
by replacing each vector e_{ijk} by its two components with respect to the
basis { u,v } of W. Classifying these $(3^{10} - 1)$ vectors is a special instance
of the more general problem of determining the orbits in a finite set S
under the action of a finite permutation group G . We could not find any
reference to an algorithmic treatement of this kind of situation. We
employed first an algorithm changing the basis at random and seeking a
relation between elements; here is its framework:

ALGORITHM 'SEARCHING FOR A RELATION'

    data: t   'element from S '

             R   'family of elements from S '

    result: possibly finds an element g in the group G

                such that g(t) belongs to R

WHILE   g is not found

           (1) one selects at random g in the group G;

           (2) one computes g(t);

      ⌐(3) IF g(t) is in R

               THEN   ' g is found '

      ᴸENd_IF

    END_WHILE

Obtaining the whole orbit of t in AT(5,2,GF(3)) can be dealt with by restricting oneself to changes of basis in V . One has to pick up an arbitrary element g of the group GL(V), and further on to compute its composition with t (namely the three-argument mapping: t(g(x),g(y),g(z)). One chooses g as some product of transvections, by making first a random choice of the product-length (upperbounded by a fixed maximum), then by choosing at random each transvection and finally by computing their product.

According to the current stage of the knowledge of the orbits, one may use this algorithm for diffent purposes by making suitable choices for the data:(i) at the beginning *one may look for a simpler representative t'* *for a given element t* (for instance R can be defined by a maximum weight w of vectors: the algorithm search then for t' having at most w non vanishing components $t'(e_i, e_j, e_k)$); (ii) as a second step one may *try to know whether* R *is a partial system of representatives* of orbits of elements from S; one suppresses temporarily an element t of R and one tests whether its orbit contains an element of the reduced family (R \ t) .

For S = AT(5,2,GF(3)) we have found, after various trials, a family R of six elements that seemed to be an eventually redundant system of representatives.

One has to keep in mind that the preceding algorithm shows *possibly* that an element t of S AT(5,2,GF(3)) has an orbit that meets a family $\underline{R}$ of elements of AT(5,2,GF(3)); but if one wants to show that its orbit **does not** intersect $\underline{R}$, the failure of this algorithm to provide an element in the intersection can only be viewed as a presumption: several facts can explain that eventually an existing relation between t and $\underline{R}$ cannot be reached. Among others possible causes of failure let us mention: (i) the time required for the computation that can be too long for testing whether an element belongs to $\underline{R}$; or :(ii) the maximum length of the product of transvections fixed *too low* (while the transformation of t in an element of $\underline{R}$ needs a longer product), or *too high* (then it takes too much time to compute the g's and the number of attempts is not large enough)... Consequently one has to make use of another algorithm that etablishes by computing some invariants that elements of $\underline{R}$ are *not* related. The underlying algebraic principle is classical: the classes the elements t of $\underline{R}$ are shown to be pairwise disjoint by computing, for each t of $\underline{R}$, a given invariant $\mu(t)$ whose values are distint. Our choice of $\mu(t)$ is specific to case S = AT(5,2,GF(3)): designate as $\mu(t)$ *the number of hyperplanes H of V whose image is of rank 1.*

Convention : Let us say that *t is defined by:* (e_{ijk}; e_{lmn}) (*or by* (e_{ijk} = $e_{i'j'k'}$; e_{lmn}) if one has the relations e_{ijk} = $e_{i'j'k'}$;), if $t(V,V,V)$ = <e_{ijk}, e_{lmn}> (of rank 2), and if the "others" e_{pqr} (those which are not mentioned in the spelling of the definition) are assumed to vanish.

<u>Theorem of the system of representatives:</u> Each element of AT(5,2,GF(3)) admits a unique representative in the family of the six elements t_i defined as follows: (t_1)– (e_{123}; e_{234}); (t_2)– (e_{123} = e_{145}; e_{234}); (t_3)– (e_{123} = e_{245}; e_{234}); (t_4)– (e_{123}; e_{145}); (t_5)– (e_{123} = e_{245}; e_{234} = e_{145}); (t_6)–(e_{123} = e_{245}; e_{234} = e_{125})

The corresponding values of $\mu(t_i)$ are: 36,16,12,24,4 and 0.

The explicit proof of this theorem can be found in [6] with the listings of the programs that were used. The proof can be checked independently of the algorithms that suggested the results (and that provided us with changings of basis that are not necessarily as simple as possible). For future investigations, the most difficult problem to be solved is the one of the choice of "good invariants" that, in each AT(n,m,K), would allow us to distinguish the orbits.

<u>Remark:</u> <u>In the one–to–one correspondence of the transfer theorem, (t_1) is sent to the Hall system described in the example of section II.</u> Besides, modifying this example by setting the definition of d to be: d = $(x_1y_2-x_2y_1)(y_3-x_3)u$ $-(x_1y_4-x_4y_1)(y_5-x_5)u$ $+(x_2y_3-x_3y_2)(y_4-x_4)v$,

yields the Hall system associated to (t_2). By similar adaptations , one gets explicit descriptions of the 6 Hall systems whose rank and 3–order are 7. since:

 1 $+a(6,1)+a(5,2)+a(4,3)$ = 13, one derives the:

Theorem: There are just 13 Hall systems of order 3^7 .

Taking into account [1] [3], all these systems are now explicitly known.

BIBLIOGRAPHY

[1] L.BENETEAU : Contribution à l'étude des boucles de Moufang et des espaces apparentés (Algèbre, Combinatoire, Géométrie), Thèse d'Etat, Université de Provence, Aix-Marseille,1981

[2].L.BENETEAU: Topics about Moufang loops and Hall triple systems,"Simon Stevin", Vol 54, n°2,(1980),107-127

[3] L.BENETEAU,Problèmes de majorations dans les quasigroupes distributifs et les groupes de Fischer,Actes Colloque "Algèbre Appliquée et Combinatoire, Univ. Scientifique et Médicale de Grenoble,(1978),pp. 22-34

[4] L.BENETEAU, J.LACAZE : Symplectic trilinear forms and related designs and quasigroups, Communications in Algebra, 16 (5), 1035-1051 (1988).

[5] L.BENETEAU, G.RAZAFIMANANTSOA : Boucles de Moufang k-nilpotentes minimales, C.R.Acad.Sci.Paris,tome 306,Série I (1988),p 743-746

[6] G.RAZAFIMANANTSOA : La k-nilpotence minimale dans les boucles de Moufang commutatives; classification partielle des applications trilinéaires alternées Thèse de troisième cycle,(Université de Toulouse III, 1988,n°3511).

[7] P.REVOY : Trivecteurs de rang 6, Bulletin S.M.F. mémoire 59,141-155, (1979)

[8] P.REVOY: Formes trilinéaires symplectiques de rang inférieur ou égal à 7, 110 ième Congrès national des Sociétés savantes, Montpellier, Sciences, fascicule III,(1985)189-194

[9] J.P. SOUBLIN : Etude algébrique de la notion de moyenne, Journ. Maths Pures et Appl. Série 9, 50, (1971), pp53-264

[10] E.B. VINBERG & A.G. ELASVILI: Classification of trivectors of a nine-dimensional space, Trudy Sem. Vekt. Tenz. Analizu, n° XVIII,(1978), pp. 197-223

[11] R.WESTWICK : Real trivectors of rank seven,
Linear and Multilinear algebra (1980) MR 82j: 15024

A Gröbner Basis and a Minimal Polynomial Set
of a Finite nD Array

Shojiro SAKATA

Department of Knowledge-Based Information Engineering

Toyohashi University of Technology

Tempaku, Toyohashi 441, JAPAN

Abstract: In this paper, the relationship between a Gröbner basis and a minimal polynomial set of a finite nD array is discussed. A minimal polynomial set of a finite nD array is determined by the nD Berlekamp-Massey algorithm. It is shown that a minimal polynomial set is not always a Gröbner basis even if the uniqueness condition is satisfied, and a stronger sufficient condition for a minimal polynomial set to be a Gröbner basis is presented. Furthermore, a simple test whether a given set of polynomials is a Gröbner basis is proposed based on the theory of nD linear recurring arrays. The observations will be important in applying the nD Berlekamp-Massey algorithm to decode some kinds of nD cyclic codes and algebraic geometry codes.

1. Introduction

It has been shown [1] that the nD Berlekamp-Massey algorithm can be applied efficiently to decoding some kinds of nD cyclic codes and algebraic geometry codes. More specifically, the algorithm finds a Gröbner basis related to the error locations from a given nD syndrome array as far as the errors are correctable. In general, it is an iterative procedure which produces a set of characteristic polynomials for a finite nD array [2, 3]. During the iterations, the set, which is called a minimal polynomial set of the array, is not necessarily a Gröbner basis of a certain ideal of the n-variate polynomial ring.

For a finite array, a minimal polynomial set is unique on a certain condition of the degrees of the polynomials in it compared to the size of the array [2, 3]. The situation is somewhat similar to that of the Berlekamp-Massey algorithm for a finite 1D array (sequence) [4], i.e., for a finite sequence, a shortest (minimal) characteristic polynomial is unique, provided that the length of the sequence is greater than or equal to twice the degree of the polynomial. It has been conjectured [3] that a minimal polynomial set is a Gröbner basis if the uniqueness condition is fulfilled. On the other hand, Theorem 2 of [5] gives a sufficient condition for a minimal polynomial set to be a Gröbner basis, which seems too weak.

Furthermore, the above considerations lead us to another conjecture on the relationship between the Gröbner basis theory [6] and the algebraic theory of linear recurring arrays [7, 8]. The conjecture is that it is possible to judge whether a given set of polynomials is a Gröbner basis by trying to generate a finite array of a certain size by using iteratively the linear recurring relations whose characteristic polynomials compose the set. It is motivated by the fact (Lemma 10 of [2]) that a minimal polynomial set F is a Gröbner basis iff it is possible to generate a perfect (infinite) array by using the linear recurring relations represented by F. If the given set is not a Gröbner basis, there will occur at a point a conflict between a pair of polynomials in the set, i.e., the value of the array determined by one polynomial does not coincide with that determined by the other polynomial. Thus, we hit on a simple test of Gröbner basis which is different from the famous algorithm by B. Buchberger [6]. If the test has less complexity of computation, it will probably help in a stage of the Gröbner basis algorithm.

In this paper, we discuss the two conjectures about the relationship between Gröbner bases and nD linear recurring arrays. In Section 2, as preliminaries, several concepts in the theory of the nD Berlekamp-Massey algorithm are reviewed and a counterexample to the conjecture in [3] is presented. In Section 3 a Gröbner basis condition and its relationship with the uniqueness condition are discussed, and in Section 4 a method of checking a Gröbner basis is presented. Some concluding remarks are given in Section 5.

2. Preliminaries

In this paper, we restrict ourselves to the 2D case ($n = 2$) and we use the same terminology as in [2] (unless otherwise explicitly stated). Thus, we consider a 2D array $u = (u_p)$ over a finite field K with support Γ which is a subset of the 2D integral lattice $\Sigma_0 = Z_+^2 = \{(i,j)|i,j$ are nonnegative integers$\}$, and a minimal polynomial set $F \in \tilde{F}(u^q)$, where $u^q = (u_p|p \in \Sigma_0^q)$ is a truncate of u whose support is $\Sigma_0^q = \{p \in \Sigma_0|p <_T q\}$ ($<_T$ is the total degree order defined over Σ_0), and $\tilde{F}(u^q)$ is the class of all minimal polynomial sets of u^q. The degree $Deg(f)$ of a (bivariate) polynomial $f \in K[z] := K[z_1, z_2]$

is defined w.r.t. the total order, and the set $S = \{Deg(f)|f \in F\}$ defines a *nondegenerate* subset of Σ_0: $\Sigma_S = \cup_{s \in S}\Sigma_s$, where $\Sigma_s = \{p \in \Sigma_0|s \leq p\}$ ($\leq$ is the partial order defined over Σ_0), and 'nondegenerate' implies that there is no distinct elements s,t s.t. $s \leq t$ in the set. The class $\tilde{F}(u^q)$ is restricted so that any polynomial f in $F \in \tilde{F}(u^q)$ with $Deg(f) = s$ has the leading coefficient $lc(f) = 1$ and the set Γ_f of exponents of nonzero terms of f s.t. $\Gamma_f \setminus \{s\} \subseteq \Delta := \Sigma_0 \setminus \Sigma_S$. Every polynomial f in F has a *minimal* (w.r.t. the partial order) degree $s = Deg(f)$ among the polynomials $f = \sum_{t \in \Gamma_f} f_t z^t$ which are *valid* for $u = u^q$, i.e., for which the linear recurring relation represented by f:

$$f[u]_p := \sum_{t \in \Gamma_f} f_t u_{t+p-s} = 0, \ p \in \Sigma_s^q,$$

holds, where z^t is an abbreviation of $z_1^{t_1} z_2^{t_2}$ for $t = (t_1, t_2)$ and $\Sigma_s^q = \{p \in \Sigma_0|s \leq p <_T q\}$. Associated to F and S, there is another pair of a set G of polynomials and a nondegenerate subset C of Σ_0 s.t. $C = \{c = r - t|$ for each $g \in G, t = Deg(g),$ and r is the first point at which g fails to be valid for $u\}$ and $\Gamma_C = \cup_{c \in C}\Gamma_c = \Sigma_0 \setminus \Sigma_S$, where $\Gamma_c = \{p \in \Sigma_0|p \leq c\}$.

The discrepancy between a Gröbner basis and a minimal polynomial set of a finite array u has an origin in the fact that the set $V(u)$ of all valid polynomials for u is not an ideal of the bivariate polynomial ring $K[z]$. Now, we recall that Theorem 4 of [2] which states that, for $F \in \tilde{F}(u^q)$, F is unique iff the condition holds:

$$\bigcup_{s \in S}(s + \Delta) \subseteq \Sigma_0^q,$$

where $s + \Delta = \{s + r|r \in \Delta\}$. It has been conjectured [3] that, if $F \in \tilde{F}(u^q)$ satisfies the above uniqueness condition, F is a Gröbner basis of a certain ideal of $K[z]$. However, we have a counterexample to the conjecture as follows.

Example 1. *(adapted from [2])* For the following 2D array u over $K = GF(2)$:

```
0  1  0  1  0
1  1  0  0
0  1  0
0  0
0  *
0
```

it can be shown that, for $q = (4, 1)$, $F = \{z_1^3, z_1 z_2 + z_1 + 1, z_2^2 + z_1^2\} \in \tilde{F}(u^q)$, $S = \{(3, 0), (1, 1), (0, 2)\}$, and $\Delta = \Gamma_{(2,0)} \cup \Gamma_{(0,1)}$, and that $\cup_{s \in S}(s + \Delta) = \Gamma_{(5,0)} \cup \Gamma_{(3,1)} \cup \Gamma_{(2,2)} \cup \Gamma_{(0,3)} \subset \Sigma_0^q$. Thus, F is unique, i.e., $\#\tilde{F}(u^q) = 1$. But, F is not a Gröbner basis, since it turns out that there occurs a conflict at $(3, 2)$ when we try to generate an infinite array by using F, or in more details, at q we have $u_q = 0$ either by $f^{(1)} = z_1^3$ or by $f^{(2)} = z_1 z_2 + z_1 + 1$, but at the next point $p = (3, 2)$ to q (w.r.t. the total order) we have $u_p = 0$ by $f^{(1)}$ and $u_p = 1$ by $f^{(2)}$. We note that the above array is different from the array of Fig. 2 of [2], which is characterized by a Gröbner basis $\{z_1^3 + z_2 + z_1 + 1, z_1 z_2 + z_1 + 1, z_2^2 + z_1^2\}$, in the value of u_p at $p = (5, 0)$.

3. Gröbner basis and uniqueness condition

In the previous section, we have observed that a minimal polynomial set $F \in \tilde{F}(u^q)$ of a finite array u^q is not necessarily a Gröbner basis even if the uniqueness condition is satisfied. Now, we consider a sufficient condition for a minimal polynomial set F of u^q to be a Gröbner basis of a certain ideal of $K[z]$, which is stronger than Theorem 2 of [4]. Our starting point is the following lemma which is directly derived from Lemma 4 of [2].

Lemma 1. For $F \in \tilde{F}(u^q)$, let $f^{(1)}, f^{(2)} \in F$ be distinct polynomials with $s^{(1)} = Deg(f^{(1)}) \neq s^{(2)} = Deg(f^{(2)})$, and $s^{(1)} + s^{(2)} \leq p$, where $p \geq_T q$. If $f^{(1)}$ is valid at p, then $f^{(2)}$ also is. In other words, if $f^{(1)}$ is not valid at p, then $f^{(2)}$ also is not.

Motivated by Lemma 1, we introduce the following definition: For $F \in \tilde{F}(u^q)$, two distinct $s^{(i)}, s^{(j)} \in S := \{Deg(f) | f \in F\}$ satisfying $s^{(i)}, s^{(j)} \leq p$ are called *directly correlated* at p and denoted as $s^{(i)} \leftarrow_p \to s^{(j)}$ iff $s^{(i)} + s^{(j)} \leq p$. Furthermore, $s^{(i)}, s^{(j)} \in S := \{Deg(f) | f \in F\}$ satisfying $s^{(i)}, s^{(j)} \leq p$ are called *correlated* at p and denoted as $s^{(i)} \leftarrow_p^+ \to s^{(j)}$ iff there exist distinct elements $s^{(i_1)}, \cdots, s^{(i_k)} \in S$ s.t. $s^{(i_1)}, \cdots, s^{(i_k)} \leq p$ and $s^{(i)} \leftarrow_p \to s^{(i_1)}$, $s^{(i_1)} \leftarrow_p \to s^{(i_2)}, \cdots, s^{(i_k)} \leftarrow_p \to s^{(j)}$. Then, we have

Theorem 1. If it holds at any point $p \geq_T q$ that, for any distinct $s^{(i)}, s^{(j)} \in S$ satisfying $s^{(i)}, s^{(j)} \leq p$, $s^{(i)} \leftarrow_p^+ \to s^{(j)}$, then $F \in \tilde{F}(u^q)$ is a Gröbner basis.

Proof: By Lemma 10 of [2], we have only to prove that we can generate a perfection v of $u = u^q$ which is an infinite extension of the finite array u and for which any $f \in F$ is valid, i.e., $v_r = u_r$ for $r <_T q$ and $f[v]_r = 0$ for $r \geq_T q$. By the mathematical induction w.r.t. the total order, it is only required to prove that we can find the value v_p at $p \geq q$ which satisfies $f[v]_p = 0$ for any $f \in F$. We choose any $f^{(i)} \in F$ with $s^{(i)} = Deg(f^{(i)}) \leq p$. Then, we define

$$v_p := - \sum_{t \in \Gamma_{f^{(i)}} \setminus \{s^{(i)}\}} f_t^{(i)} v_{t+p-s^{(i)}}, \tag{*}$$

where, by the assumption of mathematical induction, we know the values $v_{t+p-s^{(i)}}, t \in \Gamma_{f^{(i)}} \setminus \{s^{(i)}\}$, since $t + p - s^{(i)} <_T p$. (If there exists no $f^{(i)} \in F$ with $s^{(i)} \leq p$, then we can choose any value $v_p \in K$.) The above-defined value is consistent with that determined by any other $f^{(j)} \in F$ with $s^{(j)} \leq p$, since otherwise, by using Lemma 1 repeatedly, we have $f^{(i)}[v]_p \neq 0$, which contradicts the definition (*) of v_p. Q.E.D.

Combined with Example 1, the following example shows that Theorem 1 gives a tight bound of q s.t. $F \in \bar{F}(u^q)$ is a Gröbner basis.

Example 2. *(adapted from [2]) For the following array $u = u^q$ over $K = GF(2)$, $q = (2, 3)$,*

$$
\begin{array}{ccccc}
0 & 1 & 0 & 1 & 0 \\
1 & 1 & 0 & 0 & \\
0 & 1 & 0 & * & \\
0 & 0 & 1 & & \\
0 & 0 & & & \\
1 & & & & \\
\end{array}
$$

$F = \{z_1^3 + z_2 + z_1 + 1, z_1 z_2 + z_1 + 1, z_2^2 + z_1^2\} \in \bar{F}(u^q)$ *is a Gröbner basis. In fact, at any point $p \in \Sigma_{(1,3)}$, $s^{(2)} = (1,1) \leftarrow_p \rightarrow s^{(3)} = (0,2)$; at any point $p \in \Sigma_{(4,1)}$, $s^{(1)} = (3,0) \leftarrow_p \rightarrow s^{(2)}$; at any point $p \in \Sigma_{(3,2)}$, $s^{(1)} \leftarrow_p \rightarrow s^{(3)}$. Therefore, at any $p \geq_T (2,3)$, the assumption of Therem 1 is satisfied, and so we have the following perfection of u^q which is a doubly periodic array.*

$$
\begin{array}{ccccccc}
0 & 1 & 0 & 1 & 0 & 0 & 0 \\
1 & 1 & 0 & 0 & 1 & 1 & \cdot \\
0 & 1 & 0 & 0 & 0 & \cdot & \\
0 & 0 & 1 & 1 & \cdot & & \\
0 & 0 & 0 & \cdot & & & \\
1 & 1 & \cdot & & & & \\
0 & \cdot & & & & & \\
\cdot & & & & & & \\
\end{array}
$$

In connection with Theorem 1, we explore a subset Λ of Σ_0 which satisfies the following condition:

For any point $p \in \Lambda$, if any distinct $s, s' \in S$ satisfy $s, s' \leq p$,

then $s \overset{+}{\underset{p}{\leftarrow\!\!\rightarrow}} s'$.

We can prove that, if $p \in \Lambda$ and $p \leq r$, then $r \in \Lambda$, and so Λ is written as $\Lambda = \Sigma_T = \cup_{t \in T}\Sigma_t$, where T is the set of minimal points of Λ, which is nondegenerate. Let $S = \{s^{(1)}, \cdots, s^{(k)}\}$, where the elements of S are ordered as follows:

$$s_1^{(1)} > \cdots > s_1^{(k)}, \quad s_2^{(1)} < \cdots < s_2^{(k)}.$$

Since $2max(s^{(1)}, s^{(k)}) \geq 2max(s^{(i)}, s^{(j)}) > s^{(i)} + s^{(j)}$ for $s^{(i)} \neq s^{(j)}$, any point of T is contained in the subset $\Gamma_{(2s_1^{(1)}, 2s_2^{(k)})} \setminus \Delta$, where, for $s = (s_1, s_2), t = (t_1, t_2) \in \Sigma_0$, $max(s, t) := (max\{s_1, t_1\}, max\{s_2, t_2\})$. In general, $2s^{(i)} \notin \Lambda, i = 2, \cdots, k - 1$, and $\Sigma_0 \setminus \Lambda^+ \supseteq \Delta^+ := \Delta \cup (\cup_{s \in S}(s + \Delta))$, where $\Lambda^+ := \Lambda \cup \Sigma_{2s^{(1)}} \cup \Sigma_{2s^{(k)}}$. For example, in case of $k = 3$, we have $T = \{t^{(1)} = s^{(1)} + s^{(2)}, t^{(2)} = s^{(2)} + s^{(3)}\}$ as shown in Fig. 1. In Fig. 2, we have an example of T in case of $k = 5$, where $S = \{(8, 0), (6, 1), (5, 3), (2, 7), (0, 9)\}$ and $T = \{(14, 1), (13, 4), (11, 8), (8, 10), (7, 12), (2, 16)\}$. In Fig. 1, Λ^+ is almost equal to the complement of Δ^+, i.e., $\Lambda^+ \cap \Delta^+ = \emptyset$ and $\Sigma_0 \setminus (\Lambda^+ \cup \Delta^+) = \Gamma_{s^{(2)} + max(s^{(1)}, s^{(3)})} \setminus \Sigma_{2s^{(2)}}$ is a small subset.

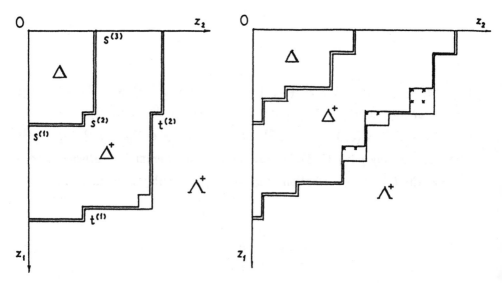

Fig. 1   Fig. 2

The situation is very similar also in Fig. 2, where, for $\Lambda^+ := \Lambda \cup \Sigma_{2s^{(1)}} \cup \Sigma_{2s^{(5)}}$

and $\Delta^+ = \Delta \cup (\cup_{1 \leq i \leq 5}(s^{(i)} + \Delta))$, $\Sigma_0 \setminus (\Lambda^+ \cup \Delta^+) = \{(5, 15), (5, 14), (6, 15),$
$(6, 14), (7, 11), (7, 10), (10, 9), (10, 8)\}$. Both examples suggest that the Gröbner
basis condition is approximate to the uniqueness condition, which might have
given an impetus to the previous conjecture. It seems difficult to give an explicit
general expression of Λ.

In terms of Λ^+, we have a corollary of Theorem 1.

Corollary. If $\Sigma_0^q \supseteq \Sigma_0 \setminus \Lambda^+$, $F \in \tilde{F}(u^q)$ is a Gröbner basis.

By the way, we recall that Theorem 5 of [3] gives a complete class of $\tilde{F}(u^q)$.
But, $F \in \tilde{F}(u^q)$ is not always a Gröbner basis. For the purpose of applying
the 2D Berlekamp-Massey algorithm, it is desirable to modify the algorithm
so as to ensure that, at every point $p \leq_T q$, $F \in \tilde{F}(u^p)$ is a Gröbner basis,
where the iterations are halted at the point q. As we have observed above,
it is possible that there exists no Gröbner basis in $\tilde{F}(u^q)$. Furthermore, the
following example shows that, when F is not unique, there can exist plural
Gröbner bases F in $\tilde{F}(u^q)$.

Example 3. *(Example 2 revisited) For the subarray $u^r, r = (1, 2)$, of the array
u of Example 2, all Gröbner bases F in $\tilde{F}(u^r)$ are as follows:*

a) $\{z_1^3 + z_1^2 + z_2 + z_1, z_1 z_2 + z_1 + 1, z_2^2 + z_1^2 + z_2 + z_1 + 1\}$,

b) $\{z_1^3 + z_2 + z_1 + 1, z_1 z_2 + z_1 + 1, z_2^2 + z_1^2\}$,

c) $\{z_1^3 + z_1^2, z_1 z_2 + z_2, z_2^2 + z_1^2\}$,

d) $\{z_1^3 + 1, z_1 z_2 + z_2, z_2^2 + z_1^2 + z_2 + z_1 + 1\}$,

e) $\{z_1^3 + z_2 + z_1, z_1 z_2 + z_1^2 + z_1 + 1, z_2^2 + z_2 + z_1\}$,

f) $\{z_1^3 + z_2 + z_1 + 1, z_1 z_2 + z_1^2 + z_1 + 1, z_2^2 + 1\}$,

g) $\{z_1^3 + z_1^2 + z_2 + z_1, z_1 z_2 + z_1^2 + z_2, z_2^2 + z_2 + z_1\}$,

h) $\{z_1^3 + z_1^2 + z_2 + z_1 + 1, z_1 z_2 + z_1^2 + z_2, z_2^2 + 1\}$,

which correspond to the perfections of u^r shown in Fig. 3a–h, respectively. It
can be shown that there exists no other Gröbner basis in $\tilde{F}(u^r)$. (The array of
Fig. 3b coincides with that of Exmaple 2.)

a)
```
0 1 0 1 1 0
1 1 0 0 1
0 1 0 0
0 0 1
1 1
0
```

b)
```
0 1 0 1 0 0
1 1 0 0 1
0 1 0 0
0 0 1
0 0
1
```

c)
```
0 1 0 1 0 1
1 1 0 1 0
0 1 0 1
0 1 0
0 1
0
```

d)
```
0 1 0 1 1 0
1 1 0 1 1
0 1 0 1
0 1 0
1 1
0
```

e)
```
0 1 0 1 0 0
1 1 1 0 1
0 1 1 0
0 1 0
1 0
1
```

f)
```
0 1 0 1 0 1
1 1 1 1 1
0 1 0 1
0 0 0
0 1
1
```

g)
```
0 1 0 1 0 0
1 1 1 0 1
0 1 1 1
0 0 1
1 0
0
```

h)
```
0 1 0 1 0 1
1 1 1 1 1
0 1 0 1
0 1 0
0 0
1
```

Fig. 3

According to Theorem 5 of [3], there are k_s alternative polynomials f of the same degree $Deg(f) = s$ in the union of all $F \in \tilde{F}(u^q)$, where $k_s = \#((s + \Delta) \setminus \Sigma_0^q)$, and so there exist $k = \prod_{s \in S} k_s$ alternative $F \in \tilde{F}(u^q)$, among which only a small number of minimal polynomial sets are Gröbner bases. Obviously, the number of these Gröbner bases is less than or equal to $2^{\#(\cup_{s \in S}(s+\Delta) \setminus \Sigma_0^q)}$. In Example 3, the value $2^{\#(\cup_{s \in S}(s+\Delta) \setminus \Sigma_0^r)} = 2^6$ is much greater than the true value 8.

4. Check of Gröbner basis

The considerations in the previous section lead us to a simple method of checking by trying to generate a perfect array whether a given set F of polynomials is a Gröbner basis. Let $S := \{Deg(f)|f \in F\}$ and $\Delta := \Sigma_0 \setminus \Sigma_S$, where we assume that S is nondegenerate. (Otherwise, F is not a Gröbner basis.) Given an appropriate choice of the initial values $u_r, r \in \Delta$, let us try to generate an infinite array $u = (u_p)$ by using the linear recurring relations represented by F iteratively for p in the total order. We know that, if it is proven to be impossible, then F is not a Gröbner basis.

The present problems are what initial values we must choose and how far we must proceed in the total order. The latter question has been answered by Theorem 1.

Example 4. Let $F = \{z_1^3, z_1 z_2 + z_1 + 1, z_2^2 + z_1^2\}$, and the initial values $u_r, r \in \Delta$, be as follows,

$$\begin{matrix} 0 & 1 \\ 0 \\ 1 \end{matrix}$$

where $\Delta = \{(0,0), (1,0), (0,1), (2,0)\}$. Then, we can calculate the values $u_r, r \in \Sigma_0^p \setminus \Delta$, consistently up to $p = (1,2)$ as follows by using F iteratively in the total order.

$$\begin{matrix} 0 & 1 & 1 \\ 0 & 0 & * \\ 1 & 1 \\ 0 \end{matrix}$$

But, at the point p, $f^{(2)} = z_1 z_2 + z_1 + 1$ and $f^{(3)} = z_2^2 + z_1^2$ yield $u_p = 1$ and $u_p = 0$, respectively. Therefore, F is not a Gröbner basis. We note that the condition of Theorem 1 has not been satisfied yet at the point $(1,2)$.

For the former question, the following example warns us that we must be careful in choosing the initial values $u_r, r \in \Delta$.

Example 5. Let $F = \{z_1^3 + z_2 + z_1 + 1, z_1^2 z_2 + z_1^2 + z_1, z_2^2 + z_1^2\}$, which is not a Gröbner basis, since the S-polynomial of $f^{(1)}$ and $f^{(2)}$ is reduced to a nonzero normal form $z_1 z_2 + z_1 + 1$. (Compare it with the Gröbner basis of Example 2.) If we start by the initial values, $u_r, r \in \Delta := \Gamma_{(2,0)} \cup \Gamma_{(1,1)}$, as follows,

$$\begin{matrix} 1 & 1 \\ 1 & 1 \\ 0 \end{matrix}$$

then we can generate the following finite array up to $p = (3,1)$, at which we have a conflict.

$$
\begin{array}{cccc}
1 & 1 & 0 & 1 \\
1 & 1 & 1 & \\
0 & 1 & & \\
1 & * & & \\
0 & & & \\
\end{array}
$$

However, from the following initial values,

$$
\begin{array}{cc}
0 & 1 \\
1 & 1 \\
0 & \\
\end{array}
$$

we can generate the perfect array shown in Example 2.

In general, we have two alternative cases when we choose the initial values arbitrarily except for the all-zero initial pattern and try to generate an infinite array by using F:

Case 1: we have a conflict at a point, which implies that F is not a Gröbner basis;

Case 2: we can generate a perfect array, which does not always imply that F is a Gröbner basis.

In Case 2, we can find a minimal polynomial set F' with $S' = \{Deg(f')|f' \in F'\}$ of the perfect array u' by the 2D Berlekamp-Massey algorithm. If $S' = S$, then F is a Gröbner basis. Otherwise, F is not a Gröbner basis (F' is a Gröbner basis of the ideal $I(u')$, where the set $V(v)$ of all valid polynomials for any perfect array v is an ideal and it is denoted as $I(v)$ [2]). Thus, even in Case 2, we can judge whether F is a Gröbner basis with an excess of computation. To dispense with the excess of computation, we need a careful choice of the initial values.

Given F, the following theorem gives a finite set of choices of the initial values at least one of which ensures us that, if F is not a Gröbner basis, we certainly have a conflict at a point in trying to generate a perfect array by using F.

Theorem 2. *Let F be a finite set of polynomials with $S = \{Deg(f)|f \in F\}$ nondegenerate, and $\Delta = \Sigma_0 \setminus \Sigma_S = \Gamma_C$. For each $c \in C$, let $u_r^{(c)} := 1$ ($r = c$), $u_r^{(c)} := 0$ ($r \in \Delta \setminus \{c\}$). Then, F is a Gröbner basis iff, for every $c \in C$, there exists a perfect array $v^{(c)}$ s.t. (1) $v_r^{(c)} = u_r^{(c)}$, $r \in \Delta$; (2) $F \subset I(v^{(c)})$.*

Proof: The necessity part is trivial by Lemma 10 of [2]. For the sufficiency part, we assume that there exists a perfect array $v^{(c)}$ for each $c \in C$ as above. Then, we have $F \subset I(v^{(c)})$, and so $F \subset \cap_{c \in C} I(v^{(c)}) = I(\{v^{(c)}|c \in C\})$, which is the characteristic ideal I_C of the set $\{v^{(c)}|c \in C\}$. (In the terminology of [8], $v^{(c)}$, $c \in C$, are the *representative arrays* of the ideal I_C.) Let F_C be a Gröbner basis of I_C with $S_C = \{Deg(f)|f \in F_C\}$, $\Delta_C = \Sigma_0 \setminus \Sigma_{S_C}$. Then, $\Delta_C \supseteq C$, since $Deg(f) \not\leq c$ for $f \in I(v^{(c)})$. (f with $Deg(f) \leq c$ fails to be valid for $v^{(c)}$ at c.) Therefore, $\Delta_C \supseteq \Delta$, which implies that each $f \in F$ is minimal in I_C, and so F is a Gröbner basis of I_C ($\Delta_C = \Delta$).   Q.E.D.

5. Concluding remarks

We have considered a condition for a minimal polynomial set $F \in \tilde{F}(u^q)$ of a finite array u^q to be a Gröbner basis of a certain ideal of the bivariate polynomial ring in connection with the uniqueness condition of F. Furthermore, we have proposed a method of checking whether a given set of polynomials is a Gröbner basis based on the theory of linear recurring arrays.

For the purpose of applying the nD Berlekamp-Massey algorithm to decoding some kinds of error-correcting codes, it is desirable to modify the algorithm so that $F \in \tilde{F}(u^p)$ is a Gröbner basis (if it exists) at every point $p \in \Sigma_0^q$. We can select a Gröbner basis F among $\tilde{F}(u^p)$ by checking whether F is a Gröbner basis. But, it will be a future problem to give an efficient method of updating a Gröbner basis at each iteration of the algorithm, i.e. finding a Gröbner basis $F^\oplus \in \tilde{F}(u^{p\oplus 1})$ from $F \in \tilde{F}(u^p)$, where $p \oplus 1$ is the next point to p w.r.t. the total order.

Acknowledgement: This work was supported in part by the Science Foundation of the Japanese Education Ministry under Grants # 02650262.

References

[1] S. Sakata, "Decoding binary 2D cyclic codes by the 2D Berlekamp-Massey algorithm, presented at the 1990 International IEEE Information Theory Symposium in San Diego and submitted for publication in *IEEE Transactions on Information Theory*.

[2] S. Sakata, "Finding a minimal set of linear recurring relations capable of generating a given finite two-dimensional array", *J. of Symbolic Computation*, vol. 5, pp. 321–337, 1988.

[3] S. Sakata, "Extension of the Berlekamp-Massey algorithm to n dimensions", *Information and Computation*, Vol. 84, No. 2, pp. 207–239, 1990.

[4] J. L. Massey, "Shift-register synthesis and BCH decoding", *IEEE Transactions on Information Theory*, vol.IT-15, pp.122–127, 1969.

[5] S. Sakata, "Synthesis of two-dimensional linear feedback shift registers", *(L. Huguet and A. Poli, ed.) Applied Algebra, Algebraic Algorithms and Error-Correcting Codes: Proceedings of AAECC-5, Menorca, 1987, Lecture Notes on Computer Science No. 356, Springer, Berlin*, pp. 394–407, 1989.

[6] B. Buchberger, "Gröbner bases: An algorithmic method in polynomial ideal theory", *(N. K. Bose, ed.) Mutlidimensional systems theory, Reidel, Dordrecht*, pp. 184–232, 1985.

[7] S. Sakata, "General theory of doubly periodic arrays over an arbitrary finite field and its applications", *IEEE Transactions on Information Theory*, vol. IT-24, pp. 719–730, 1978.

[8] B S. Sakata, "On determining the independent point set for doubly periodic arrays and encoding two-dimensional cyclic codes and their duals", *IEEE Transactions on Information Theory*, vol. 27, pp.556–565.

BOUNDS FOR DEGREES AND NUMBER OF ELEMENTS IN GRÖBNER BASES *

Lorenzo Robbiano

Dipartimento di Matematica dell'Universitá di Genova
Via L.B. Alberti 4   16132 Genova ITALY
E-mail: ROBBIANO @ IGECUNIV.BITNET

This paper is devoted to the study of upper bounds for the degrees and the number of elements in reduced Gröbner bases of ideals. The reader is assumed to be familiar with the basic concepts of Commutative Algebra (see for instance Kunz (1980) and Matsumura (1970)), as well as with the theory of Gröbner bases (see for instance Buchberger (1985) and Robbiano (1988)).

The first section is introductory. To make the paper more self-contained, some basic facts are recalled (see Proposition 1.3, Corollary 1.4); the most relevant result is Theorem 1.5, which describes a remarkable property of certain Gröbner bases of ideals. It generalizes to degrevlex-type term-orderings (see Definition 4) a result by Giusti (1988); his proof cannot be extended in this context, so some techniques are used from the theory of associated graded rings to ideals. A first easy consequence is Corollary 1.6, which generalizes a result by Lazard (1983).

In the second section we work out upper bounds for the number of elements in Gröbner bases depending on the Hilbert function. Main result is Theorem 2.4, from which many applications are derived in particular to complete intersection ideals. Also a conjecture about the sharpness of the bounds is stated at the end.

In the final and most relevant section we combine the preceding results with some theorems of Elias-Robbiano-Valla (1989) and we get upper bounds for the number of elements in Gröbner bases depending on the multiplicity and the initial degree (see Theorem 3.2). Our inequalities force many classes of ideals to be minimally generated by Gröbner bases, in particular ideals of minimal multiplicity (see Corollary 3.3) and some other determinantal ideals. Finally a degree bound is obtained for ideals with given multiplicity and initial degree (see Proposition 3.4), from which an algorithm is devised for computing the ideal of a finite set of points (with multiplicity) in the projective space.

All the work necessary to state the conjecture of section 2 and to give a computational support to the results was carried on with a systematic use of CoCoA, a COmputational COmmutative Algebra system developed at the University of Genova, Italy (see Giovini-Niesi (1990)).

This paper was written while I was visiting Queen's University (Kingston, Ontario Canada). It is a pleasure for me to express my gratitude to the Department of Mathematics, Queen's University, in particular to Prof. Geramita.

---

* The paper was written while the author was visiting Queen's University, during the academic year 1989/90. It was partly supported by the Natural Sciences & Engineering Research Council of Canada, Queen's University (Kingston, Canada) and Consiglio Nazionale delle Ricerche.

1. Upper bounds for degrees: preliminaries

Let $R := k[X_1, \ldots, X_n]$, the polynomial ring in n indeterminates over a field k, with $degree(x_i) = 1$, $i = 1, \ldots, n$ and let I be a homogeneous ideal of R. Then $A := R/I = \bigoplus_{i=0}^{\infty} A_i$ is called a standard k-algebra (see Stanley (1978)); it is a graded ring and the Hilbert function of A is the function from $\mathbb{N}$ to $\mathbb{N}$ which associates to every natural number r the dimension of A_r as a k-vectorspace. The Hilbert function H_A of A can be encoded in the power series $\mathbf{P}_A := \sum_{r=0}^{\infty} H_A(r)\lambda^r \in \mathbb{Z}[\lambda]$, the so called Hilbert-Poincare series *. It is well known that $\mathbf{P}_A$ is rational of type $Q_A(\lambda)/(1-\lambda)^d$, where Q_A is a polynomial with rational coefficients and $Q_A(1) \neq 0$; then $dim(A) := d$ is the dimension of A and $mult(A) := Q_A(1)$ is the multiplicity of A. It is also well known that there exists a suitable polynomial P_A with rational coefficients such that $H_A(r) = P_A(r)$ for $r >> 0$.

Definition 1. We recall that $reg(A)$ or $reg(I)$ is the regularity of R/I i.e. the minimum m such that $H_{R/I}(r) = P_{R/I}(r)$ for $r \geq m$.

Remark. If $dim(I) = 0$, then $reg(I) = deg(Q_A) + 1$. Namely $\mathbf{P}_A(\lambda) = Q_A(\lambda)$ and $H_A(r) = 0$ for $r \geq deg(Q_A) + 1$. In general it is easy to see that $reg(I) = deg(Q_A) + 1 - dim(R/I)$.

Definition 2. If I is a homogeneous ideal of R, it is well known that all its minimal sets of homogeneous generators have the same cardinality and the same degree-sequence. Therefore we may associate to I a sequence of positive integers $(d_1, \ldots, d_s)$ with $d_1 \geq \ldots \geq d_s$, where the d_i's are the degrees of every minimal set of generators. The number d_1 is called $Maxdeg(I)$, while the number d_s is called $indeg(I)$.

Definition 3. We recall that $grade(I)$ is the maximum length of a regular sequence contained in I and we say that I is perfect if $grade(I) = projdim(R/I)$. For our purposes it is enough to know that this is equivalent to R/I being Cohen-Macaulay. We also recall that in general, if $I = (f_1, \ldots, f_r)$, then $grade(I) \leq n - dim(R/I) \leq r$.

Proposition 1.1. Let $f_1, \ldots, f_r$ be homogeneous polynomials and let $d_i := deg(f_i)$, $i = 1, \ldots, r$ with $d_1 \geq \ldots \geq d_r$. Let $I := (f_1, \ldots, f_r)$ and let $g := grade(I)$. Then, if k is infinite, there is a regular sequence $a_1, \ldots, a_g$ in I of homogeneous elements of degrees $d_1, \ldots, d_g$ respectively, which is part of a minimal system of generators.

Proof. The well-known proof is done by induction and uses the following classical

Lemma 1.2. Let k be an infinite field and let A be a finitely generated graded standard k-algebra. Let I be an ideal of A generated by the homogeneous elements $f_1, \ldots, f_r$ of degrees $d_1, \ldots, d_r$ respectively, with $d_1 \geq \ldots \geq d_r$ and assume that $grade(I) > 0$.
Then the generic element of I_{d_1} is a non zero-divisor in A

Remark. The assumption k infinite is essential in Lemma 1.2, as the following example shows.

---

* For a description of an efficient algorithm for computing $\mathbf{P}_A$ see for instance Bigatti-Caboara-Robbiano (1990)

Example 1. Let $A := \mathbb{Z}_2[X,Y]/(X^2Y + XY^2) = \mathbb{Z}_2[x,y]$. Let $I := (x,y)$. Then $I_1 = \{x, y, x+y, 0\}$ and every element is a zero-divisor. The smallest (in degree) non zero-divisor is $x^2 + xy + y^2$. So I , whose grade is 1, cannot be generated by two homogeneous elements, such that one is a non zero-divisor. However $I = (y + x^2, x)$ and $y + x^2$ is a non zero-divisor, but not homogeneous .

A first consequence is the following result by Lazard(1981, 1983) and Giusti(1984)

Proposition 1.3. *Let k be infinite, $I = (f_1, \ldots, f_r)$ a homogeneous perfect ideal of $R := k[X_1, \ldots, X_n]$ such that $\{f_1, \ldots, f_r\}$ is a minimal system of homogeneous generators of I , $d_i := deg(f_i)$, $d_1 \geq \ldots \geq d_r$. Let $g := grade(I)$. Then*
a) $d_1 = Maxdeg(I) \leq reg(I) + dim(R/I) = reg(I) + n - g$
b) $reg(I) \leq \sum_{i=1}^{g} d_i - n + 1$
In particular if I is 0 -dimensional, then it is necessarily perfect with $g = n$, hence $Maxdeg(I) \leq reg(I) \leq \sum_{i=1}^{n} d_i - n + 1$.

Remark. If I is not perfect, there are worse upper bounds for $reg(I)$, for instance in Giusti (1988).

Definition 4. Given a term-ordering σ , and a polynomial f , we denote by $Lt_\sigma(f)$ the leading term of f w.r. to σ ; given an ideal I , we denote by $Lt_\sigma(I)$ the leading term ideal of I w.r. to σ .

Corollary 1.4. *(Lazard(1983)) With the same assumptions as in Proposition 1.3, let I be a 0 -dimensional ideal; let σ be a term-ordering and let G be any minimal Gröbner basis of I w.r.to σ . Then $Maxdeg(G) \leq \sum_{i=1}^{n} d_i - n + 1$*

Proof. Clearly $Maxdeg(G) = Maxdeg(Lt_\sigma(I)) \leq reg(Lt_\sigma(I))$ by Proposition 1.3. On the other hand I and $Lt_\sigma(I)$ have the same Hilbert function. Therefore $reg(Lt_\sigma(I)) = reg(I) \leq \sum_{i=1}^{n} d_i - n + 1$ again by Proposition 1.3 •

Definition 5. Let $R := k[\underline{X}, \underline{Y}]$, where $\underline{X} := X_1, \ldots, X_r$ and $\underline{Y} := Y_1, \ldots, Y_s$ are two independent sets of indeterminates and let σ be a degree-compatible term-ordering on R such that every term in $X_1, \ldots, X_r$ is bigger than every term containing some of the Y 's. For instance degrevlex with $X_1 > \ldots > X_r > Y_1 > \ldots > Y_s$ meets the requirements. We say that σ is of degrevlex type with respect to X and Y .

Theorem 1.5. *Let $R := k[\underline{X}, \underline{Y}]$, where $\underline{X} := X_1, \ldots, X_r$ and $\underline{Y} := Y_1, \ldots, Y_s$ are two independent sets of indeterminates and let σ be a term-ordering on R of degrevlex type with respect to $\underline{X}$ and $\underline{Y}$. Let I be a homogeneous ideal of R such that $\underline{Y}$ is a regular sequence modulo I . Let $G := \{g_1, \ldots, g_t\}$ be the reduced Gröbner basis of I w.r.to σ . Then*
a) $Lt_\sigma(g_i) \in k[\underline{X}]$
If we denote by $'$ the reduction modulo the ideal $(\underline{Y})$, by σ' the restriction of σ to $k[\underline{X}]$, and we identify $R/(\underline{Y})$ with $k[\underline{X}]$, then
b) $G' := \{g_1', \ldots, g_t'\}$ is the reduced Gröbner basis of I' . In particular G and G' have the same degree sequence, hence the same cardinality.

Proof. a) Let $g \in G$ and let us consider it as a polynomial in $\underline{Y}$ with coefficients in $k[\underline{X}]$. Then $g = g_{a_1} + g_{a_2} + \cdots + g_{a_m}$ with $a_1 < a_2 < \cdots < a_m$, $deg(g_{a_i}) = a_i$ and the degrees are taken with respect to the $\underline{Y}$'s indeterminates only. Assume for contradiction that $Lt_\sigma(g) \notin k[\underline{X}]$; by the very definition of σ there is no term in $Supp(g)$ which belongs to $k[\underline{X}]$ hence $a_1 > 0$. Moreover

(1) $g_{a_1} \in (\underline{Y})^{a_2} \, mod(I)$ hence $g_{a_1} \in (\underline{Y})^{a_1+1} \, mod(I)$

Let us denote by $B := R/I$ and by $\mathbf{y} := (I + (\underline{Y}))/I$. Then

(2) $gr_\mathbf{y}(B) := \bigoplus_n \mathbf{y}^n/\mathbf{y}^{n+1} \cong B/\mathbf{y}[Y_1, ..., Y_s] \cong R/(I + (\underline{Y}))[Y_1, ..., Y_s]$

where $\cong$ means " isomorphic to ". This fact can be seen for instance in Kunz (1980) p. 158 or Robbiano (1983). From (1) and (2) we infer that the coefficients of all the terms of g_{a_1} (meaning terms in $\underline{Y}$) belong to $I + (\underline{Y})$. Let c be such a non zero coefficient; it is a homogeneous polynomial in $k[\underline{X}]$, which can be expressed as $\sum_{h=1}^s r_h Y_h + f$ where $f \in I$. Since $(\underline{Y}) \cap k[\underline{X}] = (0)$ and $Lt_\sigma(\sum_{h=1}^s r_h Y_h) \in (\underline{Y})$, it follows that $Lt_\sigma(c) \neq Lt_\sigma(\sum_{h=1}^s r_h Y_h)$. Hence by the very nature of σ, $Lt_\sigma(c - \sum_{h=1}^s r_h Y_h) = Lt_\sigma(c)$. But $c - \sum_{h=1}^s r_h Y_h \in I$ hence c reduces to $\sum_{h=1}^s r_h Y_h$ modulo I. This contradicts the fact that G is reduced.

b) By a) and the definition of σ we get that G' is a Gröbner basis of I'. It is clearly minimal and it is also clearly reduced $\bullet$

Example 2. Let $I := (XU - Y^2, XV - Z^2)$ in $k[X, Y, Z, U, V]$. Then U, V is a regular sequence modulo I. If σ is the *deglex* ordering with $X > Y > Z > U > V$, then the reduced Gröbner basis of I w.r. to σ is $\{XU - Y^2, XV - Z^2, Y^2V - Z^2U\}$ and we see that Theorem 1.5 fails with respect to σ. On the other hand $\{Y^2 - XU, Z^2 - XV\}$ is the reduced Gröbner basis with respect to *degrevlex* and $\{Y^2, Z^2\}$ is indeed the reduced Gröbner basis of I' with respect to σ'.

Corollary 1.6. Let k be an infinite field, $I = (f_1, ..., f_r)$ a homogeneous perfect ideal of $R := k[X_1, ..., X_n]$, such that $\{f_1, ..., f_r\}$ is a minimal system of homogeneous generators of I, $d_i := deg(f_i)$, $d_1 \geq ... \geq d_r$. Let $g := grade(I)$ and suppose that $X_{g+1}, ..., X_n$ is a regular sequence modulo I (in particular this is true in generic coordinates). Let G be a minimal Gröbner basis of I w.r.to a term-ordering σ, which is of degrevlex type w.r.to the two sets of indeterminates $\{X_1, ..., X_g\}$, $\{X_{g+1}, ..., X_n\}$. Then $Maxdeg(G) \leq \sum_{i=1}^g d_i - g + 1$.

Proof. By Theorem 1.5 we get $Maxdeg(G) = Maxdeg(G')$ where G' is a minimal Gröbner basis of I' with respect to σ', so we conclude by Corollary 1.4. Of course in generic coordinates the requirement that $X_{g+1}, ..., X_n$ is a regular sequence modulo I is satisfied $\bullet$

2. Upper bounds for number of elements in a Gröbner basis

We recall some facts about binomial representations of integers (see for instance Robbiano (1990)). We make the convention that $\binom{m}{0} = 1$ for $m \geq 0$ and $\binom{m}{k} = 0$ for $0 \leq m \leq k$. Let n, i be positive integers. Then n can be uniquely written as

$$n = \binom{n(i)}{i} + \binom{n(i-1)}{i-1} + \cdots + \binom{n(j)}{j} \quad \text{where} \quad n(i) > n(i-1) > \cdots > n(j) \geq j \geq 1$$

This is called the binomial expansion of n in base i. In the following we denote by n_i such expansion. We also use the same symbol for representing its value.

Definition 1. If $n = \binom{n(i)}{i} + \binom{n(i-1)}{i-1} + \cdots + \binom{n(j)}{j}$ is the binomial expansion of n in base i, then we define

$$(n_i)_+ := \binom{n(i)}{i+1} + \binom{n(i-1)}{i} + \cdots + \binom{n(j)}{j+1}$$

$$(n_i)^+_+ := \binom{n(i)+1}{i+1} + \binom{n(i-1)+1}{i} + \cdots + \binom{n(j)+1}{j+1}$$

In the following, if $H : \mathbb{N} \longrightarrow \mathbb{N}$ is a function (typically a Hilbert function), to avoid overburdening the notations we are going to use expressions like $H(d)^+_+$ and $H(d)_+$ instead of $(H(d)_d)^+_+$ and $(H(d)_d)_+$ respectively.

We now recall some classical facts. Let $H : \mathbb{N} \longrightarrow \mathbb{N}$ be a function, $h := H(1)$. Let $R := k[X_1, \ldots, X_n]$ with $deg(x_i) = 1$. Then $R = \bigoplus R_d$ is a graded ring. Let $W_d \subset R_d$ be the subvectorspace of R_d generated by the first $H(d)$ terms in the lexicographic ordering with $x_1 > \cdots > x_n$. Then

Theorem 2.1. *(Macaulay, Stanley) The following conditions are equivalent*
1) $\bigoplus_{d \geq 0} W_d$ is an ideal of R
2) $H(0) = 1$, $H(d+1) \leq H(d)^+_+$ for every $d > 0$
3) $H = H_{R/I}$ for some homogeneous ideal I, i.e. H is the Hilbert function of R/I

Proof. See Stanley (1978) and Robbiano (1990) •

Definition 2. The ideal $\bigoplus_{d \geq 0} W_d$ is called the lex-segment ideal associated to H. If I is a homogeneous ideal, we call $S(I)$ the lex-segment ideal associated to the Hilbert function $H_{R/I}$, and by $D := Maxdeg(S(I))$.

Definition 3. If I is a homogeneous ideal, we call $\nu(I)$ the number of elements in every minimal system of generators of I.

Proposition 2.2. *With the above notations, the following holds true*

$$\nu(I) \leq \nu(S(I)) = \sum_{r=1}^{D-1} (H(r)^+_+ - H(r+1)) = \sum_{r=1}^{\infty} (H(r)^+_+ - H(r+1))$$

where H denotes the common Hilbert function of R/I and $R/S(I)$.

Proof. $\nu(I) = \sum_{r=0}^{\infty} dim(I_{r+1}/R_1 I_r) =$
$= \sum_{r=0}^{\infty} (dim(R_{r+1}/R_1 I_r) - dim(R_{r+1}/I_{r+1})) \leq$
$\leq \sum_{r=0}^{\infty} (dim(R_{r+1}/R_1 S(I)_r) - dim(R_{r+1}/I_{r+1})) =$
$= \sum_{r=0}^{\infty} (dim(R_{r+1}/R_1 S(I)_r) - dim(R_{r+1}/S(I)_{r+1})) =^{(1)}$
$= \sum_{r=0}^{\infty} dim(S(I)_{r+1}/R_1 S(I)_r) = \nu(S(I))$.
The inequality follows from Macaulay (1927). From an elementary (but complicated) computation (see for instance Robbiano (1990)) it also follows that $dim(R_{r+1}/R_1 S(I)_r) = dim(R_r/S(I)_r)^+_+$ and of course if $r > D$, then $S(I)_{r+1} = R_1 S(I)_r$, hence $H(r)^+_+ = H(r+1)$.
The conclusion follows from equality (1) •

In the following we use the notation $\#(\cdots)$ to mean "cardinality of $(\cdots)$".

Proposition 2.3. *Let I be a homogeneous ideal of $R := k[X_1, \ldots X_n]$; let σ be a term-ordering and G a minimal Gröbner basis of I w.r.to σ. Then*

a) $\#(G) \leq \sum_{r=1}^{D-1} \left(H(r)_{\ddagger}^+ - H(r+1) \right) = \sum_{r=1}^{\infty} \left(H(r)_{\ddagger}^+ - H(r+1) \right)$

If moreover I is 0-dimensional, then

b) $\#(G) \leq H(1) + \sum_{r=1}^{R-1} H(r)_+ = H(1) + \sum_{r=1}^{\infty} H(r)_+$ *   where $R := reg(I)$*

Proof. a) Of course $\#(G) = \nu(Lt_\sigma(I))$. But we know that $H_{R/Lt_\sigma(I)} = H_{R/I}$, hence the lex-segment ideal associated to $Lt_\sigma(I)$ is $S(I)$. The conclusion follows from Proposition 2.2.

b) The equality is clear since $H(r) = 0$ hence $H(r)_+ = 0$ for $r \geq reg(I)$. Let N be bigger than D and $reg(I)$. Then from a) we deduce that

$\#(G) \leq H(1)_{\ddagger}^+ - H(2) + H(2)_{\ddagger}^+ - H(3) + \cdots + H(N)_{\ddagger}^+ - H(N+1) =$

$= H(1) + \sum_{r=1}^{N} \left(H(r)_{\ddagger}^+ - H(r) \right) - H(N+1)$

But $H(N+1) = 0$ and $H(r)_{\ddagger}^+ - H(r) = H(r)_+$ by elementary properties of the Pascal triangle •

Definition 4. If H is a numerical function, we denote by $\Delta(H)$ the function defined by $\Delta(H)(n) := H(n) - H(n-1)$ and
by $\Delta^r(H)$ the function defined recursively by $\Delta^r(H) := \Delta(\Delta^{r-1}(H))$

We recall that if H is the Hilbert function of a homogeneous ideal I in R, $\underline{y} := y_1, \ldots y_s$ is a regular sequence modulo I of elements of degree 1, $R' := R/(\underline{y})$ and $I' := (I + (\underline{y}))/(\underline{y})$, then $H_{R'/I'} = \Delta^s H_{R/I}$

Theorem 2.4. *Let k be an infinite field, I a homogeneous perfect ideal of $R := k[X_1, \ldots X_n]$ of grade g and suppose that $X_{g+1}, \ldots, X_n$ is a regular sequence modulo I (in particular this is true in generic coordinates). Let G be a minimal Gröbner basis of I w.r.to a term-ordering σ, which is of degrevlex type w.r.to the two sets of indeterminates $\{X_1, \ldots, X_g\} \{X_{g+1}, \ldots, X_n\}$. Then*

$\#(G) \leq g + \sum_{r=1}^{M} (\Delta^{n-g} H(r))_+ = g + \sum_{r=1}^{\infty} (\Delta^{n-g} H(r))_+$
where $M := reg(I) + n - g - 1$

Proof. Let $\underline{Y} := X_{g+1}, \ldots, X_n$; we denote by " $'$ " the reduction modulo the ideal $(\underline{Y})$, by σ' the restriction of σ to $k[\underline{X}]$ and we identify $R/(\underline{Y})$ with $k[\underline{X}]$. Then by Theorem 1.5 we know that $\#(G) = \#(G')$, so we can apply Proposition 2.3 to I' in $k[X_1, \ldots, X_g]$ and we get the conclusion, since $H_{R'/I'} = \Delta^{n-g} H_{R/I}$ and $reg(I') = reg(I) + n - g$ •

Of course Proposition 2.3 and Theorem 2.4 can be fully used when we know *a priori* the Hilbert function of I. This is for instance the case of complete intersections.

Definition 5. An ideal I is called a complete intersection (c.i) ideal if it is generated by a regular sequence. If I is a c.i. homogeneous ideal generated by g elements of degrees $d_1, \ldots, d_g$, then we say that I is of type $(d_1, \ldots, d_g)$.

Remark. If I is a homogeneous c.i. ideal of type $(d_1, \ldots, d_g)$ of $R := k[X_1, \ldots, X_n]$, then the upper bounds of Proposition 2.3 and Theorem 2.4 only depend on $(d_1, \ldots, d_g)$. Namely they are expressed in term of the Hilbert function, which is well-known to depend only on $(d_1, \ldots, d_g)$

To illustrate how better are the bounds of Theorem 2.4 with respect to the bounds of Proposition 2.3, let us consider the following

Corollary 2.5. *Let* f, g *be a regular sequence of homogeneous polynomials in* $k[X_1, \ldots, X_n]$ *of degree* 2, d *respectively and* $I := (f, g)$. *Then*
a) *If* σ *is a term-ordering,* $n = 3$ *and* G *a minimal G- basis of* I *w.r.to* σ, *then*
$\#(G) \leq d + 2$
b) *If* $X_3, \ldots, X_n$ *is a regular sequence modulo* I *(this happens for instance in generic co-ordinates),* σ *is a degrevlex type term-ordering with respect to* $\{X_1, X_2\} \{X_3, \ldots, X_n\}$ *and* G *is a minimal G- basis of* I *w.r.to* σ, *then* $\#(G) \leq 3$

Proof. a) The Hilbert-Poincaré series of I is
$\frac{(1-t^2)(1-t^d)}{(1-t)^3} = \frac{(1+t)(1+t+\cdots+t^{d-1})}{(1-t)} = 1+3t+5t^2+\cdots+(2d-1)t^{d-1}+2dt^d+\cdots+2dt^n+\cdots$
An easy computation using Proposition 2.3 yields a).
b) $\Delta^{n-2}(H) = 1, 2, 2, \ldots, 2, 1, 0$ and the conclusion follows immediately from Theorem 2.4 ●

Example 3. Let f, g, h be a regular sequence of homogeneous polynomials of degree d in $k[X, Y, Z]$. Let σ be any term-ordering, G a minimal Gröbner basis of $I := (f, g, h)$ w.r.to σ.

Then $\#(G) \leq \begin{cases} 13 & if\ d = 3 \\ 20 & if\ d = 4 \\ 32 & if\ d = 5 \end{cases}$

Moreover if f, g, h are generic and σ is the deglex ordering, the bounds are achieved. We also observe that these numbers are exactly the bounds of Proposition 2.3 b).
This kind of examples and a lot of computational evidence lead to the following

Conjecture. The bounds of Proposition 2.3 b) are sharp for c.i. ideals and they can be achieved when $\sigma = deglex$ and the generators of I are generic.

3. Upper bounds for number of elements and degrees: applications

We start this section by recalling some results from Elias-Robbiano-Valla (1989)

Definition 1. Given two positive integers e, g, with $e \geq g+1$, we define $t = t(e, g)$ to be the unique integer such that $\binom{g+t-1}{t-1} \leq e < \binom{g+t}{t}$ and $r = r(e, g) := e - \binom{g+t-1}{t-1}$. Given three positive integers e, g, i, with $e \geq g+1$ and $2 \leq i \leq t(e, g)$, we define $s = s(e, g, i)$ to be the unique integer such that $\binom{g+s-1}{s-1} - \binom{g+s-i-1}{s-i-1} \leq e < \binom{g+s}{s} - \binom{g+s-i}{s-i}$ and $r = r(e, g, i) := e - \binom{g+s-1}{s-1} + \binom{g+s-i-1}{s-i-1}$

Definition 2. Given two positive integers e, g, with $e \geq g+1$, we define CM(e,g) to be the family of homogeneous perfect ideals I of grade g in $R := k[X_1, \ldots, X_n]$ such that $I \subseteq (X_1, \ldots, X_n)^2$ and $mult(I) = e$. We put $B(e, g) := \binom{g+t-1}{t} - r + (r_t)^+_+$
Given three positive integers e, g, i, with $e \geq g+1$ and $2 \leq i \leq t(e, g)$, we define CM(e,g,i) to be the family of homogeneous perfect ideals I of grade g in R, such that $I \subseteq (X_1, \ldots, X_n)^2$, $mult(I) = e$ and $indeg(I) = i$. We put $B(e, g, i) := 1 + \binom{g+s-1}{s} - \binom{g+s-i-1}{s-i-1} - r + (r_s)^+_+$

Theorem 3.1. *Let k be a field and $R := k[X_1,\ldots,X_n]$. Then*
a) $\nu(I) \leq B(e,g)$ *for every ideal I in* **CM**(e,g)
b) $\nu(I) \leq B(e,g,i)$ *for every ideal I in* **CM**(e,g,i).

Proof. See Elias-Robbiano-Valla (1989) •

Definition 3. We say that a homogeneous perfect ideal $I \subseteq (X_1,\ldots,X_n)^2$ of grade g is (e,g)-extremal if $I \in$ **CM**(e,g) and $\nu(I) = B(e,g)$.
We say that it is (e,g,i)-extremal if $I \in$ **CM**(e,g,i) and $\nu(I) = B(e,g,i)$.

Theorem 3.2. *Let k be an infinite field, I a homogeneous perfect ideal of $R :=$ $k[X_1,\ldots,X_n]$ of grade g and suppose that I is (e,g)-extremal (or (e,g,i)-extremal). Moreover let us assume that one of the following conditions holds*
i) *I is 0-dimensional and G is a minimal G-basis of I w.r.to a term-ordering σ*
ii) *$X_{g+1},\ldots,X_n$ is a regular sequence modulo I and G is a minimal Gröbner basis of I w.r.to a term-ordering σ, which is of degrevlex type w.r.to the two groups of indeterminates $\{X_1,\ldots,X_g\}$ $\{X_{g+1},\ldots,X_n\}$. Then*
a) $\#(G) = B(e,g)$ *(or $\#(G) = B(e,g,i)$)*
b) *After interreduction, every minimal basis of I becomes a G-basis of I w.r.to σ.*

Proof. By Theorem 1.5 we may assume I to be 0-dimensional. Then also $Lt_\sigma(I)$ is 0-dimensional and it has the same multiplicity and initial degree. Therefore $B(e,g) = \nu(I) \leq \nu(Lt_\sigma(I)) \leq B(e,g)$ (and a similar statement with $B(e,g,i)$ replacing $B(e,g)$). Since $\nu(Lt_\sigma(I)) = \#(G)$, the conclusion follows immediately •

To better illustrate Theorem 3.2 let us consider the following

Example 4. Let A be the matrix $\begin{pmatrix} T & X & Z^2 \\ X & Z & Y^2 \end{pmatrix}$ and let I be the ideal generated by the 2×2 minors of A.

I is a perfect ideal of grade 2 and $mult(I) = 5$. Now $B(5,2) = 3$, so our ideal meets the requirements of Theorem 3.2. Let σ be degrevlex with $X > Y > Z > T$. The reduced G-basis of I w.r.to σ is $\{X^2 - ZT, XZ^2 - Y^2T, XY^2 - Z^3, Z^5 - Y^4T\}$. We do not have the conclusion of Theorem 3.2, simply because Z, T is not a regular sequence modulo I. But if we let σ be degrevlex with $X > Z > Y > T$ or if we perform a generic change of coordinates, than any minimal Gröbner basis has exactly 3 elements.

Definition 4. A perfect homogeneous ideal I of grade g in $R := k[X_1,\ldots,X_n]$ is termed *ideal of minimal multiplicity* (or *of minimal degree*), if $I \subseteq (X_1,\ldots,X_n)^2$ and $mult(I) = g+1$. The projective varieties whose defining ideals are of minimal multiplicity are called *varieties of minimal degree* and they are very much studied in classical and modern geometry.

Corollary 3.3. *Let k be an infinite field, I a homogeneous perfect ideal of $R := k[X_1,\ldots,X_n]$ of grade g and minimal multiplicity. Then*
a) *I is minimally generated by $\binom{g+1}{2}$ elements of degree 2.*
b) *If $X_{g+1},\ldots,X_n$ is a regular sequence modulo I (this happens for instance in generic coordinates) and G is a minimal Gröbner basis of I w.r.to a term-ordering σ, which is of*

degrevlex type w.r.to the two groups of indeterminates $\{X_1,\ldots,X_g\}$, $\{X_{g+1},\ldots,X_n\}$, *then every minimal Gröbner basis of I w.r.to* σ *has exactly* $\binom{g+1}{2}$ *elements.*

Proof. a) This is a classical result; an easy proof is the following.
Let $\underline{X} := X_1,\ldots,X_g$, $\underline{Y} := X_{g+1},\ldots,X_n$, be such that $\underline{Y}$ is a regular sequence modulo I of linear forms. Then $\nu(I) = \nu(I')$, where $I' := (I+(\underline{Y}))/(\underline{Y})$ and as usual we identify I' with its isomorphic image in $k[X_1,\ldots,X_g]$. Now $mult(I) = mult(I') = g+1$. But I' is 0-dimensional and $I' \subseteq (X_1,\ldots,X_g)^2$, hence $I' = (X_1,\ldots,X_g)^2$, which implies that $\nu(I) = \binom{g+1}{2}$.
b) By a) and Theorem 3.2 it is sufficient to show that I is $(g+1,g)$-extremal. But $B(g+1,g)$ is clearly $\binom{g+1}{2}$ and we are done •

Remark. It is well-known that perfect ideals of grade 2 in the polynomial ring are generated by the maximal minors of $r \times r+1$ matrices. If the entries are generic linear forms, then it follows again from Theorem 3.2 that the minors become a Gröbner basis after interreduction.

Definition 5. Let H, H' be two numerical functions, $\mathbf{P}_H$, $\mathbf{P}_{H'}$ the corresponding *Poincaré* series. We say that $H >_{lex} H'$ (or $\mathbf{P}_H >_{lex} \mathbf{P}_{H'}$) if there exists $r \in \mathbb{N}$ such that $H(n) = H'(n)$ for $n < r$ and $H(r) > H'(r)$.

Definition 6. Given three positive integers e, n, i with $e \geq g+1$ and $2 \leq i$, we define $\mathbf{H}(e,n,i)$ to be the class of Hilbert functions of 0-dimensional homogeneous ideals I in $R := k[X_1,\ldots,X_n]$, such that $I \subseteq (X_1,\ldots,X_g)^2$, $indeg(I) = i$, $mult(I) = e$.

Proposition 3.4. *If* $e \geq \binom{n+i-1}{i-1}$, *then* $\mathbf{H}(e,n,i)$ *is non-empty and it has a minimum function w.r.to* $>_{lex}$. *The regularity of such function is* $e - \binom{n+i-1}{i-1} + i$; *it is the maximum among the regularities of elements of* $\mathbf{H}(e,n,i)$.

Proof. If $e \geq \binom{n+i-1}{i-1}$ and if I is a homogeneous 0-dimensional ideal of $k[X_1,\ldots,X_n]$ such that $indeg(I) = i$, then $H_{R/I}(s) = \binom{n+s-1}{s}$ for $s = 1,\ldots,i-1$. This gives already a contribution of $\binom{n+i-1}{i-1}$ to the multiplicity. If $H_{R/I}(s) = 0$ for some $s \geq i$, then $H_{R/I}(t) = 0$ for every $t \geq s$. Therefore the only way to minimize the values of $H_{R/I}(s)$ from i on, is to put them equal to 1, until e is reached. Now Theorem 2.1 guarantees that an ideal with such features exists and of course in such way we maximize the regularity. Moreover the last 1 in the sequence is in degree $e - \binom{n+i-1}{i-1} + i - 1$, hence the conclusion follows •

Example 5. The ideal described in the proof of Proposition 3.4 is the ideal generated by all the terms of degree i, with X_n^i replaced by X_n^d, where $d := e - \binom{n+i-1}{i-1} + i$. For instance if $e = 7$, $n = 3$, $i = 2$, then $I = (X^2, XY, XZ, Y^2, YZ, Z^5)$.

Corollary 3.5. *Let* I *be a homogeneous perfect ideal of grade* g *in* $k[X_1,\ldots,X_n]$. *Let* $e := mult(I)$, $i := indeg(I)$ *and assume that* $2 \leq i$. *Then*

$$Maxdeg(I) \leq e - \binom{g+i-1}{i-1} + i$$

Proof. By a standard argument one may assume that I is a 0-dimensional ideal of $k[X_1,\ldots,X_g]$. Then we conclude by Proposition 1.3 a) and Proposition 3.4 •

Lemma 3.6. Let $I_1 := (f_{11}, \ldots, f_{1a_1}), \ldots, I_r := (f_{r1}, \ldots, f_{ra_r})$ be ideals of $R :=$ $k[X_1, \ldots, X_n]$, given by generators. Let M be the submodule of R^{r-1} generated by
$\{(f_{11}, f_{11}, \ldots, f_{11}), (f_{12}, f_{12}, \ldots, f_{12}), \ldots, (f_{1a_1}, f_{1a_1}, \ldots, f_{1a_1}),$
$(f_{21}, 0, \ldots, 0), (f_{22}, 0, \ldots, 0), \ldots \ldots, (0, \ldots, 0, f_{r1}), (0, \ldots, 0, f_{r2}), \ldots, (0, \ldots, 0, f_{ra_r})\}$
Let $S := Syz(M) \subset R^{a_1 + \cdots + a_r}$ be the module of syzygies of M and let $\Phi : S \longrightarrow R$ be the homomorphysm defined by $\Phi(g_1, g_2, \ldots, g_{a1}, \ldots \ldots) = g_1 f_{11} + g_2 f_{12} + \cdots + g_{a_1} f_{1a_1}$.
Then
a) $I_1 \cap I_2 \cap \cdots \cap I_r = Im(\Phi)$
b) $I_1 \cap I_2 \cap \cdots \cap I_r$ can be computed by the computation of a single Gröbner basis of the module M, w.r.to any term-ordering.
c) Moreover, if the ideals I_i's are homogeneous and a bound is known for the degrees of generators of $I_1 \cap I_2 \cap \cdots \cap I_r$, then it is enough to compute a single truncated Gröbner basis of the module M w.r.to any term-ordering.

Proof. Easy exercise •

Let $\{Q_1, \ldots, Q_s\}$ be a set of homogeneous primary ideals of *height n* in $R :=$ $k[X_0, \ldots, X_n]$. If we put $P_i := \sqrt{Q_i}$, $i = 1, \ldots, s$, then the P_i's are maximal relevant ideals of R. Let $e_i := mult(Q_i)$ and let $I := Q_1 \cap Q_2 \cap \ldots \cap Q_s$. Of course $dim(R/I) = 1$ and $mult(I) = e := e_1 + \cdots + e_s$. By the very definition, the ring R/I is Cohen-Macaulay, hence the ideal I is perfect of grade n. So we can apply the preceding results and device an algorithm for the computation of I, which improves the obvious one stemming from Lemma 3.6. Let me describe it very informally.

INPUT $Q_1, \ldots, Q_s$ homogeneous primary ideals of R of height n.
OUTPUT $I := Q_1 \cap Q_2 \cap \ldots \cap Q_s$

STEP 1 Let M be the module described in Lemma 3.6; choose any term-ordering σ and begin the computation of a Gröbner basis G of M w.r.to σ (for description of term-orderings and Gröbner bases of modules see for instance Möller-Mora (1986)) by means of a smart algorithm, i.e. an algorithm which processes the critical pairs in ascending order of degree.

STEP 2 While computing G, compute syzygies, hence generators of I degree after degree in ascending order.

STEP 3 If no generator of I is found in degree 1, then set $g := n$, else let $r := \#$ of linearly independent generators of degree 1 and set $g := n - r$

STEP 4 When the first generator in degree bigger than 1 is found, set i to be such degree, then continue the computation of G, hence of syzygies of M up to degree $N := e - \binom{g+i-1}{i-1} + i$

STEP 5 Use Φ of Lemma 3.6 to get the result.

Termination is obvious and correctness follows from Corollary 3.5; we only observe that if some linear generator is found in I, then we have to redefine the number g in the formula of Corollary 3.5 (see STEP 3) because there we have $2 \leq i$.

Remark 1. A variant of the algorithm is to compute a sequence of pairwise intersections. This has the disadvantage of computing $s-1$ truncated Gröbner bases, but it has the advantage of computing with ideals instead of modules. Namely in Lemma 3.6, if $r=2$ we get $M \subset R^1$. This avoids the big redundant representation of M with many repetitions and many zeroes.

Remark 2. The upper bounds used in Step 4 are sharp. In fact we already know that the bounds stated in Corollary 3.5 are sharp, since we constructed 0-dimensional lex-segment ideals attaining them. Then we can use the fact that every 0-dimensional monomial ideal can be lifted to an ideal of simple points, if k is infinite (this is a classical result by Hartshorne. For an easy proof see Geramita-Gregory-Roberts (1986) or Carrá-Robbiano (1990)).

For example the ideal $I := (X^2, XY, XZ, Y^2, YZ, Z^5)$ that we considered before (see Example 5) can be lifted to the ideal $\mathbf{I} = (X(X-T), XY, XZ, Y(Y-T), YZ, Z(Z-T)(Z-2T)(Z-3T)(Z-4T)$ of the points $(0,0,0,1)$, $(0,0,1,1)$, $(0,0,2,1)$, $(0,0,3,1)$, $(0,0,4,1)$, $(0,1,0,1)$, $(1,0,0,1)$ in $\mathbb{P}_k^3$.

Remark 3. If I is not perfect, the bounds of Corollary 3.5 are no more correct, hence we cannot use the above described algorithm. For instance let I be an ideal of $R := k[X,Y,Z,T,V,W]$ with $mult(I) = 4$ and $dim(R/I) = 3$. The only $i \geq 2$ such that $4 \geq \binom{3+i-1}{i-1}$ is $i=2$ and $e - \binom{3+2-1}{2-1} + 2 = 2$. Let $I := (X,Y,Z) \cap (X,Z,V) \cap (Y,T,W) \cap (Z,T,W)$. Then $mult(I) = 4$, $dim(R/I) = 3$ and $indeg(I) = 2$. But $I = (XT, XW, YZ, ZT, ZW, YTW, YVW)$. On one hand this proves that I is not perfect, on the other one this proves that we cannot extend our algorithm, as it is, to non perfect ideals.

REFERENCES

Bigatti, A. Caboara, M. Robbiano, L. (1990). *On the computation of Hilbert-Poincaré series.* Preprint.

Buchberger, B. (1985). *Gröbner bases: an algorithmic method in polynomial ideal theory.* Recent Trends in Multidimensional Systems Theory. (Bose N.K. Ed.) Reidel.

Carrá, G. Robbiano, L. (1990). *On superG-bases.* To appear on J. Pure Appl. Algebra.

Elias, J. Robbiano, L. Valla, G. (1989), *Number of generators of ideals.* Preprint.

Geramita, A. Gregory, D. Roberts, L. (1986) *Monomial ideals and points in the projective space.* J. Pure Appl. Algebra **40** 33–62.

Giovini, A. Niesi, G. (1990) *CoCoA: a user-friendly system for Commutative Algebra.* To appear in the Proceedings of the DISCO conference (Capri 1990).

Giusti, M. (1984) *Some effective problems in polynomial ideal theory.* Proceedings of EUROSAM 84 Springer Lecture Notes in Computer Science **174** 159–171.

Giusti, M. (1988) *Combinatorial dimension theory of algebraic varieties.* Computational Aspects of Commutative Algebra. Academic Press.

Kunz, E. (1980) *Einführung in die kommutative Algebra und algebraische Geometrie* Vieweg Studium Bd **46** Aufbaukurs Mathematik Braunschweig/Wiesbaden.

Lazard, D. (1981) *Résolution des systèmes d'équations algébriques.* Theoretical Computer Science, **15** 77–110.

Lazard, D. (1983) *Gröbner bases, Gaussian elimination and resolution of systems of algebraic equations.* EUROCAL 83 Springer Lecture Notes in Computer Science **162.**

Macaulay, F. S. (1927) *Some properties of enumeration in the theory of modular systems.* Proc. London Math. Soc. **26** 531–555.

Matsumura, H. (1970) *Commutative Algebra.* Benjamin, New York.

Möller, M. Mora, T. (1984) *Upper and lower bounds for the degree of Gröbner bases.* Proceedings of EUROSAM 84 Springer Lecture Notes in Computer Science **174** 172–183.

Möller, M. Mora, T. (1986) *New constructive methods in classical ideal theory.* J. Algebra **100** 138–178.

Robbiano, L. (1983) *On normal flatness and some related topics* Commutative Algebra: Proceedings of the Trento Conference. Lecture notes in Pure and Applied Mathematics. **84** Marcel Dekker.

Robbiano, L. (1988) *Introduction to the theory of Gröbner bases,* Queen's Papers in Pure and Applied Mathematics N. **80 Vol V.**

Robbiano, L. (1990) *Introduction to the theory of Hilbert functions,* To appear in Queen's Papers in Pure and Applied Mathematics.

Stanley, R.P. (1978) *Hilbert functions of graded algebras,* Adv in Math. **28** 57–83.

Standard Bases of Differential Ideals [1]

François Ollivier

Laboratoire d'Informatique de l'X (LIX) [2]
École Polytechnique
F-91128 Palaiseau Cedex (France)
cffoll@frpoly11.BITNET
ollivier@cmep.polytechnique.fr

Abstract: The aim of this paper is to introduce a new definition of standard bases of differential ideals, allowing more general orderings than the previous one, given by Giuseppa Carrá-Ferro, and following the general definition of standard bases, given in [O3], valid for algebraic ideals, canonical bases of subalgebras, etc.

Differential standard bases, as canonical bases, suffer a great limitation: they can be infinite, even for ideals of finite type. Nevertheless, we can sometimes bound the order of intermediate computations, necessary to make some elements of special interest appear in the basis.

As an illustration, we consider a differential rational map $f: A^n_{\mathcal{F}} \mapsto A^n_{\mathcal{F}}$, and show that if f is birational, then $\operatorname{ord} f^{-1} \leq n \operatorname{ord} f$. Partial standard bases computations provide then two algorithms to test the existence of f^{-1}. The first one is also able to determine the inverse, if any. The second only determines existence, but we can provide a bound of complexity depending only of n, $\operatorname{ord} f$ and the number of derivatives.

0. Introduction

The theory of standard bases introduced here is not a new variant of the standard bases of $\mathcal{D}$-modules introduced by Castro [Cas]. We will deal with commutative differential rings, not rings of differential operators.

Effective—or almost effective—methods for solving systems of differential algebraic equations go back to the work of Riquier and Janet (cf. [Ja] 1920). Their results have been then improved by Ritt ([R1] 1932, [R2] 1950) who gets an effective method only

(1) Partially supported by GDR G0060 *Calcul Formel, Algorithmes, Langages et Systèmes* and PRC *Mathématiques et Informatique*.

(2) Équipe *Algèbre et Géométrie Algorithmiques, Calcul Formel*, SDI CNRS n° 6176 et Centre de Mathématiques, Unité de Recherche Associée au CNRS n° D0169

if the ground field allows effective factorization. This drawback has been removed by SEIDENBERG ([S] 1956), whose method has been recently implemented by Sette DIOP [D]. The original method of Ritt has also been studied by WU, first in the algebraic case, and then in the differential one. It is particularly interesting for automatic theorem proving in elementary geometry (see [Ch]). Another point of view on this matter may be found in the work of POMMARET [P2], who uses the language and results of the formal theory of partial differential equations, initiated by SPENCER.

Nevertheless, for the best of my knowledge, no method has been developed yet to answer the membership problem for a differential ideal. The computation of a characteristic set only gives partial results: the polynomial in the ideals are reduced to 0, but the reciprocal is false except for prime ideals. Furthermore, no general method has been given to compute a genuine characteristic set, and not only a coherent and autoreduced set.

We can hope that a generalization of standard bases will give a satisfactory answer. Indeed, in [Car] (1987), Giuseppa CARRA'-FERRO introduced a definition for differential standard bases. We provide here a more general one, allowing a wider class of orderings, and underlying the connections with the theory of standard bases, canonical bases, etc, following the abstract definition given in [O3]. The main trouble is that differential standard bases are in general infinite. This was already the case for canonical bases of subalgebras (see [KL], [RS], [O2]).

We are still able to prove that a completion process converges to a standard basis, meaning that after a finite number of steps we will get a basis up to a given order of derivation. We have no way yet to determine the complexity of a partial computation, nor to check it has been performed without using explicit information on the structure of the ideal, and theoretical results of differential algebra.

Anyway, this is not so far from the algebraic situation, which may be intractable, except for "well behaved" ideals. We provide an illustration of this point of view by giving algorithms to test whether a differential rational map admits an inverse. We had already proved complexity bounds for algebraic rational maps and polynomial ones (see [O1], [O2] and [O3]), using a theorem of O. GABBER. We extend this theorem to differential maps, proving that $\operatorname{ord} f^{-1} \leq n \operatorname{ord} f$, if n is the number of variables. This work is a by-product of our interest in control theory and modeling, where the search for effective and efficient tests for abstract properties of structures, as identifiability, requires such theoretical investigations (see [O1] and [O3]).

1. Standard bases

We will denote by $\mathcal{F}$ a differential field of characteristic zero.

1.1. Preliminary results of differential algebra

We limit ourselves to the essential results and definitions needed in the following. Details may be found in the classical books of RITT [R1] and [R2], or in KOLCHIN [Ko], KAPLANSKY [Ka], and POMMARET [P1]. I will mostly follow the exposition and notations of [Ko].

DEFINITION 1. *A differential ring is a ring with a finite set* $\Delta = \{\delta_1, \ldots, \delta_m\}$ *of differential operators, i.e. internal mappings* δ *satisfying*

$$\delta(ab) = \delta a\, b + a\, \delta b$$
$$\delta(a+b) = \delta a + \delta b,$$

and such that $\delta_i\, \delta_j = \delta_j\, \delta_i$.

A differential field is a field which is a differential ring.

DEFINITION 2. *A differential ideal of a differential ring is an ideal* $\mathcal{I}$, *such that* $\forall \delta \in \Delta\ \delta \mathcal{I} \subset \mathcal{I}$.

Following classical notations, we denote by (S), the ideal generated by S, and by $[S]$ the differential ideal generated by S.

DEFINITION 3. *A differential ideal* $\mathcal{I}$ *is said to be perfect if* $a^n \in \mathcal{I}$ *implies* $a \in \mathcal{I}$

The perfect ideal generated by Σ is denoted by $\{\Sigma\}$.

DEFINITION 4. *We will denote by* Θ *the abelian free monoïd generated by* Δ. *For any set* X, *we denote by* ΘX *the set of derivatives* $\Theta \times X$. *The element* (θ, x) *will by written* $x_{(\theta)}$. *There is an action of* Θ *on* ΘX *defined by* $\tau x_{(\theta)} = x_{(\tau\,\theta)}$.

The algebra of differential polynomials $\mathcal{F}\{X\}$ will be the algebra $\mathcal{F}[\Theta X]$ with the unique set of derivations Δ extending derivations on $\mathcal{F}$ and such that $\delta x_{(\theta)} = x_{(\delta\,\theta)}$. $\mathcal{F}\{X\}$ is a differential algebra over $\mathcal{F}$. Those derivations also extend to the field of fractions of $\mathcal{F}\{X\}$, denoted by $\mathcal{F}(X)$. We call a monomial of $\mathcal{F}\{X\}$, a polynomial which is a product of derivatives or 1, and a term the product of a monomial by a non-zero element of $\mathcal{F}$.

Lemma 5. *If* Σ *is a subset of a differential ring* R (*resp. a differential* R-algebra A), *then the differential ideal* (*resp.* R-algebra) *generated by* Σ *is equal to* $(\Theta \Sigma)$ (*resp.* $R[\Theta \Sigma]$). ■

In general, if $\mathcal{G}$ is a differential field extension of $\mathcal{F}$, and η a subset of $\mathcal{G}$, we denote by $\mathcal{F}\langle\eta\rangle$, the differential field extension of F generated by η. If R is a differential ring and Σ a subset of a R-algebra A, we denote by $R\{\Sigma\}$ the differential R-algebra generated by Σ. From now on, we will only consider differential polynomials over a field $\mathcal{F}$ of characteristic zero, with finite set of variables $X = \{x_1, \ldots, x_n\}$. The set of derivatives of the field $\mathcal{F}$ will be $\Delta = \{\delta_1, \ldots, \delta_m\}$.

THEOREM 6. *If* $\mathcal{I}$ *is a perfect differential ideal of* $\mathcal{F}\{X\}$, *then there exists a finite set* Σ *of differential polynomials such that* $\mathcal{I} = \{\Sigma\}$.

PROOF. See [Ko]. ■

This is the best we can do. The set of differential polynomials is not noetherian.

THEOREM 7. *If* $\mathcal{I}$ *is a perfect ideal of* $\mathcal{F}\{X\}$, *then there exists a unique set* $\{\mathcal{I}_1, \ldots, \mathcal{I}_r\}$ *of prime ideals such that*

$$\mathcal{I} = \bigcap_{i=1}^{r} \mathcal{I}_i,$$

and $\mathcal{I}_i \not\subset \mathcal{I}_j$; $i \neq j$. Those prime ideals are said to be the components of $\mathcal{I}$.

DEFINITION 8.   Let $\mathcal{I}$ be a prime differential ideal of $\mathcal{F}\{X\}$, $(\eta_1, \ldots, \eta_n)$ a n-uple of elements in a differential field extension $\mathcal{G}$ of $\mathcal{F}$. η will be said to be a generic zero of $\mathcal{I}$ over $\mathcal{F}$ if $\{P \in \mathcal{F}\{X\} | P(\eta) = 0\} = \mathcal{I}$. The generic zeroes of $[0]_{\mathcal{F}[x]}$ are called generic elements of $\mathcal{G}$ over $\mathcal{F}$.

An extension $\mathcal{U}$ of $\mathcal{F}$ is said to be universal if for all finite extension $\mathcal{F} \subset \mathcal{G} \subset \mathcal{U}$, all finite set X and all prime differential ideal $\mathcal{I} \subset \mathcal{G}\{X\}$, there exists a generic zero of $\mathcal{I}$ over $\mathcal{G}$ in $\mathcal{U}$.

All differential fields admit a universal extension (see [Ko]). This notion is in fact the same as the universal domains of Weil, if the set of derivatives is empty. It allows to throw away some logical difficulties in the definition of differential algebraic varieties given by Ritt. They may be defined as the sets of zeroes of differential ideals in a universal extension $\mathcal{U}$, chosen once and for all. The variety associated to an ideal $\mathcal{I}$ is denoted by $V(\mathcal{I})$. The differential affine space of dimension r over $\mathcal{F}$, $\mathbf{A}^r_{\mathcal{F}}$, is the set of zeroes of $[0]\ \mathcal{U}^r$. We refer to [Ko] for more details on differential algebraic geometry and conclude this short introduction by a powerful result first proved by RITT in the ordinary differential case, and latter extended by KOLCHIN.

DEFINITION 9.   The order of $\theta = \prod_{i=1}^r \delta_i^{\alpha_i}$ is $\sum_{i=1}^r \alpha_i$, and the order of a derivative θx is the order of θ. We denote by Θ_r the set of derivation operators of order less than or equal to r and by $\mathrm{ord}\,v$ the order of a derivative. The order of a differential polynomial is the maximal order of its derivatives.

PROPOSITION 10.   Let $\mathcal{I}$ be a prime ideal of $\mathcal{F}\{X\}$, η a generic zero of $\mathcal{I}$, the function $H : \mathbb{N} \mapsto \mathbb{N}$ such that $H(r)$ is the (algebraic) transcendance degree of $\mathcal{F}(\Theta_r \eta)$ over $\mathcal{F}$ is equal to a polynomial $\omega_{\eta/\mathcal{F}}$ for r great enough. Furthermore

$$\omega_{\mathcal{I}}(r) = \sum_{i=1}^m a_i \binom{r+i}{i},$$

where a_m is the differential dimension of $\mathcal{I}$, i.e. the differential transcendance degree of $\mathcal{F}\langle \eta \rangle$ over $\mathcal{F}$. ∎

The greatest i such that $a_i \neq 0$ will be called the differential type of $\mathcal{I}$, $\tau_{\mathcal{I}}$, and $a_{\tau_{\mathcal{I}}}$ the typical differential dimension of $\mathcal{I}$. As $\omega_{\eta/\mathcal{F}}$ does only depend of $\mathcal{I}$, we can also denote it by $\omega_{\mathcal{I}}$. If V is an irreducible algebraic differential variety defined by a prime ideal $\mathcal{I}$, we extend to it the definitions given above.

THEOREM 11.   (Ritt–Kolchin) Let $\mathcal{I} = \{P_1, \ldots, P_r\}$, where the maximal order of the P_i is e, and $\mathcal{J}$ a component of $\mathcal{I}$, whose differential type is $m - 1$, then the typical differential dimension of $\mathcal{J}$ is less than or equal to $n\,e$.

PROOF. See [Ko chap. IV § 17 p. 199]   ∎

1.2. Admissible orderings. Reduction

We need to define suitable orderings to allow reductions in $\mathcal{F}\{X\}$. This implies to strengthen the definitions valid in the pure algebraic case to take derivations into account.

DEFINITION 1. *Let $<$ be a total ordering on the set $\mathcal{M}$ of monomials of $\mathcal{F}\{X\}$. We extend derivations to $\mathcal{M}$ by taking δM to be the maximal monomial involved in the polynomial δM. By convention, $\delta 1 = 1$. The order $<$ is said to be admissible if*

a) $M > 1 \quad M \neq 1$,

b) $M > M'$ *implies* $M''M > M''M'$,

c) $\delta M > M \quad M \neq 1$,

d) $M > M'$ *implies* $\delta M > \delta M'$.

If $<$ is admissible, we denote by $\mathrm{lm}P$ the leading monomial of P, by $\mathrm{lc}P$ its leading coefficient. We call reductum of P the polynomial $P - \mathrm{lc}\,P\,\mathrm{lm}\,P$.

So we define δM in $\mathcal{M}$ to be $\mathrm{lm}(\delta M)$. I think no misunderstanding can result of this notation, which will be useful later.

We now need to describe some admissible orderings. For this, we first define admissible orderings, i.e. rankings in the words of Ritt, on the set of derivatives ΘX. They are orderings which satisfy c) and d) in the definition above. Considering elements of Θ as monomials, e.g. in $\mathbf{Q}[\Delta]$, we take an admissible ordering on Θ. We extend it to ΘX with the following definitions.

DEFINITION 2. *The ordering on ΘX defined by $x_{i,(\theta)} < x_{i',(\theta')}$ if $i < i'$ or $i = i'$ and $\theta < \theta'$ is said to be the lexicographical ordering extending $<$. The ordering defined by $x_{i,(\theta)} < x_{i',(\theta')}$ if $\theta < \theta'$ or $\theta = \theta'$ and $i < i'$ is the derivation ordering extending $<$.*

It is easily seen that those orderings are admissible (see [Ko chap. 0 § 17 p. 50]).

REMARK 3. If $<$ on Θ respects the order, then the derivation ordering $<$ on ΘX respects the order too, it is said then to be *orderly*.

Let $<$ be an admissible ordering on derivatives, we can extend it to monomials of $\mathcal{F}\{X\}$ in the following way. Consider two monomials $M = \prod_{i=1}^{r} v_i^{\alpha_i}$ and $M' = \prod_{i=1}^{s} \nu_i^{\beta_i}$, where the v_i and ν_i appear in strictly decreasing order. We take $M < M'$ if there exists $j \leq r, s$ such that $v_i = \nu_i \; i < j$, $\alpha_i = \beta_i \; i < j$, $v_j < \nu_j$ or $v_j = \nu_j$ and $\alpha_j < \beta_j$.

PROPOSITION 4. *The ordering $<$ defined above is an admissible well ordering on monomials. If $<$ is orderly, its extension to monomials is also orderly, i.e. $\mathrm{ord}\,P > \mathrm{ord}\,Q$ implies $P > Q$.*

PROOF. It is immediate that a) and b) are satisfied. In order to prove c) and d), we only have to remark that $\delta m = \delta v_1 v_1^{\alpha_1 - 1} \prod_{i=2}^{r} v_i^{\alpha_i}$. If $<$ is orderly on derivatives, then $\mathrm{ord}\,P < \mathrm{ord}\,Q$ implies that the leading derivative of P is smaller than that of Q, so that $P < Q$.

We now show that $<$ is a well ordering. It is known that all admissible orderings on variables are well orderings (see [Ko]). Consider now an infinite sequence $M_0 > M_1 > \cdots$ of monomials. The leading derivatives of these monomials appear in decreasing order, so that for some integer r the chain they form will become stationary. Let v be the leading derivative of M_i for $i > r$. The degree in v of M_i $i > r$ will be decreasing too, so that for $i \geq s \geq r$ this degree becomes a constant integer d. Dividing M_i by v^d, for $i \geq s$, we secure a new strictly decreasing sequence of monomials, whose leading derivatives are smaller than v. Repeating the argument, we build an infinite strictly decreasing sequence of derivatives: a contradiction. ∎

So admissible orderings on monomials actually exist. It will be useful to consider other orderings than those coming from the previous propositions. We may first remark that if P is a differential polynomial of degree d, then θP is also of degree d, moreover if P is homogeneous, θP is homogeneous too. We shall need some more convenient grading on $\mathcal{F}\{X\}$, defined by taking the weight of a monomial $\prod_{i=1}^{r} v_i^{\alpha_i}$ equal to $\sum_{i=1}^{r} \alpha_i \operatorname{ord} v_i$. A polynomial whose monomials are of the same weight is called *isobaric*. The maximal weight of monomials of a polyomial P is called the weight of P (wt P). The derivative δP of an isobaric polynomial is not in general isobaric, except if the coefficients of P lie in the field of constants of $\mathcal{F}$, but for all polynomial $P \notin \mathcal{F}$ wt $\theta P = $ wt $P + $ ord θ—we only consider characteristic zero!

Lemma 5. *If $<$ is an admissible ordering on monomials, we get a new admissible ordering $\prec$ by taking $M \prec M'$ if $\deg M < \deg M'$ or if $\deg M = \deg M'$ and $M < M'$. The same applies when considering the weight, or the partial degree according to some subset of X.*

If $<$ is a well ordering, then $\prec$ is also a well ordering ∎

REMARK 6. More generally, we can use all the admissible gradings defined in [Ko chap I § 7 p. 72].

Recursive use of this lemma allows to build a wide class of orderings, for example elimination orderings. In the following, we will suppose that such an ordering $<$ has been chosen once and for all.

We now come to reduction.

DEFINITION 7. *We say that a polynomial P is elementarily reduced by Q to R if there exist a monomial M and a derivation operator θ such that $\operatorname{lm} P = M \operatorname{lm} \theta Q$ and $R = P - (\operatorname{lc} P / \operatorname{lc} Q) M \theta Q$. We write it $P \xrightarrow{Q} R$. We say that P is elementarily reduced to R by a set of polynomials Σ if there exists $Q \in \Sigma$ such that $P \xrightarrow{Q} R$. P will be said to be reduced to R by Σ if there exists a chain of elementary reductions*

$$P = P_0 \xrightarrow{\Sigma} P_1 \xrightarrow{\Sigma} \cdots \xrightarrow{\Sigma} P_r = R.$$

We denote it by $P \xrightarrow{\Sigma_*} R$.

We say that P is totally reduced to R by Σ if P is reduced to R by Σ or if the reductum of P is totally reduced to R' by Σ and $R' = \operatorname{lc} P \operatorname{lm} P + R'$. P is irreducible by Σ if there is no Q such that $P \xrightarrow{Q} Q$.

REMARK 8. If we use the fact that $\theta \operatorname{lm} P = \operatorname{lm}(\theta P)$, for $P \notin \mathcal{F}$, with the extension of derivations to monomials made above, it becomes obvious that the reducibility of P by Q only depends of the leading monomials of P and Q. It is then easily that, if P is reducible by Q, the weight (or degree) of the leading monomial of P is not less than that of Q. It is also obvious that $P \xrightarrow{Q} R$ implies $\operatorname{lm} R < \operatorname{lm} P$.

Lemma 9. $P \xrightarrow{\Sigma_\bullet} 0$, iff $P = \sum_{i=1}^r M_i\,\theta_i\,P_i$, where the M_i are terms, and the P_i polynomials in Σ, with $\mathrm{lm}(M_i\,\theta_i\,P_i) > \mathrm{lm}(M_j\,\theta_j\,P_j)\ i < j$. ∎

We can build an effective reduction process which takes a polynomial P and a finite list of polynomials Σ and returns a polynomial R such that $P \xrightarrow{\Sigma_\bullet} R$ and R is irreducible by Σ. We begin by reduction with respect to a single polynomial. We use the syntax of the IBM computer algebra system Scratchpad II for the algorithms.

REDUCTION ALGORITHM

reduction(P, Q) == reduction$(P, Q, 1)$

reduction(P, Q, r) ==
  $\deg\mathrm{lm}P > \deg\mathrm{lm}Q$ or $\mathrm{wt}\,\mathrm{lm}\,P > \mathrm{wt}\,\mathrm{lm}\,Q$ => return P
  $\mathrm{lm}\,Q\backslash\mathrm{lm}\,P$ => return reduction$(P - (\mathrm{lc}\,P/\mathrm{lc}\,Q)\,(\mathrm{lm}\,P/\mathrm{lm}\,Q)\,Q, Q)$
  for $i \in [r, \dots, m]$ repeat
    if $(P_2 := \text{reduction}(P, \delta_i\,Q, i)) \neq P$ then return reduction(P_2, Q)
  P

PROOF. We first prove that the process stops and returns P if it is irreducible by Q. If we can apply the remark above, it stops on the first line. If not, the process is recursively repeated with derivatives of P. As their weight increases by 1 at each new step, the remark will necessarily apply after a finite number of steps. Now, if P is reducible, its leading monomial needs to be a multiple of the leading monomial of some $\theta\,Q$. A solution will to be found by trying all successive derivatives of Q, whose leading monomials have weight less or equal to the weight of P, which is done. We perform then an elementary reduction, and repeat the process. It needs to stop, for $<$ is a well ordering, and so there is no infinite sequence of elementary reductions. ∎

It is now simple to get a reduction algorithm for a list of polynomials, or for total reduction.

1.3. Definitions

DEFINITION 1. *Considering the multiplicative monoïd $\mathcal{M}$ of monomials in $\mathcal{F}\{X\}$, with the derivations acting on it as in def. 2.1, we call a subset E a differential monoïdeal if it is a monoïdeal—i.e. if $\mathcal{M}E \subset E$—, and if $\Delta\,E \subset E$.*

REMARK 2. Obviously, the set of leading monomials of a differential ideal is a differential monoïdeal. Of course the "derivations" defined on $\mathcal{M}$ are not real ones, but the mere reflect of derivations acting on polynomials. Indeed, the mapping δ_i themselves do not need to be derivations. We only need that $\mathrm{lm}\delta\,P = \mathrm{lm}\delta(\mathrm{lm}\,P)$ and that $\delta(P+Q) = \delta\,P + \delta\,Q$, so that we could use more general differential operators, say $d = \delta_1^2 - \delta_2^3$ and define standard bases for d-ideals, i.e. ideals $\mathcal{I}$ such that $d\mathcal{I} \subset \mathcal{I}$, but for this we would need a more complicated definition of reduction, and a wider class of syzygies (see [O3]).

Using derivations, we are indeed able to restrict the set of syzygies to consider, for given a product of monomials $M\,M'$, $\delta(M\,M')$ equals $\delta\,M\,M'$ or $M\,\delta\,M'$, so that the differential monoïdeal generated by a subset E of $\mathcal{M}$ is equal to $\mathcal{M}\,\Theta\,E$ (see subsection 4. bellow).

DEFINITION 3. *A subset G of a differential ideal $\mathcal{I}$ is said to be a standard basis if $\operatorname{lm} G$ generates $\operatorname{lm} \mathcal{I}$ as a differential monoïdeal.*

THEOREM 4. *Let G be a set of polynomials, $\mathcal{I}$ a differential ideal. Then the following propositions are equivalent:*
i) G is a standard basis of $\mathcal{I}$,
ii) $G \subset \mathcal{I}$ and there is no non-zero element of $\mathcal{I}$ reduced with respect to G,
iii) $G \subset \mathcal{I}$ and all the elements of $\mathcal{I}$ are reduced to 0 by G,
iv) a differential polynomial is in $\mathcal{I}$ iff it is reduced to 0 by G.

PROOF. $i) \implies ii)$. If G is a standard basis of $\mathcal{I}$ it is a subset of $\mathcal{I}$. Now, the leading monomial of any non-zero polynomial in $\mathcal{I}$ is in $\mathcal{M} \ominus \operatorname{lm} G$ using remark 2 above, so that it is reducible by G.

$ii) \implies iii)$. As $G \subset \mathcal{I}$, if $P \xrightarrow{G} Q$ with $P \in \mathcal{I}$, then $Q \in \mathcal{I}$, so that we can perform repeated reductions using ii). As chains of reductions are finite, *the result of any reduction process* is 0, which is more than iii).

$iii) \implies iv)$. $\Rightarrow$ is immediate from iii). $\Leftarrow$ Again, as $G \subset \mathcal{I}$, if $P \xrightarrow{G_*} 0$, P needs to be in $\mathcal{I}$.

$iv) \implies i)$. All polynomials in G are reduced to 0 by G, so that $G \subset \mathcal{I}$. As all polynomials in $\mathcal{I}$ are reduced to 0 by G, they are reducible, so that $\operatorname{lm} \mathcal{I} \subset \mathcal{M} \ominus \operatorname{lm} G$. Using the first part of the proof, we have indeed equality. ∎

DEFINITION 5. *A standard basis G of $\mathcal{I}$ is said to be minimal if $\operatorname{lm} G$ is a minimal set of generators of $\operatorname{lm} \mathcal{I}$. A minimal standard basis G is called reduced if all polynomials $P \in G$ are totally reduced by $G \setminus \{P\}$.*

PROPOSITION 6. *Any ideal admits minimal standard bases and a unique reduced standard basis. An ideal admits a finite standard basis iff it admits a finite minimal standard basis. In this case, all the minimal standard bases are finite.* ∎

1.4. Characterization

We have completed the easiest part with definitions. The completion process will rely on more tedious results.

DEFINITION 1. *Let P and Q be two differential polynomials, we call a syzygy between P and Q a 2-uple $(M \theta P, M' \theta' Q)$, where $M, M' \in \mathcal{M}$, $\theta, \theta' \in \Theta$, of polynomials with the same leading monomials. An essential syzygy is a syzygy with M and M' minimal and such that there is no other syzygy $(N \tau P, N' \tau' Q)$ satisfying $\vartheta(N \tau \operatorname{lm} P) = M \theta \operatorname{lm} P$ and $\vartheta(N' \tau' \operatorname{lm} Q) = M' \theta' \operatorname{lm} Q$ for some ϑ, the derivations being taken in $\mathcal{M}$.*

We call S-polynomial associated to the syzygy (U, V), the polynomial $\operatorname{lc} V\, U - \operatorname{lc} U\, V$. The rank of the syzygy will be the common leading monomial of U and V.

EXAMPLE 2. Consider ordinary differential polynomials in $\mathcal{F}\{x\}$. There is only one admissible ordering on Θ and Θx. We choose the ordering on monomials coming from prop. 2.4. Take $\mathcal{I} = \{x^2\}$. There is an essential syzygy $(\delta x\, x^2, x\, \delta(x^2))$. The syzygy $(\delta^2 x\, x^2, x\, \delta^2(x^2))$ is not essential. The only essential syzygies different from that already given are of the form $(\delta^{n+1} x\, \delta^n(x^2), \delta^n x\, \delta^{n+1}(x^2))$ $n \geq 1$. This shows that syzygies may involve twice the same polynomial, and that there is in general an infinite number of essential syzygies.

DEFINITION 3.   Let Σ be a set of polynomials, P a polynomial in $[\Sigma]$. We call rank of P with respect to Σ the smallest monomial M such that (1) $P = \sum_{i=1}^{r} Q_i\,\theta_i\,P_i$, where the P_i belong to Σ, the Q_i are terms and $\operatorname{lm} Q_i\,\theta_i\,P_i \le M$.

REMARK 4.   The rank of P is greater than or equal to the leading monomial of P. If P is reduced to 0 by Σ, it is equal to $\operatorname{lm} P$. We may consider, e.g. $\Sigma = \{\delta_1 x + \delta_3 x, \delta_2 x + \delta_3 x\}$ and $P = \delta_1\delta_3 x - \delta_2\delta_3 x$, assuming pure lexicographical ordering on Θ with $\delta_1 > \delta_2 > \delta_3$. Then, P is of rank $\delta_1\delta_2 x > \operatorname{lm} P$ with respect to Σ. If P is the S-polynomial associated to a syzygy between elements of Σ, then the rank of P is less than or equal to the rank of the syzygy. We can further notice that if P is of rank M, Q of rank N, then QP is of rank at most NM, and that θP is of rank at most θM.

THEOREM 5. G is a standard basis of the differential ideal $\mathcal{I}$ iff G generates $\mathcal{I}$ and all the S-polynomials associated to the set of essential syzygies between elements of G are reduced to 0 by G.

PROOF. $\Longrightarrow$ is obvious since S-polynomials are in $\mathcal{I}$.

The reciprocal is the consequence of the following more precise theorem.   ∎

THEOREM 6.   Let M be a monomial, Σ be set of polynomials, such that all S-polynomials associated to the set of essential syzygies between elements of Σ of rank less than or equal to M are reduced to 0 by Σ. Then, if P is of rank less than or equal to M with respect to Σ, P is reduced to 0 by Σ.

PROOF. Suppose it is not so. Among the P of minimal rank N which do not satisfy the conclusion, we choose one with smallest r in formula (1) of def. 3. The integer r is greater than 1. If not, P would be reduced to 0 by P_1. Now, we may decompose the sum (1) in two parts, e.g. $P = R_1 + R_2$ with $R_1 = Q_1\,\theta_1\,P_1$ and $R_2 = \sum_{i=2}^{r} Q_i\,\theta_i\,P_i$. Obviously, R_1 and R_2 need to be reducible, for they admit a decomposition (1) with a sum of at most $r-1$ polynomials with leading monomials at most N. This implies that R_1 and R_2 have the same leading monomial and opposite leading coefficients: if not P would be reducible.

We first prove that r is greater than 2. If $r = 2$, the polynomial $P = Q_1\,\theta_1\,P_1 + Q_2\,\theta_2\,P_2$ is the product of a S-polynomial, by a non zero element of $\mathcal{F}$. Without loss of generality we may suppose it is a S-polynomial. If this syzygy is essential, P is reducible: a contradiction. If not, suppose Q_1 and Q_2 are not minimal. They admit a proper common factor L, and P/L is of rank smaller than N, so that it is reducible and so is P: another contradiction.

The last case is when there exists a syzygy (U, V) between P_1 and P_2 such that $N = \operatorname{lm}\vartheta U = \operatorname{lm}\vartheta V$ for $\vartheta \neq 1$. The rank of (U, V) is less than N, so that the S-polynomial S associated to (U, V) is reduced to 0. This implies that the rank of ϑS is $\vartheta\operatorname{lm}S$, strictly less than N. Now, we may develop:

$$\vartheta S = aP + \text{a sum (1) of rank less than } N,$$

where $a \in \mathcal{F}\ a \neq 0$. Hence P is of rank less than N: a final contradiction to $r = 2$.

Using lemma 2.9, we may now decompose R_2 as a sum (1) $\sum_{i=1}^{s} Q_i'\,\theta_i'\,P_i'$, with

$$\operatorname{lm}(Q_i'\,\theta_i'\,P_i') > \operatorname{lm}(Q_j'\,\theta_j'\,P_j')\ i < j.$$

Let $T = Q_1\,\theta_1\,P_1 + Q_1'\,\theta_1'\,P_1'$, R_1 and R_2 having opposite leading terms $\operatorname{lm}T < N$. Furthermore $r > 2$ implies that T is reducible, so that T is of rank less than N. If we write P as $T + \sum_{i=2}^{s} Q_i'\,\theta_i'\,P_i'$, we conclude that P is of rank less than $N = \operatorname{rank} P$. ∎

The main idea is very general and follows a scheme for the proof of analogous theorems in other generalizations of standard bases (see [O3] where the proof of prop. 2.1.13 is very similar).

1.5. Completion process

We now have enough material for investigating a completion process. The first step is to build, or rather to enumerate a set of essential syzygies. Differential syzygies between elements of Σ are algebraic syzygies between elements of $\Theta\,\Sigma$. So we can use the criteria detecting unuseful syzygies valid in the algebraic case. We will mostly use two of them, as an illustration.

CRITERION 1. If $(M\,\theta\,P, N\,\tau\,Q)$ is an essential syzygy such that $M = \operatorname{lm}\tau\,Q$, then the associated S-polynomial is reduced to 0 by the set $\{P, Q\}$. ∎

COROLLARY 1. If P and Q are polynomials whose leading monomials are linear, i.e. are mere derivatives $\theta\,x_i$ and $\tau\,x_j$, then if $x_i \neq x_j$ all syzygies between P and Q are reduced to 0 by $\{P, Q\}$. If $x_i = x_j$, then we only have to consider the syzygy $(\tau'\,P, \theta'\,Q)$, where τ' and θ' are such that $\tau'\,\theta = \theta'\,\tau = \gcd(\theta, \tau)$. ∎

CRITERION 2. If $P, Q, R \in \Sigma$, $S = (U, V)$ is an algebraic syzygy between $\theta\,P$ and $\tau\,Q$, $\operatorname{lm}\vartheta R$ divides the rank of S and the algebraic syzygies between $\theta\,P$ and ϑR, $\tau\,Q$ and ϑR are both reduced to 0 by Σ, then S is reduced to 0 by Σ. ∎

CRITERION 3. If some derivative $\theta\,P$ is reduced to 0 by Σ, no syzygy involving a derivative $\tau\,\theta\,P$ needs to be considered. ∎

This simply rephrases well known results for algebraic standard bases (see [Bu]). More details on this mater may be found in [O3].

In the following completion process, G is the list which tends to a standard basis as the process goes. It will be indeed a standard basis if it stops. L_1 is the list of polynomials or derivatives of polynomials already considered, and L_2 is the list of newly appeared polynomials or derivatives, which should be used to try new syzygies. L_3 is the list of polynomials coming from the reduction of S-polynomials.

We suppose that $buildSyz(L_1, L_2)$ is a procedure which returns all algebraic syzygies between two derivatives in the list L_2, or a derivative in L_1 and one in L_2; it uses criteria 1 and 2 to discard useless syzygies, when possible. The procedure $isRed(S)$ returns P if the syzygy corresponds to the algebraic reduction of the derivative P and 0 otherwise.

We can also use cor. 1 to test if there is no more syzygies to consider. Except if the ideal is [1], this is the only way I know to reduce to a finite set of syzygies—we may imagine cases where the basis is finite and there is still an infinite number of syzygies to consider. Indeed the main example of ideals with finite standard bases are linear ones (see [Car cor. 5 p. 138]).

The procedure $linTestY(L_1, L_2)$ returns $true$ if the two following conditions are satisfied:

a) there is no more syzygies between elements of L_2 to consider, using cor. 1,

b) the leading derivatives of polynomials in L_2 are all strictly greater than the derivatives appearing in the leading monomials of polynomials in L_1.

Of course, we are sometimes lucky enough to build a finite standard basis and finish the completion process even in non-linear cases (see bellow ex. 6.5).

COMPLETION PROCESS

```
completionProcess(Σ) ==
    -- First suppress 0 and remove duplicate polynomials
    Σ := removeDuplicates delete(0, Σ)
    -- If there is a constant polynomial it is finished
    for P ∈ Σ repeat if P ∈ F then return [1]
    G := Σ; L₁ := Σ; L₂ := Σ; L₃ := []
    while L₂ ≠ [] repeat
    -- We use cor. 1 to test if all remaining syzygies may be discarded
        if linTest(L₁, L₂) then return G
        -- We construct new syzygies between "old" polynomials in L₁ and "new" ones in L₂,
        -- or two new polynomials in L₂
        lSyz := buildSyz(L₁, L₂)
        for S ∈ lSyz repeat
            -- If the syzygy is the algebraic reduction of a derivative,
            -- all syzygies involving this derivative may be removed
            P := isRed(S); delete(P, G); delete(P, L₁); delete(P, L₂)
            if (R := reduction (sPol(S), L)) ≠ 0 then
                -- If non-zero, the reduction of the S-polynomial is kept in L₃
                L₃ := cons(R, L₃)
                -- If R ∈ F it is finished
                if R ∈ F then return [1]
        G := append(G, L₃)
        -- Derivatives already considered are appended to L₁
        L₁ := removeDuplicates append(L₁, L₂)
        -- New polynomials coming from the reduction of S-polynomials
        -- and new derivatives are collected in L₂
        L₂ := append(L₃, [δ P|(δ, P) ∈ Δ × L₂])
        L₃ := [] output(G)
    G
```

THEOREM 2.   If the process stops it returns a minimal standard basis G of Σ. Otherwise, let G_i denote the set of polynomials, which is returned by the process at the end of the ith loop, then:

a) $G = \bigcup_{i=1}^{\infty} G_i$ is a standard basis of Σ,

b) $G' = \bigcap_{i=1}^{\infty} \bigcup_{j=i}^{\infty} G_j$ is a minimal standard basis.

PROOF. At the beginning, $G = \Sigma$, so G generates $[\Sigma]$. During the process, if a polynomial is removed from G, then its reduction is added to G. So G still generates $[\Sigma]$. In both cases, all the S-polynomials coming from syzygies between elements of G, which are not thrown away using the criteria are reduced to 0 by G, so that is a standard basis using theo. 4.5.

For the same reason, $\bigcup_{j=i}^{\infty} G_j$ is a standard basis for all i, so that G' is also a standard basis. As a polynomial $P \in G'$ is irreducible by $G' \setminus \{P\}$, G' is minimal. ∎

REMARK 3. If we use an orderly ordering, or a ordering which respects the weight, we can modify this process to make it stop if there is no more syzygies to compute, with order or weight less than or equal to a given integer.

If think a few words are necessary to stress on the difference on the completion process given there, and the approach in [Car]. G. Carrá-Ferro proceeds by repeated computations of algebraic standard bases, so that the same work may be done many times. We only have here one process based on reduction of differential syzygies, which do not appear in her paper.

This allows sometimes to prove we have secured a finite basis, simply because the process stops (ex. 6.5 bellow), as she needs in all cases to rely on some a priori mathematical knowledge. Of course, those improvements are far to solve everything.

1.6. Examples

Before considering examples, first a few remarks.

REMARK 1. The completion process only uses the operations of the ground field, so that the polynomials in the standard basis have coefficients in the subfield generated by the coefficients of the input polynomials.

REMARK 2. If $\mathcal{I} = [P_1, \ldots, P_r]$, where the P_i are homogeneous, the standard basis, which is the limit of our construction process will be homogeneous, as well as the reduced standard basis of $\mathcal{I}$. The same apply with isobaric polynomials, if all their coefficients are constants. In such cases, the weight, or degree of the polynomials in any basis cannot be less than the minimal weight or degree of the generators. So, considering a finite set Σ of isobaric polynomials with constant coefficients, we only have to run the completion process up to wt P in order to test if P belongs to $[\Sigma]$.

REMARK 3. Suppose we are given an ordinary differential ideal generated by a system of state or pseudo state equations :

$$x_{1,(r_1)} = P_1(x_{1,(r_1-1)}, \ldots, x_1, \ldots, x_{n,(r_n-1)}, \ldots, x_n)$$

$$\vdots$$

$$x_{n,(r_n)} = P_n(x_{1,(r_1-1)}, \ldots, x_1, \ldots, x_{n,(r_n-1)}, \ldots, x_n).$$

For any orderly ordering $\{x_{i,(r_i)} - P_i\}$ is already the reduced standard basis of the generated ideal, and the procedure given above will stop. It it also a characteristic set.

EXAMPLE 4. We consider the ideal $\mathcal{I} = [x^2]$ already given in [Car], using the same ordering as in example 4.2. RITT has shown that $(u')^{2^{p}-1}$ belongs to $[u^p]$, so that for all r, $x_{(r)}^q \in \mathcal{I}$ for some integer q, which is greater than 1, using remark 2 above. Furthermore, $x_{(r)}^q$ can only be reduced by a polynomial in the basis with leading monomial $x_{(r)}^s$ $s \leq q$. As $x_{(r)}^s$ is the smallest monomial of weight $r s$, it is in the reduced basis. So $[x^2]$ has no finite standard basis.

This shows that standard bases may be actually infinite, and even worse that it may be indeed the general case, for this example is very simple.

EXAMPLE 5. We now consider $\mathcal{I} = [P]$, where $P = x^2 + x + 1$. The first syzygy which appears is $(x' P, x P')$. The associated S-polynomial is $x x' + 2 x'$ which is reduced to $3/2 x'$, using P'. We add x' to the basis. P' is reduced to 0 by x'. Using crit. 3, all syzygies involving $P^{(s)}$ $s \leq 1$ may be discarded. P and x' are mutually totally irreducible, and using crit. 1, there is no syzygy involving only x'. Hence, the reduced standard basis of $\mathcal{I}$ is finite and equal to $\{x^2 + x + 1, x'\}$. In our process, P' is deleted from L_2. The only polynomial in L_2 is x', and $L_1 = \{P\}$. So the process stops using lin Test.

As shown by this example there also exist non-trivial ideals with finite standard bases, *which may be found in a finite number of steps by our completion process.*

Standard bases are often used to perform elimination of a set of variables. If we are lucky enough to secure a finite standard basis for a suitable ordering, this also works in the differential case.

PROPOSITION 6. *Let $\mathcal{I}$ be a differential ideal of $\mathcal{F}\{X\}$, Y a subset of X. Using lemma 2.5, we take any ordering $<$ on monomials and build a new ordering $\prec$ by considering first the degree of polynomials in the variables Y. If G is a standard basis of $\mathcal{I}$ for $\prec$, then the subset $G' = \{P \in G | \operatorname{lm} P \in \mathcal{F}\{X \setminus Y\}\}$ is a standard basis of $\mathcal{I} \cap \mathcal{F}\{X \setminus Y\}$.*

PROOF. Due to the properties of $\prec$, all polynomials in G' are in $\mathcal{F}\{X \setminus Y\}$, and polynomials in this subring cannot be reduced by the elements of $G \setminus G'$. So a polynomial in $\mathcal{F}\{X \setminus Y\}$ is in $\mathcal{I}$ iff it is reduced to 0 by G'. We conclude using th. 4.5. ∎

2. Application to birational mappings

2.1. A bound on the order of the inverse

We consider here a rational differential mapping $f : \mathbf{A}_{\mathcal{F}}^n \mapsto \mathbf{A}_{\mathcal{F}}^n$, defined by n differential fractions $f_1, \ldots, f_n$ in $\mathcal{F}\langle x_1, \ldots, x_n \rangle$. We will develop algorithmic methods to test whether f admits a rational inverse and to find it.

In the purely algebraic case, there is a theorem, that allows to bound the degree of f^{-1} knowing the degree of f. Its exact origin is not known, but a proof, due to O. GABBER may be found in [Ba]. Following the definition in [Ba], the degree of f is the maximal degree of polynomials P_i and $Q_1 \cdots Q_n$ if f is defined by the fractions P_i/Q_i.

THEOREM 1. *Let k be an algebraic field of arbitrary characteristic, $f : \mathbf{A}_k^n \mapsto \mathbf{A}_k^n$ be a birational mapping of degree d, then $\deg f^{-1} \leq (\deg f)^{n-1}$.* ∎

Our aim is to prove an analogous theorem for differential birational mappings. The proof of Gabber uses Bézout's theorem. In the differential case, we can substitute to it th. 1.1.11, which Ritt called indeed a differential analog of Bézout's theorem. The analogy is in fact very strong, for despite a few more technicalities due to the differential stuff, the proof mostly follows the algebraic one.

DEFINITION 2. *The order of a rational differential mapping is the maximal order of the fractions that define it.*

This definition does obviously not depend on the choice of coordinates.

DEFINITION 3. *Let P be an irreducible differential polynomial. Using some admissible ordering on derivatives, we denote by v_P the leading variable of P. The initial of P will be the leading coefficient of P, considered as a polynomial in $\mathcal{F}[\nu < v_P][v_P]$, and the separant of P is the polynomial $\frac{\partial P}{\partial v_P}$.*

PROPOSITION 4. *Let $P \in \mathcal{F}\{X\}$ be irreducible, the set $\{Q \in \mathcal{F}\{X\} | \exists (a,b) \in \mathbf{N}^2 \ Q S_P^a I_P^b \in [P]\}$ is a prime ideal, which is a component of $\{P\}$. It is called the general component of P.*

PROOF. See [Ko]. ∎

Lemma 5. Let P be an irreducible polynomial of order r in $\mathcal{F}\{X\}$, V the variety defined by the general component of P, $H_1, \ldots, H_{n-1}$ be generic hyperplanes of $A^n_{\mathcal{F}}$, i.e. varieties defined by polynomials $L_i = \left(\sum_j^n \epsilon_{i,j} x_j \right) - \epsilon_{i,0}$, where the $\epsilon_{i,j}$ are generic over $\mathcal{F}$. Then $V \cap \bigcap_{i=1}^{n-1} H_i$ is an irreducible variety of differential type $m-1$ and of typical differential dimension r over $\mathcal{F}\langle \epsilon \rangle$.

PROOF. There is a characteristic set Σ of $\bigcap_{i=1}^{n-1} H_i$ for some orderly ordering, with $x_n > \cdots > x_1$, which is of the form $\{x_2 - a_2 x_1 - b_2, \ldots, x_n - a_n x_1 - b_n\}$. We can reduce P by Σ by replacing in P x_i by $a_i x_1 + b_i$. The result of this reduction is an irreducible polynomial $S(x_1)$, of the same order as P. We claim that $\{S, x_n - a_n x_1 - b_n, \ldots, x_2 - a_2 x_1 - a_2\}$ is a characteristic set of the prime ideal defining $W = V \cap \bigcap_{i=1}^{n-1} H_i$. This is true, using [Ko chap. IV § 9 lemma 2 p. 167 and discussion of Problem (a) p. 169–170], because S is irreducible in $\mathcal{F}\langle \epsilon \rangle$ and the other polynomials are all absolutely irreducible.

From the proof of [Ko chap. II § 12 th. 6 p. 115], we deduce that $\omega_V(r) = \binom{r+m}{m} - \binom{r+m-\text{ord } P}{m}$. By an elementary calculation, we can see that the type of W is $m-1$ and its typical differential dimension $\text{ord } P$. ∎

Dealing with a rational mapping, we denote by fV the Zariski closure of the set theoretical image $f(V)$. Any generic point in fV is obviously in the set theoretical image.

THEOREM 6. *Let $\mathcal{F}$ be a differential field of characteristic zero with set of derivations $\Delta = \{\delta_1, \ldots, \delta_m\}$, $f : A^n_{\mathcal{F}} \mapsto A^n_{\mathcal{F}}$ be a birational differential mapping, then $\text{ord } f^{-1} \le n \, \text{ord } f$.*

PROOF. Take generic hyperplanes $H_0, H_1, \ldots, H_{n-1}$ over $\mathcal{F}$ in $A^n_{\mathcal{G}}$. $f H_0$ is an irreducible variety which is the general component of an irreducible polynomial P of order $\text{ord } f^{-1}$. Indeed, let H_0 be defined by the linear polynomial L_0 as in lemma 5, and R_i/S_i be the fractions defining f^{-1}, then, dividing the numerator of $L(R/S)$, considered as a polynomial in the $\epsilon_{i,j}$, by its content in $\mathcal{F}\{X\}$, we secure a suitable polynomial.

So, using the lemma, $f H_0 \cap \bigcap_{i=1}^{n-1} H_i$ is an irreducible variety of differential type $m-1$ and typical dimension $\text{ord } f^{-1}$ over $\mathcal{G}$. Let η be a generic zero of that variety.

We now consider the extension $\mathcal{G}\langle f^{-1}\eta \rangle$. It is $\mathcal{G}$-isomorphic to $\mathcal{G}\langle \eta \rangle$, for f is birational. So using [Ko chap. II § 12 prop. 15 p. 117], $\omega_{\eta/\mathcal{G}}(r-h) \le \omega_{f^{-1}\eta/\mathcal{G}}(r) \le \omega_{\eta/\mathcal{G}}(r+h)$, for some integer h. So those extensions have the same type and the same typical dimension $\text{ord } f^{-1}$.

Using birational equivalence, $f^{-1}\eta$ is a generic point of the irreducible variety $V = H_0 \cap \bigcap_{i=1}^{n-1} f^{-1} H_i$. Consider the set of polynomials

$$\Sigma = \{L_0, \text{denom } L_i(P/Q) \; 1 \le i \le n-1\},$$

where L_i is a linear equation defining H_i, and $f_i = P_i/Q_i$. The set $\{T \in \mathcal{F}\{X\} | \exists a \in \mathbf{N}^n \; T \prod Q_i^{a_i} \in [\Sigma]\}$ is the prime ideal defining V (this is almost the situation of th. 2.1). This ideal is then a component of $\{\Sigma\}$, which is defined by a set of equations of maximal order $\text{ord } f$. We can now apply theo 1.1.11 to show that $\text{ord } f^{-1} \le n \,\text{ord } f$. ∎

2.2. Algorithms

We still denote by P_i and Q_i the numerator and denominator of the fractions f_i

THEOREM 1.   *Let $f_1, \ldots, f_p$ be rational differential fractions in $\mathcal{F}\langle X \rangle$. Then, the ideal*

$$\mathcal{J} = [P_i(x) - Q_i(x)T_i \; 1 \le i \le p; Q_1(x) \cdots Q_p(x)u - 1]_{\mathcal{F}\{x,T,u\}}$$

is prime. $\mathcal{J} \cap \mathcal{F}\{x,T\}$ is the ideal defining the graph of the mapping $f\colon \mathbf{A}^n_{\mathcal{F}} \mapsto \mathbf{A}^p_{\mathcal{F}}$ induced by the f_i. A fraction U/V is in $\mathcal{F}\langle X \rangle$ iff there exists in $\mathcal{J}$ a polynomial of the form $S(T)U(x) - R(T)V(x)$ such that $S(T) \notin \mathcal{J}$. Furthermore, $U/V = (R/S)(f)$.

PROOF. The proof is the same as in the algebraic case (See [SS], or [O1]). It may be found in details in [O3]. ∎

By luck, some results of commutative algebra remain true, without any modification, in the differential case!

COROLLARY 2.   *The mapping f is birational iff $p = n$ and for all $x_i \in X$ there exists in $\mathcal{J}$ a polynomial of the form $S_i(T)x_i - R_i(T)$, where $S_i \notin \mathcal{J}$. In this case, the fractions R_i/S_i define f^{-1}.* ∎

COROLLARY 3.   *There exists an algorithm using a standard basis computation to test if f is birational and find its inverse.*

PROOF. Using classical arguments on orderings (see [O1]), we know that polynomials of the wanted form will appear during the computation of the reduced standard basis of $\mathcal{J}$, for an ordering which eliminates u and then X.

Let Σ be the set of generators of $\mathcal{J}$. Using th. 1.6, $S_i(T)x_i - R_i(T) \in \mathcal{F}(\Theta_{\text{nord} f}\Sigma)$, so that we have no use to consider syzygies involving polynomials of order greater than $(n+1)\text{ord} f$. We can simply compute an algebraic standard basis of the ideal $\mathcal{J}' = (\Theta_{\text{nord} f}\Sigma)$, or use a slightly modified version of the procedure given above by discarding derivatives of order greater than $(n+1)\text{ord } f$ which makes it an algorithm. ∎

THEOREM 4.   *If $f_1, \ldots, f_p$ are rational differential fractions equal to P_i/Q_i, then the ideal*

$$\mathcal{I} = [Q_i(y)P_i(x) - P_i(y)Q_i(x) \; 1 \le i \le n; \; u \,\text{lcm}(Q_1(x) \cdots Q_i(x)) - 1]_{\mathcal{F}\langle f \, y \rangle [x,u]}$$

is prime. A fraction $R/S \in \mathcal{F}\langle x \rangle$ belongs to $\mathcal{F}\langle f \rangle$ iff $R(x) - \frac{R(y)}{S(y)}S(x)$ belongs to $\mathcal{I}$.

PROOF. The proof in the differential case is exactly similar to the algebraic one, which may be found in [O1]. A detailed proof is given in [O3] ∎

We need to remark that $\mathcal{I}$ is prime only as an ideal in $\mathcal{F}\langle f(y)\rangle[x,u]$. For example $[x^2 - y^2]_{\mathcal{F}\langle y^2\rangle[x]}$ is prime, but of course $[x^2 - y^2]_{\mathcal{F}\langle y\rangle[x]}$ is not.

COROLLARY 5. *Let* $f\colon \mathbf{A}_{\mathcal{F}}^n \mapsto \mathbf{A}_{\mathcal{F}}^n$ *be a rational differential mapping defined by* $f_1,\ldots,f_n$ *as above, then* f *is birational iff* $\forall\, 1 \le i \le n\ x_i - y_i \in \mathcal{I}$.

Moreover, the rank of $x_i - y_i$ *over the generating polynomials of* $\mathcal{I}$, *with respect to some orderly ordering, is of order less than or equal to* $(n+1)\,\mathrm{ord}\, f$.

PROOF. The first part is immediate from the theorem. The second is a consequence of theo. 1.6. ∎

This result gives another algorithm to test if f is birational. Indeed, we only have to compute the standard basis of $\mathcal{I}$ up to order $(n+1)\,\mathrm{ord}\, f$, and test if $x_i - y_i$ reduces to 0. But under this form, we have lost the expression of the inverse and still have no control of the degree of computations, without a bound on the degree of f^{-1}.

Denote by Σ the set of generators of $\mathcal{I}$, by r the bound $(n+1)\mathrm{ord}\, f$, and by X' the set $X \cup \{u,v\}$ We compute a standard basis of $\mathcal{I}_i = [\Sigma;\ v(x_i - y_i) - 1]_{\mathcal{F}\langle f\,y\rangle[X']}$. f is birational iff $\mathcal{I}_i = [1]$ for all i. Still using theo. 1.6, this means that

$$1 \in (\Theta_r\, \Sigma \cup \{v(x_i - y_i) - 1\})_{\mathcal{F}(\Theta_r f(y))[\Theta_r X']}\, .$$

So we only have to consider $N = O((n+2)((n+1)\mathrm{ord}\, f)^m)$ derivatives. Using the effective nullstellensatz of KOLLÁR, we may bound the degree of intermediate computations by $d = (\deg f)^N$. Using a classical argument, we may reduce to the triangulation of a linear system of size at most $M \times (n+2)M$, where $M = O\left((\deg f)^{N^2}\right)$ is the maximal number of monomials. The coefficients are in $\mathcal{F}\{y\}$, and of degree $\deg f$ at most. The number of elementary operations in $\mathcal{F}[y]$ is polynomial in $(n+2)M$. If we use the method of BAREISS, the intermediate coefficients are minors of the matrix, so that their degree may be bounded by $D = M\deg f$ and their size by $S = O(D^N)$. The cost of any elementary operation in $\mathcal{F}[y]$ in term of elementary operations in $\mathcal{F}$ is polynomial in S. We get then the following theorem.

THEOREM 6. *We can test that* f *is birational with a complexity in elementary operations in* $\mathcal{F}$ *polynomial in* $(n+2)MS$, *i.e. polynomial in*

$$O\left((n+2)(\deg f)^{(n+2)^3((n+1)(\mathrm{ord}\, f))^{3m}}\right).$$

∎

Of course, we cannot use exactly the completion process described above to prove this theorem, because we would not be able to control the size of coefficients.

3. Conclusion

By itself, the definition of differential standard bases introduced here provides puzzling algorithmic and combinatoric problems. For example the enumeration of essential syzygies is non trivial, and one could ask whether there exists a more efficient way than trying all algebraic syzygies between derivatives.

The problem remains open of finding a satisfactory method to answer the membership problem for differential ideals. A careful study of the structure of differential ideals with finite, or infinite bases could be an inspiration to develop better and always finishing methods. But one of the main issues would be to provide an effective differential nullstellensatz, i.e. if $[P_1, \ldots, P_k] = [1]$, $\text{ord} P_i \leq e$, $\deg P_i \leq d$ to secure a bound $r(n, m, k, e, d)$ such that $(\Theta_r P) = (1)$, of which we would deduce many results of complexity using differential standard bases computations.

4. References

[Ba] H. BASS et al. *The jacobian conjecture: reduction of degree and formal expansion of the inverse*, Bulletin of the A.M.S. vol. 7, n° 2, 1982.

[Bu] B. BUCHBERGER, *A criterion for detecting unecessary reductions in the construction of Groebner bases*, proceedings of EUROSAM'79, Marseille, Lect. Notes in Computer Science 72, 2–31, Springer Verlag, 1979.

[Car] G. CARRA'-FERRO, *Gröbner Bases and Differential Ideals*, proceeding of AAECC'5, Lect. Notes in Computer Science 356, 129–140, Spinger Verlag, 1987.

[Cas] F. CASTRO, *Théorèmes de division dans les opérateurs différentiels et calculs des multiplicités*, Thèse de troisième cycle, Université Paris VII, 19 Octobre 1984.

[Ch] CHOU Shang-Ching, *Mechanical geometry theorem proving*, D. Reidel pub. co., 1988.

[D] Sette DIOP, *Théorie de l'élimination et principe du modèle interne en automatique*, thèse de doctorat, université Paris-Sud, 1989.

[Ja] M. JANET, *Sur les systèmes d'équations aux dérivées partielles*, Journ. de Math. (8ᵉ série), tome III, 1920.

[Ka] I. KAPLANSKY, *An introduction to differential algebra*, Hermann, Paris, 1957.

[KM] Deepak KAPUR and Klaus MADLENER, *A Completion Procedure for Computing a Canonical Basis of a k-Subalgebra*, Computers and Mathematics, E. Kaltofen and S. M. Watt editors, Springer, 1989.

[Ko] E. R. KOLCHIN, *Differential algebra and algebraic groups*, Academic Press, 1973.

[Koll] J. KOLLÁR, *Sharp effective nullstellensatz*, J. Am. Math. Soc. 1, (963-975), 1988.

[O1] F. OLLIVIER, *Inversibility of rational mappings and structural identifiability in automatics*, proc. of ISSAC'89, Portland, Oregon, ACM Press, 1989.

[O2] F. OLLIVIER, *Canonical bases: relations with standard bases, finiteness conditions and application to tame automorphisms*, to appear in the proceedings of MEGA'90, Castiglioncello.

[O3] F. OLLIVIER, *Le problème de l'identifiabilité : approche théorique, méthodes effectives et étude de complexité*, Thèse de Doctorat en Sciences, École Polytechnique, Juin 1990.

[P1] J. F. POMMARET, *Differential Galois theory*, Gordon and Breach, New-York, 1983.

[P2] J. F. POMMARET, *Effective method for systems of algebraic partial differential equations*, preprint, 1989.

[R1] J. F. RITT, *Differential equations from the algebraic standpoint*, A.M.S. col. publ. vol. XIV, 1932.

[R2] J. F. RITT, *Differential algebra*, A.M.S. col. publ. vol. XXXIII, 1950.

[RS] L. ROBBIANO and M. SWEEDLER, *Subalgebra Bases*, preprint, Cornell Univ., 1989.

[S] A. SEIDENBERG, *An elimination theory for differential algebra*, Univ. California Publications in Math., (N.S.), 3, n° 2, 31–65, 1956.

[SS] D. SHANNON and M. SWEEDLER, *Using Groebner bases to determine algebra membership, split surjective algebra homomorphisms and determine birational equivalence*, preprint 1987, appeared in J. Symb Comp. 6 (2-3).

COMPLEXITY OF STANDARD BASES IN PROJECTIVE DIMENSION ZERO II

Marc Giusti *
SDI CNRS 6176 "Calcul formel, Algèbre et Géométrie algorithmiques"
(associée au Centre de Mathématiques, URA CNRS D.0169)
Laboratoire d'Informatique
Ecole Polytechnique
91128 PALAISEAU CEDEX
giusti@cmep.polytechnique.fr
cfmagi@frpoly11.BITNET

Abstract

This paper is the continuation of a previous one devoted to the complexity of standard bases in projective dimension zero (see Giusti, 1989). We improve the upper bound on the maximal degree of elements in a standard basis, with respect to any choice of coordinates and any compatible ordering, given in *loc.cit.* This bound is sharp, being attained for complete intersections and lexicographic ordering.

INTRODUCTION

Let us consider an algebraic projective subvariety of $\mathbf{P}^n$, defined by a homogeneous ideal I of the polynomial algebra $R = k[x_0, \ldots, x_n]$ over some ground field k. A standard basis of I, w.r.t. a total compatible ordering of the monomials of R, depends strongly on the chosen order and on the coordinates, and can vary a lot.

However in low dimension, namely less or equal to zero, we can give a sharp upper bound for the maximal degree of the generators of a standard basis, with respect to any choice of coordinates and any compatible ordering. It is first expressed with intrinsic quantities (depending only on the ideal itself) associated to the Hilbert function. It improves the result given in Giusti (1989, Proposition 1.2.).

---

*Partially supported by a grant from GDR G0060 "Calcul Formel, Algorithmes, Langages et Systèmes" and from PRC "Mathématiques et Informatique".

Eventually, if d is the maximal degree of the input generators of I, this bound is d^n in dimension zero. This result was obtained by I. Bermejo (1990), in the following particular case : the ideal defines an arithmetically Cohen-Macaulay projective variety of dimension zero, the coordinates are rather pleasant ("commodes" in the sense of Kouchnirenko), and the order on the monomials is the pure lexicographic one. It is sharp, being attained for complete intersections and pure lexicographic ordering, as already noticed by I. Bermejo.

The proof is essentially combinatorial, using stable subsets of $\mathbf{N}^{n+1}$, and has practically nothing to do with commutative algebra ...

1 STABLE SUBSETS

1.1 Some preliminary facts

Let E be a *stable* subset of $\mathbf{N}^{n+1}$, i.e. such that for every a in E and every b in $\mathbf{N}^{n+1}$, $a + b$ is still in E. We shall recall in this section different results on stable subsets.

1.1.1 Classical definitions

A stable subset E is minimally generated by a finite subset $B(E)$ ("Dickson's lemma"). The *degree* of a point $a = (a_0, \ldots, a_n)$ of $\mathbf{N}^{n+1}$ being the integer $|a| = a_0 + \cdots + a_n$, we shall denote by $D(E)$ the maximal degree of the elements of $B(E)$.

The function HF_E, which associates to every integer s the number of elements of degree s not belonging to E, is called the *Hilbert function* of E. For the argument s large enough, this function is equal to a polynomial $HP(E)$, the so-called *Hilbert polynomial* of E. The *regularity* of the Hilbert function is the smallest integer $H(E)$ after which the function coincides with the polynomial.

The degree of the Hilbert polynomial is then the *dimension* of E, and the leading coefficient multiplied by $dim(E)!$ is the *degree* of E (noted $deg(E)$). Finally by convention the degree -1 will be associated to the zero polynomial.

1.1.2 Induction on the ambient dimension

To study stable subsets, a natural idea is to proceed by induction on n, so we introduce the sections $E_i^{(j)}$ of E by the hyperplane $\{x_j = i\}$. Again by a noetherianity argument (Dickson's lemma), this section is constant for large i. Let us denote by $e^{(j)}(E)$ (or simply $e(E)$ or e if there is no ambiguity on the axis x_j) the smallest integer after which this happens. We shall speak of the "section at infinity" $E_\infty^{(j)}$, or simply E_∞ instead of this constant section $E_i^{(j)}$ ($i \geq e^{(j)}$). As an application, we have the following :

1.1.3 Lemma :

Considering sections orthogonal to any fixed axis, we obtain :

$$H(E) \leq Max\{i + H(E_i) \mid i = 0, \ldots, e\}$$

Proof : it is given in Giusti (1988, proof of Proposition 3.4).

1.1.4 Saturation

For every index j $(j = 0, \ldots, n)$, let us consider the cartesian product of the section at infinity $E_\infty^{(j)}$ by the corresponding axis x_j. This product is a stable subset of $\mathbf{N}^{n+1}$ containing E. The intersection of all these cylinders $(j = 0, \ldots, n)$ is still stable, and contains E. We call it the *saturation* of E, or the *saturated* stable subset $G(E)$ associated to E. A useful property of the couple $E \subseteq G(E)$ is :

1.1.5 Lemma :

The Hilbert polynomials of a stable subset and its saturation coincide, since the difference $F(E) = G(E) - E$ is a finite subset.

Proof : Let us discard first the trivial case where $F(E)$ is empty. Then let $A = (A_0, \ldots, A_n)$ be a point of $F(E)$, and let us draw from A the $n+1$ lines $A + \mathbf{N}^{\{0, \ldots, 0, j, 0, \ldots, 0\}}$. Any of this lines intersects E, because if not it should also intersect the complementary of the corresponding section at infinity, and A should not belong to the saturation. So there is a point of E on each line, which implies that $F(E)$ is finite.

This notion of saturation is general, but was already used by Giusti (1989, Lemma 1.2.1.) in the case of 0-dimensional stable subsets for the same purpose.

1.2 Negative dimensional case

In this situation, the complement $\mathbf{N}^{n+1} - E$ is finite, say of volume $vol(E)$ (we say also that this last integer is the covolume of E). In particular, the integers $vol(E_i)$ are zero for i greater or equal to e, and not before.

1.2.1 Lemma :

Let E be a stable subset of negative dimension. Then :

$$vol(E) \geq Max\{i + vol(E_i) \mid i = 0, \ldots, e\}$$

Proof : The integers $vol(E_i)$, $(i = 0, \ldots, e-1)$ form a decreasing sequence of integers at least 1. Thus :

$$
\begin{aligned}
vol(E) &= \sum_{i=0}^{e-1} vol(E_i) \geq vol(E_0) + (e-1) \\
&= Max\{vol(E_i) \mid i = 0, \ldots, e-1\} + Max\{i \mid i = 0, \ldots, e-1\} \\
&\geq Max\{vol(E_i) + i \mid i = 0, \ldots, e-1\}
\end{aligned}
$$

Furthermore, E_e being the whole ambient space is of covolume zero. As the point $(0, \ldots, 0, e)$ is a generator of E, there are e points on the x_n-axis not belonging to E, and $vol(E)$ is at least e. Thus the proof is complete.

1.2.2 Lemma :

Let E be a stable subset of negative dimension. Then :

$$D(E) \leq H(E) \leq vol(E)$$

Proof : the first inequality is proved in Giusti (1988, Proposition 3.6.), the second by induction on n. The assertion is clear when n is equal to 0. Then :

$$
\begin{aligned}
vol(E) \quad &\geq \quad Max\{vol(E_i) + i \mid i = 0, \ldots, e\} \quad (1.2.1) \\
&\geq \quad Max\{H(E_i) + i \mid i = 0, \ldots, e\} \quad \text{(induction hypothesis)} \\
&\geq \quad H(E) \quad (1.1.3)
\end{aligned}
$$

1.3 Zero dimensional case

1.3.1 Structure of saturated subsets

We are first interested in a saturated stable subset G of dimension zero, which is in this case an intersection of cylinders whose section is of zero or finite covolume :

$$G = \bigcap_{j=0}^{n} (G_\infty^{(j)} \times \mathbf{N}^{\{0, \ldots, 0, j, 0, \ldots, 0\}})$$

Let us call *trivial* a cylinder corresponding to the whole ambient space. The Hilbert polynomial is a non zero constant, whose value is the degree of the saturated subset, and :

$$deg(G) = \sum_{j=0}^{n} vol(G_\infty^{(j)})$$

1.3.2 Regularity of the sections

Unfortunately a section G_i is in general no more saturated, which prevents us from straightforward inductions. Nevertheless, an upper bound of the regularity is the following :

Let G be a stable saturated subset of dimension zero. Considering sections orthogonal to some fixed axis, we get :

$$H(G_i) \leq Max(H(G(G_i)), vol(G_\infty))$$

Proof : we can assume, without loss of generality, that we are looking at sections orthogonal to the last axis x_n. From :

$$G = \bigcap_{j=0}^{n} (G_\infty^{(j)} \times \mathbf{N}^{\{0, \ldots, 0, j, 0, \ldots, 0\}})$$

we can deduce, intersecting by $\{x_n = i\}$:

$$G_i^{(n)} = G_\infty^{(n)} \bigcap_{j \neq n} (G_\infty^{(j)} \cap \{x_n = i\}) \times \mathbf{N}^{\{0,\ldots,0,j,0,\ldots,0\}} \tag{1}$$

By abstract non-sense, we see that the second part of the previous intersection is the saturation of G_i, and that $F(G_i)$ is a subset of the complementary of $G_\infty^{(n)}$. In particular, the Hilbert functions of G_i and its saturation coincide for any argument larger or equal to $vol(G_\infty^{(n)}) \geq H(G_\infty^{(n)})$ by 1.2.2. Hence the regularity of G_i is at most $Max(H(G(G_i)), vol(G_\infty))$.

1.3.3 Constancy of the sections

Let G be a stable saturated subset of dimension zero. Considering sections orthogonal to the last axis, we get :

$$e^{(n)}(G) \leq Max\{vol(G_\infty^{(j)}) \mid j \neq n\}$$

Proof : let j be an index not equal to n. For i larger or equal to $vol(G_\infty^{(j)})$, 1.2.2 implies that $G_\infty^{(j)} \cap \{x_n = i\}$ is equal to its whole ambiant space, hence for i larger or equal to $Max\{vol(G_\infty^{(j)}) \mid j \neq n\}$, the decomposition (1) yields that G_i is constant equal to G_∞.

Let us turn now to the degree and the regularity of G itself. We know already the following fact established in Giusti (1989, Lemma 1.2.2.) :

1.3.4 Lemma :

Let G be a stable saturated subset of dimension zero. Then :

$$D(G) \leq deg(G)$$

The behaviour of the regularity is the same :

1.3.5 Lemma :

$$H(G) \leq deg(G)$$

Proof : The proof goes by induction on the number of the non trivial cylinders.

The first case is when this number is one, where it is clear since G is the cylinder $G_\infty \times \mathbf{N} = G_0 \times \mathbf{N}$: by 1.2.2, the regularity of G_0 is at most its covolume, which is nothing else than the degree of G. Hence $H(G) = H(G_0) - 1$ is at most $deg(G) - 1$.

In the general case, the saturated subset is an intersection of more than two cylinders, and assume without loss of generality that the last axis x_n corresponds to one of them. So using successively 1.1.3 , 1.3.2 and the induction hypothesis, :

$$
\begin{aligned}
H(G) \;\leq\; & Max\{i + H(G_i) \mid i = 0, \ldots, e\} \\
\leq\; & Max(Max\{i + H(G(G_i)) \mid i = 0, \ldots, e\}, Max\{i + vol(G_\infty) \mid i = 0, \ldots, e\}) \\
\leq\; & Max(Max\{i + deg(G(G_i)) \mid i = 0, \ldots, e\}, e + vol(G_\infty)) \\
\leq\; & Max(Max\{i + deg(G_i) \mid i = 0, \ldots, e\}, e + vol(G_\infty))
\end{aligned}
$$

By 1.3.3 , the second term is at most $Max\{vol(G_\infty^{(j)}) \mid j \neq n\} + vol(G_\infty^{(n)})$, itself at most $deg(G)$. Now :

$$
\begin{aligned}
i + deg(G_i) &= i + \sum_{j \neq n} vol(G_\infty^{(j)} \cap \{x_n = i\}) \\
&\leq \sum_{j \neq n} i + vol(G_\infty^{(j)} \cap \{x_n = i\})
\end{aligned}
$$

Taking the maximum of all these quantities, for i running from 0 to e, yields the lemma using 1.2.1 and 1.3.3.

2   The theorem :

Let E be a stable subset of $\mathbf{N}^{n+1}$ of dimension 0. Then :

$$
D(E) \leq Max(deg(E), H(E))
$$

Proof : it follows the proof given in Giusti (1989, 1.2.3), in a slightly different way. For the convenience of the reader, we give it completely.

We divide the basis $B(E)$ into two classes of generators : the first ones belong to $B(G)$, the others do not. If $A = (A_0, \ldots, A_n)$ is in the last class, being a generator of E there exists an index i such that $(A_0, \ldots, A_i - 1, \ldots, A_n)$ is no longer in E but stays in G, hence is in $F(E)$. Either the degree of any element in the second class is at most $deg(E)$, and the conclusion holds using the lemma 1.3.4 ; or not, and then using again the lemma 1.3.4, we see that the maximal degree $D(E)$ is attained for an element A not in $B(G(E))$, so $D(E) - 1$ is at least $deg(E)$. Notice that there exists a point in $F(E)$ of degree $D(E) - 1$, so let us compute the Hilbert function in this degree :

$$
\begin{aligned}
HF_E(D(E) - 1) &= HF_F(D(E) - 1) + HF_G(D(E) - 1) \\
&> HF_G(D(E) - 1)
\end{aligned}
$$

The regularity of HF_G is bounded above by $deg(G(E)) = deg(E)$ (1.3.5, 1.1.5) hence by $D(E) - 1$. So :

$$
HF_E(D(E) - 1) > HP_G(D(E) - 1) = HP_E(D(E) - 1)
$$

We conclude that $H(E)$ is at least $D(E)$, and in all cases we proved the proposition.

2.1   Corollary

Let k be a field, R the polynomial algebra $k[x_0, \ldots, x_n]$ and I a homogeneous ideal of R. Let us choose any compatible total ordering on the monomials of R. Let us consider the projective subvariety $V(I)$ of $\mathbf{P}^n$ defined by I, and let us assume that its dimension is zero.

With the notations and under the assumptions above, the maximal degree of a standard basis, w.r.t. any choice of the coordinates and of the ordering, is at most the maximum of the degree of the variety $V(I)$ and the regularity of the Hilbert function of I.

2.2 Corollary

Assume in the corollary 2.1 that the ideal I is given by generators of degree at most d. Under the same assumptions, the maximal degree of a standard basis, w.r.t. any choice of coordinates and compatible total ordering, is at most d^n.

Proof : the degree of the variety is at most d^n by Bézout's theorem. The regularity is at most $1 + (n+1)(d-1)$ by a result of Lazard (1981). After treating the limit cases, the corollary follows.

2.3 An example

In the following example the upper bound of the theorem is attained, by the regularity and not the degree, despite of the quite different asymptotic behaviours.

Let $d \geq 2$ be an integer, and let us consider the ideal $(z^d - y^d, yz - xy, xz)$ of $k[x, y, z]$. Geometrically, it defines a unique point $(1, 0, 0)$ of $\mathbf{P}^2$. A standard basis with respect to the lexicographic order induced by $z > y > x$ is easily computed, and is $(z^d - y^d, yz - xy, xz, y^{d+1}, xy^d, x^2y)$. So it is easily seen that the degree is 1, the regularity $d+1$, this last integer beeing the maximal degree of the standard basis.

REFERENCES

I. BERMEJO (1990), Sur les degrés d'une base standard minimale pour l'ordre lexicographique d'un idéal dont la variété des zéros est Cohen-Macaulay de dimension 1, *Notes aux Compte-Rendus de l'Académie des Sciences de Paris*, t. **310**, Série I, 591-594.

M. GIUSTI (1988), Combinatorial dimension theory of algebraic varieties, in "Computational Aspects of Commutative Algebra", special issue, *Journal of Symbolic Computation, Academic Press* **6**, 115-131.

M. GIUSTI (1989), Complexity of standard bases in projective dimension zero, Proceedings of EUROCAL 87 (European Conference on Computer Algebra, Leipzig, RDA), *Lecture Notes in Computer Science 378, Springer Verlag*, 333-335.

D. LAZARD (1981), Résolution des systèmes d'équations algébriques, *Theoretical Computer Science* **15**, 77-110.

SYSTOLIC ARCHITECTURES FOR MULTIPLICATION OVER FINITE FIELD GF(2$^{\mathbf{m}}$)

Menouer Diab

Laboratoire AAECC/LSI, IRIT, Université Paul Sabatier
118 Route de Narbonne, 31062 Toulouse cedex, FRANCE

ABSTRACT: *In this paper, we develop two systolic architectures for the product-sum computation $P = AB + C$ in the finite field $GF(2^m)$. The multipliers consist of m basic cells arranged into a serial-in, serial-out one-dimensional systolic array. They need only one control signal. The first multiplier is semi-serial (coefficient B is input in parallel), and performs simultaneously two product-sum computations $P = AB + C$ and $P' = A'B + C'$. The bits of the coefficients A, C, A', C' are received serially. The bits of the results P and P' are generated serially. The second multiplier is serial (coefficients A, B, and C are input serially), and performs one product-sum computation at a time. The bits of the coefficients A, B, and C are received serially. The bits of the result P are generated serially. In all the cases, the architectures are simple, regular, and possess the properties of concurrency and modularity. As a consequence, they are well suited for VLSI design.*

1   INTRODUCTION

Finite fields play a central role in several practical and important domains. They can be applied to error-correcting codes [7-2], digital signal processing [11], and cryptography [13]. In particular, the encoding and decoding algorithms of a binary REED-SOLOMON (*RS*) code require algebraic operations in some field $GF(2^m)$. Among these operations, they need particulary the product-sum operation $P = AB + C$ [8-6-12]. Addition in $GF(2^m)$ is bit-independent and can be performed using two-input XOR gates. However, multiplication in $GF(2^m)$ is a more difficult task.

Several circuits have been proposed to realize multiplication in $GF(2^m)$ [7-2-1-4]. Unfortunately, they have some disadvantages such as irregular wire routing, complicated control problems, nonmodular structure or lack of concurrency [5]. To solve these problems, some systolic multipliers have been proposed in [14-8-15]. The multipliers proposed in [14-8] are semi-serial (B is input in parallel). Whereas the multiplier proposed in [15] is serial (all the coefficients are input serially). The multiplier proposed in [8] is less complex than the one proposed in [14]. Only one control signal is needed rather than two in [14]. However, its elementary computation time increases with m.

In this paper, two systolic multipliers are proposed. The first one is semi-serial and performs simultaneously two product-sum computations, rather than one computation at a time in [8-14]. It requires only one control signal and can operate at high rates. A less complex multiplier than the one presented in [8] can be derived from this multiplier. It performs one product-sum computation at a time. However, as in [8], its elementary computation time increases with m. The second multiplier is serial and performs one product-sum computation at a time. It requires one control signal and can operate at high rates. It is less complex than the one proposed in [15]. Contrary to [15], it avoids the use of an external XOR gate to perform the product-sum computation. Moreover, to perform this computation, the multiplier proposed in [15] needs $2m$ additional flip-flops to delay the bits of C. To perform the same computation no additional flip-flops are needed for our multiplier.

2 A SEMI-SERIAL MULTIPLIER FOR $GF(2^m)$

Let $F(x) = \sum_{k=0}^{m} f_k x^k$ $(f_m = f_0 = 1)$ be a primitive irreducible polynomial, and α a root of $F(x)$, then the elements of $GF(2^m)$ can be represented as polynomials of α with degree less than m. That is $GF(2^m) = \left\{ \sum_{k=0}^{m-1} a_k \alpha^k \text{ with } a_k \text{ in } GF(2) \right\}$. Let $A = \sum_{k=0}^{m-1} a_k \alpha^k$, $B = \sum_{k=0}^{m-1} b_k \alpha^k$, and $C = \sum_{k=0}^{m-1} c_k \alpha^k$ be three elements of $GF(2^m)$. Suppose $P = \sum_{k=0}^{m-1} p_k \alpha^k$ is the product of A and B, ie $P = AB$.

P can be written as:
$$P = \sum_{j=0}^{m-1} a_{m-1-j} B \alpha^{m-1-j} \qquad (1)$$

Suppose that: $\quad 0 \le i \le m-1 \quad P(i) = \sum_{j=0}^{i} a_{m-1-j} B \alpha^{i-j} \qquad (2)$

From (1) and (2), it follows that $P = P(m-1)$. From (2), one obtains:

$$0 \le i \le m-1 \quad P(i) = \alpha P(i-1) + a_{m-1-i} B \quad ; \quad P(-1) = 0 \qquad (3)$$

Suppose $P(i) = \sum_{j=0}^{m-1} p(i,j)\alpha^j$, then from (3) and the fact that $\alpha^m = \sum_{j=0}^{m-1} f_j \alpha^j$, the following relation is obtained:

$0 \le i \le m-1 \ ; \ m-1 \ge j \ge 0$
$p(i,j) = p(i-1,j-1) + a_{m-1-i}b_j + f_j p(i-1,m-1) \ ; \ p(-1,j) = p(i,-1) = 0$

Using the function $start(i)$ defined as: $start(i) = \begin{cases} 0 & \text{if } i \equiv -1 \pmod{m} \\ 1 & \text{else} \end{cases}$

a more general relation can be established:

$-1 \leq i \leq m - 1 \; ; \; m - 1 \geq j \geq 0$

$$p'(i,j) = p(i-1, j-1) + a_{m-1-i}b_j + f_j p(i-1, m-1) \tag{4}$$

$$p(i,j) = p'(i,j).start(i) \; ; \; p(i,-1) = 0 \; ; \; p_j \equiv p'(m-1,j)$$

Using relation (4) several times one obtains:

$m - 1 \geq j \geq 0 \quad p'(m-1,j) = p(m-j-2,-1) + p(m-j-2, m-1)f_0 +$

$$\ldots + p(m-2, m-1)f_j + a_j b_0 + \ldots + a_0 b_j$$

with $\quad p(m-j-2, -1) = 0$

If the term $p(m - j - 2, -1)$ is replaced by c_j, rather than 0, the term $p'(m - 1, j)$ computed by (4) will be the term p_j of $P = AB + C$. This yields to the following relation which performs the product-sum operation $P = AB + C$:

$-1 \leq i \leq m - 1 \; ; \; m - 1 \geq j \geq 0$

$$p'(i,j) = p(i-1, j-1) + a_{m-1-i}b_j + f_j p(i-1, m-1) \tag{5}$$

$$p(i,j) = p'(i,j).start(i) \; ; \; p(m-j-2,-1) = c_j \; ; \; p_j \equiv p'(m-1,j)$$

In (5), note that for a fixed value of i, the same terms a_{m-1-i}, $p(i-1, m-1)$ and $start(i)$ are used to compute the coefficients $p(i,j)(m-1 \geq j \geq 0)$. Similary, for a fixed value of j, the same terms b_j and f_j are used to compute the coefficients $p(i,j)(0 \leq i \leq m-1)$. Using this remark, relation (5) can be transformed to obtain the following *uniform recurrent equation* [9-10]:

$-1 \leq i \leq m - 1 \; ; \; m - 1 \geq j \geq 0$

$$p'(i,j) = p(i-1, j-1) + a(i,j)b(i,j) + f(i,j)p^*(i,j)$$

$$p(i,j) = p'(i,j).start(i,j) \; ; \; p(m-j-2,-1) = c_j \; ; \; p_j \equiv p'(m-1,j)$$

$$start(i,j) = start(i,j+1) \; ; \; start(i,m) = start(i)$$

$0 \leq i \leq m - 1 \; ; \; m - 1 \geq j \geq 0 \tag{6}$

$$a(i,j) = a(i,j+1) \; ; \; a(i,m) = a_{m-1-i}$$

$$p^*(i,j) = p^*(i,j+1) \; ; \; p^*(i,m) = p(i-1, m-1)$$

$$b(i,j) = b(i-1,j) \; ; \; b(-1,j) = b_j \; ; \; f(i,j) = f(i-1,j) \; ; \; f(-1,j) = f_j$$

From (6), one can use the *dependence mapping procedure*, presented in [9-10], to design a systolic multiplier. This procedure can be described briefly as follows:

The terms used in equation (6) are indexed by two integers i and j. To the term $p(i,j)$ is associated the point (i,j) of the Euclidean space. When the values of i and

j change, a set of points S is obtained: $S = \{(i,j)/ -1 \leq i \leq m-1 \, ; \, m-1 \geq j \geq 0\}$. According to equation (6), the computation of $p(i,j)$ at point (i,j) depends on points $(i-1,j-1)$, $(i,j+1)$, and $(i-1,j)$. Joining these points to (i,j), for (i,j) in S, yields to the *dependence graph* corresponding to equation (6). This graph is shown in Fig.1.(for more details about this procedure see [9-10]).

Let us denote as $t(i,j)$ the time at which computation at point (i,j) is done. If we assume that every computation takes one unit of time, and using the fact that a computation is performed only when its operands are ready, we obtain (Fig.1):

$$\forall \ (i,j) \in S \qquad t(i,j) = (m-1-j) + 2i \qquad (7)$$

If the graph of Fig.1 is projected along $vector(1,0)$, a set of m points is obtained: $\{(0,j)/m-1 \geq j \geq 0\}$. $(0,j)$ is the projection of points $\{(i,j)/ -1 \leq i \leq m-1\}$. To the point $(0,j)$ a given cell $cell_j$ is associated. It carries out the computations corresponding to the terms $p(i,j)(-1 \leq i \leq m-1)$. In the resulting multiplier (obtained by projection along $vector(1,0)$), the bits corresponding to A and C will move serially through the cells. The bits corresponding to B and $F(x)$ remain at the same positions and then will be input in parallel. The same situation occurs to the bits p_j of P. Indeed, from equation (6) and the graph of Fig.1, it follows that p_j is computed by $cell_j$ ($p_j = p'(m-1,j)$) and is not transfered to the neighbouring cells. In order to ensure a serial output for the bits of P, another variable psr (product-sum result) is introduced (see Fig.1). It ensures the serial transfer of bits p_j. The final *uniform recurrent equation* is obtained by adding to (6) the following equation:

$$-1 \leq i \leq 2m-2 \ ; \ m-1 \geq j \geq 0$$
$$psr(i,j) = p'(i,j).\overline{start(i,j)} + psr(i-1,j-1).start(i,j) \qquad (6')$$
$$p_j \qquad \equiv psr(2m-2-j, m-1)$$

To see that equation (6)-(6') gives the same results p_j as (6), note that:

$$psr(2m-2-j, m-1) = psr(2m-3-j, m-2) = \ldots = psr(m-1,j) = p'(m-1,j)$$

thus: $\qquad m-1 \geq j \geq 0 \qquad psr(2m-2-j, m-1) = p_j$

The algorithm and the structure of $cell_j$ are shown in Fig.2. According to relation (7), the terms $p(i,j)$ and $p(i+1,j)$ are computed respectively at times $t(i,j) = (m-1-j)+2i$ and $t(i+1,j) = (m-1-j) + 2(i+1)$, thus $t(i+1,j) - t(i,j) = 2$. It follows that $cell_j$ is active one cycle every two clock cycles (clock cycle=unit of time). To permit more parallel processings, one solution consists in using the cycle where the cell is inactive to compute another product-sum operation $P' = A'B + C'$.

The multiplier architecture for the field $GF(2^4)$ is shown in Fig.3. It consists of m basic cells. The bits f_j and b_j ($m-1 \geq j \geq 0$) of $F(x)$ and B respectively are stored in m-bit registers. The interleaved bits a_{m-1}, a'_{m-1}, a_{m-2}, $a'_{m-2}, \ldots,$ a_0, a'_0 of A and A' are received serially at the input a_{in} of the leftmost cell. At the same

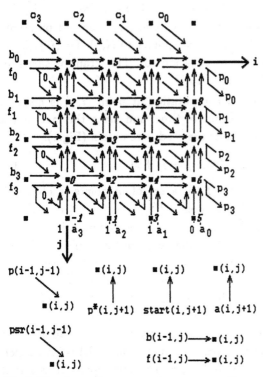

Fig.1 The dependence graph corresponding
to equation (6) $(m = 4)$

```
Algorithm :
  begin
    if  start_in = 0
        then
          begin
            p_out := 0;
            psr_out := p_in + a_out b_in + f_in p*_out
          end
        else
          begin
            p_out := p_in + a_out b_in + f_in p*_out;
            psr_out := psr_in
          end;
    a_out := a_in ;  p*_out := p*_in ;  start_out := start_in
  end.
```

Fig.2.a The algorithm of a basic cell

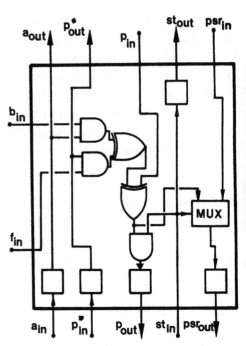

Fig.2.b The circuit of a basic cell

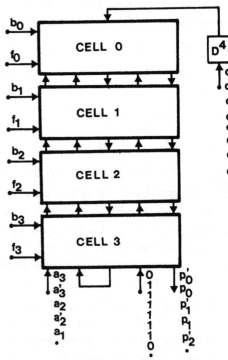

Fig.3 A semi-serial multiplier
for the field $GF(2^4)$

time, the interleaved bits c_{m-1}, c'_{m-1}, c_{m-2}, $c'_{m-2}, \ldots$, c_0, c'_0 of C and C' are received serially at the input c_{in} of the m-bit shift register D^m. This register is used for the purpose of synchronization. All the bits arrive at a rate of one bit per clock cycle. Only one control signal is needed in this multiplier. The leftmost-cell input $start_{in}$ is always set to 1 except when the bits a_{m-1} and a'_0 are received at the input a_{in}. The control signal ensures the initialization of the multiplier for a new product-sum computation, and the output of the bits p_j and p'_j ($m - 1 \geq j \geq 0$) of the respective results P and P'. The interleaved bits p_{m-1}, p'_{m-1}, p_{m-2}, $p'_{m-2}, \ldots$, p_0, p'_0 of P and P' are generated, serially through the leftmost-cell output psr_{out}, at a rate of one bit per clock cycle. According to equation (7), the first bit p_{m-1} of P is computed in $cell_{m-1}$ at the time unit $t(m - 1, m - 1) = 2(m - 1)$. At the same time unit, it is available at the multiplexer output of $cell_{m-1}$ (see Fig.2). After being loaded in the flip-flop corresponding to psr_{out} in $cell_{m-1}$, it will be generated at the leftmost-cell output psr_{out} at the time unit $2(m - 1) + 1$. Since the first bit a_{m-1} of A enters the leftmost-cell input a_{in} at the time unit $t(0, m) = -1$, $2m$ time units are then needed for the generation of the first bit p_{m-1} of P.

Compared with the multiplier presented in [14], our multiplier is less complex. Indeed, it contains $6m$ flip-flops while the multiplier developed in [14] uses $10m + 2$ flip-flops. They contain the same number of gates. Instead of two control signals in [14], only one control signal is used. Another advantage of this multiplier over those presented in [14-8] is that it preserves the bits a_i of A. This is interesting for the design of systolic RS encoders [8]. The use of m supplementary flip-flops to preserve the bits a_i is avoided. On the other hand, it computes two product-sum operations $AB+C$ and $A'B+C'$ which offers more parallel processings than the multipliers presented in [8-14]. If the flip-flops corresponding to p_{out} and psr_{out} are eliminated (see Fig.2), a multiplier which performs one product-sum computation at a time is obtained. Its gate complexity is better than the one proposed in [8]. Indeed, this last one contains in addition m inverters. Its flip-flop complexity is also better when it is integrated in the RS systolic encoder developed in [8]. The integration of the multiplier presented in [8] in the encoder requires m additional flip-flops per encoder cell. Their function is to preserve the bits a_i of A.

3   A SERIAL MULTIPLIER FOR $GF(2^m)$

In this section, we develop a serial-systolic multiplier based on the idea of E.R.BERLEKAMP used in [3].

Let $B_c = \{1, \alpha, \alpha^2, \ldots, \alpha^{m-1}\}$ be a conventional basis of $GF(2^m)$ and $B_d = \{\lambda_0, \lambda_1, \ldots, \lambda_{m-1}\}$ its dual basis. Suppose that the coefficients A, C and P are expressed in B_d, and the coefficient B is expressed in B_c. That is: $A = \sum_{k=0}^{m-1} a_k \lambda_k$, $B = \sum_{k=0}^{m-1} b_k \alpha^k$, $C = \sum_{k=0}^{m-1} c_k \lambda_k$, and $P = \sum_{k=0}^{m-1} p_k \lambda_k$

The trace of an element x in $GF(2^m)$ is defined as follows:

$$Tr : GF(2^m) \longrightarrow GF(2) , \quad x \longmapsto Tr(x) = \sum_{k=0}^{m-1} x^{2^k}$$

The useful properties of Tr are briefly reviewed as follows [8-3]:

P1. $\forall \, \lambda, \mu \in GF(2)$; $\forall \, x, y \in GF(2^m) : Tr(\lambda x + \mu y) = \lambda Tr(x) + \mu Tr(y)$

P2. $\forall \, x \in GF(2^m) :$ if $\quad x = \sum_{k=0}^{m-1} x_k \lambda_k \quad$ then $\quad \forall \, i : 0 \leq i \leq m-1 \quad Tr(\alpha^i x) = x_i$

P3. $\forall \, x, y \in GF(2^m) :$ if $\quad x = \sum_{k=0}^{m-1} x_k \lambda_k \; , \; y = \sum_{k=0}^{m-1} y_k \alpha^k \quad$ then $\quad Tr(xy) = \sum_{k=0}^{m-1} x_k y_k$

Suppose that in B_d : $\qquad 0 \leq i \leq m-1 \qquad \alpha^i A = \sum_{j=0}^{m-1} a(i,j) \lambda_j \qquad (8)$

from (8), P1, P2, P3 and the fact that $P = AB + C$, it follows that:

$$0 \leq i \leq m-1 \qquad p_i = \sum_{k=0}^{m-1} a(i,k) b_k + c_i \qquad (9)$$

from (8) and P2, one obtains: $\quad a(i,j) \quad = \quad Tr(\alpha^j (\alpha^i A)) \quad = \quad Tr(\alpha^{j+1}(\alpha^{i-1} A))$
thus: $\qquad 0 \leq i \leq m-1 \; ; \; 0 \leq j \leq m-2 \qquad a(i,j) = a(i-1,j+1)$
for $j = m-1$, $\quad a(i,m-1) = Tr(\alpha^m (\alpha^{i-1} A))$. From (8), P1, P2 and the fact that
$\alpha^m = \sum_{k=0}^{m-1} f_k \alpha^k$, it follows that:

$$1 \leq i \leq m-1 \qquad a(i,m-1) = \sum_{k=0}^{m-1} f_k a(i-1,k) \qquad \text{then}$$

$1 \leq i \leq m-1 ; 0 \leq j \leq m-1$

$$a(i,j) = a(i-1,j+1) \; ; \; a(i-1,m) = \sum_{k=0}^{m-1} f_k a(i-1,k) \qquad (10)$$

from (9) , (10) one obtains : $\quad 0 \leq i \leq m-1 \quad p_i = \sum_{k=0}^{m-1} a(i-1,k+1) b_k + c_i \qquad (11)$

The terms $\quad a(i,1), a(i,2), \ldots, a(i,m) \quad$ used to compute p_{i+1} can be determined, from (10) and (11), as follows:

$0 \leq i \leq m-1 \; ; \; 1 \leq j \leq m-1 \quad a(i,j) = a(i-1,j+1) \; ; \; a(i,m) = \sum_{k=0}^{m-1} f_k a(i-1,k+1)$

$0 \leq k \leq m-1 \qquad a(-1,k+1) = a_k \qquad (12)$

suppose that : $0 \leq i \leq m-1 \; ; \; 0 \leq j \leq m-1 \quad p(i,j) = \sum_{k=0}^{j} a(i-1,k+1) b_k \qquad (13)$

then : $\quad 0 \leq i \leq m-1 \; ; \; 0 \leq j \leq m-1 \quad p(i,j) = p(i,j-1) + a(i-1,j+1) b_j$

if we choose c_i as the initial value $p(i, -1)$, the result $p(i, m-1)$ will be equal to p_i given in (11).Thus:

$$0 \leq i \leq m-1 \ ; \ 0 \leq j \leq m-1 \tag{14}$$
$$p(i,j) = p(i, j-1) + a(i-1, j+1)b_j \ ; \ p(i, -1) = c_i \ ; \ p_i \equiv p(i, m-1)$$

Suppose that : $\quad 0 \leq i \leq m-1 \ ; \ 0 \leq j \leq m-1 \quad a^*(i,j) = \sum_{k=0}^{j} a(i-1, k+1)f_k \tag{15}$

from (15), it follows that:

$$0 \leq i \leq m-1 \ ; \ 0 \leq j \leq m-1$$
$$a^*(i,j) \ = a^*(i, j-1) + a(i-1, j+1)f_j \tag{16}$$
$$a^*(i, -1) = 0 \ ; \quad a^*(i, m-1) = \sum_{k=0}^{m-1} a(i-1, k+1)f_k$$

using the function $start(j)$ defined as: $\quad start(j) \ = \begin{cases} 0 \text{ if } j \equiv 0 \pmod{m} \\ 1 \text{ else} \end{cases}$

a more general equation is obtained from (12),(14) and (16):

$$0 \leq i \leq m-1 \ ; \ 0 \leq j \leq m-1$$
$$p(i,j) \ = c_i.\overline{start(j)} + p(i, j-1).start(j) + a(i-1, j+1)b_j;$$
$$a^*(i,j) = a^*(i, j-1).start(j) + a(i-1, j+1)f_j \ ; \quad a(-1, j+1) = a_j$$
$$0 \leq i \leq m-1 \ ; \ 1 \leq j \leq m \tag{17}$$
$$a(i,j) = a(i-1, j+1).start(j) + a^*(i, m-1).\overline{start(j)} \ ; \quad p_i \equiv p(i, m-1)$$

In equation (17), note that for a fixed value of j, the same terms b_j, f_j and $start(j)$ are used to compute $p(i,j)$, $a^*(i,j)$ and $a(i,j)$ for $0 \leq i \leq m-1$. Similarly, for a fixed value of i, the same term c_i is used to compute $p(i,j)$ for $0 \leq j \leq m-1$. Using this remark, (17) can be transformed to obtain the following *uniform recurrent equation*:

$$0 \leq i \leq m-1 \ ; \ 1 \leq j \leq m$$
$$a(i,j) = a(i-1, j+1).start(i,j) + a^*(i, m-1).\overline{start(i,j)}$$
$$0 \leq i \leq m-1 \ ; \ 0 \leq j \leq m-1$$
$$p(i,j) \ = c(i,j).\overline{start(i,j)} + p(i, j-1).start(i,j) + a(i-1, j+1).b(i,j) \tag{18}$$
$$a^*(i,j) = a^*(i, j-1).start(i,j) + a(i-1, j+1).f(i,j) \ ; \quad a(-1, j+1) = a_j$$
$$b(i,j) \ = b(i-1, j) \ ; \ b(-1, j) = b_j \ ; \ f(i,j) = f(i-1, j) \ ; \ f(-1, j) = f_j$$
$$start(i,j) = start(i-1, j) \ ; \ start(-1, j) = start(j)$$
$$c(i,j) = c(i, j-1) \ ; \ c(i, -1) = c_i \ ; \ p_i \equiv p(i, m-1)$$

In order to obtain a serial systolic-multiplier, where the bits corresponding to the coefficients A, B, C, P and the polynomial $F(x)$ are processed serially, the *dependence graph* associated with equation (18) must be projected along *vector* $(0,1)$. Indeed, from equation (18) it follows that the bits corresponding to A, B and $F(x)$ will move serially. However, with this projection, it follows from (18) that the bits of C and P do not move. Thus in the resulting multiplier (obtained by projection along *vector* $(0,1)$) the coefficients C and P are respectively input and output in parallel. In order to ensure a serial input for the bits of C and a serial output for the bits of P, another variable p^* is introduced. It ensures the serial transfer of bits corresponding to C and P. The final *uniform recurrent equation* is obtained by affecting to (18) the following modifications:

$$0 \leq i \leq m - 1 \; ; \; 0 \leq j \leq m - 1$$
$$p(i,j) = p^*(i-1,j+1).\overline{start(i,j)} + p(i,j-1).start(i,j) + a(i-1,j+1).b(i,j)$$
$$p^*(-1,j+1) = c_j \; ; \; p_j \equiv p^*(m-1,j+1) \tag{18'}$$
$$0 \leq i \leq m-1 \; ; \; 1 \leq j \leq m \quad p^*(i,j) = p^*(i-1,j+1).start(i,j) + p(i,m-1).\overline{start(i,j)}$$

To see that equation (18)-(18') gives the same results p_i as (18), note that:

$$p(i,0) = p^*(i-1,1) + a(i-1,1).b(i,0)$$

from $(18) - (18')$, it follows that : $p^*(i-1,1) = p^*(i-2,2) = \ldots = p^*(-1,i+1) = c_i$

then : $\qquad\qquad 0 \leq i \leq m-1 \qquad p(i,0) = c_i + a(i-1,1).b(i,0)$

on the other hand : $\quad 0 \leq j \leq m-1$

$$p^*(m-1,j+1) = p^*(m-2,j+2) = \ldots = p^*(j,m) = p(j,m-1) = p_j$$

The *dependence graph* associated with equation (18)-(18') is shown in Fig.4. The time at which computation at point (i,j) is performed is given by:

$$0 \leq i \leq m-1 \; ; \; 0 \leq j \leq m-1 \qquad t(i,j) = 2i + j \tag{19}$$

If the graph of Fig.4 is projected along *vector*$(0,1)$, a set of m points is obtained. These points are $\{(i,0)/ \; 0 \leq i \leq m-1\}$. The point $(i,0)$ is the projection of points $\{(i,j)/ \; 0 \leq j \leq m\}$. To the point $(i,0)$ a given cell $cell_i$ is associated. It carries out the computations corresponding to the terms $p(i,j)(0 \leq j \leq m-1)$. The algorithm and the structure of $cell_i$ are given in Fig.5.

The multiplier architecture for the field $GF(2^4)$ is shown in Fig.6. It consists of m basic cells arranged into a one-dimensional systolic array. The bits f_j, a_j, b_j and c_j corresponding respectively to $F(x)$, A, B and C are received serially at the respective inputs f_{in}, a_{in}, b_{in} and p^*_{in} in the leftmost cell. These bits enter the systolic array in the same order and arrive at a rate of one bit per clock cycle. The leftmost cell input $start_{in}$ is always set to 1 except when the bits f_0, a_0, b_0 and c_0 are received in the leftmost cell. The function of the control signal is to initialize the multiplier for a new product-sum

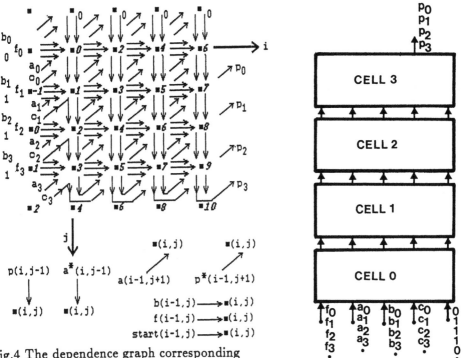

Fig.4 The dependence graph corresponding
to equation (18)-(18')($m = 4$)

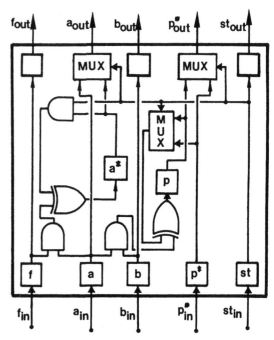

Fig.6 A serial multiplier for the field $GF(2^4)$

Algorithm :
```
begin
    if  start = 0
        then
            begin
                p*out := p ;  p := ab + p*;
                aout := a* ;  a* := af
            end
        else
            begin
                p*out := p* ;  p := p + ab;
                aout := a ;  a* := a* + af
            end;
    fout := f ;  f := fin ;  a := ain;
    bout := b ;  b := bin ;  p* := p*in;
    startout := start ;  start := startin
end.
```

Fig.5.a The algorithm of a basic cell

Fig.5.b The circuit of a basic cell

computation, and to output the bits p_j of P via connexions p_{in}^*, p_{out}^*. The bits p_j of P are generated serially through the rightmost-cell output p_{out}^*. They are generated at a rate of one bit per clock cycle. According to equation (19), the first bit p_0 of P is computed in $cell_{m-1}$ and generated at the rightmost cell output p_{out}^* at the time unit $t(m-1,1) = 2(m-1) + 1$. Since the first bits a_0, b_0, c_0 and f_0 enter the leftmost cell at the time unit -1, $2m$ clock cycles are then needed for the generation of the first bit p_0 of P.

Compared with the multiplier proposed in [15], our multiplier is less complex. Indeed, it contains $10m$ flip-flops while the multiplier developed in [15] contains $12m$ flip-flops. The additional flip-flops are used for the purpose of synchronization. Their function is to delay the bits c_i of C. Moreover, to perform the product-sum computation the multiplier presented in [15] needs an additional external XOR gate. No additional gates are needed in our design. If the flip-flops f, a, b, p^* and $start$ are eliminated from the cell shown in Fig.5, another serial systolic-multiplier is obtained. Its flip-flop complexity is better than the previous one. Indeed, it contains 5 flip-flops per cell while the previous one contains 10 flip-flops per cell. m clock cycles are needed for the generation of the first bit p_0 of P. However, its elementary computation time increases with m.

4   CONCLUSION

Two systolic architectures have been proposed in this paper for the product-sum computation $P = AB + C$ in $GF(2^m)$. The first multiplier performs two product-sum computations at the same time. In a later paper, it will be used in the design of systolic RS encoders with interleaving level $2I$. In the second multiplier, the product-sum computation is performed without addition of any external flip-flop or XOR gate. The property of this multiplier, to operate in a bit-serial manner, is very attractive for the design of a universal bit-serial systolic RS encoder. We are currently studing its integration in the systolic RS encoder developed in [8]. In [8], the encoding is performed by polynomial division. The use of the polynomial and dual basis for the representation of A, B, C and P is not inconvenient. In our encoder, B will correspond to the coefficients g_i of the generator polynomial $G(x)$, A to the quotient coefficients and P, C to the coefficients of the intermediate remainders (for more details see [8]). Consequently, the coefficients g_i and the generated codeword symbols will be respectively expressed in the polynomial and the dual basis. Compared with the one developed in [3], our encoder will take more chip area. This is generally the case for systolic architectures since they consist of several cells [15]. However, contrary to [3], our design will be much more expandable and programmable, since it doesn't focus on one particular m, $F(x)$ or $G(x)$ each time. Moreover, it has the advantages of regularity, modularity, concurrency and can operate at high rates.

ACKNOWLEDGMENT

The author would like to thank Professor A.POLI for his valuable comments.

REFERENCES

[1]   T.C.Bartee and D.I.Schneider, Computation in finite fields, Inform. contr. vol.6, (1963) 79–98.

[2]   E.R.Berlekamp, Algebraic coding theory (New York:Mc Graw-Hill, 1968).

[3]   E.R.Berlekamp, Bit-Serial Reed-Solomon Encoders, IEEE Trans. Inform. Theory, vol.IT-28, no.6 (1982) 869–874.

[4]   R.G.Gallagher, Information theory and reliable communication, (New York:Wiley, 1968).

[5]   H.T.Kung, Why systolic architectures?, IEEE Trans. on comp., vol.C-15, (1982) 37–48.

[6]   K.Y.Liu, Architecture for VLSI design of Reed-Solomon Encoders, IEEE Trans. on comp., vol.C-31, no.2, (1982) 170–175.

[7]   P.J.Mac Williams and N.J.A Sloane, the theory of error-correcting codes, (North-Holland, Amsterdam, 1978).

[8]   A.Poli et L.Huguet, CODES CORRECTEURS Théorie et applications, (MASSON 1989).

[9]   P.Quinton, INTRODUCTION AUX ARCHITECTURES SYSTOLIQUES, IRI-SA, Publication interne no.319 (1986) 1–11.

[10]   P.Quinton et Y.Robert, ALGORITHMES ET ARCHITECTURES SYSTOLI-QUES, (MASSON 1989).

[11]   I.S.Reed and T.K.Truong, The use of finite fields to compute convolutions, IEEE Trans. Inform. Theory, vol.IT-21, no.2, (1975) 208–213.

[12]   H.M.Shao, T.K.Truong, L.J.Deutsch, J.H.Yuen, and I.S.Reed, A VLSI design of a pipeline Reed-Solomon Decoder, IEEE Trans. on comp., vol.C-34, no.5, (1985) 393–403.

[13]   C.C.Wang, T.K.Truong, H.M.Shao, L.J.Deutsch, J.K.Omura, and I.S.Reed, VLSI architectures for computing multiplications and inverses in $GF(2^m)$, TDA progress report (1983).

[14]   C.S.Yeh, I.S.Reed and T.K.Truong, Systolic multipliers for finite fields $GF(2^m)$, IEEE Trans. on comp., vol.C-33, no.4, (1984) 357–360.

[15]   B.B.Zhou, A New Bit-serial systolic Multiplier over $GF(2^m)$, IEEE Trans. on comp., vol.C-37, no.6, (1988) 749-751.

PARSAC-2: A Parallel SAC-2 Based on Threads*

Wolfgang Kuechlin
Department of Computer and Information Science
The Ohio State University
Columbus, OH 43210-1277, U.S.A.
⟨Kuechlin@cis.ohio-state.edu⟩

Abstract: We describe the design of PARSAC-2, a parallel version of the SAC-2 Computer Algebra system. In PARSAC-2, parallelism is based on multiple *threads* (lightweight processes) executing on a shared memory multiprocessor. The *S-threads* subsystem provides threads which are capable of parallel list processing on a shared heap. The S-threads heap memory is designed to allow concurrent list cell allocation by multiple threads with minimal synchronization overhead. S-threads may also perform parallel garbage collection, and a slightly weaker form of storage management called preventive garbage collection. We present an example of algorithm development in PARSAC by parallelizing the SAC-2 algorithm IPRODK, an integer multiplication routine based on Karatsuba's method. Using empirical data from this experiment, we demonstrate that S-threads permit a parallelization of SAC-2 down to the lowest algebraic level. Finally, we show how a key parameter of the S-threads memory design influences parallel performance.

1 Parallelizing SAC-2

Our overall goal being the parallelization of the SAC-2 Computer Algebra system [CL] on shared memory multiprocessors, a key problem was to develop a suitable parallel programming environment. In doing so, we had to achieve a careful balance between ease of use and portability through high-level tools on the one hand, and efficiency, low overhead, and small parallel grainsize on the other hand. Given the ongoing uncertainty about the "right" parallel programming paradigm, we made it our main objective to set benchmarks for performance while not ignoring ease of programming. In particular, we wanted to find the "lowest" algebraic level at which the system could profitably be parallelized in practice, and we wanted to measure actual speed-ups at that level.

We chose a shared memory multiprocessor architecture for its ease of use, potential for relatively fine grained parallelism, and significance for future desktop systems. We chose a relatively low-level, threads based parallelization approach with the intention to maximize performance, but also because we wanted to study the impact of the parallel environment on system performance before relying on more complex parallelization tools.

---

*This material is based upon work supported by the Ohio State University Office of Research and Graduate Studies under Award No. 221152, and by the National Science Foundation under Award No. CCR-9009396.

Threads of control are *lightweight processes* which can be forked with low overhead to execute a function in parallel with the caller. Since threads are provided by the Mach operating system [ABG+86], there is a portable, stable, high quality implementation that delivers operating system grade performance. However, PARSAC-2 can be ported to any UNIX machine which provides a minimal threads package.

PARSAC-2 consists of three major bodies of code. The S-threads system [Kue90] provides the environment for parallel list processing based on some basic threads package such as C Threads [CD87]. Standard SAC-2 [CL] provides the algebraic core, executing essentially unmodified on S-threads. Finally, there is a growing number of parallel algorithms written in C; parallel constructs are furnished by the threads system, and sequential algebraic manipulation is provided by SAC-2.

For inclusion in PARSAC, the SAC-2 system was translated from its native ALDES [Loo76] into C. The functionality of the modules FP (Fortran Primitives) and BS (Basic System) is now taken over by the C environment (plus a few macros). The only other omissions are the LP (List Processing) level algorithms COMP (list cell allocation), GC (garbage collection) and BEGIN1 (LP initialization).

These algorithms are replaced by the S-threads system, which provides enhanced threads that can allocate list cells concurrently with very low synchronization overhead. All other SAC-2 functions can be executed unmodified on S-threads, thus providing parallel list processing and all higher-level parallel symbolic computation. An S-thread may be thought of as a C thread extended by a local environment containing a heap fragment and additional information. However, PARSAC-2 relies on the C Threads system not only for forking/joining of basic threads but also for additional synchronization primitives like mutex and condition variables.

A simple example of a parallel PARSAC algorithm is given below. It consists of a small C program forking two concurrent S-threads, each of which executes an unmodified sequential SAC-2 algorithm. Parallelization of a divide-and-conquer based bubble-merge sort for lists is shown in [Kue90]. Parallel isolation of real roots of integral polynomials with PARSAC-2 is discussed in [CJK90].

Example 1 *The SAC-2 function* J:=INEG(I) *negates an (arbitrary precision) integer* I. *In PARSAC-2,* I1 *and* I2 *may be negated in parallel as follows.*

```
sthread_t st;
st=sthread_fork(INEG,I2,NULL);
J1=INEG(I1);
J2=sthread_join(st);
```

The call to sthread_fork *creates a new thread of control executing* INEG(I2) *concurrently with the parent; it also returns an ID of the child. The result is passed back to the parent when it joins the child with* sthread_join. *The* NULL *parameter in the fork indicates the result passing mode (cf. Section 2.2).*

In the following, we will first look at relevant parallel architectures and software environments. We will then discuss the design of PARSAC in more detail in Section 2. Finally we will present empirical results with a parallel implementation of IPRODK in Section 3.

1.1 Parallel Architectures and Software Environments

Current research in parallel Computer Algebra (cf. [DDF89]) focuses on both *shared memory* and on *distributed memory* machines consisting of multiple independent standard microprocessors. From an implementation point of view, main problems are ease of programming (leaving much of the existing CA systems intact and requiring little additional code), parallel grainsize, and portability.

Distributed memory architectures, as e.g. the Intel Hypercube, employ communication by passing (slow) messages, but they scale to many thousands of processors. Shared memory architectures either use a common bus (e.g. Encore Multimax, Sequent Balance) or a network switch (e.g. BBN Butterfly) for fast communication. The former scale only to a few dozen processors, while the latter may have hundreds, even up to a few thousand processors.

Distributed memory machines are relatively hard to program (messages have to be passed, data-structures have to be distributed), and provide only for coarse grained parallelism. On shared memory machines, parallel programming can be done using lightweight processes (*threads of control*), which consist essentially only of a program counter and a private stack, and which share common resources such as memory (page tables) and files (buffers). Threads may be provided by an operating system such as Mach or OS/2, or by a threads user library. A threads system at the application level typically allocates some number of *workers*, each on its own processor. The workers then execute micro-tasks specified by only a function- and a data-pointer, much like Universal Turing Machines working on $\langle M, w \rangle$ pairs from the universal language.

C Threads [CD87] are a minimal, though extendable, threads abstraction which is accessible from C as a system of library calls. The C Threads package provides functionality for manipulating threads of control through primitives to *fork*, *join*, *exit*, or *detach* a thread; it also provides for synchronization between threads through primitives to *signal*, or *wait* for, a condition, and to *set*, and *reset*, binary semaphores (mutex variables). Portability is achieved through Mach; due to their simplicity, C threads can also be implemented directly with reasonable effort (cf. [Sam89]).

For wider portability, especially over distributed memory machines, the Linda programming environment [ACG86, CG89] has been proposed. Linda adds a level of abstraction in that it provides a uniform system-wide *tuple space* as a communication medium. Processes send messages by placing data tuples into Linda space, and they receive messages by reading or removing tuples; a process M is created if a special $\langle M, w \rangle$ tuple is sent. On shared memory machines, a Linda like system may be implemented on top of a threads system, thus adding abstraction but also extra overhead.

There are many other high-level parallel environments, such as the Chare Kernel [KS88] or ALPS [Vis88]. Similar questions have to be answered for all of them: Do they allow *symbolic* computation? Are their parallel constructs adequate for symbolic computation? Are they stable and well enough implemented to support programming in the large? Do they provide *useful* portability across architectures, i.e. do they preserve efficiency of an application? Most importantly, how much computational overhead do they introduce, and is the added abstraction worth that price? We have chosen to begin the parallelization of SAC-2 by working directly with threads to establish a benchmark of parallel grainsize, speed-up, and efficiency, achievable under this approach. This should later help determine the trade-off between speed and abstraction when compared to a more high-level approach.

1.2 Parallel Computer Algebra

A survey of current work on parallel algorithms and systems for algebraic manipulation was conducted by Ponder [Pon88a, Pon88b, Pon88c]. His findings on up-and-running systems have generally been disappointing. In particular, no systems using shared memory architectures were reported. Work reported in [DDF89] centers on hypercube architectures and vector processing supercomputers. Seitz [Sei89] discusses a parallel SAC-2 implementation of the modular polynomial resultant algorithm on an (Ether-)net of Sun Workstations, but without giving empirical results.

The work most closely related to ours is the parallel implementation of Collins' CAD algorithm reported by Saunders *et al.* [SLA89]. There the extension phase of CAD was executed in parallel on 7 worker processes mforked with OS multitasking primitives onto 7 processors of a Sequent Symmetry. This work has achieved a speed-up efficiency of nearly 50% and provided valuable empirical data on the potential for parallel execution on the highest algebraic level of SAC-2.

There were two significant limitations to that approach, both of which we address in the design of PARSAC. First, the usual global list AVAIL of available list cells was used, shared between all processes. Saunders *et al.* noted a significant parallel overhead of their system, which they presume "may be explained almost entirely by the cost of the lock on AVAIL" at every list allocation.

Second, programming in the large was not possible, so that only one task (CAD cell extension) of very large grainsize was performed in parallel. In some examples, however, a few cells took very long to extend, causing other processors to idle. Adding more processors would not help on these examples without using finer grained parallelism. Fine grained parallelism however will not significantly help as long as the system depends on one global available cell list which becomes a sequential bottleneck.

Saunders *et al.* suggested using the Linda system to achieve architecture independence and to facilitate large-scale parallelization. On shared memory machines, however, hardware and architecture independence can also be achieved through the use of Mach and C-threads directly. It can even be argued that all *useful* architecture independence can be realised this way, because shared memory and distributed memory architectures are too different to allow ported software to remain *efficient* without substantial reorganization.

2 On the Design of PARSAC-2

In [SLA89], Saunders *et al.* conclude "... subprocesses need greater independence and lower overhead. The challenge seems to be to achieve greater independence of the subprocesses (less dependence on the single shared SPACE) and adaptive granularity."

PARSAC-2 addresses this challenge through its design based on threads and a reorganization of the heap of list cells (SAC-2 SPACE). Times for forking and joining a thread are so low that in many cases there may be little need to provide for adaptive granularity. Instead, given enough fine-grained tasks, it becomes possible (within the limits of the threads implementation) to simply saturate the system with threads of all grain sizes so that all workers are busy all the time (cf. Section 3).

Another main concern of ours was ease of programming in the system, and the ability to *gradually* parallelize based on existing code. The implementation language of the SAC-2 system is ALDES [Loo76], which is translated to highly portable FORTRAN.

Since there is no way to manipulate a C pointer in ALDES, and hence no way to fork a C thread, all our parallel programming is done in C.

After experimenting with an interface between C and a FORTRAN based SAC-2 library it became apparent that programming under this approach would be feasible, but very cumbersome. A new translator, written by Hoon Hong, made it possible to convert the entire SAC-2 library to C, and hence to call SAC-2 algorithms from C in a natural and efficient way, without any interface. Most importantly, it became possible to parallelize a SAC-2 algorithm simply by modifying its C code, e. g. by replacing a function call by a thread fork (cf. Example 1).

As a welcome side-effect, the C system stack is now used for marking lists for garbage collection, so that the ALDES/SAC-2 system stack is no longer necessary. Therefore, an S-thread does not have to allocate an extra stack partition, and the C programmer does not have to push ALDES *unsafe* variables onto the SAC-2 stack. In fact, the distinction between *safe* and *unsafe* variables disappears entirely.

2.1 Memory Organization

As we shall confirm in Section 3 below, locked access to a global AVAIL may become a sequential bottleneck due to contention, especially as the effective parallelism in the system increases. This effect is well known from global run-queues in threads packages, cf. [Sam89, ALL88].

Motivated by the distributed thread run-queues of [ALL88], PARSAC-2 therefore employs a flexible scheme by which AVAIL can be dynamically distributed over the active threads. The SAC-2 list cell memory SPACE is now divided into *pages* of user defined size; these pages are organized into a global list page_avail under a global lock. Each thread maintains its own set of SPACE pages on which it organizes a private AVAIL list with unlocked access. If the thread runs out of cells, it performs locked access to page_avail and obtains a fresh new page (through passing of a single pointer). Each locked access therefore provides as many new cells as fit on a page, at the same time lowering synchronization cost and distributing it over many list cells.

Page size is a system parameter under user control. Choosing the pages as large as possible is equivalent to a complete distribution of AVAIL, with no synchronization cost but with the disadvantage of assuming equal memory needs for all threads. Choosing the minimal page size of one cell lets each thread have just as many cells as it needs, with the disadvantage of a global lock acquisition for each cell allocation. In practice, page size can be gradually increased until algorithm timings do no longer improve. We found sizes of 32 cells or above sufficient for IPRODK (cf. Section 3), while other experiments with polynomial GCD calculations required page-sizes in excess of 100 cells.

The reorganization of memory into pages is completely transparent to existing SAC-2 algorithms, with the exception of COMP, BEGIN1, and GC. Under some restrictions, garbage collection can now be carried out on a *per thread* basis without the need to suspend all other threads first. A sufficient condition for independent garbage collection is to not have cross pointers between local thread memories; this is realized e. g. in pure functional programming with pass by value parameters in parallel function calls.

Although the page structure can remain transparent, the programmer can also use it explicitly to significantly reduce the amount of garbage collections. If a function is executed on its own thread, it continues to allocate local space pages for its list cells until its result is computed. Then the result, which is typically small w. r. t. the space consumed in intermediate computation, can be copied onto a few new pages, which are transferred to the caller (by passing a pointer). Now all other pages are known to contain garbage; they may be transferred back to the global page pool. In effect, compacting garbage collection is thus carried out at selected points, using knowledge about location of data and garbage, which therefore does not have to be computed at random times by marking from the stack. Thus we trade marking for a small amount of copying.

This scheme is inspired by the memory organization of Prolog systems (see [MW88]) and could be viewed as an attempt to discover a stack structure on the heap. All our empirical data has been obtained in this way; general garbage collection is not implemented yet in PARSAC-2. Preventive garbage collection can be utilized independent of parallel calls through a small set of functions to manipulate sets of space pages, such as to detach pages from thread memory, to explicitly return pages to page_avail, or to attach pages to thread memory.

Example 2 *The SAC-2 function* F:=IFACT(I) *computes the list* F *of prime factors of the integer* I. *In PARSAC-2,* I1 *and* I2 *may be factored in parallel as follows, while reclaiming intermediate list memory used by both calls to* IFACT.

```
sthread_t st;
st=sthread_fork(IFACT,I2,lcopy);
M=space_detach();   /* M = thread memory up to now. */
F1=IFACT(I1);
Mp=space_detach(); /* M' = list memory used by IFACT(I1). */
F1=(*lcopy)(F1);    /* Copy result onto new pages. */
space_free(Mp);     /* Free M' for further use. */
space_attach(M);    /* Re-attach old environment. */
F2=sthread_join(st);
```

lcopy *is a pointer to a copy function for non-cyclic lists, provided by the S-threads system. As a parameter in* st=sthread_fork(IFACT,I2,lcopy) *it causes the result of* IFACT(I2) *to be copied to new pages before being transferred back to the parent in the join; this allows the local list memory of* st *to be freed after the thread exits. Essentially the same actions have been programmed explicitly for the sequential call; of course, a generic routine could also be provided. At the end, thread memory* M *has only been increased by the cells used to copy* F1 *and* F2.

2.2 The S-Threads Library

S-threads are made available to the application programmer as a PARSAC-2 system library supporting the SAC-2 list processing system. As seen in Section 2.1, a thread which allocates list cells has a local environment of SPACE pages; we call this a SPACE-using thread, short *S-thread*. Consequently, an S-thread must receive an environment upon start-up, and it must be able to receive lists as input parameters, and pass lists as results. All this increases thread handling overhead—it makes the process heavier—both at run-time and at programming time. It is the objective of library design and implementation to keep this overhead at a minimum while providing useful functionality and portability. As a rule of thumb, we set parallel grainsize to some 10 times parallel overhead, so that even when the system executes on just one processor it becomes only about 10% slower.

Our current implementation builds upon a C Threads library for basic thread handling, and for synchronization primitives. C threads are provided here as an interface to the Encore Parallel Threads library [Doe87, Enc88], while under Mach they would be provided as an interface to Mach threads. The extra overhead required by S-threads over C threads is around 300 machine instructions each per fork or join, which makes a fork/join pair about 50% heavier than under our C threads alone. With our rule of thumb, we thus arrive at a parallel grainsize of around 20,000 instructions (or 10ms on our machine). However, thread overhead depends on subtle implementation details (cf. [ALL88]); it may vary significantly between different C thread libraries, and in particular it may degrade under increasing workload. Also, there is additional overhead to be performed by the forked thread before its main function can start and after it has finished (see below). Therefore some caution is in order and we usually only assume a parallel grainsize of 20–30ms. For more timings see Section 3.1.

The S-threads library consists of two parts, providing support for manipulating threads of control and list memory, respectively. Main thread manipulation functions are sthread_fork and sthread_join. st=sthread_fork(func,arg,copyfunc) creates a new S-thread st as follows: first, an S-thread control block is obtained into which the parameters are entered; then, a C thread is forked whose main function is a generic S-thread shell routine that takes the S-thread control block as a parameter. The shell first executes r=(*func)(arg) and then performs finalization of the thread; finalization includes copying r onto new pages by copyfunc, if a copy-function has been provided. Finally, the underlying C thread is caused to exit with the S-thread control block as result. A parent absorbs a child thread st by sthread_join(st). First, the C thread underlying st is joined and the S-thread control block is retrieved. Then the result is extracted and the control block is freed.

Both the S-threads system and the C threads system accept only one pointer-size argument for the top-level function of a forked thread. Multiple arguments must therefore be packaged in a data-structure whose life-time is guaranteed to be as long as that of the forked thread. Likewise, packaged parameters must be unpacked by the forked function. When parallelizing an existing SAC-2 algorithm, it was found convenient to change it as little as possible. Therefore a separate small "shell" function was usually created for each parallelized algorithm which unpacks parameters and calls the algorithm proper.

The memory subsystem provides manipulation functions for thread list memory which is organized in pages of list cells. These functions are used in the threads subsystem, but they may also be used by the application programmer to implement preventive garbage collection (cf. Section 2.1). Every S-thread contains a private (possibly empty) set of memory pages with an available cell list. Every SAC-2 COMP allocates a list cell from this list on one of these pages. If the list is exhausted, a new page is allocated by locked access to the global page list. A list structure can be copied to a fresh set of pages by first detaching the existing page set from the thread environment, and then executing a copy function such as lcopy (provided by the memory subsystem). The empty list environment will then cause COMP to allocate new pages. The detached environment must be freed or re-attached before the S-thread exits, or it will be permanently lost.

348

3 Empirical Results

Our work was done on an Encore Multimax with 12 processors of type NS32332, rated at 2 MIPS each, and 64 MB main memory. The operating system was UMAX 4.3, and C-thread calls were emulated by Encore Parallel Threads (EPT) [Doe87, Enc88]. In all our timings we used only 8 of the 12 processors to be free of side-effects from other processes in a time-sharing environment. Times were obtained from the Encore's microsecond wall-clock and rounded to milliseconds.

3.1 S-Thread System Timings

System overhead introduced by S-threads consists of a (larger) part for the underlying C threads system and a (smaller) additional overhead of the extra S-threads code.

On our machine, a call to cthread_fork takes about $450\mu s$, and a call to cthread_join takes about $350\mu s$ to return (assuming no waiting for the join). According to [Enc88], the EPT system is not optimized, and an improvement by as much as 50% should be possible. As noted above, system timings are also affected considerably by implementation details, load, caching (warm vs. cold start), etc. An extensive performance analysis, however, is useful only for an optimized portable threads system, e. g. under Mach.

The following table summarizes times for key operations of the S-threads system. For fork and join, we measured both the S-thread overhead, and the total time including the time for the underlying C thread fork or join. Time for space page allocation consists of an overhead part that is independent of page size, and a part linear in page size due to initializing the available cell list on the page. The latter cost is just under $4\mu s$ per cell. Page deallocation is independent of page size. All timings assume no contention for shared resources and no wait on synchronization.

operation	#calls	mean[μs]	std. error
get S-thread control block	1000	99	23
fork S-thread (overhead)	1000	136	18
fork S-thread with C thread	1000	600	106
free S-thread control block	1000	92	12
join S-thread (overhead)	100	172	19
join S-thread and C thread	100	408	56
get space page (overhead)	100	143	18
free space page	1000	120	23

3.2 Application Timings

The SAC-2 function C:=IPRODK(I,J) computes the product of the SAC-2 (long) integers I and J by Karatsuba's method (see [CMW82, BB88]). We parallelized the algorithm by forking the recursive calls in parallel. Our main objective here is to measure parallel grainsize and also the influence of the size of space pages on parallel performance. We do not claim to have solved the question of how to best multiply integers in parallel. For timings of integer multiplication on a distributed memory machine see [Roc89].

The sequential SAC-2 algorithm IPRODK makes 3 recursive calls at each level of recursion, except for inputs smaller than 16 SAC-2 β-digits, or 448 bit, in length. Hence the sequential run-time is in $O(n^{\log_2 3})$ (cf. [BB88]). In algorithm piprodk we forked two of the recursive calls in parallel. Therefore, the number of threads created is in $O(n^{\log_2 3})$ and with enough processors the run-time is in $O(n)$. In algorithm `piprodk1` we only forked one of the recursive calls in parallel. The number of active threads is then in $O(n)$, and the run-time is in $O(n \log n)$.

phase	input length [bit]	piprodk threads	piprodk1 threads
1	< 448	1	1
2	< 896	3	2
3	< 1792	9	4
4	< 3584	27	8
5	< 7168	81	16

In the following graphs, each data point represents an average of 3 runs on the same random inputs. The computation phases of the above table have been indicated in all graphs.

Figure 1 shows the run-time of IPRODK *vs.* `piprodk`. Time growth for `piprodk` appears linear in phases 1–4, although up to 27 threads are mapped to only 8 processors in phase 4. Due to the divide-and-conquer nature of the algorithm, however, fewer than the maximum number of threads are active during most of the *divide* and the *combine* computations, and each of the 27 micro-tasks involving base cases of the recursion are very short ($30ms$ at maximum). During the period when there are about 3 threads for each processor, the system must behave as if it had 27 processors of one third the speed. Obviously this period of oversaturation is insignificant in phase 4, but it does become a factor in phase 5 when up to 10 threads are mapped to each processor.

Figure 2 shows parallel speed-up (sequential time / parallel time) and speed-up efficiency (speed-up / number of processors). Since we did not measure actual processor utilization, efficiency figures are lower bounds based on the maximum number of processors used.

We then looked at `piprodk1`, in the hope of getting better speed-up efficiency, since fewer processors are used. Indeed we found that the efficiency trough in phase 3 could be smoothened. Efficiency is now around 100% in phase 1, 60–70% in phase 2, 40–47% in phase 3, some 30% in phase 4, and around 40% in phase 5. Since efficiency for `piprodk` in phase 5 is 60–70%, when 81 threads are mapped onto 8 processors, it is obvious that the processors are underutilized by `piprodk1`, so that our efficiency figures are very conservative. Figure 1 also shows that `piprodk` is significantly faster than `piprodk1`, obviously because it creates more threads faster which can make use of idle processors. Finally, we note that if two of `piprodk`'s threads are executed sequentially due to lack of processors, we effectively convert algorithm `piprodk` into `piprodk1`, plus a small amount of thread-handling overhead.

This graceful degradation of performance provides for an important measure of robustness in a parallel system. It also enables the programmer to focus on algorithmic issues because process placement is determined by the threads system at run-time, depending on actual processor configuration and utilization. This is especially important when building portable systems whose actual use is hard to foresee.

In the last set of experiments we measured the impact of page size on the performance of `piprodk`. Note how the different phases show up in the graphs of Figure 3: page size has to be larger if there are more active threads competing for memory pages. At a page size of 2 cells or less, parallel performance is essentially equal to sequential

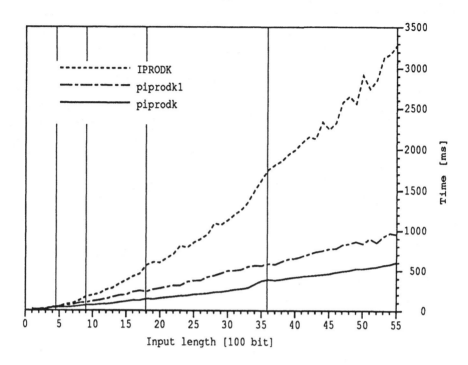

Figure 1: IPRODK *vs.* **piprodk1** and **piprodk**, 64 cells/page

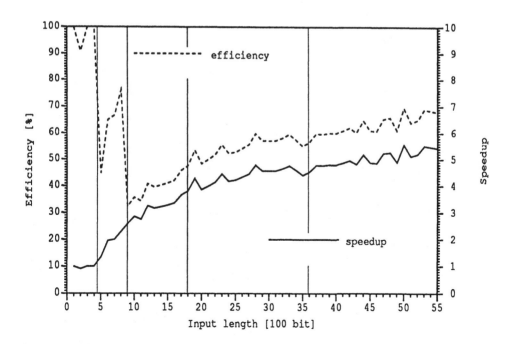

Figure 2: Speed-up and speed-up efficiency for IPRODK *vs.* **piprodk**

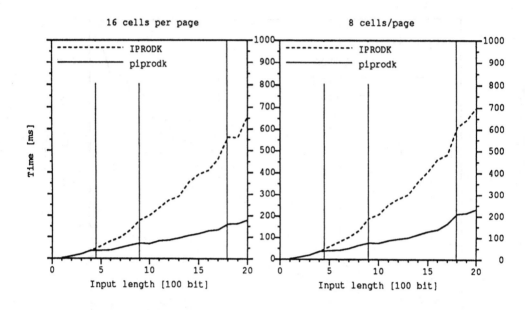

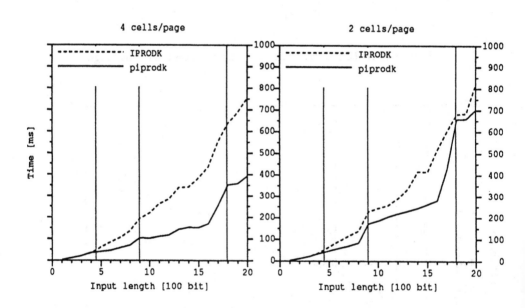

Figure 3: Influence of page sizes 16, 8, 4, 2 on **piprodk**.

performance. Hence IPRODK is not parallelizable (at this grainsize) without some degree of distribution of the available cell list. By analogy to distributed run-queues in threads systems, this also shows that subtle implementation details of the system underlying a parallel Computer Algebra package can profoundly influence the parallelizability of algorithms.

4 Conclusion

We have seen that by carefully redesigning the SAC-2 systems levels with S-threads, with particular attention to the parallel list processing system, parallelization of SAC-2 becomes profitable at the lowest algebraic level, *viz.* arithmetic in basic algebraic domains (see [CMW82]), as exemplified by Karatsuba integer multiplication. This suggests that under this approach an even higher efficiency of parallelization can be achieved on the coarse-grained parallelism afforded by higher system levels.

Acknowledgements

I wish to thank George Collins, Jeremy Johnson, and Jeff Ward, for many important discussions during design and implementation of PARSAC-2. Mike Ulm provided vital help with implementation, timings, and criticism, of an earlier version of the S-threads system; the final system timings are due to Nick Nevin. The system would not exist in its current form without the new ALDES-to-C translator written by Hoon Hong.

References

[ABG+86] Mike Accetta, Robert Baron, David Golub, Richard Rashid, Avadis Tevanian, and Michael Young. Mach: A new kernel foundation for UNIX development. Technical report, Computer Science Department, Carnegie Mellon University, Pittsburgh, PA 15213, August 1986.

[ACG86] S. Ahuja, Nicholas Carriero, and David Gelernter. Linda and friends. *Computer*, 19(8):26–34, August 1986.

[ALL88] Thomas Anderson, Edward Lazowska, and Henry Levy. The performance implications of thread management alternatives for shared memory multiprocessors. Technical Report TR 88-09-04, Department of Computer Science, University of Washington, September 1988.

[BB88] Gilles Brassard and Paul Bratley. *Algorithmics: Theory and Practice.* Prentice-Hall, Englewood Cliffs, New Jersey, 1988.

[BCL82] Bruno Buchberger, George E. Collins, and Rüdiger Loos. *Computer Algebra: Symbolic and Algebraic Computation*, volume 4 of *Computing Supplementum*. Springer Verlag, Vienna, 2nd edition, 1982.

[CD87] Eric C. Cooper and Richard P. Draves. C threads. Technical report, Computer Science Department, Carnegie Mellon University, Pittsburgh, PA 15213, July 1987.

[CG89] Nicholas Carriero and David Gelernter. How to write parallel programs: A guide to the perplexed. *ACM Computing Surveys*, 21(3):323–357, September 1989.

[CJK90]   George E. Collins, Jeremy Johnson, and Wolfgang Kuechlin. Parallel real root isolation using the coefficient sign variation method. To appear in: Proc. CAP-90, Ithaca, N.Y., Academic Press, June 1990.

[CL]   G. E. Collins and R. G. K. Loos. SAC-2 system documentation. On-line documentation and program documentation. In Europe available from: R. G. K. Loos, Universität Tübingen, Informatik, D-7400 Tübingen, W-Germany. In the U.S. available from: G. E. Collins, Ohio State University, Computer Science, Columbus, OH 43210, U.S.A.

[CMW82]   George E. Collins, Maurice Mignotte, and Franz Winkler. Arithmetic in basic algebraic domains. In *Computer Algebra: Symbolic and Algebraic Computation* [BCL82], pages 189–220.

[DDF89]   J. Della Dora and J. Fitch, editors. *Computer Algebra and Parallelism*. Computational Mathematics and Applications. Academic Press, London, 1989.

[Doe87]   Thomas W. Doeppner. Threads, a system for the support of concurrent programming. Technical Report CS-87-11, Department of Computer Science, Brown University, June 1987.

[Enc88]   Encore Computer Corp. *Encore Parallel Threads Manual*, January 1988.

[Gon89]   Gaston H. Gonnet, editor. *Proc. International Symposium on Symbolic and Algebraic Computation*, Portland, Oregon, July 1989. ACM, ACM Press.

[KS88]   L. V. Kalé and Wennie Shu. The Chare-Kernel language for parallel programming: A perspective. Report UIUCDCS-R-88-1451, Department of Computer Science, University of Illinois at Urbana-Champaign, Urbana, IL, August 1988.

[Kue90]   Wolfgang W. Kuechlin. The S-threads environment for parallel symbolic computation. To appear in: Proc. CAP-90, Ithaca, N.Y., Academic Press, June 1990.

[Loo76]   R. G. K. Loos. The algorithm description language ALDES (Report). *ACM SIGSAM Bull.*, 10(1):15–39, 1976.

[MW88]   David Maier and David S. Warren. *Computing with Logic: Logic Programming with Prolog*. Benjamin Cummings, Reading, MA, 1988.

[Pon88a]   Carl G. Ponder. *Evaluation of "Performance Enhancements" in Algebraic Manipulation Systems*. PhD thesis, Computer Science Division, University of California, Berkeley, CA 94720, U.S.A., August 1988.

[Pon88b]   Carl G. Ponder. Parallel processors and systems for algebraic manipulation: Current work. *ACM SIGSAM Bull.*, 22(3):15–21, July 1988.

[Pon88c]   Carl G. Ponder. Parallelism and algorithms for algebraic manipulation: Current work. *ACM SIGSAM Bull.*, 22(3):7–14, July 1988.

[Roc89]   Jean-Louis Roch. *L'Architecture du Systeme PAC et son Arithmetique Rationnelle*. PhD thesis, Institut National Polytechnique de Grenoble, Grenoble, France, December 1989.

[Sam89]   Ioannis Samiotakis. A thread library for a non-uniform memory access multiprocessor. Master's thesis, The Ohio State University, 1989.

[Sei89]   Steffen Seitz. Parallel algorithm development. In Della Dora and Fitch [DDF89], pages 223–232.

[SLA89]   B. D. Saunders, H. R. Lee, and S. K. Abdali. A parallel implementation of the cylindrical algebraic decomposition algorithm. In Gonnet [Gon89], pages 298–307.

[Vis88]   Prasad Vishnubhotla. Synchronization and scheduling in ALPS objects. In *Eighth International Conference on Distributed Computing Systems*, San Jose, CA, 1988.

EXPONENTIATION IN FINITE FIELDS USING DUAL BASIS MULTIPLIER

Masakatu MORII † and Yuzo TAKAMATSU †

† Department of Computer Science, Ehime University
Matsuyama, 790 JAPAN
Tel. +81-899-24-7111,   Tel. Facsimile +81-899-23-0672

ABSTRACT: Implementing finite fields arithmetic is very important, when realizing error control systems and cryptosystems. Recenty several algorithms for implementing multiplication in $GF(2^m)$ have been proposed. When using the polynomial (or standard) basis representation, it is also important that efficient squaring algorithm is improved.

In this paper we present an efficient bit-serial squarer in polynomial basis representation for $GF(2^m)$. First, we give an interesting relation between exponentiation and maximum length feedback shift register sequences(m-sequences) in $GF(q^m)$. Secondly, we present an efficient sequarer in $GF(2^m)$ based upon Berlekamp's bit-serial multiplier (also called dual basis multiplier) architecture. The squarer has very simple structure and can compute the square in $\lceil \frac{m}{2} \rceil$ steps.

1. INTRODUCTION

Implementing finite fields arithmetic is very important, when realizing error control systems and cryptosystems. Recenty several algorithms for implementing multiplication and inversion in $GF(2^m)$ have been proposed [1]. It is well known that the inversion and exponentiation in $GF(2^m)$ can be generally decomposed into squaring and multiplying. The squares can be computed easily by cyclic shift when using normal basis representation in $GF(2^m)$. However it is important that an efficient squarer is improved when only using polynomial (or standard) basis representation. The transformation from a polynomial basis to a normal basis is a linear operation, but the computation is not easy. Especially it is well-known that dual basis multiplier, which is based on polynomial basis representation, is in fact the efficient method of requiring minimum circuitry for implementation. It has the advantage that multiplication by a fixed constant can be hard-wired, for example, the case of encoding Reed-Solomon codes. Furthermore a polynomial basis representation is familiar to us in the fields of algebraic coding theory and the techniques.

In this paper we present an efficient squarer based on bit-serial architecture in polynomial basis representation for $GF(2^m)$. The squarer has very simple structure and can compute the square in $\lceil \frac{m}{2} \rceil$ steps. Furthermore it is shown that we can derive more efficient circuits for the squaring when using trinomial as feedback polynomial.

2. SQUARING AND DUAL BASIS MULTIPLIER

2.1 Dual basis multipler

E.R.Berlekamp has developed a bit-serial multiplication algorithm over $GF(2^m)$ for the concatenated Reed-Solomon /Viterbi channel encoding system for deep-space down-link [2]. The multiplier is called *dual basis multiplier* since the algorithm is based on the relation between a polynomial basis and its dual basis. Berlekamp's multiplication technique minimizes the area on the chip required by the multipliers. However, his representation is in fact the only possible method that requires minimum circuitry for implementation; it has the additional advantage that multiplication by a fixed constant can be hard-wired. In general we should required an extra AND/OR-gate for the multiplier when multiplying any two elements. In other words, we have to make a transformation from a polynomial basis to its dual basis. Afterwards, the method for multiplying without basis transformation [3] and a new interpretation of the multiplying algorithm [4] have been presented.

2.2 Finite fields operations and the discrete-time Wiener-Hopf equation[4]

In this section we shall present a relationship between dual basis multiplier and discrete-time Wiener-Hopf equations (DTWHE) over finite fields. The extension of this relationship will give an efficient bit-serial squarer in this paper.

The discrete-time Wiener-Hopf equation is defined as a system of linear inhomogeneous t equations with t unknowns a_i $(i = 0, 1, \cdots, t-1)$, $2t-1$ constant coefficients s_i $(i = 0, 1, \cdots, 2t-2)$, that are not all zero, and t constants b_i $(i = 0, 1, \cdots, t-1)$ such that

$$
\begin{bmatrix}
s_{2t-2} & s_{2t-3} & \cdots & s_{t-1} \\
s_{2t-3} & s_{2t-4} & \cdots & s_{t-2} \\
\cdots & \cdots & \cdots & \cdots \\
s_{t-1} & s_{t-2} & \cdots & s_0
\end{bmatrix}
\bullet
\begin{bmatrix}
a_0 \\
a_1 \\
\cdots \\
a_{t-1}
\end{bmatrix}
=
\begin{bmatrix}
b_{t-1} \\
b_{t-2} \\
\cdots \\
b_0
\end{bmatrix}
\tag{1}
$$

The problem of solving DTWHE arises in a great many applications, especially in autoregressive filter design.

Let $\mathbf{u}$ and $\mathbf{v}$ be any elements over $GF(q^m)$. Then they can be represented by following polynomials:

$$u(z) = u_{m-1}z^{m-1} + u_{m-2}z^{m-2} + \cdots + u_0 \tag{2}$$

and

$$v(z) = v_{m-1}z^{m-1} + v_{m-2}z^{m-2} + \cdots + v_0 \tag{3}$$

where u_k, v_k $(k = 0, 1, \cdots, m-1) \in GF(q)$. We represent $\mathbf{W} = \mathbf{u} \cdot \mathbf{v}$ as the polynomial:

$$w(z) = w_{m-1}z^{m-1} + w_{m-2}z^{m-2} + \cdots + w_0 \tag{4}$$

where w_k $(k = 0, 1, \cdots, m-1) \in GF(q)$. We then have

$$w(z) = u(z) \cdot [v(z)] \quad mod\, f(z) \tag{5}$$

where $f(z)$ is an irreducible monic polynomial of degree m over $GF(q)$. From the above preliminaries, we have the following theorem.

Theorem 1:

$$
\begin{bmatrix}
\sum_{j=0}^{m-1} u_j tr(\beta\alpha^j), & \sum_{j=0}^{m-1} u_j tr(\beta\alpha^{j+1}), & \cdots, & \sum_{j=0}^{m-1} u_j tr(\beta\alpha^{j+m-1}) \\
\sum_{j=0}^{m-1} u_j tr(\beta\alpha^{j+1}), & \sum_{j=0}^{m-1} u_j tr(\beta\alpha^{j+2}), & \cdots, & \sum_{j=0}^{m-1} u_j tr(\beta\alpha^{j+m}) \\
\cdots & \cdots & \cdots & \cdots \\
\sum_{j=0}^{m-1} u_j tr(\beta\alpha^{j+m-1}), & \sum_{j=0}^{m-1} u_j tr(\beta\alpha^{j+m}), & \cdots, & \sum_{j=0}^{m-1} u_j tr(\beta\alpha^{j+2m-2})
\end{bmatrix}
$$

$$
\bullet
\begin{bmatrix}
v_0 \\
v_1 \\
\cdots \\
v_{m-1}
\end{bmatrix}
=
\begin{bmatrix}
\sum_{j=0}^{m-1} w_j tr(\beta\alpha^j) \\
\sum_{j=0}^{m-1} w_j tr(\beta\alpha^{j+1}) \\
\cdots \\
\sum_{j=0}^{m-1} w_j tr(\beta\alpha^{j+m-1})
\end{bmatrix}
\tag{6}
$$

where α is the root of $f(z)$ over $GF(q^m)$, and β is an any non-zero element of $GF(q^m)$.

Theorem 1 shows the relationship between the DTWHE over $GF(q)$ and the multiplication. From the diferent point of view, we consider the following equation:

$$v(z) = \frac{w(z)}{u(z)} \quad mod\, f(z). \tag{7}$$

From (7), Theorem 1 also shows that solving DTWHE over $GF(q)$ with m unknowns is equivalent to performing the division of $GF(q^m)$.

2.3 Main theorem

We can derive a relation between squaring and dual basis multiplication algorithm from the extension of Theorem 1. In this section we present the generalized relation between raising the elements in $GF(q^m)$ to qth power and linear feedback shift-register (LFSR) sequences.

We represent $\mathbf{w} = \mathbf{u} \cdot \mathbf{v}^{q^r}$ as the polynomial:

$$w(z) = w_{m-1}z^{m-1} + w_{m-2}z^{m-2} + \cdots + w_0 \tag{8}$$

where $w_k \quad (k = 0, 1, \cdots, m-1) \in GF(q)$. We then have

$$w(z) = u(z) \cdot [v(z)]^{q^r} \quad mod \; f(z) \tag{9}$$

From (9), we have the following theorem.

Theorem 2:

$$
\begin{bmatrix}
\sum_{j=0}^{m-1} u_j tr(\beta \alpha^j), & \sum_{j=0}^{m-1} u_j tr(\beta \alpha^{j+q^r}), & \cdots, & \sum_{j=0}^{m-1} u_j tr(\beta \alpha^{j+q^r(m-1)}) \\
\sum_{j=0}^{m-1} u_j tr(\beta \alpha^{j+1}), & \sum_{j=0}^{m-1} u_j tr(\beta \alpha^{j+q^r+1}), & \cdots, & \sum_{j=0}^{m-1} u_j tr(\beta \alpha^{j+q^r(m-1)+1}) \\
\cdots & \cdots & \cdots & \cdots \\
\sum_{j=0}^{m-1} u_j tr(\beta \alpha^{j+m-1}), & \sum_{j=0}^{m-1} u_j tr(\beta \alpha^{j+q^r+m-1}), & \cdots, & \sum_{j=0}^{m-1} u_j tr(\beta \alpha^{j+(q^r+1)(m-1)})
\end{bmatrix}
$$

$$
\bullet
\begin{bmatrix} v_0 \\ v_1 \\ \cdots \\ v_{m-1} \end{bmatrix}
=
\begin{bmatrix}
\sum_{j=0}^{m-1} w_j tr(\beta \alpha^j) \\
\sum_{j=0}^{m-1} w_j tr(\beta \alpha^{j+1}) \\
\cdots \\
\sum_{j=0}^{m-1} w_j tr(\beta \alpha^{j+m-1})
\end{bmatrix}
\tag{10}
$$

where α is the root of $f(z)$ over $GF(q^m)$, and β is an any non-zero element of $GF(q^m)$.

Theorem 2 gives the relation between squaring and dual basis multiplier when $q = 2$ and $r = 1$.

2.4 The proof of Theorem 2

To prove Theorem 2, we shall use the following two lemmas.

Lemma 1 *(Mattson and Solomon* [5]*)* : Let the LFSR sequence over $GF(q)$ with characteristic irreducible polynomial $f(z)$ be

$$m_0, m_1, m_2, \cdots.$$

Then

$$m_k = tr(\beta\alpha^k) \tag{11}$$

for $k = 0, 1, \cdots$, where α and β are elements of $GF(q^m)$.

Let the minimum positive integer λ be satisfying

$$z^\lambda \equiv 1 \quad mod\ f(z) \tag{12}$$

λ is called *order* of the characteristic irreducible polynomial $f(z)$. Then we have the following lemma.

Lemma 2: Let

$$z^d \equiv R(z) \quad mod\ f(z)$$
$$= \gamma_{m-1} z^{m-1} + \gamma_{m-2} z^{m-2} + \cdots + \gamma_0. \tag{13}$$

Then

$$m_{q^r d+k} = \sum_{i=0}^{m-1} \gamma_i tr(\beta\alpha^{k+q^r i}) \tag{14}$$

where $0 \le d \le \lambda - 1$ and $\gamma_i \in GF(2)$ for $i = 0, 1, \cdots, m-1$.

Proof: It is almost same as that of Lemma 2 in Ref.(4) when using the relationship in finite fields with the characteristic q as follows:

$$[R(\alpha)]^{q^r} = R(\alpha^{q^r}) \tag{15}$$

Q.E.D.

Proof of Theorem 2: From (2), (3) and (8), we have

$$z^\tau \equiv u_{m-1} z^{m-1} + u_{m-2} z^{m-2} + \quad \cdots \quad + u_0, \tag{16}$$

$$z^v \equiv v_{m-1} z^{m-1} + v_{m-2} z^{m-2} + \quad \cdots \quad + v_0 \tag{17}$$

and

$$z^\omega \equiv w_{m-1} z^{m-1} + w_{m-2} z^{m-2} + \quad \cdots \quad + w_0, \tag{18}$$

where τ, v and ω are positive integers and satisfying $0 \le \tau, v, \omega \le \lambda - 1$. From (9) and (15), we have

$$z^{\omega-\tau} \equiv z^{q^r v} \quad mod\ f(z) \tag{19}$$

Using Lemmas 1 and 2, we have

$$tr(\beta\alpha^{w-\tau+k}) = \sum_{i=0}^{m-1} v_i tr(\beta\alpha^{k+q^{\tau}i}). \tag{20}$$

Then we have m linear equations with m unknowns v_i for $i = 0, 1, \cdots, m-1$ from (20). Moreover we consider the case of $k = \tau, \tau+1, \cdots, \tau+m-1$, and be substituting the relation in Lemma 2, yielding the proof.

Q.E.D.

2.5 Relationship between squaring and Theorem 2

Suppose that $q = 2$ and $r = 1$. Let $\tilde{w}_i$ be defined as follows:

$$\tilde{w}_i = \sum_{j=0}^{m-1} w_j tr(\beta\alpha^{j+i}) \tag{21}$$

where $i = 0, 1, \cdots, m-1$. It should be noted that the sequences:

$$\sum_{j=0}^{m-1} u_j tr(\beta\alpha^{2k+j}), \quad k = 0, 1, \cdots, \lfloor \frac{m-1}{2} \rfloor + m - 1$$

and

$$\sum_{j=0}^{m-1} u_j tr(\beta\alpha^{2k+j+1}), \quad k = 0, 1, \cdots, \lfloor \frac{m}{2} \rfloor + m - 2$$

are LFSR sequences which are generated from the LFSR based on the feedback polynomial given by $f(z)$. Therefore using two dual basis multipliers, we can generate the sequences as follows:

$$\tilde{w}_0, \tilde{w}_2, \tilde{w}_4, \cdots, \tilde{w}_{2\lfloor \frac{m}{2} \rfloor}$$

and

$$\tilde{w}_1, \tilde{w}_3, \tilde{w}_5, \cdots, \tilde{w}_{(2\lfloor \frac{m}{2} \rfloor)-1}$$

That means we can compute $[v(z)]^2$ using the dual bais multipliers if we can generate LFSR sequences as given in the matrix of left-hand side of (10). However it seems that we cannot generate the feedback shift-register sequences without increasing not a lettle AND/OR gate area. This is not advantageous for our purpose which aims to decrease the gate area as possible.

3. SQUARER BASED ON HARD-WIRED DUAL BASIS MULTIPLICATION

3.1 Bit-serial squarer

From a different point of view, we shall pay attention to computing only square .

Let us consider the following equation:

$$w(z) = [v(z)]^2 \cdot u(z) \quad mod\, f(z)$$

It is noted that $u(z)$ changes places with $[v(z)]^2$ and $u(z) = 1$. From the situation and Theorem 1 , we can generate the sequence $\tilde{w}_0, \tilde{w}_2, \tilde{w}_4, \cdots, \tilde{w}_{2\lfloor \frac{m}{2} \rfloor}$ is generated by hard-wired dual basis multiplier. We will discuss this point in detail.

Let us consider two systems of linear inhomogeneous equations. One is that of the linear inhomogeneous m equations with m unknowns $a_i\,(i = 0, 1, \cdots, m-1)$ such that

$$\begin{bmatrix} tr(\beta), & tr(\beta\alpha), & \cdots, & tr(\beta\alpha^{m-1}) \\ tr(\beta\alpha), & tr(\beta\alpha^2), & \cdots, & tr(\beta\alpha^m) \\ \cdots & \cdots & \cdots & \cdots \\ tr(\beta\alpha^{m-1}), & tr(\beta\alpha^m), & \cdots, & tr(\beta\alpha^{2m-2}) \end{bmatrix} \bullet \begin{bmatrix} a_0 \\ a_1 \\ \cdots \\ a_{m-1} \end{bmatrix} =$$

$$\begin{bmatrix} \sum_{j=0}^{m-1} u_j tr(\beta\alpha^j) \\ \sum_{j=0}^{m-1} u_j tr(\beta\alpha^{j+2}) \\ \cdots \\ \sum_{j=0}^{m-1} u_j tr(\beta\alpha^{j+2(m-1)}) \end{bmatrix} \tag{22}$$

The other is that of the linear inhomogeneous m equations with m unknowns $b_i\,(i = 0, 1, \cdots, m-1)$ such that

$$\begin{bmatrix} tr(\beta), & tr(\beta\alpha), & \cdots, & tr(\beta\alpha^{m-1}) \\ tr(\beta\alpha), & tr(\beta\alpha^2), & \cdots, & tr(\beta\alpha^m) \\ \cdots & \cdots & \cdots & \cdots \\ tr(\beta\alpha^{m-1}), & tr(\beta\alpha^m), & \cdots, & tr(\beta\alpha^{2m-2}) \end{bmatrix} \bullet \begin{bmatrix} b_0 \\ b_1 \\ \cdots \\ b_{m-1} \end{bmatrix} =$$

$$\begin{bmatrix} \sum_{j=0}^{m-1} u_j tr(\beta\alpha^{j+1}) \\ \sum_{j=0}^{m-1} u_j tr(\beta\alpha^{j+3}) \\ \cdots \\ \sum_{j=0}^{m-1} u_j tr(\beta\alpha^{j+2m-1}) \end{bmatrix} \tag{23}$$

We can solve them since we have α, β and

$$u_i = \begin{cases} 1 & i = 0, \\ 0 & i \neq 0. \end{cases} \tag{24}$$

From these solutions and Theorem 2, we have the relations as follows:

$$\begin{bmatrix} \sum\limits_{j=0}^{m-1} v_j tr(\beta\alpha^j), & \sum\limits_{j=0}^{m-1} v_j tr(\beta\alpha^{j+1}), & \cdots, & \sum\limits_{j=0}^{m-1} v_j tr(\beta\alpha^{j+(m-1)}) \\ \sum\limits_{j=0}^{m-1} v_j tr(\beta\alpha^{j+1}), & \sum\limits_{j=0}^{m-1} v_j tr(\beta\alpha^{j+2}), & \cdots, & \sum\limits_{j=0}^{m-1} v_j tr(\beta\alpha^{j+m}) \\ \cdots & \cdots & \cdots & \cdots \\ \sum\limits_{j=0}^{m-1} v_j tr(\beta\alpha^{j+m-1}), & \sum\limits_{j=0}^{m-1} v_j tr(\beta\alpha^{j+m}), & \cdots, & \sum\limits_{j=0}^{m-1} v_j tr(\beta\alpha^{j+2m-2}) \end{bmatrix}$$

$$\bullet \begin{bmatrix} a_0 \\ a_1 \\ \cdots \\ a_{m-1} \end{bmatrix} = \begin{bmatrix} \tilde{w}_0 \\ \tilde{w}_2 \\ \cdots \\ \tilde{w}_{2m-2} \end{bmatrix} \tag{25}$$

and

$$\begin{bmatrix} \sum\limits_{j=0}^{m-1} v_j tr(\beta\alpha^j), & \sum\limits_{j=0}^{m-1} v_j tr(\beta\alpha^{j+1}), & \cdots, & \sum\limits_{j=0}^{m-1} v_j tr(\beta\alpha^{j+(m-1)}) \\ \sum\limits_{j=0}^{m-1} v_j tr(\beta\alpha^{j+1}), & \sum\limits_{j=0}^{m-1} v_j tr(\beta\alpha^{j+2}), & \cdots, & \sum\limits_{j=0}^{m-1} v_j tr(\beta\alpha^{j+m}) \\ \cdots & \cdots & \cdots & \cdots \\ \sum\limits_{j=0}^{m-1} v_j tr(\beta\alpha^{j+m-1}), & \sum\limits_{j=0}^{m-1} v_j tr(\beta\alpha^{j+m}), & \cdots, & \sum\limits_{j=0}^{m-1} v_j tr(\beta\alpha^{j+2m-2}) \end{bmatrix}$$

$$\bullet \begin{bmatrix} b_0 \\ b_1 \\ \cdots \\ b_{m-1} \end{bmatrix} = \begin{bmatrix} \tilde{w}_1 \\ \tilde{w}_3 \\ \cdots \\ \tilde{w}_{2m-1} \end{bmatrix} \tag{26}$$

where $\tilde{w}$, $(i = m, m-1, \cdots, 2m-1)$ is defined as

$$\tilde{w}_i = \sum_{j=0}^{m-1} w_j tr(\beta\alpha^{j+i}) \tag{27}$$

Since both a_i and b_i are fixed constants for $i = 0, 1, \cdots, m-1$, we can compute squares in $\lceil \frac{m}{2} \rceil$ steps using hard-wired dual basis multiplier as shown in Fig.1, where we specify the following initial values:

$$R_i = \tilde{v}_i, \qquad i = 0, 1, 2, \cdots, m-1 \tag{28}$$

and

$$\tilde{v}_i = \sum_{j=0}^{m-1} v_j tr(\beta\alpha^{j+i}) \tag{29}$$

In general the input: $\tilde{v}_i$, needs transforming from v_i. In the same way it is required transforming to the true solution: w_i from the output: $\tilde{w}_i$. However it should be noted that the basis transformation is not necessary for us when using irreducible trinomial as $f(z)$ [3][4] .

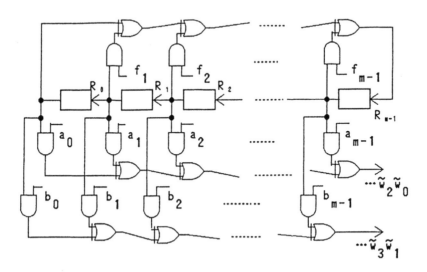

Fig.1 A squarer based on hard-wierd dual basis multiplication algorithm for $GF(2^m)$ with $f(z) = z^m + f_{m-1}z^{m-1} + f_{m-2}z^{m-2} + \cdots + f_1 z + 1$

3.2 Example

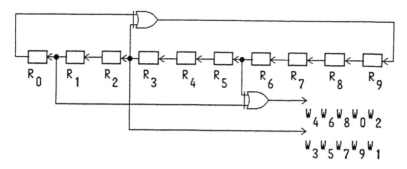

Fig.2 A squarer based on hard-wierd dual basis multiplication algorithm for $GF(2^{10})$ with $f(z) = z^{10} + z^3 + 1$

We give the squarer in $GF(2^{10})$ with irreducible trinomial as follows:

$$f(z) = z^{10} + z^3 + 1.$$

Suppose that $\beta = \alpha^{1018}$ in (22) and (23), and that satisfying $f(\alpha) = 0$. Then we have

$$
\begin{bmatrix}
0 & 0 & 1 & 0 & 0 & 0 & 0 & 0 & 0 & 0 \\
0 & 1 & 0 & 0 & 0 & 0 & 0 & 0 & 0 & 0 \\
1 & 0 & 0 & 0 & 0 & 0 & 0 & 0 & 0 & 0 \\
0 & 0 & 0 & 0 & 0 & 0 & 0 & 0 & 0 & 1 \\
0 & 0 & 0 & 0 & 0 & 0 & 0 & 0 & 1 & 0 \\
0 & 0 & 0 & 0 & 0 & 0 & 0 & 1 & 0 & 0 \\
0 & 0 & 0 & 0 & 0 & 0 & 1 & 0 & 0 & 0 \\
0 & 0 & 0 & 0 & 0 & 1 & 0 & 0 & 0 & 0 \\
0 & 0 & 0 & 0 & 1 & 0 & 0 & 0 & 0 & 0 \\
0 & 0 & 0 & 1 & 0 & 0 & 0 & 0 & 0 & 0
\end{bmatrix}
\bullet
\begin{bmatrix}
a_0 \\
a_1 \\
\cdots \\
a_{m-1}
\end{bmatrix}
=
\begin{bmatrix}
0 \\
1 \\
0 \\
0 \\
0 \\
0 \\
1 \\
0 \\
0 \\
0
\end{bmatrix}
\tag{30}
$$

and

$$
\begin{bmatrix}
0 & 0 & 1 & 0 & 0 & 0 & 0 & 0 & 0 & 0 \\
0 & 1 & 0 & 0 & 0 & 0 & 0 & 0 & 0 & 0 \\
1 & 0 & 0 & 0 & 0 & 0 & 0 & 0 & 0 & 0 \\
0 & 0 & 0 & 0 & 0 & 0 & 0 & 0 & 0 & 1 \\
0 & 0 & 0 & 0 & 0 & 0 & 0 & 0 & 1 & 0 \\
0 & 0 & 0 & 0 & 0 & 0 & 0 & 1 & 0 & 0 \\
0 & 0 & 0 & 0 & 0 & 0 & 1 & 0 & 0 & 0 \\
0 & 0 & 0 & 0 & 0 & 1 & 0 & 0 & 0 & 0 \\
0 & 0 & 0 & 0 & 1 & 0 & 0 & 0 & 0 & 0 \\
0 & 0 & 0 & 1 & 0 & 0 & 0 & 0 & 0 & 0
\end{bmatrix}
\bullet
\begin{bmatrix}
b_0 \\
b_1 \\
\cdots \\
b_{m-1}
\end{bmatrix}
=
\begin{bmatrix}
0 \\
0 \\
0 \\
0 \\
0 \\
0 \\
0 \\
0 \\
0 \\
1
\end{bmatrix}
\tag{31}
$$

Solving (30) and (31), we have

$$
a_i = \begin{cases} 1 & i = 1 \text{ or } 6, \\ 0 & otherwise. \end{cases}
$$

$$
b_i = \begin{cases} 1 & i = 3, \\ 0 & otherwise. \end{cases}
$$

Let us consider computing $[v(z)]^2$, where

$$
v(z) = v_9 z^9 + v_8 z^8 + v_7 z^7 + v_6 z^6 + v_5 z^5 + v_4 z^4 + v_3 z^3 + v_2 z^2 + v_1 z + v_0,
$$

and

$$
[v(z)]^2 \bmod f(z) = w_9 z^9 + w_8 z^8 + w_7 z^7 + w_6 z^6 + w_5 z^5 + w_4 z^4 + w_3 z^3 + w_2 z^2 + w_1 z + w_0,
$$

The sequence, $w_0, w_1, \cdots, w_9$, can be generated by a shift-register circuit as shown in Fig.2, where we specify the following initial values:

$$
R_0 = v_2, \quad R_1 = v_1, \quad R_2 = v_0, \quad R_3 = v_9, \quad R_4 = v_8,
$$
$$
R_5 = v_7, \quad R_6 = v_6, \quad R_7 = v_5, \quad R_8 = v_4, \quad R_9 = v_3.
$$

It does not seem that we will be able to construct a squarer more simply without based on our idea since we have the relationship between v_i and w_i as follows:

$$
\begin{array}{llll}
w_0 = v_0 + v_5 & w_1 = v_9 & w_2 = v_1 + v_6 & w_3 = v_5 \\
w_4 = v_2 + v_7 + v_9 & w_5 = v_6 & w_6 = v_3 + v_8 & w_7 = v_7 \\
w_8 = v_9 + v_4 & w_9 = v_8 &&
\end{array}
$$

4. IRREDUCIBLE TRINOMIALS OVER GF(2)

When using polynomial basis representation, it is very important problem that efficient squaring algorithm is improved since inversion and exponentiation in $GF(2^m)$ can be generally decomposed into squaring and multiplying algorithm. For computing square, we need to have about $\frac{m^2}{4}$ Ex-OR (exclusive or) gates, supposing that $f(z)$ has $\frac{m}{2}$ terms. If we can be fortunate enough to use the irreducible trinomial as $f(z)$, we need to have about m Ex-OR gates.

In our proposed squarer, the complexity of implementation, including the number of Ex-OR gates, is dependent upon the solution of two linear inhomogeneous m equation as given in (22) and (23). When m is fixed, there is a great possibility of constructing very simple squarer by a suitable choice of $f(z)$, α and β carefully. In fact, we can derive the simple circuit when using trinomial as feedback polynomial as shown in the following example. It should be preferable that these solutions for (22) and (23) have low weights if possible, since we hope to construct the squarer as simple as possible. Generally these solutions are not all simple. However we make them simple in general when using irreducible trinomial over $GF(2)$ as $f(z)$. Furtheremore we can generate $[v(z)]^2$ directly without basis transformation [3],[4].

Let us use the following irredusible trinomial:

$$f(z) = z^m + z^k + 1 \tag{32}$$

and consider each case [1] as follows:

Case 1: m is odd and k is even.
Case 2: m is odd and k is odd.
Case 3: m is even and k is odd.

---

[1] In the case that m is even and k is even, we know that no irreducible trinomial exists.

For some β, we have their solutions as following theorem:

Theorem 3:

(Case 1)

$$a_i = \left\{ \begin{array}{ll} 1 & if \quad i = \frac{m+k-1}{2}, \\ 0 & otherwise, \end{array} \right. \qquad b_i = \left\{ \begin{array}{ll} 1 & if \quad i = k \quad or \quad \frac{k}{2}, \\ 0 & otherwise, \end{array} \right.$$

(Case 2)

$$a_i = \left\{ \begin{array}{ll} 1 & if \quad i = \frac{k-1}{2}, \\ 0 & otherwise, \end{array} \right. \qquad b_i = \left\{ \begin{array}{ll} 1 & if \quad i = k \quad or \quad \frac{m+k}{2}, \\ 0 & otherwise, \end{array} \right.$$

(Case 3)

$$a_i = \left\{ \begin{array}{ll} 1 & if \quad i = \frac{k-1}{2} \quad or \quad \frac{m+k-1}{2}, \\ 0 & otherwise, \end{array} \right. \qquad b_i = \left\{ \begin{array}{ll} 1 & if \quad i = k, \\ 0 & otherwise, \end{array} \right.$$

Proof: We can understand this theorem easily from the following lemmas:

Lemma 3: When using irreducible trinomial given in (13) as feedback polynomial, we have the finite sequence in part of the LFSR sequence as follows:

$$m_i = \left\{ \begin{array}{ll} 1 & i = k-1, \ k+m-1 \ or \ 2m-1 \\ 0 & otherwise. \end{array} \right. \tag{33}$$

where $i = 0, 1, 2, \cdots, 2m-1$.

From Lemma 1 and Lemma 3, we have

$$tr(\beta \alpha^i) = \left\{ \begin{array}{ll} 1 & i = k-1, \ k+m-1 \ or \ 2m-1 \\ 0 & otherwise. \end{array} \right. \tag{34}$$

where $i = 0, 1, 2, \cdots, 2m-1$.

Lemma 4: For $tr(\beta \alpha^i)$ given in (34) , we have

$$\left[\begin{array}{cccc} tr(\beta), & tr(\beta\alpha), & \cdots, & tr(\beta\alpha^{m-1}) \\ tr(\beta\alpha), & tr(\beta\alpha^2), & \cdots, & tr(\beta\alpha^m) \\ \cdots & \cdots & \cdots & \cdots \\ tr(\beta\alpha^{m-1}), & tr(\beta\alpha^m), & \cdots, & tr(\beta\alpha^{2m-2}) \end{array} \right]^2 = I \tag{35}$$

where I is the $m \times m$ unit matrix.

When using hard-wired dual basis multiplier based on theorem 2, as the example shown in previous chapter, we can generate $[v(z)]^2$ in $\lceil \frac{m}{2} \rceil$ steps as following sequence:

$$w_i = \tilde{w}_{((k-i-1))},$$

where $i = 0, 1, \cdots, m-1$ and

$$((i)) \equiv i \quad mod\ m.$$

It should be noted that we need to have only two Ex-OR gates independant of m .

5. CONCLUSION

We have presented a bit-serial squarer in $GF(2^m)$ when using polynomial basis for the representation of elements. Especially, Theorem 2 in this paper is very interesting since the relationship between dual basis multiplication and squaring is given. Furthermore we have shown that we can derive much more simple bit-serial squarer when using irreducible trinomial as feedback polynomial.

ACKNOWLEDGMENT

One of authors, M.Morii, would like to thank Prof. M.Kasahara of Kyoto Institute of Technology for his continuing guidance and warm-hearted encouragement.

REFERENCES

(1) T.Beth and D.Gollmann: "Algorithm engineering for public key algorithms", IEEE Journal on Selected Areas in Commumn., **7**, 4, pp.458-466 (1989)

(2) E.R.Berlekamp: "Bit-serial Reed-Solomon encoder", IEEE Trans., Inform. Theory, **IT-28**, 6, pp.869-874 (1982)

(3) R.J.McEliece: Finite fields for computer scientist and engineers, Kluwer Academic (1987)

(4) M.Morii, M.Kasahara and D.L.Whiting: "Efficient bit-serial multiplication and the discrete-time Wiener-Hopf equation over finite fields", IEEE Trans., Inform. Theory, **IT-35**, 6, pp.1177-1183 (1989)

(5) H.F.Mattson and G.Solomon: "A new treatment of Bose-Chaudhuri codes", SIAM J. Appl. Math., **9** pp.654-669 (1961)

APPLICATIONS OF CAYLEY GRAPHS

G. Cooperman, L. Finkelstein and N. Sarawagi

College of Computer Science
Northeastern University
Boston, Ma. 02115, U.S.A.

Abstract

This paper demonstrates the power of the Cayley graph approach to solve specific applications, such as rearrangement problems and the design of interconnection networks for parallel CPU's. Recent results of the authors for efficient use of Cayley graphs are used here in exploratory analysis to extend recent results of Babai et al. on a family of trivalent Cayley graphs associated with $PSL_2(p)$. This family and its subgroups are important as a model for interconnection networks of parallel CPU's. The methods have also been used to solve for the first time problems which were previously too large, such as the diameter of Rubik's $2 \times 2 \times 2$ cube. New results on how to generalize the methods to rearrangement problems without a natural group structure are also presented.

1. Introduction

Each finite group G, together with a generating set Φ, determines a directed graph called a Cayley graph. Once a Cayley graph has been constructed for G, it is possible to obtain algorithmic solutions to the following problems: describe a complete set of rewriting rules for G relative to some lexicographic plus length ordering on the words of Φ [9]; obtain a set of defining relations for G in terms of Φ [6]; and find a word in Φ of minimal length that represents a specified element of G. The last problem is called the *minimal word problem* for G. The solution of the minimal word problem provides an optimal strategy for many rearrangement problems, where the elements of the generating set have some physical significance. These include problems in communications, which can be viewed as token movements on graphs [11], as well as such popular puzzles as Rubik's cube.

There has been a great deal of interest recently in Cayley graphs and their generalization, Schreier coset graphs, for their exceptionally nice characteristics both as models for traditional parallel network architectures and as a potential source of new networks for parallel CPU's [1, 3, 5, 7, 8]. Using Cayley graphs, researchers have discovered new regular graphs with more nodes for a given diameter and for a given number of edges per node than were previously known. This allows construction of larger networks, while meeting design criteria of a fixed number of nearest neighbors and a fixed maximum communication time between arbitrary nodes.

This paper uses theoretical techniques originally developed for designing parallel networks of CPU's [9], and applies them to applications requiring a (sub)optimal solution to the minimal word problem. The purpose is as much to demonstrate the power of the Cayley graph approach, as to solve specific applications. Using these techniques, it has been possible to make empirical observations and computational progress which would not have been possible in a more traditional approach, such as the rewriting system approach described by the authors in an earlier work [4]. An immediate example is the computation of the full Cayley graph of Rubik's $2 \times 2 \times 2$ cube (3,674,160 nodes) on a SUN-3 with 8 megabytes of storage in less than 60 CPU hours.

In understanding the significance of this work, it is important to observe that in applications which involve large groups, one is most often limited by memory resources rather than time. The theoretical tools described in [9] provide both space-efficient data structures and CPU-efficient algorithms for computing with Cayley graphs and Schreier coset graphs. In particular, we are able to implicitly store a minimal spanning tree for a Cayley graph (or Schreier coset graph) using only $\log_2(3)$ bits per node, independant of the size of the generating set, plus additional storage which is small in comparison to the total. Routing depends on the nature of the representation of the underlying group, but can always be performed efficiently for the class of permutation groups. Similar ideas have appeared elsewhere, (see for example [12]), although this is the first time that these techniques have been successfully applied to Cayley and Schreier coset graphs.

These techniques greatly increase the range of problems that can be solved. To see this, it is necessary to first review the definition of a Cayley graph.

A *Cayley graph* $\mathcal{G}$ is a directed graph associated with a group G, and set of generators Φ. The nodes of $\mathcal{G}$ are the elements of G and the edges are labelled by generators in Φ. We will always assume that Φ is closed under inverses. If α and β are two nodes connected by a directed edge (α, β) and the edge is labelled by $\phi \in \Phi$, then $\beta = \alpha\phi$ as an element of G. A *Schreier coset graph* is similarly defined, but requires the additional specification of a subgroup H of G. A Schreier coset graph $\mathcal{G}_H$ is defined to be a directed graph whose nodes are the right cosets of H in G and whose edges are labelled by generators in Φ. If $H\alpha$ and $H\beta$ are two nodes connected by the directed edge $(H\alpha, H\beta)$ with label ϕ, then $H\alpha\phi = H\beta$.

Memory, rather than CPU time, represents the limiting resource for constructing both Cayley graphs and Schreier Coset graphs within the computer technology of today and the near future. This can be shown informally by examining the requirements for using simple breadth-first search to construct a spanning tree for a Cayley graph. This clearly requires space proportional to $|G|$, the order of the group G. The corresponding time requirements are proportional to $|\Phi||G|$, assuming hashing takes constant time. $|\Phi|$ is usually small for many group generating sets of interest, while $|G|$ is on the order of thousands, millions, or more. A unit memory operation may require 8 bytes (one 4-byte word to store a node representation, and one 4-byte word to store a pointer to a parent). The corresponding time to examine the $|\Phi|$ neighbors of a known node (to find new nodes) is usually significantly less than a CPU millisecond on a SUN-3 workstation.

Under the assumptions of 8 bytes and 1 millisecond per node in G and assuming $|G| = 10,000,000$, the computation will require 80 megabytes of storage (excluding storage for the hash table) and approximately 14 CPU hours on a workstation. Since hashing of new nodes represents random accesses throughout data memory, efficient execution requires that all data be stored in semiconductor memory. Otherwise, frequent random disk accesses for virtual memory would make the program unacceptably slow. Hence, the requirements of 80 megabytes of semiconductor memory and 14 SUN-3 CPU hours clearly show the memory to be the limiting resource. In today's technology, the use of a supercomputer would show memory to be even more of a limiting resource. A future generation of parallel CPU's only strengthens further the argument that memory is the critical resource for constructing Cayley graphs.

There are many interesting applications that are outgrowths of the ability to compute with large examples. In section 3, these techniques are applied to an important family of trivalent Cayley graphs associated with the parametrized family of groups $PSL_2(p)$, for p a prime. Babai, Kantor and Lubotzky [2] describe an elegant routing algorithm for these graphs with the property that a path between any two nodes has length at most $45\log(|G|)$. Although this provides an upper bound on the diameter, these graphs are substantially more dense than indicated by the work of Babai et al., and therefore more interesting as possible interconnection networks. This has reduced the worst case estimate of the diameter to $22.5\log(|G|)$. Furthermore, in these groups short relations have been discovered that hold for all primes p.

In section 4, we present some results about the Cayley graph for Rubik's $2 \times 2 \times 2$ cube. To the authors' knowledge, this is the first time that the diameter of Rubik's $2 \times 2 \times 2$ cube has ever been computed. Yet, it was carried out in LISP on a SUN-3 with 8 megabyte of memory, and used included 1 megabyte for the main data structure. Estimates of space required to map out other well-known groups are presented. In particular, in the implementation of an interconnection network for parallel CPU's, a graph with 64,000 nodes could be mapped out and stored at each node of the network, yet consuming only 13 kilobytes for each instance.

In section 5, the methodology is extended to Schreier coset graphs. This allows one to model certain rearrangement problems in which the composition of two legal moves is not always possible. This is the case, for example, in the popular 15-puzzle or more generally to certain token movement problems on graphs [11]. In the token movement problem, labelled tokens are placed on the nodes of a graph. At least one token is designated a blank token. A *legal* move is one which interchanges a blank token with any token currently residing on a neighbor node. The object is to see if a goal configuration can be reached from some initial configuration through a sequence of legal moves.

This methodology has a wide range of applications including problems in operations research and the management of memory in totally distributed systems. Kornhauser et al. [11] developed an approach to this problem in which the question of whether the tokens can be arranged in a specified configuration is reduced to the group membership problem [13], which is solvable in polynomial time. Unfortunately their methodology for finding solutions will never yield an optimal length solution, because each of their group generators is composed of several token moves.

In order to remedy this, we describe a previously unpublished technique [10] and show how it can be extended so that many of these pebble moving problems can be solved optimally by finding a path of shortest length in a Schreier coset graph from the identity coset to a specified coset. This means that each generator, for a certain group associated with the graph, corresponds to a legal move and that the cosets for a suitably chosen subgroup are in 1-1 correspondence with the set of states of the problem. Using techniques in [9], we may now achieve substantially shorter solutions than those using a direct group theoretic approach. We illustrate these ideas in section 5, for the 8-puzzle.

2. Space-Efficient Data Structures for Cayley Graphs

Given a finite group G and generating set Φ, there are two keys to our construction of a space-efficient data structure for the Cayley graph $\mathcal{G}$. First, an easily computable function *count* is used which assigns to each element $g \in G$, a unique integer in the range 0 to $|G| - 1$. This allows us to store information for a node of the graph in an array of length $|G|$ by using *count* as an index into the array, instead of storing an explicit node representation. Second, the distance from a node to the identity modulo some base is used instead of pointers to the parent or other neighboring nodes. Similarly for a Schreier coset graph C_H defined by the subgroup H of G, a function $count_H$ can be used which assigns to each right coset Hg of H, a unique integer in the range 0 to $[G : H] - 1$.

The functions *count* and $count_H$ assume the existence of some concrete representation for G, such as a group of permutations or matrices. In the case where G is an arbitrary permutation group, *count* is defined using standard ideas from computational group theory. The function $count_H$ is far more subtle and depends on a delicate counting argument given in [9]. The general description of *count* and $count_H$ is omitted, since this work does not depend on the details of computation of those functions. A method for defining *count* in the case where G is the set of unimodular 2×2 matrices over $GF(p)$, p a prime, is given in the next section.

In order to simplify the discussion, we restrict our attention to Cayley graphs. The case of Schreier coset graphs is a straightforward generalization. Given G, Φ and $\mathcal{G}$, allocate a bit vector D of length $2|G|$, and associate with each pair of bits a unique address from 0 to $|G| - 1$.

If $count(g) = i$, then we store in $D[i]$, the distance modulo 3 in $\mathcal{G}$ from g to the identity node e. Note that the distance from g to e, is the minimal length of any word in Φ which represents g. D will sometimes be referred to as the *2-bit data structure* for $\mathcal{G}$. We define a *parent* of node g to be any neighbor which has distance to the identity one less than that of g. Note that g need not have a unique parent. Similarly, we define a *child* of g to be any neighbor which has distance one more than that of g. A *sibling* of g is any neighbor which has the same distance as g. Note that only the identity node does not have a parent.

Finding a minimal word representation for $g \in G$ is simple once D has been constructed. The idea is to create a path of minimal length from g to e by choosing an arbitrary parent node as the successor of each node along the path. Each time a new node is selected, the distance to e is diminished by one unit. Since e has no parent, the path eventually must terminate at e. The length of the path is equal to the length of a minimal word representation for G. In order to find a parent of g, it suffices to check the values of $D[count(g\phi)]$ for each $\phi \in \Phi$. If g is not equal to e, then there exists a parent node $g\phi_1$ of g. We then continue the process with g replaced by $g\phi_1$. If $\phi_1, \phi_2, \ldots, \phi_k$ is the sequence of edge labels along the path from g to e, then $g\phi_1 \cdots \phi_k = e$. Since Φ is closed under inverses, it then follows that $\phi_k^{-1} \cdots \phi_1^{-1}$ is a word of minimal length which represents g.

The time for computing a minimal word representation for $g \in G$ is $O(d|\Phi||count|)$, where d is the diameter of $\mathcal{G}$, and $|count|$ is the cost of invoking the function $count$. In the case where G is represented as a permutation group of degree n, then $|count| = O(m^2)$ where $m < n$ is the size of a *base* for G. A base is a subset of $\{1, 2, \ldots, n\}$ with the property that only the identity of G fixes every point of the set (see [13] for related concepts). Many interesting groups have bases with a small number of points. For these cases, the cost of computing $count$ will be small in proportion to n. In the family of groups $PSL_2(p)$ of section 3, the cost of computing $count$ is $O(1)$.

The above scheme leads to a simple routing algorithm for $\mathcal{G}$. Suppose a path of minimal length between two arbitrary nodes g and h is desired. If $\phi_1 \cdots \phi_k$ are the labels along a path of minimal length from e to $g^{-1}h$, then the path from g specified by the sequence $\phi_1, \ldots, \phi_k$ will terminate in h and have minimal length. (We identify the group element gh^{-1} with the name of the corresponding node.)

Storing the distance of each node to the identity modulo 3 clearly requires at least $\log_2(3) \approx 1.58$ bits per node. One can use 8 bits to store such data for 5 nodes, leading to 1.6 bits per node, which is close to the theoretical optimum. To efficiently compute the data structure initially requires additional temporary storage. In [9], a more complex scheme is described to compute the data structure using 2 bits per node of storage in time proportional to $d|G||\Phi||count|$. That scheme requires that $count^{-1}$ be computed within the same time bounds as $count$.

The construction from start of the 2-bit array D from the generators of a group follows. First, initialize D so that $D[i] = 3$, for $1 \leq i \leq |G|-1$ and $D[0] = 0$ (assuming that $count(e) = 0$). The value 3 for an entry in D is a marker which indicates that the final value, which must be 0, 1, or 2, has not yet been entered. Assume that for $\ell \geq 0$, we have filled in the value of $D[count(g)]$ for all nodes g with distance at most ℓ from e. The initialization takes care of the case when $\ell = 0$. To compute the correct D values for nodes at distance $\ell + 1$, for each j with $D[j] = 3$ check if $g = count^{-1}(j)$ has a parent h at distance ℓ from e. This is a necessary and sufficient condition for g to have distance $\ell + 1$ from e. To check this, compute $D[count(g\phi)]$ for each $\phi \in \Phi$. If any has value $\ell \bmod 3$, then enter $\ell + 1 \bmod 3$ in $D[j]$.

The time to construct the 2-bit data structure can be reduced to $|G|(d + |\Phi||count|)$ by using $\log_2(5)$ bits per node [9]. The method also generalizes to Schreier coset graphs.

3. Exploring $PSL_2(p)$

This section is concerned with an exploratory analysis of rewrite rules and short word algorithms for $PSL_2(p)$, p a prime, $p > 2$. The purpose of this section is to show the advantages of using our Cayley graph techniques to explore an interesting class of trivalent graphs discovered by Babai, Kantor and Lubotsky [2] associated with the groups $PSL_2(p)$, p a prime. We present some computational results on the actual diameters of these graphs together with the discovery of some new general relations for these groups, which lead to a shorter routing algorithm than the one described by the authors.

The trivalent graphs for $PSL_2(p)$ which we have studied are only a special case of a more general theory developed by Babai et. al. Their main result shows that every finite simple group G has a set of at most 7 generators so that every element $g \in G$ can be written as a word of length $O(\log_2(|G|))$ in these generators. Furthermore, they give an algorithm for finding a short, but not minimal, word representation for each element of G in terms of these generators. The implication for Cayley graphs is clear. The case where $G = PSL_2(p)$ is the core case and appears to be the one for which the corresponding Cayley graph will be the most dense.

The existence of a short word algorithm is very significant because it leads to a host of possible dense Cayley graphs as models for interconnection networks with good built in routing algorithms. Some very dense Cayley graphs have recently been discovered (see section 1). However, many of these do not have "natural" routing algorithms. The hypercube is an example of an interconnection network with a particularly nice routing algorithm, but which is relatively sparse. A node of the hypercube can be represented as a vector of bits (0's and 1's). A step to another node can be represented as toggling a single bit.

The group $PSL_2(p)$ is the quotient group of $SL_2(p)$ (the set of 2×2 matrices over $GF(p)$ of determinant 1) by its center which is the cyclic group generated by $-I$, where I is the 2×2 identity matrix. Our investigation begins with the generating set $S = \{x(1), x(1)^{-1}, r' \equiv h(1/2)r\}$, for $PSL_2(p)$, where

$$x(t) = \begin{pmatrix} 1 & t \\ 0 & 1 \end{pmatrix}, h(b) = \begin{pmatrix} b^{-1} & 0 \\ 0 & b \end{pmatrix} \text{ for } b \neq 0, \text{ and } r = \begin{pmatrix} 0 & -1 \\ 1 & 0 \end{pmatrix}.$$

It was shown in [2] that the diameter of the Cayley graph $\mathcal{G}(p)$ for $PSL_2(p)$ with respect to S is $O(\log_2 |PSL_2(p)|)$. In fact it was shown to be bounded by $45 \log_2(|PSL_2(p)|)$. This is a direct consequence of a clever short word algorithm. Since $|PSL_2(p)| = p(p^2 - 1)/2$, this leads to the upper bound on the diameter of $\mathcal{G}(p)$ of $\lceil 135 \log_2(p) - 45 \rceil$. A closer reading of [2] shows that the diameter of $\mathcal{G}(p)$ is actually bounded by $45 \log_2(p)$.

We found the true diameters by generating the 2-bit data structure for $\mathcal{G}(p)$ for various p up to $p = 131$. The generating set used was S in all cases. The encoding function *count* that was used for for the 2-bit data structure was defined from G to the set of positive integers $< p^3$. For any $g = \begin{pmatrix} a & b \\ c & d \end{pmatrix} \in PSL_2(p)$, since $ad - bc = 1$, $count(g) = bp^2 + cp + d$ if $d \neq 0$. Otherwise, $count(g) = ap^2 + cp$. The true diameters are compared with the theoretical estimate below. Most graphs $\mathcal{G}(p)$ were generated in a matter of minutes, and the largest ($p = 131$) required a few hours and half a megabyte of data, using the above data encoding resulting in a density of half the optimal density.

Group $(PSL_2(p))$	Number of Nodes	Bound on Diameter $(45 \log_2(p))$	True Diameter
$PSL_2(3)$	12	72	3
$PSL_2(5)$	60	105	6
$PSL_2(7)$	168	127	9
$PSL_2(11)$	660	156	11
$PSL_2(13)$	1,092	167	12
$PSL_2(17)$	2,448	184	14
...		...	
$PSL_2(107)$	612,468	303	25
$PSL_2(109)$	647,460	305	28
$PSL_2(113)$	721,392	307	26
$PSL_2(127)$	1,024,128	315	27
$PSL_2(131)$	1,123,980	317	26

This large difference in the actual diameters of these Cayley graphs from the estimated diameters led to a closer study of the short word algorithm in [2]. Instead of the theoretical upper bound of $45 \log_2(p)$, the table shows an empirical fit of the diameter to $4 \log_2(p)$.

The original short word algorithm is reproduced here, followed by a discussion of the improvements that were made. Let $g = \begin{pmatrix} a & b \\ c & d \end{pmatrix} \in PSL_2(p)$ with $ad - bc = 1$ and $c \neq 0$ then

(A)
$$g = x(c^{-1}(a-1))rx(-c)rx(c^{-1}(d-1))$$

(B)
$$r = h(2)r'$$

Moreover if $0 \leq n < p$, $m + 1 = \lceil \log_4(p-1) \rceil$ and $n = \sum_{i=0}^{m} a_i 4^i$ is the base 4 representation of n, then

(C)
$$x(n) = h(2)^{-m}x(a_m)h(2)x(a_{m-1})h(2)\ldots x(a_1)h(2)x(a_0)$$

where $a_i \in \{0,1,2,3\}$. (A) implies that the length of g with respect to the generating set S is $3 \max_{1 \leq t < p}(length(x(t))) + 2 \; length(r)$. If $c = 0$ then $rg = \begin{pmatrix} a' & b' \\ c' & d' \end{pmatrix}$ with $c' \neq 0$. Therefore the diameter of $\mathcal{G}(p)$ is at most $3 \max_{1 \leq t < p}(length(x(t))) + 3 \; length(r)$. Since $r = h(2)r'$ and by (C) the length of any $x(t)$ and r with respect to S depends upon the length of $h(2)$, it is crucial to get a short word representation for $h(2)$. Babai et al. gave a word for $h(2)$ in S of length 13,

$$h(2) = x(1)^{-2}r'x(1)^2r'x(1)r'x(1)^{-4}r'.$$

This formula is true for all $PSL_2(p)$. Using the fact that the length of $h(2) \leq 13$, the diameter of $\mathcal{G}(p)$ was estimated to be $\leq 45 \log_2(p)$.

Since the constant 45 in the estimate of the diameter depended on the length of the word representation of $h(2)$, it was natural to try to find a minimum length word representation for $h(2)$. Our approach was to use the 2-bit data structure for $\mathcal{G}(p)$ to compute the shortest word for $h(2)$ in each $PSL_2(p)$ as p was increased.

It was soon observed experimentally that for a number of values $p \geq 11$ a minimum word for $h(2)$ was given by the same word of length 9 in $x(1)$ and r'.

$$h(2) = x(1)r'x(1)^4 r'x(1)r', \quad p \geq 11$$

This led to the hypothesis that the relation would be true for all p, $p > 2$. This fact can be directly verified by simply transforming the above elements to matrices over $GF(p)$ and performing the required multiplication.

The shorter formula for $h(2)$ lowered the estimate from $45\log_2(p)$ to $32\log_2(p)$. Further study of the algorithm then resulted in a decrease of the diameter to $22.5\log_2(p) - 33$, using symbolic manipulation.

The three identities

(i)
$$r'h(2)^{-m} = h(2)^m r',$$

(ii)
$$x(t)h(2)^m = h(2)^m x(4^m t) \ \forall m \geq 0, \text{ and}$$

(iii)
$$rx(t)r = r'x(4t)r'$$

were employed to find a shorter word for g in equation (A) as follows.

Using identity (iii) in equation (A) we get,

(A')
$$g = x(c^{-1}(a-1))r'x(-4c)r'x(c^{-1}(d-1))$$

If $c = 0$ then $r'g = \begin{pmatrix} a' & b' \\ c' & d' \end{pmatrix}$ with $c' \neq 0$. Since r' is a generator, the diameter of $PSL_2(p)$ is at most $1 + length(g)$, where g is as in equation (A').

Let $u = c^{-1}(a-1) = \sum_{i=0}^m u_i 4^i$, $t = -4c = \sum_{i=0}^m t_i 4^i$, and $v = c^{-1}(d-1) = \sum_{i=0}^m v_i 4^i$ be their base 4 representations, $(u_i, t_i, v_i \in \{0, 1, 2, 3\})$. Then rewriting (A') we see, $g = x(u)r'x(t)r'x(v)$.

Using (C), (i) and (ii) successively in the above equation, we can obtain the following equation.

$$g = h(2)^{-m} x(u_m)h(2)x(u_{m-1})h(2)x(u_{m-2})h(2)\ldots x(u_1)h(2)x(u_0)$$
$$r'x(t'_m)h(2)x(t'_{m-1})h(2)x(t'_{m-2})h(2)\ldots x(t'_1)h(2)x(t'_0)$$
$$r'x(v_m)h(2)x(v_{m-1})h(2)x(v_{m-2})h(2)\ldots x(v_1)h(2)x(v_0)$$

By the above expansion, $length(g) \leq 4m \ length(h)+3(m+1) \ length(x(3))+2$. Since $length(h) \leq 9$, and $length(x(3)) \leq 3$, so $length(g) \leq 45m + 11$. Moreover, $m + 1 = \lceil \log_4(p-1) \rceil$ implies that $length(g) \leq 22.5\log_2(p) - 34$. This proves the following result.

Proposition 3.1. $diameter(\mathcal{G}(p)) \leq 22.5\log_2(p) - 34.$

As a step to further improve the theoretical estimate of the diameter for $\mathcal{G}(p)$, the minimum words for $x(t)$ were studied experimentally, using the precomputed 2-bit data structure for $\mathcal{G}(p)$. Since $x(t)^{-1} = x(p-t)$ in $PSL_2(p)$, only the minimum words for $x(t)$ for $1 \leq t \leq (p-1)/2$ needed to be examined. This was done for $PSL_2(p)$ for all prime p from 3 to 131. The minimum words either fell into the following four patterns, or were well-determined concatenations of these four patterns. We would expect that new patterns would emerge for significantly larger p, but the range up to 131 seems to be sufficient for currently envisioned applications.

(a) For sufficiently small t depending on p, $x(t)$ had a minimal word of the form $x(1)^t$.

(b) For each $p > 33$, $x((p-3)/2)$ had the same minimum word of length 15,

$$x((p-3)/2) = r'x(1)^4 r'x(1)^{-1} r'x(1)^{-1} r'x(1)^4 r'.$$

When this word was multiplied out over $GF(p)$, it was found to be equal to $x(-3/2)$. This provided a word, $W_{3/2}$, in the elements of S for $x(-3/2)$ of length 15. Concatenating $W_{3/2}$ with $x(1)^i$ forms minimal words for $x(i + (p-3)/2)$ for some i. For example $x((p-1)/2)$ had minimal word $x(1)W_{3/2}$ of length 16, for $p > 33$.

(c) For many large p, $x(20)$ had the same minimal word W_{20} of length 19. For example $W_{20}x(1)$ and $W_{20}x(1)^2$ were the minimal words for $x(21)$ and $x(22)$ of lengths 20 and 21 respectively. We also observed that the minimal word for $X(23)$ was not $W_{20}x(1)^3$ of length 22, but was $W_{24}x(1)^{-1}$ of length 21, where W_{24} was the minimal word for $x(24)$ of length 20. This lead to a set of formulas for $x(4i) = W_{4i}$ where

$$W_{4i} = r'x(1)^{-1} r'x(1)^{-4} r'x(1)^{i-2} r'x(1)^{-4} r'x(1)^{-1} r'$$

is a word in S of length $14 + i$, for all $i > 2$. W_{4i} concatenated by $x(1)$, $x(1)x(1)$ or $x(1)^{-1}$ gives words for $x(4i+1)$, $x(4i+2)$ or $x(4i-1)$ of length $15+i$, $16+i$ or $15+i$ respectively.

(d) Finally there were several values of t for which the shortest word, when multiplied out over the rationals yielded formulas for $x(1/4)$, $x(7/4)$, $x(9/4)$, $x(3/8)$, $x(13/8)$, $x(11/8)$, $x(9/8)$.

Using the above four patterns and the formulas discovered, the shortest word for any $x(t) \in PSL_2(p)$ can easily be found for $p \leq 131$. Disregarding the minimal words of type (d), a simple short word algorithm for $x(t)$ is described below.

Short-Word-Algorithm *Input:* An arbitrary t, $1 \leq t < p$, where $PSL_2(p)$. *Output:* A short word for $x(t)$.

If $t \leq (p-1)/2$
    If $t = (p-1)/2$
        Then if $t < 16$ Output $x(1)^t$
        Else Output $x(1)W_{3/2}$
    Else write $t = 4q + r$, where $0 \leq r < 4$
    $d \leftarrow$ minimum $\{t, 15 + (p-3)/2 - t, 14 + q + r, 14 + q + 2\}$
        if $d = t$ output $x(1)^t$
        if $d = 15 + ((p-3)/2) - t$ output $W_{3/2}x(1)^{t-(p-3)/2}$
        if $d = 14 + q + r$ and $r < 3$ output $W_{4q}x(1)^r$
        if $d = 14 + q + 2$ and $r = 3$ output $W_{4(q+1)}x(1)^{-1}$
Else $t > (p-1)/2$ output the inverse word of Short-Word-Algorithm$(p-t)$

Using the above algorithm to find a short word for $x(t)$, and using the factorization of an arbitrary $g \in G$ in terms of the $x(t)$'s and r', we can show that the diameter of $\mathcal{G}(p)$ is at most $3(p/10 + 16.3) + 3$. For $p < 349$ this leads to a better estimate of the diameter than the best known asymptotic theoretical bound, $22.5 \log_2(p) - 33$.

Another interesting application of the 2-bit data structure, is its use to find all possible rewrite rules in the elements of S which are true for all $PSL_2(p)$. It was hoped to find a family of rewrite rules which would reduce an algorithmically derived short word for a given group element into a still shorter word. The idea is as follows. If an element $g \in G$ has two different minimal word representations, say w_1 and w_2, then both w_1 and w_2 can be found easily from 2-bit data structure. Therefore $w_1 w_2^{-1} = e$ is a relation of even length in G. Further, if an element $g \in G$ and the element sg where $s \in S$, have respective minimal words w_1 and w_2 of the same length then again w_1 and w_2 can be found easily, and $sw_1 w_2^{-1} = e$ is a relation of odd length in G. In this way all such relations of odd and even length in G can be found.

All such relations in $PSL_2(131)$ up to length 20 were generated as described above, and then checked to see which relations were also true for all p. Three such universal relations were found of length 2, 9, and 21.

(i)
$$(r')^2 = 1$$

(ii)
$$(r'x(1)^2)^3 = 1$$

(iii)
$$r'x(1)^{-1}r'x(1)^4r'x(1)r'x(1)^4r'x(1)^{-1}r'x(1)^{-4} = 1$$

4. Large Problems

In addition to efficiently exploring many smaller cases, as in section 3, the space-efficient version of Cayley graphs is especially important in solving larger problems which would formerly have either been infeasible or required much larger computers. As an example, we find that on a SUN-3 workstation using LISP we are able to process 1,000,000 nodes of a Cayley graph for $PSL_2(p)$ every 4 CPU hours. The rate is roughly independent of p.

Our largest example to date is finding the diameter of the Cayley graph for the group Rubik2 associated with Rubik's $2 \times 2 \times 2$ cube. Singmaster [14, p. 60] poses this as an outstanding problem. This cube consists of the corners of the traditional Rubik's $3 \times 3 \times 3$ cube, while ignoring all other sub-blocks. The entire Cayley graph for Rubik2 was mapped out in place and stored, using $\log_2(5)$ bits/node $\times 3,674,160$ nodes = 1 megabyte of space. The task used the more general function *count* instead of the PSL-specific encoding and required 60 CPU hours on a SUN-3 workstation.

The more traditional method of breadth-first search, as described in the introduction, would have required 8 bytes/node $\times 3,674,160$ nodes = 30 megabytes of memory for data, or a ratio of 30 times more data storage. If a hashing scheme was used the ratio would have been proportionately increased due to overhead of the hash table. While this represents only an argument with respect to order of magnitude, the ratio of 30 illustrates the power of the proposed methodology.

The traditional generating set for Rubik2 consists of nine elements. If an orientation of the $2 \times 2 \times 2$ cube is fixed, there are three basic moves: u = rotation of the upper face by 90 degrees clockwise; f = rotation of the front face by 90 degrees clockwise; and l = rotation of the left face by 90 degrees clockwise. This, along with their inverses and their squares form the nine element generating set.

The diameter of the Cayley graph for Rubik2 with these nine generators is 11. This was found as a result of constructing the 2-bit data structure for the Cayley graph. If the generating set is restricted to the six independent generators and inverses $\{u, f, l, u^{-1}, f^{-1}, l^{-1}\}$, then the diameter of the Cayley graph is 14. This computation took 100 CPU hours.

5. Extension to Schreier Coset Graphs

Many applied rearrangement problems have invertible state transitions (legal moves), but do not have a natural group structure. This prevents one from applying the Cayley graph methodology. An interesting (and difficult) example of a rearrangement problem without a natural group structure is the classic 15-puzzle. This has 15 tiles and a blank arranged in a 4×4 rectangle. A legal move consists of interchanging a tile and the blank. The goal is to achieve a specified configuration of the tiles.

One would like to impose a group structure on the 15-puzzle in which the generators are the legal moves and the binary operator is composition. Where legal moves exist, associativity holds and there is an inverse. The identity is the null move. However, arbitrary moves cannot be composed, and so there is not a well-defined binary operator for the group. For example, a legal move interchanging a tile in position 7 with a blank in position 8 cannot be followed by a legal move interchanging a tile in position 2 with a tile in position 3.

A new method is given for re-formulating the 15-puzzle as a coset problem, in which each legal move corresponds to a generator of a group, and the goal state corresponds to a certain coset of a specified subgroup of the group. This allows the use of Schreier coset graphs as described in section 2.

Kornhauser et al. [11] described a formulation which has the defect of mapping a product of legal moves into a generator for a specified group. They also show that any state accessible as a product of arbitrary legal moves in the original formulation will also be accessible as a product of generators in the new formulation. Thus the existence problem is formulated as a group membership problem [13] in a group-theoretic setting. However, a word in the group-theoretic setting will in general correspond to a longer word in terms of the original legal moves, and so their approach cannot be used to find optimal solutions. The approach described here does not suffer that defect.

For purposes of exposition, the 8-puzzle is discussed. This is the 3×3 analogue of the 15-puzzle above. The extension of the technique to the 15-puzzle and other problems will be obvious.

The Eight Puzzle. The 3×3 board of the 8-Puzzle can be thought of as being in one of three *configurations* based on the location of the blank space. Each of the configurations has a set of four *orientations*: up, right, down, and left. We label the cells in each configuration and orientation as shown below. The configuration appears above the *configuration label* (A, B, or C), and four cells for the orientation appear below. The blank space of a configuration is shown in square brackets.

1	2	3		10	11	12		19	20	21
4	5	6		13	14	15		22	[23]	24
7	8	[9]		16	[17]	18		25	26	27
	A				B				C	

28 29 30 31	32 33 34 35	36 37 38 39

Given a state of the 8-puzzle in configuration A and orientation 28, there are two possible legal moves: sliding the tile in cell 8 into cell 9, and sliding the tile in cell 6 into cell 9. Both of these result in a board position in Configuration B, although the latter move requires a built-in rotation of the board by 90 degrees so that the blank ends up in the center cell of the bottom row.

For the moment, we ignore the orientation cells, and describe the action on the 9 tiles of a given configuration. Consider the move $(6 \rightarrow 9)$. This is represented by mapping the cells of configuration A into configuration B with a 90 degree rotation. Thus, applying $(6 \rightarrow 9)$ in configuration A followed by a 90 degree rotation yields the following configuration.

7	4	1
8	5	2
6	[9]	3

This arrangement must then be mapped into the cells of configuration B. Doing so yields the permutation given below. (Cells of configuration C are fixed).

$$\begin{pmatrix} 1 & 2 & 3 & 4 & 5 & 6 & 7 & 8 & 9 & 10 & 11 & 12 & 13 & 14 & 15 & 16 & 17 & 18 & 19 & 20 & 21 & 22 & 23 & 24 & 25 & 26 & 27 \\ 12 & 15 & 18 & 11 & 14 & 16 & 10 & 13 & 17 & 7 & 4 & 1 & 8 & 5 & 2 & 6 & 9 & 3 & 19 & 20 & 21 & 22 & 23 & 24 & 25 & 26 & 27 \end{pmatrix}$$

Finally, the orientation cells must be mapped. Since the previous move involved a 90 degree rotation, the four orientation cells of configuration A would be mapped by a cyclic rotation into the orientation cells of configuration B.

In this manner we can represent each of the nine basic moves of the 8-puzzle. Three examples are given below. The right arrow indicates which tile is being moved into which blank square. The two configurations that are interchanged are also indicated, along with the mapping of cells. Those cells which are mapped to themselves are not shown. For the 8-puzzle the generators are transpositions, although this is not a requirement of this methodology.

$(8 \to 9)$ $(A \leftrightarrow B)$ (0 degree rotation)

$$\begin{pmatrix} 1 & 2 & 3 & 4 & 5 & 6 & 7 & 8 & 9 & 10 & 11 & 12 & 13 & 14 & 15 & 16 & 17 & 18 & 28 & 29 & 30 & 31 & 32 & 33 & 34 & 35 \\ 10 & 11 & 12 & 13 & 14 & 15 & 16 & 18 & 17 & 1 & 2 & 3 & 4 & 5 & 6 & 7 & 9 & 8 & 32 & 33 & 34 & 35 & 28 & 29 & 30 & 31 \end{pmatrix}$$

$(14 \to 17)$ $(B \leftrightarrow C)$ (0 degree rotation)

$$\begin{pmatrix} 10 & 11 & 12 & 13 & 14 & 15 & 16 & 17 & 18 & 19 & 20 & 21 & 22 & 23 & 24 & 25 & 26 & 27 & 32 & 33 & 34 & 35 & 36 & 37 & 38 & 39 \\ 19 & 20 & 21 & 22 & 26 & 24 & 25 & 23 & 27 & 10 & 11 & 12 & 13 & 17 & 15 & 16 & 14 & 18 & 36 & 37 & 38 & 39 & 32 & 33 & 34 & 35 \end{pmatrix}$$

$(24 \to 23)$ $(B \leftrightarrow C)$ $(-90$ degree rotation)

$$\begin{pmatrix} 10 & 11 & 12 & 13 & 14 & 15 & 16 & 17 & 18 & 19 & 20 & 21 & 22 & 23 & 24 & 25 & 26 & 27 & 32 & 33 & 34 & 35 & 36 & 37 & 38 & 39 \\ 25 & 22 & 19 & 26 & 24 & 20 & 27 & 23 & 21 & 12 & 15 & 18 & 11 & 17 & 14 & 10 & 13 & 16 & 39 & 36 & 37 & 38 & 33 & 34 & 35 & 32 \end{pmatrix}$$

Let G be the group generated by the obvious 9 generators corresponding to legal moves of the 8-puzzle, and let H be the point stabilizer subgroup of G which stabilizes all the cells of configuration A (1 through 9 and 28 through 31). A scrambled puzzle is then represented by a permutation, where the cells of configuration A are mapped according to how the puzzle is scrambled, with the cells of the configuration in which the scrambled puzzle lies. In fact a coset of G in H corresponds to a unique scrambled puzzle. Solving this scrambled puzzle, in group theoretic terms, is to find a word in the 9 nine generators for the coset representative of H in G that contains the permutation representing the scrambled puzzle.

We generated the 2-bit data structure for the for the Schreier coset graph of the 8-Puzzle and found that the diameter of the graph was 31. This signifies that given any scrambled 8-puzzle it can be unscrambled in less than 32 moves. In the construction of the 2-bit data structure for this Schreier coset graph, we used a special purpose encoding function, assigning to each coset a unique integer in the range $[0, 12 * 8^7 = 25165824]$. This is 100 times larger than optimal, but it can be computed very efficiently.

Once the 2-bit data structure is constructed for the cosets of H in G, it is used to to find a minimal word representing any coset in terms of the generators (moves of the 8-puzzle) of G. Hence to solve a scrambled 8-Puzzle, first represent the given board X as a permutation σ, which interchanges the cells of configuration A (i.e. $A_1 \cup A_2$) with the cells of the configuration in which X lies, according to the scrambled board X, then find a minimal word w in the generators for the coset $H\sigma$. Applying the moves in the word w^R (the reverse of the word w), in order, to the scrambled puzzle will unscramble it.

Pebble Motion on Graphs. The above representation of the 8-puzzle as a group in which a legal move corresponds to one generator of the group, can be generalized to the pebble coordination problem [11]. If the graph has n nodes and if k is the number of configurations of the graph based on the position(s) of the blank node(s). Label the nodes of each configuration consecutively $(1 \ldots, nk)$ starting with the nodes of the configuration (A) corresponding to the goal state. Then each legal move and each scrambled state can be represented as a permutation of $\{1, \ldots, nk\}$. We can find a word in the generators, which represents the initial (scrambled) state by finding an appropriate coset representative of a coset of the group representing the problem in an appropriate point stabilizer subgroup (the subgroup that stabilizes all the cells of configuration A).

References

1. F. Annexstein, M. Baumslag, A.L. Rosenberg, "Group Action Graphs and Parallel Architectures", COINS Technical Report 87-133, Computer and Information Science Department, University of Massachusetts (1987).

2. L. Babai, W.M. Kantor and A. Lubotsky, "Small diameter Cayley graphs for finite simple groups", European Journal of Combinatorics 10 (1989), pp. 507-522.

3. J-C. Bermond, C. Delome, and J-J. Quisquater, "Strategies for interconnection networks: Some methods from graph theory", *Journal of Parallel and Distributed Computing* 3 (1986), pp. 433-449.

4. C.A. Brown, G. Cooperman, and L. Finkelstein, "Solving Permutation Problems Using Rewriting Rules", *Symbolic and Algebraic Computation* (Proc. of International Symposium ISSAC '88, Rome, 1988), Springer Verlag Lecture Notes in Computer Science **358**, 364–377.

5. L. Campbell, G.E. Carlsson, V. Faber, M.R. Fellows, M.A. Langston, J.W. Moore, A.P. Mullhaupt, and H.B. Sexton, "Dense Symmetric Networks from Linear Groups", preprint.

6. J.J. Cannon, "Construction of Defining Relators for Finite Groups", *Discrete Math.* **5** (1973), pp. 105-129.

7. G.E. Carlson, J.E. Cruthirds, H.B. Sexton, and C.G. Wright, "Interconnection networks based on a generalization of cube-connected cycles", *I.E.E.E. Trans. Comp.* **C-34** (1985), pp. 769–777.

8. D.V Chudnovsky, G.V. Chudnovsky and M.M. Denneau, "Regular Graphs with Small Diameter as Models for Interconnection Networks", *3rd Int. Conf. on Supercomputing*, Boston, May, 1988, pp. 232–239.

9. G. Cooperman and L. Finkelstein, "New Methods for Using Cayley Graphs in Interconnection Networks", to appear in *Discrete Applied Mathematics*, Special Issue on Interconnection Networks, (also Northeastern University Technical Report NU-CCS-89-26).

10. M. Frydenberg, A. Riel, N. Sarawagi, Unpublished manuscript.

11. D. Kornhauser, G. Miller and P. Spirakis, "Coordinating Pebble Motion on Graphs, the Diameter of Permutation Groups and Applications", *Proc. 25th IEEE FOCS* (1984), pp. 241–250.

12. T. Ohtsuki, "Maze-Running and Line-Search Algorithms", article in *Advances in CAD for VLSI*, 4, North Holland, Amsterdam (1986).

13. C.C. Sims, "Computation with Permutation Groups", in *Proc. Second Symposium on Symbolic and Algebraic Manipulation*, edited by S.R. Petrick, ACM, New York, 1971.

14. D. Singmaster, *Notes on Rubik's Magic Cube*, Enslow Publishers, Hillside, N.J., 1981.

Duality between Two Cryptographic Primitives

Yuliang Zheng

Tsutomu Matsumoto

Hideki Imai

Division of Electrical and Computer Engineering
Yokohama National University
156 Tokiwadai, Hodogaya, Yokohama, 240 JAPAN

Abstract

This paper reveals a duality between constructions of two basic cryptographic primitives, *pseudo-random string generators* and *one-way hash functions*. Applying the duality, we present a construction for *universal one-way hash functions* assuming the existence of one-way permutations. Under a stronger assumption, the existence of distinction-intractable permutations, we prove that the construction constitutes a *collision-intractable hash function*. Using ideas behind the construction, we propose *practical* one-way hash functions, the fastest of which compress nearly $2n$-bit long input into n-bit long output strings by applying only *twice* a one-way function.

1 Introduction

Pseudo-random string generators and one-way hash functions are two basic cryptographic primitives. Informally, a *pseudo-random string generator* is a function that on input a random string called a *seed* outputs a longer string which can not be efficiently distinguished from a truly random one. In contrast, a *one-way hash function* outputs a short string on input a long one, with the property that it is computationally difficult to find a pair of strings that are compressed into the same one. In a sense, pseudo-random string generators and one-way hash functions behave in a *dual* fashion.

It has been proved that pseudo-random string generators can be constructed under the assumption of the existence of one-way functions [ILL89]. However, at the time when this paper was written, the best known result on the construction of (universal) one-way hash functions was based on the assumption of the existence of one-way quasi-injections [ZMI90a]. (See also [ZMI90b].) The aim of this research is to explore the intuition that there is a duality between pseudo-random string generators and one-way hash functions, and to apply techniques developed for the former to the construction of the latter under weaker assumptions. Though our aim has not been achieved yet, we obtain a new construction for (universal) one-way hash functions assuming the existence of one-way permutations. Using the construction, we design hash functions that are very efficient and seem to be also one-way.

The paper is organized as follows. In Section 2, we introduce notions and notation. In Section 3, we discuss a duality between the construction of pseudo-random string generators and that of one-way hash functions. Applying the duality, we present in Section 4 a construction for *universal one-way hash functions* assuming the existence of one-way permutations. Under a stronger assumption, the existence of distinction-intractable permutations, we prove in section 5 that the construction constitutes a *collision-intractable hash function*. In Section 6, we use ideas behind the construction to design *practical* (supposed) one-way hash functions. The fastest of these functions compress nearly $2n$-bit long input into n-bit long output strings by applying only *twice* a one-way function.

2   Terminology and Preliminaries

The set of all positive integers is denoted by N. Let $\Sigma = \{0,1\}$ be the alphabet we consider. For $n \in N$, denote by Σ^n the set of all strings over Σ with length n, by Σ^* that of all finite length strings including the empty string, denoted by λ, over Σ, and by Σ^+ the set $\Sigma^* - \{\lambda\}$. The concatenation of two strings x, y is denoted by $x \diamond y$, or simply by xy if no confusion arises. When $x, y \in \Sigma^n$, the bit-wise mod 2 addition, also called the exclusive-or (XOR), of x and y is denoted by $x \oplus y$. The length of a string x is denoted by $|x|$, and the number of elements in a set S is denoted by $\sharp S$.

Let ℓ be a monotone increasing function from N to N, and f a (total) function from D to R, where $D = \bigcup_n D_n, D_n \subseteq \Sigma^n$, and $R = \bigcup_n R_n, R_n \subseteq \Sigma^{\ell(n)}$. D is called the *domain*, and R the *range* of f. In this paper it is assumed, unless otherwise specified, that $D_n = \Sigma^n$ and $R_n = \Sigma^{\ell(n)}$. Denote by f_n the restriction of f on Σ^n. We are concerned only with the case when the range of f_n is $\Sigma^{\ell(n)}$, i.e., f_n is a function from Σ^n to $\Sigma^{\ell(n)}$. f is an *injection* if each f_n is a one-to-one function, and is a *permutation* if each f_n is a one-to-one and onto function. f is (deterministic/probabilistic) *polynomial time computable* if there is a (deterministic/probabilistic) polynomial time algorithm (Turing machine) computing $f(x)$ for all $x \in D$. The composition of two functions f and g is defined as $f \circ g(x) = f(g(x))$. In particular, the i-fold composition of f is denoted by $f^{(i)}$.

A (probability) *ensemble* E with length $\ell(n)$ is a family of *probability distributions* $\{E_n | E_n : \Sigma^{\ell(n)} \to [0,1], n \in N\}$. The *uniform ensemble* U with length $\ell(n)$ is the family of *uniform probability distributions* U_n, where each U_n is defined as $U_n(x) = 1/2^{\ell(n)}$ for all $x \in \Sigma^{\ell(n)}$. By $x \in_E \Sigma^{\ell(n)}$ we mean that x is randomly chosen from $\Sigma^{\ell(n)}$ according to E_n, and in particular, by $x \in_R S$ we mean that x is chosen from the set S uniformly at random. E is *samplable* if there is a (probabilistic) algorithm M that on input n outputs an $x \in_E \Sigma^{\ell(n)}$, and *polynomially samplable* if furthermore, the running time of M is polynomially bounded.

2.1 Pseudo-random String Generators and One-Way Functions

Let ℓ be a polynomial. A *statistical test* is a probabilistic polynomial time algorithm T that, on input a string x, outputs a bit 0/1. Let E^1 and E^2 be ensembles both with length $\ell(n)$. E^1 and E^2 are called *indistinguishable* from each other if for each statistical test T, for each polynomial Q, for all sufficiently large n, $|\Pr\{T(x_1) = 1\} - \Pr\{T(x_2) = 1\}| < 1/Q(n)$, where $x_1 \in_{E^1} \Sigma^{\ell(n)}, x_2 \in_{E^2} \Sigma^{\ell(n)}$. A polynomially samplable ensemble E is *pseudo-random* if it is indistinguishable from the uniform ensemble U with the same length.

Now we further assume that ℓ is a polynomial with $\ell(n) > n$. A *string generator* extending n-bit input into $\ell(n)$-bit output strings is a deterministic polynomial time computable function $g : D \to R$ where $D = \bigcup_n \Sigma^n$ and $R = \bigcup_n \Sigma^{\ell(n)}$. g will be denoted also by $g = \{g_n \mid n \in \mathbf{N}\}$. Let $g_n(U)$ be the probability distribution defined by the random variable $g_n(x)$ where $x \in_R \Sigma^n$, and let $g(U) = \{g_n(U) \mid n \in \mathbf{N}\}$. Clearly, $g(U)$ is polynomially samplable. The following definition can be found in [Yao82] (see also [BM84], [GGM86] and [ILL89]).

Definition 1 $g = \{g_n \mid n \in \mathbf{N}\}$ *is a* (cryptographically secure) *pseudo-random string generator* (PSG) *if* $g(U)$ *is pseudo-random.*

One-way function is the basis of most of modern cryptographic functions and protocols [IL89]. The following definition is from [ILL89].

Definition 2 *Let* $f : D \to R$, *where* $D = \bigcup_n \Sigma^n$ *and* $R = \bigcup_n \Sigma^{\ell(n)}$, *be a polynomial time computable function, and let* E *be an ensemble with length* n. *We say that (1)* f *is one-way with respect to* E *if for each probabilistic polynomial time algorithm* M, *for each polynomial* Q *and for all sufficiently large* n, $\Pr\{f_n(x) = f_n(M(f_n(x)))\} < 1/Q(n)$, *when* $x \in_E D_n$. *(2)* f *is one-way if it is one-way with respect to the uniform ensemble* U *with length* n.

We note that a function f is one-way (with respect to the uniform ensemble U with length n) iff f is one-way with respect to *all* pseudo-random ensembles with the same length. This fact will be used in the proof for Theorem 2. Next we introduce the concept of (simultaneously) hard bits.

Definition 3 *Assume that* $f : D \to R$ *is a one-way function, where* $D = \bigcup_n \Sigma^n$ *and* $R = \bigcup_n \Sigma^{\ell(n)}$. *Also assume that* $i_1, i_2, \ldots, i_t$ *are functions from* $\mathbf{N}$ *to* $\mathbf{N}$, *with* $1 \leq i_j(n) \leq n$ *for each* $1 \leq j \leq t$. *Denote by* E_n^1 *and* E_n^2 *the probability distributions defined by the random variables* $x_{i_t(n)} \cdots x_{i_2(n)} x_{i_1(n)} \diamond f(x)$ *and* $r_t \cdots r_2 r_1 \diamond f(x)$ *respectively, where* $x \in_R \Sigma^n$, $x_{i_j(n)}$ *is the* $i_j(n)$-*th bit of* x *and* $r_j \in_R \Sigma$. *Let* $E^1 = \{E_n^1 \mid n \in \mathbf{N}\}$ *and* $E^2 = \{E_n^2 \mid n \in \mathbf{N}\}$. *We say that (1)* $i_1(n)$ *is a hard bit of* f *if for each probabilistic polynomial time algorithm* M, *for each polynomial* Q *and for all sufficiently large* n, $\Pr\{M(f_n(x)) = x'_{i_1(n)}\} < 1/2 + 1/Q(n)$, *where* $x \in_R \Sigma^n$ *and* $x'_{i_1(n)}$ *is the* $i_1(n)$-*th bit of an* $x' \in \Sigma^n$ *satisfying* $f(x) = f(x')$. *(2)* $i_1(n), i_2(n), \ldots, i_t(n)$ *are simultaneously hard bits of* f *if* E^1 *and* E^2 *are indistinguishable from each other.*

2.2 One-Way Hash Functions

There are basically two kinds of one-way hash functions: *universal one-way hash functions* and *collision-intractable hash functions* (or shortly UOHs and CIHs, respectively). In [Mer89] the former is called *weakly* and the latter *strongly*, one-way hash functions respectively. Naor and Yung gave a formal definition for UOH [NY89], and Damgård gave for CIH [Dam89]. The definition for UOH to be given below is from [ZMI90a] [ZMI90b] in which many other results, such as a construction for UOHs assuming the existence of one-way quasi-injections, are presented.

Let ℓ and m be polynomials with $\ell(n) > m(n)$, H be a family of functions defined by $H = \bigcup_n H_n$ where H_n is a (possibly multi-)set of functions from $\Sigma^{\ell(n)}$ to $\Sigma^{m(n)}$. Call H a *hash function* compressing $\ell(n)$-bit input into $m(n)$-bit output strings. For two strings $x, y \in \Sigma^{\ell(n)}$ with $x \neq y$, we say that x and y collide under $h \in H_n$, or (x, y) is a collision pair for h, if $h(x) = h(y)$.

H is *polynomial time computable* if there is a polynomial (in n) time algorithm computing all $h \in H$, and *accessible* if there is a probabilistic polynomial time algorithm that on input $n \in N$ outputs uniformly at random a description of $h \in H_n$. All hash functions considered in this paper are both polynomial time computable and accessible.

Let H be a hash function compressing $\ell(n)$-bit input into n-bit output strings, and E an ensemble with length $\ell(n)$. The definition for UOH is best described as a three-party game. The three parties are S (an *initial-string supplier*), G (a *hash function instance generator*) and F (a *collision-string finder*). S is an oracle whose power is un-limited, and both G and F are probabilistic polynomial time algorithms. The first move is taken by S, who outputs an *initial-string* $x \in_E \Sigma^{\ell(n)}$ and sends it to both G and F. The second move is taken by G, who chooses, independently of x, an $h \in_R H_n$ and sends it to F. The third and also final (null) move is taken by F, who on input $x \in \Sigma^{\ell(n)}$ and $h \in H_n$ outputs either "?" (I don't know) or a string $y \in \Sigma^{\ell(n)}$ such that $x \neq y$ and $h(x) = h(y)$. F wins a game iff his/her output is *not* equal to "?". Informally, H is a universal one-way hash function with respect to E if for any collision-string finder F, the probability that F wins a game is negligible. More precisely:

Definition 4 *Let H be a hash function compressing $\ell(n)$-bit input into n-bit output strings, P a collection of ensembles with length $\ell(n)$, and F a collision-string finder. H is a universal one-way hash function with respect to P, denoted by UOH/P, if for each $E \in P$, for each F, for each polynomial Q, and for all sufficiently large n, $\Pr\{F(x, h) \neq ?\} < 1/Q(n)$, where x and h are independently chosen from $\Sigma^{\ell(n)}$ and H_n according to E_n and to the uniform distribution over H_n respectively, and the probability $\Pr\{F(x, h) \neq ?\}$ is computed over $\Sigma^{\ell(n)}$, H_n and the sample space of all finite strings of coin flips that F could have tossed.*

If E is an ensemble with length $\ell(n)$, UOH/E is synonymous with UOH/$\{E\}$. In this paper we only consider one version of UOH that is denoted by UOH/$EN[\ell]$, where $EN[\ell]$ is the collection of all ensembles with length $\ell(n)$. Other versions such as UOH/$PSE[\ell]$ and UOH/U are also of interest, where $PSE[\ell]$ is the collection of all polynomially samplable ensembles and U is the uniform ensemble, all with length $\ell(n)$. Relationships among various versions of one-way hash functions including UOH/$EN[\ell]$, UOH/$PSE[\ell]$, UOH/U

and CIH are discussed in [ZMI90a] [ZMI90b]. Of the results obtained in the two papers the most important is that *one-way hash functions in the sense of UOH/EN[ℓ] exist iff those in the sense of UOH/U exist.*

We end this section with a definition for CIH that corresponds to *collision free function family* given in [Dam89]. Let A, a *collision-pair finder*, be a probabilistic polynomial time algorithm that on input $h \in H_n$ outputs either "?" or a pair of strings $x, y \in \Sigma^{\ell(n)}$ with $x \neq y$ and $h(x) = h(y)$.

Definition 5 *H is called a* collision-intractable hash function (CIH) *if for each A, for each polynomial Q, and for all sufficiently large n,* $\Pr\{A(h) \neq ?\} < 1/Q(n)$, *where* $h \in_R H_n$, *and the probability* $\Pr\{A(h) \neq ?\}$ *is computed over* H_n *and the sample space of all finite strings of coin flips that A could have tossed.*

3 Extending and Compressing Methods

In this section we discuss a duality between the construction of pseudo-random string generators and that of one-way hash functions. Throughout this section, t and s are integers, $\ell_0, \ell_1, \ldots, \ell_s$ are polynomials in n with $\ell_0(n) = n$ and $\ell_i(n) < \ell_{i+1}(n)$, and k is a polynomial in n such that $t \cdot k(n) > n$. Denote by ℓ the polynomial $t \cdot k$.

3.1 Serial Versions

Two extending and two compressing methods which are serial in nature are introduced below. Lemma 1 (serial-extending 1) is the dual of Lemma 3 (serial-compressing 1), and Lemma 2 (serial-extending 2) is that of Lemma 4 (serial-compressing 2).

3.1.1 Serial Extending Lemmas

For each $0 \leq i \leq s - 1$, let $g^i = \{g_n^i \mid n \in \mathbf{N}\}$ be a PSG extending $\ell_i(n)$-bit input into $\ell_{i+1}(n)$-bit output strings. The following lemma is a direct consequence of the definition for PSG.

Lemma 1 (serial-extending 1) *Let* $g = \{g_n = g_n^{s-1} \circ g_n^{s-2} \circ \cdots \circ g_n^0 \mid n \in \mathbf{N}\}$. *Then* g *is a PSG extending n-bit input into* $\ell_s(n)$-bit output strings.

A PSG extending n-bit input into $(n + t)$-bit output strings is called a *t-extender*. Let y be a finite length string. Denote by $\text{head}_i(y)$ the first i bits and by $\text{tail}_i(y)$ the last i bits of y. The following lemma is due to Boppana and Hirschfeld [BH89].

Lemma 2 (serial-extending 2) *Let* $e = \{e_n \mid n \in \mathbf{N}\}$ *be a t-extender. For an n-bit string x, let* $b_i(x) = \text{head}_t \circ e_n \circ (\text{tail}_n \circ e_n)^{(i-1)}(x)$, *where* $1 \leq i \leq k(n)$. *Let* g_n *be the function defined by* $g_n(x) = b_{k(n)}(x) \diamond \cdots \diamond b_2(x) \diamond b_1(x)$. *Then* $g = \{g_n \mid n \in \mathbf{N}\}$ *is a PSG extending n-bit input into* $\ell(n)$-bit output strings.

3.1.2 Serial Compressing Lemmas

For each $1 \leq i \leq s$, let $H^i = \bigcup_n H_n^i$ be a UOH/$EN[\ell_i]$ (or CIH) compressing $\ell_i(n)$-bit input into $\ell_{i-1}(n)$-bit output strings. Naor and Yung proved the following serial compressing lemma [NY89].

Lemma 3 (serial-compressing 1) *Let $H = \bigcup_n H_n$, where $H_n = \{h = h_1 \circ h_2 \circ \cdots \circ h_s \mid h_i \in H_n^i, 1 \leq i \leq s\}$. Then H is a UOH/$EN[\ell_s]$ (or CIH) compressing $\ell_s(n)$-bit input into n-bit output strings.*

Let $\ell'(n) = n + t$, and let $C = \bigcup_n C_n$ be a UOH/$EN[\ell']$ (or CIH) compressing $\ell'(n)$-bit input into n-bit output strings. Such a UOH/$EN[\ell']$ (or CIH) is called a *t-compressor*. Then we have the following lemma that is a restricted version of Theorem 3.1 of [Dam89]. The main idea behind the lemma also appeared in [Mer89], where it was called the "metamethod".

Lemma 4 (serial-compressing 2) *Let $C = \bigcup_n C_n$ be a t-compressor. For each $c \in C_n$ and each $\alpha \in \Sigma^n$, let $h_{c,\alpha}$ be the function defined by $h_{c,\alpha}(x) = c(\cdots(c(\alpha \circ b_{k(n)}(x)) \circ b_{k(n)-1}(x)) \cdots \circ b_1(x))$, where $x = x_{\ell(n)} \cdots x_2 x_1$ is an $\ell(n)$-bit string and $b_i(x) = x_{t(i-1)+t} \cdots x_{t(i-1)+2} x_{t(i-1)+1}$. Let $H_n = \{h_{c,\alpha} \mid c \in C_n, \alpha \in \Sigma^n\}$, and $H = \bigcup_n H_n$. Then H is a UOH/$EN[\ell]$ (or CIH) compressing $\ell(n)$-bit input into n-bit output strings.*

3.2 Parallel Versions

In their nice paper [GGM86], Goldreich et al. presented a method for constructing *pseudo-random functions* from pseudo-random string generators. Their construction provides us a configuration for generating pseudo-random strings in parallel, when given polynomially many processors. This observation is the very basis of Micali and Schnorr's parallel PSGs [MSc88]. On the other hand, a parallel hashing method was considered in several papers such as [WC81], [NY89] and [Dam89]. Duality between these two methods is clear. Details are omitted here.

4 PSGs and UOHs from One-Way Permutations

Throughout this section, it is assumed that f is a one-way permutation on $D = \bigcup_n \Sigma^n$, and that $i(n)$ has been proved to be a hard bit of f.

4.1 PSGs from One-Way Permutations

Denote by $\text{extract}_i(x)$ a function extracting the i-th bit of $x \in \Sigma^n$. The following theorem is due to Blum and Micali [BM84].

Theorem 1 *Let ℓ be a polynomial with $\ell(n) > n$, and let g_n be the function defined by $g_n(x) = b_{\ell(n)}(x) \cdots b_2(x) b_1(x)$ where $x \in \Sigma^n$ and $b_j(x) = \text{extract}_{i(n)}(f_n^{(j)}(x))$. Then under the assumption that f is a one-way permutation, $g = \{g_n \mid n \in \mathbf{N}\}$ is a PSG extending n-bit input into $\ell(n)$-bit output strings.*

The efficiency of g can be improved by changing extract$_{i(n)}()$ to a function that extracts all known simultaneously hard bits of f.

4.2 UOHs from One-Way Permutations

For $b \in \Sigma$, $x \in \Sigma^{n-1}$ and $y \in \Sigma^n$, define ins$_i(x, b) = x_{n-1}x_{n-2}\cdots x_i b x_{i-1}\cdots x_2 x_1$, and denote by drop$_i(y)$ a function dropping the i-th bit of y. As the dual of Theorem 1, we have the following result.

Theorem 2 *Let ℓ be a polynomial with $\ell(n) > n$, $\alpha \in \Sigma^{n-1}$ and $x = x_{\ell(n)}\cdots x_2 x_1$ where $x_i \in \Sigma$ for each $1 \leq i \leq \ell(n)$. Let h_α be the function from $\Sigma^{\ell(n)}$ to Σ^n defined by: $y_0 = \alpha$, $y_1 = \text{drop}_{i(n)}(f_n(\text{ins}_{i(n)}(y_0, x_{\ell(n)})))$, $\cdots$, $y_j = \text{drop}_{i(n)}(f_n(\text{ins}_{i(n)}(y_{j-1}, x_{\ell(n)-j+1})))$, $\cdots$, $h_\alpha(x) = f_n(\text{ins}_{i(n)}(y_{\ell(n)-1}, x_1))$. Let $H_n = \{h_\alpha \mid \alpha \in \Sigma^{n-1}\}$ and $H = \bigcup_n H_n$. Then under the assumption that f is a one-way permutation, H is a UOH/EN$[\ell]$ compressing $\ell(n)$-bit input into n-bit output strings.*

Proof : Assume that E is an initial-string ensemble and F a collision-string finder. Let $x \in_E \Sigma^{\ell(n)}$. We show that if F finds with probability $p(n)$ a string x' such that $h(x) = h(x')$ where $h \in_R H_n$, then there is a probabilistic polynomial time algorithm M that finds, with probability greater than $p(n)/(\ell(n) - 1)$, the inverse $f_n^{-1}(w)$ of an n-bit string $w \in_T \Sigma^n$, where T is a pseudo-random ensemble.

The proving procedure consists of three parts. In the first part, we show that every execution of f_n in h defines two pseudo-random ensembles S^j and R^j, if $\alpha \in_R \Sigma^{n-1}$. From each S^j we construct another pseudo-random ensemble T^j. In the second part, we construct a probabilistic polynomial time algorithm M. M first obtains $w \in_T \Sigma^n$, where T is an ensemble chosen uniformly at random from all ensembles T^j. Then it computes the inverse $f_n^{-1}(w)$ by the use of the collision-string finder F. In the third part we estimate the probability that M outputs $f_n^{-1}(w)$ correctly.

Let S_n^1 be the probability distribution defined by $f_n(\text{ins}_{i(n)}(\alpha, x_{\ell(n)}))$, where $\alpha \in_R \Sigma^{n-1}$ and $x_{\ell(n)}$ is the last bit of $x \in_E \Sigma^{\ell(n)}$. Let $S^1 = \{S_n^1 \mid n \in \mathbf{N}\}$. Clearly S^1 is polynomially samplable (when $x \in_E \Sigma^{\ell(n)}$ is given). Now we show that S^1 is indistinguishable from the uniform ensemble.

Let A be a probabilistic polynomial time algorithm. For $v \in \Sigma$, denote by $\Pr^v$ the probability that A outputs 1 on input $f_n(\text{ins}_{i(n)}(\alpha, v))$. Since $\alpha \in_R \Sigma^{n-1}$ and $i(n)$ is a hard bit of f (by assumption), we can think of $z = f_n(\text{ins}_{i(n)}(\alpha, x_{\ell(n)}))$ as a probabilistic encryption of $x_{\ell(n)}$ [GM84]. Thus for any probabilistic polynomial time algorithm A, for any polynomial Q, for all sufficiently large n, we have $|\Pr^0 - \Pr^1| < 1/Q(n)$. Now let $v \in_R \Sigma$. Then $\Pr^v = \Pr^0 \cdot \Pr\{v = 0\} + \Pr^1 \cdot \Pr\{v = 1\} = (\Pr^0 + \Pr^1)/2$, and hence $|\Pr^v - \Pr^0| = |\Pr^v - \Pr^1| = |\Pr^0 - \Pr^1|/2 < 1/2Q(n)$. This implies that $|\Pr^v - \Pr^{x_{\ell(n)}}| < 1/2Q(n)$, no matter how $x_{\ell(n)}$ is chosen from Σ. Note that when $\alpha \in_R \Sigma^{n-1}$ and $v \in_R \Sigma$, we have $f_n(\text{ins}_{i(n)}(\alpha, v)) \in_R \Sigma^n$, since f is a permutation. From these discussions, we see that S^1 is indeed pseudo-random.

Let $R^1 = \{R_n^1 \mid n \in \mathbf{N}\}$, where R_n^1 is defined by $\text{drop}_{i(n)}(f_n(\text{ins}_{i(n)}(\alpha, x_{\ell(n)})))$. Then R^1 is also pseudo-random.

Let $S^2 = \{S_n^2 \mid n \in N\}$, where S_n^2 is defined by $f_n(\text{ins}_{i(n)}(\beta, x_{\ell(n)-1}))$, $\beta \in_{R^1} \Sigma^{n-1}$ and $x_{\ell(n)-1}$ is the second last bit of $x \in_E \Sigma^{\ell(n)}$. Then S^2 is also pseudo-random, for otherwise there would be a statistical test distinguishing R^1 from the uniform ensemble with length $n-1$. Similarly, for each $2 < j \leq \ell(n)$, we can define S^j, R^j and prove that they are both pseudo-random.

For each S^j, $1 \leq j \leq \ell(n)-1$, define T^j as follows: $w \in_{T^j} \Sigma^n$ is produced by first generating a string $w' \in_{S^j} \Sigma^n$, then reversing the $i(n)$-th bit of w', i.e., $w = w' \oplus 0^{n-i(n)} 10^{i(n)-1} = f_n(\text{ins}_{i(n)}(y_{j-1}, x_{\ell(n)-j+1})) \oplus 0^{n-i(n)} 10^{i(n)-1}$, where 0^i denotes the all-0 string of length i and $\oplus$ the bit-wise exclusive-or operation on two strings. Obviously, T^j is also pseudo-random.

We are ready to describe the algorithm M. Let O be an oracle that on input $n \in N$ outputs a string $x \in_E \Sigma^{\ell(n)}$.

Algorithm M:

1. Choose $\alpha \in_R \Sigma^{n-1}$.

2. Query the oracle O with n. Let the answer by O be x. Note that $x \in_E \Sigma^{\ell(n)}$.

3. Choose at random a $1 \leq k \leq \ell(n) - 1$, and let $T = T^k$.

4. Let $w = f_n(\text{ins}_{i(n)}(y_{k-1}, x_{\ell(n)-k+1})) \oplus 0^{n-i(n)} 10^{i(n)-1}$, i.e., choose a $w \in_T \Sigma^n$.

5. Query the collision-string finder F with n, h and x. If F finds a string x' such that $h(x) = h(x')$, then output $\text{ins}_{i(n)}(y'_{k-1}, x'_{\ell(n)-k+1})$, where y'_{k-1} and $x'_{\ell(n)-k+1}$ are defined similarly to y_{k-1} and $x_{\ell(n)-k+1}$ respectively. Otherwise, output a $z \in_R \Sigma^n$.

The running time of M is clearly bounded by a polynomial in n. Now we estimate the probability that the algorithm M outputs $f_n^{-1}(w)$ correctly. Note that when F finds an x' such that $h(x) = h(x')$, then there is an $1 \leq m \leq \ell(n) - 1$ such that $\text{ins}_{i(n)}(y_{m-1}, x_{\ell(n)-m+1}) \neq \text{ins}_{i(n)}(y'_{m-1}, x'_{\ell(n)-m+1})$ and $\text{ins}_{i(n)}(y_{j-1}, x_{\ell(n)-j+1}) = \text{ins}_{i(n)}(y'_{j-1}, x'_{\ell(n)-j+1})$ for each $m < j \leq \ell(n)$. Since k is chosen independently of F, the probability that $k = m$ is $1/(\ell(n) - 1)$. So the probability that M outputs $f_n^{-1}(w)$ correctly is $\Pr\{M \text{ outputs } f_n^{-1}(w)\} > \Pr\{M \text{ outputs } f_n^{-1}(w) \mid F \text{ finds } x'\} \cdot \Pr\{F \text{ finds } x'\} = p(n)/(\ell(n) - 1)$.

When $p(n) \geq 1/Q(n)$ for some polynomial Q, i.e., H is not a one-way hash function in the sense of UOH/$EN[\ell]$, we have $\Pr\{M \text{ outputs } f_n^{-1}(w)\} > 1/Q'$ where Q' is the polynomial defined by $Q'(n) = Q(n)(\ell(n) - 1)$. In other words, f is not one-way with respect to the pseudo-random ensemble T, hence not one-way with respect to the uniform ensemble U (see the note following Definition 2). This is a contradiction and the proof is completed. □

Now let $I(n) = \{i_1, i_2, \ldots, i_{t(n)}\}$ be known simultaneously hard bits of f with $t(n) = O(\log n)$. Let $b = b_{t(n)} \cdots b_2 b_1 \in \Sigma^{t(n)}$, $x \in \Sigma^{n-t(n)}$ and $y \in \Sigma^n$. Define $\text{ins}_{I(n)}(x, b) = x_{n-t(n)} \cdots x_{i_{t(n)}} b_{t(n)} x_{i_{t(n)}-1} \cdots x_{i_1} b_1 x_{i_1-1} \cdots x_2 x_1$, and denote by $\text{drop}_{I(n)}(y)$ a function dropping the i_1-th, i_2-th, $\ldots$, $i_{t(n)}$-th bits of y. Then by changing $\text{drop}_{i(n)}()$ to $\text{drop}_{I(n)}()$, and x_i to $x_{t(n)(i-1)+t(n)} \cdots x_{t(n)(i-1)+2} x_{t(n)(i-1)+1}$, the above constructed H is improved to a hash function that compresses $t(n)\ell(n)$-bit input into n-bit output strings.

5 CIHs from Distinction-Intractable Permutations

We were not able to prove that the hash function H constructed in Section 4.2 is also a CIH. Under a stronger assumption to be stated below, H can be proved to be indeed a CIH.

Assume that $f : D \to R$ is a polynomial time computable function. f is *distinction-intractable* at the $i(n)$-th bit if it is computationally difficult to find a pair of strings $x, y \in D_n$ such that $f_n(x)$ and $f_n(y)$ differ only at the $i(n)$-th bit. More precisely, f is distinction-intractable at the $i(n)$-th bit if for each probabilistic polynomial time algorithm M, for each polynomial Q, for all sufficiently large n, $\Pr\{f_n(x) \neq_{i(n)} f_n(y)\} < 1/Q(n)$, where $(x, y) = M(f)$ and $x_1 \neq_{i(n)} x_2$ means that x_1 and x_2 differ only at the $i(n)$-th bit. It is not hard to verify that distinction-intractableness implies one-wayness.

Theorem 3 *Assume that f is a permutation that is distinction-intractable at the $i(n)$-th bit. Then the hash function H constructed in Section 4.2 is a CIH.*

Proof : Assume for contradiction that H is not a CIH. Then there are a polynomial Q, an infinite subset $\mathbf{N}' \subseteq \mathbf{N}$ and a probabilistic polynomial time algorithm M such that M on input $h \in_R H_n$ finds with probability $1/Q(n)$ a collision-pair (x, x'), for all $n \in \mathbf{N}'$.

Since $h(x) = h(x')$, there is an $1 \leq m \leq \ell(n) - 1$ such that $\mathrm{ins}_{i(n)}(y_{m-1}, x_{\ell(n)-m+1}) \neq \mathrm{ins}_{i(n)}(y'_{m-1}, x'_{\ell(n)-m+1})$ and $\mathrm{ins}_{i(n)}(y_{j-1}, x_{\ell(n)-j+1}) = \mathrm{ins}_{i(n)}(y'_{j-1}, x'_{\ell(n)-j+1})$ for each $m < j \leq \ell(n)$. Here x_i, x'_i, y_i and y'_i are defined in the same way as in the proof for Theorem 2. It is not hard to see that $f_n(\mathrm{ins}_{i(n)}(y_{m-1}, x_{\ell(n)-m+1}))$ and $f_n(\mathrm{ins}_{i(n)}(y'_{m-1}, x'_{\ell(n)-m+1}))$ differ only at the $i(n)$-th bit, i.e., $f_n(\mathrm{ins}_{i(n)}(y_{m-1}, x_{\ell(n)-m+1})) \neq_{i(n)} f_n(\mathrm{ins}_{i(n)}(y'_{m-1}, x'_{\ell(n)-m+1}))$. Thus for each $n \in \mathbf{N}'$, M can be used to find, with the same probability $1/Q(n)$, a pair of strings $w(= \mathrm{ins}_{i(n)}(y_{m-1}, x_{\ell(n)-m+1}))$ and $w'(= \mathrm{ins}_{i(n)}(y'_{m-1}, x'_{\ell(n)-m+1}))$ such that $f_n(w) \neq_{i(n)} f_n(w')$. This contradicts the assumption that f is distinction-intractable at the $i(n)$-th bit, and the theorem follows. $\qquad\Box$

In [Dam89] a CIH is constructed under the assumption of the existence of *claw-free pairs of permutations*. Let f^0 and f^1 be permutations over $\bigcup_n \Sigma^n$. Intuitively, (f^0, f^1) is a *claw-free pair of permutations* if for all sufficiently large n, it is computationally infeasible to find a pair of strings (x, y) such that $x, y \in \Sigma^n$, $x \neq y$ and $f_n^0(x) = f_n^1(y)$. From a claw-free pair of permutations (f^0, f^1), one constructs a function $h : \bigcup_n \Sigma^{n+1} \to \bigcup_n \Sigma^n$ as follows: For each n, let $h_n(x' \diamond x_1) = f_n^{x_1}(x')$, where $x' \in \Sigma^n$ and $x_1 \in \Sigma$. Let h_n be an instance of H_n, and let $H = \bigcup_n H_n$. In [Dam89] H was proved to be a CIH.

Now we show a relationship between distinction-intractable permutations and claw-free pairs of permutations:

Theorem 4 *Assume that f is a permutation that is distinction-intractable at the $i(n)$-th bit. Let f' be the permutation defined by $f'_n(x) = f_n(x) \oplus 0^{n-i(n)}10^{i(n)-1}$, where $x \in \Sigma^n$. Then (f, f') is a claw-free pair of permutations.*

Proof: Assume that we can find two strings $x, y \in \Sigma^n$ such that $x \neq y$ and $f(x) = f'(y)$. Then $f(x) = f(y) \oplus 0^{n-i(n)}10^{i(n)-1}$, i.e. $f(x) \neq_{i(n)} f(y)$. This is a contradiction. $\qquad\Box$

It is not clear whether or not the inverse of the above theorem is also true, i.e., whether or not we can construct a distinction-intractable permutation from a claw-free pair of permutations.

6 Practical One-Way Hash Functions Are Easy to Find

In this section we show that ideas underlying Theorems 2 and 3 can be used to design *practical* one-way hash functions. The fastest of these hash functions compress nearly $2n$-bit long input into n-bit long output strings by applying only *twice* a one-way function.

Let $f : D \rightarrow R$ be a one-way function where $D = \bigcup_n \Sigma^n$, $R = \bigcup_n \Sigma^{m(n)}$ and m is a polynomial. Let k be a real with $k > 1$, s an integer with $s \geq 2$ and $\ell(n) = s(n - \lfloor n/k \rfloor)$. Typically, we choose $1 < k \leq 10$ and $2 \leq s \leq 5$. For each $\alpha \in \Sigma^{\lfloor n/k \rfloor}$, associate with it a function h_α defined by: $y_0 = \alpha, y_1 = \text{tail}_{\lfloor n/k \rfloor}(f_n(y_0 \diamond x^s)), \ldots y_j = \text{tail}_{\lfloor n/k \rfloor}(f_n(y_{j-1} \diamond x^{s-j+1})), \ldots h_\alpha(x) = f_n(y_{s-1} \diamond x^1)$. where $x = x^s \cdots x^2 x^1 \in \Sigma^{\ell(n)}$ and $x^1, x^2, \ldots, x^s \in \Sigma^{n-\lfloor n/k \rfloor}$. Let $H_n = \{h_\alpha | \alpha \in \Sigma^{\lfloor n/k \rfloor}\}$ and $H = \bigcup_n H_n$.

In practice, we first choose, uniformly at random, a string α from $\Sigma^{\lfloor n/k \rfloor}$, and fix it. Then by using the function h_α as is defined above, we can compress $\ell(n)$-bit input into n-bit output strings. This procedure is called the *Hashing Method*, and can be used as a basic step of the serial compressing method defined in Lemma 4 and the parallel compressing method mentioned in Section 3.2.

We were not able to prove that the hash function H is a CIH or a UOH/$EN[\ell]$, even under the assumption that 1 up to $n - \lfloor n/k \rfloor$ are all simultaneously hard bits of f. A sufficient condition for H to be a CIH is that *it is computationally difficult to find two distinct strings* $x, y \in \Sigma^n$ *such that* $\text{tail}_{\lfloor n/k \rfloor}(f_n(x)) = \text{tail}_{\lfloor n/k \rfloor}(f_n(y))$. A (secure) public-key encryption function can be viewed as a function satisfying the condition. For a common-key block cipher, the function from its key space to ciphertext space induced by a randomly chosen plaintext can also be viewed as a function satisfying the condition.

It seems that, when f is carefully chosen, H is *strong* enough and also *efficient* enough for practical applications. In the remaining portion of this section, we present two concrete examples. One is based on the Rabin encryption function, and the other on a common-key block cipher called xDES. There is another good example based on the RSA encryption function with low exponents. Discussions for it are analogous to the first example, and hence omitted here.

6.1 Compressing via the Rabin Encryption Function

Let $M_n = pq$ where p and q are $n/2$-bit long randomly generated primes. Denote by Z_{M_n} the residue classes of integers modulo M_n. The Rabin encryption function *rabin* is defined by $rabin_n(x) = x^2 \bmod M_n$ where $x \in Z_{M_n}$. For Blum integers M_n, i.e., $p \equiv q \equiv 3 \pmod 4$, it was proved that 1 up to $O(\log n)$ are simultaneously hard bits of *rabin*. Now let $k = 10$. When n is large (say ≥ 500), as the authors know, no currently existing algorithms can efficiently find two distinct elements $x, y \in Z_{M_n}$ such that the last $n/10$ bits of $x^2 \bmod M_n$ coincide with that of $y^2 \bmod M_n$.

By the use of the Rabin encryption function and the Hashing Method, we can compress 900-bit input to 500-bit output strings by performing only *twice* the multiplication of two 500-bit integers modulo an integer of the same length. This procedure can be implemented very efficiently, even by software.

6.2 Compressing via xDES

Let m be a polynomial. Informally, a common-key block cipher with length $m(n)$ is a pair of polynomial time computable functions (*encrypt, decrypt*), where *encrypt* and *decrypt* are functions from $\bigcup_n \Sigma^n \times \Sigma^{m(n)}$ to $\Sigma^{m(n)}$ that have the following properties:

1. $ptxt = decrypt_n(key, encrypt_n(key, ptxt))$ for all $key \in \Sigma^n$ and all $ptxt \in \Sigma^{m(n)}$.

2. It is computationally difficult to find $ptxt$ from $encrypt_n(key, ptxt)$ for any $ptxt \in \Sigma^{m(n)}$, without knowing key.

Let f be the function from $\bigcup_n \Sigma^n$ to $\bigcup_n \Sigma^{m(n)}$ that is defined by $f_n(x) = encrypt_n(x, \alpha)$, where $\alpha \in_R \Sigma^{m(n)}$. Then each f_n can be used to compress strings by the Hashing Method. To prevent the compressing method from *meet-in-the-middle attacks*, n should be chosen in such a way that $m(n)$ is sufficiently large, say > 120. A rigorous treatment of this subject can be found in [NS90].

Consider the perhaps most widely used modern encryption algorithm DES. According to our definition, DES is the restriction of *some* common-key block cipher on $\Sigma^{56} \times \Sigma^{64}$. DES should not be directly used to compress strings by the Hashing Method, for its key length is too short. Now we use DES as bricks to build a common-key block cipher called xDES. Our building method is based on a theory on the construction of secure block ciphers developed in [ZMI89].

Let r be a polynomial with $r(i) \geq 2i + 1$. xDES is defined by $xDES^0$, $xDES^1$, $xDES^2$, $xDES^3$, $\cdots$, where $xDES^0$ is the same as DES and, for each $i \geq 1$, $xDES^i$ is a function from $\Sigma^{56r(i)i} \times \Sigma^{128i}$ to Σ^{128i} consisting of $r(i)$ rounds of *Type-2 transformations* [ZMI89]. More details follow.

1. The definition for $xDES^0$: Same·as DES.

2. The definition for $xDES^i$ where $i \geq 1$: Let $key = key_{r(i),i} \diamond \cdots \diamond key_{r(i),2} \diamond key_{r(i),1} \diamond \cdots \diamond key_{2,i} \diamond \cdots \diamond key_{2,2} \diamond key_{2,1} \diamond key_{1,i} \diamond \cdots \diamond key_{1,2} \diamond key_{1,1}$ and $ptxt = ptxt_{2i} \diamond \cdots \diamond ptxt_2 \diamond ptxt_1$, where $key_{i_1,i_2} \in \Sigma^{56}$ and $ptxt_{i_3} \in \Sigma^{64}$ for all $1 \leq i_1 \leq r(i)$, $1 \leq i_2 \leq i$ and $1 \leq i_3 \leq 2i$. Then $ctxt = xDES^i(key, ptxt)$ is computed as follows:

   Step 0: Let $c_{0,1} = ptxt_1, c_{0,2} = ptxt_2, \cdots, c_{0,2i-1} = ptxt_{2i-1}, c_{0,2i} = ptxt_{2i}$.

   Step j, for each $1 \leq j \leq r(i)$: Let $c_{j,1} = c_{j-1,2i}, c_{j,2} = c_{j-1,1} \oplus DES(key_{j,1}, c_{j-1,2}), \cdots$, $c_{j,2i-1} = c_{j-1,2i-2}, c_{j,2i} = c_{j-1,2i-1} \oplus DES(key_{j,i}, c_{j-1,2i})$.

   Step $r(i) + 1$: Let $ctxt = c_{r(i),2i} \diamond \cdots \diamond c_{r(i),2} \diamond c_{r(i),1}$.

Let $r(i) = 2i + 1$ and $k = 3$. Using the Hashing Method, we can compress 224-bit input into 128-bit output strings by performing only twice $xDES^1$, i.e., 6 times DES.

Finally, we note that xDES can also be used in normal encryption/decryption operations, and DES can be replaced by any other secure common-key block encryption algorithm.

References

[BM84] M. Blum and S. Micali, How to generate cryptographically strong sequences of pseudo-random bits, SIAM J. on Comp. 13 (1984) 850-864.

[BH89] R. Boppana and R. Hirschfeld, Pseudorandom generations and complexity classes, in: S. Micali, ed., Randomness and Computation, (JAI Press Inc., 1989) 1-26.

[Dam89] I. Damgård, A design principle for hash functions, Presented at Crypto'89 (1989).

[GGM86] O. Goldreich, S. Goldwasser and S. Micali, How to construct random functions, J. of ACM 33 (1986) 792-807.

[GM84] S. Goldwasser and S. Micali, Probabilistic encryption, J. of Comp. and Sys. Sci. 28 (1984) 270-299.

[ILL89] R. Impagliazzo, L. Levin and M. Luby, Pseudo-random generation from one-way functions, Proc. of the 21-th ACM STOC (1989) 12-24.

[IL89] R. Impagliazzo and M. Luby, One-way functions are essential for complexity based cryptography, Proc. of the 30-th IEEE FOCS (1989) 230-235.

[Mer89] R. Merkle, One way hash functions and DES, Presented at Crypto'89 (1989).

[MSc88] S. Micali and C.P. Schnorr, Super-efficient, perfect random number generators, in: S. Goldwasser, ed., Proc. of Crypto'88, (Springer-Verlag, 1990) 173-198.

[NY89] M. Naor and M. Yung, Universal one-way hash functions and their cryptographic applications, Proc. of the 21-th ACM STOC (1989) 33-43.

[NS90] K. Nishimura and M. Sibuya, Probability to meet in the middle, J. of Cryptology 2 (1990) 13-22.

[WC81] M. Wegman and J. Carter, New hash functions and their use in authentication and set equality, J. of Comp. and Sys. Sci. 22 (1981) 265-279.

[Yao82] A. Yao, Theory and applications of trapdoor functions, Proc. of the 23-th IEEE FOCS (1982) 80-91.

[ZMI89] Y. Zheng, T. Matsumoto and H. Imai, On the construction of block ciphers provably secure and not relying on any unproved hypotheses, Presented at Crypto'89, (1989).

[ZMI90a] Y. Zheng, T. Matsumoto and H. Imai, Connections among several versions of one-way hash functions, Proc. of IEICE of Japan E73 (July 1990).

[ZMI90b] Y. Zheng, T. Matsumoto and H. Imai, Structural properties of one-way hash functions, Presented at Crypto'90, (1990).

Lecture Notes in Computer Science

For information about Vols. 1–420
please contact your bookseller or Springer-Verlag

Lecture Notes in Computer Science

Lecture Notes in Computer Science

Edited by G. Goos and J. Hartmanis

322

S. Gjessing   K. Nygaard   (Eds.)

ECOOP '88
European Conference on
Object-Oriented Programming

Oslo, Norway, August 15–17, 1988
Proceedings

Springer-Verlag

Berlin Heidelberg NewYork London Paris Tokyo

Editors

Stein Gjessing
Kristen Nygaard
Department of Informatics, University of Oslo
P.O. Box 1080 Blindern, N-0316 Oslo 3, Norway

CR Subject Classification (1988): D.2.2, D.3.3, H.2.1, F.3.3

ISBN 3-540-50053-7 Springer-Verlag Berlin Heidelberg New York
ISBN 0-387-50053-7 Springer-Verlag New York Berlin Heidelberg

Printing and binding: Druckhaus Beltz, Hemsbach/Bergstr.
2145/3140-543210

Preface

> " ..... *object oriented* seems to be becoming in the 1980s what
> *structured programming* was in the 1970s. "
>
> Brian Randell and Pete Lee

This quotation is from the invitation to the annual Newcastle University Conference on Main Trends in Computing, September 1988. It seems to capture the situation quite well, only that the object orientation is being materialised in languages and language constructs, as well as in the style of programming and as a perspective upon the task considered.

The second European Conference on Object Oriented Programming (ECOOP'88) was held in Oslo, Norway, August 15-17, 1988, in the city where object oriented programming was born more than 20 years ago, when the Simula language appeared. The objectives of ECOOP'88 were to present the best international work in the field of object oriented programming to interested participants from industry and academia, and to be a forum for the exchange of ideas and the growth of professional relationships.

The richness of the field was evidenced before the conference: the conference had 22 slots for papers, and the Program Committee was faced with the very difficult task of selecting these papers from 103 submissions. Many good papers had to be rejected. We believe that the papers presented at the conference and in these proceedings contain a representative sample of the best work in object oriented programming today. When the names of the authors were made known to the Program Committee, it turned out that it had selected the 22 papers from 13 different countries, another indication of the widespread interest in object oriented programming.

Without the help of the Norwegian Computer Society in general, and Kersti Larsen in particular, ECOOP'88 would not have been possible. We would also like to thank the Program Committee members and all the other referees. Special thanks go to Birger Møller-Pedersen, who took on many additional duties during the Program Chairman's stay abroad.

Oslo, May 1988   Stein Gjessing

Kristen Nygaard

Conference Chairman

Stein Gjessing, University of Oslo, Norway

Program Chairman

Kristen Nygaard, University of Oslo, Norway

Program Committee Members

G. Attardi, DELPHI SpA, Italy

J. Bezivin, Lib UBO/ENSTRBr. France

A. Borning, University of Washington, Seattle, USA

P. Cointe, Rank Xerox, France

S. Cook, University of London, UK

C. Hewitt, MIT, USA

S. Krogdahl, University of Oslo, Norway

B. Magnusson, University of Lund, Sweden

J. Meseguer, SRI International, USA

B. Møller-Pedersen, Norwegian Computing Center, Norway

J. Vaucher, University of Montreal, Canada

A. Yonezawa, Tokyo Institute of Technology, Japan

Exhibition Chairman

Oddvar Hesjedal, Enator A/S, Norway

Organization

DND, The Norwegian Computer Society

Table of Contents

What object-oriented programming may be - and what it does not have to be

Ole Lehrmann Madsen,
Dept. of Computer Science, Aarhus University,
Ny Munkegade, DK-8000 Aarhus C, Denmark
email: olm@daimi.dk

Birger Møller-Pedersen
Norwegian Computing Center,
P.O.Box 114 Blindern, N-0314 Oslo 3, Norway
ean: birger@vax.nr.uninett

Abstract

A conceptual framework for object-oriented programming is presented. The framework is independent of specific programming language constructs. It is illustrated how this framework is reflected in an object-oriented language and the language mechanisms are compared with the corresponding elements of other object-oriented languages. Main issues of object-oriented programming are considered on the basis of the framework presented here.

1. Introduction

Even though object-oriented programming has a long history, with roots going back to SIMULA [SIMULA67] and with substantial contributions like Smalltalk [Smalltalk] and Flavors [Flavors], the field is still characterized by experiments, and there is no generally accepted definition of object-oriented programming.

Many properties are, however, associated with object-oriented programming, like: "Everything is an object with methods and all activity is expressed by message passing and method invocation", "Inheritance is the main structuring mechanism", "Object-oriented programming is inefficient, because so many (small) objects are generated and have to be removed by a garbage collector" and "Object-oriented programming is only for prototype programming, as late name binding and run-time checking of parameters to methods give slow and unreliable ("message not understood") systems".

There are as many definitions of object-oriented programming as there are papers and books on the topic. This paper is no exception. It contributes with yet another definition. According to this definition there is more to object-oriented programming than message passing and inheritance, just as there is more to structured programming than avoiding gotos.

While other programming perspectives are based on some mathematical theory or model, object-oriented programming is often defined by specific programming language constructs. Object-oriented programming

Part of this work has been supported by NTNF, The Royal Norwegian Council for Scientific and Industrial Research, grant no. ED 0223.16641 (the Scala project) and by the Danish Natural Science Research Council, FTU Grant No. 5.17.5.1.25.

is lacking a profound theoretical understanding. The purpose of this paper is to go beyond language mechanisms and contribute with a conceptual framework for object-oriented programming. Other important contributions to such a framework may be found in [Stefik & Bobrow 84], [Booch 84], [Knudsen & Thomsen 85], [Shriver & Wegner 87], [ECOOP 87] and [OOPSLA 87,88].

2. A conceptual framework for object-oriented programming

Background

The following definition of object-oriented programming is a result of the BETA Project and has formed the basis for the design of the object-oriented programming language BETA, [BETA87a].

Many object-oriented languages have inherited from SIMULA, either directly or indirectly via Smalltalk. Most of these languages represent a line of development characterized by everything being objects, and all activities being expressed by message passing. The definition and language is build directly on the philosophy behind SIMULA, but represent another line of development.

SIMULA was developed in order to describe complex systems consisting of objects, which in addition to being characterized by local operations also had their own individual sequences of actions. In SIMULA, this lead to objects that may execute their actions as coroutines. This has disappeared in the Smalltalk line of development, while the line of development described here, has maintained this aspect of objects. While SIMULA simulates concurrency by coroutines, the model presented here incorporates real concurrency and a generalization of coroutines.

The following description of a conceptual framework for object-oriented programming is an introduction to the basic principles underlying the design of BETA. In section 3 the framework is illustrated by means of the BETA language. It is not attempted to give complete and detailed description of the basic principles nor to give a tutorial on BETA. Readers are referred to [DELTA 75], [Nygaard86] and [BETA 87a] for further reading. This paper addresses the fundamental issues behind BETA, but it has also a practical side. The Mjølner [Mjølner] and Scala projects have produced a industrial prototype implementation of a BETA system, including compiler and support tools on SUN and Macintosh.

Short definition of object-oriented programming

In order to contrast the definition of object-oriented programming, we will briefly characterize some of the well-known perspectives on programming.

Procedural programming. A program execution is regarded as a (partially ordered) sequence of procedure calls manipulating data structures.

This is the most common perspective on programming, and is supported by languages like Algol, Pascal, C and Ada.

Functional programming. A program is regarded as a mathematical function, describing a relation between input and output.

The most prominent property of this perspective is that there are no variables and no notion of state. Lisp is an example of a language with excellent support for functional programming.

Constraint-oriented (logic) programming. A program is regarded as a set of equations describing relations between input and output.

This perspective is supported by e.g. Prolog.

The definitions above have the property that they can be understood by other than computer scientists. A definition of object-oriented programming should have the same property. We have arrived at the following definition:

> *Object-oriented programming.* A program execution is regarded as a *physical model*, simulating the behavior of either a real or imaginary part of the world.

The notion of physical model shall be taken literally. Most people can imagine the construction of physical models by means of e.g. LEGO bricks. In the same way a program execution may be viewed as a physical model. Other perspectives on programming are made precise by some underlying model defining equations, relations, predicates, etc.. For object-oriented programming the notion of physical models have to elaborated.

Introduction to physical models

Physical models are based upon a conception of the reality in terms of *phenomena* and *concepts*, and as it will appear below, physical models will have elements that directly reflects phenomena and concepts.

Consider accounting systems and flight reservation systems as examples of parts of reality. The first step in making a physical model is to identify the relevant and interesting phenomena and concepts. In accounting systems there will be phenomena like invoices, while in flight reservation systems there will be phenomena like flights and reservations. In a model of an accounting system there will be elements that model specific invoices, and in a model of a flight reservation system there will be elements modelling specific flights and specific reservations.

The flight SK451 may belong to the general concept of flight. A specific reservation may belong to the general concept of reservation. These concepts will also be reflected in the physical models.

Figure 1. In the object-oriented perspective *physical* models of part of the real world are made by choosing which phenomena are relevant and which properties these phenomena have.

In order to make models based on the conception of the reality in terms of phenomena and concepts, we have to identify which aspects of these are relevant and which aspects are necessary and sufficient. This depends upon which class of physical models we want to make. The physical models, we are interested in, are models of those part of reality we want to regard as information processes.

Aspects of phenomena

In information processes three aspects of phenomena have been identified: *substance, measurable properties* of substance and *transformations* on substance. These are general terms, and they may seem strange at a glance. In order to provide a feeling for what they capture, we have found a phenomenon (Garfield) from everyday life and illustrated the three aspects by figures 2- 4.

Substance is physical matter, characterized by a volume and a position in time and space. Examples of substances are specific persons, specific flights and specific computers. From the field of (programming) languages, variables, records and instances of various kinds are examples of substance.

Figure 2. An aspect of phenomena is substance. Substance is characterized by a volume and a position in time and space.

Substance may have *measurable properties* . Measurements may be compared and they may be described by types and values. Examples of measurable properties are a person´s weight, and the actual flying-time of a flight. The value of a variable is also an example of the result of the measurement of a measurable property.

Figure 3. Substance has measurable properties

A *transformation on substance* is a partial ordered sequence of events that changes its measurable properties. Examples are eating (that will change the weight of a person) and pushing a button (changing the state of a vending machine). Reserving a seat on a flight is a transformation on (the property reserved of) a seat from being free to being reserved. Assignment is an example of a transformation of a variable.

Figure 4. Actions may change the measurable properties of substance

Aspects of concepts

Substance, measurable properties and transformations have been identified as the relevant aspects of phenomena in information processes. In order to capture the essential properties of phenomena being modelled it is necessary to develop abstractions or concepts.

The classical notion of a concept has the following elements: *name*: denoting the concept, *intension*: the properties characterizing the phenomena covered by the concept, and *extension*: the phenomena covered by the concept.

Concepts are created by *abstraction*, focussing on similar properties of phenomena and discarding differences. Three well-known sub-functions of abstraction have been identified. The most basic of these is *classification*. Classification is used to define which phenomena are covered by the concept. The reverse sub-function of classification is called *exemplification*.

Concepts are often defined by means of other concepts. The concept of a flight may be formed by using concepts like seat, flight identification, etc. This sub-function is called *aggregation*. The reverse sub-function is called *decomposition*.

Concepts may be organized in a *classification hierarchy*. A concept can be regarded as a *generalization* of a set of concepts. A well-known example from zoology is the taxonomy of animals: mammal is a generalization of predator and rodent, and predator is a generalization of lion and tiger. In addition, concepts may be regarded as *specializations* of other concepts. For example, predator is a specialization of mammal.

Elements of physical models: Modelling by objects with attributes and actions

Objects with properties and actions, and patterns of objects

Up till now we have only identified in which way the reality is viewed when a physical model is constructed, and which aspects of phenomena and concepts are essential. The next step is to define what a physical model itself consists of.

A physical model consists of *objects*, each object characterized by *attributes* and a sequence of *actions*. Objects organize the substance aspect of phenomena, and transformations on substance are reflected by objects executing actions. Objects may have part-objects. An attribute may be a reference to a part object or to a separate object. Some attributes represent measurable properties of the object. The *state* of an object at a given moment is expressed by its substance, its measurable properties and the action going on then. The state of the whole model is the states of the objects in the model.

> In a physical model the elements reflecting concepts are called *patterns*. A pattern defines the common properties of a category of objects. Patterns may be organized in a classification hierarchy. Patterns may be attributes of objects.

Notice that a pattern is not a set, but an abstraction over objects. An implication of this is that patterns do not contribute to the state of the physical model. Patterns may be abstractions of substance, measurable properties and action sequences.

Consider the construction of a flight reservation system as a physical model. It will a.o. have objects representing flights, agents and reservations. A flight object will have a part object for each of the seats, while e.g. actual flying time will be a measurable property. When agents reserve seats they will get a display of the flight. The seat objects will be displayed, so that a seat may be selected and reserved. An action will thus change the state of a seat object from being free to become reserved.

Reservations will be represented by objects with properties that identifies the customer, the date of reservation and the flight/seat identification. The customer may simply be represented by name and address, while the flight/seat identification will be a reference to the separate flight object/seat object. If the agency have a customer database, the customer identification could also be a reference to a customer object.

As there will be several specific flights, a pattern Flight defining the properties of flight objects will be constructed. Each flight will be represented by an object generated according to the pattern Flight.

A travel agency will normally handle reservations of several kinds. A train trip reservation will also identify the customer and the date of reservation, but the seat reservation will differ from a flight reservation, as it will consists of (wagon, seat). A natural classification hierarchy will identify Reservation as a general reservation, with customer identification and date, and Flight Reservation and Train Reservation as two specializations of this.

Actions in a physical model

Many real world systems are characterized by consisting of objects that perform their sequences of actions *concurrently*. The flight reservation system will consist of several concurrent objects, e.g. flights and agents. Each agent performs its task concurrently with other agents. Flights will register the reservation of seats and ensure that no seats are reserved by two agents at the same time. Note that this kind of concurrency is an inherent property of the reality being modelled; it is not concurrency used in order to speed up computations.

Complex tasks, as those of the agents, are often considered to consist of several more or less independent activities. This is so even though they constitute only one sequence of actions and do not include concurrency . As an example consider the activities "tour planning", "customer service" and "invoicing". Each of these activities will consist of a sequence of actions.

A single agent will not have concurrent activities, but *alternate* between the different activities. The shifts will not only be determined by the agents themselves, but will be triggered by e.g. communication with other objects. An agent will e.g. shift from tour planning to customer service (by the telephone ringing), and resume the tour planning when the customer service is performed.

The action sequence of an agent may often be decomposed into *partial action sequences* that correspond to certain routines carried out several times as part of an activity. As an example, the invoicing activity may contain partial action sequences, each for writing a single invoice.

> Actions in a physical model are performed by objects. The action sequence of an object may be executed *concurrently* with other action sequences, *alternating* (that is at most one at a time) with other action sequences, or as *part* of the action sequence of another object.

The definition of physical model given here is valid in general and not only for programming. A physical model of a railroad station may consist of objects like model train wagons, model locomotives, tracks, points and control posts. Some of the objects will perform actions: the locomotives will have an engine and the control posts may perform actions that imply e.g. shunting. Patterns will be reflected by the fact that these objects are made so that they have the same form and the same set of attributes and actions. In the process of designing large buildings, physical models are often used.

Figure 5. States are changed by objects performing actions that may involve other objects.

Object-oriented programming and language mechanisms supporting it

The notion of physical models may be applied to many fields of science and engineering. When applied to programming the implication is that the *program executions* are regarded as physical models, and we rephrase the definition of object-oriented programming:

> *Object-oriented programming*. A program execution is regarded as a *physical model*, simulating the behavior of either a real or imaginary part of the world.

The ideal language supporting object-orientation should be able to prescribe models as defined above. Most elements of the framework presented above are represented in existing languages claimed to support object-orientation, but few cover them all. Most of them have a construct for describing objects. Constructs for describing patterns are in many languages represented in the form of classes, types, procedures/methods, and functions.

Classification is supported by most existing programming languages, by concepts like type, procedure, and class. Aggregation/ decomposition is also supported by most programming languages; a procedure may be defined by means of other procedures, and a type may be defined in terms of other types.

Language constructs for supporting generalization/specialization (often called sub-classing or inheritance) are often mentioned as the main characteristic of a programming language supporting object-orientation. It is true that inheritance was introduced in Simula and until recently inheritance was mainly associated with object-oriented programming. However, inheritance has started to appear in languages based on other perspectives as well.

Individual action sequences should be associated with objects, and concurrency should be supported. For many large applications support for persistent objects is needed.

Benefits of object-oriented programming

Physical models reflect reality in a natural way

One of the reasons that object-oriented programming has become so widely accepted and found to be convenient is that object orientation is close to the natural perception of the real world: viewed as consisting of object with properties and actions. Stein Krogdahl and Kai A. Olsen put it this way:

"The basic philosophy underlying object-oriented programming is to make the programs as far as possible reflect that part of the reality, they are going to treat. It is then often easier to understand and get an overview of what is described in programs. The reason is that human beings from the outset are used to and trained in perception of what is going on in the real world. The closer it is possible to use this way of thinking in programming, the easier it is to write and understand programs."
(translated citation from "Modulær- og objekt orientert programming", DataTid Nr.9 sept 1986).

Physical model more stable than the functionality of a system

The principle behind the Jackson System Development method (JSD, [JSD]) also reflects the object-oriented perspective described above. Instead of focussing on the functionality of a system, the first step in the development of the system according to JSD is to make a physical model of the real world with which the system is concerned. This model then forms the basis for the different functions that the system may have. Functions may later be changed, and new functions may be added without changing the underlying model.

3. A language based on this model and comparisons with other languages

The definition above is directly reflected in the programming language BETA. The following gives a description of part of the transition from framework to language mechanisms. Emphasis is put on conveying an understanding of major language mechanisms, and of differences from other languages.

Objects and patterns

The BETA language is intended to describe program executions regarded as physical models. From the previous it follows that by physical model is meant a system of interacting objects. A BETA program is consequently a description of such a system. An object in BETA is characterized by a set of attributes and a sequence of actions. Attributes portray properties of objects. The syntactic element for describing an object is called an *object descriptor* and has the following form:

```
(#
    Decl₁; Decl₂; ...; Declₙ
do
    Imp
#)
```

where Decl$_1$; Decl$_2$; ... Decl$_n$ are declarations of the attributes and Imp describes the actions of the objects in terms of imperatives.

In BETA a concept is modelled by a *pattern*. A pattern is defined by associating a name with an object descriptor:

```
P:(#    Decl1; Decl2; ...; Decln
   do
      Imp
   #)
```

The intension of the concept is given by the object descriptor, while the objects that are generated according to this descriptor, constitute the extension. So, while patterns model concepts, the objects model phenomena.

The fact that pattern and object are two very different things is reflected in their specification. An object according to the pattern P has the following specification

```
aP: @ P
```

where aP is the name of the object, and P identifies the pattern. The fact that some objects model singular phenomena is reflected in BETA: it is possible to describe objects that are not generated according to any pattern, but are singular. The object specification

```
S:@ (# Decl1; Decl2; ...; Decln
    do
        Imp
    #)
```

describes a singular object s. The object s is not described as belonging to the extension of a concept, i.e. as an instance of a pattern. Singular objects are not just a convenient shorthand to avoid the invention of a pattern title. Often an application has only one phenomenon with a given descriptor, and it seems intuitively wrong to form a concept covering this single phenomenon. A search for a missing person in the radio includes a description of the person. This description is singular, since it is only intended to cover one specific phenomenon. From a description of a singular phenomenon it is, however, easy to form a concept covering all phenomena that match the description.

The framework presented here makes a distinction between phenomena and concepts, and this is reflected in the corresponding language: objects model phenomena and patterns model concepts. A pattern is not an object. In contrast to this distinction between objects and patterns, Smalltalk-like languages treat classes as objects. Concepts are thus both phenomena and used to classify phenomena. In the framework presented here, patterns may be treated as objects, but that is in the *programming process*. The objects manipulated in a programming environment will be descriptors in the program being developed.

Delegation based language do not have a notion corresponding to patterns. They use objects as prototypes for other objects with the same properties as the prototype object.

References as attributes

An attribute of an object may be a *reference* that denotes another object. A referenced object may be either a part of the referent (the object containing the reference) or separate from the referent.

In the flight reservation system each flight is characterized by a number of seat objects. Each seat object will have a reference to a separate object representing the reservation, one reference to a part object representing the class of the seat and one reference to a part object representing whether the seat is a

Smoking seat or not.

Given the pattern Seat,

```
Seat:(#
      Reserved:^ Reservation;
      Class: @ ClassType;
      Smoking: @ Boolean
   #)
```

a flight reservation system will contain the pattern `Flight` defined locally to pattern `FlightType`:

```
FlightType: (#
              source, destination: ...;

              Flight:
              (#
                Seats: [NoOfSeats] @ Seat
                ...
              #);

              DisplayTimeTableEntry: (# ... #);
              ...
           #)
```

For each entry in the time-table there will be a `FlightType`-object. `SK451` will be a `FlightType`-object. The actual flights on this route will be represented by `Flight`-objects. Scheduled departure time, flying time, and arrival time will be attributes of `FlightType`-objects, while actual departure time, flying time, and arrival time will be attributes of `Flight`-objects.

`Reserved, Class, Smoking` are references to objects, while `Seats` is a repetition of references to `Seat` objects.

Each `Flight`-object will consist of `NoOfSeats` objects of the pattern `Seat`. The lifetime of these `Seat`-objects will be the same as the lifetime of the `Flight`-object. Every `Seat`-object will consist of part objects that represent its class and whether it is a smoker´s seat or not. In addition it will have a reference `Reserved,` to a `Reservation`-object representing the reservation of the seat (with customer identification, date, agent, etc.). The object referenced by `Reserved` will change if a reservation is cancelled or changed to another reservation. `Class` and `Smoking` are part objects as they represent properties that may be changed.

The fact that the substance of a phenomenon may consist of substances of part-phenomena is reflected in BETA by the possibility for objects to have part objects. Part objects are integral parts of the composite object, and they are generated as part of the generation of the composite object.

Most languages support part-objects of pre-defined types, or they model it by references to separate and often dynamically generated objects. One exception is composite objects of Loops. [Stefik & Bobrow 82].

References in BETA are *qualified* (typed). A reference that may only denote `Reservation` objects, will thus be specified by

```
Reserved:^ Reservation
```

The reference `Reserved` may then not by accident be set to denote e.g. an `Invoice` object.

Patterns as attributes

Objects may be characterized by pattern attributes. A Smalltalk method is an example of a pattern attribute in the BETA terminology. In the example above Flight is a pattern attribute of FlightType. In addition FlightType instances have the attribute destination and the method DisplayTimeTableEntry. For the different instances SK451 and SK273 of the pattern FlightType, the attributes SK451.destination and SK273.destination are different attributes. In the same way SK451.DisplayTimeTableEntry and SK273.DisplayTimeTableEntry are different patterns, since they are attributes of different instances. In the same way the "classes" SK451.Flight and SK273.Flight are different. For further exploitation of "class attributes" see [Madsen86].

Actions are executed by objects

Another consequence of the definition above is that every action performed during a program execution is performed by an object. For example, if Seats are to be displayed, then the display action must either be described as the actions of Seat-objects or as the actions of a local object.

In

```
Seat:(#...
        Display:
        (#   (* display Reserved, Class, and Smoking *)   #)
      #)
```

Display is a local pattern (here just described by a comment). In order to display a Seat-object (as part of the display of the Flight) a Display-object is generated and executed. In BETA this is specified by:

```
do ... ; Seats[inx].Display; ...
```

Most object-oriented languages has a construction like this. Lisp-based languages may use the form Display(Seats[inx]). From Smalltalk it has become known as "message passing", even though concurrent processes are not involved. It has the same semantics as a normal procedure call, the only difference is that the procedure is defined in a remote object and not globally. SIMULA introduced the notion of "remote procedure call" for this construction.

In BETA Display is a pattern attribute. While Seat is a pattern defining objects with attributes only, Display has an action-part, describing how Seats are displayed on the screen. The objects in BETA will thus have different functions (or missions) in a program execution, depending on their descriptor . No objects are a priori only "data objects with methods" and no objects are a priori only "methods".

Measurable properties

As an example of a measurable property, consider the percentage of occupied seats of a flight. A flight has parts like seats, but is has no part representing the percentage of occupied seats. This is a property that has to be measured. The value "85 %" is not an object, but is rather a denotation of the *value* of some measuring object.

Actions producing measurements

In BETA a measurable property is reflected by an object that produces a value as a result of executing the object. An object may therefore as part of its actions have an *exit*-part. This consists of a list of evaluations that represents the value of the object.

```
Flight:(#   Seats: [NoOfSeats] @ Seat;
```

```
                Occupied:
                (# ...
                do (* compute NoOfReservedSeats *)
                exit NoOfReservedSeats/NoOfSeats*100
                #)
        #)
```

While a `Flight` object will have `Seat` part objects, its `Occupied` will be a pattern attribute. In this example the specification of `Occupied` is just indicated.

Measurement of percentage of occupied seats is represented by execution of an object generated according to the `Occupied` pattern. This object will not be a part of the `Flight` object, but will temporarily exist when measuring the percentage of occupied seats.

In other languages this aspect is to some degree covered by function attributes.

Actions resulting in state changes

Change of state is usually associated with assignment to variables. In physical models there is a duality between observation of state (measurement) and change of state. A measurement is reflected in BETA by the execution of an object and production of a list of values that represents the value of the measurement. Correspondingly a change of state is reflected by reception of a list of values followed by execution of the actions of an object.

In order for an `OccupiedRecord` object (e.g. of a `Statistics` object) to receive the value of the percentage of occupied seats of a flight, the `OccupiedRecord` object will have an enter-part:

```
    OccupiedRecord:@ (# Occ: @ Real enter Occ do ... #);
```

The percentage of occupied seats of a `Flight` object may then be measured and assigned to this object by

```
    SK451.Occupied → Statistics.OccupiedRecord
```

The main actions of `SK451.Occupied` are executed (the actions described after **do**), its exit-part is transferred to the enter-part of `Statistics.OccupiedRecord` and the main actions of `Statistics.OccupiedRecord` are executed. As a side-effect the main action of `Statistics.OccupiedRecord` may e.g. count the number of assignments.

The association of reception of values with actions are also found in other languages (active, annotated values).

Classification hierarchies

As mentioned above a travel agency will normally handle reservations of several kinds. Classification of reservations into a general Reservation and two specializations Flight Reservation and Train Reservation will be reflected by corresponding patterns.

```
    Reservation:
    (#
      Date: (# ... #);
      Customer: (# ... #);
    #)
```

```
FlightReservation: Reservation
                   (#
                       ReservedFligt: ^ Flight;
                       ReservedSeat: ^ Seat;
                   #)

TrainReservation:  Reservation
                   (#
                       ReservedTrain: ^ Train;
                       ReservedWagon: ^ Wagon;
                       ReservedSeat: ^ Seat;
                   #)
```

We will say that FlightReservation and TrainReservation are *sub-patterns* of Reservation and that Reservation is the *super-pattern* of FlightReservation and TrainReservation.

Besides supporting a natural way of classification, this mechanism contributes to make object-oriented programs compact. The general pattern has only to be described once. A revision of the pattern Reservation will have immediate effect on both sub-patterns. It also support re-usability. All object-oriented languages, except delegation based languages, has this notion of class/sub-class.

Given this classification of reservations, the Reserved attribute (qualified by Reservation) of each Flight object may now denote Reservation, FlightReservation and TrainReservation objects. In order to express that it may only denote FlightReservation objects, it is qualified by FlightReservation:

```
Reserved:^ FlightReservation
```

Virtuals

When making a classification hierarchy, some of the pattern attributes of the super-pattern are completely specified, and these specifications are valid for all possible specializations of the pattern. The printing of date and customer of a reservation will be the same for both kinds of reservation. Other attributes may only be partially specified, and first completely specified in specializations. The printing of reservations will depend upon whether a FlightReservation or a TrainReservation is to be printed.

The descriptor of PrintDateAndCustomer in Reservation will be valid for all kinds of reservations. The descriptor of Print will, however, depend upon which kind of reservation is to be printed. So this pattern may not be fully described in the pattern Reservation. It will perform PrintDateAndCustomer, but in addition it must print either flight/seat or train/wagon/seat. It is, however, important to be able to specify (as part of Reservation) that all Reservation objects have a Print, and that it may be specialized for the different kinds of Reservations. This is done by declaring Print as a *virtual pattern* in Reservation.

Declaring Print as a virtual in Reservation implies that

• in every sub-pattern of Reservation, Print can be specialized to what is appropriate for the actual sub-pattern, and

• execution of Print of some Reservation object, by

```
do ...; SomeReservation.Print; ...
```

where SomeReservation denotes some Reservation object, means execution of the Print, which is defined for the Reservation object *currently* denoted by SomeReservation is executed.

Execution of a virtual pattern implies late binding (to the `Print` pattern of the actual object denoted by `SomeReservation`), while qualification of `SomeReservation`, so that it may only denote objects of pattern `Reservation` or of sub-patterns of `Reservation`, assures that `SomeReservation.Print` will always be valid (`SomeReservation` will not be able to denote objects that do not have a `Print` attribute).

In languages like Smalltalk and Flavors all methods are virtuals, while in BETA, C++ [Stroustrup86] and SIMULA it must be indicated explicitly. This means, that message-passing in Smalltalk and Flavors always implies late binding, while non-virtuals in C++, BETA and SIMULA may be bound earlier and thereby be executed faster. It also has the implication, that with non-virtual methods it is possible to state in a super-pattern, that some of the methods may *not* be specialized in sub-classes. This is useful when making packages of patterns. In order to ensure that these work as intended by the author, some of the methods should not be re-defined by users of the packages.

As methods in BETA are represented by pattern attributes, the ordinary pattern/sub-pattern mechanism is also valid for these. The virtual concept and specialization of methods are further exploited in [BETA 87b].

It is well-known that object-oriented design greatly improves the re-use of code. The main reason for this is sub-classing combined with virtuals [Meyer87]. For many people this is the main issue of object-orientation. However, as pointed out above, modelling that reflects the real world is an equally important issue.

Individual action sequences

As mentioned above all actions in a BETA program execution are executed by objects. Each object has an individual sequence of actions. The model identifies three ways of organizing these sequences: as concurrent, alternating or partial sequences. These are in BETA reflected by three different *kinds* of objects: *system* objects, *component* objects and *item* objects.

Concurrent action sequences

System objects are *concurrent* objects and they have means for synchronized communication: a system object may *request* another system object to execute one of its part-objects, and the requester will wait for the acceptor to do it. When the requested object *accepts* to execute this part-object, possible parameters will be transferred and the part-object executed. If parameters are to be returned, then the requesting object must wait until the part-object is executed.

In the BETA model of the flight reservation system mentioned above, agents and flights will be represented by concurrent objects, reflecting that there will be several agents, each of which at some points in time tries to reserve seats in the same flight. Seat reservations will take place by synchronized communication (the flight object will only perform one reservation at a time), so double reservation of the same seat is avoided.

For example, an Agent object may perform the following request

```
(date,3,window) → SK451 >? ReserveSeat
```

in order to reserve a window seat in 3th row.

The object SK451 may at this point in time be performing `ReserveSeat` for another Agent object, but when it performs the accept-imperative

```
<? ReserveSeat
```

then it will accept to perform ReserveSeat. The descriptor of ReserveSeat may contain a specification telling that it may be requested by *some* Agent object or only by one specific Agent object.

As each object may have their individual action sequence it may at different stages in this sequence accept different requests. When the flight is fully booked it has come to a stage in its action sequence where it does not accept ReserveSeat requests.

The underlying model of a language determines to a certain degree which kind of concurrency is supported. While languages supporting objects as the main building blocks will have objects executing actions concurrently (even if this may only be accomplished by concurrent execution of methods as in ConcurrentSmalltalk), Lisp-based languages will have concurrency based on concurrent evaluation of expressions (futures).

Alternating action sequences

Component objects in BETA are alternating objects, that is objects where at most one object is executing at a time and where the shift of control from object to object is non-deterministic.

In a BETA model of the system above the activities Tour Planning, Invoicing and Costumer Service will be represented by component part-objects of Agent objects, and each Agent will be an object, that executes these component objects alternating:

```
Agent: (#   ...
        do...;
           (| TourPlanning | Invoicing | CostumerService |);
           ...
        #)
```

The activity TourPlanning may consist of planning a series of tours that are bought earlier, and just waits to be planned. Correspondingly the activity Invoicing may consist of writing invoices for a series of tours. The activity CostumerService consists of waiting for customer requests and fulfilling them. A shift from TourPlanning or Invoicing to CostumerService will thus only take place, when there is a request for it, so it will not be part of the descriptor of neither TourPlanning nor Invoicing when this shall happen.

Partial action sequences

The example above with the execution of an object according to the Print attribute of some Reservation object is an example of a partial action sequence, represented by an item object. The action sequence of Print is executed as part of the "calling" object.

All languages have a notion of partial action sequences. The notion of procedure and function cover this aspect of actions. Method invocation as a result of message passing is a special case of a partial action sequence, where the procedure to perform is defined in an object different from the invoking object.

One of the characteristics of most other object-oriented languages is that everything is an object with methods and all activity is expressed by message passing and method invocation. Objects in these languages do not have any individual action sequence; they are only "executing methods" on request. Objects in these languages may thus not support alternating or concurrent action sequences. One exception is the Actor model of execution [Agha86], where the objects are concurrent, but sub-classes are not supported in the common sense of the word. Work has been initiated to make concurrent Smalltalk, but this work does not include giving the objects their own sequence of actions.

4. What object-oriented programming does not have to be

As mentioned in the introduction many properties are associated with object-oriented programming. In this section we will comment on some of the misunderstandings (according to the definition given here) of object-oriented programming.

Everything are objects with methods and all actions are message passing

A property common to most object-oriented programming languages is that everything has to be regarded as objects with methods and that every action performed is message passing. The implication of this is that even a typical functional expression such as

```
6+7
```

gets the unnatural interpretation

```
6.plus(7)
```

Even though 6 and 7 are objects (integer so), and they are also in the definition of object-oriented programming presented here, then there is no reason that + may not be regarded as an object that adds two integer objects:

```
plus(6,7)
```

Thinking object-oriented does not have to exclude functional expressions when that is more natural. Functions, types and values are in fact needed in order to describe measurable properties of objects.

Object-oriented programming and automatic storage management

According to the definition of object-oriented programming given here it has not necessarily anything to do with *dynamic generation of objects*. This is one of the properties often associated with object-oriented programming. In many object-oriented languages it is only possible to generate objects dynamically. But whether objects are parts of other objects or generated independently and dynamically, is not crucial for whether program executions are organized in objects or not. It is demonstrated above that in BETA it is possible to specify part-objects. Program executions are still organized in objects with attributes and actions, but some of the objects are allocated as part of other objects. As shown above a Flight object consists of Seat objects, and these are constituent parts of a Flight object. When objects are generated dynamically, as e.g. Reservation objects will be in the example above, it is, however, important that the implementation includes an automatic storage management system.

Late (and unsafe) binding of names gives slow execution (and unreliable) systems

Object-oriented programming does not necessarily imply *late and unsafe binding of names*. As mentioned above, pattern attributes of BETA objects and procedures in C++ objects may be specified as non-virtual, which means that late binding is not used when invoking them.

When Smalltalk or Flavors objects reacts on a message passed to it with "message not understood", it has nothing to do with Smalltalk or Flavors being object-oriented, but with the fact that they are untyped languages.

The combination of qualified (typed) references and virtuals in BETA implies that it may be checked at compile time that expressions like "aRef.aMethod" will be valid at run time, provided of course that aRef denotes an object and not **none**. And still a late binding determines which aMethod (of which sub-pattern) will be executed. Which aMethod to execute depends upon which object is currently denoted by aRef.

In the example above the reference SomeReservation will be qualified by Reservation. This means, that SomeReservation may denote objects generated according to the pattern Reservation or sub-patterns of Reservation. As Print is declared as a virtual in Reservation, it is assured that

```
SomeReservation.Print
```

always is valid and that it will lead to the execution of the appropriate Print. However, the use of untyped references in Smalltalk-like languages have the benefit, that recompilation of a class does not have to take into consideration the rest of the program.

What makes late binding slow is not only the method look-up. If a method in Smalltalk has parameters, then the correspondence between actual and formal parameters must be checked at the time of execution. Print will e.g. have a parameter telling how many copies to print. This will be the same for all specializations of Print, and should therefore be specified as part of the declaration of Print in Reservation.

In BETA this is obtained by *qualifying virtuals*. The fact that Print will have a parameter is described by a pattern PrintParameter:

```
PrintParameter:(# NoOfCopies: @ Integer; enter(NoOfCopies) do ... #)
```

Qualifying the virtual Print with PrintParameter implies that all specializations of Print in different sub-patterns of Reservation must be sub-patterns of PrintParameter, and thus have the properties described in PrintParameter. This implies that Print in all sub-patterns to Reservation will have a Integer NoOfCopies input-parameter.

If object-oriented programming is to be widely used in real application programming, then the provision of typed languages is a must. As Peter Wegner says in "Dimensions of Object-Based Language Design":

"..., the accepted wisdom is that strongly typed object-oriented languages should be the norm for application programming and especially for programming in the large."

As demonstrated above it does not have to exclude flexibility in specialization of methods or late binding.

Inheritance/code sharing

Since inheritance has been introduced by object-oriented languages, object-oriented programming is often defined to be programming in languages that support inheritance. Inheritance may, however, also be supported by functional languages, where functions, types and values may be organized in a classification hierarchy.

In most object-oriented languages classes are special objects, and inheritance is defined by a message forwarding mechanism. Objects of subclasses send (forward) messages to the super-class "object" in order to have inherited methods performed. This approach stresses *code sharing*: there shall be only one copy of the super-class, common to all sub-classes. With this definition of inheritance it is not strange that "distribution is inconsistent with inheritance" [Wegner] and that "This explains why there are no languages with distributed processes that supports inheritance" [Wegner].

In the model of object-oriented programming presented here the main reason for sub-classing (specialization) is the classification of concepts. The way in which an object inherits a method from a super-class is - or rather should be - an implementation issue, and it should not be part of the language definition.

According to the definition of patterns and objects in BETA given above, patterns are not objects, and in principle every object of pattern P will have its own descriptor. It is left to the implementation to optimize by having different objects of P share the descriptor. Following this definition of patterns and objects there is no problem in having two objects of the same sub-pattern act concurrently and even be distributed. The implementation will in this case simply make as many copies as needed of the pattern, including a possible super-pattern. This does not exclude that a modification of the super-pattern will have effect on all sub-patterns.

Multiple inheritance

Multiple inheritance has come up as a generalization of single inheritance. With single inheritance a class may have at most one super-class, whereas multiple inheritance allows a class to have several super-classes. Inheritance is used for many purposes including code sharing and hierarchical classification of concepts. In the BETA language inheritance is mainly intended for hierarchical classification. BETA does not have multiple inheritance, due to the lack of a profound theoretical understanding and also that the current proposals seem technically very complicated.

In existing languages with multiple inheritance, the code-sharing part of the class/sub-class construct dominates. Flavors has a name that directly reflects what is going on: mixing some classes, so the resulting class has the desired flavor, that is the desired attributes. For the experience of eating an ice cone it is significant whether the vanilla ice is at the bottom and the chocolate ice on top, or the other way around. Correspondingly, a class, that inherits from the classes (A, B) is not the same as a class that inherits from the classes (B, A).

If, however, multiple inheritance is to be regarded as a generalization of single inheritance and thereby as a model of multiple concept classification (and it shall in the model presented here), then the order of the super-classes should be insignificant. When classifying a concept as a specialization of several concepts then no order of the general concepts are implied, and that should be supported by the language.

Single inheritance is well suited for modelling a strict hierarchical classification of concepts, i.e. a hierarchy where the extensions of the specializations of a given concept are disjoint. Such hierarchies appear in many applications, and it is often useful to know that the extensions of say class predator and class rodent are disjoint.

By classifying objects by means of different and independent properties, several orthogonal strict hierarchies may be constructed. A group of people may be classified according to their profession leading to one hierarchy, and according to their nationality leading to another hierarchy. Multiple inheritance is often used for modelling the combination of such hierarchies. It may, however, be difficult to recognize if such a non-strict hierarchy is actually a combination of several strict hierarchies.

5. Conclusion

The programmers perspective on programming is perhaps more important than programming language constructs. Object-oriented programming should not be defined only by specific language constructs. It is absolutely possible to think and program object-oriented even without a language that directly supports it.

It is a great advantage to use an object-oriented language that directly supports object-orientation. Such a language should support:

- Modelling of concepts and phenomena, i.e. the language must include constructs like class, type, procedure.
- Modelling classification hierarchies, i.e. sub-classing (inheritance) and virtuals.
- Modelling active objects, i.e. concurrency or coroutine sequencing, combined with persistency.

The benefits of object-oriented programming may be summarized as follows:

- Programs reflect reality.
- Model is more stable than functionality.
- Sub-classing and virtuals improves re-usability.

The above statements are of course not objective, in the sense that it is arguable what is natural, easy and stable. For a mathematician it may be more natural to construct a model using equations.

Finally we would like to stress that a programming language should support other perspectives than object-orientation. There are many problems that may be easier to formulate using procedural, functional or constraint-oriented programming. BETA supports procedural programming and has good facilities for functional programming. Work is going on to improve the support for functional programming and to include support for constraint-oriented programming. The overall perspective will still be object-oriented, but transitions may be described as functions without intermediate states. Support for constraint-oriented programming will allow for expressing constraints and for a more high level description of transitions. Loops is an example of a language supporting several perspectives.

Acknowledgement. The framework presented here is mainly a result of the authors participation in the BETA project. In addition to the authors, Bent Bruun Kristensen, Aalborg University Centre, Denmark and Kristen Nygaard, University of Oslo, Norway, have been members of the BETA team. Jørgen Lindskov Knudsen, Kristine Stougård Thomsen, Jon Skretting and Einar Hodne have contributed with useful discussions and by commenting the paper. A very early version (in Danish) appeared in the special December 1987 issue of Nordisk Datanytt edited by Stein Gjessing.

References

[Agha86] G. Agha: *An overview of Actor Languages*. Sigplan Notices Vol.21 No.10 October 1986.

[BETA87a] B.B. Kristensen, O.L. Madsen, B. Møller-Pedersen, K. Nygaard: *The BETA Programming Language*. In: [Shriver & Wegner 87].

[BETA87b] B.B. Kristensen, O.L. Madsen, B. Møller-Pedersen, K. Nygaard: *Classification of Actions or Inheritance also for Methods*. Proceedings of the Second European Conference on Object Oriented Programming, Paris, June 1987.

[Booch 86] G. Booch: *Object-Oriented Development*, IEEE Trans. on Software Engineering, Vol. SE-12, No. 2, Feb. 1986.

[DELTA] E. Holbaek-Hanssen, P. Haandlykken, K. Nygaard: *System Description and the DELTA Language*, Publication no. 523, Norwegian Computing Center, 1975.

[ECOOP 87] *Proceedings of European Conference on Object-Oriented Programming*. BIGRE+GLOBULE No.54, June 1987

[Flavors] H. Cannon: Flavors, A Non-Hierarchical Approach to Object-oriented Programming. Draft 1982

[JSD] M. Jackson: *System Development*. Prentice Hall 1983.

[Knudsen&Thomsen85] J. Lindskov Knudsen and K. Stougård Thomsen: *A Conceptual Framework for Programming Languages*. DAIMI PB-192, Aarhus University, April 1985

[Madsen86] O.L. Madsen: *Block Structure and Object Oriented Languages*. In [Shriver & Wegner 87].

[Meyer87] Reusability: *The Case for Object-Oriented Design*. IEEE Software, Vol.4, No.2, March 1987.

[Mjølner] *MJØLNER, A highly efficient Programming Environment for industrial use*. Mjølner Report No.1.

[Nygaard86] K.Nygaard: *Basic Concepts in Object Oriented Programming*. Sigplan Notices Vol.21 No.10 October 1986.

[OOPSLA 87,88] *OOPSLA, Object oriented Programming Systems, Languages and Applications*. Conference Proceedings, 1986 and 1987.

[Shriver & Wegner 87] B. Shriver, P. Wegner: *Research Directions in Object-Oriented Languages*, MIT Press, 1987.

[SIMULA67] O.J. Dahl, B. Myhrhaug & K. Nygaard: *SIMULA 67 Common Base Language*, Norwegian Computing Center, February 1968,1970,1972,1984

[Smalltalk] A. Goldberg, D. Robson: *Smalltalk 80: The Language and its Implementation*. Addison Wesley 1983.

[Stefik & Bobrow 82] D.G. Bobrow and M. Stefik, : *Loops: An Object-Oriented Programming System for InterLisp*, Xerox PARC 1984.

[Stefik & Bobrow 84] M. Stefik, D.G. Bobrow: *Object-Oriented Programming: Themes and Variations*, The AI Magazine, 1984.

[Stroustrup86] B. Stroustrup: *The C++ Programming Language*. Addison Wesley 1986

[Wegner] P. Wegner: *Dimensions of Object-Based Language Design*. Tech. Report No. CS-87-14, July 1987. Brown University

Teaching Object-Oriented Programming is more than teaching Object-Oriented Programming Languages

Jørgen Lindskov Knudsen and Ole Lehrmann Madsen

Computer Science Department, Aarhus University,

Ny Munkegade 116, DK-8000 Aarhus C, Denmark.

E-mail: jlk@daimi.dk -and- olm@daimi.dk

Abstract

One of the important obligations of an expanding research area is to discuss how to approach the teaching of the subject. Without this discussion, we may find that the word is not spread properly, and thus that the results are not properly utilized in industry. Furthermore, discussing teaching the research area gives additional insight into the research area and its underlying theoretical foundation. In this paper we will report on our approach to teaching programming languages as a whole and especially teaching object-oriented programming.

The prime message to be told is that working from a theoretical foundation pays off. Without a theoretical foundation, the discussions are often centered around features of different languages. With a foundation, discussions may be conducted on solid ground. Furthermore, the students have significantly fewer difficulties in grasping the concrete programming languages when they have been presented with the theoretical foundation than without it.

Introduction

Most text books on programming languages describe the *technical differences* between various language constructs. This implies that emphasis is often concentrated around *features* of one language compared to features of another language. This makes it difficult to discuss the *qualitative difference* between languages. The well-known "Turing Tarpit"* states the fact that, on theoretical basis, any computation which can be expressed in one of the familiar programming languages can also be expressed in any of the others — including Turing machines. Thus comparison of features should be more than a discussion about whether or not a given construct may be *simulated* in another language. Furthermore, "a technical discussion" of programming languages is often lacking arguments about the programmers *perspective*† on programming. (One illustrative example of this approach can be found in [17].)

---

*According to W.A. Wulf[41] the "Turing Tarpit" was originally formulated by Alan Perlis.

†Please note, that others use the phrase *paradigm* instead of perspective here, but the use of paradigm in computer science has been questioned from several different sources; see e.g. [26].

Instead of technical details it is often much more fruitful to discuss requirements for supporting one or more perspectives. However, there are books that discuss languages relative to one perspective. The perspectives are usually based on mathematical models. (One illustrative example of this approach can be found in [37].) Few books are devoted to the object-oriented perspective. This may be due to the fact that the foundation/basis of object-oriented programming has not yet been very well formulated.

The purpose of this paper is to describe how programming languages are being taught at the Computer Science Department, Aarhus University. This teaching is highly influenced by 15–20 years of research in programming languages and system development in Scandinavia, mainly in Oslo and Aarhus. For more than 10 years, the teaching of programming languages at the Computer Science Department, Aarhus University has been heavily influenced by the object-oriented perspective. The approach to teaching object-oriented programming as well as the structure of the present courses will be described.

First a description of the overall approach to teaching Computer Science at the Department will be given. This is followed by a description of the objectives for teaching the subject of programming languages, leading to a description of the approach to teaching object-oriented programming. Finally, the courses that have been given over the last few years, both in the department and to industry are presented.

1 Background

When planning s course it is important to be conscious of the prerequisites of the students in order to design the most effective course. Our courses have primarily been given to students at the Computer Science Department and the overall approach to teaching computer science at the department will therefore be described.

The Computer Science Department at Aarhus University has grown out of the Institute of Mathematics. This has resulted in a strong influence of theoretical approaches to subjects. That is, the prime emphasis in teaching computer science is put on teaching theories and perspectives. Teaching concrete techniques and methods are considered as utilizations of the theories and perspectives. This implies that the students are trained in handling abstract notions besides being able to utilize these abstract notions in approaching concrete problems. That is, they are taught the abstract notions in order to make them capable of (relatively easy) learning techniques and methods for applying the abstract notions on concrete problems.

A study for the Master's degree is supposed to take 5 years. Most students, however, take considerable longer time to complete their degree. During the first 3 years approximately half of the time is devoted to computer science. The other half is usually mathematics and statistics. After 3 years the students are at the level of a Bachelors degree. The last 2 years are full time computer science, including a thesis.

With respect to programming languages, the students are trained during the first 2 years of study in using traditional procedural languages, such as Pascal, Modula-2, and Concurrent Pascal. This implies that their perspective on programming is highly influenced by the procedural programming perspective.

The courses described here are given on year 3–5 of the study.

2 Objectives

The fundamental principle is that teaching concrete programming languages should be a subordinate objective in teaching the subject of programming languages. There is a number of reasons for this:

- It is very difficult to predict which programming languages will be the most influential in industry 10–20 years ahead (unless we settle with the good old workhorses Cobol and Fortran). Furthermore, we *have to* make sure that the students of today are able to access the programming languages in the 21'st century (it's only 11 years ahead). By teaching them concrete languages of today, we are liable to make it difficult for them to access the languages of the 21'st century.

- In any teaching situation, it is most important to emphasize the principles and utilize this insight to access concrete examples of the principles. If you e.g. teach people object-oriented programming by just giving a course on Smalltalk, they may have difficulties in understanding the basic principles of object-oriented programming. They will very likely equalize object-oriented programming with programming in Smalltalk. Furthermore they often have difficulties in actually learning Smalltalk.

- By learning principles, techniques and concepts, the student will be able to evaluate different programming languages on basis of the principles, and not on basis of more or less important concrete differences (such as syntax). Furthermore, with well-chosen principles, there is a better chance of the evaluation being fair to all languages under consideration, and not being in favor of one specific language. For any concrete programming language, it may be difficult for the student to distinguish the important and general constructs of the language from the always present idiosyncrasies. State-of-the-art in programming language design has not yet reached a level where it is possible to design a language that does not end up having some poorly designed features, even if the overall principles are good and sound. Simula 67[3] and Smalltalk-80 are good examples of this. The basic principles behind these languages were excellent at the time of invention. Still a user of these languages is confronted with a large number of poorly designed edges.

- The students must be made able to consider using different languages for different programming tasks. In this case it is important that the student (when he later acts as a system developer) is aware of the perspectives which underlie the specific languages (or rather, is able to identify the underlying perspective of different languages) since the underlying perspective of a programming language in a sense outlines the borders of the application areas for which the particular programming language is well-suited, and therefore will have an impact on the programming process.

We also strive towards avoiding the discussion of *features* of programming languages, and stress that in order to make languages accessible it is very important that the concepts *simplicity*, *consistency* and *orthogonality* are the primary guidelines — features will never be fully utilized or understood if they are nothing but features. Simplicity, consistency and orthogonality of language constructs are what makes a language accessible, irrespective of which programming perspective the language supports.

The quantitative approach to evaluating programming languages has some serious defects, the "featurism" mentioned above being one of them. Without more abstract notions of what constitutes important aspects of a programming language, one is seriously in danger of the "Turing tarpit". This may stop any serious discussion of different programming languages, since it does not make any distinction between *supporting* and *simulating* a particular language construct. One very good example of not making this distinction clear is the discussion by Per Brinch Hansen of selecting language constructs to be included in the Edison programming language[6]. One of these discussions is about whether it is necessary to include both the **repeat**- and the **while**-statements. P.B. Hansen argues that since **repeat** may be expressed in terms of **while** there is no need for the **repeat**. This is the "Turing tarpit" since applying the argument repeatedly reduces any control statement to being only specific **goto** structures, and since not having the **repeat** in the language places the burden on the programmer to implement the **repeat** *each* time he finds a need for it. In this case we will say that the Edison language *simulates* the **repeat** concept.

The question is now: What do we demand of a language in order for it to *support* a given concept? Let us use the **repeat** example again. If the **repeat** were present in a language, we would of course state that the language supports the **repeat** concept, but more generally we would say that a language supports the **repeat** concept if there exists a mechanism in the language that makes it possible to state the **repeat** concept as an abstraction which may then be used on equal terms with the built-in concepts. In this way it is possible to create new abstractions that can be safely implemented once, and then securely utilized over and over. That is, the consistency of the abstraction is expressed once in the implementation and not scattered all over the program as with simulated concepts discussed above.

Returning to object-oriented programming, it is important to be aware that object-oriented programming is a lot more than inheritance, objects, and message passing or member function calling, very much the same way as structured programming is a lot more than goto-less programming.

3 Approach

The approach to teaching programming languages and especially object-oriented programming is very much influenced by the perspective you have on the role of the programming language in the system development process. In fact this role is a three-way role: as a means for expressing concepts and structures (*conceptual modeling*), as a means for *instructing* the computer, and as a means for *managing* the program description. Just focusing on the role as a means for instructing the computer is far to narrow. In the role for conceptual modeling, the focus is on constructs for describing concepts and phenomena. In the role for instructing the computer, the focus is on aspects of the program execution such as storage layout, control flow and persistence. Finally, in the role for managing the program description, focus is on aspects such as visibility, encapsulation, modularity, separate compilation, library facilities, etc.

Some of the success in teaching programming languages can be traced back to the emphasis that is put on using these roles as the foundation of the approach. Here the roles as means for conceptual modeling and prescription have proven very effective, and to some extent this makes the approach to teaching programming languages novel. It has been found that restricting the discussion of programming languages to the role of instruction (or coding) is far to restrictive, primarily because the end-product of a programming process (the program) cannot (and should not) be viewed in isolation from the programming process and thereby the application domain.

3.1 Perspectives

Teaching the perspective of object-oriented programming cannot (or should not) take place in isolation from other perspectives. Extensive parts of the courses are therefore devoted to programming perspectives as such, and presentation of various different programming perspectives.

Procedural programming[‡] is taken as the starting point for the discussion. Functional/logical programming and object-oriented programming are then described as two different reactions to several problems related to the concept of state in procedural programming. In functional programming, the approach has been to eliminate the concept of state, whereas the approach taken in object-oriented programming has been to treat the concept of state as a first-class citizen. In addition various other perspectives such as the process perspective, the type system perspective and the event perspective are treated. The latter three perspectives are not treated extensively but primarily in the context of the other perspectives.

Below a short formulation of the perspectives are given.

Procedural Programming

A program execution is regarded as a (partially ordered) sequence of procedure calls, manipulating

---

[‡] To ease the writing we will use the phrase "... programming" interchangeable with the phrase "the ... programming perspective".

data structures. This perspective is the most common and supported by languages like Algol, Pascal, C and Ada. Procedural programming has the prime focus on the instructive role of the programming language and very little support for the other roles of the programming language, and is therefore not sufficient.

The courses we give do only treat procedural programming on the level of perspective since the students in their previous courses have been trained extensively in procedural programming. We do, however, cover Ada as a representative of state-of-the-art within procedural programming languages.

Functional Programming

A program is regarded as a mathematical function, describing a relation between input and output. In functional programming, the concept of state or variable is eliminated entirely. I.e functional programming is variable free programming.

Lisp is often mentioned as the most prominent "functional programming" language. It is well-known that most Lisp variants also have variables and thereby state. For this reason it is important to stress that instead of classifying a given programming language as either a "functional programming language", a "procedural programming language", etc., it is often more useful to discuss to what extent a given programming language has support for functional programming, object-oriented programming, etc. There are few programming languages that are based purely on one perspective.[§]

In the courses, functional programming is treated on the level of perspective. The students are trained in functional programming using the Scheme programming language[27]. In another part of the department, a course is devoted entirely to the subject of functional programming using the Miranda language[39]. This course is complementary to the course described here, since its main emphasis is on the theoretical foundation for functional programming. Students are advised to take that course if they have special interest in functional programming.

Constraint-Oriented (logic) Programming

A program is regarded as a set of equations, describing relations between input and output. As in functional programming, the concept of state is eliminated in constraint-oriented programming. Prolog is the most dominant example of a language supporting the constraint-oriented perspective.

In the courses we treat constraint-oriented programming on the level of perspective and exercises the constraint-oriented programming perspective using the Prolog language. In another part of the department, a course is devoted entirely to the subject of logic programming using the Prolog language and again students are advised to take that course if they have special interest in constraint-oriented programming.

[§]To ease the writing, we will however use the phrase "... programming language" to mean "programming language with primary support for ... programming".

Object-Oriented Programming

A program execution is regarded as a physical model, simulating the behavior of either a real or imaginary part of the world. The object-oriented perspective on programming is in contrast to the above perspectives that are focusing either on manipulations of data structures or on mathematical models. The object-oriented perspective is closer to physics than mathematics. Instead of describing a part of the world by means of mathematical equations, a *physical model* is literally constructed. This means that elements of the program execution are regarded as models of *phenomena* and *concepts* from the real world. The part of the world being modeled is described by the program. Some of the well-known examples of languages supporting this perspective are Smalltalk-80, Beta and C++.

This "definition" cannot be seen in isolation but must be understood in a broader context (this applies for the other perspectives as well.) In the courses we elaborate extensively on this broader context as described in section 3.2.

The Process Programming Perspective

A program execution is regarded as consisting of a set of processes, each involved in their own activities, and communication with other processes.

The process perspective is focusing on structuring the transformations on state. Some of the well-known examples of languages supporting this perspective are CSP[7], Concurrent Pascal, and Ada.

In the courses we treat the process perspective at the level of perspective and study CSP and the process aspects of Ada.

It is also discussed to what extent the process perspective is at the same level as some of the other perspectives. You may e.g. view the process perspective as a development of the procedural perspective or as subordinate to the object-oriented perspective.

Concurrent programming in languages like Concurrent Pascal and Ada is in our view mainly carried out as a generalization of a "sequential" procedural perspective.

The modeling of objects with individual action sequences is a fundamental part of the Scandinavian tradition for object-oriented programming. Simula 67 has support for coroutines and Beta[16]. has support for coroutines and concurrency. For this reason it is natural to view the process perspective as subordinate to the object-oriented perspective.

Again we must stress (and this is of course also done in the courses) that there is no objective way of defining what is right and wrong with respect to the different perspectives on programming.

The Type System Perspective

The type system perspective may be viewed as a subordinate perspective of the constraint-oriented perspective, where the relations are described by means of type structures. Since type systems are an integrated part of many procedural languages, these perspectives co-exist harmonically in the same programming language. Many type systems have been proposed in the past, most notably the Pascal type system, the Ada type system, the ML type system (incorporating type inference and polymorphism), and the Cardelli and Wegner type system (incorporating hierarchical types).

Type systems are treated at the perspective level, and are related to the object-oriented perspective by studying the relations between hierarchical type systems and classification hierarchies. We focus on the Ada type system and on the Beta type system.

The Event Perspective

A program execution is regarded as a (partially ordered) set of events. The event perspective is a theoretical approach to the process perspective. The most notable representatives of the event perspective are Petri net models[28], Calculus for Communicating Systems (CCS)[24], CSP-85[8].

The event perspective is only treated at the perspective level, and as with the functional and constraint-oriented perspective, students with special interest in the event perspective are advised to follow the Petri net course, given by others in the department.

As indicated above we do not pretend to be objective in our teaching in the sense that we make it clear that our perspective on programming is mostly object-oriented. This implies that we (in the interest of honesty) inspire the students to take the specific courses directed towards the other perspectives if they find special interest in them.

3.2 Teaching Object-Oriented Programming

Having described the context of object-oriented programming, we will now present the actual subjects treated in our courses on object-oriented programming.

3.2.1 Theoretical Foundation for Object-Oriented Programming

As stated above, the object-oriented perspective must be accessed on basis of a theoretical foundation and not on basis of specific language constructs. The theoretical understanding of object-oriented programming which will be outlined in the following is among others a result of research activities that the authors have carried out together with a number of other people. It is important to stress that the teaching has influenced the research too. A large number of students have treated many of the subjects in their thesis work. A more detailed description of the issues discussed may be found in [13] and [16]. The foundation is highly influenced by the work reported in [9,22,32,25].

Modeling

In order to clarify the different roles that the programming language plays in the programming process, we have to look more closely at that process. The programming process may be described as a modeling process in which several sub-processes take place. The figure illustrates the

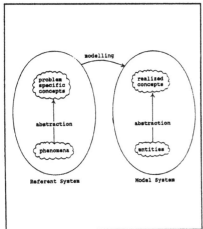

programming process as a modeling process between a referent system and a model system. The *referent system* is part of the world that we are focusing on in the programming process, and the *model system* is a program execution modeling a part of the referent system on a computer. The referent system is the concrete physical world or some imagination of a future physical world, and as such it consists only of phenomena. As a characteristic human activity, we create concepts in order to capture the complexity of the world around us — we make abstractions. That is, in the referent system, both phenomena and concepts are important. In the model system, we find elements that model phenomena and concepts from the referent system. Objects in a Smalltalk-80 program execution are typically models of physical phenomena in the referent system and the sequence of events generated by the execution of a method is typically a model of a sub-process going on in the referent system. Concepts in the referent system are modeled by abstractions such as classes, types, procedures and functions. The program text is a *description* of the referent system and in addition it is a *prescription* that may be used to generate the model system.

The programming process can now be described in terms of this figure. During the programming process, three sub-processes are taking place: abstraction in the referent system, abstraction in the model system, and modeling. Please note that intentionally we do not impose any ordering among the sub-processes. *Abstraction in the referent system* is the process where we are perceiving and structuring knowledge about phenomena in the referent system with particular emphasis on the problem domain in question. We say that we are creating *problem specific concepts* in the referent system. This process is an integrated part of the system development process. *Abstraction in the model system* is the process where we build structures that should support the model we are intending to create in the computer. We say that we create *realized concepts* in the model system. Finally, *modeling* is the process where we connect the problem specific concepts in the referent system with the realized concepts in the model system.

Concepts and Abstraction

As discussed above, concepts and abstraction are the key notions in our understanding of the programming process. It is therefore necessary to discuss subjects like the notion of concepts and their relations to phenomena, concept understanding, and important aspects of the abstraction process.

A *phenomenon* is something in the world that has definite, individual existence in reality or the mind; anything real in itself. What constitutes a phenomenon is to some degree dependent on the view of the observer. A *concept* is a generalized idea of a collection of phenomena, based on knowledge of common properties of the phenomena in the collection. Concepts may be characterized by three aspects: the designation, extension and intension. The *designation* refers to the collection of names under which the concept is known. The *extension* refers to the collection of phenomena that the concept somehow covers, and the *intension* refers to the collection of properties that in some way characterize the phenomena in the extension of the concept.

These definitions are deliberately somewhat vague since there are (at least two) different ways to understand concepts: the Aristotelian view and the prototypical (or fuzzy) view. Space does not allow an extensive discussion of these two views — just a short characterization. In the *Aristotelian* view, the concepts are rigidly defined, leading to *sharp concept borders* and *relatively homogeneous* phenomena in the extension. The Aristotelian view is the view that can be mechanized without human interaction. The *prototypical* view, on the other hand, is characterized by *blurred concept borders*, phenomena of *varied typicality* in the extension, and *decision-making/judgement* when a phenomenon is considered for inclusion in the extension. The prototypical view is the view that best describes human concept understanding.

As it can be seen above, the programming process is faced with the problem that not only do we restrict the precision of our model by only considering a part of the world (this is a problem studied in system development courses), but equally important, the modeling process *has* to take into account the restrictions imposed by modeling a possible prototypical concept structure in the referent system into an Aristotelian concept structure in the model system.

In both the referent system and the model system, concept structures are created. This implies that we have to discuss the process of producing and using knowledge, i.e. issues related to epistemology. Part of this discussion includes an introduction to some of the work of Marx, who has split the process of knowledge into three levels:

1. *The level of empirical concreteness.* At this level we conceive reality or individual phenomena as they are. We do not realize similarities between different phenomena, nor do we obtain any systematic understanding of the individual phenomena. We notice what happens but does neither understand why it happens nor relations between phenomena. In the programming process this corresponds to a level where we are trying to understand the single objects that constitute the system. We have little understanding of the relations between the objects, e.g. how to group them into classes.

2. *The level of abstraction.* In order to understand the complications of the referent system, we have to analyze the phenomena and develop concepts for grasping the relevant properties of the phenomena that we consider. In the programming process this corresponds to designing the classes and their attributes and to organize the classes into a class/sub-class hierarchy. At this level we obtain a simple and systematic understanding of the phenomena in the referent system.

3. *The level of thoughtconcreteness.* The understanding corresponding to the abstract level is further developed to obtain an understanding of the totality of the referent system. By having organized the phenomena of the referent system by means of concepts we may be able to

understand relations between phenomena that we did not understand at the level of empirical concreteness. As well as we may be able to explain why things happen and may be able to predict what will happen.

In the process of creating concepts it is useful to identify the three well-known sub-processes of abstraction: classification, aggregation and generalization. To *classify* is to form a concept that covers a collection of similar phenomena. To *aggregate* is to form a concept by describing the properties of the phenomena by means of other concepts. And finally, to *generalize* is to form a concept that covers a number of more special concepts based on similarities of the special concepts. All three sub-processes have an inverse process, called *exemplification, decomposition* and *specialization*, respectively.

In general the process of creating new concepts cannot just be explained as consisting of the above sub-functions. In practice the definition of concepts will undergo drastic changes. This is similar to the situation with top-down and bottom-up programming. It is realized by most people that pure top-down or bottom-up development of programs is rarely possible. The understanding obtained during the development process will usually influence previous steps. It is however useful to be aware whether a problem is approached top-down or bottom-up. In the same way it is useful to be aware of the above mentioned sub-functions of abstraction.

The word abstraction may be used to characterize a process, and the sub-functions of abstraction were explained as processes going on with the aim of creating concepts. On the other hand the word abstraction may also be used in a static or descriptive way. A concept is an abstraction. Given a number of concepts, their structure may be described in terms of classification, aggregation and generalization. It is e.g. possible to describe a given concept as a generalization of a number of other concepts.

In teaching it is important that the students are aware of this distinction. When evaluating a given language they might consider to what extent the language support abstraction and its sub-functions as a process and to what extent the language supports abstraction and its sub-functions as a means for describing concept structures.

Information Processes and Object-Oriented Programming

Having discussed concepts and abstraction we turn our attention towards characterizing the part of the world we are interested in creating model systems for, and then characterize object-oriented programming in greater detail.

The kind of model systems we are interested in, are those that model information processes. An *information process* is regarded as a system, developing through transformations of its state. The *substance* of the process is organized as objects. The *state* of the substance may be measured upon through *measurable* properties, and the state of the substance may change as an effect of *transformations* on the substance. Substance is physical matter, characterized by a volume and a position in time and space. Substance have certain properties that may be measured. E.g.

measurements may be compared with other measurements. Transformations are partially ordered sequences of events that change the substance and thereby its properties. Note that by focusing on information processes, concepts exist that cannot be captured, e.g. "God", "good", "bad", etc.

In object-oriented programming, an information process is modeled by organizing the substance of the program execution as a number of *objects*. The measurable properties are modeled as *state of objects*, and transformations are organized as *action sequences* performed by objects. An object is furthermore characterized by a set of *attributes* that may be either *measurable properties*, *part-objects*, *references to objects*, *procedures*, or *classes*. Finally, an object may have an *action-sequence* associated with it. Every object has at any given point in time a state. States are changed by objects performing actions that may involve other objects. Actions may in addition be involved in the production of measurements. A *program execution* consists of a collection of objects. Objects are classified into classes, and classes may be specializations of more general classes.

3.2.2   Study of Object-Oriented Languages

Having set the scene for object-oriented programming, we turn to the study of concrete examples of programming languages supporting object-oriented programming. Here we focus on the languages Simula 67, Smalltalk-80, Beta, C++ and LOOPS, and discuss hierarchical type systems[2] and delegation systems[18]. The study of the languages Smalltalk-80 and Beta is extensive and includes training in actual programming using the systems, whereas Simula 67, C++ and LOOPS are only evaluated theoretically. Here we have found that the theoretical foundation for approaching programming languages really has paid off. The students have very little problems in grasping the concrete language constructs presented in the different languages when they utilize the theoretical foundation as the basis for the learning process. They find that the notions that are handed to them in the theoretical part of the courses are in fact useful (although they might doubt it in the beginning of the courses).

It is not possible to discuss the application of the foundation in full details in this paper but to illustrate the issues, we discuss classes as models of concepts, class/subclass hierarchies as models of generalization hierarchies, and classes as aggregations.

3.2.3   Applying Object-Oriented Theory to Traditional Languages

Besides utilizing the theoretical foundation for studying object-oriented programming languages, we apply the foundation to traditional programming languages, such as Pascal, Modula-2 and Ada. In this way, we are able to gain additional insight into the foundation, but also to study the limitations of the support for object-oriented programming in the traditional languages. As an example, we discuss the relations between (generic) packages in Ada and abstraction (especially generalization).

3.2.4 Persistency

One of the major drawbacks of present object-oriented systems is that they do not support multi-user usage very well — they are essentially single-user systems (e.g. the Smalltalk-80 system). In order to support multi-user projects, persistent objects and shared program libraries must be supported within the object-oriented framework (i.e. object-oriented databases). The subject of object-oriented databases and persistence is presently not discussed in detail. Only a discussion of the underlying ideas and the reasons for the presently growing interest in the area is included. It is however mandatory that this subject is included in our courses in the near future.

3.2.5 Integration of Perspectives

The discussions of the various perspectives on programming give rise to discussions of possible ways for integration of the perspectives in one language. As already indicated, we find that some integration is both possible and fruitful. Since our perspective on programming is centered around object-oriented programming, we discuss how the other perspectives can be integrated in an otherwise object-oriented programming language. For this discussion the Beta programming language has been chosen.

As mentioned in section 3.1, the Beta language integrate the procedural, process, object-oriented and type system perspectives in one unified language. The limitations of these perspectives, and the elegance of certain solutions using the functional and constraint-oriented perspectives are constant inspirations to the discussions. We find that integration can be utilized to specify purely functional transformations on the states of objects, and to specify constraints on the interrelations between objects. Although not implemented, it seems to be possible to specify a purely functional sub-part of Beta such that parts of a running Beta system is specified using the functional perspective. With respect to integration with the constraint-oriented perspective, various different approaches are being considered. The proof-of-existence can be found in the Smalltalk/V system[43] that contains a Prolog subsystem, but this system has not been found sufficient. With respect to the event perspective, we have found that the process perspective should be chosen as the pragmatic approach to multi-sequential programming, using the event perspective as the inspiring theoretical foundation.

3.2.6 Integration with Related Subjects

Traditionally the study of programming languages have been integrated with aspects of compiler construction, formal language theory, machine architecture and mathematical semantics. The courses described here are also concerned with integrating with aspects of system development. As indicated above, the approach to system development and the approach to programming languages are related in such a way that selecting a programming perspective will have an impact on the system development process as a whole and to some degree on the resulting product. We discuss those relations and their impacts as an integrated part of the courses.

4 Courses

Several different courses based on the above premises have been given. The present line of courses at the department is described together with two courses given to people from the industry.

At the Computer Science Department we are offering two courses on programming languages. The two industrial courses are one on object-oriented programming and one on Smalltalk/V.

Bachelor level

The first course is at the third (and last) year of the Bachelor level program. The course is partly on systems development and partly on programming languages. The course is occupying 1/3 of the student program for a whole year of which the programming language part is apx. 1/2.

The part on systems development, includes the Jackson System Development Method(JSD)[11] and various approaches to prototyping. The programming language part covers the following topics: The programming process and conceptual modeling. Presentation of different perspectives on programming. Introduction to and practical training in Smalltalk-80, Scheme and Beta. The introduction to JSD is related to object-oriented programming where it is emphasized that JSD in fact is very close to an object-oriented methodology. It is discussed to what extent JSD may be strengthen by using an object-oriented language. Smalltalk-80 is also used as an example of an environment that supports fast prototyping.

As mentioned, the important issues in teaching is not teaching the actual languages. This makes it difficult to find good textbooks on the subject. Take for example Scheme. In the courses we want to demonstrate to what extent Scheme supports the various perspectives on programming. That is the teaching is concerned with features for supporting procedural programming, features for functional programming and features for object-oriented programming. The book by Abelson and Sussman[1] is an excellent book for a course on introduction to programming. However it is far to big to be used by students already familiar with programming. Other books on Scheme (and most books on Lisp in general) introduces the language feature by feature, and does not relate it to perspectives. The material used in the course includes (parts of) [4,19,16,42,30,13,20].

The reasons for using Smalltalk-80, Scheme and Beta are: Smalltalk-80 and Scheme are representatives of flexible, dynamic languages without static typing. This make them well-suited for exploratory programming. Beta on the other side is a language with a static type system and intended for production programming. Smalltalk-80 and Beta are representatives of the two major directions in object-oriented programming. Finally Scheme and Beta are languages that are not solely based on one perspective. This is in contrast to Smalltalk-80 that has little support for other perspectives than object-orientation.

Master's level

The second course is at the Master's level with a bachelor degree (including the above course) as the only explicit requirement. The course is on advanced features of programming languages with most emphasis on programming language support of object-oriented programming. The course is occupying 1/3 of the student program for one semester.

The course covers the following topics:

- The programming process and conceptual modeling[13].

- Types, packages, generics, tasks from Ada.

- Various definitions of "object-oriented programming"[20,40,36].

- Inheritance, delegation and enhancement[5,18,29,23,10].

- Multiple inheritance[35,38,14].

- Modularization[21,33].

Furthermore, the Beta programming language is used extensively throughout the entire course. The references are indications of material used. The course is organized mainly as a seminar course where the students are giving oral presentations of selected topics with the purpose of opening discussions on the topics. At the end of the course, the students are asked to write a small report on a selected topic within the course.

At the end of the course it is evident that the students are very able in programming perspectives, their relative merits and application areas, and object-oriented design and programming. Finally, they are able to evaluate particular languages with respect to their relation to specific application domains. Their ability to actually construct programs using some specific programming language is not the subject of this course, but experience has shown that the students are becoming accustomed to learning new programming languages and use them effectively after a short learning period. We find this to be a contribution of our theoretical approach to learning the subject of programming languages.

Industrial Courses

In the industrial environment we have been given two courses on object-oriented programming. The first industrial course was called "Object-Oriented Programming" and has been taught twice. The courses were arranged by "Datalogforeningen" (an association of Danish computer scientists, mainly in industry). The courses were limited to members of the association, and in effect this meant that the attendants were all having a Master's in computer science from Aarhus University. They had at least 4 years experience in industrial settings and had only very limited previous experiences with object-oriented programming.

The course was two-days with lectures and discussions. The course material was journal and conference papers, and language descriptions (almost identically to the material used in the department courses). The subjects discussed were object-oriented programming as discussed above and the Smalltalk-80 and Beta programming languages with a brief discussion of the C++ language. The second course had attached to it a one-day workshop one week after the course, covering exercises in actual programming using the Smalltalk-80 and Beta systems.

Since the attendants were computer scientists with extensive training in handling theoretical approaches to problems, we found that stressing the theoretical approach to object-oriented programming was very fruitful. The attendants were very active during the course and very many of the discussions were centered around applying object-oriented programming in real-life industrial settings. The following workshop showed that they were able to handle object-oriented design very well, and their primary problems could be traced down to problems in expressing these designs in the concrete languages and to problems in handling the systems.

The second industrial course was given as an in-house course. The course was on object-oriented programming using the Smalltalk/V system on IBM personal computers. The attendants were mostly engineers with no previous experiences with object-oriented programming but experienced in traditional procedural programming languages (Pascal and Modula-2) and assembler programming.

The course was two-days with lectures, discussions and class-room problem solving followed by an one-day workshop with one week delay, covering exercises in actually using the Smalltalk/V system working on relative small programming tasks in groups. The course material was the tutorial and reference book accompanying the Smalltalk/V system plus some articles from the Byte issue on Smalltalk-80[44].

Having no previous experience in object-oriented programming and most importantly not being used to extensive theoretical approaches, the attendants were in the beginning rather confused (Question: *When do we start learning something about Smalltalk/V programming?*). In the last part of the course they realized the importance of taking the broader view (Reaction: *Oh, that's why I need to think differently!*). Finally, in the workshop the attendants had their major problems in handling the Smalltalk/V system (they had nearly no previous experience using window/mouse based interaction), whereas they were able to take the initial steps in the direction of creating classification hierarchies. In time of writing it is known that at least two of the companies having representatives at the course are in the process of experimenting with object-oriented programming using either Smalltalk-80, Smalltalk/V or C++.

5   Final Remarks

Thus ended the story about teaching object-oriented programming at the Computer Science Department at Aarhus University. Computer scientists in the 21'st century will be forced to master several programming perspectives and a magnitude of languages supporting those perspectives. Without the theoretical understanding of the underlying perspectives they may run into difficulties. Mastering concrete programming languages and literacy in features will not be sufficient.

The most important single pay-off of the theoretical approach to the subject has been the ease with which the students have been able to access the concrete programming languages. When approaching new languages they have proven to be able to characterize the language in question on basis of the theoretical foundation, and thus avoiding discussing features of one language compared with features in another language. That is, they make qualitative evaluations on basis of the theoretical foundation and not quantitative evaluations on basis of features.

Looking back, we have found that we as teachers (and to some respect as researchers) are lacking some profound knowledge within other research areas. Some of our work have strong connections to philosophy, linguistics, philosophy of science, and psychology. The danger is that we in those disciplines are somewhat amateur researchers. We have therefore been very conscious to tell the students that we are not experts in all these disciplines, and that our approach might have some defects if seen from those research disciplines. However, we hope to improve our knowledge in those areas. We are also aware that many of the interdisciplinary issues mentioned here are well-known to people working with knowledge representation within artificial intelligence and data bases, where modeling of real world phenomena are central issues. Our approach is not primarily directed towards issues of knowledge representation, but on issues of software construction. This implies, that we are primarily discussing the various aspects of utilizing object-oriented design principles in the context of software construction and not knowledge representation. This is the reason why we are discussing the object-oriented perspective in this very broad context of system development, program description, and program prescription as well as in the context of traditional languages like C, Pascal, Modula-2 and Ada.

6 Acknowledgements

The motivation to write this paper came from discussions in an ad-hoc working group at ECOOP'87 in Paris. Here it became evident that there is a great need for discussing how people teach object-oriented programming.

This paper reports on the experiences of teaching programming languages at the Computer Science Department at Aarhus University for more than 10 years. Many people have been involved in this process. Furthermore, the approach have been influenced by the last 15–20 years of research in system development and programming languages in Scandinavia. Proper acknowledgement of all these people is impossible. However, Kristine Stougård Thomsen must be mentioned for taking very actively part in the design and implementation of several of the courses. A special thanks must go to the large number of students who have actually taken the courses.

7 References

1. H. Abelson, G.J. Sussman & J. Abelson: *The Structure and Interpretation of Computer Programs*, MIT Press, 1985.

2. L. Cardelli & P. Wegner: *On Understanding Types, Data Abstraction, and Polymorphism*, Computing Surveys, 17(4), 471–522 (December 1985).

3. O.-J. Dahl, B. Myhrhaug & K. Nygaard: *Simula 67, Common Base Language*, Norwegian Computing Center, 1970.

4. A. Goldberg & D. Robson: *Smalltalk-80: The Language and its Implementation*, Addison-Wesley Publishing Company, 1983.

5. D.C. Halbert & P.D. O'Brian: *Using Types and Inheritance in Object-Oriented Programming*, IEEE Software, 4(5), 71–79 (September 1987).

6. P. Brinch Hansen: *The Design of Edison*, Software — Practice & Experience, 11, 363–396 (1981).

7. C.A.R. Hoare: *Communicating Sequential Processes*, Comm. of the ACM, 21(8), 666–677 (August 1978).

8. C.A.R. Hoare: *Communicating Sequential Processes*, Prentice-Hall, Inc., 1985.

9. E. Holbaek-Hanssen, P. Haandlykken & K. Nygaard: *System Description and the DELTA Language*, publication no. 523, Norwegian Computing Center, February 1977.

10. C. Horn: *Conformance, Genericity, Inheritance and Enhancement*, Proceedings of the European Conference on Object-Oriented Programming (ECOOP'87), Paris, France, June 1987.

11. M.A. Jackson: *System Development*, Prentice-Hall Inc., 1983.

12. J. Lindskov Knudsen & K. Stougård Thomsen: *A Taxonomy for Programming Languages with Multisequential Processes*, Journal of Systems and Software, 7(2) (June 1987).

13. J. Lindskov Knudsen & K. Stougård Thomsen: *A Conceptual Framework for Programming Languages*, Computer Science Department, Aarhus University, DAIMI PB-192, 1985.

14. J. Lindskov Knudsen: *Name Collision in Multiple Classification Hierarchies*, Proceedings of the European Conference on Object-Oriented Programming (ECOOP'88), Oslo, Norway, August 1988.

15. B. Bruun Kristensen, O. Lehrmann Madsen, B. Møller-Pedersen & Kristen Nygaard: *Syntax Directed Program Modularization*, in P. Degano & E. Sandewall (eds.): *Integrated Interactive Computing Systems*, North-Holland Publishing Company, 1983.

16. B. Bruun Kristensen, O. Lehrmann Madsen, B. Møller-Pedersen & K. Nygaard: *The Beta Programming Language*, in [31].

17. H. Ledgard & M. Marcotty: *The Programming Language Landscape*, Science Research Associates, Inc., 1981.

18. H. Liebermann: *Using Prototypical Object to Implement Shared Behavior in Object Oriented Systems*, Proceedings of Conference on Object-Oriented Programming Systems, Languages and Applications (OOPSLA'86), Portland, Oregon, September 1986.

19. B.J. MacLennan: *Principles of Programming Languages: Design, Evaluation, and Implementation*, CBS College Publishing, 1983.

20. O. Lehrmann Madsen & B. Møller-Pedersen: *What Object-Oriented Programming may be — and what it does not have to be !*, Proceedings of the European Conference on Object-Oriented Programming (ECOOP'88), Oslo, Norway, August 1988.

21. O. Lehrmann Madsen: *Block Structure and Object-Oriented Languages*, in [31].

22. L. Mathiassen: *Systemudvikling og systemudviklingsmetode* (in Danish), Computer Science Department, Aarhus University, DAIMI PB-136, 1981.

23. B. Meyer: *Genericity versus Inheritance*, Proceedings of Conference on Object-Oriented Programming Systems, Languages and Applications (OOPSLA'86), Portland, Oregon, September 1986.

24. R. Milner: *A Calculus of Communicating Systems*, Springer Lecture Notes in Computer Science, Vol. 92, Springer Verlag, 1980.

25. K. Nygaard: *Basic Concepts in Object Oriented Programming*, Sigplan Notices, 21(10), 128–132 (October 1986).

26. K. Nygaard & P. Sørgaard: *The Perspective Concept in Informatics*, in G. Bjerknes, Pelle Ehn & Morten Kyng (eds.): *Computers and Democracy — A Scandinavian Challenge*, Avebury, 1987.

27. J. Rees & W. Clinger (eds.): *Revised³ Report on the Algorithmic Language Scheme*, Sigplan Notices, 21(12), 37–79 (December 1986).

28. W. Reisig: *Petri Nets — An Introduction*, Springer Verlag, 1985.

29. D. Sandberg: *An Alternative to Subclassing*, Proceedings of Conference on Object-Oriented Programming Systems, Languages and Applications (OOPSLA'86), Portland, Oregon, September 1986.

30. B. Sheil: *Power Tools for Programmers*, Datamation, 29(2) (February 1983).

31. B.D. Shriver & P. Wegner (eds.): *Research Directions in Object-Oriented Programming*, MIT Press, 1987.

32. J.M. Smith & D.C.P. Smith: *Database Abstractions: Aggregation and Generalization*, ACM TODS, 2(2) (June 1977).

33. A. Snyder: *Inheritance and the Development of Encapsulated Software Components*, in [31].

34. B. Stroustrup: *An Overview of C++*, Sigplan Notices, 21(10) (October 1986).

35. B. Stroustrup: *Multiple Inheritance for C++*, Proceedings of EUUG Spring '87 Conference, 1987.

36. B. Stroustrup: *What is "Object-Oriented Programming"?*, Proceedings of the European Conference on Object-Oriented Programming (ECOOP'87), Paris, France, June 1987.

37. R.D. Tennent: *Principles of Programming Languages*, Prentice-Hall Inc., 1981.

38. K. Stougård Thomsen: *Inheritance on Processes, Exemplified on Distributed Termination Detection*, International Journal of Parallel Programming, **16**(1), 17-52 (1987).

39. D. Turner: *An Overview of Miranda*, Sigplan Notices, **21**(12), 158–166 (December 1986).

40. P. Wegner: *Dimensions of Object-Based Language Design*, Proceedings of Conference on Object-Oriented Programming Systems, Languages and Applications (OOPSLA'87), Orlando, Florida, October 1987.

41. W.A. Wulf: *Languages and Structured Programs*, in R.T. Yeh (ed.): *Current Trends in Programming Methodology*, Vol. I, Prentice-Hall Inc., 1977.

42. *Scheme Manual (Seventh Edition)*, MIT, September 1984.

43. *Smalltalk/V: Tutorial and Programming Handbook*, Digitalk Inc., 1986.

44. *Special Issue on Smalltalk-80*, Byte, Aug. 1981.

The Mjølner Environment: Direct Interaction with Abstractions

Görel Hedin & Boris Magnusson
Department of Computer Science, Lund Institute of Technology
Box 118, S-221 00 Lund, Sweden
email: gorel@dna.lth.se, boris@dna.lth.se

Abstract

This paper presents the user interface to programs and their execution in the Mjølner Programming Environment. The key idea is to present the programming language abstractions, such as classes and procedures, as individual windows which the user can interact with directly. This approach is used consistently to visualize both a program and its execution. The windows are arranged hierarchically reflecting the static nesting of blocks. The window hierarchy gives powerful support for interaction and navigation in a program. Incremental compilation techniques are used to make a high level of interaction and integration possible.

1 Introduction

In this paper we present the user interface to programs and their execution in the Mjølner Incremental Programming Environment. The objective of the Mjølner project is to provide highly interactive programming environments for industrial use. Mjølner is primarily intended to support strongly typed object oriented languages in the Simula tradition [DMN72]. There seems to be a very rapidly growing interest in this area with many new languages emerging, e.g. Beta [KMMN87], Eiffel [Mey87], C++ [Str86] and Trellis/Owl [Sch86].

The major contribution in the design of the Mjølner user interface is the consistent focus on the abstractions used by the programmer. Thus the language abstraction mechanisms *class* and *procedure* ("method" in Smalltalk terminology) are brought out and presented as windows. A hierarchical window system is used to represent nesting of these constructs. The effect is comparable to the effect of the Macintosh finder for hierarchical file systems with the advantage of direct manipulation of objects on the screen and ease of navigation. We claim that the user interface is *object-oriented*.

The Mjølner environment includes an incremental compiler and a very flexible run-time system which allows modification of a running program. This results in a system that offers unusually close integration of program modification and execution. The object oriented user interface style is used throughout the programming and execution process. The environment is under development and will initially support programming in Standard SIMULA [Sim87].

The Mjølner project [DLMM86] is a Nordic effort in which companies and universities in Norway, Denmark and Sweden participate. The incremental programming environment presented here is only one of several activities within the project. Other activities focus on program databases, an implementation of the Beta language, syntax-directed editors, and object-oriented specification languages [MBD87].

The rest of this paper is organized as follows: In section 2 we describe the concept of hierarchical windows on which the user interface is built. Section 3 describes the environment from the user point of view and the facilities available. In section 4 some brief comments on the implementation are given. Section 5 concludes the paper.

2 Object-Oriented Use of Hierarchical Windows

The use of personal workstations with high resolution graphics, mice, and window techniques has had a revolutionary effect on the development of programming environments. Windows allow the user to view and interact with several things at the same time, thus getting rid of the "modes" that are inherent in traditional terminal-based mini-computer environments [Tes81]. Windows can be used in different ways. We differ between *activity-oriented* use of windows and *object-oriented* use of windows.

2.1 Activity-Oriented Use of Windows

The Smalltalk environment [Gol84] is a pioneer project in its utilization of windows for observing and manipulating both the program source and objects that are created during program execution. The use of windows in the Smalltalk environment can be characterized as *activity-oriented*, i.e. each kind of window enables the programmer to perform a certain activity. A typical example is the Smalltalk Browser window. The browser serves as a viewport into the underlying source code structure. The user can browse through this structure, looking at a single class or procedure ("method") at a time. In order to look at two procedures at the same time, two browser windows are needed. There is no particular relation between the browser windows. It is e.g. possible to let them show the same procedure body. Other kinds of windows in Smalltalk are Inspectors (used to inspect the contents of an object) and Workspaces (used to type in commands to the system). Magpie [DMS84] and Trellis [BHK87] are other examples of environments using windows this way.

2.2 Object-Oriented use of windows

A contrasting way of utilizing windows can be characterized as *object-oriented*. In this approach each window has a one-to-one correspondence with an object, making it possible for the user to *identify* the window with the object itself. The user performs all activity relevant to the object directly on or in the window. The Macintosh Finder [Wil84] is an example of a system using windows this way. Files and directories are shown as individual windows (or icons). This object-oriented use of windows was pioneered in the Star user interface [SIKVH82]. It lends itself to natural and powerful interaction mechanisms since the objects (windows) can be manipulated directly rather than by commands [Sch83]. E.g. a file can be moved to another directory in the Macintosh Finder by simply dragging its icon to the new directory window.

2.3   Hierarchical Windows

In Mjølner, windows are used in an object-oriented way, primarily to represent abstractions in programs but also on the top level as an interface to the file system. *Hierarchical windows* are used to express containment relations between objects (windows). The Mjølner hierarchical window system [HNRR88], is implemented on top of a kernel of X-window primitives and currently runs on both Sun and Vax workstations. The hierarchical window system allows any window to contain a local set of full functionality windows. Windows can be moved, resized, and iconized. Sibling windows and icons may overlap freely.

Traditional window systems, with a single set of overlapping windows, follow the metaphor of overlapping papers on a desktop. The hierarchical window system implies a generalized metaphor; a paper on the desktop may act as a viewport into a new desktop. When a "desktop" is moved on the screen, all its "papers" follow with it. If a "desktop" is made larger, more of its "papers" come into view.

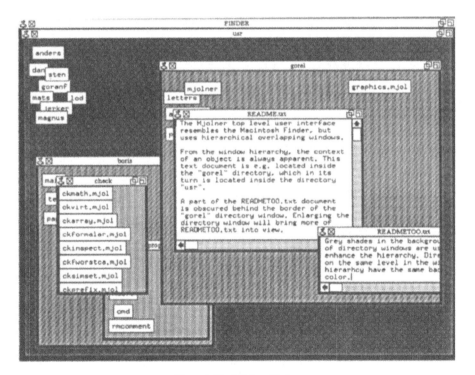

Figure 1. The Mjølner Finder

2.4   Use of Windows in Mjølner

Figure 1 shows an example of how windows are used in Mjølner. The top level window (the Mjølner Environment window) contains the file system. Each directory is shown as a window containing windows and icons for each of the files in it, similar to the Macintosh Finder. Clicking on a file icon expands it to a window displaying the file in some appropriate way (depending on the file type) and allowing appropriate interaction. E.g. clicking on a text-file icon expands it to a text window with editing facilities. Contrary to the Macintosh Finder however, a Mjølner window which is expanded from an icon keeps its place in the window hierarchy. An object (e.g. a file) is thus always presented in its context (e.g. its surrounding directories) regardless of its current icon/window state.

The full advantage of hierarchical windows is seen when windows are used within applications to present substructures. In figure 2, a program file "graphics.mjol" has been expanded to a Mjølner programming window. This window contains the program represented as a hierarchy of nested blocks (classes and procedures), each shown as an individual window or icon. In this case, the window hierarchy is more than just an organizational device since it actually reflects the static scope rules of the programming language.

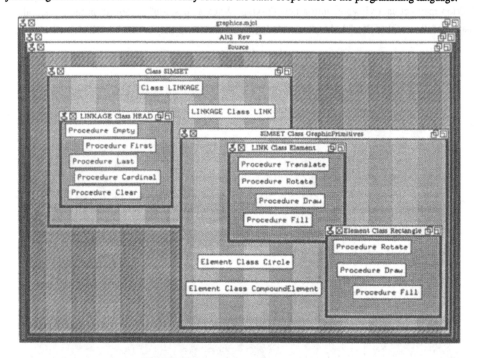

Figure 2. A Mjølner Programming Window

GraphicPrimitives is an application class containing classes for different graphic elements. Class Element describes the general behavior of a graphics element; translating, rotating, drawing, and filling. This behavior is overridden as needed in the subclasses. GraphicPrimitives inherits classes for handling double linked lists from the standard Simula class SIMSET. The elements are to be organized into lists, so class Element is a subclass to the standard class LINK in SIMSET.

It is important that the window hierarchy is used for "the right thing". The graphical layout of windows inside windows suggests using the window hierarchy to model containment relations. E.g. in the file hierarchy a directory *contains* its files. In the program block hierarchy, the class GraphicPrimitives *contains* the class Element and its subclasses Rectangle, Circle, and CompoundElement . The class Rectangle in its turn *contains* the procedures Rotate, Draw, and Fill. In an object-oriented programming language there are many other relations that are important as well, e.g. subclassing, call stacks, and references between objects. In section 3 we will return to how these relations may be modeled in Mjølner.

2.5 Interaction Advantages and Problems

There are several advantages of the object-oriented use of hierarchical windows:

Objects are easy to find, since each object has a natural place in the window hierarchy. The window hierarchy can be used as a built-in browser of objects, and makes it easy to navigate in large systems.

Object context is always shown. The context is often crucial to identify an object, e.g. to differ between procedures Draw in different classes. When interacting with an object, it is often valuable to view parts of its context at the same time. E.g. when editing a procedure one usually wants to see the instance variables of the class at the same time. In the hierarchical window system the contextual relation between objects is always apparent.

Good utilization of screen space. Overlapping windows increase the virtual screen space considerably. The usage of hierarchical windows is an aid in better utilization of this space. A single mouse click brings a window and its inner windows to the top of the screen, or iconizes a window freeing the screen space of it and all its inner windows. This allows whole contexts with interior window structure to be brought in or out of view by a single mouse click. It is easy to arrange the windows so that only information of current interest is shown, leaving less interesting information as icons behind the expanded windows.

Interaction is more complex in a hierarchical window system than in a traditional window system. One problem is that the inner windows tend to become small. In order to enlarge an inner window, its enclosing windows often need to be enlarged as well, which sometimes becomes cumbersome. Another problem occurs when the user is temporarily uninterested in the full context of a window. The enclosing window borders then take up screen space which could be used for better purposes. Further experimentation is needed to come up with smooth interaction mechanisms which solve these problems.

2.6 Related Interaction Styles

The idea to use hierarchies of graphical objects to represent programs is also used in e.g. the BOXER environment [SA86]. In BOXER the key graphical structure is a box containing text which in its turn can contain inner boxes. A box is not a window, rather it is considered a special sort of character which is moved by cut-paste operations on the text. The boxes can be iconized and expanded by the user. They can, however, not be moved around freely because of their position in a text. This gives heavy restrictions on what parts of the system can be viewed simultaneously. BOXER is mainly intended to be used in education and to give novice programmers a "what you see is what you have" interface. We agree with the BOXER designers in the merits of such an interface, and we think it is very useful also for a professional programmer.

Fisheye viewing [Fur86] is a viewing strategy which "can show places nearby in great detail while still showing the whole world - simply by showing the more remote regions in successively less detail". Fisheye viewing thus share some properties of hierarchical windows, enabling the user to see both overview and detail at the same time. However, fisheye systems and hierarchical window systems have completely different styles of interaction. Fisheye viewing can be seen as an advanced scrolling technique; when the focus is moved, the context will be changed automatically as a function of the focus point. The hierarchical window system does not have a similar concept of focus. Opening a window does not change the status of other windows.

3 Programming and Execution in Mjølner

3.1 The block hierarchy

A program in Mjølner is primarily considered to be a hierarchy of blocks, where a block is a unit of abstraction such as a class or a procedure. The main characteristics of a block are that it has a local name scope and that it can be dynamically instantiated during execution: classes are instantiated to objects, and procedures are instantiated to procedure activations. The representation of the block hierarchy using a window hierarchy makes it possible for the user to navigate in the program in terms of its abstractions. A new class or procedure is created simply by creating a new window in the appropriate father window. The two-dimensional layout of son-block icons in a block window frees the user from the unnatural ordering between sibling blocks enforced by the traditional representation of a program as linear text. Since the block is the primary unit of abstraction used by the programmer, we find it very natural that blocks play the main role in the user interface.

3.2 Editing and Incremental Compilation

The local variables and the statement body of a block are presented in special subwindows of the block window. This makes it easy to e.g. view the variable declarations of a class at the same time as the body of one of its procedures, as in figure 3. The declarations and bodies are internally represented as abstract syntax trees which are unparsed as text. The editing is done directly on the tree using a syntax-directed editor. Syntax errors are thus impossible to make.

Incremental static semantic checking is performed in response to every editing operation. Static semantic errors such as type conflicts, undeclared variables, etc. are marked on the screen by underlining the erroneous construction. E.g. in figure 3, the name anElement is underlined. The user may ask for an explanation of the error, when the erroneous construction is the current focus of attention in the syntax tree. As shown in figure 4, the explanation reveals that anElement is not declared. When the declaration is added, the error-markings automatically disappear. The philosophy in Mjølner concerning errors, is that they should be visible but not intrusive. Errors are simply marked on the screen and may be corrected at any time by the user.

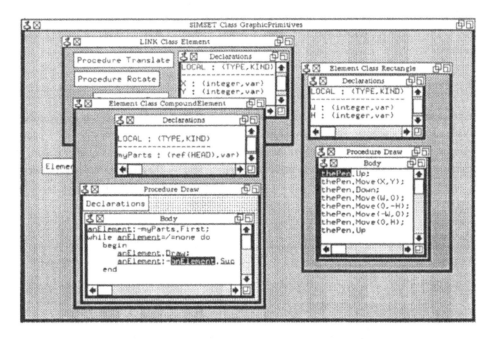

Figure 3. Declarations and Statement Body of Blocks

The class CompoundElement has a variable, myParts, which is a list of its constituents. It draws itself by walking through the list and drawing each constituent. Class Rectangle is represented by its the upper left corner (inherited from class Element) and its width and height. It uses a plotter pen to draw itself.

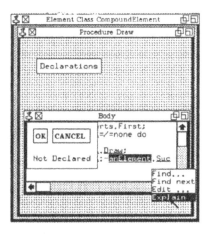
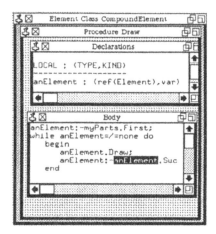

Figure 4. Correcting a Static Semantic Error

The variable anElement is underlined because it is not declared. When the user corrects the error, by inserting a local declaration, the underlining disappears.

Code is generated incrementally on a block-by-block basis, and is incrementally loaded into the executing program. This is done automatically as needed by the system, so the user does not have to issue any explicit command.

3.3 Execution

The executing program is presented to the user as a system of block instances (objects and procedure activations), each represented by a window of its own. In analogy to the program definition, the windows are organized hierarchically according to the static scope. E.g. in figure 5, the Draw procedure of CompoundElement has called the Draw procedure of a Rectangle. The procedure activation windows are shown inside the appropriate object windows. The graphic element objects are shown inside the GraphicPrimitives object.

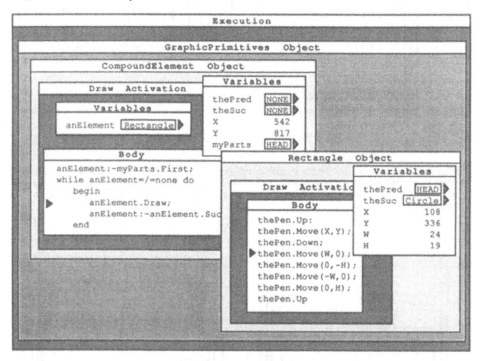

Figure 5. Objects and Procedure Activations in an Execution State.

The procedure Draw of a CompoundElement has called the Draw procedure of one of its constituent elements. This element, a Rectangle, is in the middle of drawing itself.

Each block instance window contains subwindows for data and statement body, which parallel the declaration and body windows inside a block window. The data window shows the variable values, and the statement body window shows the current execution point. An object in Simula is considered to be described by a concatenation of its class and superclasses. The data window will therefore contain inherited variables as well as variables defined in its actual class. E.g. the Rectangle object in figure 5 has thePred and theSuc variables inherited from class LINKAGE; X and Y are inherited from class Element; W and H are defined in class Rectangle.

If all existing block instances were shown on the screen, this would lead to a very cluttered view. Usually the user is interested in seeing only very few of the block instances, e.g. the currently executing instance, and some objects involved in the current computation. In Mjølner, the user can interactively open the objects and procedure activations of interest. Reference variables (pointers) in the data windows are presented as buttons. Pressing a button, using the mouse, causes the window of the referred object to appear on the screen. Windows enclosing the referred object are opened automatically when needed. If the referred object is already present on the screen, the user is given visual feedback on its position.

A call stack window presents a list of buttons for the currently active block instances. Thus, when execution is suspended, e.g. at a breakpoint, a procedure activation can be opened via a button in the call stack window. Object windows can then be opened via the reference variables in already opened windows. See figure 6.

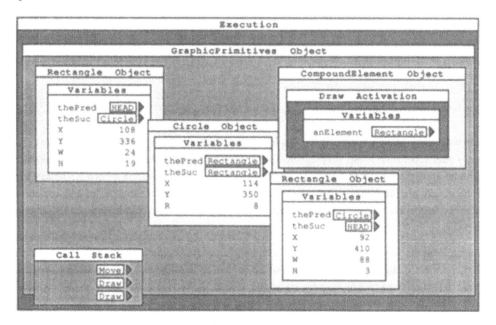

Figure 6. Following References in an Object Structure

Upon execution suspension, the user has opened the Draw activation via the lower button in the call stack. This caused the enclosing objects CompoundElement and GraphicPrimitives to appear as well. The next step was to open the Variables window in the Draw activation. The variable anElement currently refers to a Rectangle object. Pushing the Rectangle button, caused the Rectangle object to appear (in the top left corner of the picture). The user then follows theSuc references to bring the rest of the objects in the list to appear on the screen.

Execution can proceed stepwise or to a breakpoint. If desired, the contents of the block instance windows can be updated continuously during the execution. This makes it possible to e.g. monitor an object structure. More details on observation and debugging facilities are given in [THM87].

3.4 Integrated editing and execution

Although Mjølner is a compiling system, program editing and execution is very closely integrated, even more than in many interpreting systems. The program definition may freely be edited during execution. Errors can be corrected in executing programs, functionality can be added and immediately tested in an execution state with existing object structures, and experiments can be performed with executing programs.

In general, such close integration between editing and execution leads to version and consistency problems. E.g., what should happen if a variable is added to a class and there are existing objects of that class? In Mjølner, these problems are solved by providing two mechanisms; block instances of different versions may co-exist in the same execution state, and block instances may be converted from an older to a newer version. This is described in more detail in [HM86] and [HM87].

In many cases, however, the results of the close integration is very straight forward. E.g. if the implementation of a procedure is changed, Mjølner's default behavior is to let old procedure activations continue to execute according to the old version, but new procedure activations will execute the new version. New procedures and classes can be added and tried out immediately.

3.5 Presenting relations

The hierarchical window system captures the fundamental relation of static enclosure. There are, however, many other important relations in a program, e.g. the subclass-of relation, the instance-of relation (an object is an instance of a class), and the refers-to relation (for reference to objects). In section 3.3 we saw how buttons were used to follow refers-to links. They could be used also for subclass-of and instance-of relations. Buttons make it possible to easily navigate in a complex object structure in a manner similar to the ideas in hypertext systems [Con87].

An attractive extension of buttons would be to actually draw a connector between two windows when a button is pressed. Such window connectors would have to be an integral part of the window system in order to stay connected when windows are moved, resized, and iconized. The use of connectors would make it possible to show the relations between objects and their interiors at the same time, as shown in figure 7.

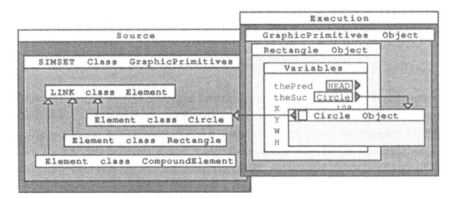

Figure 7. Presenting Relations

Window connectors could be used to show different kinds of relations. In the Source window, connectors are used to show the subclass-of relation. A connector from the Circle object to the Circle class shows the instance-of relation. A connector from theSuc variable to the Circle object shows a refers-to relation.

4 Implementation

The Mjølner environment is unusual in that it combines compilation technology with a very high degree of interactivity. Earlier incrementally compiling environments, such as the Cornell Program Synthesizer [TR81], Gandalf [MF81], and DICE [Fri84], have used a more traditional presentation of programs. In this section we will comment briefly on some important aspects of the implementation, and conclude with a report on the current status.

4.1 Incremental Compilation Techniques

To be of use in an industrial setting, an interactive environment must be able to handle large programs without noticeable increases in response times. It has therefore been of major importance in Mjølner to use incremental compilation techniques which can be *scaled up*, i.e. whose performance is relatively independent of program size. A typical bottleneck in this respect is static semantic checking. Changes to declarations may affect the static semantics in places far away from the declaration. E.g. adding a parameter to a procedure inside a class will affect all calls to that procedure. Since the static semantic checking is performed incrementally while editing, it is essential that the affected places are found quickly. In Mjølner, a method based on attribute grammars is used, but allowing side-effects when evaluating the attributes [HDM87]. The side-effects are used to build up information structures which are used to quickly find affected places after a change.

Code generation is not as time-critical as static semantic checking, since the code is needed only when execution is resumed. In Mjølner, code is generated in chunks on a block-by-block basis. To minimize the delay before execution can be resumed, we are considering to implement lazy code generation and code generation in the background, utilizing the programmer's "think time", similarly to as it is done in the Magpie environment [SDB84].

4.2 Flexible Run-time Environment

An important feature of Mjølner is that execution is allowed to continue after the program has been edited. To accomplish this, Mjølner uses a very flexible run-time architecture. The generated code-chunks are arranged in a "template tree" which parallels the block tree. A template is a data structure which contains information relevant to all instances of a particular block; activation record size, garbage collection information, link to superclass template, etc. Block calls (procedure calls and class "new"s) are performed by indirect jumps via the template tree. The code for a block call is thus independent of the position of the called block. When new code for a block is generated, it is loaded incrementally into the executing program, and linked into the template tree. The new code will be used automatically the next time the block is called.

The template tree is a generalization of the standard way of implementing message sending (virtual procedure calls) in Simula, using a virtual-table per class. The execution overhead is very small; given the static father link of a new block instance, three indirect addressing instructions are needed to find its code and template (StaticFather.Template.SonTemplate(i).Code). This can be compared to two direct addressing instructions in a traditional non-incremental Simula system.

Since the execution overhead is so small, there is no need for separate development and production environments. The executing program and the development system run as separate UNIX processes and communicate via a small kernel embedded in the run-time system. During development, the executing program is normally started via the development environment. However, the program can just as well be run "stand-alone" without the development system. If needed, e.g. at a run-time error, the development system can be attached to the executing program.

More details on the template-tree run-time architecture, and consistency aspects of mixed editing and execution, are given in [HM87].

4.3 Current Implementation Status

Mjølner is implemented in Standard Simula and executes on Sun and Vax workstations. At the time of writing (May 1988), the system is under development. Extensive work remains to be done before it can be used in practice. Editing and incremental semantic analysis supports a substantial subset of Simula. Code generation is currently being integrated into the system. The run-time system is not yet implemented, so at present, programs cannot be executed. However, a prototype run-time system for a toy language has been developed earlier, to evaluate the design [MM85].

In its current state, the system can handle programs of about the same size as the one shown in the examples. There are many rather straight-forward ways of optimizing the system, so we have good hope of being able to handle more realistic programs in a near future.

5. Conclusions

We have presented the user interface of the Mjølner incremental programming environment. The key idea is to focus on the programming language abstractions: classes and procedures. These are presented as nested windows, using a hierarchical window system. An execution is similarly presented as a structure of nested objects and procedure activations. The window hierarchy provides a powerful and natural means of navigation in terms of the abstractions. While navigating and manipulating the program in terms of his abstractions, the user is encouraged to regard the program as a physical structure of nested classes and procedures.

The user interface is object-oriented in the sense that the objects on the screen (windows) can be *identified* with the objects in the program (classes, procedures, objects, activations). This gives the user a sense of directly interacting with the abstractions in the program. Compilation and debugging tools are not explicitly visible to the programmer, but available as functionality in the relevant objects.

The high level of interaction is made possible by incremental compilation techniques. Especially important is the flexible run-time environment which supports incremental updates of the loaded executing program. This allows an unusually close integration of program modification and execution.

Acknowledgements

We are deeply indebted to the rest of our local Mjølner group for taking part in the development and implementation of the ideas presented in this paper: Magnus Taube, Sten Minör, Lars-Ove Dahlin, Anders Gustavsson, Jerker Nilsson, Dan Oscarsson, Mats Bengtsson, and Göran Fries.

The hierarchical window system was implemented by the Norwegian Mjølner group.

We also want to thank Claus Nørgaard and the referees for constructive comments on the paper.

The Mjølner project is partially funded by a grant from the Nordic Fund for Technology and Industrial Development. The Swedish sub-project has an additional grant from the Swedish Board for Technical Development (STU).

References

[BHK87]   P. O'Brien, D. Halbert, M. Kilian. The Trellis Programming Environment. OOPSLA '87 Conference Proceedings. SIGPLAN Notices, December 1987. pp 91-102.

[Con87]   J. Conklin. Hypertext: An Introduction and Survey. IEEE Computer. Sept. 1987. pp 17-41.

[DLMM86] H.P. Dahle, M. Löfgren, O.L. Madsen, B. Magnusson. Mjølner - A Highly Efficient Programming Environment for Industrial Use, Mjølner Report no. 1. Dept. of Computer Science, Lund Institute of Technology, Sweden, 1986. Also in Proceedings of the 15th Simula Users' Conference, Jersey, 1987.

[DMN72]   O.-J. Dahl, B. Myhrhaug, and K. Nygaard. Simula: Common Base Language. Norwegian Computer Center. 1972.

[DMS84]   N. Delisle, D. Menicosy, M. Schwartz. Viewing a Programming Environment as a Single Tool. Proceedings of the ACM SIGSOFT/SIGPLAN Software Engineering Symposium on Practical Software Development Environments. SIGPLAN Notices, May 1984. pp 49-56.

[Fri84]   P. Fritzson. Preliminary Experience from the DICE system, A Distributed Incremental Compiling
             Environment. Proceedings of the ACM SIGSOFT/SIGPLAN Software Engineering Symposium on
             Practical Software Development Environments. SIGPLAN Notices, May 1984. pp 113-123

[Fur86]   G. W. Furnas. Generalized Fisheye Views. In Proceedings of ACM SIGCHI Human Factors in
             Computing Systems Conference. 1986, pp 16-23.

[Gol84]   A. Goldberg. Smalltalk-80: The Interactive Programming Environment. Addison-Wesley, 1984.

[HDM87]   G. Hedin et. al. Incremental Semantic Analysis in Mjølner. Mjølner Report S-LTH-25.1, Lund
             Institute of Technology, Sweden. 1987.

[HM86]   G. Hedin, B. Magnusson. Incremental Execution in a Programming Environment Based on
             Compilation. 19th Hawaii International Conference on System Sciences, Honolulu, 1986.

[HM87]   G. Hedin, B. Magnusson. Supporting Exploratory Programming in Simula. In Proceedings of the
             15th Simula Conference, Jersey, 1987.

[HNRR88]   T. Hauge, I. Nordgard, T. Rød, G. Raeder. Gungne, Functional Specification. Mjølner Report, N-
             EB-4.2, EB Technology, Nesbru, Norway, January 1988.

[KMMN87]   B. B. Kristensen, O. L. Madsen, B. Møller-Pedersen, K. Nygaard. The BETA Programming
             Language. In B.D. Shriver, P. Wegner (ed.) Research Directions in Object Oriented Programming.
             MIT Press 1987. pp 7 - 48.

[MBD87]   B. Møller-Pederson et. al. Rationale and Tutorial on OSDL: An Object-Oriented Extension of SDL.
             Computer Networks vol. 13, No. 2, 1987, pp. 97-117.

[Mey87]   B. Meyer. Reusability: The Case for Object-Oriented Design. In IEEE Software, March 1987.

[MF81]   R. Medina-Mora, P. Feiler. An Incremental Programming Environment. IEEE Trans. on Software
             Eng. Sept 1981. pp 472-482.

[MM85]   B. Magnusson, S. Minör. III - an Integrated Interactive Incremental Programming Environment Based
             on Compilation. ACM SIGSMALL Symposium on Small Systems, May 1985.

[SA86]   A. A. diSessa & H. Abelsson, BOXER: a Reconstructible Computational Medium, CACM Sept.
             1986, pp 859-868.

[Sch83]   B. Schneiderman, Direct Manipulation: A Step Beyond Programming Languages, IEEE Computer,
             August 1983.

[Sch86]   C. Schaffert et al. An Introduction to Trellis/Owl. OOPSLA'86, SIGPLAN Notices, Nov 86.

[SDB84]   M. Schwartz, N. Delisle, V. Begwani. Incremental Compilation in Magpie. Proceedings of the ACM
             SIGPLAN Symposium on Compiler Construction, SIGPLAN Notices 19, 6, (June 1984).

[Sim87]   Data Processing - Programming Languages - SIMULA. Swedish Standard SS 63 61 14. SIS.
             Stockholm, Sweden, June 1987.

[SIKVH82]   D.C. Smith et al., Designing the Star User Interface, Byte Magazine, April 1982.

[Str86]   B. Stroustrup. The C++ Programming Language. Addison-Wesley, 1986

[Tes81]   L. Tesler, The Smalltalk Environment, Byte August 1981, pp. 90-147.

[THM87]   M. Taube et. al.,The Mjølner Observation Tool. Proceedings of the 15th Simula Users' Conference,
             Jersey, 1987.

[TR81]   T. Teitelbaum, T. Reps. The Cornell Program Synthesizer: a Syntax-Directed Programming
             Environment. CACM Sept. 1981. pp 563-573.

[Wil84]   G. Williams. The Apple Macintosh Computer, Byte Magazine. February 1984. 30-54.

Inheritance as an Incremental Modification Mechanism
or
What Like Is and Isn't Like

Peter Wegner and Stanley B. Zdonik
Department of Computer Science
Brown University, Providence, RI 02912

Abstract: Incremental modification is a fundamental mechanism not only in software systems, but also in physical and mathematical systems. Inheritance owes its importance in large measure to its flexibility as a discrete incremental modification mechanism. Four increasingly permissive properties of incremental modification realizable by inheritance are examined: behavior compatibility, signature compatibility, name compatibility, and cancellation. Inheritance for entities with finite sets of attributes is defined and characterized as incremental modification with deferred binding of self-reference. Types defined as predicates for type checking are contrasted with classes defined as templates for object generation. Mathematical, operational, and conceptual models of inheritance are then examined in detail, leading to a discussion of algebraic models of behavioral compatibility, horizontal and vertical signature modification, algorithmically defined name modification, additive and subtractive exceptions, abstract inheritance networks, and parametric polymorphism. Liketypes are defined as a symmetrical general form of incremental modification that provide a framework for modeling similarity. The combination of safe behaviorally compatible changes and less safe radical incremental changes in a single programming language is considered.

1. Introduction

Incremental modification facilitates reusing a conceptual or physical entity in constructing an incrementally similar one. It arises in incremental problem solving, incremental specification, incremental compilation, and incremental editing. Evolution of both natural and computational systems may be described and controlled by incremental modification mechanisms. Recursive specifications are dynamically incremental in that they specify the solution of a problem with parameter (n+1) in terms of the same problem with parameter n.

Inheritance is a particular kind of incremental modification mechanism that transforms a parent entity P with a modifier M into a result entity R = P+M, as in Figure 1. The parent, modifier, and result have a record structure with a finite number of attributes:

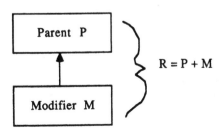

Figure 1: Incremental Modification by Inheritance

$$P = (P1,P2,...,Pp)$$
$$M = (M1,M2,...,Mm)$$
$$R = (R1,R2,...,Rr)$$

The composition operator + is asymmetrical since the parent P and modifier M play different roles in the composition process. The asymmetric role of P and M in determining R is brought out by the following notation:

result R
  inherits P;
  modified by M;

The attributes of M may be independent of or overlap with those of P:

(1) **Independent Attributes:** The attributes of M are disjoint from those of P.
    R has p+m attributes consisting of the union of those in P and M.

(2) **Overlapping Attributes:** The attribute names of M overlap those of P.
    R has the attributes explicitly added in M and those attributes of P whose names do not occur in M. The attributes of M take precedence over those of P much as identifiers declared in an inner textually nested module of a block-structure language take precedence over those declared in an outer module.

Incremental modification is clearly simpler for independent than for overlapping attributes. However, software evolution generally requires modifying existing attributes of an entity rather than merely adding new ones. We therefore investigate constraints on the incremental modification of attributes, including very strong constraints such as behavioral compatibility and very weak constraints such as cancellation of attributes.

This work is part of a wider study [We1] which examines the following attributes of inheritance:

modifiability: How should modification of inherited attributes be constrained?
granularity: Should the unit of inheritance be classes or instances, entities or attributes?
multiplicity: How should multiple inheritance be managed?
quality: What should be inherited? Specifications, code, or both?

We focus narrowly on incremental modification, recognizing that an inheritance mechanism requires design decisions for modifiability to be supplemented by decisions concerning granularity, multiplicity, and quality.

2. The Essence of Inheritance

Since our models of incremental modification depend on general assumptions about the nature of inheritance and on the kinds of entities that inherit, we characterize these notions more precisely. Inheritance is discussed first, to bring out the nature of inheritance hierarchies before committing to specific kinds of inherited entities.

What is the essence of object-oriented inheritance? It is a hierarchical incremental modification mechanism for entities defined by sets of attributes that allows any R = P + M defined as in section 1 to be further modified, say to R1 = R + M1. We assume further that inherited attributes are more essentially part of the inheriting object than attributes that are merely used or invoked, just as eye colors are more essentially part of people than the car which they drive or the house in which they live.

We adopt the view of Cook [Co1, Co2] who defines inheritance as a composition mechanism that internalizes inherited attributes by late (execution-time) binding of self-reference to the inheriting object. Self-reference in a type or class is bound to the object on whose behalf an operation is being executed, rather than to the textual module in which the self-reference occurs.

Dynamic binding of self-reference at execution time captures the essential difference between inheritance and invocation. Late binding of "self" allows an inherited entity P to assume the identity of each of the objects that uses its attributes while preserving its textual independence. Its attributes are shared by inheriting entities but behave like indigenous attributes of each inheritor. One advantage of sharing over copying of inherited attributes is that behavior modification of objects inheriting P can be realized by modular modification of the shared parent P rather than by internal surgery on entities that contain copies of P.

The execution-time binding of self-references in a parent P and modifier M to an object X with template R = P+M is illustrated in Figure 2. When an object X receives a message to invoke one of its attributes (operations) it binds self-references in both P and M to the object X.

In the example below, P defines the attributes P1,P2,...,Pp and P1 contains a textual self-reference "self A" to an attribute A. A may be a locally defined attribute Pi of P or a nonlocal attribute Mj defined in the modifier M.

P =
  P1: ... self A; ...
  P2: ...
  ... ...
  Pp: ...

When "self A" is executed, it invokes a definition of A in P or M according to the following rules:

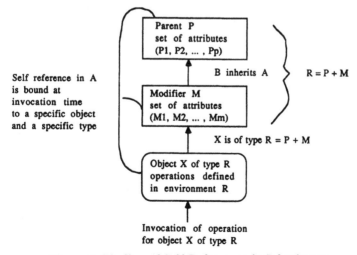

Figure 2: Binding of Self-References in Inheritance

(1) **Redefined attribute:** A is defined in both P and M (A = Pi = Mj for some i,j).

A is bound to the attribute definition Mj in M, which blocks the definition of the similarly named attribute Pi in P. The meaning of A is not the locally defined attribute of P but the nonlocal attribute of M.

(2) **Virtual attribute:** A is defined in M but not in P.

A is called a virtual attribute of P, since it is used but not defined in P. P is called a *virtual entity* (*virtual class* in Simula, *abstract type* in Smalltalk). The definition of P is incomplete in that P must rely on attributes defined in a descendant before objects that use P as a template can be instantiated. Note that if A is defined in M it is bound to the attribute Mj of M independently of whether A is also defined in P.

(3) **Recursive attribute:** A is defined in P but not in M.

A is bound to a locally defined attribute Pi in P. If the attribute A is the same Pi in which the self-reference occurs, then this is a recursive invocation. However, if A refers to another attribute of P then this is a recursive invocation of the object P but not a direct recursive invocation of the attribute definition for A.

Delayed binding of self-reference allows "self A" in P to assume the identity of a variety of different objects at execution time. It allows attributes of a parent to be internalized by the objects that use them in the sense that self-references in the parent refer to the identity of the currently invoking object. It allows parents to specify virtual resources that are not yet defined and to require the definition of such resources in descendant templates.

Attributes of an entity may be locally defined, inherited, or virtual. Inherited and locally defined attributes together determine the resources provided by an entity to its clients. Virtual attributes arise when the provided resources of an entity are insufficient ot fulfil its contract with clients. If "needed resources" > "provided resources" then attributes which are needed but not provided are virtual and must be supplied by a descendant.

The meaning of self-reference in objects, just as in recursive functions and procedures, can be defined in terms of least fixed points. Cook refers to entities with unbound self-references as generators and models binding of self-references in generators P by their least fixed point Y(P). Inheritance is realized by composing the generators P and M to obtain P+M and then taking the least fixed point Y(P+M) of this composite generator. In contrast, invocation is defined in terms of the composition Y(P) + Y(M) of fixed points. Thus Y(P + M) $\neq$ Y(P) + Y(M), and the difference between the left and right hand sides in fact captures the difference between M inheriting P and M invoking P.

In a world without self-reference, inheritance reduces to invocation and inheritance hierarchies are simply tree-structured resource-sharing hierarchies. However, recursive definitions are just as fundamental for objects as for functions and procedures. The progression from non-recursive Fortran to recursive procedure-oriented languages is being repeated for object-oriented languages. Incremental modification mechanisms should be designed from the start to properly handle self-reference.

3. Types and Classes

The terms "type" and "class" are defined to accord with their current usage. Type and class hierarchies based on these definitions are found to have very different properties.

Types are motivated by type checking and may be defined by a predicate for recognizing expressions of the type. Classes determine collections of objects and may be defined by templates for object creation. Types have a type-checking semantics while classes have an instance creation semantics, as indicated in Figure 3.

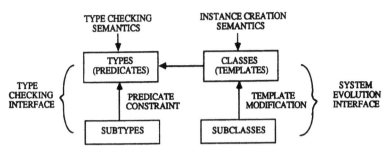

Figure 3: Relation Between Types and Classes

The definition of types as predicates and classes as templates leads to a definition of subtypes in terms of predicate modification and subclasses in terms of template modification. Subtypes are defined in terms of constraints that determine a subset of the set defined by the parent predicate. Subclasses are defined in terms of template modifications that may involve radical modification or even cancellation of template components.

Template modification is more powerful than subtyping as an incremental modification mechanism but also less tractable. Inheritance will be defined as a mechanism for template modification rather than subtyping. This reflects not only the reality of inheritance mechanisms in actual programming languages but also the fact that evolutionary processes of incremental change in the real world rarely conform to the stringent constraints of subtyping and are better modeled by subclassing.

Classes are a special kind of type, namely a type whose predicate is a template specification. For every class C there is a type predicate that characterizes the set of all potential instances of the class, namely the predicate "is an instance of the class C". However, types do not necessarily have an associated class since predicates do not necessarily specify templates.

Class-based languages automatically have an associated type system. Class declarations may be viewed as type declarations. Class hierarchies of object-oriented languages automatically have associated type hierarchies. For every statement about classes there is a corresponding statement about associated types, but the converse is not necessarily true. The condition "every object belongs to a class" implies that every object also belongs to a type. However, the converse condition, "every object belongs to a type," does not imply that every object belongs to a class.

Both classes and types serve to classify values into collections with uniform attributes. But classes are collections of created or potentially creatable instances, while types focus primarily on the predicate. Every class has two associated collections: the collection of all possible instances of the class and the collection of instances that has actually been created.

Types may be specified syntactically by signature specifications, semantically by behavior specifications, and pragmatically by implementations, as illustrated in Figure 4. Signature specifications in terms of record types are underspecifications since the semantics of typed record components is left unspecified. Behavior specifications in principle precisely determine the desired goal or task realized by an object but cannot be uniformly specified in any formal specification language. Implementations overspecify the behavior by committing to a particular implementation of the behavior. Classes are pragmatic template specifications.

We are interested in incremental modification mechanisms associated with each of the specification techniques:

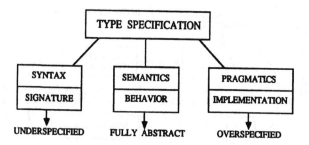

Figure 4: Syntactic, Semantic, and Pragmatic Type Specification

(1) Incremental modifications of behavior are generally behaviorally compatible with the behavior of the parent and are called subtype specifications. The subtype is related to the parent type by an is-a relation. Behavior of types and subtypes can be defined by algebras, but the relation between the algebra of a type and its subtype can be quite subtle.

(2) Incremental modifications of signatures determine compile-time checkable subclasses that are not in general behaviorally compatible with the parent type.

(3) Implementation specifications are more flexible in their support of incremental modification mechanisms than behavior or signature specifications, since template modifications of an arbitrary nature may be specified. Behavior and signature compatible modification may be supported at the level of implementation by constraints on the implementation mechanism. However, incremental modification that violates behavior or signature compatibility may also be supported.

The above analysis of types in terms of their mechanisms for specification yields mechanisms for incremental modification closely related to those that arise in actual object-oriented systems.

4. Varieties of Incremental Modification

The constraints on incremental modification imposed by our assumptions about inheritance, types, and classes leave considerable room for variation.

Four increasingly permissive properties of incremental modification are contrasted in Figure 5: behavior compatibility, signature compatibility, name compatibility, and cancellation. Each has a different conceptual model of entities in the inheritance hierarchy. Behavioral compatibility views entities as behaviors modeled by algebras, signature compatibility views entities as signatures modeled by partial ordering lattices, and name compatibility views entities as implementations modeled by an operational semantics. Cancellation views entities as sets that can be enlarged or restricted by respectively weakening or strengthening the constraints of set membership.

(1) Behavior-compatible modification (types)

The entities to be modified are types whose behavior may be modeled by many-sorted algebras. Syntax is specified by signatures and semantics by interpretations or by equational axioms. Modified behavior of subtypes is specified by behaviorally compatible subalgebras. We examine the notion of behaviorally compatibility and demonstrate that true behavioral compatibility which satisfies the principle of substitution is more restrictive than generally supposed.

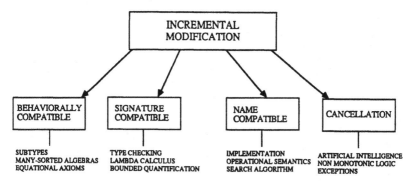

Figure 5: Varieties of Incremental Modification

(2) Signature-compatible modification (signatures)

The entities to be modified are signatures which are syntactic algebra specifications without any associated semantics. They determine a coarser equivalence relation over expressions than behavioral specifications. Subsignatures may be derived by horizontal extension (adding new components) or vertical modification (constraining existing components). Subsignatures are not behaviorally compatible with parent signatures because a given component may be replaced by a type-consistent component with different behavior.

(3) Name-compatible modification (classes)

The entities to be modified are class templates specified by their implementation. Behavior modification is realized by modifier templates which overlay their parent templates. The semantics of modification may be specified operationally by a search algorithm. To find the definition of a named attribute we first look for it in the modifier. If it is not defined in the modifier, we look for it in the ancestor hierarchy, either finding the attribute or reporting that it is undefined.

(4) Inheritance with cancellation (exceptions)

Traditional inheritance usually focuses on subtypes or subclasses defined by increasing the severity of constraints. Cancellation focuses on the relaxation of constraints, thereby allowing extension as well as restriction of sets defined by inheritance. This is reflected in our discussion of exceptions, abstract inheritance networks, and nonmonotonic reasoning systems. Cancellation may occur at the level of behavior, signatures, or names. At the level of names it may be realized by modifying the search algorithm so that search beyond the local level is blocked for cancelled names.

4.1. Behaviorally Compatible Modification

There was much debate in the 1960s whether Algol 68 and Pascal should be behaviorally compatible modifications of Algol 60 and whether IBM System/360 should be a behaviorally compatible modification of the 700 series of computers. Later it was debated whether Ada should be a behaviorally compatible modification of Pascal or simply Pascal-like. In all these cases it was decided that the constraint of behavioral compatibility was too onerous, in spite of the considerable advantages to users of such "upward compatibility".

Behavioral compatibility places strong constraints on incremental modification. When these constraints prevent us from realizing desired incremental changes, then behavioral compatibility must be discarded. But when desired incremental changes are behaviorally compatible, then great benefits can be realized by reusability of both code and concepts.

4.1.1. Specification of Behavior by Algebras

In order to define behaviorally compatible modification, we need a precise notion of behavior. Behavior may be specified by algebras with a signature and a semantics. For single-sorted algebras the signature consists of a sort S and operation symbols fi of arity ni:

Sig = (S; f1:n1, f2:n2, ... , fK:nK)

A signature determines a collection of well-formed expressions for the algebra but does not associate any semantics with expressions. A semantics of an algebra with signature Sig is given by an interpretation that associates a domain (carrier set) of values with S and a function over the domain of arity ni with every function symbol fi. Alternatively, semantics may be associated with a signature by equational axioms such as commutativity and associativity that specify that expressions transformable into each other by axioms are semantically equivalent.

A many-sorted algebra S has a set of sorts S = (S1,S2,...,SN). Function symbols have arguments and values of specified sorts. The interpretation associates a value set with each sort and associates functions from domains to ranges with each function symbol.

An order-sorted algebra is a many-sorted algebra with a partial ordering on its sorts that induces a subtype ordering relation on algebras. A partial algebra is an algebra in which functions may be partial (undefined for some arguments).

Algebras are a precise specification of behavior for classes of objects, with S representing the state and fi representing the attributes of the record structure. They provide a framework for the precise representation of substructure.

4.1.2. Three Kinds of Behavioral Compatibility

Different linguistic notions of subtype yield different notions of behavioral compatibility. We define three successively stronger notions of behavioral compatibility associated with the following three notions of subtype [BW]:

Subset subtype: Int(1..10) is a subset subtype of Int
Isomorphically embedded subtype: Int is an isomorphically embedded subtype of Real
Object-oriented subtype: Student is an object-oriented subtype of Person

4.1.2.1. Subset Subtypes and Partial Compatibility

Subset subtypes need not be closed under their operations. The result of an operation on arguments in a domain may lie outside the domain. We define a form of behavioral compatibility appropriate to subset subtypes called "partial compatibility":

Partial Compatibility: A subtype is partially compatible with its parent type if corresponding operations on corresponding arguments give corresponding results whenever the result is defined for the subtype.

Partial compatibility is not true behavioral compatibility because the result may be defined for the supertype but not for the subtype. For example, "6+7" is undefined for the subtype Int(1..10) but is certainly defined for the type Int.

4.1.2.2. Isomorphically Embedded Subtypes and Subcomplete Compatibility

Isomorphically embedded subtypes are better behaved than subset subtypes, because they are closed under their operations. They may be associated with a different notion of behavioral compatibility called subcomplete compatibility.:

Subcomplete compatibility: A subtype is subcompletely compatible with its parent type if corresponding operations on corresponding arguments of the subtype are defined in the subtype whenever they are defined in the supertype and yield corresponding results.

Subcomplete compatibility yields true compatibility when operations and arguments are restricted to the subtype, but may result in incompatibility when arguments are extended to include those of the supertype. In particular, the assignment "x := y;" where x is a variable of the supertype and y is an object of the subtype, causes y to be unprotected against inadmissible assignments of supertype values to its components.

If the operations on Int are extended to include division, then Int and Real are no longer subcompletely compatible, since Int is closed only for addition and multiplication and not for division. In mathematics, closure under division provided the motivation for extending the integers to the rationals. Closure under operations has also provided the motivation for further extensions to algebraic, real, and complex numbers. Subcomplete extension is of limited applicability since it cannot admit such partial operations in the subalgebra. Nevertheless, it is useful in practice to view subtypes as being algebras only under operations for which they are closed and to view partial operations of the subalgebra, such as division for integers, as being defined only for the algebra of the supertype.

4.1.2.3. Object-Oriented Subtypes and Complete Compatibility

In order to avoid altogether situations in which operations can be undefined on values of the subtype and defined on values of the supertype, we consider subtypes whose argument domain is precisely the same as that of the parent type. Operations defined for both the subtype and supertype have corresponding values for corresponding arguments over their identical domains. However, the subtype may have additional operations, as is the case for students who are persons with certain specialized operations such as "grade-point average". The notion of behavioral compatibility for such subtypes will be called "complete compatibility".

Complete compatibility: A subtype is completely compatible with its supertype if it has the same domain as the supertype and, for all operations of the supertype, corresponding arguments yield corresponding results.

Complete compatibility yields compatible behavior not only in the context of the subtype but also in the context of the supertype. In particular, it permits values of the subtype to be freely manipulated as values of variables of the supertype with no fear of inadmissible behavior.

The three kinds of behavioral compatibility share the property of yielding the same result when operations on corresponding arguments are defined for both the subtype and the supertype. They differ in the degree to which operations of the subtype may be undefined. Partial compatibility allows lesser definability for arguments in the domain of the subtype, subcomplete compatibility allows lesser definability only in the domain of the supertype, and complete compatibility requires equal definition for the common domain of the subtype and supertype.

4.2. Signature-Compatible Modification

When behavior is too difficult to specify precisely and completely we approximate it by a signature. In particular, type-checking algorithms usually define types in terms of signatures rather than behavior, and define the notion of subtype in terms of compile-time-checkable constraints on signatures rather than in terms of restrictions on or extensions of behavior.

A semantics-preserving signature-compatible modification is a signature-compatible modification where the attributes of the signature are guaranteed to preserve their semantics provided they remain defined. We define the notions of horizontal extension and vertical modification of signatures syntactically and then show that semantics-preserving horizontal extensions are behaviorally compatible while vertical extensions are not.

4.2.1. Horizontal Extension

A type Person1 with a name attribute of type "String" may be defined as follows:

type Person1 = (name: String);

Person is a horizontal extension Person1 with an attribute "age":

type Person = (name: String, age: Int(0..120))

Horizontal extensions are subtypes that specialize the parent type and have a richer set of attributes than the parent type. The set of objects possessing the richer set of attributes is a subset of the set of objects of the parent type. An increase in the richness of attributes goes hand in hand with a decrease in the number of objects possessing the attributes.

4.2.2. Vertical Modification

The type Person can be vertically extended to the type Retiree as follows:

type Retiree = (name: String, age: Int(65..120);

This vertical extension specializes the name component of Person to a subset of its domain. Checking that a type defined by inheritance is a vertical extension of the type from which it inherits is more difficult than checking for horizontal extension and may in general require dynamic rather than static checking.

Vertical and horizontal extension can be combined:

Let RT1 = (s1: T1)
and RT2 = (s1: T11, s2: T2)

then RT2 is a subtype of RT1
if T11 is a subtype of T1

Vertical modification is particularly common in database systems, although the concept is certainly of more general applicability. The semantic data model (SDM) [HM81] contains a rich subtype definition mechanism that is largely based on this type of relationship. The predicate-defined subclass in the SDM is an example of exactly this type of definition.

Notice that this type of definition allows the type of an object to change as the values of the attributes on which the predicate depends change. For example a person whose age increases beyond 65 suddenly becomes a Retiree.

4.2.3. The Principle of Substitutability

The definition of complete compatibility is motivated by the following principle:

Principle of substitutability: An instance of a subtype can always be used in any context in which an instance of a supertype was expected.

Of our three previously defined notions of behavioral compatibility, only complete compatibility guarantees substitutability. Complete compatibility is realized by semantics-preserving horizontal extension. The following example illustrates that, however, semantics-preserving vertical modification does not guarantee substitutability:

p: Person;
r: Retiree;

p := r;
set-age (p, 40);

The operator "set-age" works as long as the Retiree type is not defined as a subtype of Person. As soon as this relationship is established, the set-age operation given above fails. The assignment of a Retiree to a variable of type Person (which must be allowed if we wish to maintain the principle of substitutability) has caused a problem: we cannot set the retiree's age to 40. Adding the subtype has contradicted an implicit assertion made in the definition of the type Person. The type Person originally said that the age of a Person can be set to any integer value between 0 and 100. By adding the type Retiree as a subtype, we have said that there are now some persons for whom setting their age to any value between 0 and 64 is illegal.

No restriction of the domain can ever satisfy the principle of substitutability since assignment of a subtype value that is not a supertype value to a supertype variable violates the principle.

4.2.4. Subsets Are Not Subtypes

Should the term "subtype" be used only for subtypes that satisfy the principle of substitutability, or should it be used more broadly to include subset subtypes and isomorphically embedded subtypes? If we restrict the term to its narrow meaning, then subsets are not subtypes. That is, subsets such as Retiree defined by domain constraints are not subtypes because they do not satisfy the principle of substitutability. We say that the relationship between Person and Retiree is not subtype but a subset relationship.

Although the subset relationship does not conform to the principle of substitutability. it can be used for code reusability. If B is-a-subset-of A and A defines a piece of behavior (e.g., a method) that is not defined in B, this behavior can be inherited by B.

The requirement of substitutability and the associated notion of subtype and behavioral compatibility is too strong in many practical situations. It is the subset relation rather than the subtype relation that corresponds to classical is-a relation. The fact that a Retiree is always a Person is in most contexts more important than the fact that certain operations on persons (those appropriate for young persons) are inappropriate when performed on retirees. The restriction that Retirees be behaviorally indistinguishable from young persons is clearly unreasonable. The requirement of substitutability implies such indistinguishability and is therefore unreasonable too. However, for subtypes where such indistinguishability is reasonable, substitutability allows useful compile-time optimizations as well as the ability to evolve code smoothly by adding subtypes that do not break existing programs.

In practice the following weaker substitution principle is often adequate as a basis for behavioral compatibility:

Principle of read-only substitutability: An instance of a subtype can always be used in read-only mode in any context a supertype is expected.

Retirees are read-only substitutable for persons since attributes of retirees are always valid person attributes. Subtypes that are read-only substitutable may be freely used to create new instances, for comparison and discrimination, and in any other supertype context that does not involve assignment.

Note that subset subtypes as well as isomorphically embedded subtypes are read-only substitutable in the context of supertypes. Partial and subcomplete compatibility are both sufficient to ensure read-only substitutability of a subset in the context of the supertype.

4.3. Name-Compatible Modification

Name-compatible modification requires only the name and not the signature of the parent type to be preserved in the result. For strongly-typed languages name compatibility is a more permissive incremental modification mechanism than signature compatibility which allows the type as well as the behavior of arguments to be replaced. However, for non-strongly-typed languages, expressions need not have signatures and name compatibility is the only applicable mechanism.

Whereas signature-compatible modification is specified in terms of predicates for type recognition, name-compatible modification is generally specified in terms of search algorithms for names in a model of implementation:

```
procedure search (name, module)
  if (name = localname) then do localaction
  else if (inherited module = nil) then undefinedname
  else search(name, inheritedmodule)
```

This search procedure describes the basic operational mechanism of searching for a local name and following the inheritance chain if no local name exists. It may be used to implement signature compatibility by introducing a compile-time check which filters out signature incompatible modifications, ensuring that only signature-compatible hierarchies are acted on by the search algorithm at execution time. Behaviorally compatible change may in principle also be implemented by filtering out inadmissible modifications, but checks for behavioral compatibility cannot in general be effectively performed, either at compile time or execution time.

Name compatibility is the simplest form of incremental modification in practical programming languages because it requires no extra checking at compile-time or execution time.

The relation between behavioral, signature, and name compatibility may be summarized by the following characterization of their underlying models:

behavioral compatibility: algebraic and axiomatic models
signature compatibility: inclusion relations on domains
name compatibility: model of implementation, search algorithm

If we have to choose a single model of incremental modification then name compatibility may well be the best candidate because because it can be flexibly constrained to realize more restrictive forms of compatibility. The models that underlie behavioral and name compatibility cannot be as easily adapted for other purposes.

4.4. Incremental Cancellation

Inheritance with cancellation allows the modifier M to delete as well as modify attributes of the parent P. Attribute deletion is the inverse of horizontal extension and causes the result class to be a generalization of the parent class. It is a symmetrical form of incremental modification that does not distinguish between the creation of subtypes and supertypes. It is the most radical incremental change mechanism specified by inheritance. Examples of attribute deletion include the following:

(1) attribute deletion: The types Ageless-Person or Nameless-Person can be defined by respectively cancelling the attributes age or name from the type Person. Deletion of an attribute may occur when a new attribute causes an already existing attribute to become redundant. For example, adding social security to the type Person causes the name attribute to become redundant so it can be deleted.

(2) exceptions: Sets with exceptions, such as birds which cannot fly, can be handled by the mechanism of cancellation. This is done by defining a basic set of birds which can fly that excludes penguins and ostriches and adding an exception class of non-flying birds by cancellation of the fly attribute. Thus exceptions to be added to a uniformly-defined set may be specified by cancellation of attributes.

Since attributes of a class may refer to or invoke each other, deletion of an attribute may cause a problem if another attribute refers to it. The deleted attribute becomes a virtual attribute that must be defined in a subclass to provide a self-contained object interface.

The search algorithm for name compatibility can easily be adapted to cancellation by adding a test for cancelled attributes.

if (name = cancelledname) then undefined

4.4.1. Exceptions

Exception mechanisms in programming languages identify and handle undefined or abnormal arguments of a procedure, or elements of a class. We are here concerned only with the identification of exceptions, which is a form of incremental modification, and not with exception handling, which involves language design issues beyond the scope of this paper.

Exceptions may enlarge a class by including exceptions in a class even though they do not satisfy the class specification or, alternatively, shrink a class by excluding exceptions even though they do satisfy the class specification. Exceptions that enlarge a class will be called additive exceptions while exceptions that shrink a class will be called subtractive exceptions, as in Figure 6.

Additive exceptions extend a class to elements that do not satisfy the constraints of class membership. For example, [Bo] has a real-estate class for houses priced under $200,000 that is extended to higher-priced houses by exceptions. [To] views birds as mammals that fly, and extends this class to "birds that can't fly" like ostriches and penguins by exceptions. The base class is extended by admitting classes or individual elements that violate the class constraint.

Cancellation of an attribute may be viewed as the limiting case of broadening a constraint so it is always satisfied. Finer granularity for subclass extension is obtained by allowing constraints to be modified rather than altogether eliminated. An operational mechanism for handling additive exceptions by relaxation of constraints is developed in [Bo] for both exception subclasses and exception instances.

68

Subtractive exceptions define new classes by exclusion of classes (or individual elements) from a base class. For example, the class Voter may be defined from the base class Person by a variety of exclusionary exception classes like minors, felons etc. The class Retiree could be defined from the base class Person by an subtractive exception class Nonretirees.

Class exceptions provide a general framework for modeling incremental modification. Additive and subtractive incremental modification may be defined in terms of additive and subtractive class exceptions. Traditional inheritance has emphasized subtractive incremental modification while class exceptions have emphasized additive incremental modification. However, our perspective allows us to identify these two mechanisms for the management of change.

Mechanisms for additive and subtractive incremental modification of sets arise in other disciplines. For example, type-1 and type-2 errors in statistics, corresponding to errors of omission and commission, can be modeled by additive and subtractive exceptions. In logic, incompleteness may be mitigated by additive exceptions for true but unprovable formulae, while unsoundness may be mitigated by subtractive exceptions for provable but untrue formulae.

Figure 6 may be interpreted in the domain of logic. The formulae of a formal system have interpretations in which theorems denote objects in a possible world. The axioms determine a class specification and the theorems derivable from the axioms determine the elements of the class. Exceptions arise when the set of provable (valid) formulae do not correspond to the set of derivable (true) formulae. When the derivable formulae are a subset of the true formulae (incompleteness) additive exceptions may be used to augment the set of valid formulae. For example, the class of all birds is obtained from flying birds by augmenting it with the class of non-flying birds. When not all derivable formulae are true (unsoundness) subtractive exceptions may be used to eliminate provable untrue formulae. For example, the class of voters is obtained from the class of persons by eliminating minors and felons.

Cancellation of attributes may be used for specifying exceptions but is by itself a relatively inflexible mechanism. Adjusting constraints on given attributes by either weakening or strengthening them is more flexible (has finer granularity). Weakening of attribute constraints enlarges the associated class and is realized by additive exceptions, while strengthening attribute constraints determines a subclass and is realized by subtractive exceptions.

5. Abstract Inheritance and Nonmonotonicity

5.1. Inheritance as an Abstract Relation

We examine the relation between programming language inheritance and the abstract concept of inheritance developed by Brachman [Br1, Br2], Touretzky [To], Etherington [Et],

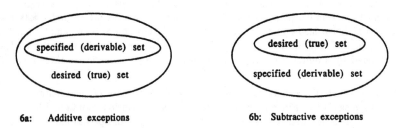

6a: Additive exceptions   6b: Subtractive exceptions

Figure 6: Additive and Subtractive Exceptions

and Reiter [Re] for semantic networks.

Programming language inheritance is an incremental modification operation on record structures with overlapping inheritable attributes. It is an imperative mechanism for creating new classes and associated objects during system evolution.

In contrast, abstract inheritance is a declarative relation among abstract entities in an inheritance network. Nodes of an inheritance network have no internal structure. The only structure is that determined by edges between neighboring nodes. Inheritance allows global relations among non-neighbors to be inferred from local relations among neighbors.

The simplest inheritance networks are tree structures with just a single kind of edge called an is-a relation. The properties of inheritance in such a network are simply the axioms of partial ordering, namely reflexivity, antisymmetry and transitivity. They allow inferences about inheritance between ancestors and descendants to be inferred from knowledge of inheritance between parents and children.

Bipolar inheritance networks with two kinds of links, called is-a and is-not links, are more expressive than unipolar nets. The semantics of bipolar nets may be characterized by the following axioms, where p, q, r represent classes, a represents instances, and x represents classes or instances [THT1]:

reflexivity of is-a: p is-a p
transitivity of is-a: p is-a q and q is-a r implies p is-a r
symmetry of is-not: p is-not q implies q is-not p
backward is-not propagation: x is-a q and q is-not r implies x is-not r
is-not inference for instances: p is-a q and a is-not q implies a is-not p

Although nodes have no internal structure, there are two kinds of nodes (class and instance nodes). "a is-a p" represents class membership (set membership) while "p is-a q" represent a subclass (subset) relation.

Bipolar semantic nets have been used to model multiple inheritance. However, when different paths of a multiple inheritance net yield contradictory conclusions more subtle analysis is required [THT2]. For example, a semantic net with the following links yields a contradiction:

Nixon is-a Quaker
Nixon is a Republican
Forall(x) Quaker(x) is-a Pacifist(x)
   Quakers are pacifists
Forall(x) Republican(x) is-not Pacifist(x)
   Republicans are not pacifists

The doubt concerning whether Nixon is a pacifist is resolved in different ways by skeptical reasoners who give up easily, refusing to reason further about apparently contradictory conclusions, and credulous reasoners, who examine further consequences of both sides of a contradiction.

The apparent contradiction may be resolved by allowing classes to have exceptions. Nixon is clearly either a nonpacifist Quaker or a pacifist Republican. There is no way of deciding between these two alternatives on the basis of the given information, but we happen to know Nixon is a nonpacifist Quaker, and therefore an exception to the rule that Quakers are pacifists.

Exceptions may be handled by default logic [Re], which allows default reasoning that may have to be revised as more information becomes available. Default reasoning is non-monotonic since adding new facts may cause previously valid inferences to become invalid.

5.2. Nonmonotonicity

Monotonicity is a desirable property of incremental modification mechanisms that guarantees preservation of properties of a parent in an incrementally modified result. It arises in many different contexts.

Monotonic functions are functions over an ordered domain that preserve ordering of arguments, guaranteeing that larger arguments can never lead to a smaller result. Scott [Sc] interprets partial ordering over arguments that are functions in terms of their degree of definition, so that monotonicity becomes the property that a more defined function can never be transformed into a less defined result.

Monotonic inference is inference in which already proved theorems and facts can never become untrue. Classical logics are monotonic. Inference systems for abstract inheritance with exceptions are nonmonotonic because general (default) inferences about inheritance may later be negated by information about specific exceptions. For example, the general inference that Nixon is a pacifist because he is a Quaker must be revised when the specific information that Nixon is not a pacifist is added to the database. The general inference that ostriches are not birds because they do not fly is negated by the specific information that ostriches are non-flying birds.

In the context of inheritance, monotonicity is the condition that properties of a parent class are preserved in the result class. Our discussion of behavioral compatibility indicated that this condition was too strong for incremental software evolution. All forms of incremental modification other than complete behavioral compatibility are in fact nonmonotonic.

Nonmonotonic systems are untidy because we cannot assume that properties of classes or validity of inferences are preserved once they have been established. But in spite of this they are necessary because incremental system evolution is in practice nonmonotonic.

5.3. Structural Versus Abstract Inheritance

The emphasis in the present paper is on structural inheritance among entities whose internal structure consists of finite sets of attributes. From the abstract point of view, this corresponds to a particular model (interpretation) of abstract inheritance that imparts an attribute structure to abstract entities. Frames are another concrete model of abstract inheritance that is in many respects similar to the object-oriented interpretation.

Our concrete model of inheritance allows us to develop algorithms for implementing inheritance that depend on the specialized structure of the model. Our structural relations go beyond is-a and is-not relations to include a variety of incremental modification mechanisms which can be defined in a variety of interesting and useful ways in terms of the structure of inherited and inheriting entities. However, the abstract point of view complements the concrete structural view in exhibiting the fundamental abstract structure of inheritance and allowing us to see clearly the differences between inheritance in abstract structures and concrete computational structures.

Programming language inheritance hierarchies have is-a relations but have not felt a need for is-not relations, although such a relation could easily be defined operationally. Is-not is a potentially useful relation. For example, if we have "Chevy is-a Car" and "Toyota is-a Car", then the relation "Chevy is-not Toyota" tells us the important fact that Chevy and Toyota are disjoint sets. Inheritance hierarchies make no provision for expressing such information because they are viewed as imperative structure for finding inherited operations rather than declarative structures for expressing relations. The integration of declarative and imperative features of inheritance hierarchies deserves further study.

The nodes of a bipolar net may be interpreted as propositions with "x is-a p" asserting p(x) and "x is-not p" asserting not(p(x)). In the domain of programming languages nodes are interpreted as types. The correspondence between the interpretation of nodes as propositions and types has an analogue in the domain of constructive type theory [ML]. In the present context, just as in constructive type theory, the interpretation of entities as propositions is simply an abstraction of their interpretation as types. The basis for the abstraction arises from the strong relation between monotonic inference and behavior-preserving type inheritance.

6. Parametric (Generic) Types

Parametric types capture a form of type similarity fundamentally different from the incremental similarity captured by inheritance. Whereas inheritance allows adding, deleting, and modifying operations, parametric types have a fixed set of generic (polymorphic) operations that may be specialized in a uniform manner, often allowing the same code to be used for a wide range of specializations. The similarity among parametric types is called "parametric similarity" in contrast to the "incremental similarity" determined by inheritance.

We are here interested primarily in incremental similarity but examine its relation to parametric similarity because it is better understood and has been widely studied in the context of ML [Mi], CLU [LSAS], and Ada [DOD]. The differences between incremental and parametric similarity have been studied in [Me] and [CW] and are illustrated by the following comparison:

(1) The dynamic binding of "self" to different objects and types at execution time has no counterpart for generic types. Instantiation of a generic type to a particular non-generic type occurs at compile time, while specialization of an inherited type occurs at execution time by binding "self" to an invoking object. Generic types are conceptually instantiated at compile time by macro expansion or an equivalent technique, while inherited types are shared and may be modified by "blocking" or "method combination" at execution time

(2) Generically similar types specialize a fixed set of polymorphic operations, while incrementally similar types add, cancel, or modify operations. Parametric and incremental similarity are complementary mechanisms for type modification that play different roles in application programming.

(3) Parametric types must be instantiated to specific types before they can be used as templates for creating objects, and instantiated types cannot in general be further instantiated. Inherited types can generally serve as both ancestors for inheritance by descendant types and templates for instances, and incrementally defined types in an inheritance hierarchy can generally be further incremented.

The ability to add new operations may be expressed by an implicit formal parameter that may optionally be instantiated to an arbitrary descendant type that in turn has an implicit formal parameter for further incremental modification. "Self" is an additional implicit parameter that is instantiated at operation invocation time to the object on whose behalf an operation is invoked.

Thus parametric types have an explicit formal parameter that parameterizes a fixed set of parametric operations, while inherited types have an implicit parameter that serves to augment the set of operations when it is instantiated by a subtype in the context of an invocation of an object of the subtype. These two forms of parameterization represent complementary forms of variability. In particular, parametric types are useful in representing common structure and behavior for a range of types, such as that of stacks or lists of arbitrary type, while inherited types are useful in incrementally augmenting the set of operations for the purpose of specialization or evolution.

Generative similarity is useful for generating multiple instances of similar objects whose differences are captured by a parameter. Incremental similarity is useful in managing change that arises during evolution or exception handling. Generative similarity and the associated mechanism of parametric types are more uniform and have more developed mathematical models than incremental similarity. The relation between generic and incremental similarity has been studied in [Me] and [CW].

Parametric similarity was first studied by Strachey [Str]' and was the basis for the development of ML [Mi, Ha]. We view parametric polymorphism as a particular kind of type similarity. Looking at polymorphism in this way in turn suggests that other forms of type similarity may correspond to other forms of polymorphism. For example signature-compatible type similarity has been called "inclusion polymorphism" [CW]. Inheritance determines a more permissive form of polymorphism that we call "incremental polymorphism". Incremental polymorphism includes inclusion polymorphism (signature compatibility) as a special case.

Signature compatibility was examined in [Ca]. A framework for combining parametric polymorphism, signature compatibility, and data abstraction was developed in [CW]. Incremental similarity is more general than signature compatibility in that it captures behavioral as well as domain-range similarity, and forms of incremental change that violate signature compatibility in order to model exceptions and system evolution. Such generalization requires sacrificing static type checking but permits more adequate modeling of evolution in real-world applications.

Figure 7 illustrates the similarities and differences between parametric and incremental notions of type similarity. In both cases a common ancestor captures the similarity of a collection of descendants. But edges connecting ancestors to their descendants have an entirely different meaning in these two cases. For parametric types, edges connect a generic type to an instantiation for a particular parameter value. For inheritance hierarchies, edges connect ancestor types to descendants that incrementally modify the set of operations of the ancestor.

The relation between parametric and incremental similarity may be illustrated by thinking of operations of a type as instruments in an orchestra. Variations of a particular melody can be realized by changing the pitch and volume of a fixed set of instruments or by adding harmonies for new instruments. Variations with a fixed set of instruments correspond to

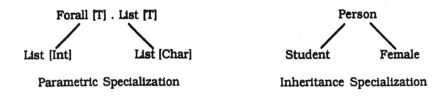

Figure 7: Common Ancestors Versus Parametric Types

parametric similarity while harmonies with new instruments correspond to incremental similarity. Mozart was able to achieve new musical effects by making use of the crescendo, a new form of musical technology that extended the range of parametric variation. Inheritance is an example of new computational technology that extends the range of incremental variation.

7. Liketypes

The four incremental modification mechanisms may be named as follows:

behavioral compatibility: R subtype P or R is-a P
signature compatibility: R subsig P
name compatibility: R subclass P
cancellation: R1 like R2

Like is the most general of the relations and subsumes all the others. Types related by a like relation may be called *liketypes* by analogy with the term subtypes for types related by a subset relation.

Whereas the subtype relation is asymmetrical, the like relation is symmetrical:

if "R1 like R2" then "R2 like R1".

Cancellation may be combined with the is-a relation to obtain a symmetrical relation strictly weaker than the general like relation which we call is-like [We2]:

R1 is-like R2 iff ((Exists P) [R1 is-a P and R2 is-a P])

This relation holds between common subtypes of a parent P. It is useful in modeling type evolution from a common parent. When we are given just the subtypes R1 and R2 then the problem of determination of the common ancestor P may be formulated as a unification problem [AN].

Since only is-a relations may obtain between components of R1 and R2 and their common ancestor P, the relation is-like is strictly weaker than the general like relation. Corresponding symmetric relations for signature and name compatibility may be defined:

R1 sig-like R2 iff ((Exists P) [R1 subsig P and R2 subsig P])
R1 name-like R2 iff ((Exists P) [R1 subclass P and R2 subclass P])

In many applications we are interested in formulating the relation between similar liketypes in terms of their differences rather than their similarities. For example, consider a military application where the type "tank" evolves through a sequence of small changes to its armor. Let T1, T2, T3 be three successive versions of the type "tank". Then T2 can be defined by a minor modification of T1 and T3 by a minor modification of type T2. We have "T1 like T2" and "T2 like T3", where the common properties C12 of T1 and T2 and C23 of T2 and T3 include practically all the properties of tanks, but are likely to be slightly different.

In these circumstances incremental evolutionary changes M12, M23 are more descriptive than common properties C12, C23:

T2 = T1 + M12
T3 = T2 + M23

74

The relation "R1 like R2" may be interpreted as an assertion that is true or false or as an incremental modification rule for specifying R2 = R1 + M in terms of R1. For purposes of application programming, the interpretation of "R1 like R2" as a rule for incrementally specifying R2 in terms of R1 is often more useful. Specification of R2 in terms of differences from R1 is often preferable to specification in terms of common attributes C.

8. Combining Like Relationships

Current languages generally select a single incremental subtyping mechanism that is built into its type system. However, each incremental type definition technique is methodologically useful. We therefore consider combining them all in a single programming language.

Such a language would have the ability to accept subtype definitions that establish any of the like relationships discussed here between a type T and a descendant type S. The type lattice could have freely mixed inheritance links representing any of the varieties of type similarity.

It is important to understand how such a heterogeneous type lattice affects our type-checking mechanism. We will do this by looking at the transitivity properties of like relationships. Consider first the interaction between is-a and name-like (abbreviated in Figure 8 to like) inheritance.

Suppose that "B name-like A", since it redefines the f method defined in A to be f'. This notation indicates that f and f' have the same name, but their bodies are different. Also, "C name-like B", since it redefines the g method that was introduced in B, as in Figure 8(a). The result is that C name-like A, since C inherits the modified f method from B.

Suppose that B name-like A by rewriting the f method, and C name-like B in that it adds one new method g. This is the case shown in Figure 8(b). Similar to the case above, A and C are related as C name-like A, since C inherits the redefined f method from B.

Suppose that B is-a A by adding one method and C name-like B in that it has the same methods but has redefined one of them, as in Figure 8(c). Thus, C is-a A because none of the methods on A have been modified in C.

From this we see that in order to determine the transitive relationship between A and C in situations like the above, we need to analyze the relationship between corresponding methods at each level of the lattice.

In the example below the four kinds of liketypes are abbreviated as in section 7. We also assume the function name(f) returns the name of the function f, and the function sig(f)

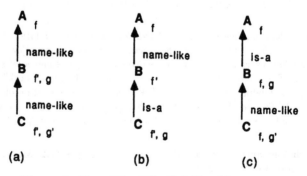

Figure 8. Transitive Like Relationships.

returns the signature of f.

The general rules for reasoning about the relationship between two types in the type lattice that are on a common path to the root can be stated as follows. If A supports the methods $f_{a1},..., f_{am}$, B supports the methods $f_{b1},..., f_{bn}$, and B is an indirect subtype of A, then

> **If** [For all f_{ai}] [There exists f_{bj}] $(f_{ai} = f_{bj})$ **then** B **is-a** A
> **else if** [For all f_{ai}] [There exists f_{bj}] (sig (f_{ai}) = sig (f_{bj})) **then** B **sig-like** A
> **else if** [For all f_{ai}] [There exists f_{bj}] (name (f_{ai}) = name (f_{bj})) **then** B **name-like** A
> **else** B **like** A

With this kind of analysis, we can determine how to handle type checking in our programming language. For example, consider the following piece of code:

```
x: A;
y: C;
x := y;
```

where C is an indirect subtype of A. In order to determine if the assignment is allowable, we can first apply the above rule to the types A and C to determine the effective relationship between them. We can then use the rules of the language for type compatibility between a type and its immediate subtypes to determine whether or not a particular assignment is legal.

Consider the example shown in Figure 9. The effective relationship between Vehicle and Toyota is name-like. If our language does not guarantee behavioral compatibility between types (as in Smalltalk), then the first assignment would be legal. Toyota supports all of the named behavior of Vehicle, allowing the substitution of a Toyota wherever a Vehicle was expected. It will always be possible to drive either, even though they might have different semantics.

The second assignment, however, would not be allowed in a language that guaranteed the kind of substitution mentioned in the previous paragraph. The rules tell us that the relationship between Vehicle and Car-Sculpture is a like. This means that we cannot use a Car-Sculpture wherever a Vehicle was expected, because someone might try to drive it.

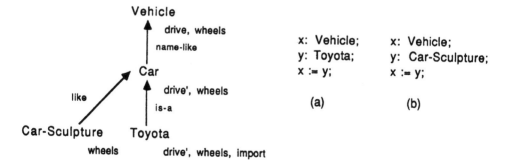

Figure 9. Example of type checking

For type checking on function application, we must simply match the compatibility rules for that function. That is, if we have a function f defined on type A and we try to apply f to a member of an indirect subtype B, we must ensure that the definition of f available on that subtype meets the compatibility rules of the language. If the language requires behavioral compatibility, then we must check that no type between A and B redefines f. Notice that the effective relationship between A and B might be a weak form of like; however, this is irrelevant since it is only the one function f that matters.

9. Conclusion

This paper contributes to the design of inheritance mechanisms by presenting major design alternatives and exploring the models, motivation, and methodology underlying each alternative.

We started with the objective of better understanding inheritance by classifying the varieties of incremental modification that are legitimate realizations of inheritance. This led us to a discussion of the essence of inheritance and of the distinction between types and classes. It led to a novel algebraic characterization of alternative forms of behavioral compatibility, to the analysis of horizontal and vertical signature compatibility, and to the distinction between additive and subtractive exceptions as a model of additive and subtractive incremental modification. It led to the distinction between concrete and abstract inheritance and between monotonic and nonmonotonic incremental modification. It led to an exploration of how multiple like mechanisms can be combined in a single language. The fact that this approach has yielded many interesting analyses suggests that viewing inheritance as an incremental modification mechanism is worthwhile.

10. Acknowledgements

The authors are indebted to William Cook for contributions to the section on inheritance and for the term "liketype", to Lynn Stein for contributions to the section on types, to Kim Bruce for contributions to the algebraic specification of behavioral compatibility, and to numerous other readers who have made helpful comments. This work was supported in part by NSF under contract DCR-8605567, by DARPA under order #4768, and by the IBM TJ Watson Research Center.

11. References

[AN] Ait Kaci H. and Nasr R., Login: A Logic Programming Language with Built-in Inheritance, Journal of Logic Programming, 1986.

[Bo] Borgida A., Exceptions in Object-Oriented Languages, Sigplan Notices, Oct 1986.

[Br1] Brachman, R., "I Lied About the Trees Or Defaults and Definitions in Knowledge Representation", AI Magazine, Fall, 1985.

[Br2] Brachman, R., "What Is-a is and Isn't", AI Magazine, Fall 1985.

[BW] Bruce K. B., and Wegner P., An Algebraic Model of Subtypes and Inheritance, Brown University Report, July 1987, also in Proc Roscoff Conference on Database Programming Languages, Sept 1987.

[Ca] Cardelli L., A Semantics of Multiple Inheritance, In LNCS Vol 173, Ed G. Kahn, 1984.

[CW] Cardelli, L. and Wegner, P., On Understanding, Types, Data Abstraction, and Polymorphism, ACM Computing Surveys, December, 1985.

[Co1] Cook, W., A Denotational Semantic Model of Inheritance, Forthcoming PhD Thesis, Brown University Summer 1988.

[Co2] Cook, W., The Semantics of Inheritance, Brown University Technical Report, March 1988.

[DOD] Ada Reference Manual, US Dept of Defense, July 1980.

[Et] Etherington D. W., Formalizing Nonmonotonic Reasoning, Artificial Intelligence, 1987.

[GR] Goldberg, A. and Robson, Smalltalk80: The Language and Its Implementation, Addison-Wesley, 1983.

[Ha] Harper R., An Introduction to ML, Edinburgh University Technical Report, 1987.

[LSAS] Liskov B., Snyder A., Atkinson R., and Schaffert C., Abstraction Mechanisms in CLU, CACM, August 1987.

[Me] Meyer B., Object-Oriented Software Construction, Prentice Hall, 1988.

[Mi] Milner R., A Proposal for Standard ML, Proc Symposium on Lisp and Functional Programming, ACM, August 1984.

[ML] Martin-Lof P., Constructive Mathematics and Computer Programming, in Mathematical Logic and Computer Programming, Hoare and Shepherdson Eds, Prentice Hall International, 1985.

[Re] Reiter R., A Logic for Default Reasoning, Artificial Intelligence, 1980.

[Sc] Scott D., Data Types as Lattices, Siam Journal of Computing, September 1976.

[Sn] Snyder, A., Encapsulation and Inheritance in Object-Oriented Languages, OOPSLA, 1986.

[Str] Strachey C., Fundamental Concepts in Programming Languages, lecture Notes for International Summer School in Computer Programming, Copenhagen, August 1967.

[THT1] Thomason, R. H., Horty, J. F., and Touretzky, D., A Calculus for Inheritance in Monotonic Semantic Nets, CMU-CS-86-138, July 1986.

[THT2] Touretzky, D. S., Horty, J. F., and Thomasson R. H., A Clash of Intuitions: The Current State of Nonmonotonic Multiple Inheritance Systems, Proc IJCAI-87.

[To] Touretzky, D., The Mathematical Theory of Inheritance, Morgan-Kaufman, 1986.

[We1] Wegner, P., Object-Oriented Concept Hierarchies, Brown University Technical Report, May 1988.

[We2] Wegner, P., The Object-Oriented Classification Paradigm, in Research Directions in Object-Oriented Programming, Edited by Shriver and Wegner, MIT Press, 1987.

[Zd] Zdonik S. B., Can Objects Change Type? Can Type Objects Change?, Proc Roscoff Conference on Database Programming Languages, Sept 1987.

GSBL: An Algebraic Specification Language Based on Inheritance

S. Clerici & F. Orejas
Facultat d'Informàtica, Universitat Politècnica de Catalunya
Pau Gargallo 5, (08028) Barcelona, SPAIN

Abstract

At the specification level, inheritance can be defined as subtyping by means of order sorted specifications [GM85]. Subtyping is, obviously, a very important notion, allowing not only to work with a non rigid type structure, but also providing an adequate basis for error handling in algebraic specifications. However, in our opinion, subtyping and order sorted specifications do not play the same rôle as inheritance in program design. In this paper, we will present a hierarchical organization for specifications, based on a different concept of inheritance which, we think, corresponds, methodologically, to the usual inheritance relation defined at the programming level. This new relation allows to work with *incomplete* specifications with several levels of detail and, as a side-effect, it may play the rôle of genericity. The use of this notion of inheritance is shown by means of the GSBL specification language built around this new concept, whose use and formal semantics are sketched.

Key words and phrases: Algebraic specification, inheritance, genericity, specification languages.

1. Introduction

Inheritance in object oriented programming languages can be defined as a "specialization" relation among classes. This specialization relation may take different forms. Conceptually, inheritance may be seen as subtyping. For instance the class of natural numbers may be considered to be a subclass of the integers. Indeed, the formal semantics of inheritance has been given in terms of subtyping [CAR84].

However, from a methodological standpoint, inheritance in program design is more than subtyping. On one hand, implementation (in the sense given in the abstract data type literature [EKMP82]) relations are usually considered as inheritance. For instance, in [MEY87] binary search trees are considered a subclass of tables. On the other hand, inheritance plays a rôle very similar to genericity in languages like Ada. The definition of a class of *lists* having as a client a class of *values* may be seen as equivalent [MEY86] to the definition of a generic package of lists in Ada. The instantiation of generic parameters would cause the same effect as defining lists over any subclass of values.

These two aspects make of inheritance a major design concept, since it provides the basis for both top-down, by deferring the description of some aspects of a superclass that will later be refined in a subclass, and bottom-up object-oriented design, by facilitating the reuse of objects through the hierarchical organization provided by the inheritance and client relations.

At the specification level, inheritance can be defined as subtyping by means of order sorted specifications

[GM85]. Subtyping is a very important notion, allowing not only to work with a non rigid type structure, but also providing an adequate basis for error handling in algebraic specifications. However, in our opinion, subtyping and order sorted specifications do not play the same rôle as inheritance in program design. In this paper, we present a hierarchical organization for specifications, based on a different concept of inheritance which, methodologically, corresponds better to the usual inheritance relation defined at the programming level.

Specification design is a process consisting in building a formal and complete description of a problem (and, possibly, of its intended solution) from an informal, often incomplete and sometimes contradictory set of requirements. Therefore, during this process, the specifier, by interacting with the customer, would have to detect the possible contradictions and to add the necessary detail to make the final specification complete.

As a consequence, specification design cannot be seen as a "linear" process as it is, simplistically, often seen, i.e. a process in which, at every time we have a full description of part of the problem that we, continuously, enlarge until the full problem is specified. Rather, when designing a specification, at any moment we have to deal with specifications describing only partially some aspects of the problem. Also, in subsequent steps we may refine these specifications by *completing* them, or we may have to backtrack and redo part of the work.

Therefore, a language aimed at giving support to the specification design process, especially for building large specifications, should offer the possibility of dealing with incomplete specifications, describing partial aspects of a problem, and of refining these specifications by completing them. Also, it would have to facilitate the design of modular specifications and the organization of its components in a hierarchy reflecting the design process, in order to enhance comprehensibility, to simplify the modification of some of its components, to facilitate its reuse and to assure the correctness of the whole specification if every component is correct.

Algebraic specification languages cover, in some sense, most of these aspects. Modularity is inherent to such languages, built over the abstract data type concept. Several forms of hierarchical organization, based on a notion of refinement, have been proposed. Broy and Wirsing [WPPDB83] based hierarchical specifications over the concept of enrichment or extension. Burstall and Goguen [GB80] proposed a hierarchical organization based on a two-dimensional structure: horizontal refinements correspond, also, to enrichments, while vertical refinements correspond to implementations.

The possibility of dealing with "incomplete" specifications, although not explicitly stated, is also present in most specification languages. However, this is not quite true for the hierarchical organization associated to the refinement relation defined by adding detail, i.e. completing a specification. In our opinion, according to the design process we foresee, the kind of hierarchical organization needed for dealing with large specifications is precisely this one, together with the usual one based on the extension concept, that may be seen as playing the same rôle as the *client* relation in object oriented programming languages [MEY87].

Incomplete specifications may be seen, within the algebraic approach, as specifications with loose semantics, i.e. the class of models defined by the given specification is not an isomorphy class. In this sense, "complete" specifications would have an isomorphic semantics (for instance, initial) and completing (refining) a specification would mean restricting its associated class of models.

Within an incomplete specification some parts may be considered to be completely defined (for instance, the booleans may be considered fully defined within a larger and incomplete specification). The semantics of specifications in which some parts are completely defined may be stated in terms of *constraints* [REI80, BG80, SAN81, EWT82].

As it happens at the programming level, genericity is embedded within our inheritance notion. Parameterized specifications may be seen as incomplete specifications with a completely defined part. For instance, *Lists_of(values)* can be seen as a specification in which the parameter, *values*, is only sketched, while *lists* (once a given set of values is given) is completely defined. In this sense, parameter passing may be seen as a refinement of the parameterized specification, since the result is "more complete".

However, the converse is not true. The use of parameterized specifications for dealing with incomplete specifications is limited. Specifically, we could have problems if we would want to write a specification of *Lists_of(values)* with an operation *choose*, that we have not decided yet which element out of a list must select: since this operation needs to be considered incompletely defined, it would have to be in the parameter specification. However, this would be technically impossible, since the resulting signature of the parameter specification would be syntactically incorrect.

In the rest of the paper, we will introduce a specification language built around this notion of inheritance, and show some small examples of its use. Also, all this concepts will be made more precise by sketching the formal semantics of the language.

2. Overview of GSBL

GSBL is an algebraic specification language based on the ideas expressed above. That is, in GSBL specifications may be seen as having some sorts and operations not completely defined. From an object-oriented point of view, these sorts and operations correspond to the generic attributes of the object, and the sorts and operations completely defined to the fixed ones.

Then new objects may be defined from old ones in two ways: by extending previously defined specifications, then we say that the new specification is defined **over** the old ones, and by consistently redefining some "incomplete" part of an old specification, then we say that the new specification is a subclass of the old one, since we may consider the former a specialization of the latter. We consider, then, the objects of GSBL hierarchically organized by this two relations. Moreover, if a class A is defined over a class B and C is a subclass of B, then a new class D, obtained by substituting, in A, B by C, is considered implicitly defined. D is considered to be over C and subclass of A.

Being more specific, the language mechanism for creating a new object is the **class definition** that has the following scheme (all the clauses are optional).

```
CLASS  Class name

    OVER  < overlist >                              --over clause

    SUBCLASS-OF  < subclasslist >                   --subclass clause

    WITH    SORTS  < sortlist >                     --with clause
            OPS  < opslist >
            EQS  < varlist > < equationlist >

    DEFINE  SORTS  < sortlist >                     --define clause
            OPS  < opslist >
            .EQS  < varlist > < equationlist >
END-CLASS
```

For instance, the following specification describes any ordered set of values.

```
CLASS  With-order
OVER Boolean
WITH  SORTS With-order
      OPS
            _ = _ ,  _ ≤ _ : With-order  ×  With-order  →  Boolean
```

```
EQS   { a, b, c: With-order }
       a  == a  = true
       a  == b  ∨ ¬( a  == b ) = true
       a  == b  ⇒ b  == a  = true
      ( a  ≤  b )  ∨  ( b  ≤  a ) = true
      ( a  ≤ b  ∧  b  ≤ c )  ⇒ a  ≤ c  = true
      ( a  ≤  b )  ∧  ( b  ≤  a ) =  ( a  == b )

DEFINE  OPS   _≥_ ,  _<_ ,  _>_   : With-order  ×  With-order  →  Boolean
        EQS   { a, b : With-order }
               a  ≥  b = b  ≤   a
               a  >  b = ¬( a  ≤  b )
               a  <  b  =  b  >  a

END-CLASS
```

The *over* clause imports the Boolean specification and establishes an **over** relation between the two specifications. The *with* clause declares the new operations and sorts, introduced by the specification, which are not completely defined. Finally, the *define* clause presents the completely defined sorts and operations.

Hence, the specification for the class *With-order* (in addition to the Boolean subspecification) will have, then, a new sort, with its very name, that is considered to be not completely defined. Also, the specification introduces two new infix operations $==$ *and* $≤$ with equations that express several properties such us symmetry, transitivity, reflexivity, but, again, without a complete definition. On the contrary the new operations $≥, <$ *and* $>$ are considered to be fully defined by the equations. Of course, this definition may be seen as depending from the precise definition of $==$ *and* $≤$.

Intuitively, we may consider the signature Σ of the specification SP for a class A as being the union of two pairs $W = <S_w, Op_w>$ and $D = <S_d, Op_d>$ corresponding to the **non completely defined** and **completely defined** sorts and operations of A, respectively. In the example, we have that *Boolean* is a basic class, with all its sorts and operations completely defined. Then, the whole resulting signature of the *With-Order* specification is $\Sigma_{With-order} = W \cup D$ where

$$W = < \{With\text{-}order\}, \{==, ≤\}> \quad \text{and} \quad D = \Sigma_{Boolean} \cup < \varnothing, \{≥, <, >\}>$$

Note that neither W nor D need to be signatures, because the arity of their operations may involve sorts belonging only to the other part.

Formally, the completely defined parts are **constraints** [REI80, BG80, SAN81, EWT82]. A constraint, as we will see in the following section, may be seen as the complete definition of some sorts and operations in terms of others. In the example, the *With-Order* specification defines two constraints: the first one is the definition of booleans and the second one is the definition of $≥, <$ *and* $>$. That is, every *define* clause establishes a constraint within the specification defining the sorts and operations, declared in the clause, in terms of the rest of the specification.

For example, *Natural* can be defined as an instance of *With-order* using the *subclass* clause as follows.

```
CLASS Natural
   SUBCLASS-OF  With-order
   DEFINE   SORTS Natural
            OPS   == , ≤
                  0 :  → Natural
                  succ : Natural  → Natural
            EQS   { a, b : Natural }
                  0  == succ(a) = false
                  succ(a)  == succ(b)  =  a  == b
                  0  ≤ succ(a) = true
                  succ(a)  ≤ succ(b)  =  a  ≤ b
END-CLASS
```

Natural has been constructed as subclass of *With-order*, the first one **inherites** the whole specification of the second one, with the **implicit renaming** of *With-order* by *Natural*. Hence, note that, to write the new specification, it has only been necessary to declare which sorts and operations, from the W part of the superclass, are now completely defined (namely *Natural*, $=$ and $\le$), and to add the necessary equations and additional operations (*0* and *succ*). That is, *Natural* has refined the specification *With-order* by completing it. Let us observe that, in the resulting specification for *Natural*, the W part will be empty, since all sorts and operations are now completely defined.

Within a specification, it is reasonable to think that the completely defined parts should be "protected" with respect to the rest of the specification, i.e. a specification is considered correct if the whole specification is *consistent* with respect to the constraints and if every constraint is *sufficiently complete* with respect to the other constraints. For instance, we should not allow an equation in the previous specification stating *true = false*, nor a completely defined term generating a *junk* value (not equivalent to *true* or *false*) of sort bool. Otherwise, probably, the set of models of the specification would be empty.

As a consequence, we ask all equations introduced by a new specification, either in the *with* or *define* clauses to be consistent with previous constraints, either coming from an *over* or a *subclass* clause. Also the operations introduced in the *define* clause should not introduce junk on any previous constraint. Finally, for methodological reasons, we also ask for consistency of new equations with respect to any subspecification declared in the *over* clause, being completely defined or not, since, otherwise, the *over* relation would not be a true extension relation. This means that our specifications extend the specifications declared in the *over* clause with a mixture of the *protecting* and *extending* enrichments of OBJ2.

In the previous examples, the *over* and the *subclass* clause have been used as a shorthand for writing specifications. However, as it has been already mentioned, this is not their only rôle. In fact, we can declare an already existing specification as subclass of another one just to establish this connection between them. Then, in this case, subclass declaration is similar to a **view** declaration in OBJ2. Indeed, this clauses are also the declarations that serve to define the over and subclass relations which are the basis for the hierarchical organization of specifications advocated in this paper. For instance, in the example, *Natural* is **subclass-related** to *With-order* and **over-related** to *Boolean*.

A class may be **subclass-related** to several other classes, i.e. **subclass** is a multiple inheritance relation. Furthermore, a class may be **subclass-related** to another one in more than one way. For example, we may want to define two different ways for ordering sequences of naturals, say comparing only the first elements, or perhaps comparing the length. We may construct this new class in the following way:

> *CLASS Natseq-with-order*
> *SUBCLASS-OF Natsequence,*
> > *With-order [eqfirst: $=$; l.e.first: $\le$; l.first: $<$; g.e.first: $\ge$; g.first: $>$] ,*
> > *With-order [eqleng: $=$; l.e.leng: $\le$; l.leng: $<$; g.e.leng: $\ge$; g.leng: $>$]*

....

The rest of the specification would include the equations (and, possibly, the new operations) needed to complete the definition of the inherited and renamed (by means of the renaming indicated inside the square brackets) operations.

On the other hand, two objects may be **over** and **subclass-related** at a time. For example, we may want to specify how to induce an order relation on a set by means of a function from this set to an already ordered one:

```
CLASS Orderable
    OVER Inductor : With-order
    SUBCLASS-OF With-order
    WITH OPS
        f : Orderable → Inductor
    DEFINE OPS   =, ≤
    EQS { a, b : Orderable }
        a  = b = f( a )  = f( b )
        a ≤ b = f( a ) ≤ f( b )
END-CLASS
```

It may be noted that in the resulting specification some operations are overloaded, namely $=, ≤, ≥, <$ *and* $>$ for *Inductor* and for *Orderable*, but, because of their arity, no confusion arises. If it does, a renaming for the *Orderable* operations would have to be made. In fact, *Orderable* has been declared over *Inductor:With-order* and not just over *With-order* to avoid name confusion.

The two relations are transitive, therefore if we would have had *Orderable* already defined, (and *Natstequence* had the operation *length*) we may have had constructed *Nqtseq-with-order* as follows

```
CLASS Natseq-with-order
SUBCLASS-OF Natsequence,
    Orderable [ Natural : Inductor; first: f; equfirst: = ; l.e.first: ≤ ; l.first: < ; g.e.first: ≥ ; g.first: > ] ,
    Orderable [ Natural : Inductor; length: f; equleng: =; l.e.leng: ≤ ; l.leng: < ; g.e.leng: ≥ ; g.leng: > ]
END-CLASS
```

Then, because of the transitivity of the subclass relation, *Natseq-with-order* will become a subclass of *With-order*, and thus it will be possible to define a new specification in which *Natseq-with-order* plays the rôle of *Inductor*, and so on.

In GSBL genericity is a consequence of the effects of the combination of the over and the subclass relations. As it was already said, if a class A is defined over a class B and C is a subclass of B, then a new class D, obtained by substituting, in A, B by C, is considered implicitly defined. Then, D is considered to be over C and subclass of A. This causes the same effect as parameter passing in usual algebraic specification languages. For instance, if a class *Ordered_sequences* is defined over *With-order*, this would, implicitly, cause the definition of a class of ordered sequences of naturals over *Naturals*.

In what follows, and to end this section, we will see a small example of generic programming [GOG84] using GSBL. Let us first consider a very frequent problem in programming, finding the maximum of a structure of elements having an order relation defined over them. Independently of the kind of structure, we only need a way for traversing it and examining its elements. Then, we first provide a specification of *Traversable* structures.

```
CLASS Traversable
    OVER   Boolean
           Element : ANY
    WITH   SORTS   Traversable
           OPS     first :   Traversable  →   Element
                   rest :    Traversable  →   Traversable
                   end? :    Traversable  →   Boolean
END-CLASS
```

where *ANY* is a predefined class whose specification is

```
CLASS ANY
    WITH SORTS ANY
END-CLASS
```

that may be seen as a superclass of any specification. For example the usual *stack of elements*, may be viewed as an instance of Traversable as follows:

Stack SUBCLASS-OF Traversable [top: first ; pop: rest ; emptystack? : end?]

Also, an example of trees as traversable structures may be seen in the appendix. Now, if we want to obtain the maximum of a traversable structure we may construct the following class:

```
CLASS    Maximizable
 OVER   Element : With-order
 SUBCLASS-OF Traversable
 DEFINE   OPS
               maxim : Maximizable    →  Element
               update: Maximizable   ×   Element   →  Element
          EQS   { m: Element;   x: Maximizable }
               maxim( x ) = update( rest( x ), first( x ) )
               update( x, m ) = IF end?( x )
                          THEN   m
                          ELSE  IF m < first( x )
                          THEN update( rest( x ), first( x ) )
                          ELSE update( rest( x ),  m )
     END-CLASS
```

Let us now suppose that the problem is to find the most useful object belonging to a set of tools, where tool is a generic type whose elements have a profit and a weight (see the full specification in the Appendix), assuming that the utility of such objects is defined by the quotient profit/weight. Then the new specification is:

```
CLASS  Useful-objects
 OVER  Object:Tool
 SUBCLASS-OF   Set [ Object : Element ] ,
                      Maximizable[   Object: Element;  the-most-useful: maxim
                                     any : first;   anyless: rest;   ∅ ?: end?   ]
     WHERE     Object SUBCLASS-OF Orderable[ Real: Inductor  ;  utility: f
                                     eq-useful: equiv ;   less-useful: <  ]
       DEFINE OPS   utility
               EQS    { a : Object }
               utility(a ) = profit(a) /  weight(a)
     END-WHERE
 END-CLASS
```

where the *WHERE* clause is a local declaration inside the *SUBCLASS* clause. It is used to induce an order relation over objects, so that *Object* is a subclass of *With-order*, and the declaration of *Useful-objects* as subclass of *Maximizable* is correct. A similar effect would have been achieved declaring, in a separate class definition, *Object* as subclass of *Orderable* and, then, *Useful-objects* over *Object*.

We could continue this example until arriving, for instance, to the specification of a solution of the knapsack problem by means of a greedy approach. The rest of the specification can be found in the Appendix.

To end, the following picture shows the connections among the specifications that we have presented up to now. The dashed lines denote the **over** relation, the arrows the **subclass-of** relation, and the full lines without head arrows specifications equivalent up to renaming. The different shapes of the boxes, used to represent classes, correspond to a graphic notation that is of no interest now.

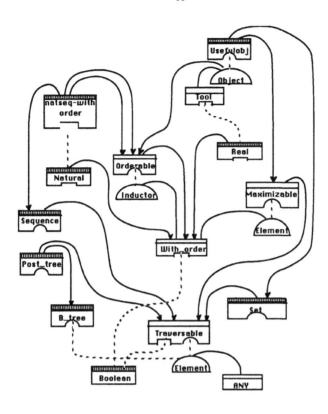

3. Semantics of GSBL

In this section we will sketch the ideas over which the formal semantics of GSBL is defined. First, we will introduce some basic definitions and results concerning algebraic specifications with constraints. Then, we will overview the formal semantics of the main constructions of the language. The reader is assumed to have some background on algebraic specification (for instance, [EM85]).

Definition 1

A **presentation** P is a tuple $< \Sigma, E >$, where Σ is a signature i.e. a pair $< S, Op >$ where S is a set of sorts and Op a set of operations, and E is a set of Σ-equations.

A **constraint** C on P is a pair of presentations (P_1, P'_1) such that $P_1 \subseteq P'_1$ and $P'_1 \subseteq P$. A constraint C is **persistent** iff for any P'_1-algebra A, we have $(A|_{P_1})|^{P'_1} \cong A$

Notation : $\_|_P$: $\text{Alg}_{P'} \longrightarrow \text{Alg}_P$ and $\_|^{P'}$: $\text{Alg}_P \longrightarrow \text{Alg}_{P'}$ denote, respectively, the forgetful functor and the free functor associated to the inclusion $P \subseteq P'$

A constraint, as we have already said, indicates that the P'_1 part of the specification P is completely defined, once the P_1 part is given. This may be made precise with the following definition:

Definition 2

An algebra $A \in \text{Alg}_P$ **satisfies** a constraint (P_1, P'_1) with $P'_1 \subseteq P$ iff $(A|_{P_1})|^{P'_1} \cong A|_{P'_1}$

This means that a P-algebra satisfies the constraint (P_1, P'_1) if the P'_1 part of the algebra is freely generated from the P_1 part. Technically, this is called an *initial or data constraint* because it "follows" the initial algebra semantics philosophy. In particular, if we want the models of a presentation to be an extension of a basic data type, for instance the booleans, then we would need to define a constraint, in which the first part is the empty specification and the second part is the boolean specification. As a consequence, an algebra would satisfy the constraint if it is an extension of the boolean initial algebra. The kind of specifications we will deal with are, therefore, specifications with constraints:

Definition 3
A **specification** SP is a pair $<P, \zeta>$ where P is a presentation and ζ is a family of constraints on P.

As we have said in the previous section, certain properties guarantying the "protection" of the constraints are needed to assure correctness of the specifications.

Definition 4
A presentation P is **compatible** with a constraint (P_1, P'_1), $P'_1 \subseteq P$ iff
$$\forall t_1, t_2 \in T'_1(X_1)_s, s \in S'_1 - S_1 \quad P \vdash t_1 = t_2 \Rightarrow P'_1 \vdash t_1 = t_2$$
A pair of constraints $C_1 = (P_1, P'_1), C_2 = (P_2, P'_2)$ are **mutually non destructive** if $\forall A \in Alg_{P1+P2}$ and for i = 1,2
$$(A|_{P_i})|^{P'_i} \cong (A|^{P'1+P'2})|_{P'_i}$$
A specification SP = $<P, \zeta>$ is **correct** if ζ is a family of persistent and mutually non destructive constraints and P is compatible with ζ.

Compatibility of constraints can be characterized proof-theoretically by means of consistency and sufficient completeness properties:

Theorem 5
Two constraints (P_1, P'_1) and (P_2, P'_2) are **mutually non destructive** iff for i=1,2 it holds:
   a. (**consistency**) $\forall s \in S'_i - S_i$ and $t_1, t_2 \in T_{\Sigma'_i}(X_i)_s$,
   $P'_1 + P'_2 \vdash t_1 = t_2 \Rightarrow P'_i \vdash t_1 = t_2$
   b. (**sufficient completeness**) $\forall s \in S'_i - S_i$ and $t \in T_{\Sigma'_1 + \Sigma'_2}(X_1 + X_2)_s$
   $\exists t' \in T_{\Sigma'_i + \Sigma 1 + \Sigma 2}(X_1 + X_2)_s$ such that $P'_1 + P'_2 \vdash t = t'$

Now, we may define the semantics of a specification:

Definition 6
The **semantics** of a specification SP is defined by the following class of models
$$Mod(SP) = \{ A \in Alg_P / P \text{ satisfies } \zeta \}$$

From now on, we will assume the correctness of all specifications we are dealing with, since that assures the existence of models. Moreover, it also assures the compatibility, between the syntactic and the semantic level, of certain constructions used in the definition of algebraic specification languages. We do not include these results since we feel that they are slightly out of the scope of this paper.

The over relation of our language is based on the following notion of extension:

Definition 7

A specification SP_1 is an **extension** of another specification SP_2 if

    a. $SP_2 \subseteq SP_1$

    b. $\forall\ t_1, t_2\ T_{\Sigma 2}(X_2)$, $SP_1 \vdash t_1 = t_2 \Rightarrow SP_2 \vdash t_1 = t_2$

We do not ask for sufficient completeness properties since they are implicit in the correctness of SP_1 and SP2. To define the subclass relation we will need a notion of specification morphism:

Definition 8

A **specification morphism** $f : <R_1, \zeta_1> \rightarrow <R_2, \zeta_2>$ is a presentation mophism $f : R_1 \rightarrow R_2$, such that $\forall\ A \in Alg_{P2}$ if A satisfies (P_2, P'_2) then $U_f\ (\ A \mid_{f(P'_1)}\ )$ satisfies (P_1, P'_1), where U_f is the forgetful functor associated to f.

The idea of this definition is that we ask SP_2 to contain "stronger" constraints than SP_1. Now, to end with this part, the following theorem is the basis for defining the result of the parameter passing-like mechanism of GSBL (combining the subclass and the over relation) and assuring its correctness.

Theorem 9

Let SP_2 be an extension of SP_1, let $f : SP_1 \rightarrow SP'_1$ be a specification morphism (wolog, see the following note, we assume that $(SP_2 - SP_1) \cap SP'_1 = \varnothing$). The result of substituting SP1 by SP1' in SP2 is the specification $SP'_2 = <P'_2, \zeta'_2>$ where $P'_2 = P'_1 + f'(P_2)$, $\zeta'_2 = \zeta'_1 + f(\zeta_2)$ and f' is the presentation morphism defined as extension of f in the usual pushout construction. Then we have:

    1. SP'_2 is a correct specification, i.e. ζ'_2 is a family of persistent and mutually non destructive constraints and P'_2 is compatible with respect to ζ'_2.

    2. f' is a specification morphism.

Note. If $(SP_2 - SP_1) \cap SP'_1 \neq \varnothing$ we consider that f' renames the common sorts and operations names.

Now, we may provide the semantics for GSBL. The basic notion for its definition is the concept of environment. An environment is a set of specifications related by extensions and morphisms defining the **over** and the **subclass-of** relations, respectively. Every language construction modify the environment adding new specifications and relationships among them. In particular, the effect of a correct *class definition* on the environment is obtained by the composition of the effects of all its constituent clauses.

    A *class definition* denotes a specification, whose presentation is obtained by putting together the presentations of the subspecification inherited by the *over* and *subclass* clauses and, then, including the sorts and operations declared in the *with* and *define* clauses. The constraints of this specification are, also, all the inherited constraints together with a new constraint (P_1, P_2), where P_2 is the whole presentation associated to the class definition and P_1 is P_2 minus the sorts, operations and equations of the *define* clause, in case there is one. Then, the effect of a class definition on the environment consists on adding to it the specification associated to the class, together with all the new relations declared in the *over* and *subclass* clauses. In addition, these clauses may contain some local declarations that also add to the environment new specifications and relations.

    Before defining more precisely the effect of all the clauses of a class declaration, let us first define the effect of a *class instantiation*, i.e. the substitution, as defined in GSBL, within a class C_1, of a class C_2 by a class C_3, when C_1 is over C_2 and C_3 is a subclass of C_2 via a specification morphism f. Substitutions of this kind may occur in a subclass declaration like:

Object SUBCLASS-OF Orderable [Real : Inductor; utility: f; eq-useful: ==; less-useful: <]

or within local declarations of the form:

Input: Traversable[Data:Element; select: first]

in the *over* and the *subclass* clauses. This definition, as it was already said, is based on theorem 9, but it is slightly more complicated. The causes for this additional complication are the following:

- The definition of the subclass C_3 may be implicitly given in the instantiation declaration. For instance, in the previous declaration Traversable is defined over *Element*, and, implicitly, a new class called *Data*, that maybe was not in the environment, is now being implicitly defined. In the example the new class *Data* would just be a renaming of the class *Element*.

- Also, the specification morphism establishing the subclass relationship between C_2 and C_3 may be implicitly given, partially or totally. In GSBL two elements (sorts, operations or classes) with the same name (and the same arity, in the case of operations, or the same components, in the case of classes) are identified. Therefore, there is no need to establish an explicit binding between two such elements. This has the advantage of an economy of writing. However, the price to be paid is a possible loss of correctness, since inconsistencies may be provoked. Later on, we shall again discuss this issue.

Now, the semantics of a class instantiation is the following. Given a class C over $C_1, ... , C_n$, and a binding b, defined by the list $[e_1: e'_1, ..., e_m: e'_m]$, where the e_i are either sorts, operations or names of classes, then if b is a correct binding for C in the current environment (i.e. if the resulting specification is correct), C [b] denotes the class obtained as follows:

- First the C_{over} subspecification is obtained from the disjoint union of $C_1, ... , C_n$ (i.e. if $C_1, ... , C_n$ contain elements with the same name they will not be yet identified). For instance, in the definition of *Input*, C_{over} would be the union of *Boolean* and *Element* (a renaming of *ANY*), since traversable is defined over these specifications.

- Next, the specification C'_{over} is defined as a subclass of C_{over} by renaming some parts of it consistently with the binding b. In the definition of *Input*, C'_{over} would be obtained renaming in C_{over} the sort *Element* by *Data* and the operation *first* by *select*.

- Then, applying the construction of theorem 9, we obtain the specification C'':

$$
\begin{array}{ccc}
C_{over} & \hookrightarrow & C \\
h \downarrow & & \downarrow \\
C'_{over} & \hookrightarrow & C''
\end{array}
$$

that is C'' would contain all the elements of C'_{over} plus the new sorts, operations, equations and constraints introduced by C.

- Finally, C[b] is obtained from C'' identifying all the elements with the same name. As it was said before, this identification can cause an inconsistency in the resulting specification. Then, one could think, a priori, that the correctness of the whole specification would have to be tested, with a consequent loss of modularity. But, in fact, if only subspecifications are identified the subclass relation guarantees correctness. And when isolated sorts and operations are also separately identified, it is possible to find conditions in which some kind of local consistency, between the affected subspecifications, is sufficient to ensure that the resulting class C [*b*] is a correct specification.

Now, we may define the semantics of all the clauses of a class definition. The semantics of an *Over* clause of the form *OVER* $C_1, ... , C_n$ in the *class definition* for A, consists in the following modifications of the environment:

1. Constructing the A_{over} subspecification of A. Also, at this moment A will denote A_{over} in the environment

2. Adding to the environment the relationships A **over** C_i, and

3. If a class variable definition B: C[*b*] is inside the *over* clause, then it causes the creation of a local class B instantiating C[*b*]. It also has as effect the following local relationships betwen B and the classes related to C. For each class D such that C **over** D, B **over** *h* (D) being *h* the morphism defined by *b*, if C **subclass-of** D via *h'* then B **subclass-of** D via *h* ∘ *h'*

The semantics of a *subclass* clause of the form *SUBCLASS-OF* $C_1, ... , C_n$ in the *class definition* for A consists in the following modifications of the environment:

1. Updating the specification associated to A in the current environment, adding all the inherited elements

2. Adding to the environment the relationships A **subclass-of** C_i, and A **over** C' for each C' such that C_i **over** C'

A *where* clause inside the *subclass* clause, causes the same effect as adding to the environment the result of the local declarations inside the *where*, and, then, evaluating the rest of the *subclass* clause.

The semantics of the *with* and *define* clauses of the form *WITH* w.sortlist w.oplist w.equlist , DEFINE sortlist oplist equlist in the *class definition* for A consist in modifying the environment, by adding to the presentation associated to A, the sorts in w.*sortlist,* operations in w.*oplist,* and equations in w.*equlist* . They also add to the ζ constraints family associated to A a new constraint (A', A"), where A" is the new presentation associated to A, and A' is A" minus the sorts, operations and equations of the *define* clause, in case there is one.

For instance, the declaration of the class *Knapsack* (see the appendix) would cause the definition in the current environment of a new specification, obtained by adding the operations and equations from the *WITH* and the *DEFINE* clauses to the union of the specifications *Real, Tool, Pair* (included in the *OVER* clause) and *Sequence* (from the *SUBCLASS* clause), transformed according to the declared substitutions. Also, a new constraint will be added establishing the definition of *remaining-cap, performance* and *put-object* in terms of the rest of the presentation. Additionally, the new **over** and **subclass-of** relations introduced by the class *KNAPSACK* would be included in the environment.

4. Conclusion

In this paper, a new approach for specification design has been introduced. The main idea consists in dealing with incomplete specifications by means of *constraints* and in introducing a new subclass relation, with the meaning of "is more detailed than", allowing for true inheritance in specification design. In particular, this new notion of inheritance embeds the concept of genericity as it is usually understood in algebraic specification.

Based in this subclass relation, the specification language GSBL has been defined, showing its use in specification design by sketching an example of parameterized programming [Gog84]. Moreover the idias used for defining its formal semantics have been introduced characterizing the correctness criteria that specifications must fulfil.

ACKNOWLEDGEMENTS

This work has been partially supported by Comisión Asesora de Investigación (ref. 2704-83).

4. References

[BG77]   Burstall, R.M.; Goguen, J.A. "Putting theories together to make specifications", Proc. V IJCAI, Cambridge Mass., 1977, pp. 1045-1058.

[BG80]   Burstall, R.M.; Goguen, J.A. "The semantics of Clear, a specification language", Proc. Winter School on Abstract Software Specification, Springer LNCS 86, pp. 292-332, 1980.

[CAR84]   Cardelli, L. "The semantics of multiple inheritance", Proc. Colloquium on the Semantics of Data Types, Sophia-Antipolis, 1984.

[CW85]   Cardelli, L.; Wegner, P. "On understanding types, data abstraction and polymorphism", Computer Surveys 17, 4 (Dec. 1985), pp. 471-522.

[EKMP82]   Ehrig, H., Kreowski, H.-J., Mahr, B., Padawitz, P. "Algebraic implementation of abstract data types", Theoret. Comp. Sc. 20 (1982), pp. 209-263.

[EWT82]   Ehrig, H., Wagner, E.G. Thatcher, J.W. "Algebraic constraints for specifications and canonical form results", Institut für Software und Theoretische Informatik, T.U. Berlin Bericht Nr. 82-09, 1982.

[EM85]   Ehrig, H., Mahr, B. "Fundamentals of algebraic specification 1", EATCS Monographs on Theor. Comp. Sc., Springer-Verlag, 1985.

[FGJM85]   Futatsugi, K, Goguen, J.A., Jouannaud, J.-P., Meseguer, J., "Principles of OBJ2", Proc. 12th POPL, Austin 1985.

[GOG84]   Goguen, J.A. "Parameterized Programming", IEEE Trans. on Soft. Eng. SE10, 5 (Sept. 1984), pp. 528 - 543.

[GB80]   Goguen, J.A., Burstall, R.M. "CAT, a system for the structured elaboration of correct programs from structured specifications", Report CSL-118, Comp. Sc. Lab., SRI Int., 1980.

[GM85]   Goguen, J.A.; Meseguer, J. "Order-sorted algebra I: partial and overloaded operators, errors and inheritance", SRI Int., Comp. Sc. Lab. Rep., 1985.

[MEY86]   Meyer B. " Genericity versus Inheritance ", Proc. ACM conf. Object-Oriented Programming Syst, Languages, and Applications, ACM, New York, 1986, pp. 391-405

[MEY87]   Meyer B. "Reusability: The Case for Object-Oriented Design", IEEE Trans. Software Eng. March 1987, pp. 50-65.

[REI80]   Reichel, H. "Initially restricting algebraic theories", Proc. MFCS 80, Springer LNCS 88 (1980), pp. 504-514.

[SAN81]   Sannella, D. "A new semantics for Clear", Report CSR - 79 - 8, Univ. of Edinburgh, 1981.

[WPPDB83] Wirsing, M., Pepper, P., Partsch, H., Dosch, W., Broy, M. "On hierarchies of abstract data types", Acta Informatica 20 (1983), pp. 1-33.

APPENDIX

CLASS Postree
SUBCLASS-OF B-tree , Traversable [empty? : end?]
DEFINE
 OPS   first, rest
 EQS   { t_1 , t_2 : Postree   ; x : Element }

  first(maketree(t_1 , x, t_2)) = IF empty?(t_1) ∧ empty?(t_2)
                                THEN x
                                ELSE IF empty?(t_1)
                                          THEN first(t_2)
                                          ELSE first(t_1)

  rest(maketree(t_1 , x, t_2)) = IF empty?(t_1) ∧ empty?(t_2)
                                THEN t_1
                                ELSE IF empty?(t_1)
                                          THEN maketree(t_1 , x, rest(t_2))
                                          ELSE maketree(rest(t_1), x, t_2)

END-CLASS

CLASS B-tree
 OVER Boolean
       Element: Any
DEFINE
 SORTS B-tree

 OPS empty : → B-tree

      maketree: B-tree × Element × B-tree → B-tree

      empty ? : B-tree → Boolean

      left , right : B-tree → B-tree

      root : B-tree → Element

 EQS { t_1 , t_2 : B-tree   x : Element }

      left, right , root (empty) = ?

      left (maketree(t_1 , x, t_2)) = t_1

      right (maketree(t_1 , x, t_2)) = t_2

      root(maketree(t_1 , x, t_2)) = x

      empty ? (empty) = true

      empty ? (maketree(t_1 , x, t_2)) = false
END-CLASS

CLASS Greedy
 OVER Data , Solution: ANY
        Input : Traversable [Data: Element , select: first]
WITH
 OPS init: Input → Solution

      stop-cond.: Input × Solution → Boolean

      feasible: Solution × Data → Boolean

      unite: Solution × Data → Solution
DEFINE
 OPS solve: Input → Solution

      apply: Input × Solution → Solution

 EQS { c: Input; s: Solution }

  solve(c) = apply(c, init(c))

  apply(c, s) = IF stop-cond.(c, s) THEN s
            ELSE
              IF feasible(s, select(c))
              THEN apply(rest(c), unite(s, select(c)))
              ELSE apply(rest(c), s)
END-CLASS

```
CLASS   Tool
OVER    Real
WITH
  SORT   Tool

  OPS    profit: Tool  →  Real

         weight: Tool  →  Real
END-CLASS
```

```
CLASS  Knapsack
OVER  Real
          Object: Tool
          Obj-quant : Pair [ Object: Elem1; Real: Elem2;  * : makepair
                              which: selec1;  quant: selec2 ]
SUBCLASS-OF Sequence [ Obj-quant : Element ]
WITH
  OPS capacity:   →  Real
DEFINE
OPS  remainig-cap, performance:  Knapsack  →  Real
     put-object:  Knapsack  ×  Object  →  Knapsack
EQS  { o : Object;   c:  Real;  m: Knapsack  p: Obj-quant }
  remaining-cap( empty )= capacity
  remaining-cap( *  (m , *  ( o, c) ) =  remaining-cap( m )- weight (o) . c
  performance ( empty )= 0
  performance( *  (m , *  ( o, c ) ) =  performance( m ) + profit (o) . c
  put-object( m, o) = IF remainig-cap( m ) ≥ weight( o )
              THEN  put( * ( o, 1 ) ,m)
              ELSE put( * ( o, remaining-cap(m)/ weight(o) ) , m)
END-CLASS
```

```
CLASS Greedy-knapsack
OVER  Objects-set: Useful-Objects
      Optim-knap:  Knapsack
SUBCLASS-OF Greedy[ Objects-set: Input;  Optim-knap: Solution;  Object: Data;
                    empty-knapsack: init;   not-full: feasible;
                    put-object: unite;   fill-knapsack: solve ]
  WHERE
      Objects-set SUBCLASS-OF Traversable[ Object: Element;    the-most-useful: first
                                    remove.optimum : rest ;  ∅ ?: end?   ]
        DEFINE
        OPS  remove-optimum
        EQS  { x : Object-set }
        remove.optimum( x ) = remove( x, the-most-useful( x ) )
    END-WHERE
DEFINE
OPS  stop-cond;  empty-knapsack;   not-full
EQS  { x : Objects-set;  a, b : Object;   m: Optim-knap }
  stop-cond( x, m ) =  ∅ ?( x ) ∨ ( remaining-cap(m) ≡ 0 )
  empty-knapsack( x ) =  ∅
  not-full( m, a ) =  remaining-cap(m) > 0
END-CLASS
```

Name Collision in Multiple Classification Hierarchies

Jørgen Lindskov Knudsen

Computer Science Department, Aarhus University,

Ny Munkegade 116, DK-8000 Aarhus C, Denmark.

E-mail: jlk@daimi.dk

Abstract

Supporting multiple classification in object-oriented programming languages is the topic of discussion in this paper. Supporting multiple classification gives rise to one important question — namely the question of inheritance of attributes with identical names from multiple paths in the classification hierarchy. The problem is to decide how these multiple classification paths are reflected in the class being defined. One of the conclusions in this paper is, that by choosing strict and simple inheritance rules, one is excluding some particular usages of multiple classification. This leads to the notion of attribute-resolution at class definition, which means that the programmer in some cases is forced or allowed to resolve the potential ambiguity of the inherited names. The concept of attribute-resolution is managed through the identification of two conceptually different utilizations of specialization (unification and intersection), and two different attribute properties (plural and singleton) to guide the attribute-resolution.

Introduction

One of the vital issues when designing programming languages or software systems using the object-oriented perspective, is multiple classification. In this paper, we will restrict ourselves to deal with programming languages in the sense that our examples and our terminology are influenced by work done within programming language design. However, we will claim that the discussion is relevant in the design process of object-oriented systems, too. In fact, we find that a major part of the design of an object-oriented system is language design, bringing the following discussion into the realm of object-oriented system design. In the past there have been many proposals for programming language support for object-oriented programming with multiple classification. Some of the most notable proposals are the object-oriented extensions to Lisp: LOOPS[3] and FLAVORS[13], the proposal for multiple classification in the class hierarchy of Smalltalk-80[5], the ThingLab system[4], and the programming languages Galileo[1], Amber[6], Eiffel[10], and C++[11]. The problem of name collision has been dealt with very differently by these proposals. Some of the proposals treat name collision in the hierarchy as illegal; others treat name collisions as separate declarations of equal right, while others treat name collisions as specialization of the attribute. As it can be seen, no general agreement of the treatment of name collision has been reached yet. This paper will examine the underlying issues in order to reach a unified understanding of name collision and also to understand why there isn't *one right* treatment of name collision in class hierarchies with multiple classification. Underlying the discussion is an aim to solve as many name collisions as possible at compile-time, and to ensure the highest degree of polymorphism.

1 Discussion of Object-Oriented Programming and Multiple Inheritance

Object-oriented programming is one of the buzzwords of the eighties of which all agree without agreeing on what it is. In the following, I will shortly discuss my view of object-oriented programming to put this paper into perspective. In object-oriented programming we have shifted our attention from the program text onto the program execution. We look at the program execution as a physical model of some part of the real world. We want to structure both the program text and the actual program execution in a way that reflects this aspect of modeling. We want to be able to identify structures of the program as models of actual phenomena in the part of the world that we want to model.* This inspires us to examine the ways in which we as humans conceive and structure our knowledge of the phenomena of the world around us. Object-oriented programming is inspired by three different ways in which humans structure their knowledge. The first structuring mechanism is *classification*, where we identify that a number of different phenomena share some common characteristics. By classifying phenomena, we create concepts. A concept can be described by its name, its intension, and its extension. The *extension* of a concept is the phenomena that can be described by the concept, and the *intension* is a description of the properties which phenomena in the extension possess. Having identified some concepts, we use these concepts to create other concepts. This can be done in two different ways. Either by aggregation or by specialization. Structuring concepts by aggregation is to form a concept by describing the properties of the phenomena by means of other concepts. Specifying a concept as being an aggregate consists of specifying the sub-components and other aspects of the intension (i.e. properties of the aggregate as a whole). Classification and aggregation are not unique to object-oriented programming. In fact, almost any programming language contains language constructs for specifying classification and aggregation (e.g. type systems and record types).

The unique aspect of object-oriented programming is the language support for the structuring mechanism *specialization*. Specialization of concepts supports the specification of concepts as variants of other concepts. When we specialize a concept, we do it either by specifying further properties in the intension or by specializing one or more of the properties in the intension. Specialization of concepts gives rise to a hierarchical structure on the concepts. In object-oriented programming we

---

*In fact, we want to be able to model parts of some imaginary world, too, e.g. during the design of a new application without any predecessors. To ease the writing, we will only use the term "part of the world" instead of "part of the world or part of some imaginary world".

utilize this hierarchical structure in the class hierarchy, and furthermore we may utilize the structure to support polymorphic programming. By polymorphic programming as part of object-oriented programming we understand the ability to utilize the hierarchical structure of concepts in the specification of e.g. parameters. Let us assume that we have a parameterized program fragment, where one of the parameters is specified as class A. Then the program fragment can be compile time checked with respect to the legal usages of this parameter, since the manipulations of the parameter is legal, as long as it is manipulated according to the specification given in class A. Since specialization is property preserving (i.e. if an instance of class A has property z, then instances of all specializations of A will have the same property z) it is now possible to use instances of all specializations of class A as actual parameter to the program fragment, since they will possess at least the properties described in class A. We do not say that specialization is semantics preserving, since most language constructs for specialization do not necessarily preserve the semantics of the attribute (especially of actions), although this is the ultimative goal. For a more detailed discussion of this approach to object-oriented programming, see [8] and [12].

Looking at existing specialization hierarchies reveals that very many of them cannot be described by tree-structured hierarchies, since they contain concepts which are specialized from at least one general concept but along two (or more) paths in the hierarchy. The language support for specialization in many programming languages allows only for tree-structured classification hierarchies, thus limiting the expressive power of the language. The concept of multiple inheritance (or multiple classification) is *one* approach to loosen this limitation.

2 Definition of Terms

In order to ease the following discussion, a few terms need to be clarified. In fact, these terms are used in the above discussion in accordance with these definitions. By the term *hierarchy*, we will mean structures that can be described by acyclic, directed graphs.

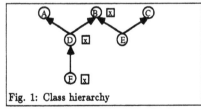

Fig. 1: Class hierarchy

The term *class hierarchy* will be used to cover the hierarchy of all the classes in a given system and their sub/super-class relationships. As described above, the class hierarchy can be used both for specifying inheritance of properties and for polymorphic programming.

The term *classification hierarchy* will be used to cover that part of the class hierarchy which is involved in the classification of one particular class. That is, there is one class hierarchy in a system, but several classification hierarchies (one for each class in the system). In fact, the class hierarchy is the union of all classification hierarchies.

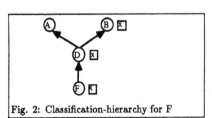

Fig. 2: Classification-hierarchy for F

96

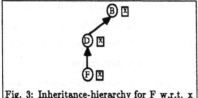

Fig. 3: Inheritance-hierarchy for F w.r.t. x

The term *inheritance hierarchy* of a class with respect to a particular attribute is the part of the classification hierarchy that covers the inheritance paths of that attribute.

Name collision can arise in two different ways depending on whether the collision is a result of the presence of more than one super-class, or whether the collision arises because of ambiguities between the class itself and its super-classes. It clarifies the discussion if these two types of name collision are separated.

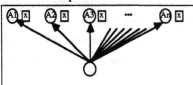

Fig. 4: Horizontal name collision

The term *horizontal name collision* will be used to cover name collisions, where a class inherits several attributes with the same name from different super-classes. Note that horizontal name collision cannot arise in tree-structured hierarchies.

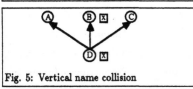

Fig. 5: Vertical name collision

The term *vertical name collision* will be used to cover name collisions, where a class defines an attribute with the same name as one (or more) attributes, inherited from one of its super-classes.

Please note, that both horizontal and vertical name collision might be involved in a particular name collision, if e.g. in fig. 5, class A and/or class C has an x-attribute, too.

3 Issues of Name Collision

Essentially, there are three different views on the consequences of a name collision. We say that a name collision is *intended* if different attributes with the same name describe the same phenomenon. We say that a name collision is *casual* if different attributes with the same name describe different phenomena. And we say that a name collision is *illegal* if the relation between attributes, names and phenomenon must be unique. In the following sections, we will discuss these views in more detail.

3.1 Intended Name Collision

When a name collision is regarded as an intended name collision, we are actually dealing with one attribute (defined by the name; i.e. the relation between attribute name and phenomenon is unique). The attribute will have several specifications (one for each inherited attribute, and possibly one in the class itself) which together must constitute the full specification of the unique attribute. In a programming language where all name collisions are regarded as intended, the phenomenon is modeled by one attribute with multiple specifications. This means that, in order to be able to ensure the polymorphic property of the classification hierarchy, the specifications must not be in conflict. If there are conflicts, it is impossible to unify these separate specifications into one specification for the phenomenon. It is not obvious in which situations we are able to ensure the polymorphic property. Let us consider an example:

Fig. 6a: Intended vertical name collision   Fig. 6b: Intended horizontal name collision

If we have an intended vertical name collision (as in fig. 6a), we can ensure the polymorphic property if we know that the specification of B.x is a specialization of the specification of A.x. This specialization property can be ensured by classification hierarchies on the specifications. This approach to specifications is taken by the Beta language[9]. Another example is type-hierarchies as exemplified in the language Amber[6]. The situation is more complex, when we consider intended horizontal name collision (as in fig.6b). At least four different situations might arise:

1. The domains of A.x and B.x might be *disjunct.*

   This might happen if A.x is a variable of type integer in the range: 1–99, and B.x is a variable of type integer in the range: 200–1000.

2. The domains of A.x and B.x might be *inconsistent.*

   This might happen if A.x and B.x both model temperature but the domain of A.x is Fahrenheit, whereas the domain of B.x is Celsius.

3. A.x and B.x might be of *different nature.*

   This might happen if A.x is a variable of some type (e.g. integer), whereas B.x is an operation.

4. The classes of A.x and B.x might have *a common superclass.*

   In this case, the two attributes are to some extent related, and it might therefore be plausible to consider them as different views on the same attribute.

3.2   Casual Name Collision

When a name collision is considered casual, we are allowing several attributes with the same name but with different, and not necessarily related specifications. In this situation it is important to be able to distinguish between the different attributes by some means other that their names. The most immediate solution is to *qualify* attribute names with the name of the class from which it is inherited (this qualified name is unique). That is, in fig. 6a it must be possible to denote both A.x and B.x, whereas in fig. 6b it must be possible to denote both A.x, B.x, and C.x. In the case of casual name collision, it is useful to use horizontal and vertical overwriting. *Horizontal overwriting* means that in class B in fig. 6a it is possible to state that the x-attribute of B is e.g. A.x. *Vertical overwriting* means that in class C in fig. 6b it is possible to state that the x-attribute of C is e.g. B.x. Horizontal and vertical overwriting does not exclude the possibility of denoting the other x-attributes by qualification — it is merely a short-hand.

98

3.3 Illegal Name Collision

When a name collision is considered illegal, the relation between names, attributes and phenomena must be unique. This means that name collisions will always give rise to compile-time errors, and not run-time errors.

3.4 Summary of Name Collision

The above discussion can be summarized by the following table:

	Intended	Casual	Illegal
Horizontal	• Disjunct • Inconsistent • Different nature • Specializations	• Qualification • Horizontal overwriting	
Vertical	• Specializations	• Qualification • Vertical overwriting	

Figure 7: Summary of issues of name collision

4 The Need for Programmer Control

By examining a selected classification hierarchy, we find that all three views on name collision are useful, and each corresponds to different aspects of programming and modeling, and that choosing one particular interpretation will result in the inability to express certain structures. That is, using one particular view in connection with either vertical or horizontal name collision is a matter of choosing to express one particular relationship between the inherited attributes, and the specified attribute (if there is such one).

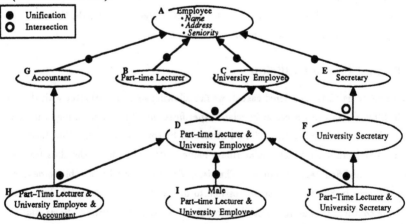

Figure 8: An example classification hierarchy

To guide the discussion of the various possibilities involved in dealing with name collisions in multiple classification hierarchies, we will examine the classification hierarchy in figure 8. Let us

assume that we are in the process of developing an accounting system for the university administration. Currently we are focusing on the structures for handling the data concerning employees
(name, address, job-category, salary, etc.). Assume that we are utilizing an object-oriented system,
and that some part of the classification hierarchy is developed outside our organization (that is,
we cannot make changes to parts of the system — only expand those parts). There is one important requirement; namely that an employee must only be represented as one employee-object in the
system.

We have only identified three attributes of the employee class to support the discussion. Further attributes may be specified in both the employee class and in the shown specializations (e.g.
accountant). In the hierarchy there are four examples of multiple classification, namely classes D,
F, H and J. Now looking at the attributes *Name*, *Address*, and *Seniority* there is no doubt that
throughout the entire hierarchy, the attributes *Name* and *Address* are singleton (i.e. any instance
of any class in the hierarchy will only have one *Name* and one *Address* attribute). The question
is more subtle when we consider the *Seniority* attribute. In that case we often find that a single
person is employed at the same university in more that one position at a time. As an example, we
may find a person who is professor at one department, but at the same time part-time lecturer at
another department. And he may even be accountant of some foundation, administrated by the
university in question. The *Seniority* attribute is concerned with the seniority of the person as
employed in the particular job-category. Since we know that any employee is employed in at least
one job-category, it is meaningful to specify that any employee-object has a *Seniority* attribute. But
the seniority of one particular person is dependent on whether we consider his/her seniority as e.g.
accountant or as e.g. part-time lecturer. This might lead to specifying that the *Seniority* attribute
should be inherited down the hierarchy with duplicates when multiple classification is involved. This
is, however, erroneous since the class F models secretaries employed at the university, and as such
secretary instances should only have one *Seniority* attribute. In fact, the following table indicates
the intended distribution of *Seniority* attributes in this small class hierarchy:

	A	B	C	D	E	F	G	H	I	J
Number of Seniority attributes	1	1	1	2	1	1	1	3	2	2

Figure 9: Table showing number of *Seniority* attributes

The problem is how do we obtain the situation where some name collisions are treated as intended,
others as casual, and yet others as illegal? It is obvious that choosing one particular view on name
collisions (e.g. casual name collision) will not result in the above table.

5 Discussion of Solutions

Looking at the classification hierarchy in figure 8 and the table in figure 9, one can see that the
specialization taking place in the specification of class D and H is different from the specialization

of class C and E to class F. If we look at (G,D) ⇒ H and (C,E) ⇒ F, we will see that the number of *Seniority* attributes in H resp. F will be either 3 resp. 2, or 1 resp. 1. This inspires to look closely at the underlying semantics of the classes H and F. Class H models employees who at the same time are employed as part-time lecturer, university employee, and accountant; that is, holding three job-positions, whereas class F models employees who are secretaries at a particular university; that is, holding only one job-position. This gives us the motivation for introducing two different specialization methods. The first specialization method (called *unification*) takes care of the kind of specialization where the specialized class is supposed to model the unification of all the classes in its classification hierarchy; that is, if a horizontal name collision should occur, it should be treated as a casual horizontal name collision, giving rise to multiple attributes with the same name. We call such a class an unification class.[†] The second specialization method (called *intersection*) takes care of the kind of specialization where the specialized class is supposed to model the intersection of all the classes in its classification hierarchy; that is, if a horizontal name collision should occur, it is treated as an intended horizontal name collision if the attribute for all the immediate superclasses is inherited from a common superclass. We call such a class an intersection class. To motivate this rule, let us look at figure 8 and assume that class D is an intersection specialization. Then the name collision of the two *Seniority* attributes from B resp. C is treated as an intended horizontal name collision, since the attribute originates from class A which is a common superclass of both B and C. If classes B and C, on the other hand, both had an x-attribute (not inherited from class A), the name collision in D would be treated as a casual horizontal name collision. Now, applying the above rule to the hierarchy in figure 8 will give us the intended distribution of the *Seniority* attribute, if we assume that classes D and H are unification classes, and class F is an intersection class. If we however look at the *Name* and *Address* attributes, we do not get the intended distribution, since there will be multiple copies in the classes D and H. This is highly undesirable, since it may give rise to inconsistencies in the contents of these different copies of this semantically identical attribute. It is therefore not sufficient to device two different specialization methods — we have to specify inheritance properties of individual attributes, too. We will therefore introduce the concept of *singleton* attributes with the semantics that they may only exist in one copy in any of the future specializations of the class. All other attributes are said to be *plural*. The singleton property is associated with the attribute in the class that initially declared the attribute. Looking at figure 8, the attributes *Name* and *Address* must be specified as singleton in class A in order to obtain the desired distribution of attributes.

Of course, it may be possible to specify the singleton property on a class as a whole implying that all attributes of the class are given the singleton property. This is merely a shorthand for the common

---

[†]Please note, that unification is conceptually different from aggregation, since a unification of classes A, B and C specifies that the unified class can be approached from three different perspectives (namely those perspectives that are defined by the classes A, B and C). This is called subtyping by combination in [7]. An aggregation of classes A, B and C specifies that the aggregated class is composed of an instance of class A, an instance of class B, and an instance of class C. This is called subtyping by composition in [7], and part hierarchy in [2].

case where the whole class is shared by all subclasses in the classification hierarchy. Singleton classes are very similar to virtual classes in the proposal for multiple inheritance in C++[11].

6 Discussion of Unification and Intersection Inheritance

The detailed properties of unification and intersection inheritance can be discussed in detail by examining the cases outlined in figure 10. These cases illustrate the various possible types of hierarchies that may arise in multiple classification hierarchies. In the following, I will give some comments on the most important cases in order to justify the formal rules for inheritance in multiple classification hierarchies that are given in section 7.

Case 1: Single Inheritance

    If class B inherits from class A using single inheritance, name collision is dealt with using the well-known rules from tree-structured classification. We will not discuss this case further in this article.

Case 2: Disjoint Multiple Inheritance

    The case of disjoint multiple inheritance is the place where we decide to consider some name collisions as being illegal.

2(a): Unification

    When disjoint hierarchies are combined using unification inheritance, we consider name collisions as being casual, and allow duplicate instances of attributes having the same name. The reason is, that we want to combine two independent hierarchies. An example is combining a hierarchy concerning job type (teacher, secretary, trucker, etc.) with a hierarchy concerning nationality (Danish, Swedish, American, etc.). If there is an attribute X in both hierarchies, then this attribute will not be considered as being the same attribute (i.e. the two hierarchies use the same name X by coincidence).

2(b): Intersection

    When hierarchies are combined using intersection inheritance, we are stressing that the involved hierarchies are considered as mutually contributing to the full specification of the new class. That is, the new class is created by merging attributes. In the case of a name collision, we have to consider whether it makes sense to merge the attributes. If the attributes are not defined in a common superclass (that is, the attributes have each their own defining statement), then there is no way to ensure, that the attributes are related in any way (see section 3.1), and any automatic rule must consider name collisions in disjoint intersection inheritance as *illegal* name collisions.

Case 3: Simple Multiple Classification

    In this simple case of multiple inheritance, the classification hierarchies of the superclasses

102

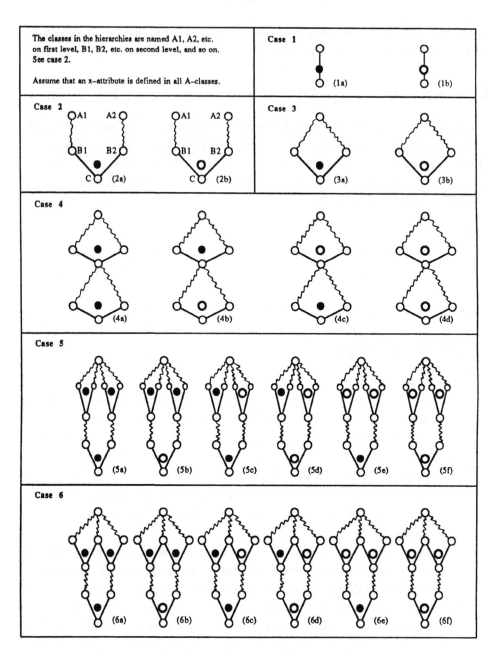

Figure 10: Important Multiple Classification Hierarchies

share a common superclass in which the X attribute is defined (and no multiple inheritance is involved in the superclass hierarchies).

3(a): Unification

The same as disjoint unification above, giving rise to two X attributes in class C.

3(b): Intersection

In this case, the superclass hierarchies share a common superclass in which the X attribute is defined, and it is therefore possible to assure that the inherited X attributes are related and thus it makes sense to merge them into one attribute.

Case 4: Chained Multiple Classification

Chained multiple inheritance is similar to simple multiple inheritance, so only two sub-cases need to be commented on:

4(a): Chained Unification

When a unification class (e.g. C) is a common super-class in another unification class (e.g. E), all attributes are inherited along all available paths in the resulting inheritance hierarchy for E w.r.t. X. In this case giving rise to four X attributes in E. It may be argued, that class E only contains two X attributes, and the formal rules in section 7 can relatively easy be modified to reflect such a decision.

4(c): Unification after intersection

Having intersected an attribute does not imply that the attribute is made singular (i.e. further specializations may contain duplicates of the attribute), but merely that at this level in the hierarchy, the attribute is unique. If this attribute is further inherited multiple times in a unification, it will give rise to multiple instances of the attribute. In this case, class E will possess two X attributes.

Case 5: Two-level Multiple Classification

The case of two-level multiple classification deals with the case where an attribute is defined in a common superclass, and that class is (independently) specialized into several classes. These specialized classes are then involved in separate multiple classifications that in turn are classified into one class using multiple classification. None of those cases needs further comments.

Case 6: Two-level Mixed Multiple Classification

This case is similar to case 5, except that the specialized classes are involved in multiple classifications that are overlapping. Here cases 6(a), 6(c), and 6(e) need to be commented on:

6(a): Two-level Mixed Multiple Unification

This case is very similar to case 4(a), leading to four X attributes in class E. However, similar to case 4(a), it may be argued, that class E only contains three X attributes, and the formal rules in section 7 can relatively easy be modified to reflect such a decision.

6(c): Merging

In this case, the X attribute which is inherited along the path A-$B2$-$C1$-$D1$ is merged into the X attribute inherited along path A-$B2$-$C2$-$D2$ since the origin (class $B2$) is involved in both class $D1$ and $D2$.

6(e): Unification after overlapping Intersection

With respect to the number of X attributes, this case is identical to case 4(c).

7 Formal Specification of Inheritance Rules

After the above discussion of the objectives for inheritance rules, it is time to state them more formally. The formal rules will be followed by the rationale for the intrinsics of the rules.

7.1 Notation

$X_B^{singleton}$ $\equiv$ The set of all paths in the inheritance-hierarchy for class B w.r.t. the singleton attribute X. If X is not a singleton attribute of B, $X_B^{singleton} = \emptyset$.

X_B^{plural} $\equiv$ The set of all paths in the inheritance-hierarchy for class B w.r.t. the plural attribute X. If X is not a plural attribute of B, $X_B^{plural} = \emptyset$.

X_B^{merged} $\equiv$ The set of merged attributes named X in class B. Each element in the set is a set containing all the paths, contributing to the particular merged X. If X is not a plural attribute of B, $X_B^{merged} = \emptyset$.

$super(B)$ $\equiv$ The set of all superclasses of class B.

$$X_{super(B)}^{singleton} \equiv \bigcup_{A \in super(B)} X_A^{singleton}$$

$$X_{super(B)}^{plural} \equiv \bigcup_{A \in super(B)} X_A^{plural}$$

$$X_{super(B)}^{merged} \equiv \bigcup_{A \in super(B)} X_A^{merged}$$

$$\bigcup X_{super(B)}^{merged} \equiv \left\{ p \in S \mid S \in X_{super(B)}^{merged} \right\}$$

Comments on Notation

The sets $X_B^{singleton}$, X_B^{plural} and X_B^{merged} contain the information necessary to decide the number of X-attributes in B, divided into three categories, depending on whether the particular X-attribute is *singleton*, *plural* or *merged*. When an instance of class B is created, the sets are used in the following way:

For each element in X_B^{plural}, an X-attribute is instantiated with the qualification given by the path in the classification hierarchy for B specified by that element. Intuitively, this rule states that a plural attribute is to be instantiated with the full qualification along its (unique) inheritance path in the classification hierarchy for its class.

For each element in X_B^{merged}, an X-attribute is instantiated with a qualification that is given by the path in the classification hierarchy for B, specified by the *longest common path* (LCP) of that element. LCP is the longest common path of the set. Intuitively, this rule states that a merged attribute cannot always be instantiated with the full qualification along its inheritance path in the classification hierarchy for its class since that path is not necessarily unique. A merged attribute may be inherited along multiple paths in the classification hierarchy (as with the *Seniority*-attribute in class J in figure 8). Since these multiple paths may have conflicting qualifications for the attribute, the only qualification that can be ensured for the attribute is the qualification which all inheritance paths for that particular instance of the attribute agree upon, namely that given by the nearest branching node in the inheritance hierarchy (defined by LCP).

If $X_B^{singleton} \neq \emptyset$, one attribute is instantiated with a qualification given by $LCP(X_B^{singleton})$.

The inheritance paths are represented in the sets in the following way: Upper-case letters indicate names of classes, lower-case letters indicate sub-paths. The classes in the path are separated by either '.', '†' or '‡' where '$A.B$' indicates that B inherits from A using single inheritance, '$A†B$' indicates that B inherits from A using unification inheritance, and '$A‡B$' indicates that B inherits from A using intersection inheritance. In path expressions, '$\star$' is taken to denote any of '.', '†' or '‡'. For any path p, $|p|$ represents the length of the inheritance path (i.e. the number of classes along the path).

7.2 Plural and Singleton definition in B

$$X_B^{plural} = \{B\} \cup X_{super(B)}^{plural}$$

$$X_B^{singleton} = \{B\} \cup X_{super(B)}^{singleton}$$

Comments on Plural and Singleton Definition Rules

The rules for plural and singleton definitions state that defining a plural attribute in a class gives rise to one more potential attribute of that name in all subclasses of that class. Defining a singleton attribute implies the presence of at least that instance of the attribute. In both cases, the definitions from the superclasses are retained.

7.3 Single inheritance in B

$$X_B^{singleton} = \left\{ s.B \mid s \in X_{super(B)}^{singleton} \right\}$$

$$X_B^{plural} = \left\{ p.B \mid p \in X_{super(B)}^{plural} \right\}$$

$$X_B^{merged} = \left\{ \{s.B \mid s \in S\} \mid S \in X_{super(B)}^{merged} \right\}$$

7.4 Unification inheritance in B

$$X_B^{singleton} = \left\{ s\dagger B \mid s \in X_{super(B)}^{singleton} \right\}$$

$$X_B^{plural} = \left\{ p\dagger B \mid p \in X_{super(B)}^{plural} \wedge \left(\not\exists a\ddagger s \in \bigcup X_{super(B)}^{merged} : p = a \star p' \wedge |a| > 0 \right) \right\}$$

$$X_B^{merged} = \left\{ \{s\dagger B \mid s \in S\} \cup \{a\ddagger p\dagger B \mid a\star p \in X_{super(B)}^{plural} \wedge (\exists a\ddagger s \in S : |a| > 0)\} \mid S \in X_{super(B)}^{merged} \right\}$$

Comments on Unification Inheritance Rules

The case of unification inheritance is complex. Let us first comment on the rule for X_B^{plural}. This rule states that the set of X-attributes in B is those attributes inherited from the superclasses that have not been merged in any of the superclasses. The exact rule states that a plural attribute is turned into a merged attribute, iff the plural attribute is inherited along the same path as one of the paths of the merged attribute up to the node in the classification hierarchy where the attribute is made merged. Intuitively this implies that name collisions are considered casual unless the name collision is between a plural and a merged attribute with a common subpath.

The first part of the rule for X_B^{merged} states that all merged attributes in the superclasses are inherited unconditionally. The second part of the rule states the actual merging of a plural attribute into a merged attribute. Exactly this is done by injecting the (slightly modified) path of the plural attribute into the element of X_B^{merged} containing the common subpath. Intuitively this implies that the specification of that particular instance of the merged attribute is extended to include the specification given by the plural attribute. This implies that the name collision is considered intended.

7.5   Intersection inheritance in B

if $\quad \exists a : \left(|a| > 0 \wedge \forall p \in X^{plural}_{super(B)} \cup \bigcup X^{merged}_{super(B)} : p = a \star p' \vee p = a \right)$

then $X^{singleton}_B = \left\{ s \ddagger B \mid s \in X^{singleton}_{super(B)} \right\}$

$\qquad X^{plural}_B \quad = \emptyset$

$\qquad X^{merged}_B \quad = \left\{ \left\{ s \ddagger B \mid s \in X^{plural}_{super(B)} \cup \bigcup X^{merged}_{super(B)} \right\} \right\}$

else illegal name collision on X in B

Comments on Intersection Inheritance Rules

The conditional part of the rule states that intersection is legal iff all inherited plural attributes share a common subpath in their inheritance path. If not, the name collision is considered illegal, since the origin of the different plural attributes is distinct parts of the hierarchy with no connections in the inheritance path of attributes.

If, on the other hand, the inherited plural attributes share a common subpath, they are all merged into one merged attribute. Intuitively this rule states that intersection is like unification except that all inherited plural attributes must be considered as all contributing to one particular attribute. If this is not possible, the name collision is considered illegal.

8   Resolving Conflicts

The above inheritance rules do not resolve the conflicts, but merely restrict the possibilities for conflicts. Now assume that we want to access the x-attribute of class B. If only one x-attribute exists in $X^{merged}_B \cup X^{plural}_B$, no ambiguity arises. But if more than one exist, the specification of the x-attribute must be unambiguous. We will propose the following rule: If no further specification is given, the x-attribute in X^{merged}_B will be selected, if the set is not empty. If the set is empty, or if one of the x-attributes in X^{plural}_B is wanted, a qualification must be given to resolve the ambiguity. This qualification can be given by specifying a subpath of the inheritance path of the attribute in the inheritance hierarchy for the attribute from the present class and to a node, where the attribute is unique (according to the above rule). This implies that the above mentioned node must be either the defining node of the attribute, or a node where the attribute has taken part of an intersection inheritance.

9 Horizontal and Vertical Overwriting

Instead of resolving conflicts (as described above) each time it may arise, it might be easier to allow for horizontal or vertical overwriting. If this is allowed, the above rule for resolving conflicts must be extended to check for the existence of an overwriting, in which case the qualification can be considered unambiguous. Naturally, further qualification is allowed, and sensible, in the case where the overwriting is not what is wanted in the present situation.

9.1 Explicit Inheritance

In section 7, the default rules for inheritance of attributes in a multiple classification hierarchy are given. The reasons for the rules are based on the recognition of the classification hierarchy as a conceptual hierarchy. There might be cases in which the expected inheritance does not conform with the conceptual hierarchy. This might be the case in situations where the classification hierarchy is slightly ill-suited for the specific application, but where reorganization for some reason is not applicable.

In these situations some restricted usage of horizontal overwriting can be allowed. In fact, in this situation we are dealing with intended name collision which cannot be dealt with by the default rules. Returning to section 3.1, four situations might arise. The case of the inherited attributes being disjunct cannot be remedied by horizontal overwriting. The next two cases can be remedied by horizontal overwriting by means of language constructs for specifying transformations between two domains, or to a lesser extent by language constructs for interchanging variable denotations with operation denotations. The case of the inherited attributes having a common superclass is the most easily remedied case, since it only requires abilities similar to the abilities already used in intersection inheritance and further described in [12].

Vertical overwriting is the usual method redeclaration in Smalltalk-80, or the virtual declaration in languages like Simula67, C++ or Beta. The most interesting version of virtual declaration is the one in Beta, since virtual declarations in Beta to some degree ensure that inherited properties are not invalidated by the vertical overwriting.

10 Conclusion

The prime result here is the recognition of the need for supporting all three views on name collision in one programming language, unless one accepts to be unable to express certain structures. As stated in the introduction, the discussion in this paper is in itself valuable as many proposals have been put forward, and in many cases the arguments and the discussion of the alternatives aren't given. This paper offers such a discussion relieved from the burden of promoting one particular solution at the same time. It will thus hopefully be a source of inspiration for future designers of programming languages with support of classification hierarchies with multiple classification.

References

1. A. Albano, L. Cardelli & R. Orsini: *Galileo: A Strongly-Typed, Interactive Conceptual Language*, ACM TODS, **10**(2), 230–260 (June 1985).

2. E. Blake & S. Cook: *On Including Part Hierarchies in Object-Oriented Languages, with an Implementation in Smalltalk*, Proceedings of the European Conference on Object-Oriented Programming (ECOOP'87), Paris, France, June 1987.

3. D.G. Bobrow & M.J. Stefik: *Loops — Data and Object-Oriented Programming for Interlisp*, Discussion papers, Proceedings of the European Conference on AI, Orsay, France, July 1982.

4. A. Borning: *The Programming Language Aspects of ThingLab, a Constraint-Oriented Simulation Laboratory*, ACM TOPLAS, **3**(4), 353–387 (October 1981).

5. A. Borning & D.H.H. Ingalls: *Multiple Inheritance in Smalltalk-80*, Proceedings of the National Conference on AI, Pittsburgh, PA, 1982.

6. L. Cardelli: *Amber*, AT&T Bell Labs Technical Memorandum, 11271–84092–410TM, 1984.

7. D.C. Halbert & P.D. O'Brien: *Using Types and Inheritance in Object-Oriented Languages*, Proceedings of the European Conference on Object-Oriented Programming (ECOOP'87), Paris, France, June 1987.

8. J. Lindskov Knudsen & K. Stougård Thomsen: *A Conceptual Framework for Programming Languages*, Computer Science Department, Aarhus University, DAIMI PB-192, 1985.

9. B. Bruun Kristensen, O. Lehrmann Madsen, B. Møller-Pedersen & K. Nygaard: *The Beta Programming Language*, in B.D. Shriver & P. Wegner (eds.): *Research Directions in Object-Oriented Programming*, MIT Press, 1987.

10. B. Meyer: *Eiffel: Programming for Reusability and Extendibility*, ACM Sigplan Notices, **22**(2), 85–94 (February 1987).

11. B. Stroustrup: *Multiple Inheritance for C++*, Proceedings of the Spring '87 EUUG Conference, Helsinki and Stockholm, May 1987.

12. K. Stougård Thomsen: *Inheritance on Processes, Exemplified on Distributed Termination Detection*, International Journal of Parallel Programming, **16**(1), 17–52 (1987).

13. D. Weinreb & D. Moon: *Flavors: Message Passing in the Lisp Machine*, MIT AI Memo No. 602, November 1980.

Reflexive Architecture:
From ObjVLisp to CLOS

Nicolas Graube*

Equipe Mixte L.I.T.P. & Rank Xerox France
Université Paris VI
Tour 45-55 Porte 209
4, Place Jussieu
75252 Paris cedex 05
France

Abstract

This paper presents the design of a minimal set of instruction for a class system embedded in Lisp: ObjVLisp. We re-use the set of postulates describing the operational behaviour of ObjVLisp to discuss and derive a new implementation based on a reduced set of functions and a more self-contained description. Then we develop the ObjVLisp experience in building metaclass architecture to propose a layered and incremental meta-object protocol for CLOS.

1 Introduction.

One goal of the ObjVlisp project is to provide the user with a class system both understandable, self-described, small (described in approximately two pages of Lisp) and extensible to other different class system paradigms.

---

*This research was partly funded by the *GRECO de programmation du CNRS*.

To reach this goal ObjVLisp is constructed with a virtual machine[1] (a set of primitive functions [7]), and a circular definition of the basic class architecture (Class and Object).

This paper attempts to discuss, clarify and minimise this virtual machine. This minimisation will be operated mainly by reinforcing the self-description, i.e. by moving knowledge from the virtual machine to ObjVLisp itself: the object world.

Then we focus our attention on Common Lisp Object System (CLOS)[2], an object-oriented paradigm for Common Lisp [11]. CLOS enhances fusion into Common Lisp through both syntactical and semantical choices. These choices are synthesized in integration of message passing in functional application via generic functions and merging structure and Common Lisp types in CLOS' classes.

With the previous virtual machine, modified in order to characterize the behaviour of CLOS, we also attempt to describe a minimal virtual machine and a layered metaclass architecture for CLOS.

Since our main concern is minimality and reflexion, we do not address the efficiency problem even if both concerns are not antinomical. Optimizations are beyond the scope of this paper.

2 ObjVLisp's Postulates.

ObjVLisp is a good synthesis of a class system based on six postulates which both fully depict the operational behaviour of the system and give a guideline to possible implementations. These postulates were descibed in [7] as follows:

P1 : An object represents a piece of knowledge and a set of capabilities.

P2 : The only protocol for activating an object is message passing: a message specifies which procedure to apply (denoted by its name, the selector), and its arguments.

P3 : Every object belongs to a class that specifies its data (attributes called fields) and its behaviour (procedures called methods). Objects will be dynamically generated from this model, they are called class instances. Following Plato, all class instances have the same structure and shape, but differ through the value of their common instance variables.

P4 : A class is also an object, instantiated by another class: its metaclass. Consequently (**P3**), to each class is associated a metaclass which describes its behaviour as an object. The initial primitive metaclass is the class Class.

---

[1]ObjVLisp stands for Object Virtual Lisp and does not rely on any particular dialect of Lisp. Different implementations have been provided for Le_Lisp (A., Deutsch, J-P., Briot), InterLisp-D (N., Graube), (Kyoto, Xerox) CommonLisp (A., Deutsch & C., Consel, N., Graube) and VLisp (P., Cointe)[4] [7].

P5 : A class can be defined as a subclass of one (or many) other class(es). This subclassing mechanism allows sharing of instance variables and methods, and is called inheritance. The class Object represents the most common behaviour shared by all objects.

P6 : If the instance variables owned by an object define a local environment, there are also class variables defining a global environment shared by all the instances of a same class. These class variables are defined at the metaclass level.

3 Discussion of the ObjVLisp Implementation.

We will now describe a new implementation of ObjVLisp where we use postulates as a guideline for the discussion of the various components of both the virtual machine and the object world. When necessary we will pinpoint some possible alternatives to the interpretations of postulates.

The first point to define is the basic class architecture as we need it in the following explanation. The architecture is defined by **P4**. It implies the presence of a self-instantiated class, Class, which will be the root of the instantiation tree. Once the self-instantiation problem has been solved the possibility of self description arises, then Class defines its own structure in such a way that it becomes both its class and its instance. We call this architecture a reflexive descriptive architecture, because while we share the same fundamental idea of meta-description with reflective languages[9], the word "descriptive" is used to express that we are only concerned with the description of the architecture, not with the operational process of *reflection*.

We use the reflexivity of the system for translating some of the instructions of the previous virtual machine[7] into the object world. Thus a common problem is infinite regression triggered by an over powering self-description.

3.1 The Link Between an Instance and its Class: Isit.

P3 represents an important problem. An object cannot live without its class, otherwise it cannot find the real associated semantic of all its embeded values.

Then a problem with class languages is to provide a way of representing the class of any object.

We have to decide the real status of the class reference. Should it be hidden from the user, thus part of the virtual machine, or accessible and thus part of the object world?

While the first proposition enhances safety and allows modifications only to be accomplished at the lower level, the second one enforces reflexivity and thus authorises any kind of manipulation at the object level.

The solution adopted by ObjVLisp is to put the class reference into an instance variable named isit[2] defined by Object thus, according to **P5**, inherited by every object of the system.

Once the status problem has been solved we need to describe the process involved in the search for this information. As isit is an instance variable, we could provide a method to access its value, triggered by message passing, which would obviously be defined in Object. Unfortunately this method leads us into an infinite regression because, in order to apply a method, along the lines of **P3** we must find the object's class, precisely what this method supposedly returns.

Then a method cannot be of any help in solving this problem, we have to define a specific function, not part of the object world, for retrieving this information. This function, Class-of, must hold some knowledge about where it can find the link in any object between itself and its class. This implies that every object will share the same basic shape, usable by Class-of.

3.1.1 The Basic Object Creation: Make-Object.

Defining Class-of means freezing both the global shape of every object and the physical position of the class reference within this structure. Then it has to be defined in conjunction with the basic object allocator, make-object, which will be the only entry-point for memory allocation[3].

```
(defun make-object (a-class)
      (let ((an-object (make-sequence 'vector 2)))
           (setf (first an-object) a-class) an-object) )

(defun class-of (an-object) (first an-object) )
```

3.1.2 Named vs Anonymous Classes.

Another point to sort out is the kind of information we expect to find in the link. Two sorts of information can be provided: the name of the mother class or a reference to the mother class.

---

[2]The name isit was chosen for historical reasons in relation to Smalltalk-72.

[3]For practical reasons, make-object will also set the link between the newly created object and its class i.e. the caller of make-object.

Storing the name of the class implies that the system can only handle named classes and thus reduce the expression power of the extension. However this scheme enables us to provide a very simple self-description for Class because defining a new class with the same name as an already existent one will automatically erase the latter.

On the other hand, storing the reference of the mother class frees us from the naming problem and enables us to use anonymous classes. While with the class name the self-description was trivially easy, this process needs to be more intricate and involves a pointer manipulation and more precise knowledge of the object structure.

3.2 Object Creation: new.

From **P3** we learn that instances are dynamically generated from the class. In combination with **P2** we know that creation is triggered through message passing and that the creation method must be located in the class of the class from which we want to make a new instance. Thus it is in the metaclass that we will find the description of the instance generator. This method will give the physical description of the instance which must be coherent with the general shape explained in the previous section, i.e. it must rely upon the use of make-object.

While the global shape cannot be customised because of class-of, the inside organisation of the new instance, i.e. where we put the associated value of the instance variables, is under the control of the creation method.

Three different schemes can be derived, each one defining different possibilities of parametrisation.

1. As generating a new object is performed by a method, we can include in it the description of the parametrisation or export it in an other method as follows:

```
new       (λ (self &rest args)
;; self is automaticaly bind to the object receiver.
          (send (send self 'basicNew (iv-value self 'iv)) 'initialize args) )
basicNew  (λ (self ivs)
          (let ( (an-object (make-object self))
;; This is the parametrisation.
                 (structure (make-sequence 'list (length ivs) )) )
;; Here is an assumption about the global shape of an object, i.e. a two field sequence.
          (setf (second an-object) structure) an-object) )
```

This first solution implies that we hold some methods which give information about this parametrisation essentially needed for retrieving values.

2. Unlike the previous solution we use an instance variable of the metaclass to store the description of the parametrisation.

```
new  (λ (self &rest args)
        (send (funcall (iv-value (class-of self) 'constructor) self (iv-value self 'iv))
              'initialize args) )
```

Then all instances of classes of the same metaclass will share the same internal structure.

3. With the same idea as the previous solution, but by storing the value of the instance variable in the class, we get the following description:

```
new  (λ (self &rest args)
        (send (funcall (iv-value self 'constructor) self (iv-value self 'iv))
              'initialize args) )
```

Allowing different classes of the same metaclass to customise their instances in a different manner.

In both of the latter solutions, constructor can be associated to:

```
#'(λ (self ivs)
     (let ((an-object (make-object self))(struct (make-sequence 'list (length ivs))))
          (setf (second an-object) struct) self))
```

We have derived three different correct definitions for the method new, which is part of the object world, but all of them must depend upon the use of the allocation primitive strictly derived from the postulates.

3.3 Finding the value of instance variables: iv-value.

One fundamental action is to ask for the associated value of an instance variable inside an object[4]. As the Object world cannot handle the basic access to the associated instance variable[5], there must be a primitive of the virtual machine to perform this task.

---

[4]In the previous code this was accomplished by the Lisp form: iv-value.

[5]Providing a method for this task will cause infinite regression because of method description under instance variables, or implies another primitive for accessing the specific value associated to the instance variable holding the method description.

116

3.3.1 The Quest for Meaning.

From **P3** we learn that all of the description of the data of any object is held by its class. Therefore there must be some information in the class about instance variables defined for its instances.

In ObjVLisp this knowledge is described in the iv instance variable which holds all instance variables' specific information, i.e. their names, initial values and positions in the related object. Finding the associated value of a specific instance variable of an object depends on finding the associated value of the iv instance variable in an object's class and extracting all the information needed for retrieving the value. In order to complete this task we must be able to find the associated value of iv in the object's class and so on from class to class. But from **P2** we know that there is a loop in the instantiation graph and while all classes share this graph, the latter process will loop endlessly. Thus, in order to avoid this infinite regression we have to provide some information at a certain point in the search. As the loop starts at Class, we must provide the description of all instance variables defined by it. This will be done via a primitive instruction from the virtual machine: class-all-iv which takes an object supposed to be the self-instantiated class and returns the description of all its instance variables. Then a possible description of this function is:

```
(defun iv-value (an-object a-name )
    (find-slota-name an-object
            (if (eq an-object (class-of an-object)) (class-all-iv an-object)
                (iv-value (class-of an-object) 'iv) ) ) )
```

In which *find-slot* is only a do-loop which finds the named *a-name* instance variable description through the list of all instance variable descriptions, and then finds the associated value in the object *an-object*.

Class-all-iv is very simple to describe but relies on the internal organisation of the class Clas

3.3.2 The Quest for Physical Structure.

Because only the general shape is defined as being the same for each object, it is at the metaclass level, or at the class level, depending on the choice of implementation, that we can find the information about the internal physical structure of an object. Therefore the previous section only defines the way to retrieve the description of a specific instance variable. We have to modify the previous definition so that it will also find the physical description of the object in order to access the associated value embedded in this structure. As before we have to provide a primitive instruction, class-organization, which will give some information about the structure of Class, to avoid infinite regression[6].

---

[6]All iv-value description must be modified if we want to handle the isit case properly. But in order to clarify the presentation we omitted this treatment.

```
(defun iv-value (an-object a-name )
    (multiple-value-bind (iv structure)
            (if (eq an-object (class-of an-object))
                (values (class-all-iv an-object) (class-organization an-object))
                (values (iv-value (class-of an-object) 'iv)
                (iv-value (class-of (class-of an-object)) 'structure) ) )
            (find-slot a-name an-object iv structure) ) )
```

3.4   Handling Multiple Inheritance.

P5 induces the necessity of handling multiple inheritance schemes. This can be part of the object world as the only implication is the fact that the path produced by this scheme can be found both by the control structure, while finding the description of the method, and during the collecting of all the inherited instance variables. Thus we can define a method located in **Class** for computing a linearisation of the multiple inheritance graph i.e. also named linear extension. This facility enables us to provide different inheritance schemes coexisting within the same system.

3.5   Control structures: send and run-super.

According to **P2** there is only one control structure in the object world which is **send**. While finding the right method can be easily parameterised via the contents of the instance variable which holds the linear extension and is deduced by the multiple inheritance treatment, the application of this method onto the arguments is not part of the object world. Then **send** has to be part of the virtual machine.

```
(defun send (an-object a-selector &rest args)
    (apply (find-method a-selector (iv-value (class-of an-object) 'linear-extension) ) )
            an-object args) )
```

A problem related to **send** is the processing of shared definitions of methods via the run-super. As we use multiple inheritance, run-super can no longer be just a static reference to the first super class of the class definition (as in Smalltalk-80[8]), handled by some processes which are part of the object world, i.e. *code walker* or others, but must be involved in a more general dynamic process, and must be described as a primitive. This also implies that some information has to be provided by **find-method** to permit this behaviour. Thus finding the method must also return the class of the description in order to go further in its search when executing a **run-super**.

Therefore we summarise the definition of a reduced virtual machine for ObjVLisp in the following figure 1.

class-of	make-object	iv-value	setf-iv-value
send	run-super	class-all-iv	class-organisation

Figure 1: ObjVLisp Reduced Instruction Set.

We made a first attempt at the description of a meta-object protocol for CLOS by using ObjVLisp's ability to describe new class taxonomy paradigms. Then we defined CLOS as an embedded world of ObjVLisp and we provided a set of classes which depicted all specific CLOS behaviour. More precisely we concentrated all of the differences in a metaclass which made the connection between the two worlds. Unfortunately even if this scheme works, the cooperation between these two worlds was unsatisfactory because some basic CLOS principles go against some basic ObjVLisp's ones[7].

Then while this experience was not, in our opinion, a complete success we decided to apply the same description process as the one for ObjVLisp to CLOS.

4 Common Lisp Object System in Eight Postulates.

Surprising as it might be, ObjVLisp and CLOS are depicted by some common postulates. Therefore we shall only pointout the differences.

P2' : The only protocol for activating an object is generic function application. A generic function is made up of different methods described for specific typed parameters. When applying a generic function the system tries to find the best defined method of this generic function, i.e. the one for which specific parameters are closer to actual arguments according to class inheritance.

P5' : A class can be defined as a subclass of one (or many) other class(es). This subclassing mechanism allows sharing of instance variables and methods, and is called inheritance. The class T represents the most common behaviour shared by all objects and the least defined type of Common Lisp.

P6' : If the instance variables owned by an object define a local environment, there are also class variables defining a global environment shared by all the instances of a same class. These class variables are specific instance variables defined in the class but with a particular allocation scheme. We call them *slots* instance variables or class variables.

P7 : As long as it is possible, elements of the extension such as slots, methods and generic functions are to be considered as objects and instances of specific classes.

---

[7]This needs to be synthesized in major modifications of the virtual machine.

P8 : Inside methods we can use inheritance via `call-next-method`. But in an orthogonal way we can use *method combination* which enables us to organize methods in a specific way.

5 Our Implementation of CLOS.

As for ObjVLisp an implementation can be depicted using postulates as guideline. In this case we do not give all the details but only focus on the major differences and on the associated reflexive descriptive architecture[8].

Note that some specific CLOS behaviour is not defined via postulates such as temporized instantiation, class changing protocol or specific method combination, mainly because these are not a fundamental aspect of CLOS behaviour but rather features which can be defined in the object world.

5.1 Basic access to Slots.

P6' and **P7** introduce important changes and imply both a generalised vision of instance variables and the consideration of some new primitive functions.

First we have to modify the `iv-value` so that access to both true instance variables and class variables is possible. In order to acces class allocated instance variables one solution is to describe these objects as instances of a specific class, `class-allocated-slot`, and instance allocated instance variables as instances of `instance-allocated-slot`. Then when we try to find the associated value of a slot in an object, we just have to apply a generic function, `get-value`, defined on both the instance and the slot description. Providing two different methods, one for each slot class, can solve the problem of where to find the associated value. Unfortunately this pretty solution leads once again to infinite regression because every class in the system describes their slots as objects, particulary generic function objects. Then we have to rely on a simple test on the class of slot description or simply on some information embedded in the description.

Secondly we have to consider the fact that slots are now true objects and that the simple action of finding the name of a slot in a slot description is a real problem. As applying a generic function leads us to infinite regression we have to provide two simple functions: `slot-name` and `slot-description`.

---

[8]Roughly all the modifications can be made by considering message passing as an application of a generic function where only the first argument is used in the method selection.

As there are no other great differences from ObjVLisp virtual machine description, we can summarise all the instructions in figure 2. Note that basic differences are limited to slots and the control structure otherwise they share the same description.

class-of	make-object	slot-value
setf-slot-value	slot-name	slot-description
class-all-slots	class-organisation	call-next-method

Figure 2: CLOS Instruction Set.

6 A Layered Architecture for CLOS.

While the external description of CLOS object behaviour can easily be found in the first two chapters of [2], it is not the case with the meta-object protocol which is still under discussion. However we try remain as faithful as possible to the first ideas explained in [1].

A basic principle is to define a reduced set of classes, thus defining a kind of *micro*-CLOS (μCLOS), which describes all basic CLOS behaviour. Then, by using μCLOS as a basis we will build all specific CLOS classes. This stratified construction will be completed using multiple bootstrapping as some classes are mutually defined. The key notion is that from bootstrap to bootstrap we get closer and closer to the final description of CLOS, coherent with all postulates. Then it is not surprising to give an incremental description of the virtual machine as well. Each description must be coherent with the postulates which have been introduced by the previous bootstrap[9].

6.1 The ObjVLisp Kernel Revisited: μCLOS.

Along the lines of **P3** we can define the very first kernel of μCLOS: the self-instantiated class Class, root of the instantiation tree and its first instance from which it inherits, the class Object. While T has been defined as being the most common class of all the components of the CLOS system, it is more like the only connection with the Common Lisp type system than Object which has to be considered as the most common class of all CLOS classes.

---

[9]As ObjVLisp was described with a single bootstrap this necessity did not appear in its description.

Defining Class and Object will need a first bootstrap and the use of an early version of make-instance[10] and of defclass. As generic functions cannot yet be provided make-instance is only a simple Lisp function using make-object.

6.2 Class, Object and Slot: The Holy Trinity.

P7 implies that slot descriptions must be instances of a class. We can then instantiate a new object from Class: the class Slot. It inherits from Object and depicts all the behaviour for slot description. While all existing slot description must be instances of this class, we must perform a second bootstrap in order to give a coherent definition to Object, Class and Slot.

Slot-value needs to be modified in order to handle object slot description. To be more precise we just have to give the final description of slot-name and slot-allocation while their corresponding postulate is introduced into the system. We have to rewrite some of the defclass description to make it handle slot instantiation and to provide a more complete definition of the initialize function so that it distinguishes between terminal objects and classes.

6.3 Types and T: A Connection with the Type System.

T is an entry point in the Common Lisp Type Lattice but is only there to be inherited and must act like an abstract class. However as Common Lisp Types have to be considered as classes we must provide a hook where we can describe their representation in terms of class. Thus we have to provide a metaclass Types, instance of Class, from which we describe all the necessary classes.

If the Types definition does not involve any particular problem it is not the same for T as it must be inherited by Object. Thus a third bootstrap is necessary enabling us to provide the right description of Object.

6.4 Methods and Generic-Functions.

Each time we had to check the class, or the type of an object to specify the correct behaviour of a function, we performed an explicit test, embedded in this Lisp definition. We will now introduce all necessary classes in order to integrate P2'. Two classes can describe basic generic functions, i.e. generic functions which handle both method definition on a cartesian product of

---

[10]make-instance is ObjVLisp's new correspondant for CLOS.

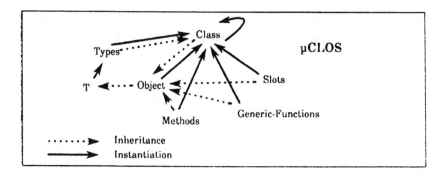

Figure 3: μCLOS.

classes or types and use of the hierarchy defined on classes or types and the code sharing scheme via the call-next-method utilisation. These classes are Methods and Generic-Functions their definitions are straightforward and do not need any other bootstrapping mechanism.

Once they are defined we can move all the Lisp functions defined on objects into generic functions. At this point μCLOS is complete, it defines both a full class system for Common Lisp and a good basis for an extented description of CLOS.

6.5 CLOS as an embedded world of μCLOS.

When we made a description of CLOS as an embedded world of ObjVLisp, we encountered some problems as some of ObjVLisp's postulates clash with some basic CLOS ideas.

Describing CLOS as an embedded world of μCLOS while sharing the same idea as the previous attempt does not cause any difficulties because none of μCLOS's postulates can differ from CLOS ideas.

Our only task is to extend μCLOS classes so that they will describe the complete behaviour of CLOS. We started this process by increasing Class and Slot so that we have complete slot description instances of Clos-Slot which inherits from Slot and is co-instantiated with Clos-Class, a subclass of Class. This description needs an ultimate bootstrap as a result of the co-instantiation process. Note that as this bootstrap only concerns these two recently defined classes, the redefinition can only be local. A complete CLOS system has been implemented along this line and runs under Xerox Common Lisp. Finally because CLOS's generic functions are more powerful than the one μCLOS defines we must describe subclasses

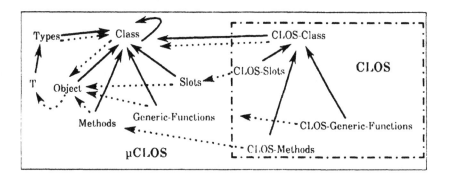

Figure 4: CLOS as embedded world of μCLOS.

of Methods and Generic-Functions respectively as Clos-Methods and Clos-Generic-Functions in order to depict their full behaviour.

7 Future Work.

The descriptions of the ObjVLisp and CLOS virtual machines were accomplished using postulates as a guideline. A second step which needs to be carried out by is an automatic derivation of a virtual machine through postulates, using denotational semantics or other formal descriptions. In the same direction as the one taken for the description of both ObjVLisp and CLOS, we will describe a complete class system for Le_Lisp Version 16[5]. This system will enhance metaclass programming and complete integration of the type system which will be defined at the same time.

Acknowledgements

We would like to thank Jean-Pierre Briot, Pierre Cointe, Alain Deutsch, Sara Dootson, Jean-François Perrot and Christian Queinnec for their encouragement and constant help.

References

[1] Bobrow, D.G., DeMichiel L.G., Gabriel R.P., Keene S., Kiczales G., Moon D.A, Common Lisp Object System Specification, Chapter 1, 2 and 3, X3J13 (ANSI COMMON LISP), March 1987.

[2] Bobrow, D.G., DeMichiel L.G., Gabriel R.P., Keene S., Kiczales G., Moon D.A, Common Lisp Object System Specification, Chapter 1 and 2, X3J13 (ANSI COMMON LISP), November 1987.

[3] Bobrow, D.G., Kiczales G., The Common Lisp Object System Metaobject Kernel A Status Report, *IWoLES 88*, Afcet, Afnor, LITP and Inria, Paris, France, February 1988.

[4] Briot, J-P., Cointe, P., A Uniform Model for Object-Oriented Languages Using The Class Abstraction, *IJCAI 87* Proceedings of the Tenth International Joint Conference on Artificial Intellingence, pp. 40-43, Milan, Italy, August 1987.

[5] J., Chailloux, M., Devin, J-M., Hullot, LeLisp, a Portable and Efficient Lisp System, *Lisp and Functional Programming*, pp 113-122, Austin, Texas, USA, August 1984.

[6] Cointe, P., Towards the design of a CLOS Metaobject Kernel: ObjVLisp as a first layer, *IWoLES 88*, Afcet, Afnor, LITP and Inria, Paris, France, February 1988.

[7] Cointe, P., Metaclasses are First Class: the ObjVlisp model, *OOPSLA '87*, Special Issue of SIGPLAN Notices, Vol. 22, No 12, pp. 156-167, Orlando, Florida, USA October 87.

[8] Goldberg, A., Robson, D., Smalltalk-80 - The Language and its Implementation, Addison-Wesley, Reading MA, USA, 1983.

[9] Maes, P., Concepts and Experiments in Computational Reflection, *OOPSLA '87*, Special Issue of SIGPLAN Notices, Vol. 22, No 12, pp. 147-155, Orlando, Florida, USA October 87.

[10] Queinnec, C., Cointe P., Types, Classes, Metatypes, Metatypes Classes: an open-ended data representation model for Eu-Lisp *to appear in the Lisp and Functional Programming conference*, Paris, LITP, January 88.

[11] Steele Jr., G., F., Common Lisp: The Language, Digital Press, 1984.

A ObjVLisp Reduced Instruction Set.

```lisp
;;; A Reduced Instruction Set for ObjVLisp.
;;; Object allocation.  [Primitive 1]
(defun make-object (a-class)
    (let ((basic-structure (make-sequence 'vector 2)) )
        (setf (elt basic-structure 0) a-class)
        basic-structure) )
;;; Find the link between an object and its class.  [Primitive 2]
(defun class-of (an-object) (elt an-object 0))
;;; Message Passing.  [Primitive 3]
(defun send (an-object selector &rest args)
    (do ((classes (iv-value (class-of an-object) 'linear-extension) (rest classes)))
        ((null classes) (send an-object 'error "Method ~S not found" selector args))
        (do ((methods (iv-value (first classes) 'methods) (rest (rest methods))))
            ((null methods))
            (and (eq selector (first methods))
                (return-from send (apply (second methods) an-object
                                         selector (rest classes) args)) ) )))
;;; run-super  [Primitive 4]
(defmacro run-super (&rest args)
    (let ((classes (gensym))
          (methods (gensym))
          (args (gensym)) )
      '(block run-super
        (do ((,args (list ,@args))
            (,classes *classes* (rest ,classes)))
            ((null ,classes) (send an-object 'error "Method ~S not found" *selector* ,args ))
            (do ((,methods (iv-value (first ,classes) 'methods) (rest (rest ,methods))))
                ((null ,methods))
                (and (eq *selector* (first ,methods))
                    (return-from run-super (apply (second ,methods)
                                            self *selector*
                                            (rest ,classes) ,args))) )))))
;;; iv-value  [Primitive 5]
(defun iv-value (an-object name)
    (multiple-value-bind (description structure)
            (if (eq an-object (class-of an-object))
                (values (class-all-iv an-object) (class-organisation an-object))
                (values (iv-value (class-of an-object) ''iv)
                        (iv-value (class-of (class-of an-object)) 'organisation)) )
            (do ;; a basic description is: (name pos initial-value)
```

```
                    ((desc description (rest desc))
                     ;; Here is an assumption, structure hold (constructor,accessor,modifier).
                      (acces (elt structure 1)) )
                    ((null desc)  (send an-object 'error
                                              "This instance variable is unknown: ~S" name))
                    (and  (eq name (elt (first desc) 0))
                         (return-from iv-value  (funcall acces
                                                      an-object
                                                      (elt (first desc) 1) )) ) )))
;;; set-iv-value, this form also can be defined inside iv-value...  [Prmitive 6]
(defun set-iv-value (an-object name value)
    (multiple-value-bind (description structure)
             (if (eq an-object (class-of an-object))
                 (values (class-all-iv an-object) (class-organisation an-object))
                 (values (iv-value (class-of an-object) 'iv)
   (iv-value (class-of (class-of an-object)) 'organisation)) )
             (do ((desc description (rest desc))
                  (acces (elt structure 2)) )
                 ((null desc)  (send an-object 'error
                                            "This instance variable is unknown: ~S" name))
                 (and  (eq name (elt (first desc) 0))
                      (return-from set-iv-value
                          (funcall acces
                                an-object
                                (elt (first desc) 1) value) ) ) ) ) ) )
;;; setf form
(defsetf iv-value set-iv-value)
;;; basic class structure
;;; #( isit [(isit iv name supers linerar-extension organisation) .... ])
;;;  [Prmitive 7]
(defun class-all-iv (an-object) (elt (elt an-object 1) 1) )
;;;  [Prmitive 8]
(defun class-organisation (an-object) (elt (elt an-object 1) 5) )
;;;; Class hand make...
 ...
;;; First bootstrap
  ...
;; First Object creation
(send class 'new
        :name 'object
        :supers ()
        :iv '(isit)
```

```
        :methods '(initialize
                  (lambda (args)
                    (mapc  #'(lambda (iv)
                              ((lambda (find)
                                (and find
                                    (setf (iv-value self (elt  iv 0))
                                          (second find))))
                              (member (intern (symbol-name (elt iv 0)) 'keyword)
                                      args) ) )
                          (iv-value (class-of self) 'iv))
                    (setf (iv-value self 'isit) (class-of self)) ) ) )
; Then we rewrite the class definition...
(send class 'new
      :name 'class
      :supers '(object)
      :iv '(iv name supers linear-extension organisation methods)
       :methods  '(basicNew (lambda ()
                              (funcall (elt (iv-value self 'organisation) 0)
                                  self
                                  (iv-value self 'iv)) )
                  new (lambda (&rest args)
                            (send (send self 'basicNew) 'initialize args) )
                  initialize (lambda (args)
                                    ;; Feed slots
                                    (run-supers args)
                                    ;; compute linear extension.
                                    (send self 'compute-linear-extension)
                                    ;; collect all slots.
                                    (send self 'collect-slots)
                                    ;; walk-methods
                                    (send self 'code-walker)
                                    ;; here we are.
                                    (setf (symbol-value (iv-value self 'name)) self)
                                    self)
                  compute-linear-extension (lambda () ... )
                  collect-slots (lambda () ... )
                  error (lambda (&rest args) ... )
                  code-walker (lambda () ... ) ) )
;; Then complete the bootstrap by redefining Object and adjust all pointers.
```

Nesting in an Object Oriented Language is NOT for the Birds

P. A. Buhr* — C. R. Zarnke**

* Dept. of Computer Science, University of Waterloo, Waterloo, Ontario, Canada, N2L 3G1
** Waterloo Microsystems Inc., 175 Columbia St. W., Waterloo, Ontario, Canada, N2L 5Z5

ABSTRACT

The notion of nested blocks has come into disfavour or has been ignored in recent program language design. Many of the current object oriented programming languages use subclassing as the sole mechanism to establish relationships between classes and have no general notion of nesting. We argue that nesting (and, more generally, hierarchical organization) is a powerful mechanism that provides facilities that are not otherwise possible in a class based programming language. We agree that traditional block structure and its associated nesting have severe problems, and we suggest several extensions to the notion of blocks and block structure that indirectly make nesting a useful and powerful mechanism, particularly in an object oriented programming system. The main extension is to allow references to definitions from outside of the containing block, thereby making the contained definitions available in a larger scope. References are made using either the name of the containing entity or an instance of the containing entity. The extensions suggest a way to organize the programming environment for a large, multi-user system. These facilities are not available with subclassing, and subclassing provides facilities not available by nesting; hence, an object oriented language can benefit by providing nesting as well.

Key words: Object-Oriented, Nesting, Block Structure, Programming-in-the-Large

1. Introduction

The notion of block structure, as in nested blocks (nesting), has come into disfavour or has been ignored over the last decade. This is reflected in the fact that many of the current object oriented programming languages use subclassing as the sole mechanism to establish relationships between classes (Smalltalk [GOLDB83], LOOPS [BOBRO83], C++* [STROU86], Objective-C [COX86]) and have no general notion of block structure.

---

*C++ allows textual nesting, but its meaning is as if the contained class were not nested at all; instead, the contained class is considered to appear at the same level as the containing class. Hence, in C++, references from the contained class to variables in the containing class are not allowed; this fails to meet the normal criterion for block structure.

However, there are a few important exceptions. Both Simula [SIMULA87], the first programming language to provide an object oriented programming style, and its successor BETA [KRIST87] support the idea of block structure, in particular, nesting of class definitions, as well as the programming language LOGLAN82 [BARTO84]. However, the authors of Simula realized that block structure in Simula was too restrictive, since a class can only be referred to in the block where it is declared, and extended the concept of block structure in BETA by allowing references to a class from outside the class in which it is declared [MADSE87].

Blocks, which are formed by several programming language constructs such as BEGIN blocks, procedures, classes, packages, etc., provide two main functions: to introduce new name spaces into the definitional structure of a programming language and to introduce a new level for the control of visibility of names within a name space. Subsequent activation of a block causes allocation of all the data items contained within it. Because block structure provides a general mechanism for introducing and structuring name spaces, each program has its own name space and nesting gives the ability to create new ones. In this paper, we wish to review criticisms against block structure and to rebut them, to discuss briefly the extensions to block structure that were introduced in Simula and extended in BETA, and to extend these ideas even further with the intent of providing a mechanism for structuring an entire system using blocks. (The framework for this discussion assumes a unilingual programming environment [WEGBR71, TEITE78, GOLDB83, TEITE84, HEERI85, ATKIN85, MEYER87, BUHR87].) Our extensions depart substantially from traditional notions in block structured languages.

2. Normal Block Structure

The normal Algol scope rules preclude references to any definitions not within the current block or one of its containing blocks. The basic reason for this restriction is to assure what we call **existence**, that is, that at execution time the block frame in which a variable is declared has been instantiated. Existence is not an issue for entities such as constants and simple types. The main criticisms against block structure given in [CLARK80] and [HANSO81], many of which are addressed in [TENNE82], focus on the problems of excessive visibility of names in containing blocks. These visibility problems can usually be handled by adding appropriate facilities to the programming language so that importing and exporting of names is under explicit control. Our major complaint is the opposite situation, that is, the inability to refer to definitions in blocks other than itself or its ancestors. It is our contention that items in such blocks can be referenced and yet still be able to guarantee *statically* that they or data they use will exist at execution time (i.e. the compiler can guarantee closure).

3. Extending References to Components of a Block

Referring to fields of a record is the simplest example of the kinds of references that we are talking about. The reason that internal references are possible for a record is that a field is selected from an instance of the record, which contains instances of all the fields of the record. Simula augments the concept of a data record to contain procedures as well as data items in the CLASS construct, collectively called **attributes** in Simula but which we prefer to call **components**, and uses the same syntax to refer to both procedures and data items, for example:

```
CLASS xxx
    VAR v : ...
    PROC p(...) ...
END CLASS

VAR x : xxx
x.v        ⟵⎺⎺ references to components inside a block from outside
x.p(...)   ⟵⎯⎯┘
```

Procedures defined in the class must execute in the environment (context) of the class. As a result, a procedure must be qualified by an instance because the scope rules allow global references from the procedure to the class instance.

The next extension to class definition is in BETA, which allows references to class components of a CLASS, for example:

```
CLASS xxx
    CLASS yyy
        ...
    ...

VAR x : xxx
VAR y : x.yyy   ⟵ qualify internal class yyy by an instance of xxx
```

Extending references to classes within a block makes them a true component by enlarging their visibility to that of other components. That is, conventionally a class is only accessible within the containing block where it is defined and now it is also accessible outside of it. Qualification is allowed in the type specification of a declaration, where normally only a simple type name is allowed. The qualification uses conventional syntax and its meaning is consistent with the meaning of qualification in executable statements, that is, to select a component from a definition. The meaning of this kind of reference is to create an instance of the contained class and to use the qualifying instance as its context. In the above example, if yyy referred to components in xxx (or even outside of xxx if this was allowed), the contexts for those components are available. Both Simula and LOGLAN82 allow class definitions to be nested within other class definitions but the contained class definition cannot be referred to outside the containing one.* BETA is less restrictive, but it does limit references to one level down in the textual (static) definition structure (see also [HOUSE86]).

---

*This is the conclusion we have drawn from the one reference manual that we have on LOGLAN82. We are open to correction on this point about LOGLAN82.

4. Multiple Dependent Inner Class Instances

As discussed in [MADSE87], references to nested classes provides the ability to have multiple instances of a contained class associated with a single instance of an containing class, for example:

```
VAR x : xxx
VAR y1, y2, y3 : x.yyy
```

Here there are three instances of class yyy (y1, y2, y3) associated with the instance of class xxx (x). This allows the children to carry out actions jointly with their common parent. Both C++ and Smalltalk have a storage class (static and classVariableNames, respectively) which allow variables to be created that are unique to a class, independent of the number of instances of the class that are created; or in the case of Smalltalk, the number of subclasses that create it implicitly. This *does* mimic having multiple children for a single parent by having the static variables be considered to be the parent. Unfortunately, this *does not* generalize to allow multiple instances of the containing class with multiple instances of the contained class in each, for example:

```
CLASS xxx
    VAR x, y, z : ...       ← local variables of xxx
    CLASS yyy ...
    . . .

VAR x1, x2: xxx             ← create 2 parent classes
VAR y1x1, y2x1, y3x1 : x1.yyy   ← create multiple children in each parent
VAR y1x2, y2x2, y3x2 : x2.yyy
```

Here there are separate instances of x, y and z created for each instance of x1 and x2, and y1x1, y2x1 and y3x1 access one set in x1 and y1x2, y2x2 and y3x2 access the other set in x2. If x, y and z had static storage class, only one instance of them would be created for both x1 and x2.

Like Madsen, we feel dependent instances are an extremely useful and powerful facility in an object-oriented programming language. Madsen gives several examples of the use of nested classes in [MADSE87]. The following are more examples which we feel are important in the construction of large systems.

4.1. User-Definable Variable-Size Data Structures

Multiple dependent instances can provide a programming technique that is a compromise between universal garbage collection and no garbage collection. The technique is as follows: a containing class can act as a separate heap for storage allocation and its internal class(es) can be the types that are allocated in the heap. This is similar to the idea of a collection in Euclid [LAMPS77] or an area in PL/I [IBM81]. The following outline is a simple example of how a variable-length string data type might be defined using dependent instances; a more general scheme has been used to implement a variable-string class in C++.

```
CLASS VstrHeap(NoOfVar, NoOfChar : INT)
    RECORD VstrDesc                          descriptor for each VstrVar
        length : 0..NoOfChar                 length of string
        posn   : 1..NoOfChar                 position in VstrText
    END RECORD
    VAR VstrVar  : [1..NoOfVar] VstrDesc      indirect pointers to strings (handles)
    VAR VstrText : [1..NoOfChar] CHAR         contains text for all strings

    CLASS Vstr
        VAR VarNo : 1..NoOfVar                subscript of descriptor

        INFIX '+'( ... ) ...                  definition of assignment
        PROC substr( ... )  ...
        PROC length( ... )  ...
        PROC concat( ... )  ...
    INITIALIZATION
        VarNo + AllocVstrDesc()               get descriptor entry
    TERMINATION
        FreeVstrDesc(VarNo)                   free descriptor entry
    END CLASS

    PROC AllocVstrDesc() ...
    PROC FreeVstrDesc() ...
END CLASS

VAR vh : VstrHeap(20, 5000)  create heap
VAR a, b, c : vh.Vstr        create variable-length string variables
...
a + 'abc'
b + 'xy'
c + a.substr(2, b.length)
```

Clearly, there are drawbacks to this approach such as having to explicitly declare the containing class, and should multiple heaps be created, instances in one are *not* readily interchangeable with those in another; however, universal garbage collection has its drawbacks, too, such as the cost in execution time, storage space and system complexity. It is felt that the compromise provided by multiple dependent instances is a reasonable one. It provides flexibility to the type definer in creating sophisticated data structures that are dynamically created at a reasonable cost in execution time, storage space, and convenience to the user of the data type and yet do not require explicit freeing, thereby eliminating the dangling pointer problem.

4.2. User-Definable File Definitions

One of our goals is to be able to define, in an object-oriented paradigm, types that correspond to files in traditional systems. This is discussed in detail in [BUHR86] where only three new programming language constructs are needed to construct a file-like definition; these definitions rely heavily on nesting. Only the OBJECT construct will be mentioned here. An OBJECT is similar to a CLASS except instances of it are not directly accessible after declaration and its storage is not managed directly by the compiler. The skeleton structure for a sequential file definition is:

```
OBJECT SeqFile(RecordType : TYPE)  ← polymorphic type definition
    VAR FirstRecord, LastRecord : REF RecordType
    declaration for the records and routines to manage them in the file
INITIALIZATION
    FirstRecord ← LastRecord ← NULL
TERMINATE
    termination code, none for SeqFile
END OBJECT
```

Unlike variables in a block that are accessible immediately after declaration, files are not accessible after they are declared because they are created in a separate memory that is not always accessible by the hardware. Files must be made accessible either implicitly (Multics, Smalltalk, IBM System 38) or explicitly. We feel that explicit access is an important facility to have. As well, information must be created and maintained about each user's access to a file. For example, when reading from a file, a pointer to the current record read must be maintained for each reader. However, this access information must not be stored in the file because it may be difficult to find and adjust it correctly after system failure.

While current object-oriented programming languages do not allow the definition of files, some mechanism exists to access the system's files and, in general, it is done in the following way. There is a CLASS for the access information which contains all the access routines and access related variables. This separates the access information from the file information and allows multiple access declarations for multiple accessors to a particular file. When creating the access information, it must be tied to the particular file that it is making accessible, for example;

```
VAR f  : SeqFile(INT)    ← or some other mechanism to create a new instance of a file
VAR fa : AccessInfo(f)   ←pass the file as a parameter to establish a connection
                           between the access information and the file

fa.write(3)
i ← fa.read()
```

Terminating access is done in the termination code for the access structure. What we dislike about this approach is that the system must still provide some explicit mechanism that can be invoked by the initialization code and termination code to cause the file to become accessible/inaccessible by the hardware (e.g. creation/freeing of page tables or system buffers). As well, there is only a coincidental connection through the parameter to the access information that establishes the fact that the file will be made accessible. The compiler cannot be aware that this connection will ultimately make the file accessible/inaccessible and hence it cannot make any inferences about this important aspect of the behaviour of file-like objects.

An alternate solution is to define the access information class in the object. Then, the way to access a file is through the nested class because it contains all the access routines. When a nested class is used in this way it is called an **access class**, for example:

```
OBJECT SeqFile(RecordType : TYPE)
    VAR FirstRecord, LastRecord : REF RecordType
    declaration for the records in the file

    CLASS SeqAccess
        ESCAPE EndOfFile
        VAR CurrentRecord : REF RecordType

        PROC read RETURNS r : REF RecordType    read a record
        PROC reset                              position to 1st record
        PROC write(r : RecordType)              write a record
        PROC update(r : RecordType)             replace a record (same length)
        PROC recreate                           destroy all records
    INITIALIZATION
        CurrentRecord ← NULL
    TERMINATION
        termination code, none for SeqAccess
    END CLASS
INITIALIZATION
    FirstRecord ← LastRecord ← NULL
TERMINATE
    termination code, none for SeqFile
END OBJECT
```

The access class would be used in the following way:

```
VAR f  : SeqFile(INT)
VAR fa : f.SeqAccess

fa.write(3);
1 ← fa.read()
```

After the instantiation of SeqAccess, the SeqFile must be made accessible so that SeqAccess and its routines can refer to data items in it. Similarly, de-allocation of the access class makes the object inaccessible. Thus, the duration of object access is the duration of the access class. The actual mechanism for making the file accessible/inaccessible is not explicit but implicit through creation and freeing of the access class, which is something that the compiler has full knowledge of and hence, can possibly take advantage of. Thus, the access class provides the necessary storage creation, duration, and scope to implement explicit accessibility and support concurrent accessors, and it is tied into the programming language. Other issues dealing with serializing concurrent access to the object's internal resources when using nesting are discussed in [BUHR86, KRIST87].

5. Controlling Visibility

Traditional blocks (BEGIN, procedure, etc.) prevent access to their data components from outside because the block (and hence its data components) has not been instantiated; on the other hand, instances of classes originally allowed access to *all* internal components except internal classes (Simula67), because a class instance contains instances of all its internal components. However, the notion of abstract data types made it desirable to restrict visibility to components of classes, but not to prevent access completely. Thus, a mechanism for selectively making components visible was needed. The simplest approach is to segregate components of a class into two categories: visible/invisible, as in Ada®

®Ada is a Registered Trademark of the U.S. Department of Defense

[ADA83], Simula (protected/hidden) and C++ (public/protected/private). The following is a brief discussion of a more complex visibility scheme (see [BUHR88] for complete details).

While components can be divided into two broad categories, those that are visible to gain access to the object's functionality, and those that are strictly internal for implementation of this functionality, there are situations where not all the visible components need to be or should be visible. In other words, the subdivision into visible and invisible is too coarse; a finer partitioning is wanted in many cases. Often there are natural groupings of visible components related to different services provided by the class. For example, a sequential file type provides a storage service and access to this storage. The visible components of a sequential file type could be grouped into the following functional groups: reading, updating, extending, modifying and recreating the information in the storage container. In general, each functional group is implemented by several components and each group contains only a subset of the visible components. Normally programming languages do not support specification of the functional groupings. However, these inherent groupings provide an extremely useful way of describing other facilities not directly related to the operation of the object, in particular security and arbitration of concurrent accessors.

5.1. Perspectives

A **perspective** is the name we give to a group of components that are logically related by the function that they jointly provide. This is like naming operation lists in ALPHARD [SHAW81]. Usually there are several perspectives defined for an object. These groupings are given names and often form a hierarchy of functionality. For example, the following diagram shows the perspectives for SeqFile, the hierarchical relationships among the perspectives, and the components that make up the perspectives.

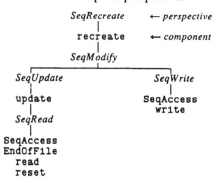

Each perspective includes all the components below it in the hierarchy. Notice that a component may be associated with several perspectives; for example, SeqAccess, which is needed to access a SeqFile no matter what type of access is desired, appears in SeqRead and SeqWrite. The exact number of perspectives depends on the implementer's intuitive notion of what constitutes an abstraction. These perspectives are declared and their relationship established by using a partially ordered set, as in:

```
PERSPECTIVE p : (SeqRead,
                 SeqWrite,
                 SeqUpdate > SeqRead,
                 SeqModify > SeqUpdate & SeqWrite,
                 SeqRecreate > SeqModify)
```

(Partially ordered sets are not peculiar to perspectives but can be used for other purposes in the programming language.) These names are then associated with components appropriate for the perspective.

```
OBJECT SeqFile
   PERSPECTIVE p : ...
   CLASS SeqAccess {SeqRead, SeqWrite}
      ESCAPE EndOfFile {SeqRead}
      PROC read {SeqRead} ...
      PROC reset {SeqRead} ...
      PROC write {SeqWrite} ...
      PROC update {SeqUpdate} ...
      PROC recreate {SeqRecreate} ...
```

Any component without a perspective is strictly internal to the object and is not part of the operations provided to the user. Hence, the perspectives define a user's interface to the object. Notice that there are now multiple interfaces for an object, which is different from most programming languages that support strong abstraction mechanisms, where there is a single interface per definition. The mechanism for granting programs and users visibility through perspectives is not discussed here, but it is similar to user definable capabilities as the perspectives are used both at compile time and execution time. For example, a perspective can be associated with a declaration, as in: VAR fa : f.SeqAccess {SeqRead}. Here the compiler can perform checks that only those components in SeqRead are used by fa. The instance can also use this information at execution time to check if a user is authorized to make the access to f and if so, schedule the access appropriately with other users depending on the functionality the perspective allows.

6. Structuring the System

The basic motivation behind much of the previous discussion is to apply the notion of strong typing across the entire system. In particular, we want to guarantee type consistency between a user's program and the system, between a user's program and conventional files (these being the sole kind of persistent data in a conventional system) and between interactive commands and the operations they invoke. (For further discussion of our motivation, see [BUHR87].) There are two important consequences of this motivation. First, all interfaces for modules that define the system, the files, and the commands must be retained and be accessible during compilation. Second, files must be objects that are completely definable in the programming language as this gives them an interface.

The interfaces for modules defined in the system and by users will be grouped appropriately into **libraries**, each containing modules that are likely related to one another in some fashion. Each user would likely have at least one library that he uses and augments. System modules, shared by many users, would likely be grouped into several libraries. These libraries are essentially the same as those described in Ada. In conventional systems, a library would have to have a representation in the file system so that it will persist, and the compiler would have to have at least some rudimentary knowledge of the file system and library structure so that it can access a library during compilation.

We propose that the libraries be defined in a hierarchy (as they undoubtedly would be in a conventional file system) that is visible within the programming language; the modules in each library then constitute an extension of this hierarchy. As well, we propose that each module node in the hierarchy be able to be extended to define all of the submodules (procedures and other scope-defining constructs) of that module. Thus, a single hierarchy defines the libraries, the modules and the blocks of user and system programs. Each node will consist of a list of components contained within that node. Although it is not essential, our intention is that these components be stored, not in textual form, but in an internal form as the symbol table that would result from the compilation of the declarations at that node. This eliminates reparsing when that node is used in another compilation. Because declarations are not in textual form, they must be maintained and modified with a special program which we call the **program editor**. The reader should understand that, although declarations are stored in symbol table form, examples in the paper will be illustrated in textual form (i.e. what the program editor might display when browsing through the hierarchy).

Although each node must contain at least the visible components defined in it, we intend that most nodes will also contain all invisible components, too, so the node will consist of all items declared within the corresponding module or block. Each component may have its source code associated with it, comprising the body (executable code) for that component; some components, such as uninitialized variables, will not have code. The source code can be stored in textual or parse tree form as either will suit our purposes. A component that defines a new scope (such as a class) will point at the node corresponding to this new scope; this produces the hierarchy we have just described.

We call this hierarchy of library, module and block definitions the **definitional hierarchy**. The hierarchy corresponding to a single module shows the traditional structure of that module in terms of nested blocks and submodules. In fact, textual nesting of blocks could be considered as an artificial way of representing this hierarchy; except for BEGIN blocks, which are not supported in the hierarchy, the two are essentially equivalent. We intend that the various kinds of nodes (libraries, modules, etc.) can be used throughout the hierarchy, except that a library node may not contain data items as components. As has been seen in Simula and BETA, concepts like libraries and modules may be able to be implemented by a single construct in the programming language. This reduces the number of different kinds of nodes in the definitional hierarchy but does not alter the hierarchy.

A brief example of structuring a system using this approach is:

```
ENVIR System
    all the system declarations
    ENVIR UserId ...   ← type defining a user

    VAR Buhr, Zarnke : UserId   ← create two new users
```

An ENVIR is a special construct (see [BUHR87]) that is similar to a class in most respects but has the following property relevant to this discussion: instantiation of an ENVIR also creates a new node in the definitional hierarchy (as if a new block definition had been made). The program editor can then be used to add definitions to it, in the same way as for conventional blocks, as in:

```
ENVIR Buhr
    CONST yyy = ...          ⎤
    CLASS xxx ...            |
    VAR i : INT              ⎬ added using the program editor
    PROC p(...) ...          |
        VAR i : INT          |
        PROC q(...) ...      ⎦
```

This results in the following definitional hierarchy:

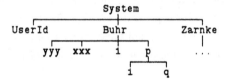

6.1. Referencing in the Hierarchy

If the conventional visibility rules applied within the hierarchy, then a component would have to be placed so that all the blocks (or programs) that needed to refer to it were descendant blocks. This would produce a structure that was top heavy because all shared components would have to be placed at or near the root. The extreme case is when all definitions are at the same level so all may reference one another (the **flat name space** approach).

Instead, we intend that a user employ the freer organizational style used at the module level — where entities are grouped by conceptual relationship rather than actual dependence — for the definition of nodes (similar to symbolic links in a file hierarchy). This, in turn, requires a way to refer to components anywhere within the hierarchy, not just to those contained in higher level nodes. We adopt the usual qualification-name syntax for such references, where the qualifiers are the names of successively lower-level nodes in the hierarchy. If one node in the hierarchy needs to refer to a component in another node, then it uses a qualified name relative to the common ancestor of the two nodes. A simple example is sharing a type or constant contained in another user's environment, as in:

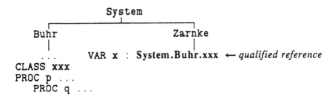

Here, the type of **x** is defined in a location which is not directly above it. It is also possible use the qualification in executable statements, such as the following procedure calls from Zarnke's code:

```
System.Buhr.p(...)
System.Buhr.p.q(...)  ← call of nested procedure
```

The above reference to the nested routine q will, in general, not be done, but it does illustrate the extent to which referencing in the definitional hierarchy is allowed. The reader may wonder why routine q is embedded inside of routine p when it needs to be used outside of p. From user Buhr's standpoint, q is never used outside of routine p and that is why he located it inside of p. If another user wishes to use routine q, this requires unfolding the existing definitions. However, the other user may not be able to make such a modification because he is not authorized to make changes to Buhr's environment. Thus, the other user would be forced to copy the code (if that is allowed) or write his own routine q; both these solutions are unacceptable because they violate the notion of reusability which is so important in programming-in-the-large. Allowing the other user to reference q directly solves these problems. In the future, if more users wish to access q, the code might be reorganized to reflect this new kind of usage. Our contention is that, if the choice is between allowing references that are not allowed by traditional scope rules and duplicating/copying the definition to another location where it will be accessible/visible, surely it is much more preferable to allow the reference to the original definition. Without these extensions, traditional block structuring is too restrictive and so the organization facilities that it provides are relegated to programming-in-the-small [KRIST87, p. 19]. With these extensions, a system can be structured according to usage relationships and yet references can be made from throughout the hierarchy, encouraging reusability [COX86]. This increase in visibility is the major factor in making block structure significantly more useful.

The main difficulty with our approach is to guarantee that the compiler can tell when a qualified reference can and cannot be allowed. In the nested procedures example of p and q, the reference to procedure q from outside procedure p may not be valid if q has a reference to any variable outside itself. Even if a block frame for p was created to act as the context for the execution of q, the lack of initialization of this block frame would make the execution of q unsafe. Because of the definitional hierarchy, the compiler can determine if q depends on its context and hence indicate when a reference to such a dependent entity is invalid. We call an entity that does not depend on its static context **independent**.

Our approach to structuring the system will now be contrasted with the structure of existing systems. Module interconnect languages are discussed as a mechanism to perform this structuring [DEREM76, PARNA85]. We agree with the benefits that might be achieved by such languages, but they likely require much of the interface information that is stored in our definitional hierarchy, and so it makes more sense to us to extent the definitional hierarchy to include module interconnect language information rather than creating a new language to do this.

In Smalltalk [GOLDB83], classes can be grouped into categories by the browser; however, this structure is not in any way related to the structure of the class definitions or the language itself. Essentially, all Smalltalk CLASS definitions are in a flat name space and the browser provides some artificial structure to this flat space but only for the purpose of examination by a user, not for references by objects. Browser names are not part of the Smalltalk language. This same flat name space is the way many Lisp [WILEN86] systems are implemented, too. In a large program development multi-user system, forcing all programs and all users to use the same name space is, in general, not acceptable. This problem is further exacerbated if the system also shares the same name space.

In C++ the structuring mechanism is the file system; a user employs it to store his programs. The classes of a C++ program are organized in a flat space within a source file and then a programmer can create some artificial structure for the source files through the UNIX® file system and preprocessors. However, the file system is outside the realm of the programming language (and hence no type checking) although its various levels may be very similar to the name spaces created through block structure in a program. Hence, while classes in these flat name space may have complex relations with one another, there is no requirement that this structure be reflected in any organizational scheme that is known in the programming language. Similar problems exist in Ada where the with statement is used to include packages but there is no connection between the name of the entity specified in the with statement and the actual library that contains it. Simula has the same problem when using a library class as a block prefix; how is the class prefix name found when it is not defined within the current source program?

Notice that in our system if an extended reference is made to an entity which also contains an extended reference to another entity, there are no problems. Each entity will be looked up in the context of the reference. This is not the case in UNIX/C where included source files are looked up in the current environment (unless absolute path names are specified) and not the included file's environment. This is because including is based on copying and our scheme does not copy but uses a definition intact as if it appears at its point of static definition. This is exactly how references are made to procedures, classes, etc. in the programming language. Hence, the same scheme is used to reference what are traditionally thought of as library routines and routines within a program.

®UNIX is a Registered Trademark of AT&T Bell Laboratories.

6.2. Separate Compilation

A node in the definitional hierarchy may have properties that are largely independent of its kind (i.e. module, procedure, etc.), such as whether it is separately compilable. We propose that any node in the hierarchy can be identified as being separately compilable, for example:

```
CLASS p COMPILABLE
    . . .
    CLASS q COMPILABLE
        . . .
        PROC r(...) COMPILABLE
            . . .
    PROC s(...) INLINE
        . . .
    PROC t(...)
```

Here each procedure that has a COMPILABLE clause is eligible to be compiled separately, and components with no compilation clause are compiled as part of the node that contains them. In the above example, t is compiled as part of the compilation of p (like Algol68C and Simula). INLINE has the standard meaning. There is no order on compilation; however, at least the interface for all referenced items must be defined before a node using them can be compiled. Normally, module nodes will be designated separately compilable but individual procedures might be declared as being separately compilable during their development. Separate compilation at this level, however, will exact a cost at execution time. Adding or removing a compilation clause can be done without a change to the structure of the program. Traditional systems often force a flattening of the nested scope into modules or packages to introduce a compilation unit.

As well, it is possible to use the definitional hierarchy to define precisely the location of object code generated by the compilation of each compilation unit. By introducing a definitional-time declaration of a code area, it is possible to have each compilation unit specify where its code will be placed, as in:

```
CLASS p COMPILABLE m ← p's code is placed in code area m
    . . .
    CLASS q COMPILABLE n ← q's code is placed in code area n
        . . .
```

It is our intention to be able to execute object code directly from these code containers and not necessarily to copy object code together to form the executable equivalent of a PROGRAM in Pascal (code containers are like Multics segments). Hence, we can largely do away with traditional linking (i.e. copying separate compilation units together) and prefer to dynamically execute from the original copy of the code. The user can control the number code containers that must be made accessible at execution time by controlling which compilation units go in which containers.

7. The Complex Situations

One anomaly that results from adopting this scheme is that two separate names might have to be specified to instantiate a class contained in another class: one for the instance that is the context for the instantiation and one for the type for the instance, for example:

```
CLASS aaa
    CLASS bbb
        VAR x :  ...
        CLASS ccc
            x ← global reference to x in bbb
        END CLASS
    END CLASS

    bbb.ccc CLASS ddd  ← prefix is a qualified reference to a type
        . . .
    END CLASS

    VAR b : bbb
END CLASS

VAR a : aaa
VAR d : a.ddd
```

Unfortunately, the declaration of d is invalid because, in order to instantiate ddd, it is necessary to provide an instance of bbb because there is a dependence between the super-class for ddd, that is, ccc and class bbb. However, a reference such as a.b.ddd is also illegal because ddd is not a component of bbb. All this can be determined by the compiler. To deal with this we imagine a specification such as:

```
                      ┌─ instance context
                     ┌─┐
        VAR d : a.b|aaa.ddd
                    └─────┘
                        └─ type
```

where the instance name to the left of the | indicates the context, and the (potentially qualified) name to the right of the | indicates the type of the instance.

The inclusion of both subclassing and nesting complicates the rule for looking up names that are not completely qualified. This is because there are now two paths that can potentially be searched, for example:

```
        CLASS www
            CLASS xxx
                . . .

        . . .
        CLASS yyy
            www.xxx CLASS zzz
                x ← reference to a variable not defined in the current scope
            . . .
```

The name x can be looked up in one of two sequences of blocks: zzz, xxx, www, and possibly the context of www; or zzz, yyy, and then the context of yyy. The first search path follows the path of the subclassing chain; the second follows the hierarchy of definition. A similar problem exists with multiple inheritance, that is, which order are the ancestors looked in to resolve an unqualified name.

The search rule we have chosen is to search the superclass chain to its end but not the context containing the superclass and then proceed back to the starting point and search the definitional hierarchy. This will require that two contexts be maintained at execution time for each class instance, including each superclass instance. These contexts can all be determined at compile time. This particular rule was chosen because of the way we are structuring persistent entities in our system [BUHR87].

8. Implementation

We have started implementation of an integrated programming environment that is structured using extended block structure. The implementation work is in C++ using nested blocks to structure the work. This requires that we simulate nested classes by passing the containing class as an explicit context. We have considering the idea of writing a preprocessor to support nested classes in C++, but C++ is rather unstable at the moment. As well, we have used the storage management technique of multiple child classes in the implementation of dynamically sized types, such as variable-length strings (which appears on the Usenix C++ examples tape). We have compiled a library of storage managers that can be used by implementors of dynamically sized types. Having a number of storage managers allows us to try different ones with a particular type to find a good one.

9. Conclusion

Our idea of extended block structure is a straight-forward generalization of block structure and its usage follows naturally from the methods of programming in object oriented programming. Our scheme does not destroy the ideas of abstract data types and encapsulation. If a programmer does not want references to internal components, they can be protected. Extended block structure provides facilities that are not available in existing object oriented programming languages; these facilities can be used to solve programming-in-the-large problems, such variable-sized data structures and file definition. But most importantly, using nesting to form the definitional hierarchy and extended references to access items within the hierarchy provides a means to achieve strong typing across the entire system for a statically-typed unilingual programming environment. Other interesting problems might also lend themselves to elegant solutions using extended block structure.

10. Acknowledgments

We would like to thank Ron Pfeifle for helping with some of the background material, and Glen Ditchfield and Lauri Brown for reading the final draft.

11. References

ADA83   United States Department of Defense. "Reference Manual for the Ada Programming Language". *ANSI/MIL–STD–1815A–1983*, February 1983, Springer–Verlag, New York

ATKIN85   Atkinson, M. P., Morrison, R. "Types, Binding and Parameters in a Persistent Environment". *Persistence and Data Types Papers for the Appin Workshop*, Persistent Programming Research Report 16, University of Glasgow, Dept. of Computing Science, Scotland, August 1985, pp. 1-24

BARTO84   Bartol, W. M., et al. "Report on the LOGLAN82 Programming Language". Polska Akademia Nauk, Instytut Podstaw Informatyki, Panstwowe Wydawnictwo Naukowe, Warszawa—Lodz, 1984

BOBRO83   Bobrow, D. G., Stefik, M. "The LOOPS Manual". Xerox Corporation, 1983

BUHR86   Buhr, P. A., Zarnke, C. R. "A Design for Integration of Files into a Strongly Typed Programming Language". *Proceedings IEEE Computer Society 1986 International Conference on Computer Languages*, Miami, Florida, October 1986, pp. 190-200

BUHR87   Buhr, P. A., Zarnke, C. R. "Persistence in an Environment for a Statically-Typed Programming Language". *Persistent Object Systems: their design, implementation and use.*, Persistent Programming Research Report 44, University of Glasgow, Scotland, August 1987, pp. 317-336

BUHR88   Buhr, P. A., Zarnke, C. R. "Protection in a Multi-User Object-Oriented Environment". in preparation.

COX86   Cox, B. J. *Object Oriented Programming*, Addison—Wesley, 1986

CLARK80   Clarke, L. A., Wilden, J. C., Wolf, A. L. "Nesting in Ada Programs is for the Birds". *SIGPLAN Notices*, vol. 15, no. 11, November 1980, pp. 139-145

DEREM76   DeRemer, F., Kron, H. H. "Programming-in-the-Large Versus Programming-in-the-Small". *I.E.E.E. Transactions on Software Engineering*, vol. SE-2, no. 2, June 1976, pp. 80-86

GOLDB83   Goldberg, A., Robson, D. *Smalltalk—80: The Language and its Implementation*, Addison—Wesley, May 1983

HANSO81   Hanson, D. R. "Is Block Structure Necessary?". *Software—Practice and Experience*, vol. 11, 1981, pp. 853-866

HEERI85   Heering, J., Klint, P. "Towards Monolingual Programming Environments". *ACM Transactions on Programming Languages and Systems*, vol. 7, no. 2, April 1985, pp. 183-213

HOUSE86   House, R. T. "Alternative Scope Rules for Block-Structure Languages". *The Computer Journal*, vol. 29, no. 3, June 1986, pp. 253-260

IBM81   IBM. *OS and DOS PL/I Language Reference Manual*, Manual GC26—3977—0, September 1981

KRIST87   Kristensen, B. B., Madsen, O. L., Moller-Pedersen, B., Nygaard, K. "The BETA Programming Language". *Research Directions in Object-Oriented Programming*, Shiver, B., Wegner, P. (eds.) MIT Press, Computer Systems Series, 1987, pp. 7-48

LAMPS77   Lampson, B. W., Horning, J. J., London, R. L., Mitchell, J. G., Popek, G. L. "Report on the Programming Language Euclid". *SIGPLAN Notices*, vol. 12, no. 2, February 1977, pp. 1-79

MADSE87   Madsen, O. L. "Block Structure and Object Oriented Languages". *Research Directions in Object-Oriented Programming*, Shiver, B., Wegner, P. (eds.) MIT Press, Computer Systems Series, 1987, pp. 113-128

MEYER87   Mayer, B. "Eiffel: Programming for Reusability and Extendibility". *SIGPLAN Notices*, vol. 22, no. 2, February 1987, pp. 85-94

PARNA85   Parnas, D. L., Clements, P. C., Weiss, D. M. "The Modular Structure of Complex Systems". *I.E.E.E. Transactions of Software Engineering*, vol. SE-11, no. 3, March 1985, pp. 259-266

SHAW81   Shaw, M. *ALPHARD: Form and Content*, Springer-Verlag, New York, 1981

SIMULA87   *Simula - Data Processing, Programming Languages*, Swedish Standard SS636114 SIS, Stockholm, Sweden, June 1987

STROU86   Stroustrup, B. *The C++ Programming Language*, Addison-Wesley, 1986

TEITE78   W. Teitelman, J. W. Goodwin, A. K. Hartley, et al., *Interlisp Reference Manual*, Xerox Palo Alto Research Center, 1978

TEITE84   W. Teitelman, *The Cedar Programming Environment: A Midterm Report and Examination*, Xerox Palo Alto Research Center, Technical Report CSL-83-11, June 1984

TENNE82   Tennent, R. D. "Two Examples of Block Structuring". *Software—Practice and Experience*, vol. 12, 1982, pp. 385-392

WEGBR71   Wegbreit, B. "The ECL programming system". *Proceedings of AFIPS 1971 FJCC*, vol. 39, AFIPS Press, Montvale, N. J., pp. 253-262

An Object-oriented Exception Handling System for an Object-oriented Language.

Christophe Dony
Laboratoires de Marcoussis, CGE Research Center,
route de Nozay, 91460 Marcoussis, FRANCE.
&
LITP, University of Paris VI,
4 place Jussieu, 75005 Paris, FRANCE.

Abstract.

We present an original exception handling system especially designed for object-oriented languages, making actual information hiding possible and taking into account specific issues of object-oriented languages. It allows association of handlers with expressions as well as with object classes, using a well defined semantics. It offers an object-oriented and extensible representation of exceptions, handlers and knowledge about exceptions. Handlers can specify both resumption and termination. There are no distinctions between system and user defined exceptions. With this system, fault tolerant programs and well specified encapsulations can be written, simple and powerful integration of new user-defined exceptions and secure as well as readable non local moves can be implemented.

In this paper, we examine object oriented specific issues related to exception handling. We discuss the exception handling mechanisms available in current object-oriented languages and explain why they do not provide the ability to define fault tolerant encapsulations. Our system description shows how the utilization of the object-oriented formalism solves, in an efficient and simple way, some well known problems related to exception handling such as : how to create exception hierarchies, how to signal fatal or continuable exceptions with the same primitive, how to pass arguments to handlers, and so on.

Key words and phrases : Exception handling, Object-oriented programming, Fault-tolerant encapsulations, Resumption model, Debugging environments.

1. Introduction.

We present an original exception handling system, designed for an object-oriented language and taking advantage of the object-oriented formalism. This system is implemented and currently used.

Our system improves program reliability and reusability [Meyer 87] by allowing the creation of fault tolerant encapsulations, making actual information hiding possible. Indeed, each procedural abstraction (method) is able to handle lower level exceptions raised by inner method activations; and is able to specify all its possible responses, including exceptional ones.

Since atypical events can be dealt with, we improve the expressiveness of the language. We allow dialogues based on exceptional situations : a method M can answer *"I don't know"* or *"impossible to do this because ..."*, by raising an exception. The method caller can anticipate and say *"if the response to this message is an impossibility then ..."*. Control structures that make these dialogues possible allow secure and clear non local moves [Testard-Vaillant 86] to be written, as they provide a way of communication between nested method invocations.

An important characteristic of the system relies on the object-oriented representation of exceptions and handlers. Some well-known exception handling problems are thus easily and efficiently solved by the possibility of defining slots and methods on the method, handler and exception classes. For example, two slots allow handlers to know precisely where the exception was raised. New possibilities arise, for example exceptions can be organized in hierarchies. The debugging possibilities of programs which use exception handling mechanisms are enhanced. For example the system can be requested to give the name of all the methods where a given exception could be raised.

We take into account the object-oriented representation of problems as well as the communication by message passing while allowing association of handlers with both expressions and classes. All handlers take into account the exception hierarchy.

Our system is based on an analysis of similar works for procedural languages such as *Pl/I* [Pl/I 78], *Clu* [Liskov 79], *Ada* [Ichbiah 79] and *Mesa* [Mitchell 79]. Basic definitions and references are provided in section 2.

In section 3, we discuss exception handling problems specific to object-oriented languages. We explain why attaching handlers to classes is insufficient and does not enhance reusability, why attaching handlers to dynamic units is also insufficient, why exceptions should be first class objects and why *unknown-selector* should not be the only fundamental exception.

Our system is also based on a study of exception handling systems in current object-oriented languages. In section 4, we review data structures within the *Lisp+Flavors* system [Moon 83]. We examine the signaling and handling possibilities of current object-oriented languages with special attention to *Smalltalk* [Goldberg 83].

Section 5 is a presentation of our system. We briefly introduce the language *Lore* for which it is implemented. We give a system overview and explain our main choices. We then describe how to create, signal and handle exceptions, how to restore environments when leaving a sequence of code, how to define default-handlers and interactive resumption propositions.

2. Exception handling basic definitions.

Research for improving verifiability, reliability and robustness of programs led to concepts such as modules [Parnas 72], data abstraction [Hoare 72], software engineering [Boehm 75] and software fault tolerance and avoidance [Liskov 79] [Christian 82]. The program structures for handling exceptional events [Goodenough 75] [Levin 77] [Berry 85] [Knudsen 87] are mainly designed to implement fault-tolerant encapsulations, i.e. software objects able to return well defined and foreseen answers, whatever may happen while they are active, even though an exceptional situation occurs.

An exception can be defined as a situation leading to an impossibility of finishing an operation. The term "exception" implies that this situation is not necessatily an error case. Three types of exceptions can be distinguished (see [Goodenough 75] for examples) : the **domain exceptions** raised when the input assertions of an operation are not verified, the **range exceptions** raised when the output assertions of an operation are not verified or will never be, and the monitoring exceptions raised to implement controlled non-local moves.

To **raise** (a procedure invocation raises an exception) or to **signal** (a procedure activation signals an exception) an exception results in interruption of the usual sequence followed by search and invocation of a handler for that exception.

Handlers are attached to (or associated with) entities for one or several exceptions (according to the language, an entity can be a program, a process, a procedure, a statement, an expression, or a data). Handlers are invoked when an exception is signaled during the execution or the utilization of one of these protected entities. They can raise new exceptions, **exit** or **terminate** the entity invocation and sometimes **resume** the signaler, i.e. transfer control to the statement following the signaling statement.

3. What an exception handling system should provide.

Because of the lack of space, we focus on exception handling problems specific to object-oriented languages. Four major issues are developed : is there a need for new exceptions? Which handler attaching possibilities should be provided? How to define fault-tolerant encapsulations? What can be expected of object-oriented representation of exceptions? We outline eight main objectives that constitute our system basis.

3.1. New exceptions.

Object-oriented design allows operations to be broken down into sub-operations defined on relevant sub-domains of the whole definition domain. Then, message sending (as in *Smalltalk*) as well as generic functions and multi-methods (as in *Common-loops* [Bobrow 86]) may convert numerous domain exceptions (and some range exceptions[1]) into exception **unknown-selector** raised[2] by the message sending primitive. For example, evaluating the message *"["abc" + 1]"* raises : *"Error: string "abc" does not understand message +"*, instead of a conventional domain exception *"Error: +, argument "abc" is not a number"*. One may think that since all domain tests are performed via message sending (or generic function invocation), all classical exceptions are converted into *unknown-selector*. However, in many cases, specific exceptions have to be raised. On the one hand, this is the case (1) when the input assertions of an operation may be based on the types of several arguments and not only on the receiver's type (this remark does not apply to generic functions); (2) when the input assertions are not based on argument types or include complex calculations. On the other hand, it is not always possible (or it is too expensive) to divide an operation into sub-operations defined on sub-domains. This last point is true when sub-domains do not exist and cannot be defined (as classes) in the language (for example "positive integers" or "[1 5] union [7 9]") or when this breakdown would lead to the creation of too many subclasses, etc.

---

[1] A range exception for an operation is sometimes nothing but a domain exception for an inner operation. E.g. *end-of-file* is a range exception for procedure *read-string* and is a domain exception for inner operation *read-char*.

[2] when no method is associated with the given selector in the given receiver class, or when there is no method the parameter type tuple of which matches the given argument tuple.

> **Objective 1 : Although there are fundamental exceptions, it should be possible for the users, to define new ones.**

3.2. Where to define handlers.

One of the characteristics of the object-oriented design is that procedural knowledge is attached to object classes. We think that such a possibility is not powerful enough to store the variety of responses to exceptional situations. Three levels of knowledge about handling exceptional events should be distinguished : (1) knowledge that is independent from any execution context as well from the object data-base state, (2) knowledge that is common to a set of objects, (3) knowledge related to a language expression the evaluation of which may signal an exception All these points depend on handler attachment capabilities.

• **Exception handlers.** The most general handlers must be valid in any cases, e.g. they can print an error message or make a general correction proposition. They are always invoked when no more specific handlers can be found. They should be associated with exceptions themselves.

> **Objective 2 : It should be possible to attach handlers to exceptions.**

• **Class handlers.** Attaching handlers to an object is not a new idea [Goodenough 75], such handlers are invoked when an exception is raised during an access to that object. From an object-oriented view, similar handlers can be attached to object classes. They allow to define, for all instances of a class, a common behavior in front of exceptional situations, which however is independent of any execution context. For this last reason, we will speak of static handlers[3].

Here is an example where static handlers are useful : encapsulation of all the messages to a particular object [Pascoe 86]. We want to perform some operations around the execution of the original message. Some applications are : debugging, concurrent access or context restoration. What is done in the handler is independent of the execution context. *Encapsulator* is a class on which a handler for exception "unknown-selector" has been defined. Each instance of this class can then encapsulate another object. When a message is sent to an encapsulator object instead of the encapsulated one, the exception *unknown-selector* is raised and handled, then additional operations are performed around the original message execution the result of which is used to resume the signaler.

> **Objective 3 : It should be possible to attach handlers to classes.**

• **Expression handlers.** Handlers attached to expressions allow the specification of the program continuation when the invocation of the expression to which they are attached raises an exception that they handle.

Here is an example where expression handlers and context-dependent responses to an exception are needed : two methods defined on the same class have to react in different ways to the same exception, which is, in both cases, raised by sending the same message to the same object. Handlers defined on

---

[3][Liskov 79] spoke of static association to enhance the fact that in *Clu* it is possible to determine, by a static analysis of the program, the handlers to be invoked. Since new classes are rarely dynamically defined, this is also true of our "class handlers".

a class would be insufficient since they neither know where the exception was raised nor in which nesting context; they cannot provide a context-dependent answer. Let two methods *divide* and *modulo* be defined on the class of *natural numbers* using method *minus* which is itself defined with primitive method *predecessor*, raising the exception *negative-number* when the message predecessor is sent to 0. Division of *x* by *y* is defined by successive subtractions (*minus*),each time a counter is incremented, no entry assertion is tested but the exception *negative-number* caught and the counter value is returned. *Modulo* is defined in the same way but there is no counter and the handler for the same exception returns the value of *x* (cf. § 5.5).

Notice that, in these definitions, handling exceptional cases is an important feature because it avoids the test "Y ≤ X" which is as costly to compute (in our example) as the operation "X minus Y". Moreover, since it is not possible (in general) to define the subclass of numbers that are lower or higher than others, this test could not be bypassed by subclass creation and message sending.

Objective 4 : It should be possible to attach handlers to dynamic entities.

3.3. Fault-tolerant encapsulation.

Two opposite solutions to information hiding can be found in object-oriented languages. In the first one, creating fault-tolerant encapsulation is viewed as the ability to handle exceptions within object classes, i.e exceptions are not propagated outside the method in which they are signaled but handled within the receiver's class. In the second one, a fault-tolerant encapsulation is an entity able to catch and hide the exceptions raised by their inner modules, but also able to propagate relevant exceptions and return exceptional answers.

Objective 5 : Exceptions should first be propagated along the invocation chain.

With such propagation rules, a method has a predefined protocol to answer "*I do not know*" or "*I cannot do this*" to its caller. Furthermore, it enhances reusability since message senders are able to give context-dependent answers to exceptional situations.

As a result of such a decision, all static handlers become default handlers since they are invoked only if no expression handlers has been found. A specific primitive then has to be designed in order to implement the encapsulator example.

3.4. Object-oriented knowledge representation.

In procedural or functional languages, exceptions are strings, symbols or variables declared to be of type: "exception". Knowledge about exceptions is scattered in the handlers. It is odd to see that most object-oriented languages do not take advantage of their data structuring possibilities to represent exceptional events. In *Smalltalk*, an exception is a selector but not a first class object. Thus it cannot own any characteristics, cannot be inspected, modified or upgraded.

Yet, exceptions are complex entities, they own not only attributes such as the context where they are raised, but also procedural characteristics that describe for example how to report an error message or how to propose solutions to the failure. As soon as exceptions are represented as data-types, following the ideas developed in the *flavors* system [Moon83] or in *Taxis* [Nixon 83], *it* becomes simple and natural to specify their static characteristics as attributes (slots) and their dynamic knowledge as methods.

> Objective 6 : Exceptions should be hierarchically organized classes.

Here are some of the main advantages of this choice.

• An exception (a concept) or an instance of an exception is, and can be used as, a first class object, with all debugging and extensibility consequences this can entail.

• Exception handling specification benefits from the object-oriented formalism advantages. Exceptions can be organized in a hierarchy based on common behaviors : exception *divide-by-zero* is naturally a subclass of *arithmetic-exception*. Properties can be shared : when *unbound-instance-variable* is raised and not handled, the following proposition "P" is made : "supply a value to store as the value of X". This proposition is not only valid for *unbound-instance-variable* but for *unbound-variable* too; for this reason, it is defined on an upper-class named *cell-content-error* as a *document-proceed-type* method. Besides, one method is defined on each exception to properly store the new value.

• All predefined properties are reusable via subclassing and method overriding. The new exceptions can be integrated as "sub-exceptions" of existing ones. As a consequence, the handlers for the new exceptions can take advantage of all the behaviors already defined on the upper exceptions. Furthermore the system and all the applications using this exception handling system will offer in the same way the same basic default debugging environment.

• Handler definition is powerful, since handlers do not only handle one (one kind of) exception but all exceptions that are sub-classes of it. E.g, handlers for *arithmetic-exception* catch *overflow*, *divide-by-zero* as well as *arithmetic-exception* itself.

> Objective 7 : All handlers should take into account the exception hierarchy.

Such ideas can still be improved : e.g., specific handling possibilities may be defined on suitable exceptions. Consequently, thanks to message passing, it is implicit that the resumption message cannot be sent to a fatal exception just as the termination message cannot be sent to a proceedable one.

> Objective 8 All handling primitives should be generic operations.

4. Object-oriented exception handling systems.

4.1. Data structures.

As we pointed out in the last section, there are few examples of object-oriented languages where exceptions are first class objects, *Flavors* being one example [Moon 83]. Main principles of this system (from the data structure viewpoint) relies on the fact that exceptions are classes, an instance of one of these classes is created when an exception is raised, all handlers for that exception receive this instance as argument. Thus handlers can access, via inspection and message sending, to the behavior defined on this exception (this flavor). Handlers are neither objects nor methods but *Lisp lambda expressions*, they can only be attached to expressions, both resumption[4] and termination are possible. *Taxis* provides similar capabilities and also allows to associate default handlers with exceptions.

Important works, based on the object-oriented representation of exceptions in the language *Taxis*, have been done in languages designed to manage data-bases and information systems [Borgida 85]. The adaptation of these techniques to object-oriented languages [Borgida 86] brings efficient solutions to the problems raised by exceptional objects and values. Exceptional objects can be exceptional instances (objects with slot values breaking constraints) or exceptional subclasses (that do not want to inherit certain properties defined on upper-classes). Exceptional data are stored as instances of exception classes and do not interfere with "normal" ones. This prevent from creating multiple "abstract" classes or modifying the range of the slots. Exceptional data can be accessed after exceptions were raised, these accesses are slower than classical ones, but standard computations are not slowed down by management of possible occurrences of exceptional cases.

4.2. Signaling and handling capabilities.

• **Standard solutions.** A classification of object-oriented languages based on their exception handling possibilities brings out three categories, where the main specificities remain in the handler attachment possibilities. None of these categories is wholly satisfactory.

Smalltalk is the main example of the first one, where handlers for all exceptions can only be statically attached to classes.

The second kind of languages consists in extensions of procedural or applicative languages. These extensions are turned towards new data type creation without inheritance (for example *Clu*) or with inheritance (for example *Simula* [Dahl 70], *C++* [Stroustrup 86] or *Flavors*). These extensions have been done without modification of the standard or existing exception handling mechanisms. Regardless of the resulting systems - there is no room here to examine all the possibilities - these do not provide solutions to associate handlers with classes or with exceptions (objectives 2 & 3).

Finally, some languages (such as Loops [Bobrow 83] or Objvlisp [Cointe 84]) also built on top of procedural or functional ones, choose a compromise: some exceptions relative to object manipulations are raised and handled as in *Smalltalk*, the other ones depend on the underlying language. As a consequence, the handling possibilities are unequal : it is impossible to attach a handler for *unknown-selector* to an expression, and impossible to attach a handler for *divide-by-zero* to a class.

• **Smalltalk.** The *Smalltalk* solution to exception handling is a model for many object-oriented languages. In order to signal run-time exceptions, the *Smalltalk* evaluator sends, to the current object, a message corresponding to the current exception. Therefore handlers are methods pointed out by exception selectors, and are attached to classes. Thus, exceptions raised by methods defined on a class are handled within that class. There are three main possible selectors : *DoesNotUnderstand* (of which *ShouldNotImplement* and *SubclassResponsability* are variants) reports the *unknown-selector* exception, *PrimitiveFailed* reports system errors and lastly the selector *Error* reports all the other exceptions.

---

[4]Handlers can be executed in the lexical environment of the caller when they are defined using the *function* special form.

This system is simple and efficient since it only uses message sending and method definition which are basic operations. Besides its expressive power seems insufficient and some of the above-mentioned objectives are not fulfilled satisfied.

• Handlers cannot be attached to expressions or statements. Exceptions cannot be propagated to operation callers; as stated earlier, this limits the reutilization possibilities.

• There is no well suited location to store knowledge (about exceptions) which is independent of the class hierarchy. In other words, all of this knowledge has to be stored in handlers attached to the hierarchy root class (object).

• Exceptions are not objects, they cannot own properties, they are not organized in a hierarchy.

• There is no simple and predefined way to create new exceptions (this must not be confused with the possibility of signaling new exceptional cases using the selector "error").

• Exceptions, except the two basic ones, are anonymous; a handler cannot know which exception it handles unless some meaningful arguments are provided.

• All of the knowledge about an exception must remain in its handlers. Then, the same method "error" defined on a class may have to handle several exceptions. This may lead to confusion and non object-oriented programming style.

5. Description of the system.

5.1. The Lore language.

Our system has been developed and implemented for the object-oriented language *Lore* [Caseau 87] [Benoit 86] dedicated to knowledge representation. Non basic parts of our system are written in *Lore* and take advantage of its possibilities. Here are some important features. Classes are sets, properties defined on classes (slot, methods, etc) are relations whose domain and range are *Lore* sets. All properties such as *slots, relations* or *methods* are objects. Among all the consequences, one is, for example, that properties can have properties. Communication is based on message passing and on the multi-method concept : methods are defined on the message receiver class but several methods with the same name can be defined on a class. Here are a few syntax examples. (Words in italic are *Lore* predefined objects, bold words are existing slots, methods or special forms)

• Message sending ≡ [receiver selector arg1 ... argN]

• Class creation ≡ [square isa *class* **superset** figure]

• Relation definition ≡ [square **has** *relation* length **range** *integer* **init-value** 10]

• Method definition ≡ [square **has** *method* grow **comment** "..."
      filter (i integer) ; Parameter name and type.
      form [oself length **is** [[oself length] + i]]] ; The body of the method.

• Contracted definition ≡ [rectangle isa *class* **superset** figure **with**
      (*slot* length **range** *integer*)
      (*method* area **form** [[oself length] * [oself width]])]

• Square instantiation and slot assignment ≡ [s1 isa square length 10]

- Assignment ≡ [value -> variable-name]
- Modifying a slot value ≡ [s1 length is 20]
- Asking for squares the length of which is 20 ≡ [length.square of? 20]

5.2. System overview.

Our exception handling system can be used in the same way by the system and the by user programs. Our specification is based on the above-mentioned objectives.

• **Status of exceptions.** Exceptions are classes (objective 6); an instance of an exception X is created each time X is raised. Defining an exception is nothing but creating a new class (objective 1), new exceptions can be raised and handled exactly in the same way as predefined ones. Two basic exception types are defined : the fatal (**error**) and the proceedable (**warning**) ones.

• **Status of handlers.** Handlers are methods (i.e objects in *Lore*) or instances of a specific class named **protect-handler**. They can be attached to expressions, to classes and to exception classes (objectives 2, 3 & 4), they are searched and invoked in that order (see § 5.4). Static handlers (attached to classes) are default handlers. All handlers are aware of the exception hierarchy (objective 7), the handlers for **exceptional-event** (the exception lattice root) will thus catch all exceptions. The mechanisms used to search and invoke handlers are the standard *Lore* mechanisms used to find (using the receiver, the selector and the arguments) and invoke methods.

• **Which model?** Both the **resumption** and the **termination** models are supplied. Termination means that the method activation that raises the exception is exited. Resumption means that after the execution of the handler, control returns to the statement following the signaling one. Although the resumption model has some drawbacks [Liskov 79], we choose it for its expressive power useful for interactive or simulation applications, where operators, users or experts may bring interactive solutions; debugging propositions are based on the same principle [Moon 83] [Wertz 83]. This model is as efficient to handle and restart long computations (the same possibilities can be obtained with a termination model assuming association of numerous handlers which can be time and place consuming and makes programs tedious to write). Lastly, the resumption model offers to all users a well defined way to access calculation history [Testard-Vaillant 86], which can be very useful to compute special applications, to explain a calculation or to show the current state of a program execution.

• **Handling possibilities.** All handlers may reference the variable xself, they can access and modify global variables and objects. Three methods are defined on exception classes (objective 8). Exit and retry defined on error and resume defined on warning can be used within handlers. Resume causes the resumption of the interrupted computation, exit leads to the execution termination of the expression to which the handler was attached, retry is a variation of exit such that, after termination, the expression is once again executed under the same protections. It is possible to signal an exception within a handler.

• **Semantics of signaling.** The method signal is designed to raise any exception. Arguments can be submitted and will be transmitted to handlers. The signaler can choose the type of the signaled exception. The handlers are responsible for resuming or exiting. These choices are consistent with the theory according to which the signaler knows the seriousness of a situation, whereas the caller knows the context of the operation invocation and can decide with full knowledge of the facts.

5.3. Exception classes.

Exceptions are instances of the class **exception-class**, itself a subclass of **class**. There are four basic instances of **exception-class**. <u>Exceptional-event</u> is the exception lattice root. <u>Error</u> is the set of exceptional-events for which resumption is impossible. <u>Warning</u> is the set of exceptional-events for which termination is impossible. For the instance of <u>exception</u> - subset of **error** and **warning** - both termination and resumption are possible. (cf. Fig. 1).

• **Creating new exceptions.** An instantiation of **exception-class** creates a new exception on which slots and methods can be defined; these slots can be assigned while signaling and accessed (modified) within handlers. This solution provides a communication mechanism between signalers and handlers. Three basic slots are defined on **exceptional-event** to contain the signaling context (**xobj xprop**) and general arguments (**xargs**). Several methods are defined, among which **print** is intended to display the basic error message.

The following example shows the definition of **unknown-selector** where the method **print** is redefined, and makes use of the slot values to report the error.

```
[unknown-selector isa exception-class superset exception with
  (slot receiver range object comment "The receiver of the wrong message.")
  (slot selector range symbol comment "The selector of the wrong message.")
  (multi-slot args range object comment "The Arguments of the wrong message.")
  (method print form .... [[oself receiver] print] ....)]
```

5.4. Signaling.

• **Syntax.** Any method can signal an exception. **Signal** is a method defined on the class **exception-class**, so signaling consists in sending the message **signal** to an instance of **exception-class**. **Signal** is nothing but a redefinition of the method **new** that first creates an instance of an exception and then searches an handler for it.

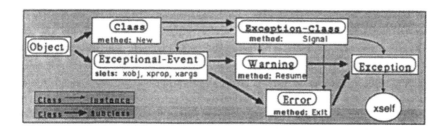

Fig. 1 : **Basic exception types and meta-type.**

It is important to understand the distinction between the generating class **exception-class** which is a meta-concept, an exception such as **error** which is an instance of it, and an instance of *error* created when *error* is raised (cf. fig. 1).

Exception instances are always named <u>xself</u>, (from now this term will be used to reference the instance of a current exception). Their slots <u>xobj</u> and <u>xprop</u> are automatically assigned to the object and to the property that are currently active when the exception is raised. All others slots defined on the exception are assigned to the given values (if not provided, default values are used). Any handler will receive xself as unique argument.

```
[unknown-selector signal  receiver 1  selector 'foo  args '(2)]
```

- **Defining method interfaces and retreiving information.** The two relations <u>signals</u> and <u>propagates</u> are defined on the set named **method** to store the exceptions that a method may raise. The following definition of the method *divide* specifies that it may signal the exception *divide-by-zero*.

```
[integer has method divide  signals divide-by-zero   filter (...)   form ...]
```

This information can be retreived by inspecting the method interface (the values of all its slots) or by searching all the methods where exception may be raised (this can be done in a simple way in *Lore* with the relation *of?* defined on the set *relation*)

```
[signals of? divide-by-zero]
= (divide.integer  modulo.integer  ....)
```

- **Looking for a handler.** To fulfil the fifth objective, the "call relation" [Levin 77] is first taken into account i.e. a handler attached to the expressions that dynamically include the signaling one is searched. The search stops as soon as a handler, the parameter type of which is an upper type of the signaled exception, is found. By default, the dynamic handler attached by the system to the top-level loop is found and invoked. It first looks for default-handlers attached to the class or upper classes of the signaling-time active object[5] (value of *[xself xobj]*); if none can be found, it looks for default-handlers attached to the signaled exception itself (*xself*'s type).

5.5. Handling.

One or several handlers can be attached to any expression or class. All handlers have one parameter named xself the type of which is the exception to be handled and which will be bound to the instance of the current exception as soon as the handler is invoked. Determining whether a handler is valid consists in matching its parameter type with the current exception one.

- **Handler bodies.** The methods **resume**, **exit** and **retry** can be used within handlers, they accept an optional argument which is evaluated in the handler environment. Of course, exceptions can be signaled within a handler.

---

[5] i.e. the last receiver of a message. For all exceptions raised by the system while analyzing a message invocation (ex. unknown-selector), this definition is ambiguous; in such cases, we consider the receiver of the wrong message as the active object.

• **Resumption** : The method **resume** is defined on the class **warning** (cf. fig. 1); thus, this message cannot be sent to an instance of **error**. The argument value becomes the value returned by the method **signal**. This implies that a method which signals a non fatal exception should foresee what it will do with the signaling result, in case the exception is resumed.

• **Termination** : The method **exit** is defined on the class **error**. The context between the signaler and the association is discarded, the argument value is returned as value of the expression to which the handler was attached.

• **Defining dynamic handlers.** Dynamic handlers are instances of the class **protect-handler**, internally attached to expressions by using the primitive __protect__, their extent is dynamic and their scope is undefined, they are executed in the lexical environment of the association operation. The definition syntax is "*[handler protect expression]*", a handler takes the following form "*(exception-name form handler-body)*". Within the handler body, the implicit parameter \self can be referenced.

• Here is an example using termination which implements the *modulo* examples, where *minus* raises the exception *negative-number* (cf. § 3.2).

```
[natural has method modulo filter (p natural)
  signals divide-by-zero  ; May signal this exception.
  form  if  [p = 0]  then  [divide-by-zero signal]  else
                ; A handler for negative-number exits the while loop
                ; and returns current oself value.
        [(negative-number form [xself exit oself])  ; The handler.
          protect  ; Attaches the handler to the while loop.
          [t  while  [[oself minus p] -> oself]]]]]
```

• Here is another example using **exit** and **retry** where two handlers are attached to the same expression. This is a very simple version of the *Lore* top-level generator. The method *loop* is designed to prevent a top-level exit after an error occurred. Signaling the exception *exit-top-level* forces the legal exit. The idea is to easily create new instances and new kinds of top-levels.

```
[exit-top-level isa exception-class superset error]  ; In order to force exit.
[top-level isa class with
  (method prompt form ...)
  (method body form [t while [oself prompt] [[[standard-io read] eval] print]])
  (method quit form [exit-top-level signal])  ; The only way to quit.
  (method loop form
    [((exit-top-level form [xself exit 'bye])  ; Catches exit-top-level and force exit.
      (exceptional-event form [xself default-handles]))  ; a simplified version of default-handling.
      protect  ; Attaches the two handlers to the following expression.
      [t while [oself body]] ] ) ]  ; Infinite loop until an exception occurs.
```

The method **default-handles** may use **retry** to cause a restart (cf. the following box). It is now very simple to create subclasses and override methods *prompt* or *body*, e.g. "*[debugger isa class superset top-level ....]*". New instances are obtained and invoked by typing for instance "*[debug isa debugger]*" and then "*[debug loop]*".

• **Automatic creation of dynamic handlers.** A problem appears when trying to implement the encapsulator example since we have to attach a handler to the class *encapsulator* (cf. § 3.2). A **class-handler** cannot be used since such handlers are default handlers and may be overriden by a dynamic one; consequently, we cannot ensure that exception *unknown-selector* will be handled as it should to fulfil the requirement.

In order to solve such problems, a primitive has been designed wich takes a class argument and ensures that an expression handler will be dynamically attached to each message toward an object of that class. Enabling this primitive costs one slot access for each message sending, to check whether a handler is defined on the receiver's class or upper classes and has to be attached to the transmission. This access can generally be computed at compilation time.

• **Defining default handlers.** All static handlers are standard methods named <u>default-handles</u>. The most general ones are defined on the exceptions and have no parameters. Those attached to classes must be defined with a parameter the type of which is the exception to be handled. To look for default-handlers, the system first sends, to the signaling-time active object, the message default-handles with the instance of the current exception (xself) as argument. The *Lore* message passing mechanism automatically selects the appropriate handler if one exists. If not, the system sends the same message to xself.

As an example, let us consider the most general default-handler defined on **exceptional-event**.

```
[exceptional-event has method default-handles
   form [oself describe-error] ; provides information about the error
        [oself display-propositions] ; displays interactive propositions (cf. § 5.7)
        ; If no proposition is chosen
        [xself retry]] ; Executes the top-level actions again, (cf. the above box)
```

5.6. When-exit or how to restore contexts whatever happened.

One problem with the resumption model is the restoration of valid contexts. Indeed, signalers as well as handlers that propagate exceptions do not know whether there will be resumption or termination. Thus, they do not know whether they have to perform restorations since resumption may occur (cf. "cleanup handlers" [Goodenough 75]). To prevent making association syntax and our model implementation more complex, we provide the primitive <u>when-exit</u> allowing to attach restoration actions to expressions. Whatever happens, these restorations will be performed when the expression execution will end. **When-exit** returns the expression value.

As an example, consider the method signal which creates an instance of the signaled exception and destroys it after it has been handled, in resumption cases as well as in termination cases.

```
[exception-class has method signal ....
   form [xself as [oself new ....] ; instanciation of the current exception
        [[xself search-and-invoke-handler] when-exit [xself kill]] ] ]
```

5.7. Interactive propositions.

We provide a *Lore* implementation of interactive propositions. These are simpler to define and to use than the *Flavor* one [Moon 83]. The principle consists in storing, within exceptions, correction propositions and correlated actions to be executed when a proposition is chosen. The propositions can be displayed one at a time or all together (as done by the default-handlers). "Sub-exceptions" inherit from propositions defined on upper exceptions.

```
> [1 + [5 foo]]
*** <Unknown-Selector> : foo is not a selector for receiver 5.
*** In the <method> <+.number> while sending + to the <integer> 1
------- 1 : Supply a value to use as result of this message.
------- 2 : Supply a new selector.
------- 3 : Supply a new receiver.
>>> 2
Enter the new selector, please : factorial
121
```

Implementation is based on the ability of defining new kinds of methods (more precisely new sets of properties), and on the ability of defining slots on properties. We thus define a subclass of *method* called **propose-method** with a slot named **when-chosen**. This slot is designed to contain the method to be invoked when the proposition is chosen. In order to display all propositions, default handlers only have to look for and invoke all the properties of type **propose-method** in the current exception dictionary. Example : here is the creation of a proposition on exception **unknown-selector**.

```
[unknown-selector has propose-method ask-new-selector when-chosen 'new-selector
    form ["Supply another selector." print] ]

[Unknown-Selector has method new-selector
    form ["Enter the new selector, please : " print]
        [xself resume .....]⁶
```

6. Conclusion.

Object oriented programming has led to many improvements in system specification, software quality and reusability. However most current object oriented languages do not take advantage of past studies about software fault tolerance.

We have presented a mechanism for exception handling in object oriented languages designed mainly to implement fault-tolerant encapsulations. Its originality lies first in the fact that we have paid attention to the specific characteristics of such languages : a method is able to react when an inner method invocation raises an exception, and a solution is provided to associate handlers with object and exception classes. Second, we provide a is a full object-oriented implementation of the main well known exception handling techniques coming both from procedural and object-oriented languages. Our system has been implemented within the *Lore* object-oriented language and is currently used in simulation and AI applications.

---

[6]Since it is possible to resume with a new receiver or a new selector, there must be a resumption protocol between the signaler and the handler.

This system has also been designed to be a powerful basis for an interactive debugging environment which is intended for detection, localization and correction of errors, i.e. of non-handled exceptions. It allows each debugging tool to handle exceptional situations. For example, the stepper handles exceptional situations so that users do not have to follow all the details of an execution (cf. [Lieberman 84]) to localize an error context. We have also taken advantage of the non-local exit and resumption possibilities to write a stack examiner in *Lore*. Any user could in the same way write his own.

Extensions are planned. It is yet possible, without significant modifications, to include, in our system, the possibilities of storing and accessing exceptional data through exception handling mechanisms, as described in [Borgida 86]. These features are a powerful alternative to "mixin" and method combination in order to describe inheritance hierarchies with exceptions. They make it possible to create *exceptional classes* or *exceptional-instances*.

Another possible extension lies in the search for handlers via the "use" relation (vs. the "call" relation) as presented in [Levin 77]. This would consist in looking for handlers attached to entities that use (vs. "are currently using")[7] the object within which the exception was raised.

Acknowledgements.

I have greatly benefited, while specifying this system, from the comments abd suggestions of Michel Bidoit. I wish to thank Francoise Carre, Yves Caseau and Olivier Danvy for commenting drafts of this paper, Francois-Xavier Testard-Vaillant for interesting discussions, Hélène Kawa and Catherine Jourdan for commenting english style.

References.

[Benoit, Caseau, Pherivong 86] C.Benoit, Y.Caseau, C.Pherivong : Knowledge Representation and Communication Mechanisms in Lore. Proc. of ECAI'86, Brighton, July 1986.

[Berry 85] D.M.Berry, S.Yemini : A Modular Verifiable Exception-Handling Mechanism. ACM Transaction on Programming Languages and Systems, Vol. 7, No. 2, pp. 213-243, April 1985.

[Bidoit 85] M. Bidoit et al. : Exception Handling: Formal Specification and Systematic Program Construction I.E.E.E. Transactions on Software Engineering, Vol. SE-11, Number 3, March 1985, pp.242-252.

[Bobrow 83] D.G.Bobrow, M.Stefik : The Loops Manual. Xerox Parc, 1983.

[Bobrow 86] D.G.Bobrow & al : Merging Lisp and Object-Oriented Programming. Proc. of OOPSLA'86, Special issue of Sigplan Notices, Vol. 21, No. 11, pp. 17-29, November 1986.

[Borgida 85] A.Borgida : Language Features for Flexible Handling of Exceptions in Information Systems. ACM Transactions on Database Systems, Vol. 10, No. 4, pp. 565-603, December 1985.

[Borgida 86] A.Borgida : Exceptions in Object-Oriented Languages. ACM Sigplan Notices, Vol. 21, No. 10, pp. 107-119, October 1986.

[Caseau 87] Y.Caseau : Etude et Réalisation d'un langage objet : LORE. *Thèse de l'université Paris-Sud*, Orsay, France, Novembre 1987.

---

[7]These entities may be inactive at raising time.

[Christian 82] F.Christian : Exception Handling and Software Fault Tolerance, IEEE Transactions on Computers, Vol. C-31, No. 6, pp. 531-540, June 1982.

[Cointe 84] P.Cointe Implémentation et interprétation des langages objets Applications aux langages Formes, Objvlisp et Smalltalk. *Thèse d'état*, Université Paris 6, IRCAM, France, Décembre 1984.

[Dahl 70] O.Dahl, B.Myhrhaug, K.Nygaard : SIMULA-67 Common Base Language. SIMULA Information, S-22 Norwegian Computing Center, Oslo, Norway, October 1970.

[Etherington 83] D.Etherington, R.Reiter : On inheritance hierarchies with exceptions. Proc. of AAAI-83, pp. 104-108, August 1983.

[Goldberg,Robson 83] A. Goldberg, D. Robson : SMALLTALK 80, the language and its implementation. Addison Wesley 1983.

[Goodenough 75] J.B.Goodenough : Exception Handling: Issues and a Proposed Notation. Communication of the ACM, Vol. 18, No. 12, pp. 683-696, December 1975.

[Ichbiah 79] J.Ichbiah & al : Preliminary ADA Reference Manual. Rationale for the Design of the ADA Programming Language. Sigplan Notices Vol. 14, No. 6, June 1979.

[Knudsen 87] J.L.Knudsen : Better Exception Handling in Block Structured Systems. IEEE Software, pp. 40-49, May 1987.

[Lieberman 84] H.Lieberman : Step Toward Better Debugging Tools For Lisp. ACM, Conference Record of the 1984 ACM Symposium on LISP and Functional Programming.

[Levin 77] R.Levin : Program structures for exceptinal condition handling. Ph.D. dissertation, Dept. Comput. Sci., Carnegie-Mellon University Pittsburg, June 1977.

[Nixon 83] B.A.Nixon : A Taxis Compiler. Tech. Report 33, Comp. Sci. Dept., Univ. of Toronto, April 83.

[Liskov 79] B.Liskov, A.Snyder : Exception Handling in CLU. IEEE Transactions on Software Engineering, Vol. SE-5, No. 6, pp. 546-558, Nov 1979.

[Meyer 87] Reusability: The Case for Object-Oriented Design. IEEE Software, pp. 51-64, Mars 1987.

[Mitchell 77] J.G.Mitchell, W.Maybury, R.Sweet : MESA Language Manual. Xerox Research Center, Palo Alto, Calif., Mars 1979.

[Moon 83] D. Moon, D. Weinreb : LISP Machine Manual, Fourth Edition. MIT Artificial Intelligence Lab., Cambridge, Massachussets, (July 1981).

[Pascoe 86] G.A.Pascoe : Encapsulators: A New Software Paradigm in Smalltalk-80. Proc. of OOPSLA'86, Special issue of Sigplan Notices, Vol. 21, No. 11, pp. 341-346, November 1986.

[PL/I 78] Multics PL/I Reference Manual, Cedoc 68, Louveciennes, Septembre 1978.

[Stroustrup 86] B.Stroustrup : The C++ Programming Language. AT&T Bell Laboratories, Murray Hill, New Jersay. Addison-Wesley, March 1986.

[Testard-Vaillant 86] F.X.Testard-Vaillant : Exceptions and Interpreters. AIMSA'86, Varna, September 1986.

[Wertz 83] H.Wertz : An Integrated, Interactive and Incremental Programming Environment for the Development of Complex Systems. in Integrated Interactive Computing Systems, pp. 235-250, ED P.Degano & E.Sandewall, North-Holland 1983.

On the darker side of C++

Markku Sakkinen

Department of Computer Science, University of Jyväskylä
Seminaarinkatu 15, SF-40100 Jyväskylä, Finland
Electronic mail: markku@jytko.jyu.fi

Abstract

We discuss several negative features and properties of the C++ language, some common with C, others pertaining to C++ classes. Remedies are proposed for most of the latter ones, most of the former ones being feared to be already incurable. The worst class-related defects claimed in present C++ have to do with free store management. Some hints are given to programmers on how to avoid pitfalls.

1. Introduction

The C++ programming language [Stro1, Stro2] is a rather new language for which, evidently, no standardising efforts are yet underway; but it has had significant influence on the draft ANSI standard of the C language, as mentioned in [Bana]. It is reportedly used quite a lot at AT&T, where it was originally developed. In addition to that, C++ seems to gain popularity in the UNIX™ community. The USENIX society has recently arranged a workshop on C++ [Caro]. There are commercial implementations available, e.g. [Gloc]. Furthermore, various software packages have been and are being implemented in C++ [Carg, Rich, Wien, Nuut, Gorl]. There does not seem to be much critique yet published on the language; in [Snyd] and [Wegn] some of its features are compared with several other languages. The paper [Nuut] does indicate rather strong discontent with C++, but does not specify it closer (although that paper comes from Finland, too, I do not suspect a general unsuitability of C++ for Finnish temperaments).

This paper tries to bring up some points in and around C++ that I think bad or problematic. Some of them are flaws in the currently available implementations only, some might be considered and ameliorated in the evolution of the language, but some others are certainly inherent and should be taken into account by programmers when deciding which language to use for a given task. The focus will be on semantics, orthogonality, compile-time detection of possible errors, and somewhat on run-time efficiency. Problems that concern concrete syntax only are bypassed, because I consider them both relatively unimportant and to a high degree matters of taste. This paper is *not* a balanced assessment of C++, the language's virtues are mostly mentioned only where they are connected to some problem. In consequence, readers are warned that the language is not as bad as would appear from the present exposition alone; read some of Stroustrup's articles to see the sunny side.

My practical acquaintance with C++ stems from an ongoing project, the purpose of which is a document database management system. In that project, the AT&T C++ Translator, Release 1.0 [AT&T] was used first. Now the work is continuing with Glockenspiel 'designer C++' [Gloc], on another computer. While the final revision of this paper was going on, our department got the Release 1.2 of AT&T's product for the first computer; there is no significant experience on it yet. It is a definite lack in my background that I have no practical experience with e.g. Simula, Ada®, or Smalltalk™. All readers can take § 2 with a grain of salt because I am no authority on object orientation.

At the request of the Program Committee, I have tried to make the text understandable to people without previous knowledge of C++ or even C. For this purpose, there is a short Appendix describing several language features that are used in the examples. Those who are already committed to C will probably either find nothing new in the criticism in § 3 and 4, or disagree with it.

2. About object orientation and language extensions

There is no consensus about what 'object' and 'object orientation' precisely mean. The paper [Stro3] approximately equates 'object-oriented programming' with '*inheritance*', trying to distinguish it clearly from 'data abstraction', which is presented as another important goal of C++. It looks to me that the issues of object *integrity* and *identity* [Khos] have not been considered important, or that the C heritage has made it impossible to take them very well into account. Interestingly, the taxonomy of [Wegn] also almost ignores the identity and integrity of objects, except in connexion with databases (persistent objects).

Wegner classifies C++ as an object-oriented language, while CLU [Lisk] qualifies as class-based but not object-oriented because there is no inheritance. Ada is classified only as object-based but not class-based because its *packages* have no class, i.e. type. It would be more appropriate to regard the *restricted private types* as classes and their instances as objects; Ada would then be class-based. In the sense of [Wegn], C++ has no data abstraction (because instance variables of objects can be directly accessible) and no virtual resources (because **virtual** functions cannot be left unimplemented in their base class). It has somewhat non-strict inheritance (operations of ancestors can be redefined in descendants, but only if they were declared **virtual** in the first place), and inheritance is by code sharing. It can also be regarded as strongly typed.

A conspicuous omission in [Wegn] is that no distinction is made between languages in which "everything is an object" (Smalltalk-80 [Gold] *et al.*) and those in which objects are just one kind of entity among others. One can write huge programmes in C++ without defining any classes at all. In Smalltalk, one must program in an object-oriented way since no other paradigm is available. This is not to say that the Smalltalk way is "good" — I don't know whether there exists any extremely object-oriented language that offers even nearly the same possibilities for structured programming and compile-time checking as C++ or Ada. Probably the "Turing tar-pit" is easily lurking whenever programming is reduced to a very small set of primitive concepts.

One of the primary goals in the design of C++ was upward compatibility with C [Harb, Bana], as far as feasible. This goal has been very well attained, too. As a consequence, previous C users can quite well upgrade *gradually* to programming in C++, in the first step just feeding their existing C code through the C++ translator and checking if some small modifications would be necessary. Unfortunately, this approach has necessarily transported several drawbacks of C to C++ as well.

If we compare C with Pascal, for instance (a language of roughly the same age), we find that the latter is more object oriented as far as concerns the integrity of data objects. The C language was originally designed with much more concern to machine registers than programmer-defined objects. Moreover, while Pascal is (even overly) strict, C is sloppy. Some features have been defined so as to be convenient for their most obvious application, but causing illogicalities in more complex combinations. (In this respect, C resembles the UNIX command interface.) On the other hand, the existence of pointers to procedures (always 'functions' in C terminology) makes C more object oriented in the sense that behaviour can be connected to data. Further, the generality of pointer expressions can often simplify the handling of complex objects in comparison to Pascal.

Extending some existing language with *lower-level* capabilities is not very difficult in general. The extreme in this direction is the escape to assembler, which exists in several languages or implementations.

But when someone sets out to enrich an existing language with object-oriented or other *higher-level* features, trying to keep totally upward compatible with the base language can be problematic. Obviously, it is easier to extend a language that seems too restricted (e.g. Pascal) than one that has very general, powerful, and accordingly error-prone facilities (e.g. C). One recent example of extending Pascal in an object-based direction (in CLU fashion) is presented in [Saje, Olsz].

Several "machine-oriented high-level languages" such as Mary [Conr] have tried to solve the dilemma of powerful low-level features and protected high-level environment by defining a *safe subset* of the language and requiring some explicit operations (e.g. compiler options) or notation for programme modules or sections of them that use *unsafe* features. This principle could perhaps be applied to C++, too, to alleviate the heterogeneity between high-level and even very low-level operations. Of course, it would be much more difficult to decide *a posteriori* which facilities should be classified as unsafe than design a new language with an eye to this classification.

3. Miscellaneous problems inherited from C

The concept of a *type* is somewhat vague already with simple types. For instance, char and short are something between full-fledged types and int crammed into a smaller space. In contrast, e.g. a *pointer* to short is a true type of its own, different from a pointer to int. Moreover, long is a true type, separate from int, although they are physically identical in typical 32-bit implementations. Because of operator and function *overloading*, type is a more important concept in C++ than C, and the vagueness thus more irritating. As an example, one cannot define an overloading of a function identifier such that there is one variant for an int parameter and another for a char parameter.

Enumerations, which possess different degrees of "typeness" in different C implementations (or are not implemented at all) [Harb], are definitely not types in C++, just another way to declare int constants and synonyms for 'int'. This is a pity, especially considering the overloading facility. Furthermore, since C++ has a general means to declare named constants (which C traditionally lacks), enumerations are completely superfluous under the present definition.

A programmer can prescribe the evaluation order of expressions by using parentheses, *except* between operators of the same precedence. The compiler is free to rearrange those operators that are regarded as associative or commutative. This stipulation in C is intended to allow more extensive optimisations. It overlooks the fact that even the basic arithmetic operations are *not* absolutely associative because of overflows, underflows, and rounding errors. This misfeature could be removed from C++ without affecting upward compatibility with C.

The language proper is not concerned with input and output; those functions are relegated to the standard library. (Ada has followed the same model.) This already tends to make them more error-prone than incorporating them into the language, because both compile-time and run-time possibilities to check function parameters are limited (although better in C++ than in most dialects of C). The facilities of the standard I/O functions are on a very low level of abstraction when compared to Pascal. However, just about anything *can* be done using them, whereas standard Pascal I/O is far too restricted for other than toy applications. The low level is probably not a big nuisance to software houses that define and build their own high-level I/O on top — it will be easily (?) *portable* across C++ implementations. — Object input/output is envisaged in [Stro5].

4. Array problems

In my opinion, the worst common feature of C and C++ ("degree of badness weighted by importance") is the handling of *arrays*. An array type, say, *atype* defined by the declaration

   typedef basetype atype [dimension];

is handled as a true type only when storage is allocated for an *atype* variable, or when arithmetic is done in a pointer expression of type *atype** (pointer to array of type *atype*). Otherwise, the name of an array just stands for the address of its first element. Continuing the above example,

   atype array1, *apointer;
   apointer = &array1;

the assignment is illegal in C++ (and in many C dialects); it would be legal if *atype* were any other kind of data type except array! (Cf. explanation in appendix.)

Array handling in C and C++ is very much prone to devious programming errors, mainly because it is equated to pointer handling. Of course, indexing *can* be regarded as just a special case of pointer arithmetic, but it is very common, and could be made essentially safer by treating it specially as most languages do. I have not seen "index checking" (what a familiar and natural thing to Fortran and Pascal programmers) mentioned in connection with any C or C++ compiler. There are no aggregate operations for arrays in the language, which means that even assignments between arrays must be programmed by writing explicit loops; this creates more chances for indexing errors.

The original main reason for the unfortunate way of handling arrays was probably a striving to pass parameters and function results in hardware registers. That caused arrays to be passed by address, whereas simple types are passed by value. The principle makes it in most cases impossible to write sensible and efficient array-valued functions. The actual array cannot be a local variable of the function, because it would be destroyed when returning. Therefore, it must be created by the new operator, and the caller of the function is responsible for explicitly deleting it later. Actually, the language specification [Stro1] *forbids* array-valued functions, but present compilers accept them gladly.

The unorthogonality of the C and C++ approach to arrays becomes even more evident if we think about embedding an array into a structure with no other components. As a parameter to a function, the structure would then be passed by value, the array by address. Moreover, an assignment statement would be legal for the structure, but not for the embedded array. Conversely, any structure can be embedded in a one-element array, with similar consequences.

A very important special case of arrays are character strings. They also serve well to illustrate the defects of not handling arrays as objects. To make general routines for handling strings of different length at all possible, there is a convention of marking the end of a string with a null byte (the compilers generate it for string literals). This means that the whole string must be traversed even when only the actual length must be found out. Also, there is no built-in way to mark the *reserved* length of a string variable, but the programmer must keep care of it separately. Accordingly, most string handling functions in the standard library come in two variants, one of them having as an additional parameter the maximal number of characters to be read or modified. — Note that there is no way to make a null byte a part of a string as interpreted by the standard functions.

The term 'string' is conventionally used in C and C++ literature to mean a *pointer* to an actual character array. Exactly speaking, the type of such a pointer is not 'pointer to character array' as one would expect, but 'pointer to character' (**char***), i.e. it points to the first character of the actual string. One reason for this convention must be the pointer assignment problem described in the first paragraph. One consequence is that the declared types of a pointer to a null-terminated string and a pointer to a single character become identical.

5. Classes

Classes, borrowed from Simula 67 [Dahl], are the vehicle C++ offers for object orientation. They have facilities comparable to classes in other languages. An equivalent for the **inspect** statement of Simula has deliberately not been included in C++, because it would be contrary to the quest for data abstraction. The possibilities of data hiding are very versatile; they have even been enhanced [Stro4] from the original. Each component of a class is either **private** (the default: accessible only to member and friend functions), **protected** (accessible also to member and friend functions of any derived class), or **public** (accessible wherever the class is defined).

The equivalent of a *method* in Smalltalk is called a *member function* in C++; it is common to all instances of a class. This is quite another thing than a pointer-to-function component, which naturally can be different in different instances. Functions and even whole classes (which means all their member functions) can also be declared **friends** of some class so that they can access its private components. Probably at the time of this conference, available implementations of the language will support even multiple inheritance [Stro4]. Until now, **friend** declarations must often have been used as a substitute for it.

Classes in C++ are defined in such a way that a **struct** becomes just a special case of a class, which is nice economy of concept. Also, variables of a class type can be defined like any other variables: they can belong to any storage class (need not be allocated by **new**) and be components of arrays and other classes. (Restrictions to this will be mentioned in the following sections.) Class declarations are further organised so as to make a very efficient run-time implementation possible. This principle causes compile-time drawbacks: in most cases, changing the declaration of a class, even the private parts, requires all modules utilising that class to be recompiled [Carg].

Objects of a given class are, in principle, all of the same size. As mentioned in [Stro1] (§ 5.5.8), there are ways to circumvent this restriction for class objects allocated on the free store, but they are not without problems. When one wants to implement classes for things such as well-behaved variable-sized character strings, or anything else of really dynamic size, in practice one has to declare two classes. One of them is the generally visible main class, and the other is an auxiliary class, which contains the actual variable-sized objects and is only used by the main class.

Contrarily to C++, many languages, e.g. CLU [Lisk] and those proposed in [Saje, Olsz], *always* implement aggregates indirectly via implicit pointers, thus increasing run-time overhead for every level of structure in comparison to direct aggregates. The previous paragraphs imply that in several cases, that method can be easier in the programme development phase. To be fair, C++ does not *prevent* a programmer from using classes in an indirect way in order to relieve the compile-time overhead where possible. It suffices to declare

class myclass;

in a source module where only pointers to *myclass* objects will be handled. However, the CLU approach would then result in simpler source code in those modules in which the C++ programmer must be concerned about both *myclass* and *myclass** values. We will return to this subject in § 12. — One can observe that one kind of indirect aggregate is very common in C and C++ programming: the pointer array. It is especially often used instead of an array of character strings, with obvious advantages.

The book [Stro1] uses the word 'member' (of a class) in a meaning that, in my opinion, is in contradiction with its connotations in set theory and everyday speech. In this paper, the word 'component' will be used instead. However, 'member function' does not sound misleading, because normally every invocation of a member function of some class is connected to an object ('member' in the ordinary sense) of the class.

Within a member function, there is always an implicit parameter **this**, which is a pointer to the class instance whose component the function is invoked as. Perhaps a little surprisingly, a member function of *myclass*, say, *can* be called even without an instance of *myclass*:

```
myclass* myc_p = 0;            // null pointer
myc_p -> myfunction ();        // call function via pointer
```

In the above invocation of *myfunction*, the current instance pointer **this** will simply be null. However, this bit of code will crash if *myfunction* is a **virtual** function (cf. § 8), because in that case the class object must really be accessed at run time to find out the appropriate variant of the virtual function. By adding an explicit class prefix like *myclass::myfunction* even a virtual function can be invoked.

6. Problems with constructors and destructors

The possibility to declare constructors for a class is very useful, indeed necessary for achieving sophisticated abstract data types. Among other things, they permit a distinction between initialisation and assignment, which is often crucial (although uninteresting for simple types). Constructors also allow the creation of auxiliary objects "behind the scenes", as mentioned in the previous section. Such constructors are typical cases which necessarily need a destructor as well, to delete the auxiliary objects.

However, constructors and destructors are not without problems, the way they are defined in C++. One obvious defect is that, although constructors will typically take parameters, there is no way to pass parameters in the definition of an *array* of class objects. More subtle difficulties may result from the fact that the order in which the destructors for the automatic variables of a block will be called is undefined.

The capability for the programmer to take care of memory allocation within a constructor is useful; it is the only way to create class objects of variable size. It can also allow a more efficient allocation of specific classes. Unfortunately, it is presently offered in a rather unstructured, "ad hoc" manner, by the appearance of an assignment to the automatically defined variable **this**. (Correspondingly, a zero value can be assigned to **this** in a destructor to bypass standard memory deallocation; this possibility is needed less often.) There is no compile-time check against using such a constructor on external, static, or automatic variables (which are necessarily allocated before the constructor can be called); the programmer must make an appropriate test at run time.

The present C++ translators do not allow the **new** and **delete** operators to be overloaded for a class, contrarily to § 6.2 of [Stro1]. According to [Stro4], the facility will be implemented in the next release. By overloading **new** and **delete** the need to assign to **this** in constructors and destructors can be obviated. Unfortunately, this solution does not work for variable-sized objects.

If one builds a large structure of class objects connected hierarchically (or otherwise) to each other, it can easily happen at the end of the programme that all those objects are destroyed laboriously one by one, to no benefit at all. That can take approximately as much time as building the structure. Fortunately, this "domino effect" can be avoided, e.g. by having enough strategic objects created by **new** and *not* deleting them at the end.

If a class *aclass* has a constructor with *one* parameter of some type *atype* (there can be additional parameters if they have default values), then that constructor will also be used automatically as a conversion function so that

```
atype tom; aclass jerry = tom;
```

will succeed (without compiler warning). This is mentioned in [Stro1], § 6.3.1 and 6.3.2. In many cases, one might not want such automatic conversions, as they can cause programming errors to pass unobserved. Then one must simply avoid defining one-parameter constructors.

7. Mistakes with derived classes

A *derived class* in C++ means a class type that possesses all components of its *base class*, and normally some additional components. A class can also have components of another class type; this is not quite the same as being a derived class, but the problems to be discussed in this section are the same for both cases. The major difficulties with derived classes, and classes with class components, occur in constructor and destructor functions when there is explicit storage allocation and deallocation. That is, we get more complicated problems in addition to those discussed in the previous section.

The "Reference Manual" part of [Stro1] says (in §8.5.5):

"If a class has a base class or member objects with constructors, their constructors are called before the constructor for the derived class. The constructor for the base class is called first."

Correspondingly, it says (in §8.5.7):

"The destructor for a base class is executed after the destructor for its derived class. Destructors for member objects are executed after the destructor for the object they are members of."

The reference manual recognises that the case is different if there is explicit storage allocation in the constructor of the derived class, by the following passage (in §8.5.8):

"Calls to constructors for a base class and for member objects will take place after an assignment to this. If a base class's constructor assigns to this, the new value will also be used by the derived class's constructor (if any)."

The C++ reference manual errs badly in the last point above: the constructors of both base and derived class should not be allowed to assign to this, or conflicting memory allocations will result. (The manual also forgets to say that if a destructor of a derived class assigns a zero value to this, then the destructors for the base class and any component classes should be called *before* that assignment.) Even when these errors are corrected, this approach is very difficult in practice, because it cannot generally be known at compile time where the assignment to this will actually take place.

At least both C++ implementations mentioned in § 1 make a gross error in the opposite direction to the manual: If there seems to be an assignment to this in the constructor (destructor) of the derived class, they simply do not call the constructor (destructor) of the base class at all! This bug must have caused a lot of trouble to people programming in C++. One way of handling the storage allocation problem for derived classes consistently will be presented in § 9.

8. A problem with virtual functions

The smaller difference between the base class of a derived class and a class component of a containing class is that there is no direct way to handle the "base object" as an entity, only its components separately. The main difference is the ability to define **virtual** functions in a base class, which can then be redefined in some derived classes if required.

Unfortunately, virtual functions are another feature, at least in the present implementations of C++, that does not mix freely with programmer-controlled memory allocation. Moreover, neither the book [Stro1] nor the compilers will warn you about the pitfall, which is the following. The "first hook" for the virtual function facility is a pointer, placed immediately after all declared data components of a base class that has at least one member function declared **virtual**. If *variable-sized* objects are allocated, the pointer will be left in the middle. Fortunately, there is a portable way to circumvent the difficulty.

The solution is best illustrated by a small example, showing part of a class declaration (further public components, denoted by the ellipsis, may be functions and variables), a constructor and another public member function. The private member function *contents* is defined within the class declaration itself (thus automatically becoming an **inline** function).

```
class flexstring {
    unsigned space, length;
    char* contents ()                          // pointer to start of actual string
        { return (char*) this + sizeof (flexstring); }
public:
    flexstring (unsigned = 20);                // constructor with default size
    void copy (char*);                         // copy ordinary string into flexstring
    virtual void put ();                       // output (somehow)
    . . .
};

flexstring::flexstring (unsigned size = 20)
{
    this = (flexstring*) new char [size + sizeof (flexstring)];
    space = size; length = 0;
}

void flexstring::copy (char* cp)
{
    length = min (strlen (cp), space);         // min is not a standard function,
    strncpy (contents (), cp, length);         // but strlen and strncpy are standard
}
```

The sizeof operator gives the size of a *flexstring* object as known to the compiler, including the virtual function pointer. The constructor allocates space for this *plus* the requested number of bytes for the actual string to be stored. All other member functions that need to access the actual string (*copy* above is a simple example) get its address by calling *contents*.

Another solution to the same problem is to declare an auxiliary class that has a virtual function, then derive *all* classes that need both variable-sized instances and virtual functions from that auxiliary class:

```
class virtual_aid {
    virtual void dummy () { }                  // do nothing
};
```

This solution is simpler (derived classes will not become as contrived as *flexstring* above), but probably has a greater risk of not working with all coming C++ releases if the virtual function mechanisms are changed. The completely straight approach to variable-sized class objects in § 5.5.8 of [Stro1] is successful because the class *char_stack* defined there has no virtual functions.

One should be aware that the sizeof operator is purely a compile-time device in all cases. It does not behave at all like virtual functions: if *p* is a pointer to *aclass* then sizeof($p*$) will always yield the declared size of an *aclass* instance although *p* may point to an instance of a derived class. It would not even be possible to offer a general "runtime-sizeof" operator without a significant change of current C++ object implementation. This is a consequence of the weak support of object identity.

9. Some suggestions to cope with the problems

We will try to sketch some amendments to the C++ language that would settle most of the difficulties described in § 6 to 8. A complete proposal with a detailed syntax would be a little beyond the scope of this conference paper.

As already mentioned in § 6, it will become possible to overload the new and delete operators for a class. Very importantly, the new operator function will get as one parameter the size of the object to be allocated. It is thus possible to write an allocator for a base class that will work correctly for any derived class also. Alternatively, one may write an overriding allocator for some derived class if needed. In either case, the appropriate version of **new** will be called before any constructor and so the need to assign to this within constructors disappears. Hence, the constructors can really be invoked in the order described in § 7.

An analogy of the previous paragraph holds for destructors. However, the **delete** operator function does not get any size parameter. Data structures to store programmer-allocated class objects must therefore be designed so that the size of each allocated object is known at deletion time.

This coming improvement in storage allocation and deallocation will only cater for classes whose all instances are of the same size, as we said in § 6. The reason is that the size passed to **new** is the compile-time (declared) size of the original or derived class. In principle, it would be possible to declare a differing object size for any constructor of a class. This would still be a compile-time matter, thus easily applicable even to automatic, static, and external class variables. With C++ as it stands, a programmer can obtain an equivalent effect by declaring a separate (typically derived) class for each object size. A proliferation of classes can then become a problem, although multiple inheritance may help a little.

An orderly solution to the problem of *run-time* determination of the size of each class instance (cf. example of § 8) would be more complicated. The following is one feasible solution: Corresponding to each constructor for which dynamic size determination is desired, a size calculation function with the same parameter signature as the constructor must be declared. This function will automatically be invoked with the initialisation parameters to yield the size parameter to **new**. After **new** has allocated the correct amount of space, the constructor will be called with the same initialisation parameters as the size calculation function. The compiler should allow such a constructor to be invoked *only* on objects created by the **new** operator. In present C++, the designer of a class has no means to enforce such a constraint, but obeying it is necessary with a class like *flexstring*. I cannot imagine other reasons than the creation of variable-sized objects, that would absolutely forbid a constructor to operate e.g. on automatic variables.

One consequence of the last proposal is that a variable-size constructor must not be used to initialise the base class of a derived class, nor a component of another class. In consequence, the whole example of the previous section cannot be written in this manner by just simplifying the constructor and adding a separate size calculation function: declaring functions **virtual** serves no purpose unless derived classes can be defined. This problem can be solved as suggested in § 5, in a way that might have been clearer in the first place (but we had to illustrate a point in *current* C++): We make the anonymous variable-sized part of *flexstring* a separate class, say *flexbytes*, and add a pointer to it as a component of *flexstring* (the *contents* function is no longer needed). Now, only *flexbytes* has variable-sized instances, and *flexstring* can have virtual functions.

From the object-oriented standpoint, it would appear beneficial if every class instance had a run-time descriptor at its beginning. At present, classes with virtual functions have a kind of descriptor (unfortunately not at the beginning, as explained earlier) but other classes have none. The associated overhead would not be unreasonable even if the descriptor included a length field. A standardised length field would facilitate the writing of constructors, destructors, and memory allocators / deallocators. The **struct** keyword would remain for declaring plain C structures without any implicit overheads.

The two minor problems mentioned in § 6 could be solved if considered worthwhile. Constructor parameters for an array of class objects could be passed by using the same syntax already invented for initialiser lists of aggregates ([Stro1] Reference Manual, §8.6.1). The order of destruction of a block's automatic variables could be defined as the inverse of their creation order; the newer translator versions already seem to work like this.

10. Operator overloading

The capability of operator overloading for class operands is such that a separate function must be written for each desired operator. This can cause some difficulties for both the implementer and the utiliser of a class. The implementer of a typical general-purpose class must write a great number of operator functions. The utiliser must learn the semantics of each operator separately, since they need not have similar relationships to each other as they have with the basic data types.

The most evident area in which the problem just mentioned could be alleviated are the six different relational operators. They could be taken care of by writing only one comparison function, which should be directly accessible, too. Indeed, for comparing *strings*, the standard library of C and C++ contains only a function that returns a negative number if the first string is lexicographically less than the second, a positive number in the converse case, and zero if the strings are equal. An expedient stipulation would be that a comparison function for a class automatically defines all relational operators in the obvious way, but if there is no comparison function then any or all relational operators can be defined explicitly. — The basic idea can be used in defining classes even though it is not built into the language. The paper [Sakk] elaborates on this subject.

The *modifying assignment operators* ('+=', '*=', etc.) are further candidates for reducing the number of functions. Probably the most useful way would *not* be to define them automatically on the basis of the corresponding "ordinary" operators, but *vice versa*. That means, if '+=' is explicitly defined in some class *aclass* for a right-hand operand of type *atype* (not necessarily the same as *aclass*), the variable *a* is of type *aclass* and the variable *b* of type *atype*, then the expression *a + b* would be automatically implemented as

(aclass temp = a, temp += b)

(This is not real C++, since declarations are not allowed within expressions.) An explicit definition of '+' would only be allowed for a class if '+=' is not defined for it (with the same type of right-hand operand). The same principle applies to all modifying assignment operators.

When the operators '++' and '--' are overloaded, the distinction between their postfix and prefix application is lost. This could be remedied, and the semantics of these operators with classes be made even otherwise analogous to that with basic types, as follows. If, for the class *aclass*, the operator '+=' is defined with a right-hand operand of some arithmetic type, and *a* is of type *aclass*, then the pre-increment expression ++*a* would automatically be defined as *a += 1*. Otherwise, an explicit definition for '++' could be written. The post-increment expression *a++* would in both cases be automatically implemented as

(aclass temp = a, ++a, temp)

(Even this is not real C++, of course.) The decrement operator would be handled similarly.

The undesirability of the rearrangement of expression evaluation order, noted in § 3, is really pronounced with overloaded operators. It is totally up to a class implementer to achieve all those commutativity and associativity properties that the compiler assumes some operators to have.

11. Constants and pointers to constants

C++ allows one to derive a constant type from any non-constant data type. This general 'constant' concept is very useful, exists in many other languages, and is unambiguously defined when applied to "pure data" types. However, when the base type is a class with member functions, there arises a problem that none of the references has observed: what member functions, if any, of a constant class instance should be callable? The present implementations appear to allow all member functions to be called, including the assignment operator if it is overloaded. A real solution to this problem would require, either an explicit declaration of those member (and friend) functions that are allowable with constant class objects, or disproportionate run-time effort, at least on typical current computers (with a truly sophisticated hardware architecture like that of Burroughs, it could be easier). We should thus only warn programmers not to trust "constants" of any class that has any modifying member or friend function.

The book [Stro1] discusses pointers to constants (in § 2.4.6). More exactly, a 'pointer to constant' is defined as a pointer through which the referenced object cannot be modified. In consequence, Stroustrup continues:

> "One may assign the address of a variable to a pointer to constant since no harm can come from that. However, the address of a constant cannot be assigned to an unrestricted pointer since this would allow the object's value to be changed."

This is logical. Unfortunately, the old C++ Translator [AT&T] turns things upside down: a pointer to constant can be assigned directly to an unrestricted pointer variable without any warning, but the reverse assignment cannot be done even with an explicit type cast (which normally allows almost anything to be assigned to any variable)! This behaviour has been corrected in the newer releases of the translator.

There are situations in programming when one would like to classify the above kind of pointer as a 'nonmodifying pointer' and have also a 'pointer to true constant' available. Then one could assign a static local *pointer to constant* once in a function and rest assured that even no other part of the programme could modify the constant between invocations of the function. Obviously, assignment of a *pointer to constant* to a *nonmodifying pointer* variable would be allowed, but not vice versa.

12. Some practical difficulties and hints

The definition and implementation of a typical general-purpose class takes quite a lot of effort. Certainly, the same goes for a typical general-purpose private type in Ada. Modules that use several classes will need to include several big header files [Carg] (some of the standard header files needed e.g. for standard I/O are already large). This costs so much in compilation time that one is inclined to write much longer source modules than would be optimal from some other viewpoint. Compilations become expensive and time-consuming also for the reason that the current C++ implementations translate the source code into ordinary C (with an appreciable increase in code size) and then invoke the C compiler. The paper [Dewh] can give interesting insight into some aspects of the language and the AT&T translator, even to persons who are not planning to build their own C++ compiler.

A general guideline for C++ programmers to minimise inclusion and recompilation overheads is as follows: Define hierarchies of derived classes only when every level must be visible to the "end user", as in the example 'employee - manager - director - vice_president - president' ([Stro1], § 7.2.5). Define classes with class components only when needed for runtime efficiency, or if the component classes are very simple. In other cases, declare just pointers to lower-level classes as components of upper-level classes. — Many of the ideas presented in [Stro5] seek to improve essentially the speed and ease of compilation, module management, and software maintenance. When such improvements get implemented, this advice will become less relevant.

Considering the problems discussed in § 8, you should regard any class whose constructor may create instances of different sizes as *abnormal*. Such classes should not be visible to the "end user" but be hidden behind normal classes. An abnormal class should be kept disjoint with all other classes in the sense that it should be neither a derived class itself nor a base class of another, derived class, nor a component of another class. However, no harm can arise from a normal class being a component of an abnormal one. Do not define functions that return abnormal values: according to [Stro6] there is no commitment to support variable-size objects. The earlier version of the C++ Translator [AT&T] handled such return values wrong; current versions handle them correctly in *many* cases, but a function that returns an abnormal class value may again cause mysterious errors with some future release.

If you define some class *aclass* with a constructor (a typical non-trivial class will mostly need it), it is not advisable to define functions returning a value of type *aclass* even if the class is normal. At least current C++ versions implement them in a very inefficient way. It is better to have an additional parameter of type *aclass** or *aclass&* (a reference, cf. appendix), but so that the object to hold the value is created before the call, not in the called function. Likewise for reasons of efficiency, you should avoid passing large class objects directly as parameters, because they must then be physically copied; again, rather use pointers or references, but now *to constant* since the effect of a value parameter is desired. This does not hold if the function really needs a local, modifiable copy or the parameter object. Also, a pointer to an allegedly constant class object is not completely safe (cf. § 11).

To my knowledge, there are at present no debugging facilities usable at the C++ level. When debugging, one must resort to the C level, except for source programme line numbers. The most awkward thing in this is that the C names generated by the translator from overloaded C++ identifiers can be extremely long and hard to type. Moreover, the debugging tools usually available in UNIX environments are rather unsophisticated and hard to use e.g. in comparison to the VAX/VMS™ debugger. Better tools are coming — *Pi* [Carg] is an example of a more advanced debugger. Even today, the personal overall assessment in [Tric] compares C++ favourably over Common Lisp as a software development tool.

13. Conclusion

If a little pun is allowed, perhaps incrementing C by 1 is not enough to make a good object-oriented language to all tastes. The existence of such ambitious object-oriented programming libraries as Gorlen's OOPS [Gorl] is some evidence of the capabilities of C++, although one cannot claim that C++ is object-oriented because OOPS is object-oriented and written in C++. One advantage of C and C++ over several other languages is that the capability to handle many machine-dependent aspects of programming *explicitly* in the language often makes it possible to write very portable code, paradoxical though this may sound. Of course, the same capability also makes it possible for bad programmers to write utterly unportable code.

The realm in which C++ may be competitive against truly high-level, object-oriented languages would be those tasks for which the low-level capabilities afforded by C++ are essential. Development plans for the document database management system mentioned in § 1 are an illustrative example. The user interface layer will probably be realised in Prolog by modifying an existing Prolog prototype [Salm] as required. Modules written in C++ will be used as Prolog primitives. Finally, an existing database management system might be added as the third, lowermost layer.

Acknowledgements

This work was supported by the Academy of Finland and the Ministry of Education (doctoral programme in information technology).

Bjarne Stroustrup (of AT&T Bell Laboratories) has been very helpful and communicative, among other things by sending me copies of some references that would otherwise have been omitted. This was notwithstanding that the first submitted version of this paper, which I sent him, was a lot more arrogant towards his language than the present one. Correspondence with Stroustrup has caused numerous changes especially in § 9 (elsewhere they concern mostly details).

174

The suggestions of the ECOOP'88 Program Committee have certainly helped to make this paper more generally interesting and understandable than the original version. Seppo Sippu (of our department) has given useful comments at more than one stage of writing.

UNIX is a trademark of AT&T. **Ada** is a registered trademark of the United States Department of Defense. **Smalltalk-80** is a trademark of Xerox Corporation. **VAX/VMS** is a trademark of Digital Equipment Corporation.

References

[AT&T] UNIX System V AT&T C++ Translator Release Notes, AT&T 1985 *(307-175 Issue 1)*.

[Bana] Mike Banahan, The C Book : Featuring the draft ANSI C Standard, *The Instruction Set Series*, Addison-Wesley 1988.

[Carg] T. A. Cargill, Pi - A Case Study in Object-Oriented Programming, *OOPSLA '86 Proceedings, ACM SIGPLAN Notices Vol. 21 No. 11 (November 1986), p. 350-360*.

[Caro] John Carolan, The Santa Fe Trail, *EUUG Newsletter Vol. 8 No. 1 (Spring 1988), p. 41-44*.

[Conr] Reidar Conradi and Per Holager, MARY Textbook, RUNIT (Trondheim, Norway) 1974.

[Dahl] Ole-Johan Dahl, Bjørn Myhrhaug and Kristen Nygaard, SIMULA 67 Common Base Language, Norwegian Computing Center 1968 *(No. S-2)*.

[Dewh] Stephen C. Dewhurst, Flexible Symbol Table Structures for Compiling C++, *Software - Practice and Experience, Vol. 17 No. 8 (August 1987), p. 503-512*.

[Gloc] designer C++ release 1.2 User Guide, Glockenspiel Ltd. of Dublin 1987.

[Gold] Adele Goldberg and David Robson, Smalltalk-80: The Language and its Implementation, Addison-Wesley 1983.

[Gorl] Keith E. Gorlen, An Object-Oriented Class Library for C++ Programs, *Software - Practice and Experience, Vol. 17 No. 12 (December 1987), p. 899-922*.

[Harb] Samuel P. Harbison and Guy L. Steele Jr., C : a Reference Manual, Prentice-Hall 1984.

[Khos] Setrag N. Khoshafian and George P. Copeland, Object Identity, *OOPSLA '86 Proceedings, ACM SIGPLAN Notices Vol. 21 No. 11 (November 1986), p. 406-416*.

[Lisk] Barbara Liskov *et al.*, CLU Reference Manual, *Lecture Notes in Computer Science 114*, Springer-Verlag 1981.

[Nuut] Esko Nuutila *et al.*, XC - A Language for Embedded Rule Based Systems, *ACM SIGPLAN Notices, Vol. 22 No. 9 (September 1987), p. 23-31*.

[Olsz] Jacek Olszewski, Capability Oriented Aliasing Language Rationale, *Technical Report No. 87/89*, Department of Computer Science, Monash University (Australia) 1987.

[Rich] John E. Richards, GKS in C++, *EUUG Newsletter, Vol. 7 No. 1 (1987), p. 53-64*.

[Saje] A. S. M. Sajeev and J. Olszewski, Manipulation of Data Structures Without Pointers, *Information Processing Letters, Vol. 26 No. 3 (November 1987), p. 135-143*.

[Sakk] Markku Sakkinen, Comparison as a Value-yielding Operation, *ACM SIGPLAN Notices, Vol. 22 No. 8 (August 1987), p. 105-110*.

[Salm] Airi Salminen, A method for designing tools for information retrieval from documents, *Proc. 4th Symp. on Empirical Foundations of Information and Software Sciences (1986), p. 261-272*, Plenum Press 1988.

[Snyd] Alan Snyder, Encapsulation and Inheritance in Object-Oriented Programming Languages, *OOPSLA '86 Proceedings, ACM SIGPLAN Notices Vol. 21 No. 11 (November 1986), p. 38-45*.

[Stro1] Bjarne Stroustrup, The C++ Programming Language, Addison-Wesley 1986.

[Stro2] Bjarne Stroustrup, An Overview of C++, *Object-Oriented Programming Workshop, ACM SIGPLAN Notices Vol. 21 No. 10 (October 1986), p. 7-18*.

[Stro3]   Bjarne Stroustrup, What is "Object-Oriented Programming"?, *Proc. 1st European Conf. on Object Oriented Programming, Paris (June 1987)*, also to appear in *IEEE Software, May 1988*.

[Stro4]   Bjarne Stroustrup, The Evolution of C++ : 1985 to 1987, *Proc. USENIX C++ Workshop, Santa Fe, New Mexico, U.S.A. (November 1987)*.

[Stro5]   Bjarne Stroustrup, Possible Directions for C++, *Proc. USENIX C++ Workshop, Santa Fe, New Mexico, U.S.A. (November 1987)*.

[Stro6]   Bjarne Stroustrup, *private communication*, 1988.

[Tric]   Howard Trickey, C++ versus Lisp: A Case Study, *ACM SIGPLAN Notices, Vol. 23 No. 6 (February 1988), p. 9-18*.

[Wegn]   Peter Wegner, Dimensions of Object-Based Language Design, *OOPSLA '87 Proceedings, ACM SIGPLAN Notices Vol. 22 No. 12 (December 1987), p. 168-182*.

[Wien]   Richard S. Wiener, Object-Oriented Programming in C++ - A Case study, *ACM SIGPLAN Notices, Vol. 22 No. 6 (June 1987), p. 59-68*.

Appendix: Some features of C++ (and C)

The languages C and C++ can be called Algol-like: they are imperative and block-structured, have largely the same complement of statement types and basic data types as any language of the Algol family, and allow recursion. Conspicuous syntactic differences from Algol 60 are an easier attitude to semicolons and the substitution of **begin** and **end** by '{' and '}' respectively. Comments in C are bracketed by '/*' and '*/'; C++ additionally allows end-of-line comments beginning with '//'.

C (and C++) has a text-substitution preprocessor facility that allows macros both with and without parameters. The capabilities most often needed in C++ are probably compile-time inclusion of secondary source files and conditional compilation. Several other things, common in C (e.g. defining symbolic constants as macros), can be better done in the C++ language proper. The preprocessor is rudimentary in comparison to any modern macro assembler.

The fundamental data types *of C* are signed and unsigned integers of several sizes (**char** is most naturally considered one of them), floating-point numbers, and **void** (an empty set of values). The most important derived data types are arrays (many-dimensional arrays are handled similarly to Pascal), structures (**struct**, like records in Pascal), and pointers. The *classes* of C++ are discussed in the main text. C++ allows the definition of constants (**const**) of any type.

The logic of a type definition is approximately inverse to that in most Algol-like languages: it tries to describe how one will get a value of the base type from the declared variable (or, in the case of a **typedef** statement, from a value of the new type). Thus, in the example of § 4, indexing an *atype* value gives a *basetype* value, the variable *array1* is of *atype*, and dereferencing *apointer* ('*' is the dereferencing or indirect addressing operator, prefixed to its operand) gives an *atype* value. The type 'pointer to *atype*' is often denoted by '*atype**'. The unary, prefix '&' is the referencing or address-of operator: when applied to an (addressable) object of type *sometype*, it yields a value of type *sometype**. The last statement in the example is erroneous because *array1* is not regarded as a variable of type *atype*, but instead as a constant of type *basetype**, namely &*array1[0]* (array indexes always start from 0).

Function declarations follow the same logic as variable declarations. In fact, the only thing that distinguishes the declaration of an *atype* variable from the declaration of a function returning *atype* is that there must be a pair of parentheses after the function name (even if there are no formal parameters). A function declaration that is also a *definition* is recognised from a block of code (in braces) immediately following the parameter parentheses. C++ accepts the attribute **inline** for a function, meaning that the function body shall be in-line expanded at every place where the function is called, thus minimising the overhead for very small functions.

Components of classes (both data and functions) are accessed by conventional dot notation. Alternatively, a right arrow can be used in conjunction with a *pointer* to the class type (as in the example of § 5); this is not essential but handy because the dereferencing operator '*' has lower priority than the dot (component selector).

Typing is strong as a rule (cf. § 3), but there are some implicit type conversions (cf. example in § 6). Furthermore, explicit type conversions or casts can be effected very generally between types; even if *atype* cannot be converted to *btype*, at least *atype** can be converted to *btype**. The example of § 8 uses traditional C cast notation in the constructor function *flexstring::flexstring* to convert a **char*** value to *flexstring**. Whenever the target type can be expressed as just a type name, functional notation can be used as well:

    this = flex_pointer (**new char**[somesize]);   // suppose type flex_pointer has been defined

In addition to pointers, *references* to any data type can be declared in C++: *atype&*. A reference is semantically almost the same as a constant pointer but can cause the automatic creation of a temporary variable to refer to, in some cases. Syntactically, declaring a formal parameter of a function to be *atype&* (instead of *atype**) makes function calls look just as if one had a reference parameter (**var** parameter in Pascal).

The most important storage classes of variables are **extern** (global), **static** (roughly equivalent to **own** in Algol 60), and **automatic** (allocated on the stack). The allocation and deallocation of variables in free store (dynamic memory, heap) is similar to Pascal. C++ has the standard operators **new** and **delete** for this purpose; in C various library functions are used (depending on the environment and implementation).

C and C++ are statement languages, not expression languages like Algol 68. However, instead of the more conventional assignment statement, the main workhorse is an *expression statement*. The assignment operator '=' (not to be confounded with the equality test operator '==') is just an operator that both has a side effect and yields a value. In addition to ordinary assignment, there are "modifying assignment" operators corresponding to most binary operators: their left-hand operand is evaluated once, then used in the binary operation, and last assigned the result of the operation (cf. § 10). As a special case, there are *unary* operators '++' and '--' for incrementing and decrementing an arithmetic variable by one. They can be used as either prefix and postfix operators; the value of the postfix expression is the *old* value of the variable. Another unconventional operator, usable within any expression, is the sequencing operator ',' — its left-hand operand is evaluated first (presumably for the sake of its side effects) and its value discarded, the right-hand operand is evaluated then and its value used for the whole expression (cf. § 10).

There is no 'main programme' nor are there 'procedures' in C or C++; all *statements* are within *functions*. A function with the name *main* will be recognised as the main programme, and functions with result type **void** (thus returning no result) can be defined. *Any* function can be invoked in the manner of a procedure call if the result is not needed. Functions cannot be lexically nested; all functions are either on the global level or within class declarations. It is possible to define **extern** and **static** variables outside of functions. The type of a function is defined by its signature, i.e. the type of result and the number and types of formal parameters; traditionally in C the type of a function has been determined solely by its result type. (We use the word 'parameter' in this paper, although the C and C++ community prefers 'argument'.)

Code and data are completely separated in principle, but pointers to functions are possible. There is a distinct pointer type corresponding to every distinct function type. If an implementation does not completely protect code segments at run time, code can naturally be mutilated, e.g. after casting a pointer to a function to another pointer type.

The programmer can declare any function name to be overloaded; names of member functions (of classes) are automatically regarded as overloaded. The C++ translator determines the correct function to apply from the types of actual parameters and result, applying standard conversions if necessary. Overloaded functions with the same name must therefore be distinguishable from each other by their signatures in the C++ type system. The class of a member function can be explicitly specified in a function call thus (the type of *anobject* must be *thatclass* or derived from it):

    z = anobject.thatclass::somefunction (x, y);

Virtual functions of derived classes will very often need this possibility to call the corresponding functions of their base classes. — Almost all *operators* can be overloaded analogously to functions.

It is possible to define a *constructor* function for a class, even several constructors if they have different parameter signatures. If this is done then it is guaranteed that any instance of the class, independently of storage class, will be automatically initialised by the appropriate constructor before first use. Likewise, it is possible to define a *destructor* function (only one) for a class. In this case, the destructor is guaranteed to be called automatically to operate on every instance of the class when it is being deleted. Note that this also happens to external and static variables of a class type at programme exit, but not to instances created by the **new** operator unless they are explicitly deleted by **delete**.

Prototyping an Interactive Electronic Book System Using an Object-Oriented Approach

Jacques Pasquier-Boltuck, Ed Grossman, Gérald Collaud
Institute For Automation and Operations Research (IAUF)
University of Fribourg
1700 Fribourg Switzerland

Abstract

An Integrated Electronic Book (IEB) represents a complex network of integrated information and "know how" on a given subject. In the design phase of WEBS (Woven Electronic Book System), we soon realized that, because we were describing the IEB in terms of "objects" and "methods," and because we wanted WEBS to be easily expandable and to offer a consistent user interface, we should use an object-oriented development system.

This paper does not fully discuss the concept of electronic book systems, but rather describes our own experiences developing a complex software system with an object-oriented language. The first part explains our choice of an object-oriented language and software framework. The rest describes the software architecture of WEBS, which is a class hierarchy of three layers: a software framework (MacApp) layer; a system-specific layer; and an application-specific layer.

Keywords : Object-oriented programming, application framework, user interface consistency, hypertext, electronic book.

1. Introduction

1.1 Background

At the beginning of 1984, the research project: *The Integrated Electronic Book* [Kohlas,1984] was launched at the Institute for Automation and Operations Research of the University of Fribourg, Switzerland. The achievements made by our small group of researchers, during the project's first four years, are briefly summarized below:

1. We first designed EBOOK3, a software environment or shell for both creating and consulting electronic books on IBM-PC compatibles [Savoy,1987a, b and c]. We were conscious from the beginning of the limitations of such machines (for example, very limited graphics capabilities), but since we wanted our system to be used by as large a group of people as possible, our choice was motivated by market availability and cost.

2. We then used the EBOOK3 shell to author a small set of electronic books on selected subjects in the areas of operations research, programming, and economic theory. Once two fully functional electronic books[1] existed, we set out to gain experience by analyzing the feedback offered by a group of endusers. We presented the books at several formal and informal meetings and used them as a supplement to our course support material.

3. Based on the experience gained with the EBOOK3 system, we created the first design document for WEBS (Woven Electronic Book System), the eventual successor to EBOOK3 [Pasquier-Boltuck and Collaud, 1987].

4. We soon realized that we were describing the IEB in terms of "objects" and "methods" (see Section 2). It seemed only natural that as we set to work programming the first prototype of WEBS in September 1987, we should choose an object-oriented environment; it reflected the basic principles of our conceptual design. In addition, it provided us with modularity, data abstraction, and reusability of code through inheritance. We therefore selected MacApp, an object-oriented software framework for the Macintosh™ [Schmucker, 1986a and b], and Object Pascal, the first available language that could be used with MacApp.

5. Four months after writing the first line of code, a beta version of the WEBS prototype is operational. This working prototype allows us to test and solidify our design ideas, rather than merely describing them.

1.2 Goal and Outline of this Paper

The goal of this paper is not to fully discuss the design issues related to the concept of electronic book systems[2], but rather to describe our own experiences developing a complex software system with an object-oriented language. With this in mind, the following issues arising from our research will be discussed:

- Why we decided to develop WEBS with the help of an object-oriented language.

- How we proceeded with this task.

- What conclusions can be drawn at the present stage of development.

---

[1] The titles of these two books are "Linear Optimization" and "The Application of Markov Chains in Reliability Theory".

[2] The interested reader is referred to the papers of [Savoy, 1987a, b and c], [Pasquier-Boltuck and Collaud, 1987], [Yankelovich and al.,1985] and [Conklin, 1987].

This paper will thus be organized as follows :

- Section 2 briefly describes the fundamental components of an integrated electronic book system.

- Section 3 presents the technological objectives we set while designing WEBS, and the reasoning behind our choice of MacApp and Object Pascal to attain them.

- Section 4 describes the software architecture of the WEBS prototype by sketching the three layers of its class hierarchy.

- Finally, Section 5 enumerates some of the conclusions we have drawn from our work, and contains some suggestions for improvement at various levels.

2. The IEB Concept

An Integrated Electronic Book (**IEB**) represents a complex network of integrated information and "know-how" on a given subject. For example, an IEB on Markov chains might be composed of:

- A set of **text objects** embodying the hierarchy of chapters, sections, and paragraphs which comprises the bulk of any textbook.

- A set of **graphical objects** containing the illustrations, figures, and other pictorial information of the book.

- A set of **modelling objects** including the data necessary to specify various kinds of Markov models.

Each of these objects includes a set of procedures or **methods** which allow for its creation, computation and management. Some, such as the text and graphical objects and their associated methods, are the basic components of any IEB. Others are specific to an IEB's subject; an IEB on linear optimization, for instance, would contain objects that modelled the tools used to perform a sensibility analysis.

A complete IEB is not merely a collection of objects and methods. We believe that an IEB management shell, such as WEBS, should also incorporate at least the following set of capabilities:

1. The IEB shell should allow the user to **distinguish** and **protect** the objects it manages. In the case of WEBS, objects belong either to an author, in which case they are public objects that can be consulted by all users but altered only by the author, or to a reader, in which case they are private objects with private access rights.

2. In order for the objects of an IEB to constitute a manageable and useful "knowledge base," the shell should provide tools for creating index objects[1]. These index objects should allow the user to **find and access** the piece of information in which s/he is interested. They can be compared to the tables of contents and indices of paper textbooks.

3. Finally, the shell should allow its users to **navigate efficiently throughout the various components of an IEB.** In order to achieve this, WEBS uses a hypertext [Conklin, 1987; Yankelovich et al., 1985] construct which we call a **web.** A web imposes a context in which connections can be made between various documents; each new web is a different context. When reading an IEB, two webs can be open at any given time. One belongs to the author, and contains connections that s/he wanted contained in the book. The other is where a reader, presumably studying the material in the text, can make his/her own connections, perhaps generating new insights beyond what the author had considered. A web is composed of a set of **links**, each of which represents a connection between two **blocks.** A block is simply a selection within an object of an IEB; a string of characters within a text document would be one example. These links allow for direct jumps between different parts of an IEB (and indeed, between different IEBs). Figures 1a and 1b illustrate WEBS's implementation of this fundamental notion by showing the effect of the command **Follow Link.** More than one link can connect to a given block; in such a case the user is given a choice of which one to follow. Blocks and links also contain other information, notably the **explainer**, a short user-defined string intended to give a user an idea as to why the block or link exists.

3. Technological Objectives

The EBOOK3 IEB shell was developed by one person in a reasonable amount of time. This was achieved by adopting a very restrictive user interface paradigm, and by totally separating the creation and the consultation processes, with only the latter being wholly interactive [Savoy, 1987a, b and c]. When confronted with the difficulties of extending EBOOK3 into an entirely interactive system with a powerful modern user interface, we decided that a completely new software design would best suit our new requirements. Our present strategy is based on the following non-exhaustive list of technological objectives:

- **Modularity** and **Prototyping.** The IEB is a set of interconnected objects, for which the shell provides creation, access and management methods. It was clearly necessary to design a basic structure for WEBS which we could later expand when new objects or functionality were needed. Section 4 will detail the design of this structure. One consideration, however, imposed itself from the beginning; the

---

[1] The integration of index objects and methods within the WEBS prototype is still in its development phase and will not be further discussed in this paper.

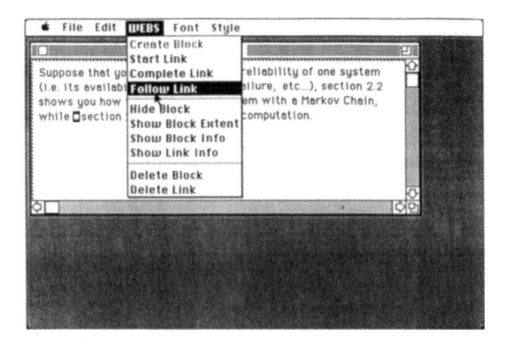

Figure 1a A Snapshot of WEBS before the Follow Link Command

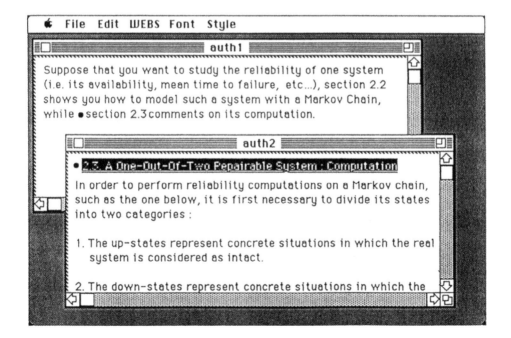

Figure 1b A snapshot of WEBS after the Follow Link Command

system had to be decomposed into modules in such a way that it would be simple both to extend its basic structure and to have several developers working in parallel. For example, while one developer used the WEBS prototype to add modules for creating and manipulating text objects, a second could do the same for modules related to graphical or modelling objects. When both developers have completed their respective tasks, the structure of WEBS should facilitate the aggregation of their modules into a single more powerful tool.

- **Understandability** and **Reusability**. When adopting a modular design, with several developers working in parallel, one must find a way of encapsulating the data and the procedures of the system into meaningful units. Thus there are essentially two goals. First, the design should facilitate the maintenance and expansion of WEBS. Second, developers should be encouraged to extensively reuse each others' code, instead of "reinventing the wheel" at each stage. The inherent modularity of Object Pascal, as well as the facility which with it allows inheritance of methods and data structures, make it an ideal choice in view of these objectives.

- **Consistency**. We felt it essential that WEBS be grounded on a sophisticated and directly manipulable user interface based upon well accepted principles such as those stated in Chapter 2 of *Inside Macintosh™* [Apple, 1985]. Particularly in a multi-developer environment, it is difficult to make such an interface consistent, so that similar commands operate similarly in all modules. We solved this crucial problem by including the user interface functionality of WEBS within a set of basic modules in such a way that a new module could inherit its appropriate behavior with a minimum of additional work on its developer's part.

- **Minimum Effort**. Since the implementation of the WEBS prototype from scratch would take too much time, we wanted a software environment which would provide us with a rich initial framework. MacApp provides such a framework.

- **Targeted Equipment** . The hardware for an IEB must provide an agreeable environment with which the user may interact. Such an environment is particularly necessary because the targeted users of IEBs are not computer specialists, and would soon become frustrated and unreceptive if their interaction with the system were hampered by the machine. The system should include a bitmapped, high resolution screen and an input device other than a keyboard, such as a mouse. This hardware would support the direct manipulation interface described above, which is easy to learn and use. Of the equipment actually available on the market, those which best respond to the above requirements are the workstations. Unfortunately, real workstations are still too expensive to be readily available to the majority of potential WEB users. We therefore selected the Macintosh™ SE as our first target machine, intending to migrate later to the more powerful Macintosh™ II.

4. The Software Architecture of WEBS

The actual objects which are manipulated by an object-oriented program are always instances of abstract objects called **classes**. For example, WEBS can instantiate and control dozens of text documents. These objects, however, all belong to the same *TTextDocument* class, which defines both the data structure and the methods associated with them. In an object-oriented design, then, the unit of modularity is the class. The WEBS prototype is based upon three layers of such classes:

1. the MacApp layer,

2. the WEBS general layer, and

3. the WEBS model layer.

This section provides a high level description of WEBS software architecture. It is divided into three subsections, each describing one of these layers.

4.1 The MacApp Layer

The hierarchy of the MacApp layer's main classes is shown in the upper part of Figure 2.

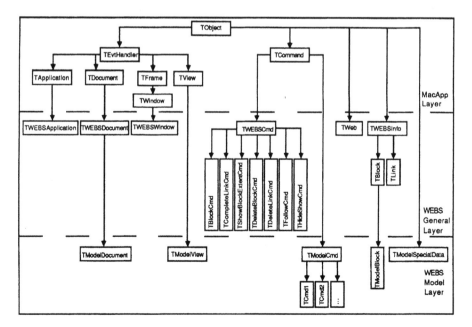

Figure 2 Class Hierarchy of WEBS Prototype

The MacApp layer provides an "expandable" application framework that encapsulates the basic behavior of the standard Macintosh™ user interface, (i.e. pull-down menus; moveable, resizeable, scrollable windows; data storage, retrieval, and printing; and the capability of reacting directly to user generated events such as keyboard or mouse input). The next paragraphs describe the most important classes that MacApp provides:

- The *TObject* class is the ancestor of all other classes. It provides utility methods for cloning and freeing objects.

- The *TEvtHandler* class defines a template method[1] and some default methods to handle several types of events arising in a Macintosh™ application (e.g. key, mouse and menu events). Since most classes must be able to react appropriately to such events, the TEvtHandler class is the ancestor of most of MacApp's other classes.

- Methods for launching an application and displaying its menu bar, and template methods for creating and initializing appropriate document objects, are contained within the *TApplication* class. This class has few fields of its own; MacApp globals, however, are generally treated as fields of the TApplication object. The TApplication class also has the responsibility of managing the main event loop. This loop, which is fully encapsulated within the MacApp layer, receives incoming events from the event queue and dispatches them to their appropriate objects; in the case of most mouse events, for example, the event is sent to the window in which the mouse pointer was located.

- The *TDocument* class is responsible for documents that belong to the application. Furthermore, the TDocument objects contain all the basic information for managing the frames, the windows and the views (see below), which allow the user to inspect and modify a document. Naturally, the methods of TDocument which allow for saving and restoring data must be overridden within the subclasses of TDocument that implement the specific types of documents that the application manages.

- The *TView* class manages the display of the data proper to a document and passes various mouse events to the appropriate objects within the view.

- The user, however, only sees a portion of a view on the screen. The *TFrame* class provides the mechanism for showing a portion of the view in a window, and for scrolling so that different parts of the view appear.

- The *TWindow* class provides methods for opening, closing, moving, resizing, activating, and deactivating windows. A TWindow object is a special type of frame, and may act as a container for other frame objects.

---

[1] Template methods basically do nothing as long as they are not overridden in a subclass of their class.

- Finally, the *TCommand* class manages most actions generated by user input that affect the data or appearance of the active document, i.e. menu commands, mouse commands, and keyboard input. A TCommand object has template methods which provide for executing a command, for undoing its effect, and for redoing its effect. Any subclass of TCommand must override these methods; its methods will act on the application documents, windows or views in order to implement the changes needed to reflect the user's action.

4.2 The WEBS General Layer

The WEBS general layer classes are illustrated in the middle part of Figure 2. The classes of this layer encapsulate the "electronic book" functionality of WEBS. This functionality includes assigning ownership to and protecting document objects, as well as a data structure and methods for dealing with the creation and use of webs, links, and blocks. Later this layer will be further extended with classes allowing for the management of index objects. The following paragraphs describe the main classes of this layer:

- The *TWEBSApplication* class is responsible both for the management of all the documents in the system and for keeping track of the hypertext objects. As such, it includes a method which allocates and initalizes a WEBS document of some requested type[1]. It also finds and installs the appropriate author's and reader's web objects.

- The *TWEBSDocument* class provides template methods which allow WEBS document objects (i.e. any object which inherits from the TWEBSDocument class) to act appropriately in a hypertext environment. A WEBS document object might, for example, be asked to hide all of its blocks. Furthermore, this class contains fields that allow it to be identified by its owner, and methods that allow it to save and restore those fields.

- Hypertext-related commands are supported by the *TWEBSCmd* class. These are basically those commands which appear in the WEBS pull-down menu presented in Figure 1. An object of this class keeps track of the web and document to which the command pertains. Any command object which implements a hypertext-related command should inherit from this object.

- Our notion of a web is implemented in the *TWeb* class. Each object of this class contains three lists: one is a list of all the links in the web; another is a list of sublists, in which each sublist is a list of blocks contained within one document; the third simply maps documents' internal identifiers to their external names and directories. The TWeb class bears the brunt of the responsibility of finding blocks and links and opening their accompanying documents; its methods reflect this.

- The *TWEBSInfo* class is the ancestor of both the block and the link classes. It contains information

---

[1]Presently, only text documents are supported by the WEBS prototype.

such as the date of creation and user-defined explanatory texts which is common to both. Its methods initialize and display this information.

- The *TBlock* class contains the fields and methods which implement WEBS block objects, providing for their creation, initialization, selection, and deletion. Since we expect that blocks will be implemented differently in each type of document (a sentence cannot be treated in the same way as a group of shapes), many of the methods are templates that will be overridden by a specific document block. All types of block objects, however, contain a list of the links attached to them; this list is therefore contained in the TBlock class description.

- A link knows its starting and ending blocks; the *TLink* class contains the fields which support this knowledge, as well as the fields which are inherited from TWEBSInfo. Most of the TLink class' power is in its methods; they allow link objects to be added to and deleted from webs, disconnected from blocks, and followed from one end to the other.

4.3 The WEBS Model Layer

The WEBS model layer contains all of the classes specific to any of the particular models which make up an IEB. These include the general text and graphic models[1] as well as individual models, such as a Markov chain simulation, which apply solely to a given electronic book. In other words, the model layer of WEBS implements the individual model objects which comprise an IEB, while the general layer of WEBS manages the classification and the connection of these objects within an IEB. Since the number of such model objects within our prototype will steadily increase, we first present a generic strategy for implementing any type of model object within the WEBS model layer and only thereafter discuss the particular case of text model objects.

The lower part of Figure 2 shows the classes which must be created to support a new type of model object. Since a model object's data are contained within or pointed to by a MacApp document, they can be saved on disk in the same way as any other MacApp document:

- The *TModelDocument* class must be a subclass of TWEBSDocument. In addition to incorporating much of the functionality that any WEBS model needs to manage its own duties (being in many ways no different than any stand-alone application), it overrides methods of its parent class that concern themselves with document-specific block operations. Amongst other things, TModelDocument objects must know how to hide and show their blocks, as well as to report whether or not a block has been selected.

- A subclass of MacApp's TView class, the *TModelView* class is a model-specific view that displays a document's data to the user. Sometimes it is instructive to show different displays of the same data;

---

[1] We call all of these objects models, thus justifying the name given to this layer.

sales figures might be exhibited as both a printed table and a bar graph. Then several different TModelViews might be defined, each being responsible for a different sort of display.

- If a new model is to do anything, command objects, subclasses of MacApp's TCommand class, will have to be defined for most of its operations. These are illustrated in Figure 2 by the *TModelCmd*, *TCmd1* and *TCmd2* classes.

- Any document-specific capabilities of a block must be implemented in the *TModelBlock* class. This class overrides those methods of its TBlock parent class that are related to the block's display; each model will have its own way to mark the existence of a block and to delimit its contents.

- Since any new model will presumably have some special data of its own, it will often be necessary to implement special data objects. Such objects will probably inherit directly from the MacApp class TObject, and are represented in Figure 2 by the *TModelSpecialData* class.

- Finally, it should be noted that the WEBS TLink object is not represented by a child class in this layer. A link only needs to know the identity of the blocks at each end. Since it does not need to know any model-specific information, it does not need to be represented in the model layer; a generic link, that connects equally to any kind of block, is all that is necessary.

Figure 3 shows what the TModelView and the TModelCmd classes look like when the model objects are text objects. In this case, the new text objects follow the class hierarchy described for the generic case above, except for the addition of several new MacApp classes which constitute a unit that supports text editing.

5. Conclusion

5.1 Achievements

Our experience with Object Pascal and MacApp has been very positive thus far. Within a four month period we have developed a prototype which integrates the WEBS general layer described in Subsection 4.2 and which fully supports text model objects. Such quick progress has been possible largely because of MacApp's support of the Macintosh™ basic user interface features. Furthermore, we have written highly modular code which, with the expandable class structure of WEBS and the inheritance capability inherent in an object-oriented language such as Object Pascal, will greatly facilitate the concurrent integration of several new model objects into our prototype.

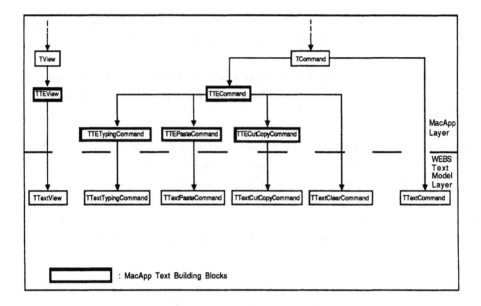

Figure 3 Partial Hierarchy of WEBS Text Model Layer

5.2 Learning Curve

Learning the syntax of Object Pascal is a matter of hours for an experienced Pascal programmer. To master the object-oriented programming style, however, is a more difficult task, which can only be achieved through ample experience, and some good examples to follow. We feel that an object-oriented language is really efficient only if it is used in conjunction with an extensive set of basic classes, such as those which make up MacApp, so that new classes developed for new applications can inherit certain fundamental behavior rather than always reimplementing it. The learning curve for a system such as MacApp is steep, and several months of experimenting with the system appear to be necessary before starting a major project. Nevertheless, the time and consistency gained by using MacApp make learning it a worthwhile effort for the serious developer.

5.3 Additional Comments

Some of the MacApp units do not offer sufficient functionality for our ultimate goals. For example, the TTEView and TTECommand classes offer only simple text editing facilities (no multiple fonts within the same document, no ruler as in MacWrite™) and MacApp does not offer a framework for organizing the display of simple graphic objects such as circles and rectangles. Although it is possible that future releases of MacApp will remedy some of these problems, we may want to develop such facilities ourselves. A major drawback to such a plan is that because of the lack of multiple inheritance capability in Object Pascal, some of this functionality might have to be included in MacApp's own classes. Such a solution would risk incompatibility with future versions of MacApp. We have already run into this problem with descendants of the TTECommand class. TTextTypingCommand, TTextPasteCommand, TTextClearCommand, and TTextCutCopyCommand are all descendants of the corresponding TTECommand classes (see Figure 3). In addition, they all have to perform certain operations on blocks. Because of the lack of multiple inheritance, and because we did not want to rewrite the parent classes, we were forced to duplicate block-handling code in each of the child classes, rather than have one copy of the code from which each of those classes could inherit.

We also believe that the MPW/MacApp system would be a better prototyping system if it were an interactive system rather than a compiled one. We have spent a lot of time over the past few months waiting for the program to compile and link each time a change is made. For a 5000 line program, compiling and linking with the MacApp units can take over 5 minutes on the Macintosh™ SE. An interactive system, or at the very least an incremental compiler, could cut out a lot of this wasted time.

6. Acknowledgements

Our research on "The Integrated Electronic Book" originated with the paper by [Kohlas, 1984] and has been and continues to be sponsored, since December 1985, by the Swiss National Science Foundation under grant number 1.018-0.84.

References

Apple *Inside Macintosh™*, Volumes I, II and III
Apple Computer, Addison-Wesley, 1985.

Conklin J. *Hypertext: An Introduction and Survey*
Computer, September, 1987.

Cox B. *Message/Object Programming: An Evolutionary Change in Programming Technology*
IEEE Software, Vol. 33, No. 1, pp. 50-61, January, 1984.

Cox B. *Software-ICs*
BYTE, June, 1985.

Cox B. *Object-Oriented Programming: An Evolutionary Approach*
Addison-Wesley, 1986.

Doyle K., Haynes B., Lentczner M. and Rosenstein L. *An Object-Oriented Approach to Macintosh™ Application Development*
Proceedings of the 3rd Working Session on Object-Oriented Languages, Paris, France, January 8-10, 1986.

Garret L. and Smith K. *Building a Timeline Editor from Prefab Parts: The Architecture of an Object-Oriented Application*
OOPSLA '86 Proceedings, Portland, Oregon, September, 1986

Goldberg A. and Robson D. *Smalltalk-80: The Language and its Implementation*
Addison-Wesley, 1983.

Goodman D. *The Two Faces of Hypercard*
Macworld, pp. 123-129, October, 1987

Haan B., Drucker S. and Yankelovich N. *An Object-Oriented Approach to Developing Consistent Integrated Applications*
IRIS Report, Institute for Research in Information and Scholarship, Providence, RI, September, 1985.

Kohlas J. *Das Integrierte Buch (eine Projektidee)*
Working Paper No 78, IAUF , April, 1984.

Meyrowitz N. *Intermedia: The Architecture and Construction of an Object-Oriented Hypermedia System and Applications Framework*
OOPSLA '86 Proceedings, Portland, Oregon, September, 1986

Pasquier-Boltuck J. and Collaud G. *The Woven Electronic Book System, (WEBS): The Enduser Model and Interface*
Working Paper No 129, IAUF, February, 1987.
A shorter version of this paper has been submitted for publication in the International Journal of Man-Machines Studies.

Savoy-a J. *Le livre électronique EBOOK3*
Proceedings of the EAO-87 congress, Cap d'Agde-France, March 23-25, 1987.

Savoy-b J. *The Electronic Book EBOOK3*
Working Paper No 137, IAUF*, July, 1987.
This paper has been submitted for publication in ACM Transactions on Office Information Systems.

Savoy-c J. *Le livre électronique EBOOK3*
Diss., Peter Lang S.A. publishers, Berne, Switzerland, 1987.
ISBN 3-2 61-03772-5.

Schmucker-a K. *Object-Oriented Programming for the Macintosh*
Hayden Book Company, Hasbrouck Heights, NJ, 1986.
ISBN 0-8104-6565-5.

Schmucker-b K. *MacApp: An Application Framework*
BYTE, pp. 189-193, August, 1986.

Tesler L. *Object-Oriented Languages: Programming Experiences*
BYTE, pp. 195-206, August, 1986.

Yankelovich N., Meyrowitz N. and van Dam A. *Reading and Writing the Electronic Book*
Computer, October, 1985.

SCOOP
Structured Concurrent Object Oriented Prolog

Jean Vaucher, Guy Lapalme & Jacques Malenfant
INCOGNITO, Dépt. d'Informatique et R.O., Université de Montréal,
CP 6128, Station "A", Montréal, CANADA H3C 3J7

ABSTRACT

SCOOP is an experimental language implemented in Prolog that tries to combine the best of logic, object-oriented and concurrent programming in a structured, natural and efficient manner. SCOOP provides hierarchies of object classes. These objects behave as independent Prolog programs with private databases which can execute goals within other objects.

SCOOP also supports parallel processes, synchronised by the exchange of messages. For simulation, a sequencing set and primitives concerned with simulated time are provided. Thus, SCOOP has the ability to describe structured dynamic systems and to encode knowledge.

The important features of SCOOP are 1) its lexical block structure designed to promote and enforce modularity and to allow verification and optimisation via a compiler, 2) its combination of familiar programming cliches: the concepts of Simula67 for macro-structuring of entities and those of standard Prolog (unification & backtracking) for local behaviour, 3) its provision for parallel activity with a clear distinction between static objects and dynamic processes and 4) its discrete simulation capability.

1. INTRODUCTION

Prolog is a relatively new programming language based on the notation of formal logic and the concepts of theorem proving[Cloc81, Colm83, Ster86]. It stresses a declarative style of programming where a program is written as a set of **facts** and **rules** pertinent to the problem at hand and execution is viewed as attempts to prove the validity of queries. Prolog has shown itself to be excellent for symbolic computation and has found many applications in areas such as natural language processing and expert systems.

Prolog is a deceptively simple language whose power derives from the systematic exploitation of two fundamental concepts: unification and backtracking. *Unification* is a pattern matching operation which is used to select applicable rules and facts and to effect parameter transmission. Backtracking means that, in trying to answer a query, the inference mechanism tries all possible combinations of facts and rules. In other words, Prolog automatically builds programs from available procedures at run-time. As a result, it is not uncommon to find that a dozen well-conceived lines of Prolog are equivalent to several pages of code in more traditional languages.

The logical elegance of Prolog and the compactness of its code makes it tempting to use for large-scale general-purpose programming. However, Prolog was not conceived with software engineering aspects in mind and there are no standard features to support hierarchical decomposition and modularity. Prolog is also weaker than traditional imperative languages in other aspects. The most glaring deficiencies lie in the following areas:

- modularity and protection,
- state changes (there is no assignment in Prolog) and
- expression of parallel activity.

These are exactly the areas where object-oriented (OO) concepts are most effective. Using the **object** methodology, systems are conceived in terms of **classes** of interacting **objects** of various types each with its own attributes, state and capabilities [Stef85]. Local procedures and state variables are usually hidden or protected so that each class may be designed and understood in isolation then assembled according to behaviour without concern for internal details. Moreover, OO languages, such as Simula [Dahl70, Dahl72], can be designed to exhibit static block-structure, so that many aspects of program behaviour can be deduced from the text. This helps in understanding, debugging and compiling.

Often, a **class** may be delared to be a specialisation or subclass of another, thereby **inheriting** all the attributes of the superclass. The inheritance mechanism facilitates incremental development and allows the creation of generic software packages whose object classes can be extended and customised. In some languages, multiple inheritance from several superclasses is allowed forming inheritance networks rather than hierarchies.

Object-oriented programming is also a good descriptive base for parallel computing [Yone87]. The notion of self-contained independent object is very close to that of parallel **process** or **actor**. Furthermore, the concept of communication by message passing between objects and local interpretation of messages maps well onto distributed environments.

In view of the complementary strengths of the logic and object-oriented paradigms, an integration would appear to be fruitful. Scoop tries to do just that: it combines what we think is the best of logic, concurrent and object-oriented programming in a natural and efficient manner. The Scoop system is implemented in Prolog but the use of a compiler and a meta-interpreter gives us great freedom to experiment in the design of the syntax and semantics of the language.

In Scoop, classes represent independent Prolog programs can operate with the full power of unification and backtracking. Inheritance between classes is also provided. Each object has its own private database: initially an exact copy of the clauses defined in its **class**. Some clauses are immutable and fixed for all objects; but others, like dynamic predicates in Prolog, can be asserted and retracted by the object. The fixed (static) clauses act as **methods** whereas the dynamic ones act as **state** variables. It is also possible to parametrise objects by inserting some dynamic clauses into the local database of the object at creation time.

An object can invoke goals within other objects. Initially, only one object, the **main process** object, is active. This object can create other objects and send them queries. To allow multiple activities to proceed in parallel, Scoop supports processes, synchronised by the exchange of messages. On a sequential machine, processes are implemented by time-sharing.

Much of our experience with object-oriented programming had been acquired through simulation with Simula, an OO language which could be extended to serve as a discrete-simulation language. As a test of the suitability of Scoop for general-purpose programming, we extended it for simulation in the same way with a sequencing set and scheduling primitives.

In the next section, we shall review related language proposals which combine some aspects of the object-oriented, concurrent and logic programming paradigms. Each is different either in its objective, its features or its base-language. Put together, the proposals cover just about all parts of Scoop. It is therefore important to underline the particular approach embodied in our proposal:

1) We take Prolog as the base language and seek to extend it in the most useful way. We do not attempt to compare Scoop to designs which go the other way and graft logic (or object, actor...) features to other base languages [Robi82, Lalo87, Yone87]

2) We believe that Scoop manages to integrate usefully more concepts than any other Prolog-based proposal.

3) Our perpective is that of *software-engineering*: we are interested in structuring Prolog so that large programs may be written, understood and assembled in modular fashion. Thus, our proposal will stress static inheritance structure to fix the meaning of names and we will not allow the user to define more flexible inheritance schemes or meta-interpreters. Similarly, we shall aim for macro-parallelism suitable for operating system processes rather than trying to exploit massively-parallel architectures

The salient features of the Scoop object/logic integration are the following:

- **block structured syntax:** designed to promote and enforce modularity and to allow verification and optimisation via a compiler.

- **familiar programming cliches**: using the concepts of Simula67 for macro-structure and those of standard Prolog for local object behaviour. Hopefully, Scoop programs should appear familiar to both object and logic programmers.

- **parallelism**: with a clear distinction between static entites (objects) and dynamic computing agents (processes)

- **discrete simulation capability.**

Scoop is not a finished language; it is an evolving experimental vehicle meant to show the feasability and desirability of the combined approach as well serve as a test-bed for new ideas. Some aspect of the language as therefore quite arbitrary and subject to evolution.

The remainder of the paper is structured as follows. First, we survey consider related language-integration proposals. Next, we describe object definition in Scoop along with remote calls and qualification of object references. After that, we consider, concurrent programming and message passing. Then, the simulation features of Scoop are described. Programming examples follow and the implementation is briefly discussed.

2. RELATED WORK

Object-logic proposals can be divided into two categories depending on wether or not they consider parallelism. In the first object-oriented language, Simula67, objects were introduced to simulate parallel dynamic entities and Simula objects operated as coroutines. Currently, over 20 years later, it is surprising that many object-oriented language proposals concentrate on modularity and neither support real or quasi-parallelism. We consider first the proposals which integrate logic and objects without considering parallelism, then those for concurrent programming.

Object-oriented logic programming

Some of the benefits of object-oriented programming, namely class-dependent methods and inheritance, can be achieved with minimal implementation by using a particular coding discipline for the methods and a SEND predicate which searches an explicit ISA hierarchy . This is the approach taken by Kahn (1981), Zaniolo (1984) and Stabler (1986). However, users must be aware of the internal representation of objects and the notation is verbose. Further study of the proposals reveals that there is no provision for changing object state and no consideration of concurrency.

Objlog [Chou87] and LOOKS [Mizo84] implement objects as part of more complex knowledge representation languages and concentrate on flexible inference schemes. LOOKS realizes a blend of objects and logic programming with class/meta-class/instances having methods defined as Horn clauses. Meta-classes here deal with control knowledge associated with a class, allowing the user to define the inference procedure to be used. Objlog is specifically oriented toward complex knowledge representation problems as architectural databases. Neither system considers parallelism and there is no structured syntax.

ESP is an extension of KL0, a Prolog like machine language developed at ICOT [Chik84]. An object in ESP is an axiom set which responds to messages by trying to refute the submitted proposition using its axiom set. Inheritance through class hierarchy is provided and "slots", representing time dependent state variables, may be associated with each object. In Scoop, we follow standard Prolog and keep state information as dynamic clauses.

SPOOL [Fuku86] is a proposal which syntactically ressembles our own. SPOOL is built on top of IBM Prolog. It a uses block-structured syntax for class declaration and supports inheritance hierarchies. Full backtracking is allowed. Objects communicate via a "send" primitive which retains the full unification capability of Prolog to allow such things as anonymous recipients. SPOOL has no parallelism and, like ESP, it uses instance variables to implement state.

Proposals with parallelism

Parallelism in logic programming is also receiving considerable attention. Two families of approaches may be identified. First, much work has been focussed on the AND/OR model of parallelism, exemplified by Parlog [Clar86], Concurrent Prolog [Shap83, Shap86] and others [Kahn86] . These languages are designed for fast execution on massively-parallel architectures and concurrency occurs at a *micro-level* of the individual term. The ability to backtrack has been seriously limited in order promote speed. The notion of object with changing state is implemented in a side-effect free manner as perpetually recursive procedures. Although logically sound, this approach is unorthodox from the point of view of traditional programming. Object-oriented inheritance can be implemented in these languages, but there is no syntax to support the methodology and the notation remains opaque [Shap83, Kahn86]. At the present, these must be considered to be machine-oriented object-logic languages.

The second family is best described by the generic title: *communicating sequential Prolog processes*. Delta-Prolog [Pere84] illustrates this type of approach. It is based on Monteiro's distributed logic and uses communication primitives similar to those of Communicating Sequential Processes. Scoop may be described as an object-oriented extension of the principles of this second family. Communicating Prolog Units [Mell86, Mell87] also are object-oriented extension of the Delta-Prolog family. However, they differ from Scoop by giving to the user more flexibility in the definition of relationship between objects (as in Actors). Scoop prefers static inheritance scheme to ease program comprehension and coding for most situation. Another example of parallel object-oriented logic language is Object-Prolog [Doma86]. In Object-Prolog, a program is divised into worlds defined as "modularised unit of knowledge" (Horn clauses). Arbitrary inheritance patterns may be coded into worlds again at the price of more programming efforts and run-time inefficiency.

Finally, one should mention T-Prolog [Futo86] which allows discrete simulation in Prolog. However, there is no attempt to provide object structure.

3. SCOOP CLASSES AND OBJECTS

Scoop modularizes Prolog programs by grouping related predicates in class definitions. A class is a template for the creation of objects of the same type. All objects of a particular class will have a common set of predicates particular to them. These predicates are separated in two sets. The first set contains static predicates which are fixed and common to all objects of the same class. These predicates can be viewed as methods. The second set contains dynamic predicates which may have initial clauses common to all objects of the same class but can also have new clauses passed as parameter at object creation and/or asserted or retracted during program execution. Dynamic predicate clauses function as individual databases private to each object that can model changes of state. To improve SCOOP's performance, dynamic predicate clauses are limited to facts whereas static predicate clauses can be either facts or rules. This has not been found to limit expressive power.

Class definition syntax

In Scoop, a class is a set of dynamic and static predicate definitions which can inherit definitions from other classes. The syntax of a class is shown below along with a typical example:

```
class <name>
    (<dynamic predicates dcl> ).
is_a <super_class name> .
dynamic.
    <dynamic facts>

static.
    <static predicate clauses>
begin
    < body >
end.
```

a) Syntax of Class

```
class person (    name/1, child/2,
                  sex/1, +needs/1 ).
dynamic.
    name (unknown).
    sex (male).
    needs (love) .
static.
    son (S)        :- child (S, male).
    daughter (D)   :- child (D,female).
end.
```

b) Typical class definition

Figure 1 - Class Declaration

The class *name* is simply an atom which identifies the class. The is_a clause indicates inheritance relationships between classes in the program. In the example, *person* has no super_class.

Dynamic predicates function as parameters and state variables. They are declared in the *dynamic declaration part* following the class name. Here, the programmer must give the mode, name and arity (number of arguments) of each dynamic predicates. In the example, four dynamic predicates are declared: name, child, sex, and needs. *Name* is a predicate with one argument, *child* has two arguments, etc...

Dynamic predicates can be of one of two "modes": either "add" or "replace". "Add" mode for a predicate is specified by preceding it with the symbol "+", otherwise default "replace" mode is assumed. The mode governs the way the parameters in the object creation will affect the local database of the object. In the example, *needs* is of mode "add". In the next section on object creation, we shall show how "mode" helps in the implementation of default attributes.

The *dynamic* part of the class definition is where the programmer gives the initial state of the objects that will eventually be created from this class. This initial state takes the form of facts which are instances of the predicates declared in the dynamic declaration list. A declared dynamic predicate need not have initial facts but every initial fact must refer to a declared dynamic predicate.

The static part lists invariant facts and rules common to all objects of the defined class. No explicit declaration of name and arity is required. The difference between dynamic and static predicates is that new dynamic facts may be passed as parameters when an object is created; they can also be asserted and/or retracted during program execution. Static predicates cannot be modified in any way during the program execution. Defining the same predicate both as static and dynamic is an error in Scoop.

The <body> defines initialisation code to be executed upon object creation. With the exception of the class name and the keywords **class** and **end**, all parts of the class definition are optional.

Object creation

Classes are merely definitions. It is the objects created from the classes that will own local databases containing dynamic clauses and have the behavior outlined by the definition. In Scoop, an object is created with the new/3 predicate:

new(*Object_ref, Class_name, Dynamic_clause_list*)

Class_name must be instantiated to the name of a defined class. The Scoop interpreter creates a new object of that class and generates a unique identifier for it. This identifier is unified with the *Object_ref* argument of **new**. The newly created object will be said to be of class *Class_name*. An object can obtain a reference to itself with the primitive **thisobject**(X).

At object creation, the object's internal database will be initialized with all the dynamic clauses defined in the dynamic part of the class. The object will also have access to all its class' static predicates. In this section, we only consider the simple case of a class without a *super_class*; inheritance will be covered later.

Dynamic_clause_list is a list of facts compatible with the dynamic predicates declared in the class. These facts will be added to the internal database of the object after the initialization. Here the mode of the predicates is taken into account. If a clause (or a set of clauses in the list) refers to a predicate with *replace* mode, all the initializing facts for this predicate will be replaced by the new fact (or set of facts). On the other hand, if it refers to a predicate with add mode, the new fact (or set of facts) will simply be added *after* the initializing facts. The two modes correspond to the familiar Prolog operations of *consulting* and *reconsulting* files.

To illustrate the difference between modes, consider the effect of the following statement creating a new person according to the class defined in Figure 1:

new(Fred, person,
          [name(fred), child(ann,female), child(joe,male), needs(money)])

The local database created for Fred would contains the following facts:

```
name(fred).          % replacing "name (unknown)."
sex(male).           % default value
needs(love).         % initial fact
needs(money).        % added to the previous default clause.
child(ann,female).   % this shows that several clauses for the same predicate
child(joe,male).     %      can be inserted at once.
```

Scoop objects function like Prolog programs. Launching the proof of a goal G within an object O is termed a *remote call* and is expressed as "O:G". In the case of the person Fred, the query "Fred:son(S)" would succeed with S=joe. Similarly, the query "Fred:needs(N)" would succeed twice with N=love then N=money. "Fred:sex(red)" would fail.

Inheritance and scope of predicates

When a class C_1 is defined by an **is_a** clause to be a subclass of another C_2, objects of class C_1 have access, not only to the predicates of C_1 but also to others inherited from C_2. This inheritance can be affected by the redefinition of predicates in sub-classes and we must clarify the scope or visibility of predicates in Scoop objects. This discussion will also be pertinent to the next section which covers the scope of predicates in *remote calls* between objects.

In the *person* example, there was no inheritance. To illustrate the scoping rules, we shall use the following contrived example where class C "**is_a** B" which "**is_a** A" .

```
class a(p1/1).
    static.
            p2(a).
            p3(a).
    end.
class b(p1/1).
    is_a a.
    static.
            p4(b).
    end.
class c.
    is_a b.
    static.
            p2(c).
            p5(c).
    end.
```

First, we introduce the concept of *inheritance chain*. The inheritance chain of an object is defined as the ordered list of classes beginning by the class from which the object was created followed by all the classes inherited through the **is_a** relationship. In the example, if X is an object of class C, the inheritance chain of X is [C, B, A]. C is said to be the lowest class in the chain and, A, to be the highest. B and C are said to be below A . With these definitions, we can proceed with the scoping rules. You should remember that the scoping rules of Scoop are static in the sense that when you know the context of a call or predicate invocation, you always can determine which clauses will be used to try the reduction (and the class where they appear) by direct inspection of the source program.

Scoping rules:

1. A predicate is said to be *defined* in a class if appears either in the static part of that class or in its dynamic declaration part.

2. A predicate is visible in its defining class.

3. A predicate visible in a class C1 is also visible in any class C2 immediately below C1, unless redefined in C2. By redefinition, we mean that a predicate with the same name and arity is defined in the class. A predicate visible in a class but not defined in this class is said to be inherited.

Briefly, a goal occuring in a given class context refers to the first definition (or redefinition) encountered starting at the class of the context and going up the inheritance chain. An example can illustrate the rules. In our example, here are the predicates visible in each class context:

in A:

| from a: | p1 | p2 | p3 | |

in B:

| from b: | p1 | | | p4 |
| from a: | | p2 | p3 | |

in C:

from c:		p2		p5
from b:	p1		p4	
from a:			p3	

Clause access: local & remote calls

An object is a context or a black box that keeps track of internal information and provides services to computing agents. Access by an object of one of its own predicates is called a *local call*. Access by an object of a predicate in another object is called a *remote call* and, in the simplest case, it is expressed as "Object:Goal" (It is worth noting here that both static and dynamic predicates can be accessed with a remote call).

In Scoop, inheritance and sub_classing means that an object can be viewed at various levels and at each level, different clauses are visible. Different forms of the calls allow access to all levels.To understand the variants of accessing, it is useful to define a computation context as a couple (Object, Class) where Class is the defining class of the Object or one of its super_classes. The couple (Object, Class) will also be referred by "Object as Class". In Simula, this notion of context is called the *qualification* of a reference. The syntax of a call takes one of the following forms:

remote calls

    X: g
    X as *class_name*: g

local calls

    self: g
    self as *class_name*: g
    super: g·
    g   % actually a local call

X: g, where X denotes an object, is the simplest remote call. It means calling the goal g in the context of the object X at its defining class, that is the one used as class_name in the *new* statement.

X as class_name: g means calling the goal g in the context of object X viewed as an object of class *class_name. Class_name* is usually a super_class of the actual class of X, but it could also be a sub_class.

The predicate invoqued for **g** is the definition of **g** visible from the*class_name*. **Self: g** and **self as class_name: g** are the same as the previous forms except that the keyword "self" designate the actual object within which the call is made (in the same fashion as Smalltalk). The use of **self** is a convenience; it is possible to to the same thing by retrieving a self-reference with **thisobject(S)** and using S instead of **self** thereafter.

Super: g means calling **g** in the context of the actual object but one class higher than the class in which "super: g" occurs.

Finally, in the body of a class C, simple use of a goal identifier G (defined in C or its super_classes) is equivalent to "self as C: G".

Assuming the previous definitons of classes **a**, **b** and **c** and the object X created with new(X, c, []), the following shows how to access some of the defined predicates (i) within the context of **b** and (ii) from a context external to X:

predicate accessed		from the context of **b**	from a context outside X
p1 of **a**	....	super : p1(_) self as **a** : p1(_)	X as **a** : p1(_)
p1 of **b**	....	p1(_)	X : p1(_)
p2 of **a**	....	p2(_)	X as **a** : p2(_)
p2 of **c**	....	self : p2(_)	X : p2(_)

Asserting and retracting dynamic clauses

In Prolog there is no assignment statement to alter state, but clauses may be added or removed from the database. In Scoop, we use the same mechanism. The predicates asserta/1, assertz/1 and retract/1 are provided for this purpose and only dynamic clauses may be asserted and retracted during program execution.

In order to protect locality and protection for an object's state, Scoop does not allow asserta/1, assertz/1 or retract/1 as goals in a remote call. This restricts modification of an object's database to its own context. Naturally, if an object wants to provide an access to the dynamic management of its clauses from other contexts, it can simply have static predicates in its class definition allowing it. This means that no object can modify another's state without its knowledge or permission.

Management and access of dynamic clauses is more complex and onerous than for static clauses. In particular, dynamic clauses are replicated in each object whereas static clauses exists in only one copy. It is good Scoop practice to declare as dynamics only the predicates that will be asserted, retracted at run-time or passed as parameters at object creation.

4. CONCURRENT PROGRAMMING IN SCOOP

In the preceding sections, Scoop classes and objects have been introduced. However, classes and objects are only static entities. The active agents in Scoop are the processes. Initially, there is only one active process: the main program. However, it is possible to create others dynamically to execute independent sequences of goals. Scoop processes execute in parallel, time-sharing the interpreter; they synchronize and exchange information by message passing.

Process definition

Process definition in Scoop is a direct extension of class definition. Here, the <body> in the **begin** part (see figure 1) is considered to be a sequence of goals that a created process will attempt to demonstrate in parallel with its creator:

The initial process of a Scoop Program is defined by a "main class" where the "begin" part is mandatory and this "main" class must appear as the very last class definition in a Scoop program.

```
main (<dynamic declaration>).
dynamic.      ...
static. ...
begin.
        <sequence of goals>
end.
```

Process creation and scheduling

Processes are created with the predicate new_process/3:

new_process(Process, Class_name, Dynamic_clauses_list).

As with the **new** primitive, this operation creates an object of class 'Class_name'. Facts in the 'Dynamic_clauses_list' act as parameters. Furthermore, an independent process, is created and launched to execute the sequence of goals in the context of the object. A unique process identifier is generated for that process (different from the object reference) and it is the process identifier which is returned in the 'Process' argument. Process identifiers can be used to effect communication via message passing (to be described later).

It is important to note that object references are distinct from process references. Objects refer to static entities whereas processes refer to dynamic executions of code. Initially, there is a strong association between a newly created process and the object serving as its initial context, but the process may move to execute in the context of other objects and several processes may be present in the same object at the same time. A process can obtain its own process reference with the primitive **thisprocess(P)** which unifies P to the unique reference of the current process. A reference to the object context where a process is currently executing is obtained via **thisobject(O)**.

Although we have chosen not to implement automatic mutual exclusion of processes within objects (as with monitors), it should be noted that the initial object of a process is private to the process and protected from any external tampering. This comes about because remote calls which could alter the object's state make use of <u>object</u> references and the <u>object</u> reference of a process is initially unknown. However, a process could change this by obtaining and broadcasting the reference of its object.

In the current implementation, Scoop processes are executed by time sharing. The interpreter repeatedly gives each process in a *ready_queue* a time slice until the queue is empty. Thus, processes in the *ready_queue* proceed in quasi-parallel fashion whereas processes not in this queue are quiescent or passive. A newly created process is deemed active and placed at the tail end of the queue. Two scheduling primitives are provided to move processes in or out of the *ready_queue*. They are:

        activate (Process)
and   passivate (Process)

            where Process must be a process reference.

Process Synchronisation

In order to perform coordinated computing, processes must be able to synchronize and exchange information. Scoop communication and synchronisation primitives are send/2 and wait/2:

            SEND (<channel> , <msg>)
and   **WAIT (<channel> , <msg>)**

The <channel> is used to select a receiver, whereas the <msg> is meant to carry the bulk of the information. For a successful message transfer to occur, both arguments in the **send** and the **wait** must match. The familiar appearance of these primitives masks some subtle points in the Scoop implementation. In particular, there is an important difference in the way that the "channel" and the "message" arguments are treated.

Send is non-blocking and it never fails. It creates a term of the form "msg(C,M)" with copies of **send**'s *<channel>* and *<msg>*parameters and places this term in a global message database. A process executing a **wait** is made to wait (if necessary) until a message with a unifiable *<channel>* parameter is available. When this occurs, the message is removed from the database and an attempt is made to unify the *<msg>* parameter. If this succeeds, the waiting process proceeds having extracted the message information via unification; if it fails, the **wait** operation is deemed to have failed and the process backtracks. Thus, a mismatch on <msg> fields can cause failure and backtracking but a mismatch on <channel> can only cause waiting. Note also 1) that a sent message can reactivate at most one waiting process and 2) that because messages are copies, the Concurrent Prolog technique of returning answers via un-instantiated variables in messages will not work.

An important aspect of the proposal is that that arbitrary terms are allowed for both arguments. In particular, although the *<channel>* could be a process reference and we are not restricted to sending to either designated objects or designated processes. In the terms of SPOOL, Scoop can have have anonymous recipients. A few examples will show the flexibility of the mechanism. Below, the <channel> is a constant denoting the type of service required. There could be multiple servers.

Sending process:	Receiving process:
send(print, int(123)),...	...wait(print, int(X)),...

In the next example, a message is sent to a known Process and the built-in predicate, *gensym*, is used to generate a unique return address for the reply. This address is sent as part of the original message.

SENDER:

RECEIVER:

```
gensym(Ret_Id)                 thisprocess ( P ),
send( Process , info(Input, Ret_Id)),   wait (P, info(X, Ret)),
                               ... < compute answer > ...
wait( Ret_Id, ans(Output)),...   send (Ret, ans(...)),...
```

For simulation, messages can also be used to implement the *seize* and *release* operations on resources:

```
seize(R) :-   wait (R,1).
release(R) :-  send(R,1).
```

5.   SIMULATION

Our model for object-oriented programming, Simula67, exhibited the power of its methodology by being a general-purpose language which could be extended easily to handle discrete-event simulation.   After implementing a first version of Scoop called POOPS [Vauc86], we decided to extend it for simulation and created SIMPOOPS [Vauc87]. The extention was found to be trivial and Scoop retains the simulation features.

In discrete-event simulation, there is implicit sequencing between activities based on the concept of simulated time for events. It is ironic that the main difficulty we encountered in this extention, was limiting the inherent concurrency of Scoop processes to ensure that only one process was active at one time.

Essentially, we extended the original scheduler with a sequencing set (SQS): that is a queue of doublets < Process, scheduled event_time > ordered by increasing event_time. Now, when the ready_queue is empty, the interpreter looks into the sequencing set and transfers the processes scheduled at the next instant of simulated time into the ready_queue. It also updates its internal clock to the new time. Program execution ends when both the ready_queue and SQS are empty. Three new primitives were also added:

- hold (DT) :

Causes the executing process to suspend itself for DT units of simulated time. This is implemented by removing the active process from the ready_queue and placing it in the sequencing set according to its scheduled reactivation time. The execution of **hold** prevents backtracking to previous goals.

- **time (T) :**

Unifies T with the current simulated time.

- **terminate :**

Cancels all current and future events and stops the simulation. This is useful when there is cyclic activity in a system and quiescence is never achieved.

Other types of synchronisation are easily implemented through the built-in SEND and WAIT primitives (i.e. SEIZE and RELEASE). In addition to these, other predicates such as **uniform(A,B,Ts)** were added to generate various random distributions. In all about 30 lines of code were required to implement all the simulation primitives.

6.   EXAMPLES

The first example below first presents a MUTEX class which implements "semaphore" operations for mutual exclusion in a concurrent environment. Next, we show the implementation of a STACK data type. A local predicate, *pile(X)*, stores the stack state and the implementation uses **assert/retract** to modify state. To ensure correct operation with parallel processes, STACK is defined "as_a" mutex object and non-atomic predicate bodies are bracketed by calls to P and V to ensure mutual exclusion.

```
class mutex.
      static.
                p :- thisobject(O), wait(O, 1).
                v :- thisobject(O), send(O, 1).
      begin.
                v.              % initialisation
      end.

class stack.
      is_a   mutex.
      dynamic.
            pile([]).
      static.
            push (X) :-  p,
                              retract( pile( S ) ),
                              assert ( pile( [X | S] ) ),
                         v.
            pop  (X) :- p,
                              retract( pile( [X | S] ) ),
                              assert ( pile ( S ) ),
                         v.
            top  (X) :-    pile( [ X | S ] ).
            empty  :-      pile ( [] ).
      end.
```

Figure 2 - Stack Definition

Stacks can be created and used as follows:

new(St,stack,nil), St.push(1), St.pop(X), write(X),...

```
class scan (margin/1,tree/1) .
static.

   trav(t(L,R)):- trav(L), trav(R) .
   trav ( X ) :-
        integer(X),
        margin(M), tab(M),
        writeln(X).
begin.
   tree(T), trav(T).
end.

main.
   begin.
      new_process(_,scan, [1 ,  t(t(1,t(3,5)),t(6,99))] ),
      new_process(_,scan, [10 , t(t(1,2),t(3,4))] ).
   end.
```

The Output:

```
              1
                   1
                   2
              3
              5
                   3
                   4
              6
              99
              *** THE END ***
              CPU  : 0.77 sec
```

Figure 3. Parallel processes

Figure 3 shows processes executing in parallel. **Scan** objects print out the leaves of binary trees they are given at creation time. The parameter *margin* serves to format the output. The main program creates two **scan** objects as processes and the output shows the parallel execution.

```
class prod (ch/1,tree/1) .
static.
        trav (t(L,R)):- trav(L).
        trav (t(L,R)):- trav(R) .
        trav ( X ) :-    X is_a number,
                         ch(C), send(C,X).

        go:- tree(T), trav(T).
        go:- ch(C), send(C,eof).
begin.
  go.
end.

class merge.
static.
        merge(eof,eof):-      !.
        merge(V1,eof):-       writeln(V1), wait(1,V1x),!, merge(V1x,eof).
        merge(eof,V2):-       writeln(V2), wait(2,V2x),!, merge(eof,V2x).
        merge(V1,V2) :-       V1<=V2, writeln(V1),
                     wait(1,V1x),!, merge(V1x,V2).
        merge(V1,V2) :-       V2<V1, writeln(V2),
                     wait(2,V2x),!, merge(V1,V2x).
begin.
        wait(1,V1), wait(2,V2), merge(V1,V2).
end.

main.
begin.
    new_process(_,prod, [ch(1),   t(t(3,7),33)] ),
    new_process(_,prod, [ch(2) ,  t(1,t(6,99))] ),
    new_process(_,merge,[]).
end.
```

The Output:
```
                1
                3
                6
                7
                33
                99
                *** THE END ***
```

Figure 4. Synchronised Merge

The next example in Figure 4 shows interprocess communication using the **send** and **wait** primitives. The example is adapted from [Dahl72]. There are two producer processes similar to the processes of Figure 3. Each generates an ordered sequence of integer values terminated by the constant **eof** and sends them along a communication channel. The merge process merges values from both channel to keep the output ordered. As in [Shap83], the merge operation is implemented as a tail recursive procedure with the cut (!) operator to prevent needless growth of the backtrack stack.

```
class client (id/1).
static.

        seize(R)        :- wait(R,1).
        release(R)      :- send(R,1).

begin.
        uniform(0,10,Ta),  hold(Ta),
        id(N), Nx is N+1,  new_process (_, client, [id(Nx)] ),
             write(N),  writeln(" Waiting"),
        seize (res),
             write(N),  writeln(" Entering resource"),
             uniform(0,8,Ts),
             hold(Ts),
        release (res),
        write(N), writeln (" Leaving system").
end.

main.
begin.
        send(res,1),
        new_process( _, client,  [id(1)] ),
        hold (20),
        writeln("Closing down system"),
        terminate.
end.
```

OUTPUT:

```
        1 Waiting
        1 Entering resource
        2 Waiting
        3 Waiting
        1 Leaving system
        2 Entering resource
        4 Waiting
        2 Leaving system
        3 Entering resource
        3 Leaving system
        4 Entering resource
        Closing down system

        *** THE END ***
        CPU  : 0.95 sec
        EVALS: 3779
        FAILS: 453
```

Figure 5.   A single server queue simulation.

Figure 5 shows a typical single server queueing simulation and the output generated. Each arriving customer generates his successor and provides him with a unique identifying number. Inter-arrival and service times are random. Seize and release operations have been implemented trivially via message passing. The main program controls the simulation. It starts the first customer and eventually shuts down the simulation. The output trace shows the interleaving of the activities of the various objects. The statistics at the end of the listing show that the execution of this program took about one second on a VAX8600.

CLASSES
  class(client)
  class(main)

STATIC PREDICATES
  static_predicate(client,seize)
  static_predicate(client,release)

DYNAMIC PREDICATES
  obj_pred(client,[id])
  obj_pred(main,[])

CLAUSES
  1: clauses(client,seize(_89),[wait(_89,1)],_90)
  2: clauses(client,release(_81),[send(_81,1)],_82)
  3: clauses(client, begin, [uniform(0,10,_70), hold(_70), id(_71), _72 is _71+1,
    new_process(_,client, [id(_72)]), write(_71), writeln(Waiting), seize(res), write(_71), writeln(
    Entering resource), uniform(0, 8,_73), hold(_73), release(res), write(_71), writeln(Leaving
    system)], _74)
  4: clauses(main, begin, [send(res,1), new_process(_,client, [id(1)]), hold(20), writeln(Closing
    down system), terminate], _45)

Figure 6. Simulation - Compiler Output

This program, which apes the Simula or GPSS style, looks familiar and is easy to understand. On the other hand, the Prolog clauses produced by the compiler to drive the interpreter are quite cumbersome. They are shown in Figure 6. Many object-oriented Prolog systems require users to program in this style.

```
class obj_parallele (id/1).
is_a obj.
static. ...
end.

class dialogue(modal/1, title/1, shape/1, rect/4).
is_a    obj_parallele.
dynamic.
        rect(50,50, 150,350).
static.
        send(Showdialog) :- id(Id), senddialog(Id, Showdialog).
        select :- id(Id), d_select(Id).
        hide :- id(Id), d_hide(Id).
        ...etc...
end.
```

Figure 7. Extract from SCOOP Graphic Expert

Finally, in Figure 7 shows an extract from the largest Scoop program written so far: about 600 lines of Scoop which combined with another 500 lines of pure Prolog implement a small expert system shell with a multi-window graphic interface. The figure shows part of the definition of a dialogue window which is defined as a parallel object to allow multiple simultaneous interactions. Of interest is the "rect(....)" dynamic clause which specifies a default size and position for dialogue windows. This work by Augustin Paar, an MSc Student in our Department, showed that Scoop was a useful and practical programming tool for large programs. By using a blend of Prolog and Scoop, acceptable interactive response could be achieved.

7. IMPLEMENTATION

Scoop is a compiler/interpreter system implemented in Prolog. Different versions have been implemented with several brands of Prologs and run on Vax, Sun and MacIntosh computers. At present the program is about 450 lines of code divided evenly between interpreter, compiler and formatted trace output. More details on the language and its implementation are available in [Vauc86 and Vauc87].

One of the main problems in a concurrent/logic integration is the combination of backtracking and parallelism. The problem lies in the implementation of backtracking to undo the variable bindings (instantiations). Typically, a meta-interpreter for Prolog undoes bindings in the reverse order to which they were made. When one tries to simulate co-routines or parallel behaviour, this is no longer the case. Our solution implements both backtrack and parallelism by exploiting a simple fact. When a clause is **asserted** and added into the database, a new copy of each of its variables is created. To backtrack, it suffices to retrieve the copy from the database to restore the state to what it was. This method of *freezing* and *melting* variables is also described by Sterling and Shapiro [Ster86]. This method of keeping track of state was forced on us by the use of C-Prolog which does not tail-recursion optimisation and forced us into a failure-driven interpreter loop.

Presently, our main concern is speed. The Scoop interpreter is about 60 times slower than the underlying Prolog interpreter, i.e. about 100 LIPS versus 6000 LIPS. Recent tests have shown that much of the overhead is due to the maintenance of backtrackable states. However, speed can be traded for granularity of parallelism. Any term which has no Scoop definition is assumed by the interpreter to be implemented in the underlying Prolog and passed on to the native Prolog interpreter. By judicious mixing of Scoop and Prolog, reasonable levels of performance, modularity and parallelism can all be achieved.

We have improved the compiler so that most method-lookup is done at compile time. Moreover, we are modifying the interpreter to take advantage of a new Prolog with tail-recursion and we plan to modify an existing Prolog engine to implement Scoop directly

8. CONCLUSIONS

Scoop integrates concurrent, logic and object-oriented programming paradigms in a natural way. Thus, Scoop has the ability to describe structured dynamic systems and to encode knowledge. The important features of Scoop are:

1) provisions of both local logic programming and global object-oriented structure
2) lexical block structure designed to promote and enforce modularity and to allow verification and optimisation via a compiler,
3) use of familiar programming cliches: the concepts of Simula67 for OOP and those of standard Prolog (unification & backtracking) for local behaviour,

4) provision for parallel activity with a clear distinction between static entites (objects) and dynamic computing agents (processes)

5) acceptable performance and

6) discrete simulation capability.

At present we are using Scoop to implement a better distributed system with a remote graphic interface server. Scoop will also serve as a vehicle for language experimentation. In particular we will experiment with alternatives to multiple-inheritance, study various ways to hide local information and enforce more protection. However, our main concern is speed and we plan to modify an existing Prolog engine to implement Scoop directly.

9. ACKNOWLEDGEMENTS

This research was supported in part by the Natural Science and Engineering Research Council of CANADA. The authors also wish to thank Charles Giguère, Director of the Computer Research Institute of Montréal (CRIM), who provided the computer facilities for the development of the early versions of Scoop.

10. REFERENCES

[Clar86] Clark K. and Gregory S. **Parlog: Parallel Programming in Logic**, *ACM Trans. on Programming Languages and Systems*, **8**, 1 (Jan. 1986), pp. 1-49.

[Chik84] Chikayama, T. **Unique Features of ESP**, *Proc. International Conference on Fifth Generation Computer Systems*, ICOT, Tokyo, pp. 292-298, (1984).

[Chou87] Chouraki, E. and Dugerdil, Ph. **The Inheritance Processes in Prolog: Multiple vertical with point of view and multiple selective with point of view**, in *GRTC technical paper #GRTC/187bis/Mars 1987* (CNRS Marseilles).

[Cloc81] Clocksin, W. & Mellish, C. **Programming in PROLOG**, Springer-Verlag, Berlin, (1981).

[Colm83] Colmerauer, A. , Kanoui H. et Van Caneghem M. **Prolog, bases théoriques et développements actuels**, *Techniques et Sciences Informatiques*, Vol. 2, N° 4, pp. 271-311, (avril 1983).

[Dahl70] Dahl, O-J., Myhrhaug, B. & Nygaard, K. (1970). **SIMULA Common Base Language**, Publication S-22, Norwegian Computing Center, Blindern, OSLO.

[Dahl72] Dahl, O-J., Dijkstra, E. & Hoare, C.A.R. (1972). **Structured Programming**, Academic Press, London.

[Doma86] Domán, A. **Object-PROLOG: Dynamic Object-Oriented Representation of Knowledge**, SzKl Comp. Research and Innovation Center, 1986, 14 p.

[Fuku86] Fukunaga, K. and Hirose, S. **An Experience with a Prolog-based Object-Oriented Language**, in *Proc. of Object-Oriented Prog. Sys., Lang. and Applic. '86 (OOPSLA)*, ACM Sigplan Notices **21**, 11 (Nov. 1986), pp. 224-231.

[Futo86] Futó, Y. and Gergely, T. **Logic Programming in Simulation**, in *Transactions of the Society for Computer Simulation* **3**, 3 (July 1986), pp. 195-216.

[Kahn82] Kahn, K., **Intermission - Actors in Prolog,** in *Logic Programming,* (Eds. Clark, K.L. & Tärnlund, S-A.), Academic Press, London, pp.213-228, (1982).

[Kahn86] Kahn, K., Tribble, E.D., Miller, M.S. and Bobrow, D.G. **Objects in Concurrent Logic Programming Languages,** in *Proc. of Object-Oriented Prog. Sys., Lang. and Applic. '86 (OOPSLA), ACM Sigplan Notices* **21,** 11 (Nov. 1986), pp. 242-257.

[Lalo87] Lalonde, W.R. **A Novel Rule-Based Facility for Smalltalk,** in *Proceedings of ECOOP'87,* Bigre+Globule **54** (Juin 1987), pp. 193-198.

[Mell86] Mello, P. and Natali, A. **Programs as Collections of Communicating Prolog Units,** in *Proc. of ESOP '86, Springer-Verlag Lecture Notes in Comp. Science* **213,** pp. 274- 288.

[Mell87] Mello, P. and Natali, A. **Objects as Communicating Prolog Units,** in *Proceedings of ECOOP'87,* Bigre+Globule **54** (Juin 1987), pp. 233-243.

[Mizo84] Mizoguchi, F., Ohwada, H. and Katayama, Y. **LOOKS: Knowledge Representation System for Designing Expert System in a Logic Programming Framework,** in *Proceedings of the International Conf. on Fifth Gen. Comp. Sys.,* 1984, pp. 606-612.

[Pere84] Pereira L.M. and Nasr R. **Delta-Prolog: a Distributed Logic Programming Language,** Proc. International Conf. on Fifth Generation Computer Systems, ICOT, Tokyo, pp, 283-291

[Robi82] Robinson, J.A. & Sibert, E.E., **LOGLISP: Motivation, Design and Implementation,** in *Logic Programming,* (Eds. Clark, K.L. & Tärnlund, S-A.), Academic Press, London, pp.299-313, (1982).

[Shap83] Shapiro E. & Takeuchi A., **Object oriented programming in concurrent Prolog,** *New Generation Computing ,* Vol 1, pp. 25-48, (1983).

[Shap86] Shapiro E. **Concurrent Prolog: a Progress Report,** *Computer,* **19,** 8, pp. 44-54 (Aug. 1986)

[Stab86] Stabler, E.P. **Object-Oriented Programming in Prolog,** *AI Expert ,* October 1986, pp. 46-57.

[Stef85] Stefik M. & Bobrow D., **Object-Oriented Programming: Themes and Variations,** *The AI Magazine,* pp. 40-62, (1985).

[Ster86] Sterling L. and Shapiro E. **The Art of Prolog,** MIT Press, 1986.

[Vauc86] Vaucher, J.G. & Lapalme, G., (1986). **POOPS: Object-Oriented Programming in Prolog,** Pub. Nº 565, Département d'informatique et de R.O., Université de Montréal.

[Vauc87] Vaucher, J.G. & Lapalme, G. **Process-oriented simulation in Prolog,** *SCS Multi-Conference on AI and Simulation,* San Diego, Jan. 1987, pp.41-46, (1987), also available as Pub. Nº 604, Département d'I.R.O., Univ. de Montréal.

[Yone87] Yonezawa, A. & Yokoro, M. (Editors) **Object-Oriented Concurrent Programming,** The MIT Press, Cambridge, Mass., (1987).

[Zani84] Zaniolo, C. **Object-Oriented Programming in PROLOG,** *Proc. International Symposium on Logic Programming,* Atlantic City, IEEE, pp. 265-270, (1984).

The Implementation of a Distributed Smalltalk

Marcel Schelvis and *Eddy Bledoeg*

Océ Nederland, P.O.box 101
5900 MA Venlo, the Netherlands
+31-77-594036

Abstract

This paper describes DistributedSmalltalk, which consists of a number of cooperating Smalltalk virtual machines distributed over a network, that provide complete distribution transparency to the image level, including transparent message passing across machine boundaries. As a result no modifications are necessary at the image level and e.g. the standard Smalltalk debugger can be used for system wide debugging. Transparent I/O is provided by means of a concept called "home objects". The performance degradation is acceptable, due to replication and the home object concept. Replication is transparent and replication consistency is guaranteed, so e.g. for replicated class objects no compatibility checking is needed. Distributed garbage, whether containing cycles or not, is collected incrementally without any synchronization being necessary.

Key words and phrases

Smalltalk, distributed processing, distribution transparency, remote procedure call, incremental distributed garbage collection, replication.

1. Introduction

This report describes the design and implementation of a distributed programming system called DistributedSmalltalk, which is completely distribution transparent. Distribution transparency implies that programmers writing distributed applications, such as multi-authoring document systems, e-mail or calendar systems, need not worry about object access, network location, replication, concurrency control, etc. DistributedSmalltalk is based upon an existing implementation of Smalltalk (Berkeley Smalltalk) and is implemented on a network of Sun workstations running Berkeley UNIX. It will serve as a vehicle for further research in distributed object-oriented computing environments and applications.

Object model in distributed systems

Distributed systems and applications are inherently more complex to program than non-distributed ones. To reduce complexity, much research work has been focused on tools that assist in the construction and programming of distributed systems. For example, message based operating systems like the V-system [11] and Mach [12] provide the communication means, but the programmer has to deal with message packaging/unpackaging and locating message targets.

With a remote procedure call facility like Sun's Rpc [10], the programmer can describe his application as a set of interacting modules via procedure calls.

Object-oriented languages like Smalltalk-80 [15] and Argus [13] can make life more easy. With these languages, the programmer can construct his applications in terms of communicating objects.

We believe that objects are an excellent way to structure a distributed system because they provide a means for data encapsulation. Data encapsulation is a powerful mechanism for controlling access to shared data.

Objects are the units of programming and distribution. Furthermore, an object can be seen as a computational unit, so therefore it has the potential for concurrent computation as exploited in POOL [14].

Regarding reliability, damage resulting from an error can be confined within an object. This property of objects is vital for fault tolerant distributed systems.

Smalltalk as a distributed system

The Smalltalk programming system can be seen as a set of objects that communicate with each other and with the user in a well defined way.

Until now, Smalltalk is a non-distributed system. Distribution of this system implies that the set of objects is divided in subsets, each subset residing on a different host. In order to call the resulting system a Smalltalk system, the semantics of the interaction between objects and between the user and the objects must not change. In other words no functionality must be changed.

The way functionality is implemented however will have to change and will depend on the location of the interacting entities (objects, users).

2. The Smalltalk system

Smalltalk is a programming system that provides functions like storage management, display handling, text and picture editing, compiling and debugging. The Smalltalk system consists of a virtual image and a virtual machine.

The virtual image

All components of the Smalltalk programming system are represented by objects. The set of all the objects in the system is called the virtual image. An object is a representation of a real world entity e.g. a display screen or an abstract entity e.g. a number. An object consists of some private data and a set of operations. The private data of an object can be manipulated only by its own operations. An object can be manipulated by other objects through its set of operations.

Objects communicate with each other by sending messages. A message is a request for an object to carry out one of its operations. A message specifies the name of the receiver object (receiver), the name of the operation (selector) and a list of object names as arguments. A message only specifies which operation has to be performed. The receiver of the message determines how the operation will be carried out.

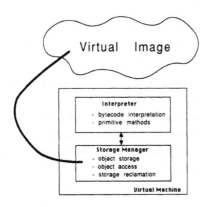

Figure 1. The Smalltalk system

A class is a set of equivalent objects. A class itself is an object. It describes the private data and the set of operations of its instances. Furthermore, a class provides the operation of creating its instances. Every object in Smalltalk is an instance of a class.

The private data of an object is described by its instance variables. An instance variable is a name which refers to one object, called its value.

Each operation of an object is described by a method. A method specifies manipulations of the object's instance variables and manipulations of other objects by means of message passing. A method specification is expressed in the Smalltalk programming language. A method specification is compiled into a bytecode representation, which is executable by a runtime kernel, called the virtual machine.

A small subset of the methods in Smalltalk are not expressed in the Smalltalk language. These are called primitive methods. Primitive methods are built into the virtual machine. Primitive methods allow the underlying hardware and virtual machine structures to be accessed.

A class can inherit methods from one other class, called its superclass. There is a class in Smalltalk, called Object, which is the ultimate superclass of all classes.

The virtual machine

The virtual machine consists of a storage manager and an interpreter.

During the lifetime of the system objects are created, live for a while and die. The storage manager provides storage for the objects that make up the virtual image and reclaims the storage occupied by objects, that have died (garbage collection). The storage manager also provides access to the memory fields of objects.

The interpreter provides a stack oriented execution environment. It accesses the following objects found in the object memory: processes, compiled methods, contexts and classes. The interpreter fetches and executes bytecodes found in compiled methods. The state of execution of a compiled method is represented by a context. Contexts correspond to stack frames or activation records. The interpreter finds the appropriate compiled method to execute by searching message dictionaries of class objects.

Some compiled methods refer to primitive methods. If a compiled method indicates a primitive method, the interpreter does a dispatch and executes the primitive. There are about a hundred primitive methods that perform arithmetic, storage management and I/O operations. Process scheduling routines are also performed by primitive methods.

3. Distributing Smalltalk

Smalltalk-80 is a single user programming system. Multiple Smalltalk programmers can exchange objects only by writing objects (or source code) into a file, transferring the file over the network and reading the file at the destination.

There has been a number of attempts to make shared access of objects easier [1][2]. They all provide a transparent message passing mechanism at the image level. The Smalltalk virtual machines (VM's) are enriched with some primitives that enable inter-VM communication, and the images with special objects that make use of these new primitives in order to send messages to each other over the network. Other attempts use a centralized data base containing the shared objects, where the database manager acts as a virtual machine and communicates with a number of VM's by means of transparent message passing [3]. Although we think the object manager of a VM indeed should support persistent storage and concurrency control, we do not believe in this centralized approach. We believe in a truly distributed system, where each VM provides this functionality. Advantages of the image level approach are:

☐ No substantial changes have to be made to the VM, and hence

☐ it is relatively easy to make VM's from different vendors work together and

☐ no performance is lost during local operation.

However there are a number of problems that are difficult to solve with an image level approach.

One problem mentioned in [2] has to do with I/O. When a remote VM is executing some method for me, and within this method a message is sent to the object Display, I want things happen on my screen and not on the screen of my colleague (who in his turn would be rather annoyed seeing my menus pop up out of the blue).

Then there is a problem with standard classes on different hosts. Is the semantics of "standard" classes everywhere the same? If not, things can go very wrong. On the other hand it would be almost impossible to keep consistency unless every user is so kind not to touch standard classes.

A third problem has to do with Smalltalk processes. With the image level approach, during a remote execution there are several Smalltalk processes involved during a remote execution, at least one sender and one receiver process. Therefore the standard Smalltalk debugger can not be used for remote debugging.

To solve these problems we adopted a different approach. We chose for the concept of distribution transparency [4] at the image level. This is the property provided by the VM's, that all consequences of distribution are concealed from the image, and therefore also from applications and users. As a result of distribution transparency, all objects in the system may be referenced in a uniform manner regardless of such factors as access, location, migration and replication. Since distribution transparency must be provided by the VM's, we call our approach the "VM level" approach. When a message is sent to some object, the VM knows whether the object is local, remote or replicated, and if it is remote, where or how to find it and if it is replicated, how to select the appropriate replica. If the receiver happens to be a local object, the local VM handles the message just like in an ordinary standalone VM. If the receiver is remote however, the message is forwarded to the remote VM. As a result objects on different VM's can work together as if they were on the same host. If there is distribution transparency, the Smalltalk process concept need not change at all. For remote execution process objects simply migrate to remote VM's, and the context chain may run across many machines. At the image level this is all transparent and therefore NO modification whatsoever is needed to e.g. the standard Smalltalk debugger or the user-interrupt code.

4. The VM level solution

In the previous chapter problems encountered when "distributing" Smalltalk are introduced. In this chapter we discuss our VM level solutions for them.

Host objects

Lets start with a number of virtual machines being capable of localizing objects and doing remote sends (and returns) and one standard Smalltalk image. In order to make a distributed Smalltalk system we can take this standard image and distribute its objects at random over the VM's (see fig.2). The resulting system however would still be a single user system, since there is only one Display object, one InputSensor, etc. and worse, they are on different machines. So the first thing we should do is making extra objects on every VM which represent the functionality of the underlying hardware, a.o. Display, Sensor, Processor, ScheduledController. We call these objects "host objects" (a host is a VM + image). Since we may now have multiple objects associated with the same name (e.g. the name Display is associated with a set of display objects, one on every host), we immediately have to deal with a new problem. How do we select the right object? In case of a message sent to Processor or

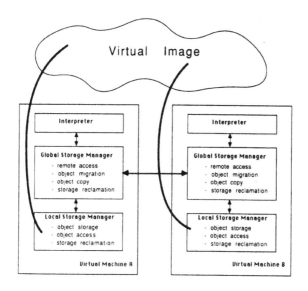

Figure 2. The DistributedSmalltalk system with two virtual machines

ScheduledControllers the local object is fine. However when e.g. Display is the receiver, the VM handling the message must find out on behalf of which user (me or my colleague?) the message is sent, and redirect it to the host of this user.

Replication

The system now can handle multiple users concurrently, but is extremely slower than its local standalone counterpart. This is because even for simple key strokes or mousebutton clicks the objects needed to handle them are scattered at random over the network. Also the method lookup process itself is very costly, as class objects, their superclasses and their method objects can reside on different hosts. To reduce the network traffic, we practically do the same as we did with the previous problem. We introduce multiple copies (replica's) of heavily used objects, one for each host.

Selection of message receivers

The receiver selection mechanism for host objects like Display is implemented in the following manner. Host objects like Display carry a flag "HOME" in their header, and Process objects have an instance variable containing the identity of the host where their process was initiated (their "home"). When a receiver is flagged "HOME" then the message sent to it is forwarded to the home of the current process. Objects that are flagged "HOME" are called "home objects".

The selection problem for replicated objects is not that urgent, in fact it seems obvious to always select the local replica in order to avoid network traffic (after all that is why we replicated in the first place), but on second thoughts it becomes clear that class objects the instances of which are of a highly interactive nature, be better selected on the host where the interaction is to take place (StandardSystem-View, Debugger, Inspector, etc.) Therefore also these objects are flagged "HOME" and the same selection mechanism is used as for host objects like Display.

Relationship between host-, home- and replicated objects

The host objects are

- □ Display, Sensor, Transcript.
- □ All class variables.
- □ Processor, ScheduledControllers.

Before we define the relation between host- and replicated objects, we have to introduce the concept "equality".

- □ Two objects are equal if they reference exactly the same objects in the same order. That is if the objects are viewed as tuples consisting of references $r1,.,rn$ and $s1,.,sm$ then $n = m$ and $r1 = s1, .. , rn = sn$.
- □ Two objects are equal if the objects they reference are equal.

An object's graph is a directed graph where the nodes are all objects that can be accessed from the root object via a path of inter-object references (the graph's branches). Two objects are called deep equal if their graphs do not share any node, other than primitive self-describing objects like SmallIntegers (Objects containing the SmallInteger "3" in this context are considered to share the object "3"). Two objects are called shallow equal if their graphs share all nodes except the roots. If the roots are shared the objects are called identical. See also [5].

A set of objects is replicated if

- □ there is exactly one object on each host and
- □ the objects are equal except for host object nodes in their resp. graphs.

From this definition follows that host objects, although not equal, are replicated. Replicated objects are

- Smalltalk (the SystemDictionary containing all global variables).
- All host objects.
- All objects that are referenced by a non-host replicated object and that are not shared.

Non-replicated objects are all remaining objects, that is objects not referenced by a non-host replicated object.

The home objects are

- Display, Sensor and Transcript (we already explained why).
- All class variables (because they often serve as a communication medium for class' instances of one user, e.g. ParagraphEditor *CurrentSelection* and *Clipboard*).
- Class objects the instances of which are of a highly interactive nature.

Restrictions on object location

To make method lookup faster (and simpler) the VM's make sure (a replica of) the class object of methods and class-instances is locally present. These restrictions are reasonable, since before a class' instance or method can be accessed, nearly always the class object itself must be accessed first. At the moment also (a replica of) the superclass object of each class object must be locally present. This restriction will be removed.

Easy implementation

A way to understand why the replication problem and the I/O problem are best handled at the VM level is to realize that within the VM there are only a few places where a check is needed. During send-bytecode interpretation one check is needed to see if the receiver is a home object and if the message should be forwarded (and this check is combined with the check for remote receivers, so there is no additional cost). In the same way there is only one place in the VM code, where a check has to be done on writing in a replicated object (this is also combined with other checks). With the image level approach checks would be necessary on numerous places. Furthermore replication by the VM can be easily implemented as an atomic action.

5. Addressing and storage management

Every VM manages its local object space, which is organized the same way as the Berkeley Smalltalk object space. In DistributedSmalltalk objects within a local object space are uniquely identified and addressable by means of their *oop* (object-oriented pointer). Most of the time objects are addressed directly, and oops are pointers to the object (instead of indices in an object table).

Forwarding objects

Sometimes objects are addressed indirectly. In this case the oop is a pointer to a "forwarding object", a Smalltalk object that contains the actual pointer. Forwarding objects are not visible from within the Smalltalk image, but for the VM they are ordinary objects. A forwarding object is used as a temporary indirection to some local object that moved to another address (and therefore changed its oop) during garbage collection or because it needed more space. In this case the forwarding object is located at the object's old address.

A forwarding object is also used to reference an object on a remote host. This kind of forwarding object is also called "proxy" [16]. A proxy contains a host identification and an index in a table on this host. This table contains the local oops of remotely referenced objects. When remotely referenced objects change oop (because of garbage collection or growing), only their table entry has to be updated.

When a forwarding object is accessed, the VM recognizes this by means of a "FORWARDING" flag in the object header and accesses the "forwarded" object instead. In case of local forwarding, the reference to the forwarding object is shortcircuited to the forwarded object.

Object spaces

A VM's object space consists of several subspaces. One subspace contains all replicated objects (ReplicaSpace), and the others contain objects according to their age (OldSpace, NewSpace, and SurvivorSpace). SmallIntegers are self-describing, their oop is at the same time their contents. All other objects have real oops that point into one of the before mentioned spaces.

Newly created objects reside in NewSpace, old objects in OldSpace. SurvivorSpace is an intermediate between New and OldSpace.

ReplicaSpace

The ReplicaSpace is the only space in DistributedSmalltalk, that does not exist in Berkeley Smalltalk. The ReplicaSpace is like OldSpace, except that it contains equal objects in the same order on every host. Since replicated objects are in a separate space, they can be recognized by their oop. Since ReplicaSpaces start at the same local address, a replicated object has the same oop on every host. For this reason remote procedure call (de)serialization routines need not translate pointers to replicated objects.

The image can be seen as a directed graph, with the Smalltalk SystemDictionary as root. ReplicaSpace is a subgraph which covers the top of the image graph. The leafs of this subgraph are either primitive data like SmallIntegers, or oops to objects in other local spaces. Host objects contain oops to local objects, that are not shared by other hosts. The remaining leafs of the graph are objects shared by all hosts. Take for example some non-replicated class object, then the Smalltalk SystemDictionary of the host where this class object resides, contains an association (which is a replicated object), the value part of which is an oop to the class object in some other local space. All other hosts have the same association except for the value part which is an oop to a proxy, which points to the same class object. The leafs of the ReplicaSpace are the roots of the other spaces.

Replication consistency control

In order to keep the ReplicaSpaces mutually consistent, every VM checks whether a store in an object affects this consistency. In case of store operations in non-host replicated objects, the modification is replicated over all other hosts by means of a broadcast message. Concurrency control is done by optimistic time-stamp ordering using Thomas' Write Rule (TWR) [6]. Creation of new replicated objects or existing ones that change oop (because they must grow) is handled in much the same way. The difference here is that no rejections are possible. Where according to TWR a modification request (for a store in a replicated object) comes "too late" and therefore is rejected, in the latter case the new object (or the object to move) simply has to be inserted between the ones with timestamps just lower and higher.

6. Garbage collection

The problem of garbage collection is that of reclaiming space occupied by "dead" objects, which is data that has become inaccessible. All data (objects) in a heap oriented system form a graph structure of objects pointing to one another. This graph contains some root objects, which are accessible by definition. Objects live when they are accessible via a path of pointers starting from a root. Otherwise they are dead.

Generation Scavenging

In DistributedSmalltalk (and Berkeley Smalltalk) the garbage collection algorithm is based on the lifetime of objects, and is called "Generation Scavenging" [7][8]. Newly created objects are stored in NewSpace. When NewSpace is filled up, NewSpace and SurvivorSpace are garbage collected with a breadth first copying graph traversal called scavenging. The roots of this graph are the set of New and Survivor objects referenced from OldSpace, ReplicaSpace or remote hosts. This root set is dynamically updated by checking on stores of pointers to New spaces. The objects in this graph are moved to a new SurvivorSpace, except for old enough objects, which are moved to OldSpace. At the end of a traversal NewSpace is empty. Since most new objects die soon (e.g. method contexts), OldSpace fills up relatively slowly, and therefore garbage collection of the much bigger OldSpace and ReplicaSpace is necessary much less frequently. This is the reason Generation Scavenging is very cheap compared to more conventional methods like mark&sweep, mark&copy and reference counting. Because the graph of living new objects is relatively small, scavenging NewSpace can be done atomically without the user noticing an interruption (non-disruptiveness).

Collecting old garbage

In Berkeley Smalltalk the OldSpace is garbage collected on user-request using a conservative mark&sweep algorithm with a file as temporary space. This is called a "reorganization".
For DistributedSmalltalk we designed a concurrent scavenging algorithm. Scavenging is done within a background Smalltalk process, that each time it is active copies a few living OldSpace objects from one side of the OldSpace to the other. The old copies become forwarding objects. Roots of the graph are the Old objects referenced from ReplicaSpace, NewSpace, SurvivorSpace and remote hosts. Objects referenced from ReplicaSpace or remote hosts are dynamically kept in a table.

Collecting distributed garbage

Figure 3 (left) shows 4 hosts containing local (cyclic) garbage, distributed non-cyclic garbage, distributed cyclic garbage and a distributed living structure (objects painted on the edge of a host are roots). Figure 3 (middle and right) shows the same hosts after a few garbage collections. All figures are snapshots from an interactive simulator called "GarbageEditor" designed to aid in the design and testing of our distributed garbage collection algorithm. It allows you to graphically construct hosts, objects, pointers and it gives visual feedback during the garbage collection of the simulated system.

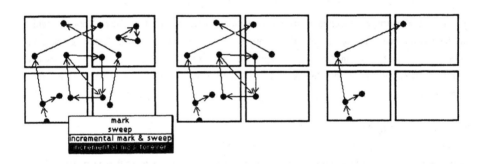

Figure 3. Four simulated hosts before, during and after some garbage collections

In order to discover dead objects in a distributed system, you should check all hosts on having pointers to a particular object. This can be accomplished by a system wide mark and sweep algorithm. The graph of living objects is traversed, the objects accessed are marked, and at the end the space of unmarked objects is reclaimed or "swept". Approaches using reference counting or weighted reference counting already suffer from the deficiency of not being capable to collect local cyclic garbage, let alone distributed cyclic garbage, so these methods are not very interesting. The global mark&sweep algorithm however, although it handles both local and distributed cycles well, does not work properly when not all hosts are able or willing to cooperate, which in an average distributed system is likely to be the case. At our R&D for example the times that all Suns are in the air will be seldom.

Incremental distributed garbage collection

Therefore we looked for an "incremental" algorithm based on a very loose way of cooperation between hosts as described in [9]. Again the problem with this incremental approach was the impossibility to collect distributed cyclic garbage. However we succeeded to solve this problem.

As in [9], garbage collection is an activity which hosts are free to apply to their local spaces whenever they want. During such a garbage collection no remote hosts are involved, only remote references are gathered. Afterwards these references are sent to the appropriate hosts. Unlike in [9], in our algorithm there is some information associated with each reference concerning the graph structure it is part of. It is not necessary that every remote host picks these references up immediately. It is even allowed that they are not received at all, e.g. when the receiving host is currently down (see Host method *sendAccessPathsAndSweepEntrance*; during *receiveAccessPaths:from:time* the exception *HostDown* is raised).

```
markAndSweep
        self unmark; markLocalLiving; sweep; sendAccessPathsAndSweepEntrance

markLocalLiving
        entrance do: [ :dgcInfo | dgcInfo prepareForDGC].
        "traversal with local roots"
        roots do: [ :aRoot |
                aRoot markLocalRecursivelyRoot: aRoot mark: #hasLocalRoot
                        backTrackWhen: [:mark | mark == #hasLocalRoot]
        ].
        "traversal with remote roots"
        entrance do: [ :aDGCInfo |
                aRoot <- aDGCInfo object.
                aRoot markLocalRecursivelyRoot: aRoot mark: aRoot
                        backTrackWhen: [:mark | mark == #hasLocalRoot | mark == aRoot]
        ]

sendAccessPathsAndSweepEntrance
        | received hostsDown |
        hostsDown <- Set new.
        accessPaths keys do: [ :aHost |
                received <- true.
                ([aHost receiveAccessPaths: (accessPaths at: aHost) from: self time: self time]
                        except when: #HostDown
                                do: [received <- false. hostsDown add: aHost]) value.
                self cleanupAccessPathsAt: aHost received: received.
        ].
         "sweep entrance"
        entrance do: [:dgcInfo | dgcInfo collectGarbageWithHostsDown: hostsDown]
```

Figure 4. Some Host methods for garbage collection

The sets of references to local objects that a host has received from remote hosts are used for the next garbage collection. When a set is not up to date this means it may refer to objects that in reality are not referenced anymore. As a result some garbage can be kept alive until some future garbage collection. In fact, as long as some host is down or inaccessible, distributed garbage part of which is on this host will not be collected. The sets are guaranteed to include every remotely referenced object, so no living objects will be garbage collected as a result of incomplete received information.

Since the information exchange is the only interaction necessary between hosts (apart from some information exchange necessary when pointers move between hosts; we managed however to incorporate this in the remote send operation itself to avoid extra network traffic) and since there are no rules prescribing some time order or any other dependency between the garbage collection activities of different hosts, no synchronization problems had to be tackled.

The simulated hosts collect local garbage using mark & sweep depth-first graph traversal as opposed to DistributedSmalltalk, where the generation scavenging technique as mentioned earlier is based on copying breadth-first graph traversal. When during graph traversal a remote reference is detected (and added to a collection of other references to the same remote host), the algorithm backtracks, so there is no remote access during traversals.

A host's *entrance* is the indirection table via which the host's objects are accessed by remote hosts. Entries in this table contain of course a pointer to the remotely referenced object, but also information for distributed garbage collection purposes (*DGCInfo*). Before a garbage collection graph traversal (see Host method *markLocalLiving*), the information received in the past from other hosts about the structure of the distributed graphs is combined and results in what is called an *AccessPath* for each remotely referenced object. This is done in *prepareForDGC*. The most interesting feature of the AccessPath mechanism is that it leads to the detection of distributed garbage whether containing cycles or not. It is also capable of handling "moving" roots in a living graph (see fig. 5). A more detailed discussion of the AccessPath mechanism is beyond the scope of this paper.

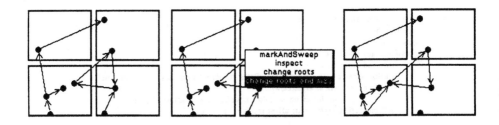

Figure 5. Living distributed graphs may have changing roots

After the AccessPaths are sent to the appropriate remote hosts, the entries in entrance that indicate "*garbage*" are removed by *collectGarbageWithHostsDown:*, that is, only if all remote hosts accessible from a particular entry received the AccessPath associated with that entry. The local objects accessible only from these removed entries will be collected the next time. Although garbage entries are detected already before the traversal (by *prepareForDGC*), they are not removed until after the traversal, because remote hosts accessible from these entries first should receive the garbage-indicating AccessPath. Absence of this information would only indicate *"not referenced (anymore) from host X"* and it would take more time than necessary for the remaining garbage to be collected (in fact the algorithm would grow a factor N in time complexity, where N is the number of remotely referenced objects in the distributed graph).

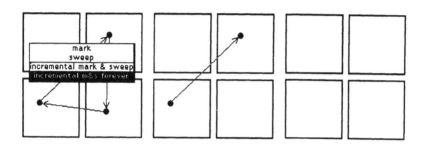

Figure 6. Simple distributed cyclic garbage before and after some garbage collections.

In fig. 6 hosts are numbered 1 to 4 from up-left to down-right. Each time a host has performed a garbage collection, it tries to contact one other host. Host 4 sends an AccessPath to host 3, host 3 to host 2 and host 2 to host 4. When there has been a sequence of garbage collections with successful sends such that a particular host (host 4 in this example) receives back the information it sent out in the past (after being processed by hosts 3 and 2 in this example) then this host will conclude that his object is not accessible anymore via the entrance. After the next successful send of the associated AccessPath, the object's entry in entrance is removed, including the last local pointer to the object. Hence the object's space will be reclaimed at the next garbage collection. All cyclic garbage will have been collected after a sequence of garbage collections on hosts *432443322* but also after e.g. *14u322142u14u32u1122u423u12321411213* ("4u" indicates unsuccessful send after garbage collection on host 4).

In fig. 7 the up-left host has three remotely referenced objects. Associated with each of the three remote pointers from this host is a set of three AccessPaths, since there exist local paths from each of the three remotely referenced objects to these pointers. In order to determine these different AccessPaths, some objects are visited more than once during graph traversal (see backtrack criterion for remote roots in *markLocalLiving*). In general the number of those objects is relatively small, so no performance degradation occurs.

As the number of remote references in an average distributed system is a small fraction of the total ("locality of reference"), the (work involved with) distributed garbage is a fraction of the (work involved with) local garbage. Therefore we could allow ourselves the luxury of implementing the major part of the algorithm at the image level (in relatively slow Smalltalk-80). The bulk of the work (collecting local garbage and remote pointers) is done in the virtual machine (written in C).

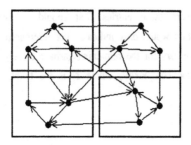

Figure 7. More complex distributed cyclic garbage

7. Message passing

Smalltalk objects communicate with each other by passing messages. A Smalltalk message can be considered to be a dynamically bound procedure call; the particular method to be bound is determined by the (super)classes of the receiver. In DistributedSmalltalk message passing is location transparent, i.e. any message may be sent to an object residing on the same or different host in a uniform way.

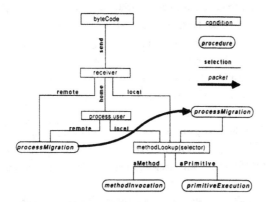

Figure 8. Send-bytecode interpretation

On a message send, if the receiver is a proxy, the host forwards the send message to the host of the remote object, that is represented by the receiving proxy. If the receiver is a home object and the host identity of the active process is not the same as the identity of the interpreting host, the send message is forwarded to the host of the active process (its home).

A send message is forwarded by means of a remote procedure call (rpc). A send rpc transfers the following objects: active process, active context, message selector, receiver and arguments. At the server host, message lookup is performed, a new context is created and the transferred process is resumed. The new context's sender is a proxy of the active context at the client host, and the resumed process has a proxy of its home process. At the client host, the active process is suspended.

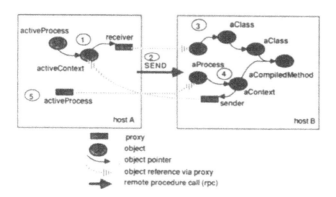

(1) if receiver is proxy (or receiver is home object ..),
     (2) do SEND(activeProcess, activeContext, send message attributes) rpc
     (3) look up compiledMethod (via class-superclass chain)
     (4) resume transferred process with new context
(5) suspend active process at client (host A), and create proxy of migrated process

Figure 9. A remote send

On a message return, if the sender of the active context is a proxy, a return rpc is initiated. A return rpc transfers the following parameters: active process, active context's sender and the result. At the server, the result is put on the stack of the suspended context. The transferred process is resumed. At the client, the active process is suspended. The same mechanism is used for remote block executions, because a block is also an object. If a block is sent as an argument to a remote receiver, a proxy of the block is transferred to the receiver. A remote block execution is in fact the result of a message send to a proxy of a block.

Smalltalk does non-preemptive scheduling for processes of the same priority. When a process migrates to a remote host, it must share the processor with other processes. In order to avoid situations where a process does not give up the processor voluntarily, periodically a Smalltalk time slicer process with adequate priority gets hold of the processor.

A process can hop through various hosts before returning to its home. It should be possible to interrupt a migrated process from its home. Therefore, after each hop (a send rpc or return rpc) the home proxy of the migrated process is updated (via an update rpc). For each particular process hopping through various hosts, there is a race condition between update rpc's. In order to solve this problem, each update rpc and each home proxy is assigned a unique timestamp according to the causal order of events. An update is performed only if the home proxy timestamp is older than the update timestamp.

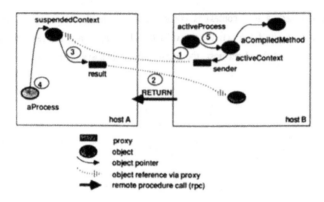

(1) if sender of active context is proxy,
(2) do RETURN(activeProcess, sender, result) rpc
       (3) put result on stack of suspendedContext
       (4) resume transferred process with suspendedContext
(5) suspend activeProcess at client (host B)

Figure 10. A remote return

During message passing, various rpc exceptions e.g. "host not available" can occur due to the distributed nature of the system. If an rpc exception occurs during a send or return, the process user will be notified by means of an exception message to a home object. If the process home is not reachable, the process is considered to be an orphan and terminated.

The default exception handling in Smalltalk opens a notifier view to the user (by means of the methods *doesNotUnderstand:* and *error:*). The notifier offers the user the opportunity to proceed or to debug. This type of exception handling is appropriate for exceptions like "message not understood" which are under control of the user. But rpc exceptions have a transient nature, and therefore we have implemented a general exception handling control structure in Smalltalk (see fig.4).

Argument access (in primitives) is read-only. When a primitive argument is a proxy, a copy is made by means of a copy rpc. Most primitives, except for BitBlt, are implemented according to the rules of data encapsulation, i.e. only the receiver is directly manipulated, not the objects it refers to nor the primitive arguments. BitBlt primitives directly access the receiver's instance variables. Therefore, when an instance variable of BitBlt is a proxy, a copy is made also.

8. Communication between hosts

Hosts communicate with each other by means of rpc. We use two forms of rpc: a synchronous unicast and a broadcast. The synchronous unicast is similar to a local procedure call, i.e. there is one caller and one callee, and during the call the caller is blocked. A broadcast has one caller and several callees. The rpc's are used for:

□ Remote message passing (send, return, update).
□ Copying read only objects (shallow copy, deep copy).
□ Initializing a host. Every new host broadcasts its network address. The receiving hosts add this address and the associated host identity to a table and return this table to the new host, which after receiving the first reply ignores the others.
□ Replication consistency control.
□ Distributed garbage collection.

The rpc parameters that are transferred are simple C integers or graph structures representing Smalltalk objects pointing to each other. Given some Smalltalk object the following can be transferred:

□ A reference to the object. If the object is replicated, just the object's oop is transferred, otherwise a proxy object.
□ A shallow copy of the object. Only the references of the objects it points to are transferred.
□ A deep copy of the object. The entire graph with the object as root is copied.
□ An n-deep copy of the object. The graph with the object as root is copied until depth n.

Most of the time only references are transferred. For performance reasons heavily used read only objects are transferred as deep copy (e.g. Strings, Rectangles and Points). Shallow copies are made for arrays, e.g. the argument array in the perform primitive.

In order to transfer Smalltalk objects across machine boundaries, it is necessary to serialize them into a sequence of bytes suitable for network transmission. The inverse operation, deserialization, converts the object back into an accessible form in a particular host address space. We have implemented a general (de)serialization procedure, that can transfer a graph of Smalltalk objects until any depth. Cycles in this graph are handled appropriately.

We use the Sun rpc library in our implementation [10]. However we had to change the Sun rpc polling mechanism in order to avoid communication deadlock, which in our application is possible because a host can be rpc client and server at the same time. When for example host A issued an rpc request to host B and is waiting for the reply, whereas B issued an rpc to A and is also waiting for the reply, we have a deadlock. Communication deadlock can be detected by polling for incoming replies and service requests simultaneously. Deadlock is broken by serving service requests first and getting the reply later.

9. Related work

Our work is related to [1] [2] [16]. While in [1] [2] the mechanisms are incorporated in the virtual image, we have chosen for the virtual machine level approach, in order to provide distribution transparency for image applications as well. Because of this approach, no modifications had to be made to existing image applications such as the debugger. As in [1] class objects can be replicated, but we have chosen for replication control to keep replicated objects consistent, instead of making run-time class compatibility checks, when instances of replicated classes have to move between machines. The I/O problem mentioned in [2] is solved with the "home object" mechanism. Our remote message passing mechanism is similar to Emerald [17]: a process moves to the destination host of the receiver of the message in order to execute the method. The problem of reclaiming cyclic distributed garbage is mentioned but not solved by [2] [16] while in [1] an expensive system-wide mark&sweep algorithm is implemented where obviously the cooperation of all machines is necessary. Instead our distributed garbage collection scheme reclaims cyclic distributed structures incrementally with a minimum of inter host communication.

1. *Current status and benchmarks*

Without any changes to existing programming tools, we now can browse on remote classes, interrupt and debug remote processes, and inspect remote objects. We promoted user interface class objects (e.g. views, controllers) to home objects in order to get an acceptable performance. Thus remote message passing is not necessary for I/O e.g. sensing input, displaying windows. For example, inspecting a remote class object, involves remote message passing for getting the model's printable representation (Strings), but not for displaying activities (menus, window clipping, text scanning). We have succeeded in keeping the system responsive.

Our incremental distributed garbage collection algorithm is implemented and tested by means of simulation, and is now being integrated in DistributedSmalltalk.

Table 1 shows the results of the micro benchmarks measured for two Sun-3/160 workstations. A comparison is made between a local send and a remote send with different types of results: self, a string, an array. A remote send is approximately 500 times slower than a local send.

Return Value	Local Send	Remote Send	(in milliseconds)
self	0.09	28	
literal	0.26	144	
array	0.20	106	

Table 1. Micro benchmarks

Table 2 shows standard benchmarks for BerkeleySmalltalk, DistributedSmalltalk with one host and DistributedSmalltalk with two hosts. In the latter case, some benchmarks (PrintDefinition, PrintHierarchy, Inspect, Compiler and Decompiler) apply to remote class objects. According to these benchmarks, a remote programming activity is 45% slower than a local one. As can be seen from table 2, additions made to BerkeleySmalltalk for testing on each message send, has made local computing about 10% slower.

	DistributedSmalltalk	DistributedSmalltalk*	BerkeleySmalltalk
BitBLT	26.6000	28.0986	27.7083
TextScanning	39.0000	36.2791	35.4545
ClassOrganizer	19.3291	15.1081	19.6429
PrintDefinition	18.1624	6.05413*	27.2436
PrintHierachy	19.6850	9.39850*	24.3902
AllCallsOn	17.0455	12.6404	21.2766
AllImplementors	21.1039	16.1692	24.2537
Inspect	31.1644	2.33813*	35.0000
Compiler	23.9785	9.65368*	27.5990
Decompiler	18.3976	1.70893*	21.2329
KeyboardLookAhead	36.3248	30.5755	40.4762
KeyboardSingle	39.9267	34.2767	44.1296
TextDisplay	28.5714	23.1884	25.1968
TextFormatting	37.8289	36.3924	42.5926
TextEditing	39.7363	31.8731	45.0855
Performance Rating	26.4813	14.2129	29.5995

(*) applied on remote class objects

Table 2. Macro benchmarks

11. Conclusions

By providing a remote message passing mechanism at the VM level, we have achieved distribution transparency for users and image applications. Therefore, no changes had to be made to existing image code, not even to the debugger. Transparent I/O is provided by means of a new concept called "home objects". Performance degradation is acceptable by means of replication and furthermore by making user interface classes like views and controllers, home objects. Local programming activity is 10% slower than in a comparable standalone system. Although remote sends are 500 times costlier than local sends, according to our macro benchmarks, remote programming activity is only 45% slower. Replication is transparent and replication consistency is guaranteed, so for replicated class objects no compatibility checking is needed. Distributed garbage, whether containing cycles or not, is collected incrementally without any synchronization being necessary.

12. References

[1] Bennett, J. K., "The Design and Implementation of Distributed Smalltalk", OOPSLA '87 Proceedings, Oct. 4-8, 1987, pp 318-330

[2] McCullough P. L., "Transparent Forwarding: First Steps", OOPSLA '87 Proceedings, Oct. 4-8, 1987, pp 331-341

[3] Purdy A. et all, "Integrating an Object Server with Other Worlds", ACM Trans. Office Information Systems, Vol. 5, No. 1, January, 1987, pp 27-47

[4] Herbert A. J. and Monk J. (editors), ANSA Reference Manual Release 00.03 (Draft), June 1987

[5] Khoshafian S. N. and Copeland G. P., "Object Identity", OOPSLA '86 Proceedings, pp 406-416

[6] Thomas R. H., "A Majority Consensus Approach to Concurrency Control for Multiple Copy Databases", ACM Trans. Database Systems, Vol. 4, No. 2, June 1979, pp 180-209

[7] Ungar D., "Generation Scavenging: A Non-disruptive High Performance Storage Reclamation Algorithm", ACM Software Engineering Notes, April 1984, pp 157-167

[8] Lieberman H. and Hewitt C., "A real-Time Garbage Collector Based on the Lifetimes of Objects", Communications of the ACM, Vol. 26, No. 2, June 1983, pp 419-429

[9] Mohamed Ali K. A-H, "Object-oriented Storage Management and Garbage Collection in Distributed Processing Systems", The Royal Institute of Technology, Dept. of Telecommunication Systems - Computer Systems, Sweden, Report TRITA-CS-8406, December 1984

[10] Remote Procedure Call Reference Manual, Sun Microsystems, Inc., Part Number 800-1177-01, Nov. 1984

[11] D.R. Cheriton, "The V Kernel: A Software Base for Distributed Systems", IEEE Software 1, April 1984, pp. 19-42.

[12] M. Accetta et al, "Mach: A New Kernel Foundation for UNIX Development", Proc. Summer 1986 USENIX Technical Conference and Exhibition, June 1986.

[13] B. Liskov, "Overview of the Argus Language and System", Programming Methodology Group Memo 40, M.I.T., Laboratory for Computer Science, February 1984.

[14] P. America, "Rationale for the design of POOL", ESPRIT Project 415, Doc. Nr. 0053.

[15] A. Goldberg and D. Robson, "Smalltalk-80: The Language and its Implementation", Addison Wesley, 1983.

[16] D. Decouchant, "Design of a Distributed Object Manager for the Smalltalk-80 System", OOPSLA '86 Proceedings, pp. 444-452

[17] E. Jul et al, "Fine-Grained Mobility in the Emerald System" ACM Transactions on Computer Systems, Vol. 6, No. 1, February 1988, pp 109-133

Implementing Concurrency Control in Reliable Distributed Object-Oriented Systems

Graham D. Parrington and Santosh K. Shrivastava
Computing Laboratory,The University of Newcastle upon Tyne,
Newcastle upon Tyne, NE1 7RU, U.K.

Abstract

One of the key concepts available in many object-oriented programming languages is that of *type-inheritance*, which permits new types to be derived from, and inherit the capabilities of, old types. This paper describes how to exploit this property in a very simple fashion to implement object-oriented concurrency control. We show how by using type-inheritance, objects may control their own level of concurrency in a type-specific manner. Simple examples demonstrate the applicability of the approach. The implementation technique described here is being used to develop *Arjuna*, a fault-tolerant distributed programming system supporting atomic actions.

Key words and phrases: Object-oriented programming, Concurrency control, Reliability, Atomic actions, Type-inheritance.

1. Introduction

A widely used technique of introducing fault-tolerance - particularly in distributed systems - is based upon the use of *atomic actions* (atomic transactions) for structuring programs [Gray 78]. An atomic action possesses the properties of *serialisability, failure atomicity and permanence of effect*. A wide variety of concurrency control algorithms have been proposed in the literature to ensure the serialisability property of atomic actions. This paper proposes a practical implementation technique for such concurrency control algorithms within the framework of an object-oriented system. The basic idea behind the approach is quite straightforward. It is assumed that the underlying operating system provides rudimentary synchronisation facilities, such as semaphores, for concurrent processes. A type can then be constructed to provide a specific concurrency control technique by exploiting these facilities. User-defined types can *inherit* this underlying concurrency control facility by making use of the type-inheritance mechanism available in object-oriented programming languages. If desired, the inherited concurrency control mechanism can also be *refined* to provide a *type-specific* concurrency control mechanism.

This paper reports on simple experiments that have been performed to show that the approach presented here does provide a very flexible means of implementing concurrency control techniques for object-oriented systems. Other papers [Dixon and Shrivastava 87, Dixon *et al.* 87] have shown how the same type-inheritance technique can be used to add the properties of failure atomicity and permanence of effect to objects with equal ease.

The use of inheritance in this fashion has a number of advantages. Firstly, it is not necessary to design and implement either a new language and run-time system, nor an operating system kernel. Secondly, it enables experimentation with different concurrency control and recovery techniques in a straightforward fashion, since the capabilities are not tied into any particular system. This approach may be contrasted with that taken by other systems in the same area including Clouds [Dasgupta *et al.* 85], Argus [Liskov and Scheifler 83], TABS [Spector *et al.* 85], and ISIS [Birman 86].

This paper is structured as follows. Section two reviews the main ideas on objects, actions and some widely used concurrency control techniques such as two-phase and type-specific locking. Section three introduces the inheritance-based concurrency control implementation technique, after first describing the relevant features of the C++ object-oriented language in which the implementations have been carried out. Sections four and five discuss some specific concurrency control implementations to illustrate the feasibility of the proposed approach. Finally, section six describes how this approach is integrated within *Arjuna* - a distributed system supporting atomic actions.

2. Objects, Actions and Concurrency Control

An object is an instance of some *type* or *class*. Each individual object consists of some variables (its *instance* variables) and a set of operations (its *methods*) that determine the externally visible behaviour of the object. The operations provided by a type have access to the instance variables and can thus modify the state of an object. Furthermore, the type of an object determines what operations may be applied to it.

In a distributed system, an operation on an object is typically invoked via a *remote procedure call* (RPC) (see [Liskov and Scheifler 83, Spector *et al.* 85, Birman 86, Dixon *et al.* 87] for some typical fault-tolerant, object-based architectures). Programs which operate on objects are executed as *atomic actions* with the properties of (i) *serialisability*, (ii) *failure atomicity*, and (iii) *permanence of effect*. The first property ensures that concurrent execution of such actions is free from interference (that is, concurrent executions can be shown to be equivalent to some serial order of execution [Eswaran *et al.* 76, Best and Randell 81]). The second property ensures that an action can either be terminated normally (*committed*) producing the intended results, or *aborted*, producing no results. This property can be obtained by the appropriate use of *backward error recovery*, which is invoked whenever a failure that cannot be masked occurs. Typical failures causing an action to be aborted are node crashes and communication failures such as lost messages. It is reasonable to assume that once an action terminates normally, the results produced are not destroyed by subsequent node crashes. This is the property of permanence of effect which ensures that state changes produced are recorded on *stable storage* which can survive node crashes with a high probability of success. A *commit* protocol is required during the termination of an action to ensure that either all of the objects updated within the action have their new states recorded on stable storage (normal or committed termination), or no updates get recorded (aborted termination).

In object-based systems, the encapsulation properties of objects make it seem natural to require that the implementation of each object type be made responsible for enforcing concurrency control and for implementing properties such as backward error recovery. A separate paper [Dixon and Shrivastava 87] has described how a type can implement backward error recovery by exploiting the type-inheritance facilities of an object-oriented programming language. This paper applies similar ideas to the problem of concurrency control.

2.1. Object-Based Concurrency Control

A very simple and widely used concurrency control technique is to regard all operations on objects to be of type *read* or *write*, which must follow the well known synchronisation rule permitting concurrent read operations but exclusive write operations. This rule is imposed by requiring that any action intending to perform an operation that is of a *read* type (*write* type) on an object must first acquire a *read lock* (*write lock*) associated with that object. To guarantee serialisability, all actions must follow a *two-phase* locking policy [Eswaran *et al.* 76] . During the first phase, termed the *growing* phase, an action can acquire locks on objects but must not release any acquired locks. The last phase of the action constitutes the *shrinking* phase, during which time held locks can be released, but no new locks can be acquired. It is also necessary to make the shrinking phase appear instantaneous by releasing all of the held locks simultaneously, to ensure that an action can be aborted unilaterally, without affecting other ongoing actions and thus avoiding the possibility of cascade-rollback.

This policy of shared read but exclusive write represents a *lock conflict rule* for an object: read-read locks do not conflict but read-write and write-write locks do. There are situations where this conflict rule can be regarded as overly restrictive, from the point of view of permissible concurrency. Consider a simple example.

Suppose there is some directory object providing the operations: *addentry(...)* (to add an entry in the directory), *rmentry(...)* (to remove a specified entry from the directory), and *lookup(...)* (to look for a given name in the directory). If *addentry* and *rmentry* operations are taking place on different entries then there is generally no reason why these two operations cannot be permitted to occur concurrently. This observation leads to the notion of *type-specific locking*, whereby a type definition includes a type-specific lock conflict rule [Schwarz and Spector 84]. Type-specific locking is a promising technique for increasing the permissible concurrency in object-based systems supporting atomic actions.

3. Utilising Type-Inheritance for Concurrency Control

Several modern programming languages support the property of *type-inheritance* - the ability for newly constructed types to acquire the properties and operations of the base types out of which they have been constructed. This section examines how this property can be used to implement object-oriented concurrency control. The approach is to construct an appropriate concurrency control base type which can then be used to derive more specific (user) types as required.

3.1. Type-Inheritance in C++

The language we will use to describe our objects is C++ [Stroustrup 86] largely because of its ease of availability and its incorporation of the features we require. C++ is an object-oriented superset of the language C, and includes facilities for type-inheritance, data abstraction, and operator overloading. The data abstraction and type-inheritance facilities are based on the *class* concept. Instances of a class are objects, with specific operations provided for their manipulation. The type-inheritance mechanism of C++ works as follows: given a base class C_1, another class C_2 - a *derived* class of C_1 - can be defined so that it inherits some or all of the operations of C_1.

Classes are defined in the manner shown in Figure 1(a) which is a skeleton declaration of a class called *baseclass*. The variable and function declarations which occur before the *public* label are private members of the class; the only operations which may access private variables or invoke private operations are the member operations of the class itself (in this example, *baseclass*, ˜*baseclass*, *op1*, *op2* and *op3*). The variable and operation declarations following the public label constitute the public interface to objects of the class (here, *op2* and *op3* in Figure 1(a); the operations *baseclass* and ˜*baseclass* are special, see below). An example of a class derived from the *baseclass* class is shown in Figure 1(b). This new class, called *derived*, inherits the public operations *op2* and *op3* from *baseclass*. In this example these inherited operations are also made public operations of the derived class by the use of the keyword *public* in the class header.

Each class may have a *constructor* which is a public operation with the same name as the class (*baseclass*() and *derived*()), and which will be invoked each time an instance of the class is created. There is also a complementary operation (˜*baseclass*() and ˜*derived*()),

```
class baseclass                          class derived : public baseclass
{                                        {
       int val1;                                int val3;
       int val2;
       op1 ();
public:                                  public:
       baseclass ();                            derived ();
       ˜baseclass ();                           ˜derived ();

       op2 ();                                  dop4 ();
       op3 ();                                  dop5 ();
};                                       };

       (a)                                      (b)
```

Figure 1: C++ syntax

called a *destructor*, which is invoked automatically when the object is deleted. The constructor allows an object to perform type-specific initialisation, and the destructor enables an object to tidy up before it is deleted. Both operations are special in that they will be automatically invoked when objects are created or deleted and even though a part of the public interface to the object, they cannot be directly invoked by a user of the object.

3.2. The Basic Concurrency Control Scheme

The basic concurrency control scheme relies on the inheritance features of the language C++. Concurrency control is achieved by first defining an appropriate class (called *LockCC*) which provides the basic concurrency control facilities that are required, and then using that class as a basis from which to derive other specific user-defined classes. Given such a class, Figure 2 shows how a user-defined type called *File* is defined so that it inherits the capabilities of *LockCC*. Use of the operations provided by *LockCC* must be explicitly coded as part of the code for the operations of the newly derived class (for example, the *open* operation in Figure 2 should contain appropriate calls on the operations of *LockCC*). This point will be explained further in a later section.

3.3 Representing Locks

In order to make the system as flexible as possible, locks are deliberately not made special immutable system types (in contrast to systems such as Clouds and Argus). Rather, in this system locks are simply another type that can be refined as required in the same fashion as any other type. This approach has several advantages. Firstly, locks can be

```
class File : public LockCC
{
        ...                 // private file data
public:
     File ();
     ˉFile ();

     open (...);        // standard file operations
     close (...);
     ...
}
```

Figure 2: The class *File*

created and manipulated in the same fashion as any other object in the system. Secondly, we do not require any new language features or modifications to the run-time environment to support them. Thirdly, the approach is very flexible, permitting different concurrency control policies to be adopted with surprising ease.

In Figure 3 we show a skeleton declaration of the *Lock* class. Instances of this class are

```
class Lock: public Object
{
     modetype lockmode;
     Uid owner;
     ...
public:
     Lock (modetype) ;
     ˉLock ();

     modetype getmode () { return lockmode; }
     Uid getowner () { return owner; }
     ...
     virtual boolean operator! = (Lock∗);
}
```

Figure 3: The class *Lock*

created as needed by the user and then passed as arguments to the *setlock* operation provided by *LockCC*.

Since instances of the *Lock* class are simply objects, they obey the usual object-oriented rules regarding encapsulation. Thus the *Lock* class provides operations to retrieve certain parts of its internal state rather than allow public access to the actual instance variables. Note that *Lock* is itself a derived class (of *Object*), and thus inherits the capabilities and operations of that class. The reasons for this will be covered in section six.

The mode of a lock object is immutable. Thus having declared a lock object to be a read lock, the object is always a read lock. If a write lock is required a new lock object with the appropriate mode must be created. This restriction relates to the way in which locks are expected to be used. It seems unlikely that having created a lock the programmer would want to change its mode from say read to write (this process of *lock conversion* can be handled in a different fashion in this system as will be explained in a later section).

The boolean operator $!=$ defined for the *Lock* class is declared to be *virtual* so that it may be redefined in any class that is derived from *Lock*. This operator is used by one of the internal (private) operations of *LockCC* to ascertain whether two locks conflict; its use will become clearer in the following section.

3.4 The Concurrency Control Class LockCC

The previous section outlined the basic *Lock* class that is supplied as a parameter to the operations of the concurrency control class *LockCC*. In this section the *LockCC* class itself is described, as is the manner in which it manipulates the locks that are passed to it.

Recall that, as mentioned in section 3.2, a user-defined class requiring concurrency control is derived from the class *LockCC* which manages the concurrency control information. So, if several instances of such a user-defined class are created, each instance will possess the capability of maintaining its own concurrency control information on a purely local basis.

The *LockCC* class provides two basic public operations; *setlock*, whose task is to set a lock upon the user-defined object; and *releaselock* whose task is to unlock the object. The private information maintained by the *LockCC* class includes a list of *Lock* objects that are currently being held, which is used to determine whether any new lock can be set. Given this information each object can determine whether a new lock request can be granted based purely on its own local information without reference to the other objects in the system. In addition, since multiple processes may be attempting to set locks upon an object concurrently this private information is updated inside a critical section protected by the semaphore *mutex*.

Given the above basic description an outline of how the *setlock* operation works is illustrated in Figure 5*. As shown here, *setlock* attempts to determine whether a conflict exists by calling the private operation *lockconflict*. If this returns the result TRUE then a conflict exists between the requested lock and (at least) one of the other locks currently set on the object, in which case the mutual exclusion semaphore is freed and the call blocks.

---

* We show a simple implementation of *setlock* where the calling process simply keeps on trying (after a brief pause) until the lock is granted. Clearly optimisations are possible, but are not discussed here.

```
class LockCC: public Object
{
      Lock__List locks__held;
      Semaphore *mutex;          // and other private concurrency
                                 // control information

      ...
      boolean lockconflict (Lock*);

public:
      LockCC ();
      ~LockCC ();
      lockstatus setlock (Lock*);
      lockstatus releaselock (Uid*);

}
```

Figure 4: The concurrency control class *LockCC*

```
lockstatus LockCC::setlock (Lock *reqlock)
{
      boolean conflict = TRUE;
      do
      {
           P(mutex);
           if (conflict = lockconflict(reqlock))
           {
                V(mutex);
                sleep();              // wait for a while
           }
      } while (conflict);
      locks__held.insert(reqlock);   // add to list of locks
      ...
      V(mutex);
      return (GRANTED);
}
```

Figure 5: The *setlock* algorithm

When the call resumes, the conflict check is again performed to see if the requested lock can now be set.

Releaselock is not intended to be called directly by the programmer, rather it is called automatically when the atomic action using the object terminates (hence the unusual parameter type). This ensures that the two-phase policy is always followed. It occurs in the public interface because in *Arjuna* atomic actions are themselves objects (instances of the class *Action*, see section 6), not part of the underlying system, and thus have no special privileges to access objects, instead they can only use the same public interface that any other object in the system could use.

The operation of checking whether two locks conflict is more sophisticated than simply allowing *lockconflict* to compare the modes of the lock objects, since in the case of type-specific locking extra information may be used to allow greater concurrency. Instead the *Lock* objects themselves are required to ascertain whether a conflict exists. This check is performed by utilising the *!=* operation provided by the *Lock* type. This operation is defined such that if *L1* and *L2* are two instances of the *Lock* class, then execution of the comparison operation *L1 != L2* returns the value TRUE if the two locks conflict, and returns FALSE otherwise.

Using this approach leads to the implementation of *lockconflict* as shown in Figure 6.

```
boolean LockCC::lockconflict (Lock *reqlock)
{
        Lock__Iterator next (locks__held);
        Lock* heldlock;

        while ((heldlock = next()) != Null)
        {
                if (*heldlock != reqlock)
                        return TRUE;
        }
        return FALSE;
}
```

Figure 6: The *lockconflict* algorithm

This implementation makes use of an iterator *next* which when called returns the next lock from the list of currently held locks. The implementation of the conflict check between locks is obviously type-specific, however, it is assumed that the conflict operator of the basic *Lock* type supports the traditional multiple reader, single writer policy, thus giving the implementation shown in Figure 7.

3.5. Extensions for Type-Specific Locking

The basic locking scheme of the previous section may be extended to take advantage of more knowledge about the semantics of individual types. Not only can derived classes inherit operations from a base class unchanged, they may also modify those operations to make them more suitable for their individual needs. This property can be utilised to allow new types of lock to be derived from the basic *Lock* type. Using this approach, the new lock type can provide its own version of the conflict operator *!=*. By doing this the user-defined type can determine what level of concurrency it will support since the programmer of the

```
boolean Lock::operator! = (Lock *otherlock)
{
    if (otherlock→getowner() ! = owner)   // no conflict if locks from same action
        switch (lockmode)
        {
            case READ:                  // holding read
                if (otherlock→getmode() ! = READ)
                    return TRUE;
                break;
            case WRITE:                 // holding write
                return TRUE;
        }
    return FALSE;
}
```

Figure 7: The basic conflict check of *Lock*

individual operations decides the type of lock that needs to be passed to the concurrency controller of the object. Thus type-specific locking is handled in this scheme in an extremely simple manner. All that is required is the derivation of a new type of lock from the basic *Lock* type and an appropriate redefinition of the conflict operation. Obviously this conflict operator can take advantage of any extra information available about the new lock type to provide additional concurrency. The examples in the next section should help to make this clear.

4. Examples of Object-Oriented Concurrency Control

In this section some simple examples are considered to illustrate the scheme in action. In the first example a simple class that implements a file type is described. It makes available the usual file operations to the client and uses the basic capabilities it inherits (from *LockCC*) to provide simple file locking.

The second example illustrates how the basic facilities can be overridden to increase concurrency using a type-specific approach. The technique adopted of deriving a new type of lock allows this to be undertaken with surprising ease.

4.1. A File Class

Consider a *File* class that allows the usual operations of *read, write, open, close*, etc. Such a class was illustrated in Figure 2. With this organisation Figure 8 illustrates how the *open* operation might be implemented. All that is required is the creation a new instance of the *Lock* class (which is automatically initialised to contain the correct information by its

```
File::open(mode)
{
        setlock(new Lock (mode));

        // now actually open file and do other housekeeping
        ...
}
```

Figure 8: The implementation of *open* for the class *File*

constructor), and to pass that instance to *setlock*. The standard implementation of the lock conflict operator is used to determine whether the requested lock can be applied and the *open* operation only proceeds when the lock has been granted.

4.2. A Directory Class

This second example illustrates a directory class that uses a type-specific locking scheme for concurrency control. A new class - a type-specific lock (*TypeSpecLock* (Figure 9))

```
class TypeSpecLock : public Lock
{
        InstanceId Id;              // some identifier that
                                    // identifies this lock
public:
        TypeSpecLock (mode, Id);    // TypeSpecLock constructor
        ~TypeSpecLock ();

        InstanceId getId () { return Id; }  // A means of accessing the Id

        virtual boolean operator! = (Lock *);
}
```

Figure 9: The class *TypeSpecLock*

is created, derived from the basic *Lock* class and it therefore inherits all of the attributes of the *Lock* class. However, this class has one important addition - a new field by which it can be identified and an operation by which that field can be interrogated.

Given this class, a directory type might be implemented as shown in Figure 10. Then in a similar fashion to the *open* operation for the *File* class, the *addentry* operation of the *Directory* class might be coded as illustrated in Figure 11 (assuming that the operations *addentry* and *rmentry* require write locks, while *lookup* requires a read lock). As was noted earlier both the *addentry* and the *rmentry* operations may proceed concurrently providing that they both manipulate different directory entries. So, to increase concurrency the implementations of *addentry* and *rmentry* construct and pass instances of the new lock type

```
class Directory : public LockCC
{
    ...                         // private directory information
public:
    Directory ();
    ~Directory ();

    addentry (char *Name ...);
    rmentry (char *Name ...);
    lookup (...);
    ...
}
```

Figure 10: The class *Directory*

```
Directory::addentry (char *Name ...)
{
    // first set an appropriate lock...
    setlock(new TypeSpecLock(mode, (InstanceId)Name));

    // then actually manipulate the directory
    ...
}
```

Figure 11: The implementation of *addentry* for the class *Directory*

(*TypeSpecLock*) to the concurrency controller via *setlock*. The implementation of the conflict operator for *TypeSpecLock* is shown in Figure 12.

```
boolean TypeSpecLock::operator! = (Lock *otherlock)
{
    if (otherlock→getowner() != owner)
        switch(lockmode)
        {
            case READ:
                if (otherlock→getmode() == WRITE)
                    return TRUE;   // Read conflicts with Write
            case WRITE:
                if((otherlock→getmode() == READ) ||
                   (Id == ((TypeSpecLock*)otherlock→getId() ))
                    return TRUE;   // RW or WW conflict on same entry
        }
    return FALSE;
}
```

Figure 12: Lock conflict check for the class *TypeSpecLock*

5. Alternative Approaches To Concurrency Control

Most concurrency controllers assume that lock requests emitted by the same action do not conflict, thus an action holding a read lock will be permitted to acquire a write lock providing no other action holds a conflicting lock. This process is often termed *lock conversion*, since its effect is to convert a weaker mode lock into a stronger mode lock. Lock conversion can be handled in the system described in this paper simply by arranging that *Lock* objects with the same owner do not conflict (as determined by the conflict operator *!* =). This simple test was included in the conflict operation *!* = in both Figures 7 and 12.

An alternative approach, adopted in ISIS, requires that conversion is only possible if the original lock had been a *promotable read lock*. Such locks are easy to implement with the approach adopted in this paper. A new type of lock - the *PLock* (derived from *Lock*) - is created and provided with an appropriate conflict operator (Figure 13) which checks all

```
boolean PLock::operator! = (Lock *otherlock);
{
    switch (lockmode)
    {
        case READ:
            if (otherlock→getmode() == WRITE)
                return TRUE;
            break;
        case PREAD:
            if (otherlock→getmode() == READ)
                break;
            if (owner ! = otherlock→getowner())
                return TRUE;
        case WRITE:
            if (owner ! = otherlock→getowner())
                return TRUE;
    }
    return FALSE;
}
```

Figure 13: The *PLock* conflict algorithm

locks for conflict regardless of ownership.

A radically different approach to locking, which enforces a *pessimistic concurrency control* policy, is the *optimistic concurrency control* policy [Kung and Robinson 81] where actions are allowed to execute without any synchronisation. At termination (commit) the action is *validated* by analyzing read/write conflicts with other ongoing actions. The validation succeeds if the committing action preserves the serialisability property, otherwise the action is aborted. Just as it is possible to define a conflict rule for type-specific locking, so is it possible to define a type-specific validation rule [Herlihy 86]. By providing another base class (*OptCC*), objects can utilise this type of concurrency control by being derived from it rather than from *LockCC* (for further details see [Parrington 88]).

6. Integration with Atomic Actions

This section describes how the concurrency control implementation technique described earlier is integrated into a reliable distributed programming system called *Arjuna* currently being developed at the University of Newcastle upon Tyne.

Not surprisingly, the type-inheritance mechanism is also employed for making user-defined types recoverable [Dixon and Shrivastava 87]. A base class *Object* provides the basic capabilities that allows a type to be recoverable. Thus a user-defined type inherits properties of recoverability from the class *Object* and concurrency control capability from the class *LockCC*. The overall class hierarchy of Arjuna is shown below as Figure 14 (see [Dixon *et al.* 87] for more details) .

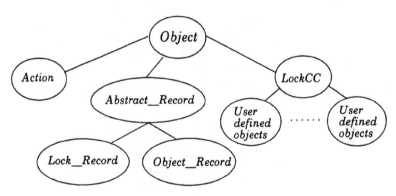

Figure 14: The basic *Arjuna* class hierarchy

Arjuna is novel with respect to other such systems in taking the approach that every major entity in the system is an object. Thus, just as locks are objects, the management of atomic actions is handled by instances of another type of object (called *Action*).

Atomic actions are available through the use of the class *Action*, which provides the operations normally associated with atomic actions, such as *Begin_Action*, *Commit_Action*, etc. *Action* manages information provided to it by *Object* and *LockCC* to ensure that the atomic action abstraction is maintained (for example, locks are released at action commit). See [Dixon 88] for more precise details.

User-defined objects which have been derived from *LockCC* are treated as *persistent*, and are normally stored in local (to a node) object stores. An object without any locks held on it (which implies that no action is currently accessing it) is treated as *passive* and its state is stored in the object store. When an action is granted a lock on an object, that object is made *active* (if it is not already so) by copying its state from the object store and associating a server process capable of receiving operation invocation requests with the object. If the client action aborts the object state held in the server process is discarded. If the action commits, the state is placed back in the object store using the capabilities provided by the class *Object* (see [Dixon *et al.* 87]).

Actions in *Arjuna* may be nested in the normal way which requires that the lock-based concurrency controller implemented by *LockCC* only finally releases locks when the top-level action commits. When a nested atomic action commits locks are propagated to the parent action [Moss 81]. Further details of this and the implementation of other concurrency controllers can be found in [Parrington 88].

7. Conclusions

The use of type-inheritance has enabled the design and implementation of a concurrency control scheme that is highly adaptable and flexible without resorting to designing a new language or system. Using this approach programmers have control over what level of concurrency a type supports. Of course, this flexibility is not without its potential penalties, since careless programming could lead to chaos as objects are manipulated without being supervised by a concurrency controller. The *Arjuna* system employs type-inheritance for incorporating the serialisability, recoverability and permanence of effect properties of atomic actions.

Of the other comparable robust object-based systems described in the literature only Avalon [Herlihy and Wing 86] is exploring using type-inheritance as opposed to constructing an entirely new language and/or system. However, the approach adopted in Avalon is different from that presented here in that control over concurrency is based on the concept of hybrid atomicity [Weihl 84] and providing user-defined operations for the commit and abort of actions. Nevertheless, its aims are similar. Their work enforces our belief that type-inheritance provides a very powerful concept for incorporating fault-tolerance in systems.

Acknowledgments

Discussions with Graeme Dixon were helpful in formulating these ideas, as were comments by Pete Lee on an earlier draft of this paper. This work was supported by an SERC/Alvey grant in Software Engineering.

References

Best and Randell 81

Best, E., and B. Randell, "A Formal Model of Atomicity in Asynchronous Systems", *Acta Informatica*, 16, pp. 93-124, 1981.

Birman 86

Birman, K.P., "Replication and Fault Tolerance in the ISIS System", Proceedings of 10th Symposium on the Principles of Operating Systems, *ACM Operating Systems Review*, Vol. 19, No. 4, pp. 79-86, 1985.

Dasgupta *et al.* 85

Dasgupta, P., R.J. LeBlanc Jr., and E. Spafford, "The Clouds Project: Designing and Implementing a Fault Tolerant, Distributed Operating System," Technical Report GIT-ICS-85/29, Georgia Institute of Technology, 1985.

Dixon 88

Dixon, G.N., "Managing Objects for Persistence and Recoverability," Ph.D Thesis, Computing Laboratory, University of Newcastle upon Tyne, in preparation.

Dixon and Shrivastava 87

Dixon, G.N., and S.K. Shrivastava, "Exploiting Type-Inheritance Facilities to Implement Recoverability in Object Based Systems", *Proceedings of 6th Symposium on Reliability in Distributed Software and Database Systems*, Williamsburg, pp. 107-114, March 1986.

Dixon *et al.* 87

Dixon, G.N., S.K. Shrivastava, and G.D. Parrington, "Managing Persistent Objects in Arjuna: A System for Reliable Distributed Computing," *Proceedings of the Workshop on Persistent Object Systems*, Persistent Programming Research Report 44, Department of Computational Science, University of St. Andrews, August 1987.

Eswaran *et al.* 76

Eswaran, K.P., *et al.*, "On the Notions of Consistency and Predicate Locks in a Database System", *Communications of the ACM*, Vol. 19, No. 11, pp. 624-633, 1976.

Gray 78

Gray, J.N.,, "Notes on Data Base Operating Systems", in Operating Systems: An Advanced Course, eds. R. Bayer, R.M. Graham, and G. Seegmueller, pp. 393-481, Springer, 1978.

Herlihy 86

Herlihy, M.P., "Optimistic Concurrency Control for Abstract Data Types", *Proceedings of the Fifth Annual ACM Symposium on Principles of Distributed Computing*, pp. 206-216, Calgary, Alberta, August 1986.

Herlihy and Wing 86

Herlihy, M.P., and J.M. Wing, "Avalon: Language Support for Reliable Distributed Systems," *Digest of Papers FTCS-17: Seventeenth Annual International Symposium on Fault-Tolerant Computing*, pp. 89-94, Pittsburgh, July 1987.

Kung and Robinson 81

 Kung, H.T., and J.T. Robinson, "On Optimistic Methods for Concurrency Control", *ACM Transactions on Database Systems*, Vol. 6, no. 2, pp. 213-226, June 1981.

Liskov and Scheifler 83

 Liskov, B., and R. Scheifler, "Guardians and Actions: Linguistic Support for Robust Distributed Programs", *ACM Transactions on Programming Languages and Systems*, Vol. 5, No. 3, pp. 381-404, 1983.

Parrington 88

 Parrington, G.D., "Management of Concurrency in a Reliable Object-Oriented Computing System," Ph.D Thesis, Computing Laboratory, University of Newcastle upon Tyne, in preparation.

Schwarz and Spector 84

 Schwarz, P.M., and A.Z. Spector, "Synchronizing Shared Abstract Types", *ACM Transactions on Computer Systems*, Vol. 2, No. 3, pp. 223-250, August 1984.

Spector *et al.* 85

 Spector, A.Z., *et al.*, "Support for Distributed Transactions in the TABS Prototype", *IEEE Transactions on Software Engineering*, Vol. SE-11, No. 6, pp. 520-530, 1985.

Stroustrup 86

 B. Stroustrup, *The C + + Programming Language*, Addison Wesley, 1986.

Weihl 84

 Weihl, W., "Specification and Implementation of Atomic Data Types," Ph.D Thesis, MIT/LCS/TR-314, MIT Laboratory for Computer Science, Cambridge, Mass., March 1984.

An Implementation of an Operating System Kernel using Concurrent Object Oriented Language ABCL/c+

Norihisa Doi   Yasushi Kodama*
Institute of Information Science, Keio University
4-1-1, Hiyoshi, Kohoku-ku, Yokohama, 223 Japan.

Ken Hirose
Department of Mathematics,
School of Science and Engineering,
Waseda University

Abstract

The ABCL/c+ is a C-based concurrent object-oriented language, designed as an extension of ABCL/1, a language developed by A.Yonezawa and others. In order to create the world of processes, a routine object is introduced which unifies procedures, functions, and objects. ABCL/c+ is then used to write an operating system kernel. The XINU operating system, developed by D. Comer and others of Bell Laboratories, is rewritten entirely in ABCL/c+. The result shows that concurrent object-oriented languages can produce a highly readable and understandable operating system kernel.

Key words and phrases: Concurrnet object-oriented language, Operating system kernel, ABCL.

1 Introduction

Object-orientation is a paradigm which enables us to produce highly reliable and reusable programs. A number of object-oriented languages have been developed and are currently being used, including Smalltalk-80 [7], Loops [1] and Esp [2]. Several **concurrent object-oriented languages**, with functions for concurrent programming, have also been developed, including ABCL/1 [10] [12], ConcurrentSmalltalk [11], Orient84/K [8], and BETA [9]. In these languages a one-to-one correspondence exists between an object definition and process, which is the unit for execution of an instance of the object. For example, in Smalltalk-80, everything is an object, and a process can be created with a block as its entity. Therefore, one of the reasons why it is difficult to write a concurrent program in Smalltalk-80 is that a process does not correspond to a class, which is the

---

*Current address: Nihon Sun Microsystems K.K., Ichibancho FS Bldg. 5F, 8 Ichiban-cho, Chiyoda-ku, Tokyo 102, Japan

unit component of a program [6]. In what are termed concurrent object-oriented languages, this correspondence is maintained, and concurrent processes can be described in a very natural way.

Since concurrent object-oriented languages can be used to describe concurrent processes easily, system programs and operating system kernels are attractive fields of application for them. However, aside from system programs, the "world of processes", i.e. an environment in which a process exists as the unit of activity, must be created to write an operating system kernel. A system kernel can be implemented as a process. This is not a good approach, however, even considering the ease of understanding and execution efficiency.

We extended the ABCL/1 language developed by A.Yonezawa et al., and developed a C-based concurrent object-oriented language, ABCL/c+ [5], which can be used to realize the world of processes, and tried to write an operating system kernel. We selected the XINU operating system [3] [4] developed by D. Comer and others of Bell Laboratories as the base, and rewrote it entirely in the ABCL/c+. As a result, it is shown that concurrent object-oriented language can produce very readable and understandable operating system kernels.

In the following, first, ABCL/c+ is described in section 2, then the relation between ABCL/c+ and C is sketched in section 3. Section 4 describes the development of the operating system kernel by using ABCL/c+, and finally the results are discussed in section 5.

2   Concurrent Object Oriented Language ABCL/c+

The ABCL/c+ is based on the language C, while the ABCL/1 language is based on Common Lisp. Therefore, ABCL/c+ is a language with types.

2.1   Object and message

Each object is in one of three **modes**: dormant, active, or waiting. Fig. 1 shows the transitions among the three modes [10]. Except for the routine object, which is discussed later, each object has two queues for retaining messages in the order of arrival. One is the queue for ordinary messages. The other is the queue for emergency messages, which can be accepted during the actions for an ordinary message, i.e. while the object is in the active mode. The actions specified for the ordinary message are temporarily suspended. The ordinary message is called an **ordinary mode message**, and the emergency message is called an **express mode message**. Their queues are called the **ordinary mode message queue** and **express mode message queue**, respectively. The message mode is explicitly identified by both the sender and receiver of the message. The pattern of acceptable messages, constraints, and the mode of the message are defined in an object.

Ordinary mode messages are processed by an object as follows (see Fig. 1). An object, when it is created, is in the dormant mode. If an acceptable message arrives, the object goes into the active mode, and starts performing the actions specified for the message. Any ordinary mode message arriving during the action is placed in the ordinary mode message queue. When the sequence of

252

actions is completed the object goes into the dormant mode. If messages are in the ordinary mode message queue, the object goes into the active mode again, and starts performing the actions. The select form (see 2.4) can be used during the actions to make the object go into the waiting mode and wait for the arrival of a specific message.

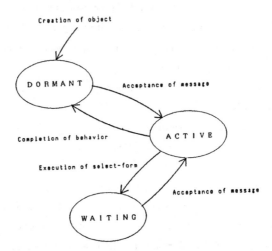

Fig. 1   Mode transition of an object

If a message arrives (or has arrived) while in the dormant mode, the object checks the ordinary mode message queue, starting from the first item, selects the first acceptable message, and deletes all the messages prior to it. If a message arrives (or has arrived) while in the waiting mode, the object checks the ordinary mode message queue, starting from the first item, selects the first acceptable message, and deletes it from the queue.

When an express mode message arrives, it is placed in the express mode message queue. It is not accepted if the object is accessing local memory, if the object is performing an action sequence specified by the atomic form (see 2.5), or if the object is performing the actions specified for an express mode message. Otherwise, if an express mode message is acceptable, the actions specified for the message are performed at once. Whether or not the actions for an ordinary mode message, which is interrupted by an express mode message, are to be resumed after the completion of the actions for express mode message can be specified by using the nonResume form (written as nonResume()). Unless the nonResume form is executed during the actions for the interrupting express mode message, the interrupted actions are resumed.

In addition to these objects, the ABCL/c+ has objects without any message queue, called **routine objects**. The routine objects are introduced to unify procedures, functions, and objects. They do not exist as a process. Routine object accepts only the call type message described later. A routine object can be defined alone, or its definition can be nested in an other routine or ordinary object. However, an ordinary object cannot be defined in a routine object. The routine object is introduced in order to write the operating system kernel for realizing concurrent objects, or the world of concurrent processes. (However, the machine-dependent part, which cannot be written in C, must be written in another language, just as in the case where C is used. In the rewriting of XINU, described in section 4, assembly language was used for the machine-dependent part. Also, since an ordinary object exists as a process, objects which receive only the now type message cannot replace the routine objects.)

If a routine object is defined in an ordinary object, the routine object acts as an internal procedure or function, and corresponds to a routine in ABCL/1 [12].

The scope of the name in the nesting follows the scope rule of the block structure. The name which is in effect in the environment directly outside is in effect unless the same name is declared as a local name in an inner routine object.

2.2 Type of message passing

The types of message passing between objects include: past type, now type, future type, and call type for routine objects. The first three types are the same as those in ABCL/1 [10]. The semantics and syntax for message passing in ordinary and express modes are as follows.

(1) Past type

The sender object sends a message to the receiver object. If it has something to be processed, it continues the processing. If a reply should be sent to some other object, the destination object can be specified as the **reply destination**. The syntax for past type message passing is as follows({...} stands for the optionals):

Ordinary mode: [target object $<=$ message {@ reply-destination}]

Express mode: [target object $<<=$ message {@ reply-destination}]

(2) Now type

The sender object sends a message to the receiver object, and waits until a reply is returned. The syntax for now type message passing is as follows:

Ordinary mode: [target object $<==$ message]

Express mode: [target object $<<==$ message]

(3) Future type

The sender object, when it sends a message to the receiver object, specifies the "reply destination" in a variable called **future variable**, and continues its actions without waiting for the arrival of the reply. The reply sent to the specified destination can be accessed only by the sender of the message. When, at a later time, the reply is accessed, the commands "ready?(future-variable)" and "dequeueFuture(future-variable)" can be used to check to see if the reply has been sent or to take out the reply. The syntax for a future type message passing is as follows:

Ordinary mode: [target object <= message $ future-variable]

Express mode: [target object <<= message $ future-variable]

(4) Call type

The sender object sends a message to the receiving routine object, and waits until a reply is returned from the receiver. The syntax for call type message passing is as follows:

Ordinary mode: [target object <- message]

Express mode: [target object <<- message]

2.3 Object definition

In the ABCL/c+ ordinary objects are defined as follows:

```
Type/composite-type-declaration;
[object object-name
definition-of-routine-object;
                             . . .
state {
    type-of-state-variable state-variable:=initial-value;
                          . . .
}
script {
(=> message-pattern @ destination-variable where constraint # type
    type-declaration-of-pattern-variable; ... ;
    {
        declaration-of-temporary-variable; ... ;
        description-of-action; ... ;
    })
                        . . .
(=>> message-pattern @ destination-variable where constraint # type
    type-declaration-of-pattern-variable; ... ;
    {
        declaration-of-temporary-variable; ... ;
        description-of-action; ... ;
    })
                        . . .
}]
```

Routine objects are defined in the same form as ordinary objects except that -> and ->> are used instead of => and =>> to specify message patterns.

The following is a detailed explanation:

(1) 'State-variables' are variables which indicate the internal state of the object.

(2) An object accepts only those messages which match with 'message-pattern' and satisfy 'constraint.' Message patterns and constraints are checked from top to bottom. 'Reply-destination' and 'future-variable' are bound to 'destination-variable.' Reply is returned if the reply form is written in 'description-of-action.'

(3) When a message is accepted, actions defined in 'description-of-action' are performed sequentially. 'Temporary-variables' are those used by the object during the action. A temporary variable can be declared anywhere before the variable is referred to.

(4) The message pattern following '=>' is for an ordinary mode message, and the message pattern following '=>>' is for express mode message. Message patterns for call type messages are specified using '->' and '->>', respectively.

(5) 'Type' is the type of the value in the reply. Reply can be in any type allowed by the language C.

(6) 'Pattern-variable' is a variable used as an element in a message pattern. When a value is accepted as a message, the value is assigned to a pattern variable only if the type of the value and the type of the pattern variable coincide. 'Reply-destination' is also a pattern variable.

(7) The state part, destination-variable, constraint, and declaration-of-pattern-variable may be omitted.

2.4 Select form

The select form is specified as follows:

```
[select
    (=> message-pattern @ reply-destination where constraint # type
    type-declaration-of-pattern-variable; ... ;
    {
        description-of-action; ... ;
    }
                    . . .
    (=>> message-pattern @ reply-destination where constraint # type
    type-declaration-of-pattern-variable; ... ;
    {
        description-of-action; ... ;
    })]
```

If a select form is encountered in the description of actions, the object goes into the waiting mode, and waits for the message which matches with the message pattern specified in the select form. The first message which matches with the pattern and satisfies the constraint is accepted, and corresponding actions are performed. As noted earlier, the message accepted by the select form may have already arrived before the object goes into the waiting mode.

2.5 Atomic form

If a sequence of actions should not be interrupted by an express mode message, enclose the action sequence as follows:

```
atomic { description-of-action, ... , description-of-action }
```

Express mode message will not be accepted while the action sequence is performed.

2.6 Reply form

A reply can be sent back by using the following notation:

```
! value-form
```

Here, 'value-form' means variables other than a future variable, object definition form, now type
message passing form, and primitive functions supplied by ABCL/c+.

2.7 Programming example

A simple example of programming in ABCL/c+ is shown in this section.

Example. A complex number is represented as an object. Fig. 2 shows an ABCL/c+ program
which computes addition and subtraction between two complex numbers, the result being given as
another object.

```
[object ComplexNumberCreator                              (=> [:Sub ComplexNumber] # int
 script {                                                     int ComplexNumber;
  (=> [:New RealPart ImaginaryPart] # int                     {
  float RealPart, ImaginaryPart;                              ![ComplexNumberCreator <== [:New
   {                                                           (RealPart -
    ![object                                                   [ComplexNumber <== :GetRealPart])
     script {                                                  (ImaginaryPart -
      (=> [:Add ComplexNumber] # int                           [ComplexNumber <== :GetImaginaryPart])]];
         int ComplexNumber;                                    })
          {
           ![ComplexNumberCreator <== [:New                (=> :GetRealPart # float
            (RealPart +                                         {
            [ComplexNumber <== :GetRealPart])                   !RealPart;
            (ImaginaryPart +                                    })
            [ComplexNumber <== :GetImaginaryPart])]];
           })                                              (=> :GetImaginaryPart # float
                                                               {
                                                                !ImaginaryPart;
                                                                })
                                                          }];
                                                          })
                                                         }]
```

Fig. 2   ComplexNumberCreator

ComplexNumberCreator creates an object which represents a complex number. When a message
:New is sent to the *ComplexNumberCreator* with *RealPart* (real part) and *ImaginaryPart* (imaginary part) as arguments, an object is created using the *RealPart* and *ImaginaryPart* as entities.
The new object accepts messages :Add, :Sub, :GetRealPart, and :GetImaginaryPart. For example,
3+4i (object name is *compl1*) and 5+7i (object name is *compl2*) are created as follows:

$$compl1 := [ComplexNumberCreator <== [:New\ 3\ 4]];$$
$$compl2 := [ComplexNumberCreator <== [:New\ 5\ 7]];$$

The object representing the result of *compl1* + *compl2* can be created as follows:

$$[compl1 <== [:Add\ compl2]];$$

2.8   Necessity of routine objects

As indicated earlier, routine objects are introduced to unify procedures, functions, and objects. Routine object does not exist as a process. Therefore, its control mechanism and semantics are quite different from those of other objects. However, what is important is conceptual consistency, such as the existence of individual objects as the single substance, and execution of the task by message passing among the objects.

Conceptual consistency is most important when systems are developed. What is most important in the software implementation of the system is the architecture on which the concept is to be based and unified notation. Given the architecture and a unified notation we can avoid changing what we are thinking about, and therefore, we can develop systems easier to understand and with less errors.

Routine objects are introduced based upon these viewpoints, and they are not a deviation from the object-oriented concept.

2.9   Necessity of the express mode

Since the express mode causes one object to interrupt its actions and another to be executed, it is not very desirable when we consider the internal state of an object. However, in order to write an operating system kernel, we adopt the express mode in ABCL/c+ because it is very effective in the realization of input/output processing.

For instance, when we rewrote the XINU, input/output interruption is processed as an express mode message to the device handler corresponding to the interruption. For example, console manager (see Fig. 5) has two objects, one manages input from console, and the other manages output. They manage the input and output buffer, respectively. For instance, the input manager object accepts the message :GetC for taking out one character at a time in the order of their arrival from the input buffer, and the message [:Tty IIn Chr] for storing input characters in the input buffer (the input manager object can accept other messages). The message [:Tty IIn Chr] is an express mode message, while the message :GetC is an ordinary mode message. This is because it is necessary to store input data in the buffer as soon as possible, lest it should be lost.

The debugging of the objects which accept the express mode message is not different from ordinary interrupt processing. However, the simplicity of the notation should be noted.

3   Relation between ABCL/c+ and C

The C syntax can be used in the type/composite-type-declaration, state, and script parts, but not the function definition, which cannot be declared in C.

3.1 Type check

In ABCL/c+, types are checked either statically or dynamically depending on where the types are defined. Types defined in the type/composite-type-declaration part, which are global, and state and script parts, which are local, are checked statically by the C compiler. The types string, future, and list, which are introduced into ABCL/c+ especially for string manipulation, future variables handling and list manipulation respectively, are checked by the ABCL/c+ preprocessor.

But, the types of the components of a message are checked dynamically, when the message is tested against the messsage pattern as to whether it is acceptable or not. As message passing is done by byte-by-byte and as we would like to accept, for example, the value having type is short into the variable which type is long, only the types corresponding to the conversion characters which are used by stream I/O in the C are checked to the messages.

The type int is used for the type of a pointer to an object, because the value of a pointer to an object is the identification number of the object.

3.2 ABCL/c+ forms

Any kinds of forms of ABCL/c+, which are called ABCL/c+ forms, and statements of the language C can be used in the state and script parts. Among other things, two ABCL/c+ forms, object defining forms and value returning (now type) message passing forms, can be written on the right hand side of the assignment statement and assign the values to variables.

3.3 Alternation of the C syntax

As ABCL/c+ forms and C statements can be used together, the syntax of C is altered as follows [5]:

(1) Use assignment operator := instead of =.

(2) Use (* and *) instead of [and] as array subscript notations.

4 Description of Operating System Kernel

In order to verify the effectiveness of ABCL/c+, we tried to write an operating system kernel. The material chosen was the operating system kernel XINU [3] developed by D. Comer and others of Bell Laboratories. XINU is written in C, and consists of more than a hundred functions. The source file of XINU contains about 5850 lines of C code and 650 lines of assembly code. Deleting comments, it has approximately 4300 lines of C code and 550 lines of assembly code.

As shown in Fi. 3, the system has a 10 layer structure, including hardware and user programs.

Fig. 4 shows the relations among modules (functions) belonging to the process management layer and their data reference.

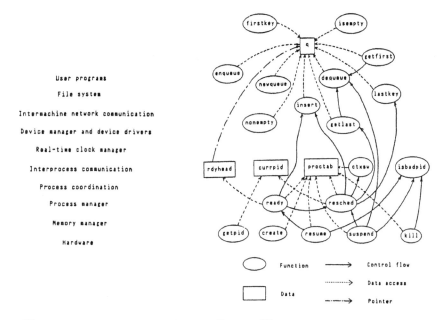

Fig. 3   The layering of components in XINU   Fig. 4   The process manager layer of XINU

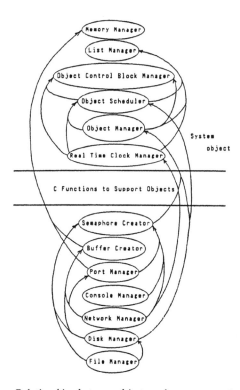

Fig. 5   Relationships between objects and message passing

Although they are structured in layers, it is not easy to understand the modules in such a complex relationship. We rewrote this system in ABCL/c+. Using basically the same data structure, the function of the new system is exactly the same as that of the old system. The objects and message passing relationships are shown in Fig. 5.

In Fig. 5 'memory manager' through 'real time clock manager' are the system objects which constitute the world of processes. Each of them is implemented as a routine object. 'Semaphore creator' and above are in the world of processes. Each of them corresponds to one or more objects (processes). 'C functions to support objects' in the middle are functions for object creation, and those for message passing. These C functions are automatically generated by ABCL/c+.

In the overall system, the command processor SHELL and ABCL/c+ compiler, which at present is the only language processor of our system, are placed on the kernel. When an object code of ABCL/c+ is executed, every time an object is created, 'C functions to support objects' are called, and they pass messages to and from system objects 'object manager', 'object scheduler', and 'object control block manager'. In so doing objects begin to exist as processes, and each process is executed in parallel by time-slicing. When an object (process) communicates with the console, it passes messages to and from 'console manager.' When it operates a file, it passes messages to and from 'file manager.'

In the following subsections, the two layers: 'process manager' and 'intermachine network communication' shown in Fig. 3 are discussed.

4.1 Process management

Generally speaking, this layer has four functions:

(1) list processing

(2) maintenance of process control block

(3) process schedule

(4) process management

List processing functions manage the ready list, semaphore list, and delta list. They use an array q which consists of three fields: priority, the predecessor, and the successor. The delta list is a list of sleeping processes in the order of the time at which they are to be awaken. These lists are bi-directional, with fields for the predecessor and successor. The second function is implemented as a table *proctab* for process management. The third function consists of the functions for scheduling and context switch using the ready list, and the variable *currpid* which maintains the identifier of the process currently being executed. The fourth function treats the creation/deletion and suspend/resume of processes.

The process control block *proctab* and the array q for lists are global data. For instance, *proctab* can be accessed from the upper layer 'interprocess communication' or from a still higher layer such as the 'real time clock manager' layer, as well as directly from this layer.

Using global data directly accessible or modifiable by other layers causes difficulties in the ease of understanding, in maintenance, and in revision.

Therefore, based upon functions and data, we divided this layer into the following routine objects:

(1) list manager

(2) object control block manager

(3) object scheduler

(4) object manager

Here, (1) (3) and (4) roughly correspond to (1) (3) and (4) above. However, the object control block manager, which consists of the object control block (process control block) and basic operations on it, did not exist explicitly in the original version.

```
[object Scheduler
 [object Schedule
  script {
   (-> :ReSched # int
      {
      int OldObjectID;

      If( ( [OCB <- [:ReferState :At CurrentObjectID]]
            == CURRENT ) &&
          ( [List <- [:LastKey ReadyTailID]]
               < [OCB <- [:ReferPriority :At CurrentObjectID]] ) )
         !OK;

      If( [OCB <- [:ReferState :At CurrentObjectID]]
            == CURRENT ) {
         [OCB <- [:PutState READY :At CurrentObjectID]];
         [List <- [:Insert CurrentObjectID ReadyHeadID
            [OCB <- [:ReferPriority :At CurrentObjectID]]]];
      }

      OldObjectID := CurrentObjectID;
      CurrentObjectID := [List <- [:GetLast ReadyTailID]];
      [OCB <- [:PutState CURRENT :At CurrentObjectID]];
      [Timer <- :Quantum];
      ctxsw( [OCB <- [:ReferObjectRegisters :At OldObjectID]],
         [OCB <- [:ReferObjectRegisters :At CurrentObjectID]] );
      !OK;
      })
 }]

 state {
   int CurrentObjectID;
   int ReadyHeadID;
   int ReadyTailID;
 }
```

```
script {
  (-> :Initialize
     {
     ReadyHeadID := [List <- :NewQueue];
     ReadyTailID := ReadyHeadID + 1;
     })

  (-> [:Ready ObjectID ReScheduleBool] # int
     int ObjectID;
     Bool ReScheduleBool;
     {
     If( Isbadpid( ObjectID ) ) !SYS_ERR;
     [OCB <- [:PutState READY :At ObjectID]];
     [List <- [:Insert ObjectID ReadyHeadID
        [OCB <- [:ReferPriority :At CurrentObjectID]]]];

     If( ReScheduleBool ) [Schedule <- :ReSched];
     !OK;
     })

  (-> :ReSchedule
     {
     [Schedule <- :ReSched];
     !OK;
     })

  (-> :ReferCurrentObjectID # int
     {
     !CurrentObjectID;
     })
}]
```

Fig. 6   Object scheduler

For instance, object scheduler is as shown in Fig. 6. This takes the form of nested routine objects. When the message :Initialize is accepted, a message is sent to the list manager, *List*, to create the ready list. The first and last entries of the list are assigned to *ReadyHeadID* and *ReadyTailID*, respectively. From the message [:Ready ...], object identifier *ObjectID* and the boolean value *ReScheduleBool*, which indicates whether or not the schedule is to be rescheduled, are obtained as arguments. After the validity of the object identifier is checked, a message is sent to object block manager, *OCB*, to put the object in the waiting state. Next a message is sent to the list manager, *List*, to insert the object into the ready list. At this time a message is sent to *OCB* to obtain the priority of this object, which is sent to *List*. Items in the ready list are in the order of their priority, from low to high. Lastly, the necessity of rescheduling is confirmed. If rescheduling is necessary, the internal routine object *Schedule* is requested to reschedule. If the message :ReSchedule is accepted, rescheduling is made by using an internal routine object. If the message :ReferCurrentObjectID is accepted, "current object" identifier *CurrentObjectID* is returned.

When the internal routine object, *Schedule*, accepts the message :ReSched, it does not perform rescheduling if the state of *CurrentObjectID* is 'CURRENT' and the priority of this object is higher than that of the object in the tail of the ready list (the ready list is in the ascending order of priority from low to high). Otherwise, the context is switched. First, if the state of *CurrentObjectID* is 'CURRENT', the state is changed to 'READY' (for example, when the real time clock manager accepts the :Sleep message, the state of the *CurrentObjectID* is set to 'DORMANT' before the :ReSchedule message is sent), and the identifier of the object and its priority are cataloged in the ready list. Then the object with the highest priority is taken out of the ready list, and its state is set to 'CURRENT'. Finally, *Timer* is set to the quantum time, and the context is switched by using ctxsw, which is implemented in the assembly language.

Thus, using nested routine objects, the scheduler can be written compactly, based upon the concepts consistent with other objects.

4.2 Network communication

Roughly speaking, this layer has the following three functions:

(1) message communication

(2) creation and management of buffer pools

(3) port management

The first function performs interprocess message communication via network, and consists of a number of system functions and several processes. A message is sent and received as a block. An outline of the transformation process is shown in Fig. 7.

The second function performs the creation of buffer pools to be used in the data link layer of (1) above and in disk management, as well as the allocation and release of buffers in the pools.

The third function performs the creation and management of "ports" which are the basic logical structure elements of frame and block in (1) above. A port is a list the basic element of which is a word.

We implemented the functions (2) and (3) each as an object, and function (1) by several objects. At present, transport, internet, and data link in Fig. 7 are implemented as one object each, and frame is implemented by 3 objects.

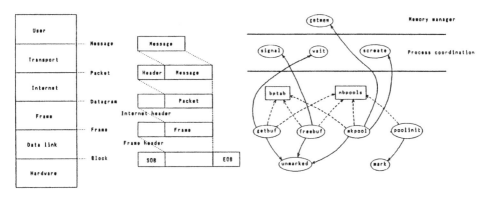

Fig. 7   Network software layers and data formats   Fig. 8   Creation and management of buffer pool
(a part of network communication)

For instance, modules (functions), data, and the reference relations of the buffer pool creation and management function in the original version are as shown in Fig. 8. The main function here is the dynamic management of the set of buffers with the size specified at buffer pool creation. The table *pbtab* is used to record, for each buffer pool, the size of buffer pool, pointer to the free buffer, and the semaphore for allocation management. The *nbpools* is used to count the number of buffer pools.

In the ABCL/c+ version, a buffer pool is created as an object, which is allocated. Each buffer has an identifier of the pool to which it belongs. Thus, no buffer is returned to a wrong buffer pool. Fig. 9 shows buffer pool creation and management object *BufferCreator*.

The object *BufferCreator* can accept :Initialize message and :New message. When the message :Initialize is accepted, state variable *NumberOfBufferPools*, which indicates the number of buffer pools currently supplied, is initialized to 0. When the message [:New ...] is accepted, the size of the buffer pool is computed, using *BufferSize* (size of a buffer) and *NumberOfBuffers* (the number of buffers), and a memory area with that size is allocated. Actually, the allocation is achieved by sending a call type message to the system object for memory management, *MemoryManager* (see Fig. 5). At this time, the pointer and pool identifier parts are added to each buffer (2*SizeOf(int)). *MemoryManager* returns the start address of the allocated area. If a memory area with the necessary size cannot be allocated, SYS_ERR will be returned.

If an area is allocated successfully, the state variable *NumberOfBufferPools* is incremented by 1. Then buffer pool is initialized by chaining all buffers using the pointer part of the buffers.

Lastly, the object *BufferPool*, which allocates and releases the buffers in the buffer pool, is created and returned to the sender of the message. The object *BufferPool* can accept the message :GetBuffer and the message :FreeBuffer. When the :GetBuffer message is accepted, the object checks to see if any buffer can be allocated in the pool. If not, it waits for the :FreeBuffer message to come in the select form and release a buffer. If a buffer is released, or one is available, the first buffer in the free buffer list, whose root is *Next*, is taken out. *Me*, that is the ID of the object *BufferPool*, is set in the pool identifier part. Then the buffer is returned to the sender of the message. When :FreeBuffer message is accepted, it checks to see if the buffer in the message belongs to it. If not, it returns SYS_ERR. If it is the buffer which belongs to the object, the returned buffer is chained to the head of the free buffer list, and OK is returned to the sender of the message.

We can see that, by making an object for each buffer pool, which manages the buffers in the pool, we can at the same time clarify and simplify the functions.

```
[object BufferCreator
 state {
   Bool InitializeFlag := TRUE;
   int NumberOfBufferPools;
 }
 script {
   (=> [:Initialize] where ( InitializeFlag ) # int
   {
       InitializeFlag := FALSE;
       NumberOfBufferPools := 0;
       !OK;
   }
   (=> [:New BufferSize NumberOfBuffers] # int
   int BufferSize, NumberOfBuffers;
   {
       short Psw;
       char *Where, *Next;

       Disable( Psw );
       if( BufferSize < BUFFER_MIN_BYTE
       || BufferSize > BUFFER_MAX_BYTE
       || NumberOfBuffers < 1
       || NumberOfBuffers > BUFFER_MAX_NUMBER
       || NumberOfBufferPools >= MAX_BUFFER_POOLS
       || ( where := [MemoryManager <- [:GetMemory
       ( ( BufferSize + 2*SizeOf(int) )*NumberOfBuffers )]] )
       == SYS_ERR ) {
       Restore( Psw );
       !SYS_ERR;
       } else {
       Restore( Psw );
       NumberOfBufferPools++;
       Next := Where;
       BufferSize += 2*SizeOf(int);

       for( NumberOfBuffers-- ; NumberOfBuffers > 0 ;
       NumberOfBuffers--, Where += BufferSize )
       *( (int *)Where ) := (int)( Where + BufferSize );
```

```
       *( (int *)Where ) := (int)NULL;

 ![object BufferPool
 script {
   (=> :GetBuffer # int
       {
         int *Buffer;

         if( Next == NULL )
         [select
           (=> [:FreeBuffer Buffer1] # int
             int *Buffer1;
             {
             *Buffer1 := (int)Next;
             Next := Buffer1;
             })];
         Buffer := Next;
         (int)Next := *Buffer;
         *( Buffer + 1 ) := Me;
         !(int)Buffer;
       }

   (=> [:FreeBuffer Buffer] # int
       int *Buffer;
       {
         if( *( Buffer + 1 ) != Me ) !SYS_ERR;
         else {
         *Buffer := (int)Next;
         Next := Buffer;
         !OK;
         }
       })
   )];
 }
)}
}]
```

Fig. 9 Buffer pool creator

5 Discussion

The routine object is introduced to unify procedures, functions, and objects. By introducing the routine object, procedures and functions for creation of the world of processes, as well as those in objects, can be constructed by the same concept as that of the objects. The example given in 4.1 clearly illustrated the advantage of this approach.

We have also shown that, in addition to its effectiveness in grouping global data along with operations on it, as seen in the case of the process control block, the object-oriented concept is very effective in the management of "objects" which exist independently as seen in the example of the buffers in 4.2.

The buffers discussed in 4.2 are used to manage data transfer, triggered by external interruption, in the data link layer and disk management. By implementing each buffer pool as a "concurrent object", data processing mechanism can be made simple and clear.

Thus, we have shown that concurrent object-oriented languages, such as our ABCL/c+, are suited to describe an operating system kernel. Although it is desirable to describe even the machine dependent part in the same paradigm, we have not yet implemented it in the ABCL/c+ language.

6   Conclusion

In this paper, we presented the outline of concurrent object-oriented language ABCL/c+ and the result of writing an operating system kernel in ABCL/c+. As far as we know, this is the first attempt to write an operating system kernel in a concurrent object-oriented language. We have demonstrated that, using a language with the routine object, which is our extension of the ABCL/1, and the ability to describe the simple and powerful concurrent objects, such as in the ABCL, a very highly understandable operating system kernel can be written.

However, we also assume that function layers within an object can lead to an even higher degree of ease in writing and understanding programs. Therefore, in order to achieve the layer of functions, we tried to make it possible to nest the definitions of ordinary objects other than routine objects in ABCL/c+. However, we have not yet been successful in finding satisfactory approaches for the method inheritance and search strategies (depth first or breadth first). This point is still under investigation.

The development of the processor of ABCL/c+ and the rewriting of XINU using ABCL/c+ are carried out on the Sun workstation and Macintosh II. The final result, the ABCL/c+ version of the XINU operating system, will run on Macintosh II.

Acknowledgements

The authors wish to thank Akinori Yonezawa, who gave them a number of useful comments in preparing this paper. Also they wish to thank the referees for their useful suggestions and comments on the first version of the manuscript.

References

[1] Borow, D.G. and Stefik, M. *The LOOPS Manual, KB-VLSI-81-13*. Xerox PARC, 1983.

[2] Chikayama, T. *ESP Reference Manual, TR-044*. Technical Report, ICOT, 1984.

[3] Comer, D. *Operating System Design : The XINU Approach*. Prentice-Hall, 1984.

[4] Comer, D. *Operating System Design - Volume II: Internetworking with XINU.* Prentice-Hall, 1987.

[5] Doi, N. and Kodama, Y. *ABCL/c+ User Manual.* Institute of Information Science, Keio University, 1987. (In Japanese.)

[6] Doi, N. and Segawa, K. "Concurrent programming in Smalltalk-80." *Computer Software*, 3(1), 1986. (In Japanese.)

[7] Goldberg, A. and Robinson, D. *Smalltalk-80 : The Language and its Implementation.* Addison-Wesley, 1983.

[8] Ishikawa, H. and Tokoro, M. "Orient 84/K: An Object-Oriented Concurrent Programming Language for Knowledge Representation." in *Object-Oriented Concurrent Programming* (ed. Yonezawa, A. and Tokoro, M.). MIT Press, 1987.

[9] Kristensen, B.B., Madsen, O.L., Moller Pedersen, B., and Nygaard, K. "Multisequential execution in the BETA programming language." *Sigplan Notice*, 20(4), 1985.

[10] Shibayama, E. and Yonezawa, A. *ABCL/1 User's Guide.* Information Science Department, Tokyo Institute of Technology, 1986.

[11] Yokote, Y. and Tokoro, M. "Concurrent Programming in ConcurrentSmalltalk." in *Object-Oriented Concurrent Programming* (ed. Yonezawa, A. and Tokoro, M.). MIT Press, 1987.

[12] Yonezawa, A., Shibayama, E., Takada, T., and Honda, Y. "Modeling and Programming in an Object-Oriented Concurrent Language ABCL/1." in *Object-Oriented Concurrent Programming* (ed. Yonezawa, A. and Tokoro, M.). MIT Press, 1987.

Debugging Concurrent Systems
Based on Object Groups

Yasuaki Honda[†]

Akinori Yonezawa

Department of Information Science,

Tokyo Institute of Technology,

Ookayama, Meguro-ku, Tokyo 152, Japan

Abstract

This paper presents a debugging method for Concurrent Object-Oriented Systems. Our method is based upon a new notion called Object Groups. An Object Group is a collection of objects which forms a natural unit for performing collective tasks. An Object Group's Task differs from C.Manning's *nested transaction* which is based on the nested request-reply bilateral message passing structures. Each Object Group's Task permits more general message passing structures. The language constructs which specify and use Object Groups have been introduced into an object-oriented concurrent language ABCL/1. The paper also describes ABCL/1's debugging tools based on Object Groups.

1 Introduction

In an object-oriented concurrent system, a number of objects are working concurrently and they communicate with each other by sending messages. Debugging programs on such a system is usually more difficult than debugging ones on a sequential system because bugs or errors may exist not only in individual behavior of each object, but also in communications and synchronization with each other.

As a basis for a debugging scheme for concurrent systems, it has been proposed to record all the events in which relevant objects are involved. The record of each object is often called a *task record* ([Manning 87]) or a *local history* ([Honda 86]). Since the data recorded in local histories are

---

[†]The author's current address is: Fuji Xerox, Co., Ltd. Akasaka, Tokyo, Japan.

low-level ones and not well-structured, it is often hard to make sense out of them directly. Based on the notion of *nested transactions*, C. Manning proposed a scheme to extract useful information for debugging from task records and implemented it as a part of his debugging system called, an Apiary Observatory[Manning 87]. A nested transaction comprises nested pairs of *request* and *reply* messages. His scheme is very powerful for debugging a system in which message passing structures are always bilateral with pairing of a request and a reply (the call/return pair).

In general, however, message passing structures in object-oriented concurrent systems[Yonezawa 87] do not necessarily conform to the request-reply bilateral protocol. To cope with such general message passing structures, the notion of *Object Groups* we propose provides a promising approach to debugging. (In this approach the data recorded in the local histories of individual objects are also the primary source of information.)

Generally, an Object Group reflects an abstraction of various problem and solution structures. When the programmer writes a large scale (object-oriented concurrent) system, s/he usually has in his/her mind a structure or an abstraction of the system. Currently proposed object-oriented concurrent languages do not provide the programmer with explicit structuring mechanisms: such structures or abstractions can only be expressed by using the implicit "knows-about" relationship among objects which is established by storing objects' names in state variables of objects.

An Object Group is a collection of objects which forms a natural unit for performing collective tasks. For example, consider to model a company in a concurrent object-oriented manner. A company basically consists of persons who are employed by the company. When the company is requested to perform some task, members of the company perform actions in cooperation and communicate with each other. Naturally, each person is modeled as an object. Actions performed by each person are simulated by procedures associated corresponding objects, and of course, communication among persons is simulated by message passing. But, should the company be modeled as an object? In fact, the company itself does not perform any task, but all the tasks are carried out by (a collection of) individual persons in the company. Thus it is often useful to model the company simply as a group of objects.

In this paper, we introduce the notion of Object Groups and its language constructs in our language ABCL/1[Yonezawa 86], which can be used to explicitly describe abstractions of structures of a large concurrent system. The rest of the paper mainly describes the applications of Object Groups to our debugging scheme and visual execution monitoring system. Also other debugging tools provided in ABCL/1 will be briefly presented. Note that the usefulness of the notion of Object Groups does not limit itself to debugging and monitoring: it is of course powerful in other activities such as designing systems and processor allocation as will be briefly discussed in Section 6.

2 The Computation Model and the Language ABCL/1

2.1 The Computation Model

The object oriented computing model assumed in this paper is a simplified version of our proposed model [Yonezawa 86]. In the model, each object has its own computing power and can perform its activity concurrently with the other objects. The computation model allows neither shared memory nor global clock. A group of objects forms a sparsely connected system.

An object has its own internal world, which consists of a local persistent memory and procedures inquiring/updating the local memory. They are called *state* and *script*, respectively. We assume that an object has its own local clock. The internal world of an object cannot be directly accessed from the other objects.

Objects interact with one another via *message passing*. In response to a message, an object executes one of the procedures in its script. Execution of a procedure (script) by an object consists of a sequence of actions of the following kinds:

- inquiring and updating the *state* of the object,
- creating new objects,
- sending and accepting messages, and
- returning a value as a reply to a received message.

Messages arriving at an object will be processed one at a time in a sequential manner.

There are two types of message passing, *asynchronous* and *synchronous*, which are called *past-type* and *now-type*, respectively. Immediately after sending a message asynchronously, an object can continue its computation. The sender object of an asynchronous message transmission expects no reply. In contrast, the sender object of a *synchronous* message transmission cannot resume its computation until the reply to this message arrives.

When a message is sent to an object to request the object to do some task, it is often useful to be able to specify where the result or reply of the requested task should be sent. This information is called *reply-destination* in our computation model (See [Yonezawa 86] for more details). Each synchronous message transmission implicitly specifies the reply-destination. The receiver object of a synchronous message transmission can forward the reply-destination to another object which is responsible for returning a reply. Note that the reply-destination specified by a synchronous message transmission can be forwarded to more than two objects. Some of them may return replies to the reply destination. In this case, the first message returned is considered as the reply for the synchronous message transmission.[1]

We assume that messages satisfy the *transmission ordering law*: Suppose that two messages M and M' have the same sender and the same receiver. If M and M' are sent in this order according

---

[1] This feature is used in the example of tree structured dictionary given in Section 3.1.

to the local clock of the sender, they are received by the receiver in the same order. In general, if *M* and *M'* are sent from the different objects and received by the same one, their arrival order cannot be determined.

2.2 The Language Constructs

In the language ABCL/1, an object is defined by the following form:

```
[object <object-name>
  (state [<state-variable> := <initial-value>] ...)
  (script
    (=> <message-pattern> @ <variable>  <behavior-description>)
                         ⋮
    (=> <message-pattern> @ <variable>  <behavior-description>))]
```

The `<state-variable>`s of an object are the variables which represent the internal *state* of the object. An object defined by the above form accepts a message which matches against some `<message-pattern>`. When a message matches against the pattern, the reply destination of the message (if it exists) is bound to the corresponding `<variable>`. After accepting a message, the object will perform a sequence of actions described in the corresponding `<behavior-description>`. The state variable declaration part and `<variable>` are optional.

In the `<behavior-description>`, sequential computation is written using Lisp-like forms and message passing and returning a reply are written in the following forms:

```
[<object> <== <message>]       ;; synchronous message passing (now-type)
[<object> <= <message>]        ;; asynchronous message passing (past-type)
[<object> <= <message> @ <reply-destination>]
                               ;; asynchronous message passing with reply destination
!<reply>                       ;; returning a reply
```

3 Object Groups in ABCL/1

3.1 Characteristics of Object Groups

In this subsection, we will discuss the characteristics of Object Groups. Possible operations on Object Groups and the tasks of Object Groups are illustrated by using a tree structured dictionary.

A tree structured dictionary is a dictionary represented by a binary tree. In a tree structured dictionary, each dictionary entry is represented as an object. This object corresponds to an entered word and remembers both its spelling and meaning. This object is called *a word object* and it is created by an object *CreateWord*.

The nodes of the tree are word objects. Each node may have at most two sub-trees. A word object, which is also a node of the tree, can accept messages [:search <word>] and [:update [<word> <meaning>]]. Upon receiving one of these messages, a word object compares <word> with its remembered spelling. If the spelling equals to <word>, then the meaning it holds is returned if the request is *:search*, or the word object updates its meaning if the request is *:update*. Otherwise, the word object sends the same message to one of its sub-trees according to the result of the comparison between the spell and <word>.

To ask the meaning of a word, a request message [:search <word>] is sent to the root of the tree. A tree search is performed inside the dictionary. If the word has already been entered, the meaning of the word is returned from the corresponding word object of the tree dictionary. When more than one *:search* request arrives, the search tasks are performed in parallel.

There is a tree-manager object which keeps the root word object of the tree structure. This dictionary also has a cache memory which remembers the last ten referred words. This cache memory is represented as an object called Cache. When the tree-manager object is requested to search a word, it sends [:search <word>] messages to both the root object of the tree structured dictionary and the cache object in parallel. (Figure 1)

The objects appeared in the tree structured dictionary example form a natural unit for performing collective tasks, such as searching or updating. In order to describe this structure explicitly, we can define an Object Group that consists of the tree-manager object, the cache object and word objects.

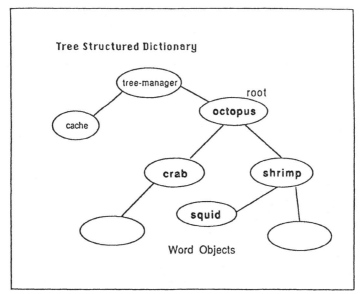

Figure 1: Objects in the tree structured dictionary

From this example, the following characteristics of the group of objects are observed.

- *Some members of the group are fixed, while some are not fixed and dynamically added to or deleted from the group.*

The tree-manager object and the cache object are the fixed members of the tree structured dictionary group. That is, these objects originally belong to the tree structured dictionary group since the group was created. In contrast, word objects are dynamically added to the tree structure when they are created. Some word objects may be removed from the tree structured dictionary group when words are removed from the dictionary.

- *Causally connected individual actions of objects in the group form a task of the group.*

- *Some of the group's fixed members represent the group as a whole and request messages sent to them often trigger the group's collective tasks.*

When a request [:search <word>] comes to the tree-manager object which represents the whole dictionary, it sends the same requests to the root node of the tree and the cache object in parallel. The meaning of the word is searched in both the cache and the tree in parallel and, if found, then it is returned. The individual actions of the objects form the searching task of the tree structured dictionary group. Note that if the cache hits, two replies are sent back, one from the cache and the other from the tree. (In this case the reply that arrived first is picked and the second one is ignored. See Section 2.1 for more details.) This kind of message passing structure cannot be conformed to the request-reply bilateral protocol.

Since these characteristics may be observed widely in various problem domains, the notion of Object Groups is useful in concurrent object-oriented systems. We incorporate the language constructs for Object Groups into our language ABCL/1.

3.2   Language Constructs for Object Groups

Object groups may be created dynamically during the execution of a program. We use the following notation to specify an *object-group* in ABCL/1:

```
[object-group <group-name>
  (fixed-members
    [<member-name> = <object-creation-form>]
                      ⋮
    [<member-name> = <object-creation-form>])
  (initialize
    <abcl-form> ...)]
```

Every object created by evaluating <object-creation-form> is a fixed member of the Object Group. Fixed members can be accessed by specifying associated <member-name>s and the name of the group they belong to. After all the fixed members are created, the <abcl-form>s are evaluated as the initialization of the group.

As an object-group form is evaluated, an object is created which manages and maintains the members of the group. The object is associated with <group-name>. Any request to an Object Group should be sent to this object.

The following is an example of a definition of an Object Group.

```
[object-group tree-dictionary
 (fixed-members
  [Cache = [Object
               (state [cache-entry := nil]
                      Manager)
               (script
                 (=> [:cooperate-with obj]
                     [Manager := obj])
                 (=> [:search word]
                     if word exists in cache-entry
                        then return its meaning in cache-entry
                        else do nothing.)
                 (=> [:update [word meaning]]
                     update the cache-entry.))]]
  [tree-manager =
     [Object
       (state [RootNode := [CreateWord <== :new]]
              Cache)
       (script
         (=> [:cooperate-with obj]
             [Cache := obj])
         (=> [:search word] @ RD
             [[Cache RootNode] <= [:search word] @ RD])
             ;; A message [:search word] and the reply destination RD
             ;; are sent to both Cache and RootNode in parallel.
         (=> [:update data]
             [RootNode <= [:update data]]))]])
 (initialize
  [Cache <= [:cooperate-with Manager]]
  [Manager <= [:cooperate-with Cache]])]
```

This object-group form defines the tree structured dictionary described in Section 3.1. When this form is evaluated, the tree-manager object and the cache object are created as the fixed members of the group. Then the initialization forms are evaluated.

Every fixed member is accessed by specifying their name and the name of the object group that it belongs to. The following functions are provided for this purpose.

```
(group-member <group-name> <fixed-member-name>)
(fixed-member-name <group-name>)
```

On evaluating the first form, a fixed member of the group <group-name> which is associated with <fixed-member-name> is returned. On evaluating the second form, a list of the names of all the fixed members are returned. Using the first form, for example, it is possible to get the meaning of a word by evaluating the following form:

```
[(group-member tree-dictionary 'tree-manager) <== [:search word]]
```

It is possible to add an object to an object group, or to delete an object from an object group by sending a request to the object group (in fact an object which manages the members of the group). Ordinary message passing forms are used with the following special message patterns:

```
[<object-group> <= [:add an-object]]
[<object-group> <= [:delete an-object]]
```

In the tree structured dictionary example, immediately after the creation of a new word object the following form should be evaluated to add it to the tree structured dictionary group. This form is evaluated in the CreateWord object, which creates the new-word object.

```
[tree-dictionary <= [:add new-word]]
```

4 Debugging Scheme based on Object Groups

4.1 Local History

As a basis for our debugging scheme, the ABCL/1 system records the local history for each object during the debug mode execution of a program. The local history of an object is a sequence of events that occur in the object. Sending or receiving a message and object creation are such events, and they are recorded as the local history as they occur. Information related to each event, such as the sender name and the message content is also recorded as the local history.

Each object keeps its local time. The local time of an event is also recorded in the local history together with the event. The local time is incremented by one unit time when the object sends a message or creates an object. When an object S sends a message M to an object R at local time Ts of S, Ts is also sent along with the message M as a time-stamp. When the object R receives this message, the sender S, the message M, and Ts are recorded in the local history of R. Suppose Tr is the local time of R immediately before it receives M. The local time of this receiving event is the maximum of $1 + Ts$ and $1 + Tr$.

By using the local time mechanism, the global history of the entire system can be re-constructed from the local event history recorded in each object, for the local history keeps a partial order of events. Note that the local history mechanism is similar to the *task record* of [Manning 87]. He used it for constructing nested transactions. But we will use the information obtained from the local history to define and construct the *Object Group's Task* described in the next section.

The ABCL/1 debugging system provides a tool called the ABCL/1 Inspector, which can be used to examine the local history of an object directly. More details of the tool will be given in Section 5.

4.2 Object Group's Task

An *Object Group's Task* is a collection of causally connected actions of individual objects in the group. When one of the object group's fixed members receives a message from outside the group, an *Object Group's Task* starts. The first message sent from outside the group to trigger the Object Group's Task is called an *entry message*. The receiver of the message causes message transmissions to other members of the group. Such other members, in turn, perform their actions, and so on. All the actions performed by the members of the group are included in the Object Group's Task. Sending messages to more than two members of the group may cause a branch of the Group's Task. Sending a message to an object outside the group or sending no messages terminates a branch of the task. An Object Group's Task is considered to be terminated when all the branches are terminated. Note that such an Object Group's Task cannot be captured by Manning's nested transactions.

Since Object Groups reflect the problem domain structure, Object Groups' Tasks reflect meaningful tasks in the domain. For example, when the tree-manager of the tree structured dictionary group receives a [:search word] message from outside, a task of the tree structured dictionary group begins. The tree-manager object sends a [:search word] message to the root node and the cache simultaneously. This causes a branch in the Object Group's Task. They start their actions upon receiving the [:search word] message. All the actions performed by the tree-manager, the root node, and the cache are parts of the group's task. If the cache does not know the meaning of the word, it does not send any message to other objects. So a branch of this group's task is considered to terminate at the cache. The root node, after receiving [:search word], sends the same message to its RightNode or LeftNode, then a tree search begins. If a word object which remembers the specified spelling exists, it returns the meaning of the word and the branch of the Group's Task terminates at the word object.

All information needed to trace an Object Group's Task is included in the local history of the members of that group.

4.3 Object Group Examiner

Object Group Examiner (OGE) is a tool to examine each task of an object group. Evaluating the following form activates the OGE:

```
(oge-start <object-group>)
```

OGE displays the following information on its sub-windows (Figure 2):

- *the fixed members* of <object-group>,
- *entry messages* of one of the fixed members,
- An *Object Group's Task* triggered by one of the entry messages, and
- *entered commands* for OGE

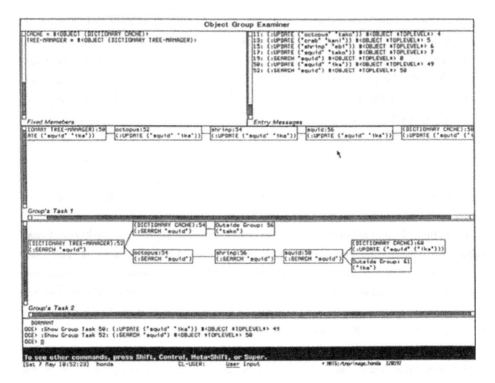

Figure 2: Object Group Examiner Window: the tree structured dictionary is examined.

On the upper left sub-window, all the fixed members of the tree dictionary group (*tree-manager* and *cache*) are displayed. These objects are mouse-sensitive and, by pointing one of them by mouse, entry messages of the selected object are displayed on the upper right sub-window which is called an *entry messages sub-window*.

Each line of the entry messages sub-window corresponds to each entry message and it includes *the received time, the message content, the sender,* and *the send time.* In Figure 2, entry messages of the tree-manager are displayed. Note that at 17, the tree-manager receives [:update ["squid" "tako"]] message. This message is wrong because "tako" does not mean a squid in Japanese.[2] To correct the wrong registration, at 50, [:update ["squid" "ika"]] is received. Each entry message is

---

[2] The word tako and ika mean octopus and squid in Japanese respectively.

also mouse-sensitive and, by pointing one of them, an object group's task invoked by the message is displayed in one of two windows which are labelled *Object Group's Task 1 and 2* in Figure 2. Two of Group's Tasks can be examined simultaneously on these windows.

On the *Object Group's Task 1* sub-window, the Object Group's Task triggered by [:update ["squid" "ika"]] received at 50 by the tree-manager is displayed. Each box appeared in the sub-window represents an individual action of an object invoked by receiving a message. For example, the left-most box represents the tree-manager's action invoked by receiving [:update ["squid" "ika"]] at 50. The tree-manager sends the same message to the root node of the dictionary. The second box represents the action of the root node (which is, in fact, a word object). The root node sends the same message to its appropriate sub-tree. The third box represents the action of the intermediate node between the root node and the node which remembers the meaning of "squid" (The action of the node is the same as the action of the root node). The fourth box represents the action of the node which remembers the meaning of "squid". It updates the meaning, then sends an [:update ["squid" "ika"]] message to the cache object. The last (right-most) box represents the action of the cache. It updates its local memory and sends no message to other objects. The *:update* task terminates at this point.

A more complicated Group's Task can be observed in the *Group's Task 2* sub-window. The Group's Task displayed on this sub-window is invoked by receiving the message [:search "squid"] by the tree-manager at 52. It is observed that after the tree-manager receives the *:search* request, it sends the same messages to both the cache and the root node in parallel. The cache sends a reply message "tako" (which is the wrong answer. See footnote.) to outside the group. Concurrently, a tree search occurs and the right answer "ika" is sent to outside the group. These two replies are sent back to the same reply destination and the first arrival is used as the reply. In this example, the wrong answer "tako" is used instead of "ika." This seems to be a bug in this program because the out-of-date meaning "tako" is returned for the *:search* request (at 52) which is sent after the *:update* request (at 50). The reason for the bug can be found by looking at the *Group's Task* sub-windows. In the *Group's Task 1* sub-window, the cache object receives the [:update ["squid" "ika"]] at 58, while the cache object receives the [:search "squid"] at 54. This *:update* action should have been completed before this *:search* action.

At this level, each action performed by each member object is not displayed in detail. To see the detail, a box which represents an individual action of an object should be pointed by mouse. The ABCL/1 Inspector is invoked to see the detail of that action which is a part of the object's local history. (See Section 5 for ABCL/1 Inspector.)

4.4 Visual Execution Monitor

We have been developing a tool called Visual Execution Monitor, which displays the changes of general status for each object in real time. The status of an object includes its name and current mode (*dormant*, *wait* or *active*). A programmer can designate additional information for each object to be displayed on Visual Execution Monitor. In the word object example, the spelling remembered by an object and its current mode can be seen on the display (Figure 3). The state variable *entered-word* of the word object, which holds the spelling of the entered word, is designated by the programmer.

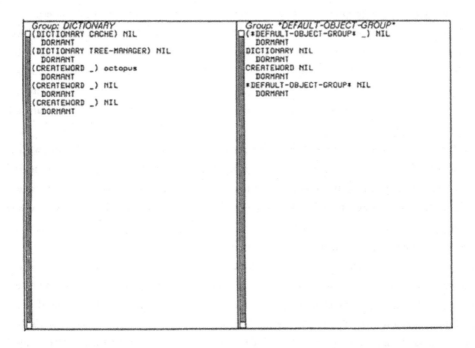

Figure 3: Visual Execution Monitor 1

Visual Execution Monitor uses Object Groups as the strategy of placing objects on the window. All the members of an object group is placed on a particular region of the window, so the dynamic changes of members during the execution such as addition and deletion can also be observed in real time. The window of the Visual Execution Monitor can be divided into more than two sub-windows to display many Object Groups simultaneously.

The left sub-window of Figure 3 shows the status of each object in the tree structured dictionary group after a word "octopus" is entered. The label displayed on top of the sub-window

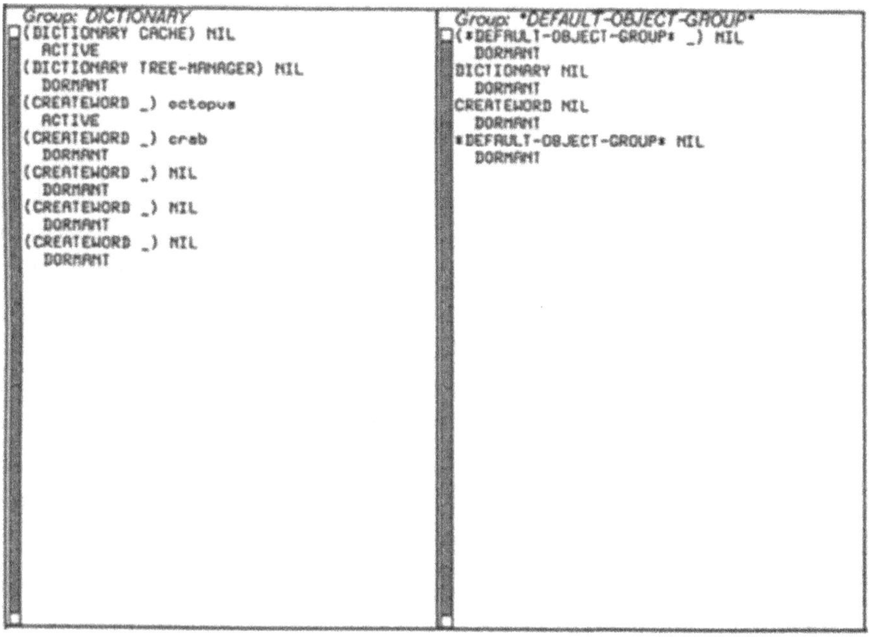

Figure 4: Visual Execution Monitor 2

shows the name of the group. Each object's information consists of two lines. The first two lines indicate the information about the cache object: the user designated information is nil (in fact, nothing is designated) and the current mode is *dormant*. The next two lines are the tree-manager's information. The information about a word object which remembers the word "octopus" is shown in the next two lines. There are two word objects which no words are entered. On Figure 4 is displayed the status of each object at a moment of carrying out the *:search* task. The cache and the root node (which corresponds to the word "octopus") are active at this moment.

J.Joyce et.al employed a process grouping technique to manage their graphical state display system called Mona [Joyce et.al 87], which is a part of their distributed system monitor. Processes are graphically displayed on a screen and, when a process changes its running state, the display is updated. A process group is created by pointing processes displayed on the screen using mouse. After a group is created, it can be repositioned, or can be incorporated into other groups as if it were an indivisible unit.

In Mona, processes are grouped by the instruction from the user monitoring system execution. In our scheme, Object Groups are specified by the program text. J. Joyce et.al use their process grouping only to improve the display, while we use Object Groups not only to improve the display but also to extract debugging information (See Section 4.3).

5 Other ABCL/1 Debugging Tools

Our simplified computation model of ABCL/1 is based on both sequentiality within an object and parallelism among objects. Thus debugging a program written in ABCL/1 requires both a scheme suitable for parallelism and that for sequentiality. We have already discussed the former one in the preceding sections. To deal with sequentiality, the current ABCL/1 system has three major tools. They are based on the ideas of traditional debugging tools.

The ABCL/1 Inspector can be used to examine the general information about an object such as the value of state variables and the local history. Objects relevant to each event are recorded in the local history. In Figure 5, the local history of the cache object is displayed in the upper sub-window. In the lower left sub-window, variable values are displayed. All the message patterns acceptable to the cache object is displayed in the lower right sub-window.

The traditional trace and break-point facilities are also provided. The trace facility simply prints the event information when an event takes place. The trace-switch can be set on selected objects or every object involved in the system so that the programmer can selectively trace objects.

The break-point facility stops the program execution when a [break-point] form is executed in an object. The object which causes the break can be examined by using the break-point top-level commands or by invoking the ABCL/1 Inspector.

6 Discussions

We have introduced the notion of Object Groups and incorporated its language constructs into ABCL/1. Its application to the debugging scheme for concurrent systems have been presented. The Object Group Examiner was presented as a tool for observing Object Group's Tasks. Also the Visual Execution Monitor was presented which can be used to observe the dynamic changes of object groups and the status of component objects in real time.

The Object Group Examiner and the Visual Execution Monitor have been incorporated into our ABCL/1 environment and we have used these tools to debug ABCL/1 programs. One of interesting example programs is a context free grammar parallel parsing algorithm proposed by Yonezawa and Ohsawa [Yonezawa 88]. Our debugging tools are very useful to debug and monitor the program itself, and the grammar rules written for the parallel parser. Also in demonstrating the parallel parser to the people who have no idea of the algorithm, these tools are very helpful because they clearly display the cooperative behavior of objects in a very structured way.

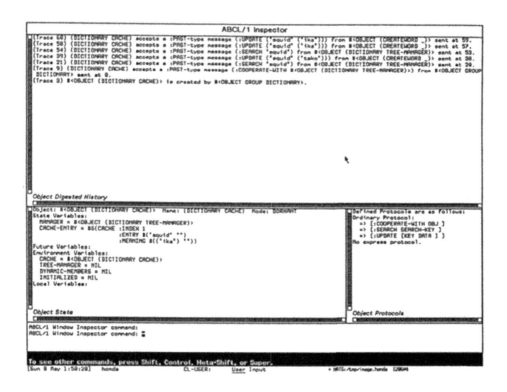

Figure 5: ABCL/1 Inspector: The cache is inspected.

Complementarily to these tools, we feel that tools are needed to present the overall behavior of a system (which might consist of multiple Object Groups). Such tools can be used to determine which Object Group's Task contains bugs. Note that the objective of these tools is the same as that of the behavioral abstraction approach given in [Bates and Wileden 83]. Bates and Wilden define an event description language. Using this language, they describe abstract events in a bottom up manner. In contrast, Object Groups' Tasks, which correspond to abstract events, can be defined more naturally by Object Groups.

The notion of Object Groups can be applied to areas other than debugging. For instance, it can be expected that objects included in an object group are so tightly connected that they communicate with each other more frequently than with objects outside the object group. This means that, in a distributed environment, to put all the members in an object group on a single processor may reduce the inter-processor communication which often degrades system performance.

We do not claim that the current design of Object Group is complete. It imposes several restrictions. For example, the current design does not allow the hierarchy of object groups. Every object must belong to only one object group. Furthermore, a member that ought to be accessed by *group-member* function must be a fixed member. The membership of an object and its access method is tightly connected. In the future design, these points need to be improved to provide more flexible and useful features.

7 Acknowledgment

We would like to appreciate various comments made by other members of the ABCL Project, E.Shibayama, T.Takada, and K.Sano.

References

[Bates and Wileden 83] P.C.Bates and J.C.Wileden.: High-Level Debugging of Distributed Systems: The Behavioral Abstraction Approach, *The Journal of Systems and Software 3*, 1983.

[Honda 86] Y.Honda and A.Yonezawa.: A Debugging Scheme for an Object-Oriented Concurrent Language, (in Japanese) *Proceedings of 3rd Annual JSSST Conference*, 1986.

[Harter et.al 85] P.K.Harter,Jr., D.M.Heimbigner and R.King.: IDD: An Interactive Distributed Debugger, *Proceedings of International Conf. on Distributed Computing Systems*, 1985.

[Joyce et.al 87] J.Joyce, G.Lomow, K.Slind, and B.Unger: Monitoring Distributed Systems, *ACM Trans. on Computer Systems*, Vol.5, No.2, 1987

[Manning 87] C.R.Manning: Traveler: An Apiary Observatory, *Proceedings of European Conf. on Object Oriented Programming*, 1987.

[Yonezawa 86] A.Yonezawa, J-P. Briot, and E. Shibayama: Object-Oriented Concurrent Programming in ABCL/1, *Proceedings of Object-Oriented Programming System, Languages and Applications*, 1986.

[Yonezawa 87] A.Yonezawa and M.Tokoro (Eds): *Object-Oriented Concurrent Programming*, The MIT Press, 1987.

[Yonezawa 88] A.Yonezawa and I.Ohsawa: Object-Oriented Parallel Parsing for Context-Free Grammars, *Proceedings of International Conf. on Computational Linguistics*, Budapest, August, 1988.

Fitting Round Objects
Into Square Databases

D.C. Tsichritzis
O.M. Nierstrasz

University of Geneva
Centre Universitaire d'Informatique
12 Rue du Lac, CH-1207 Geneva, Switzerland
UUCP: mcvax!cernvax!cui!{dt,oscar}
BITNET/EARN: oscar@cgeuge51

Abstract

Object-oriented systems could use much of the functionality of database systems to
manage their objects. Persistence, object identity, storage management, distribution
and concurrency control are some of the things that database systems traditionally
handle well. Unfortunately there is a fundamental difference in philosophy between
the object-oriented and database approaches, namely that of *object independence*
versus *data independence*. We discuss the ways in which this difference in outlook
manifests itself, and we consider the possibilities for resolving the two views, including the current work on object-oriented databases. We conclude by proposing
an approach to co-existence that blurs the boundary between the object-oriented
execution environment and the database.

1 Introduction

Consider an environment for running object-oriented applications. There are a number of *object
management* issues that should ideally be directly supported by the environment: persistence,
creation and destruction of objects, concurrency control, management of object identifiers, binding
between object instances and classes, etc. We will call the part of the environment responsible for
these issues the *object manager*. Some of these issues are operating system issues (e.g., scheduling
of activities), but many are inherently database issues. It is therefore quite natural to propose
that the object manager use a database system for dealing with database issues. An integration of
object-oriented and database systems can also be attractive for several other reasons. First, it may
provide a way for interfacing to existing databases. Next, we may get some additional database
functionality available to our objects, namely querying and transaction support. Finally, it may
be that certain object-oriented ideas will be useful for organizing databases and making databases
easier to use.

In fact, these two worlds are not so easily merged. We present the problem graphically in figure
1. Objects as they exist in object-oriented systems are depicted as round. Database instances, on
the other hand, are square. We use two different shapes to represent the fact that very different
properties of "objects" are emphasized by the two approaches. Furthermore, the object-oriented

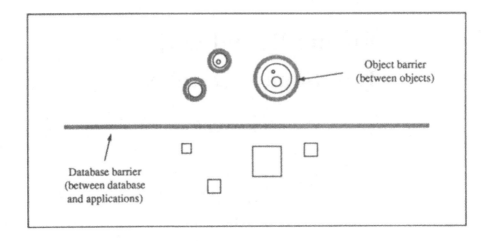

Figure 1: Object Independence vs. Data Independence

approach emphasizes what we call *object independence*, represented as an encapsulation barrier between objects, whereas the database approach emphasizes *data independence*, namely a barrier between the database and the applications. The object manager needs to provide persistence, and possibly other database functionality, for its round objects. The database system provides persistence, among other facilities, for square objects. The issue is simple. To what extent can the object manager use square database facilities to manage its round objects?

The paper consists of two parts. First, we shall discuss the main difference in outlook between object-oriented systems and database systems in terms of object independence versus data independence, and we shall see a number of specific ways in which this difference manifests itself. Second, we will illustrate the different approaches that can be taken for enabling object-oriented and database systems to co-exist. We will conclude with a proposal for merging the two by factoring out the common functionality, and permitting the two systems to each manage their own objects according to the appropriate paradigm.

2 Object Independence vs. Data Independence

Before we examine the differences between object-oriented and database approaches in detail, it is instructive to contrast their basic principles.

Object-oriented systems emphasize *object independence* by encapsulation of individual objects. Objects' contents and the implementation of their operations (i.e., their methods) are hidden from other objects. Interaction with objects is through a well-defined interface, illustrated by the paradigm of communication via message-passing. Object independence can be viewed as fundamental to *all* object-oriented concepts, including, for example, all forms of inheritance [Nierstrasz 1988]. We depict object independence as a boundary around objects, as in figure 1.

Database systems, on the other hand, emphasize *data independence* by separating the world into two independent parts, namely the data and the applications operating on them. The independence boundary is between the database and the rest of the world. This separation serves many purposes, but the most important effect is that the responsibility of the database system is well-defined, and consequently the database interface can be relatively simple.

These two principles constitute a major commitment in each field, and lead to a major cultural difference. (This has also been pointed out in [Bloom and Zdonik 1987].) Object independence is fundamental to object-oriented systems, just as data independence is fundamental to database systems. If database techniques are to be relevant to object management, we must ask whether these two principles can co-exist, and, if so, how?

Consider, for example, the design of a large application. If we believe in data independence then we must split our application into persistent, uninterpreted data, and programs that manipulate and share them. An object-oriented approach, however, would lead to a design composed entirely of objects, some of which are persistent. Manipulation of persistent objects is via their interfaces, and sharability is implicit in the message-passing paradigm. It seems clear that the two approaches lead to very different designs.

Can the two paradigms be reconciled? Can we support both data independence and object independence for the same set of objects, or must we always put our objects into the one world or the other? Maybe some objects, and therefore applications, can be handled more effectively in one way or the other. Maybe objects at some level (round objects) must be handled in one way and other objects (square objects) at a different level must be handled in a different way. Maybe the same objects can be either round or square depending on the context or the operations performed on them.

In order to develop some intuition to answer these questions, we shall try to evaluate precisely how deep this cultural difference runs. Briefly, we shall look at the following issues:

1. Are classes objects?

2. What relationships may exist between classes?

3. Should navigation be supported?

4. What operations exist on objects?

5. How are objects identified?

6. How are objects selected?

7. What is the role of classification?

8. Can object classes evolve?

9. Should the network be visible in a distributed system?

10. How are active/passive objects handled?

11. Is the object world closed (complete) or open?

2.1 Instance/class separation

Traditionally databases make a very strong distinction between instances and classes. Instances are in the database, whereas class information, i.e., the schema, is stored in the data dictionary. Many database systems have two different languages to deal with instances and classes. The Data Manipulation Language (DML) deals with operations on instances. The Data Definition Language (DDL) deals with operations, mainly creation, on classes.

It is obvious that object-oriented systems need to manage both instances and classes. In some object-oriented languages, notably Smalltalk-80 [Goldberg and Robson 1983], classes are themselves genuine objects, and can be manipulated as such. With the emergence of relational systems the DML-DDL separation was blurred, at least conceptually. After all, data dictionary tables could be viewed as relations and they could be manipulated with relational operations. However, most database systems retain a strong separation between schema and database.

The evolution from databases to knowledge bases forced a reconsideration of the instance/class separation. In conceptual models for knowledge bases, classes and instances are dealt with together and operations on classes are allowed [Mylopoulos and Levesque 1983; Brachman 1988].

The database research community has already accepted that classes can be manipulated, and that they can be structured with PART-OF and IS-A relationships, etc. Existing commercial database systems do not provide such facilities. They do provide, however, extensive facilities for class definitions in the database dictionary. It is conceivable that these facilities can be made available, and integrated as database operations. However, in doing so database systems will lose some of the simple user interfaces. The great advantage of relational systems is based on the relative few, very basic and very clear operations.

2.2 Relationships between classes

The relationships supported between database classes, whether they be relations or record types, etc., are quite restricted. They may be statically defined between classes, as in entity-relationship schemas. Relational systems, on the other hand, allow many relationships, but they are completely syntactic, based on contents and operations like joins.

In object-oriented systems we need the ability to deal with many relationships, some defined only at the instance level. Objects which know each other, or communicate with each other, are somehow related. It is not easy to model such relationships in databases. In many cases the relationships are explicit and not implied by the object's contents. Explicit relationships of this sort were forbidden in relational systems to emphasize relationships in terms of contents. Contents, however, are generally hidden in objects due to encapsulation.

Research in object-oriented databases tries to deal with that problem. Unfortunately relational systems, on the other hand, which are by now well-established, do not seem to offer the appropriate capabilities. Mapping object relationships into relational tuples or joins is not an easy task. On the other hand, going back to "information-bearing" relationships between database instances is considered a step backward.

2.3 Navigation

After many years of debate, database systems have de-emphasized point-to-point navigation between instances in the database. Such a facility was present in some older systems (notably, network and hierarchical databases) but it is now considered at worst harmful and at best old-fashioned. Many database specialists believe that reintroducing navigation in databases would be a step backward.

It is not easy, however, to see how else one should access objects with many instance-to-instance relationships when they are stored in a database. Relational systems give very powerful set-oriented operations but they are not appropriate for navigating from one object instance to another.

Fortunately people working in object-oriented databases are aware of the problem and they will probably come up with a solution. They will probably either have to utilize the relational interface in some innovative way (!) or they will have to adopt a different functional data manipulation language (e.g., Daplex).

2.4 Operations on objects

Database systems traditionally provide very few generalized types (i.e., record types, relations, etc.). As a result they can provide a small number of very general operations for queries and updates on the database objects. The operations are the same regardless of the semantics of the object involved. Queries and updates on employees, cars, accounts etc. utilize the same operations. In addition, the operations are simple. More sophisticated operations are encapsulated in transactions, which are treated separately from the database objects.

Object-oriented systems require that all objects provide their own set of operations, with some sharing through object classes and inheritance mechanisms. In addition, the methods can be logically complex. Most of the work in object-oriented databases deals with extending database operations to accommodate particular object types [Bancilhon 1988]. The extensions take two forms. First, complex objects can be defined, thus dealing with structural complexity within objects. Second, operations specific to object classes can be defined. Multiple inheritance can be used to define new classes that share operations and attributes (i.e., visible instance variables) with existing classes.

2.5 Object identifiers

Database systems utilize object identifiers internally for implementation purposes. These identifiers used to be visible and available for manipulation by the user in older database systems. The identifiers had a connotation of physical location, and they were used to physically place the database object in the system files. For this reason, they were considered harmful, and they were therefore removed from the database interface. In the relational model, and in some relational systems, tuples do not have a visible identifier. They are identified by their contents, via primary or secondary keys.

In object-oriented systems object identifiers are very important for two reasons. First, identifiers provide a permanent handle for objects that may evolve or move, in much the same way that file names hide the fact that a file's contents and physical location may change. Second, if an object's contents are properly encapsulated, they cannot be expected to provide a means for identification.

We need, therefore, to reintroduce identifiers into databases. These identifiers should be purely for identification purposes and should not, of course, be related to the physical location of objects in the database. In addition, allowable operations on identifiers should be strictly limited, since they cannot be treated in the same way as other attributes of objects. Although there is some reluctance in the database area for introducing identifiers, it should not be very difficult, at least conceptually. For particular systems, identifiers which were always present should become available and visible in some form through the database interface.

2.6 Content addressability

Databases traditionally provide operations based on selection by contents. This is especially true in relational systems, where all relationships between entities are represented by contents, and all operations are based on contents.

In object-oriented systems object contents are typically encapsulated, i.e., hidden. We are not supposed to know the values of an object's variables. Even when objects advertise visible attributes, we may not know whether they are "real" attributes, or virtual attributes computed by the object upon request. This situation presents a double dilemma. First, how can we take advantage of existing indexing mechanisms and content-oriented selection in database systems for object selection? Second, what mechanisms are appropriate for object selection in an object-oriented system?

In the first case we should try to represent some of the behaviour and the characteristics of objects in terms of attribute values visible to the other objects and to the object manager. Such external attributes play the same role as keywords for text retrieval. They are supposed to be representative for retrieval purposes. The problem is that these attributes may not ideally capture the information we need to select the objects. The work on databases for multimedia documents points out some of the problems and solutions.

In the second case we need other mechanisms to select objects which are not based on contents. Since objects encapsulate behaviour, they should also be selectable in terms of their behavioural aspects. Unfortunately databases offer very poor facilities for such selection. Very few database systems offer even simple facilities such as, "Get me the last updated record." In object-oriented systems we need to select objects in terms of where they have been, what operations they have launched on other objects, and what operations they have performed for others. What other objects a particular object knows, or has previously communicated with, may also be relevant. There is a need for behavioural selection methods and their implementation on traditional database content selection.

We should also mention that, since the selectivity for an object oriented system is not particularly high, we have to accept that browsing facilities become very important. Ideas from multimedia document browsing can be used and extended to browse through objects in an object-oriented system. This implies that we need to represent both the objects and their relationships in ways that reflect the user's model of what these objects do.

2.7 Classification

Databases traditionally have very few classes, with large numbers of instances per class. Classification in databases serves mainly to provide a means to efficiently manage and present large amounts of highly regular data. The differentiation between entities is represented by attribute contents and not by subdividing or creating extra classes. Classification in object-oriented systems serves a very different function, namely to support instantiation, encapsulation and class inheritance.

It is debatable whether object-oriented systems, or, for that matter, the real world, can be classified to such an extent. Databases were able to exploit rampant classification because they left the interpretation of classes to the manipulation programs and transactions. For certain object-oriented systems it may be difficult to force such extensive classification. It is not uncommon to find object-oriented applications with many objects that are the sole instance of their object class.

On the other hand, if instances are only one or two orders of magnitude more numerous than classes then databases have a difficulty dealing with them. There will probably be a need to treat object classes as instances as far as the database is concerned. In any case we will need a mapping between object classes and instances to database classes and instances.

2.8 Schema evolution

Traditional databases allow very little flexibility for evolution of their classes. Schema evolution is very restricted. Relational systems are better than other systems in that they sometimes permit adding attributes. However, dropping attributes or moving them to other relations is seldom permitted.

In object-oriented systems object classes should be able to change to accommodate software evolution [Skarra and Zdonik 1987]. If object classes correspond to database classes the different approaches will certainly create problems. We are again tempted to propose that object classes should be treated as database instances. This differentiation should not be a surprise to database people. After all, logical database instances are not always in correspondence with physical database instances. If object instances and classes are mapped to database instances, this will facilitate object instances changing their class. Since both the object instance and the two object classes are all treated as database instances, there should be fewer problems in representation. The problem of maintaining database consistency in the face of schema evolution has already been addressed in the object-oriented database field [Banerjee et al. 1987b].

2.9 Distribution

Traditional databases deal with distribution, if at all, by hiding it. A distributed database is a logically integrated, physically distributed database. The network is not visible, and we seldom have a notion of context, either as a geographic location, i.e., a workstation, or as a logical context.

The physical notion of context was present before in databases as the notion of an area (i.e., physical volume area where data was stored) but it was taken out. The logical context does not exist except as a logical view, which implies a partition, and perhaps transformation, of the data.

Object-oriented systems need a strong definition of context. First, we believe that objects should be aware of where they are. Physical location in the network may affect their behaviour. Second, objects, or collection of objects, may encapsulate beliefs, and we therefore need a context to define a boundary. (See, for example, [Tsichritzis et al. 1987].) Third, objects' behaviour may be affected by their context. Sometimes they should even directly inherit methods from their context. A simple example is a *text* object that inherits formatting characteristics from its surrounding scope. Finally, it seems extremely difficult to implement an environment of globally coordinated object managers where objects are managed by local object managers but in a completely integrated and transparent manner.

We are therefore faced with an interesting dilemma. On one hand distributed databases strive to provide a uniform globally integrated database. On the other hand object-oriented systems seem to require a strong notion of context. To what extent the two can co-exist depends a lot on how objects are mapped into databases.

2.10 Passive/active objects

Databases have always fundamentally viewed data objects as being passive. Operations performed on the database are issued from outside, and cause the database to be modified from one consistent state to the next. There is no real notion of the contents of the database being potentially active in the same way that processes managed by an operating system are active.

In certain cases databases have been extended with automatic triggers, but in an ad hoc manner. Triggers are low level alarm facilities for handling exceptions or for chaining operations together. Transactions encapsulate activities, but they are treated separately. Most data models and many systems completely separate the data and the transactions on them.

One of the most interesting aspects of objects is that they can be viewed as active agents [Agha 1986; Nierstrasz 1987]. They not only respond to requests from other objects, but they can trigger themselves. We do not separate objects into active and passive except in degree of activity.

Extensions of databases into object-oriented databases do not handle active objects. Other extensions in terms of automatic procedures and office tasks offer better support for activities [Zloof 1981]. We need, however, a general model of active and passive objects. Databases, as they are today, can only treat objects as passive entities.

2.11 Closed world assumption

Traditional database systems implicitly make the assumption that all information not in the database is not true. This assumption has been attacked for quite some time now by researchers who have extended databases with an inference engine. There is a solid basis for combining logic and databases, and for introducing support for recursive queries and inference.

This issue may seem irrelevant for object-oriented systems but it is important. Objects may incorporate rules. They should be able to augment the knowledge they have using inference. Since objects provide a clear context for inference, the knowledge they manipulate can represent a belief local to the object. We therefore have two effects. First, inference augments the information present in the database or in the objects. Second, objects can give us a context for partitioning knowledge into independent and potentially inconsistent beliefs.

2.12 Overloading, message passing, etc.

There are a host of other issues that seem, at first glance, to be treated differently by databases and object-oriented systems. After some reflection, however, we see that these issues have never been handled by databases and probably do not need to be.

As an example, operator overloading was not needed since databases did not deal with abstract data types, and their operations were very simple. As another example, message passing is a very appropriate paradigm for communication between objects, yet it is absent in databases. Since databases assume a passive view of data, a shared memory model is far more appropriate than a message-passing model. On the other hand, the database itself can be viewed as a large object with which one communicates through an abstract interface (i.e., via "message passing").

Finally, object-oriented systems emphasize reusability. So do databases, in an extreme manner. Passive objects are shared, and transaction definitions can be reused. Whatever notions the databases capture can be reused. Unfortunately they do not capture very many.

3 Co-existence Approaches

Now that we have examined in detail many of the apparent incompatibilities between object-oriented systems and database systems, we shall consider some of the ways in which the two viewpoints can be reconciled in order to provide some database support for objects, and possibly to provide some object-oriented functionality for databases. The possibilities we shall consider are summarized as follows:

1. Provide an interface between independent object-oriented and database systems.

2. Transparently store dormant objects in a database.

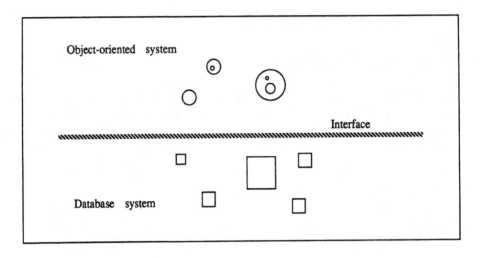

Figure 2: Separate Object-Oriented and Database Systems

3. Add object-oriented functionality to a database.

4. Add attributes and database functionality to an object-oriented system.

5. Integrate the common functionality required by both systems, and provide additional support for querying, transactions etc., for "database objects".

3.1 Separate co-existence

The simplest approach is depicted in figure 2. The application is divided into two clear parts: programs and data. Programs are encapsulated as objects using the object-oriented system. Data are managed by the database. Objects can issue direct database commands to the database. The object manager can also use the database for its own needs. There is a very clear distinction between objects and database instances.

Persistence for objects is provided directly by the object manager through, for example, a large, persistent virtual memory. Support for object creation, destruction, concurrency, etc., is dealt with by the object manager independently of the database system. This is analogous to the traditional distinction between operating systems and database systems.

This approach does not require major changes to either database systems or object-oriented systems. In addition, it allows existing applications implemented in a traditional manner to co-exist with object-oriented systems. Furthermore, the principles of object independence and data independence are generally reconciled. There are, however, two ways for objects to communicate with one another, either directly by message-passing, or indirectly through the database. If there is a need for objects to encode part of their contents in the database, then object independence can be compromised through sharing of database instances.

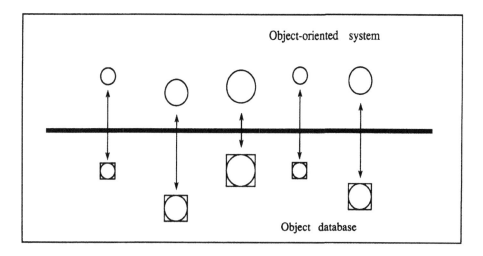

Figure 3: Active and Dormant Objects

There are two other drawbacks to this approach. First, there is a lack of uniformity. Some application entities become objects, others become database instances. Second, there will be some duplication of effort. The object manager will require some mechanisms similar to those used by the database system.

3.2   Active and dormant objects

A second approach is depicted in figure 3. Objects can be active or dormant. Active objects exist in a large, persistent workspace managed by the object manager. Objects become "dormant" either by putting themselves asleep, or by being retired by other privileged objects. Various policies can be used for deciding when to retire objects. The problem is similar to that of managing a virtual memory. When objects are dormant they are labeled using their identifier, and perhaps some other attributes, and they are stored in the database. As far as objects are concerned, there is no such thing as a database.

Dormant objects can be re-awakened by the object manager, for example, when a message is sent to them. Re-activated objects have precisely the same state as they did when they were put to sleep. Creation of a new object implies initialization of variables and a new identifier. Re-activation implies the old identifier and the same variable contents as at the time the object died.

This approach has the advantage that it is completely object-oriented. The database does not exist except as a facility for storing dormant objects. This allows objects to maintain the illusion of a purely object-oriented environment where object independence is emphasized. On the other hand, it does not address the problem of access to existing databases, nor does it provide a means to exploit other database functionality from within the object-oriented system. The database is used like a file system. Querying and transaction management are unavailable except to the object manager.

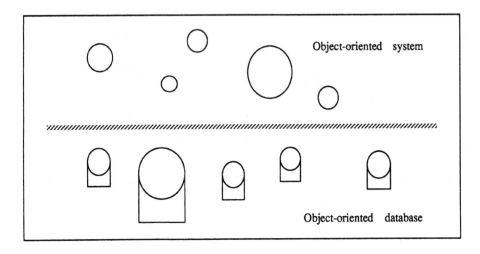

Figure 4: Object-Oriented Databases

3.3 Object-oriented databases

Object-oriented databases are depicted in figure 4. This term is *not* used to mean "databases for supporting object-oriented systems". It refers, rather, to databases that have been extended by incorporating various object-oriented concepts, i.e., abstract data types, complex objects, multiple inheritance, etc. [Bancilhon 1988; Maier and Stein 1987; Banerjee et al. 1987a; Fishman et al. 1987].

Reconciling the independence principles poses no problem. Database "objects" are clearly separated from the rest of the objects, thus providing data independence. The other objects emphasize object independence between themselves.

Due to the inclusion of object-oriented features, the interface to object-oriented databases resembles that of object-oriented systems. Interaction with database objects is similar to communication with other objects. The object manager is relieved of the responsibility of managing the vast collection of relatively docile and well-structured objects. It should be able to do a better job of catering to the needs of the other objects. Furthermore, by extending database systems with object-oriented concepts, we greatly enhance the functionality and usability of databases (though object-oriented databases will not have the simple set of generic operations that current databases have).

There is, nevertheless, a clear distinction between database instances as objects and the other objects. First of all, it is not clear that the particular object model adopted by the object-oriented system will match that of the object-oriented database, unless they have been designed with that purpose in mind. Next, object-oriented databases have been developed to support data-intensive applications, like CAD/CAM, that have to deal with abstract "objects", not to support object-oriented applications in general. As a consequence, there is an emphasis on large numbers of similar,

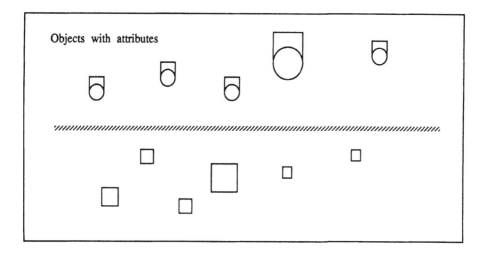

Figure 5: External attributes

well-structured objects with external attributes. It is not clear that object-oriented databases have much to offer for objects that do not satisfy these assumptions. Finally, it is not clear how to use object-oriented databases to store active objects. The execution state of an object involved in a long-term activity is not something that databases are normally expected to deal with.

3.4   Objects with external attributes

This approach is depicted in figure 5. In a large, object-oriented system, there can be a severe selection problem. (This is analogous to the problem of posing an ad hoc query on a file system.) One approach for dealing with this problem is to abstract an object's behaviour and properties in terms of a number of external attributes. These attributes, possibly including text descriptions, represent the salient properties of the object to the outside world. An object can be selected by posing a query in terms of the values of these attributes. The attribute values may be set by the object itself, or, in some cases, by the object manager (e.g., the last update time). We can view this approach as providing another (square) database-oriented interface to objects.

If we follow this approach, it would be reasonable to use the same attributes to store and retrieve the objects. All communication is addressed via the object identifier or the values of the external attributes. The object manager can therefore use these attributes to store objects in a database, and to re-activate them when required.

This approach deals well with both the addressing and the persistence problems. Existing database instances can be treated as special objects that have only external attributes. The disadvantage is that objects cannot be purely encapsulated. We introduce a notion of visible attributes. Objects are accessed not only through their identifiers but through the values of the attributes. Furthermore, it does not really solve the problem of object selection, since static attributes cannot fully capture all of the interesting dynamic properties of objects. How do you

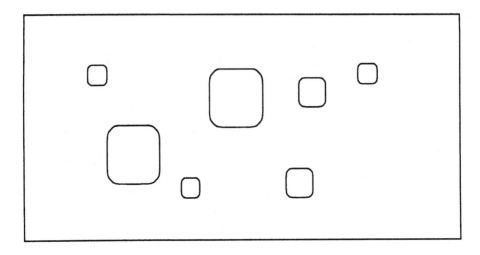

Figure 6: "Plastic" objects

decide what visible attributes your objects should have? Nevertheless, this is a nice compromise between the object-oriented and database paradigms.

3.5 "Plastic" objects

What we would really like is to merge the object-oriented and database approaches. "Plastic" objects would have it all (figure 6). They would behave like objects, obeying object independence, but they could also look like database objects, if necessary. They would be round and square at the same time!

The problem with all of the approaches we have described up till now is that they do not successfully break through the database barrier. The object manager can use some database support for nearly all objects, namely for persistence and for maintaining object identifiers. Other issues, dealing with querying, indexing, efficient storage and transaction management, may be relevant only for a large number of well-structured objects belonging to few object classes. Highly active objects can be handled in more efficient ways than by keeping track of all their changes in a database.

What we propose as a solution is to factor out the common database and object management support into a shared subsystem that is not seen by objects. Highly active objects can be stored and managed using the low-level object management subsystem. More passive objects with visible attributes exploit the same object management subsystem, but are provided with additional support for querying, etc. Given support for schema evolution, it is even possible for objects to migrate between the two parts of the object environment, either becoming more like database objects, or becoming more active. For example, redefining an object class to provide visible attributes for its instances will cause those objects to migrate to the database part of the object environment.

This approach would also accommodate existing databases or object-oriented databases in the same way as we have described above. Since external databases constitute independent object worlds, all that is required is an interface for communicating with such objects.

The advantage of this approach is that it is completely object-oriented, retaining all the claimed advantages of object-orientation. The obvious disadvantage is that the object manager is asked to solve many difficult problems. However, the real difficulty is that the application designer will be offered a choice within the same system. Applications can be designed in such a way as to emphasize a separation of activities and data, or, alternatively, they can be designed in a completely integrated manner. Sometimes the choice will be obvious. In some cases it may be very difficult.

4   Conclusions

We have discussed the difficult issues arising from a co-existence of object-oriented and database systems. Some of these issues are a natural follow-up to differentiation of approaches between programming languages and database systems. Programming languages emphasize operations while databases emphasize information representation.

There are many possible attitudes we can take.

1. The object-oriented and database fields can ignore each other. This will not be realistic and fortunately is not happening.

2. The object-oriented and database fields can develop, but each using the facilities of the other in a decoupled way.

3. The object-oriented and database fields can borrow concepts from each other, each trying to duplicate the other's efforts.

4. The object-oriented and database fields can try to merge their capabilities to arrive at systems which can smoothly integrate the facilities for both, without prohibiting either a purely database-oriented approach or a purely object-oriented approach to the problem.

We obviously prefer the fourth solution, and we believe that it is a promising direction to pursue. This presupposes, however, an open approach, accepting ideas from a different area, and respecting the limitations and constraints that each area poses. The limitations and constraints of object-oriented systems and database systems did not arise through chance or oversight. They arose because other principles and other ideas were heavily emphasized.

Finally, we should note that a discussion on whether databases should be extended, or another field should redevelop its capabilities is not new to databases. For instance, the same difficulties arose with knowledge bases. Should knowledge bases be implemented on top of databases? Should databases be extended to incorporate knowledge base ideas? Or should knowledge bases provide their own storage and retrieval facilities? This controversy has not been settled completely and we are now embarking on similar discussions concerning object-oriented systems. In the end, it is not important whether object-oriented or database ideas prevail. Rather, it is important what facilities the final system offers.

References

[Agha 1986] G.A. Agha, *ACTORS: A Model of Concurrent Computation in Distributed Systems*, The MIT Press, Cambridge, Massachusetts, 1986.

[Bancilhon 1988] F. Bancilhon, "Object-Oriented Database Systems", Proceedings 7th ACM SIGART/SIGMOD/SIGACT Symposium on Principles of Database Systems, Austin, Texas, March 1988.

[Banerjee, et al. 1987a] J. Banerjee, H. Chou, J.F. Garza, W. Kim, D. Woelk, N. Ballou and H. Kim, "Data Model Issues for Object-Oriented Applications", ACM TOOIS, vol. 5, no. 1, pp. 3-26, Jan 1987.

[Banerjee, et al. 1987b] J. Banerjee, W. Kim, H-J Kim and H.F. Korth, "Semantics and Implementation of Schema Evolution in Object-Oriented Databases", Proceedings ACM SIGMOD '87, vol. 16, no. 3, pp. 311-322, Dec 1987.

[Bloom and Zdonik 1987] T. Bloom and S.B. Zdonik, "Issues in the Design of Object-Oriented Database Programming Languages", ACM SIGPLAN Notices, Proceedings OOPSLA '87, vol. 22, no. 12, pp. 441-451, Dec 1987.

[Brachman 1988] R.J. Brachman, "The Basics of Knowledge Representation and Reasoning", AT&T Technical Journal, vol. 67, no. 1, pp. 7-24, Jan/Feb 1988.

[Fishman, et al. 1987] D.H. Fishman, D. Beech, H.P. Cate, E.C. Chow, T. Connors, J.W. Davis, N. Derrett, C.G. Hoch, W. Kent, P. Lyngbaek, B. Mahbod, M.A. Neimat, T.A. Ryan and M.C. Shan, "Iris: An Object-Oriented Database Management System", ACM TOOIS, vol. 5, no. 1, pp. 48-69, Jan 1987.

[Goldberg and Robson 1983] A. Goldberg and D. Robson, *Smalltalk 80: the Language and its Implementation*, Addison-Wesley, May 1983.

[Maier and Stein 1987] D. Maier and J. Stein, "Development and Implementation of an Object-Oriented DBMS", in *Research Directions in Object-Oriented Programming*, ed. B. Shriver, P. Wegner, pp. 355-392, The MIT Press, Cambridge, Massachusetts, 1987.

[Mylopoulos and Levesque 1983] J. Mylopoulos and H. Levesque, "An Overview of Knowledge Representation", in *On Conceptual Modelling: Perspectives from Artificial Intelligence, Databases and Programming Languages*, ed. M. Brodie, J. Mylopoulos, pp. 3-17, Springer-Verlag, New York, 1983.

[Nierstrasz 1987] O.M. Nierstrasz, "Active Objects in Hybrid", ACM SIGPLAN Notices Proceedings OOPSLA '87, vol. 22, no. 12, pp. 243-253, Dec 1987.

[Nierstrasz 1988] O.M. Nierstrasz, "A Survey of Object-Oriented Concepts", in *Object-Oriented Concepts, Applications and Databases*, ed. W. Kim and F. Lochovsky, Addison-Wesley, 1988, (to appear).

[Skarra and Zdonik 1987] A.H. Skarra and S.B. Zdonik, "The Management of Changing Types in an Object-Oriented Database", in *Research Directions in Object-Oriented Programming*, ed. B. Shriver, P. Wegner, pp. 393-415, The MIT Press, Cambridge, Massachusetts, 1987.

[Tsichritzis, et al. 1987] D.C. Tsichritzis, E. Fiume, S. Gibbs and O.M. Nierstrasz, "KNOs: KNowledge Acquisition, Dissemination and Manipulation Objects", ACM TOOIS, vol. 5, no. 1, pp. 96-112, Jan 1987.

[Zloof 1981] M.M. Zloof, "QBE/OBE: A Language for Office and Business Automation", IEEE Computer 14, pp. 13-22, May 1981.

Database Concepts
Discussed in an Object Oriented Perspective

Yngve Lindsjørn
Norwegian Computing Center
P.O. Box 114 Blindern
N-0314 OSLO 3, Norway

Dag Sjøberg
The Central Bureau of Statistics
P.O. Box 8131 Dep.
N-0033 OSLO 1, Norway

Abstract

The terminology in the area of databases and data models is inconsistent and inaccurate and thus often confusing. Some fundamental database concepts are described in this paper. The description of these concepts are based on general concepts related to the development of the object oriented languages SIMULA, DELTA and BETA. A database is defined as an information system providing information about a referent system (the modelled part of the world), and a data model is defined as "having an inherent structure and a set of tools and techniques used in the process of designing, constructing and manipulating model systems (in particular, databases)". The connection between databases, DBMS and data models and other concepts related to the development and use of databases are described in a process/structure hierarchy.

1. Introduction

The term "object oriented" is derived from the SIMULA67 programming language [*Dahl, Myhrhaug, Nygaard, 68*]. Later, other object oriented programming languages appeared. In particular, Smalltalk [*Goldberg, Robson, 83*], normally associated with an (object oriented) graphical environment, has contributed to the popularity of the object oriented paradigm.

As will be illustrated in this paper, concepts associated with the object oriented paradigm are also suitable for describing databases. The paper aims at describing a *conceptual basis* for object oriented databases and environments where ideas from the object oriented programming style are integrated with facilities provided by modern database management systems (DBMS).

Most literature concerning object oriented databases follow the Smalltalk tradition (e.g. [*Copeland, Maier, 84*], [*Maier, Stein, 87*]). Some literature do not follow any specific tradition, but are influenced by several object oriented languages (e.g. [*Zdonik, Wegner, 86*], [*Skarra, Zdonik, 87*]). This paper follows the tradition of SIMULA, DELTA and BETA [*Dahl, Nygaard, 65*], [*Dahl, Myhrhaug, Nygaard, 68*], [*Delta,*

75], [BETA, 83a], [BETA, 83b], [BETA, 85], [BETA, 87], and the development of the basic concepts presented in Section 2 is related to the development of these languages.

Following this tradition, the *object oriented perspective* can briefly be interpreted as follows:

Information processes may be understood, organized, constructed and operated in terms of a system concept in the sense that the system is "the part of the world" that develops through the process. Program executions and operations of a database are examples of such processes. The substance of the process is organized as the components of the system, called objects. Any measurable property of the substance is a property of an object. Any transformation of state is regarded as action by objects.

Some of the concepts in this paper are more extensively discussed in [Lindsjørn, Sjøberg, 87].

2. Basic Concepts

Using concepts from the object oriented perspective, we present definitions of basic concepts which constitute a basis for specialized database concepts. Most of the concepts defined here are presented earlier, see e.g. [Nygaard, 86a], [Nygaard, 86b], [Delta, 75].

The first concept defined is *process*:

A **process** is a development of a part of the world through *state transitions* during a time interval.

An example of a process is the economic development of Norway after the Second World War. The concept process comprises its sequences of changing states.

A process is always restricted in its behaviour, and the *structure of a process* is defined as:

The **Structure** of a process is the limitation of its set of possible states and state transitions.

When regarding the economic development of Norway, the general economic policy of the Norwegian government, the laws, the total manpower, etc. are *structuring* the process.

In informatics, the processes dealt with are *information processes*:

A process is regarded as an **information process** when the qualities considered are:
- its *substance*, the physical matter that it transforms,
- *measurable properties* of its substance, represented by values,
- *transformations* of its substance and thus its measurable properties.

An essential concept in this paper is the *system* concept with its close relation to the object oriented perspective.

A **system** is a part of the world that a person (or group of persons, during some time interval and for some reason) chooses to regard as a whole, consisting of
- *components*, each component characterized by
- *properties* that are selected as being relevant and by
- *state transitions* relating to these properties and to other components and their properties.

According to this definition, no part of the world is a system as an inherent property. It is a system if we choose a system perspective.

In informatics, the systems dealt with are information systems:

An **information system** is a system where the components are organized as objects, and the properties of the objects are *attributes* that are either *quantities*, *references* or *categories*.

A **state** of an information system is expressed by describing
- the *moment* at which the state is recorded,
- its *objects*, the *states of attributes*, and *transitions* going on, all at that moment.

The transitions are actions that result in changes of state of the information system. The update of objects in a system (e.g. accounts in a bank system) is an example of a transition. The assignment   x+1 -> x   is another.

Properties of components are captured by attributes:

An **attribute** of some object is the association of the object and:
- a name,
- a set of elements (the domain),
- (at any given time) one of the elements of the set.

The attributes (of the objects) are divided into three kinds - quantities, references and categories.

A **quantity** is an attribute for which:
- a way of *measuring* a state of the attribute is defined,
- there exists a *mapping* from the set of measurements of every possible state to the elements of the attribute's associated set, called the *value set*,
- the associated element, called the attribute's *value*, corresponds to the measurement of the   property at the given moment.

A *quantity* is used to denote values of measurements and is either a *variable* or a *function*.

A **reference** is an attribute for which the associated set has objects as elements. The associated element is called the attribute's *referent*.

As it appears from the definition, the referent of a reference is the object referred to at a given time. The referent could be denoted (indicated) by a name or an address. However, we emphasize that the attribute does not have a "value" being the name or the address, but denotes the actual (referred) object - the referent. Just as a value can be assigned to a quantity-variable, an object can be assigned to a reference-variable.

A **category**[1] is an attribute for which the associated set consists of prescriptions for *common structure* of categories of objects.

Procedures and functions are examples of category-attributes.

---

[1] In BETA the concept *pattern* is used instead of *category*.

Analogous to the definition of structure of a process, we define the *structure of an information system*:

The **Structure** of an information system is the limitation of its set of possible states and state transitions.

The structure is given by the category descriptions of the objects and their attributes with associated sets.

2.1 Model System - Referent System

We understand the concept "model" as an analogue device - a model is *similar* to "something". We find another use of the term in the concept "data model". A data model specifies rules for modelling and manipulating objects in an information system.

We now define the concepts *model* and *referent*:

A phenomenon M is a **model** of a phenomenon R, according to some perspective P, if a person regards M and R as similar according to P. We call R the *referent phenomenon* (or simply *referent*) and M the *model phenomenon* (or simply *model*). We call P the perspective of the model M.

As mentioned, our object oriented perspective implies that we focus upon the *system* concept, and the phenomena we discuss in this paper are regarded as (information-) systems. When we choose to regard "a part of the world" as a system, we always produce another, structurally similar system - mental or manifest - as a vehicle for relating to that "part of the world". The "part of the world" system will be called the *referent system*, the mental or manifest system produced, will be called the *model system*.

An example:
A database, containing information about employees, departments, offices and items (and their selected attributes and constraints), is a *model system* of some part of an enterprise. The selected part of the enterprise is the *referent system* and the *database* constitutes the (manifest) model system.

2.2 Categories and Abstraction Mechanisms

One way of structuring and presenting the information contents of objects, their attributes and actions is through the use of abstraction. Abstraction provides the ability to hide details and concentrate on describing the common structure of similar phenomena. We briefly present the three commonly known abstraction mechanisms - *classification*, *generalization* and *aggregation*.

The category construct unifies the constructs (abstractions) class, type and procedure by means of a category description which prescribes common structure of a category of objects[2]. These objects are said to be *category specified* (as opposed to *singular objects* which are specified together with their descriptors - as for instance inner block instances in ALGOL).

---

[2] In this context, quantities and actions (procedure/ function activations) are regarded as objects *per se*.

Below we present three different kinds of classification:

1. A *class* is intended to abstract *substance* (the substance is focused). Substance is organized as objects, and when we talk about objects, we normally mean instances of classes.
2. A *type* is intended to abstract *measurable properties* (quantities) of substance (values are focused).
3. A *procedure* is intended to abstract *actions* (actions are focused).

We refer to these categories as *class-category* (we also use *object-category*), *type-category* and *procedure-category*, respectively. We will also introduce a fourth kind of category - the *relationship-category*.

Categories can be organized in a hierarchy. This is referred to as *generalization/specialization*. A category can be a generalization of two or more categories. Conversely, categories are specialized from other categories (this can be done successively). If category B is a specialization of category A, B *inherits* all attributes (and actions) of A.

Most object oriented programming languages support generalization/specialization for class-categories only. As in BETA, this abstraction mechanism should be supported in all three kinds of categories (class-category, type-category and procedure-category).

Aggregation is the abstraction mechanism by which a category can be constructed (aggregated) from other categories (a category can also be regarded as aggregation of its constituent attributes). In BETA (and SIMULA) aggregation can be done successively. In most data models aggregation can be done at one level only.

In practise, there is no clear distinction between the process of aggregation and the process of classification. Normally, aggregation takes place as part of the process of classification (or vice versa). For instance, when classifying persons into the category PERSON, the (aggregation-) process of deciding the relevant attributes of the persons takes place at the same time.

3. Database Concepts Described in an Object Oriented Perspective

A database is an information system intended to represent (model) some referent system (see Fig. 1), and this fact is reflected in our definition:

A **database** is
- an information system that
- provides information about the state of another system, the referent system, and the database is designed to be a model of (to represent) the referent system.
- The desired correspondence between actual *database states* and referent system states, at any given time, is obtained through a set of routines reflecting an *update strategy*, and by a set of *mappings* from the database, extracting database state information in the form of *maps*.

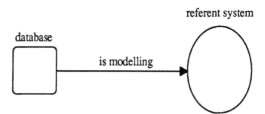

Fig. 1. A database is modelling a referent system.

An example:

A database of the employees in an enterprise is an information system used in personal administration. The enterprise and employees, with their attributes, constitute the referent system. In order to keep the database state in correspondence with the state of the referent system, a change in the referent system (e.g. a new person employed in the enterprise) causes a change in the database (a new employee-object inserted in the database). The change in the database is done in accordance with an update strategy (e.g. once a day, when "required"). Information about the employees (e.g. *name* and *salary*) could be presented as a table (the map) by a mapping from the actual database state.

The structure of the objects is specified by means of category descriptions in a *database schema*. The selected categories and their structure specifications reflect a *perspective* on the referent system (the model perspective).

3.1 Database State - Representative State

A **state of a database** is expressed by describing:
- The moment at which the state is recorded.
- The *objects* (e.g. employees, departments), the *relationships* (e.g. relationship between an employee and a department), the values of the *attributes* (e.g. *date of birth* of an employee, *name* of a department), and the ongoing *transformations* (actions on the objects - e.g. execution of update-procedures).

The information obtained by using a database does usually not include ongoing transformations. For instance, a database is *not intended to give* information about an ongoing move of an employee from one department in the enterprise to another - the purpose of the database is to give information about which department the employee actually belongs to (and perhaps former departments). This fact is reflected in our definition of *representative states* (see Fig. 2):

Representative states of a database are states concerning the *objects*, *relationships* and *attributes* that have counterparts in corresponding states in the associated *referent system*.

Relationships between objects are expressed either as references or as objects in their own right.
Working files, index tables, the execution of constraint checks, etc. are not part of the representative state.

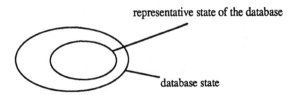

Fig. 2. Representative state of the database as part of the database state.

In a database system, the ongoing actions *per se* are not of interest from an end-user's point of view - it is the results of the actions that are of interest. End-users are interested in representative states *between* transitions.[3] A control system (e.g. a system monitoring a nuclear power plant or an auto-pilot system) is an example where information about ongoing actions is of interest. Similarly, information about ongoing actions of a simulated referent system is central when regarding a simulation system.

A special example of a description of a representative state of a database is a *back-up*. The back-up shows (is a picture of) a database in a certain (representative) state - it preserves that state. A back-up is not itself a database, but it can be used to initialize a database with the same structure as the database for which the back-up was made.

3.2 Update of the Database

In order to provide information about the actual state of the referent system, the database should (in theory) be updated at the same point of time as the changes occur in the referent system. In practise, however, there is often a period of time (a delay) between the change in the referent system and the corresponding update of the database. This phenomenon depends essentially upon the update strategy, which we divide into four main strategies:

1. The update of the database system is executed *before* the corresponding action in the referent system (e.g. room-reservation, flight-reservation).
2. When any change in the referent system occurs, a corresponding change in the database is made immediately (e.g. bank account system).
3. Batch update - the changes in the referent system that have occurred during a certain time interval (since last update), will be correspondingly changed in the database - all at the same time (once a day, once a week, etc.).
4. Update when "required" - the changes are made each time somebody needs the database to get information about the referent system. The time intervals between the updates according to this strategy are thus not equal.

In practise, the update procedure is often a combination of the strategies outlined here.

---

3 The ongoing actions are, of course, of interest when regarding *concurrency*. The interleaving of concurrent operations can produce incorrect (not representative) states of the database. In order to prevent this situation, the control mechanism known as *locking*, is introduced.

3.3 Mapping

In order to *present* the information of a database, we introduce the notions of *mappings* and *maps*. A map is a presentation of information and may be given in the forms of screen displays, statistical tables, graphical displays, etc. A map is produced as a result of a mapping from the *database space* to the *presentation space*. The database space is the space of all possible (representative) database states of a certain database, and the presentation space is the space of all possible presentations of a corresponding database.

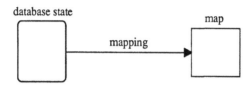

database state

map

mapping

Fig. 3. The mapping.

4. Data Model

The concept data model is frequently used within informatics - in data modelling in general and in the domain of databases in particular. In this domain, the development of the three classical data models for "database design" - the relational data model [*Codd, 70*], the network data model [*DBTG, 71*] and the hierarchical data model (e.g. [*Tsichritzis, Lochovsky, 76*]) - has taken place.

The model concept, as it is used in this context, does not correspond to our definition of the concept. From our model definition, a data model should be a *model* of a part of the world (the referent system). Rather, a data model is a collection of *principles and guidelines for the process of constructing and manipulating model systems* (e.g. databases), and a data model always reflects a perspective on this process. Instead of talking about different data models, one could thus talk about different perspectives - e.g. *the hierarchical perspective, the network perspective, the relational perspective.*

We define the concept as follows:

A **data model** is characterized by having an *inherent structure* and a set of *tools* and *techniques* used in the process of designing, constructing and manipulating model systems (e.g. databases). The data model reflects a *perspective* on this process.

A model system has a *structure*. Some part of the structure is given by general principles - *independent of the actual referent system being modelled.* We refer to this part of the structure as the *inherent structure*.[4] In fact, it would be more correct to say that a data model supports *general principles for imposing* the inherent

---

4 This concept is a modification of the concept *inherent constraint* used in e.g. [*Tsichritzis, Lochovsky, 82*].

structure of model systems. However, we find it more convenient (in accordance with common convention) to talk about *inherent structure of a data model* when discussing data models.

Examples of inherent structure of data models are the representation (structure) in the form of *hierarchies* in the hierarchical data model and in form of *networks* in the network data model. The selected categories and their attributes (including the constraints on these) give structure to a database which is *dependent of the actual referent system*, and such imposed structure is, of course, not based on principles of a data model.

Techniques of a data model could be modelling methodologies as diagrammatic techniques, techniques for schema design, query design, etc. Tools could be languages, concepts, constructs, abstraction mechanisms, automatic design aids, etc.

4.1 Database Language

The language used to describe the referent system and to describe and prescribe the structure and use of a database, we call a *database language* (DBL). A data definition language (DDL), a data manipulation language (DML) and a query language (QL) are subsets of a database language. We refer to DDL, DML and QL as *aspects* of the database language, though they today often are languages in their own right. They are used for defining, manipulating and querying databases. The distinction between the languages (aspects) is not clear. For example, should a declaration of a function be written in DDL or DML? This is one of the reasons why all these languages (aspects) should - in our opinion - be unified into one language[5].

The basic actions or operations (by means of predefined procedures or operators) that can be performed on a database, are:

retrieve (make available objects in the database)
insert (add new objects to the database)
delete (remove objects from the database)
update (change existing values of attributes in objects in the database)

Such operations transform a database state into another database state. If the operation is insert, delete or update (also referred to as *altering operations*), the operation also transforms the *representative state* of the database into another representative state.

The *structure of a database* is the limitations of its set of states and state transitions. The database states and the database state transitions are limited by two aspects of structure: the *permanent structure* and the *transient structure*.

The permanent structure is written in the DDL and DML aspects of the database language and is contained in *database schemas*. The DDL-described structure is the category descriptions of objects and their attributes together with the relationships between the objects. This structure is described in the *basic schema*

---

[5] SQL (Structured Query language), which has become an ANSI standard for the relational data model, is an example of such a unifying language.

(corresponding to the *conceptual schema* in the "ANSI/SPARC three schema architecture" [*ANSI/SPARC, 75*]). The DML-described structure is contained in *external schemas* (cf. ANSI/SPARC). The main purpose of the external schemas is to make it possible to produce maps adapted to special users (user groups). Another purpose is to protect parts of the database from different user groups. This concerns both updates and maps. A database has *one* associated basic schema, while there (normally) are *several* external schemas. When external schemas are made, the basic schema is used as basis.

The transient structure is provided by the QL aspect of the database language. A query is formed ad hoc - as opposed to a procedure (permanent structure). The QL may, however, specify the invocation of existing permanent structure, described for instance in the DML.

As mentioned, a *data model* is characterized by having an *inherent structure* and a set of *tools* and *techniques* used in the process of designing, constructing and manipulating databases. Another way to express the data model concept, is to regard a data model as a concise reflection of the *perspective* of one or more database languages. This perspective and the perspective reflected in the actual process of *using* the database language (and data model), are reflected in the modelled database:

If a referent system with a specific structure is modelled (e.g. an enterprise with hierarchical structure), it would be convenient, of course, to use a data model with an inherent structure adapted to the structure of the referent system (e.g. the hierarchical data model). In turn, when the data model is selected, a database language, among the set of associated database languages of this data model, must be chosen. The perspective reflected in the database is further specialized by the actual use of the database language (mainly by the DDL aspect) which might be restricted by e.g. economy and time limits.

4.2 Classification of Data Models

Data models can be classified into four categories (cf. [*Brodie, 84*]):

1. Primitive data models (traditional file systems, where objects are represented in records grouped into files, and the operations provided are primitive read and write operations on records).
2. Classical data models (the relational, the network and the hierarchical data model).
3. Semantic data models (data models designed to provide richer and more expressive concepts in which to capture more "meaning" than the classical data models).
4. Special purpose (application oriented) data models (semantic modelling theory applied to particular applications such as office and factory automation, VLSI, CAD/CAM, etc.).

Data models of category two and three concern databases, and they are also referred to as *database models* (e.g. [*Skarra, Zdonik, 87*]). There is, however, no clear distinction between the categories. For example, semantic data models can be used in the domain of databases, and they can be used in other domains (e.g. the *semantic network data model* in the domain of AI).

4.3 Concept Usage

Data models differ to a great extent in terminology. The domain of databases and data models would profit greatly by a standardization of concept usage. This problem concerns, in particular, the concepts *object* and *attribute* which have a lot of synonyms. The most common used synonyms are *record* and *field*. These concepts (and concepts *tuple* and *segment*) are - in our opinion - less general than the concepts object and attribute. *Record* and *field* are only suitable when considering *physical representation* - as opposed to when considering the referent system.

The concept usage in the three classical data models and the two semantic data models (the entity relationship model (ER) [*Chen, 76*], [*Chen, 77*] and the extended relational model RM/T [*Codd, 79*]) is presented in Fig. 4. The concept usage is compared with our proposed terms given in the leftmost column. For relations (associations) between objects we use the term *relationship* (as in ER). A relationship is classified into a *relationship category*.

4.4 A Comparison of some Data Models

We will now briefly compare the three classical data models and the two semantic data models ER and RM/T with respect to concept usage, abstraction mechanisms, inherent structure and specification of attributes.

The use of concepts in the data models differs significantly. This is reflected in Fig. 4.

proposed terms	corresponding terms				
	Entity Relationship	Relational	RM/T	Network	Hierarchical
object	entity	tuple, row	entity	record	segment, record
object-category	entity set	heading, relational scheme	entity type	record type	reord type
relationship	relationship	tuple, row	associative entity	set (DBTG-set), set occurence	link
relationship category	relationship set	heading, relational scheme	associative entity type	set, set type (DBTG-set type)	link
attribute	attribute	attribute, column	attribute	data item	field (field occurence)
type-category	value set	domain	domain	?	field

Fig. 4. Concept usage in some specific data models.

Today, the attributes supported are mainly quantities. In future, attributes should also include references and categories (procedures and functions, in particular).

Abstraction Mechanisms

In the data models, there are different constructs for abstracting (*classifying*) objects, relationships and attributes (see the rows *object-category*, *relationship-category* and *type-category* in Fig. 4).

Generalization is supported in the RM/T only.[6] The lack of generalization often results in an unnatural and awkward way of modelling the referent system. This abstraction mechanism is well known from object oriented languages as SIMULA, BETA and Smalltalk.

In all the data models, *aggregation* is supported at one level only - a category can be regarded as an aggregation of its constituent attributes (e.g. PERSON is aggregated from *name*, *address*, etc.). It would be convenient to model object-categories as *containing* other object-categories (*nesting*). For example, an object-category INSTITUTE could be specified as an aggregation of the object-categories STUDENT, LECTURER, MACHINERY, etc. However, in the data models discussed, it is not possible to aggregate object-categories in this way.

Inherent Structure and Relationships

The main difference between data models concerns the inherent structure - mainly in how relationships between objects are expressed.

The ER, the network, and the hierarchical data model have specific constructs for expressing relationships. The ER data model (like the relational data model) supports direct representation of many-to-many relationships, while so-called *virtual* object-categories must be introduced in the network and the hierarchical data model in order to express such many-to-many relationships.

In the relational data model, there is no specific construct for expressing relationships - both objects and relationships are represented as *relations* (perceived by its users as a collection of tables). The users (database designers and application programmers, in particular) are thus not limited by a strict structure (e.g. network and hierarchical structure) when modelling relationships.

There is no difference in the representation of one-to-one, one-to-many or many-to-many relationships in the relational data model, and this data model is the only one supporting direct representation of relationships between *more than two* object-categories. The existence of the associated objects of a relationship must be explicitly checked (cf. the problem of *referential integrity*, e.g. [*Date, 86*]). In the network and hierarchical data model, it is impossible to insert a "child-object" without the existence of a "parent-object". Correspondingly, all the "child-objects" of a relationship are automatically removed when removing the "parent-object".

---

[6] Generalization is, of course, supported in other semantic data models not discussed in this paper (e.g. the *semantic network data model* (the *IS-A* construct)).

The languages of a data model are adapted to the inherent structure. Since the inherent structure of the relational data model is reflected by the fact that the objects and relationships in relational databases are represented as relations (sets), access and manipulation of these databases are *set-oriented*. Relational DBMSs are thus said to have a *set-interface*. In network and hierarchical databases, objects and relationships are accessed and manipulated one at a time. Hierarchical and network DBMSs are thus said to have a "one-record-at-a-time" interface.[7]

Attributes

The data models support quantity attributes (in fact, quantity-variables) only - the ability to specify procedures and functions as attributes in the category-descriptions is not supported. This fact tells us that the object oriented perspective has not (yet) significantly influenced the field of databases and data models.

The ER and the relational data model give the ability to specify user-defined *value sets* (*domains*) - used as type-categories. For example, the type-category AGE could be specified with value set integers from 0 to 150. In our opinion, the "set" concept in ER is not convenient - a set should denote a set (value set or extension of categories). As the concepts are used in ER, however, they denote both the *intention* and the *extension* of categories. The concept *value set* in ER is used to abstract attributes (quantity-variables) with their values. Similar to *entity set* and *relationship set*, the concept *value set* is used both as *category* (*type-category*) - and as *set*. We use *type-category* instead of *value set*, and we use *value set* to denote the set of legal values of the associated attribute.

The network and the hierarchical data model do not support the ability to specify user-defined type-categories - all type-categories are predefined.

5. DBMS

A *database management system* (DBMS) is a package of software which makes it possible to define and manipulate databases in computers. In addition to the implementation of a database language (compiler, interpreter, run-time system, etc.), a DBMS often includes facilities as full screen editor, report generator, *system catalog*, etc.

The simplest example of a DBMS is just a file system in which the database is viewed by the users as a group of files, each file consisting of records with fields. Another example is a relational DBMS as INGRES[8] in which the database is viewed by the users as relational tables. Loosely speaking, the users interact at a higher conceptual level. INGRES is more advanced and has facilities as menu and form driven interface, report generator, system catalog, etc.

---

7 Many *non-relational* DBMSs have (or are currently extended to have) "relational" ("set") front-end. However, problems often occur because of the dependency on the underlying structure.

8 INGRES [*Stonebraker, 76*] is developed at the University of California at Berkeley.

Information about structure of a database is supported by the DBMS by means of the system catalog (or *data dictionary*). In addition to structure information, it contains information as number of objects of different categories, and information not concerning the referent system - such as indexes, update-frequencies, users, access privileges, etc.

A DBMS can be regarded as an *implementation of tools* (e.g. languages) *associated with a data model*. The perspective of the associated data model is thus reflected in the DBMS. The associated data model is also referred to as the *inherent data model* of a DBMS (e.g. [*Tsichritzis, Lochovsky, 82*]). A DBMS might have several associated data models, and it can thus be difficult to classify DBMSs according to the associated data model(s). For example, ADABAS[9] is denoted both as a *network* DBMS (e.g. [*Ullman, 82*]) and as an *inverted list* DBMS[10] (e.g. [*Date, 86*]).

The connections between data model, DBMS and database can be explained with respect to structure. A data model imposes structure on a database via (by means of) a DBMS (see Fig. 5).

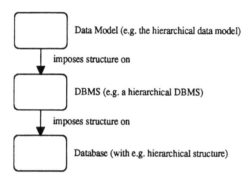

Fig. 5. Structure diagram.

In addition to the implementations of tools associated with a data model, a DBMS includes facilities independent of the associated data model (e.g. tools concerning user-interface).

6. Database Concepts in a Process/Structure Hierarchy

In this section, we describe the development and use of databases in a process/structure hierarchy, and we introduce the notions of "database process", "actors" and "database machine".

---

9   ADABAS is a DBMS from Software AG.
10 "An inverted list database is similar to a relational database - but a relational database at the low level of abstractions, a level at which the stored tables themselves *and also certain access paths to those stored tables* (in particular, certain indexes) are directly visible to the user" [*Date, 86*].

6.1 The Database Process

In a database process, the *database* is the "part of the world" being developed (cf. the process definition):

A database process is a development of a database through state transitions during a time interval.

The *subprocesses* of a database process can be described with respect to how the database is transformed:

1. Subprocesses causing transitions of the representative state of the database - the database is updated (e.g. the address of a person is changed).
2. Subprocesses exclusively producing maps by means of mappings (for instance, the process of making statistical tables concerning the age distribution of persons).

In practice, a subprocess may belong to both categories.

The execution of the database process can be classified into the following categories:

1. *Uni-sequential execution.* All subprocesses are organized in one single sequence (no alternation or concurrency between sequences occurs).
2. *Alternate execution*[11]. Every subprocess is organized (at a given time) in *one* of a finite set of sequences (no concurrency between sequences occurs).
3. *Concurrent execution.* Every subprocess is organized as a finite set of sequences (concurrency between sequences may occur).

6.2 Actors

An **actor** is a participant in a process, transforming the process states through its actions.

A **system development actor** is a participant in a process (the system development process) generating and/or modifying permanent structure for database processes.

Designers and programmers are typical system development actors. A system development actor takes part in the system development process by specifying a referent system and by describing and prescribing permanent structure for a database process (and thereby the permanent structure of the database) in database schemas by means of a database language.

A **database actor** is a participant in a database process, transforming database states through its actions.

---

11 Alternate execution is described in [*BETA, 85*] by the specific concepts *coroutine sequencing* and *alternation* which denote deterministic and non-deterministic alternate execution, respectively.

End-users are typical database actors. Other information systems (e.g. simulation systems, control systems) interacting with databases, are also examples of database actors.

Database actors take part in the database process by describing and prescribing transient structure for the process (and thereby transient-structure of the database) by means of the QL aspect of a database language. The set of possible descriptions and prescriptions is limited by the permanent structure specified by the system development actors.

System development actors, participating in the system development process, modify and produce permanent structure, while database actors, participating in the database process, produce transient structure. The actual occurrences of transient structure are not of interest to a system development actor. The system development actor is, however, interested in creating powerful tools for creation of transient structure by database actors.

A database actor might want transient structure to become permanent (structure). For example, an end-user who repeatedly forms the same query in order to produce a specific map, might want a permanent structure for that mapping.
The end-user might then change his role from database actor to system development actor (the end-user acts as system development actor).

6.3 Database Machine

A **database machine** is a computer that carries out a database process according to (permanent and transient) structure communicated (and/or described) by system development actors and database actors in an associated database language.

The concepts *database, database machine, database language* and *database process* are interrelated:

To every database, there is an associated database process carried out by an associated database machine and with structure given in an associated database language. To every database machine, there are an associated database, a database process and a database language. To a database language there exists a data model and there may exist a set of (database, database machine) couples.

6.4 Summary of Concepts

Fig. 6 summarizes the connections between some database concepts. The figure shows that both system development actors and database actors use a database language when producing and modifying permanent and transient structure description, respectively. These structure descriptions are used as prescriptions for the database machine which carries out the database process, which, in turn, transforms the database. The system development actors produce and modify structure description in accordance with the actual referent system being modelled. In addition to the transient structure produced by forming queries, the database actors produce transient structure as results of changes in the referent system (updates).

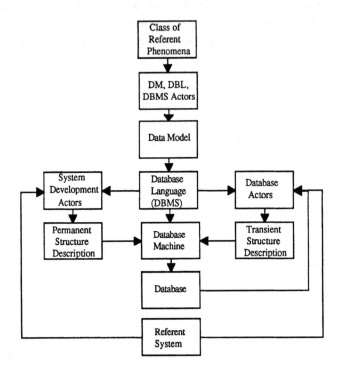

Fig. 6. Some interrelated database concepts.

DBMS (with associated data model and database language), system development and database actors are participants in, and are carrying out, processes. These processes produce structure which, in turn, limits a process (with participant actors) at lower levels. Processes and structures concerning development and use of databases can be described in a hierarchy. When applying this strict actor/process view (perspective), we identify the levels of the following process/-structure hierarchy:

Level 1:

        Process: The database process (a set of subprocesses causing state transitions of the database) carried out by a database machine.

      Structure: Transient structure (and permanent structure from level 2) limiting the database process.

Level 2:

        Process: The user processes by database actors (processes producing transient structure - at level 1).

      Structure: Permanent structure limiting the user processes (and database processes).

Level 3:

    Process: The system development process by system development actors (processes producing and modifying permanent structure - at level 2).

    Structure: Structure of the system development process (DBMS with data model and database language and, e.g., laws, existing knowledge, available resources).

Level 4:

    Process: The research and tool development process by DBMS development actors.

    Structure: .....

    ⋮
    ⋮

In general, structure at level n inherits structure at level n+1. For example, the database process is limited by transient structure (at level 1) and by permanent structure (at level 2) and by structure (at level 3) given by the actual DBMS (database language).

Acknowledgement

We would like to thank Kristen Nygaard for valuable comments and discussions, which, in particular, have contributed to the general perspective reflected in this paper. Thanks also to Birger Møller-Pedersen for useful hints and comments.

References

[*BETA, 83a*]: Kristensen, B. B., Madsen, O. L., Møller-Pedersen, B. and Nygaard, K.: "Abstraction Mechanisms in the BETA Programming Language". Proceedings of The Tenth ACM Symposium on Principles of Programming Languages, January 24-26 1983, Austin, Texas.

[*BETA, 83b*]: Kristensen, B. B., Madsen, O. L. and Nygaard, K.: "Syntax Directed Program Modularization". In "Interactive Computing Systems" (Ed. Degano, P. and Sandewall, E.), North-Holland 1983.

[*BETA, 85*]: Kristensen, B. B., Madsen, O. L., Møller-Pedersen, B. and Nygaard, K.: "The BETA Programming Language." Norwegian Computing Center, Report no 805, November 1987.

[*BETA, 87*]: Kristensen, B. B., Madsen, O. L., Møller-Pedersen, B. and Nygaard, K.: "Multisequential Execution in the BETA Programming Language." ACM Sigplan Notices, Vol. 20, No. 4, April 1985.

[*Brodie, 84*]: Brodie, M. L.: "On the Development of Data Models". Topics in Information Systems, On Conceptual Modelling (Perspectives from Artificial Intelligence, Databases, and Programming Languages) 1984 by Springer-Verlag New York Inc. pp. 19-47.

[*Chen, 76*]: Chen, P. P.: "The entity-relationship model: Toward a unified view of data". ACM Transactions on Database Systems, Vol. 1, No. 1, March 1976, pp. 9-36.

[*Chen, 77*]: Chen, P. P.: "The entity-relationship model: A basis for the enterprise view of data". Proc. AFIPS NCC No. 46 pp. 77-84.

[*Codd, 70*]: Codd, E. F.: "A Relational Model of Data for Large Shared Data Banks". Communications ACM, Vol. 13, No. 6, June 1970, pp. 377-387.

[*Codd, 79*]: Codd, E. F.: "Extending the data base relational model to capture more meaning." ACM Transactions on Database Systems, Vol. 4, No. 4, December 1979, pp. 397-434.

[*Copeland, Maier, 84*]: Copeland, G. and Maier, D.: "Making Smalltalk a Database System". Proc. ACM Sigmod, Boston, MA, June, 1984.

[*Dahl, Nygaard, 65*]: Dahl, O.-J. and Nygaard, K.: SIMULA - a Language for Programming and Description of Discrete Event Systems. Norwegian Computing Center, Oslo 1965.

[*Dahl, Myhrhaug, Nygaard, 65*]: Dahl, O.-J., Myhrhaug, B. and Nygaard, K.: SIMULA 67 Common Base Language. Norwegian Computing Center, Publ. S-2, Oslo, 1968.

[*Date, 86*]: Date, C. J.: An Introduction to Database Systems. Volume I, fourth edition, Addison-Wesley Publishing Company.

[*DBTG, 71*]: Codasyl Data Base Task Group Report, April 1971.

[*DELTA, 75*]: Holbæk-Hanssen, E., Håndlykken, P. and Nygaard, K.: System Description and the DELTA Language. Norwegian Computing Center, Publ. 523, Oslo 1975.

[*Goldberg, Robson*]: Goldberg, A. and Robson, D.: "Smalltalk-80, the language and its implementation". Addison-Wesley, 1983.

[*Lindsjørn, Sjøberg, 87*]: Lindsjørn, Y. and Sjøberg, D.: "Database Concepts. A Discussion in an Object Oriented Perspective". Master Thesis. University of Oslo, Norway. July, 1987.

[*Maier, Stein, 87*]: Maier, D. and Stein, J.: "Development and Implementation of an Object-Oriented DBMS". Research Directions in Object-Oriented Programming. Shriver, B., Wegner, P. (eds.) The MIT Press. Cambride, Massachusetts. London, England, 1987. pp. 355-392.

[*Nygaard, 86a*]: Nygaard, K.: "Proceedings of the IFIP 10th World Computer Congress". Dublin, Ireland, September 1-5, 1986.

[*Nygaard, 86b*]: Nygaard, K.: "Basic concepts in object oriented programming". (Opening lecture at the Conference on Object Oriented Programming in Wiesbaden, 24-25 September 1985).

[*Skarra, Zdonik, 87*]: Skarra, A. H. and Zdonik, S. B.: "Type Evolution in an Object-Oriented Database". Research Directions in Object-Oriented Programming. Shriver, B., Wegner, P. (eds.) The MIT Press. Cambride, Massachusetts. London, England, 1987. pp. 393-415.

[*Stonebraker, 76*]: Stonebraker, M., Wong, E., Kreps, P. and Held, G.: "The Design and Implementation of INGRES". ACM Transactions on Database Systems, September, 1976.

[*Tsichritzis, Lochovsky, 76*]: Tsichritzis, D. C. and Lochovsky, F. H.: "Hierarchical Data Base Management: A Survey". ACM Comp. Surv. 8, No. 1 (March 1976).

[*Tsichritzis, Lochovsky, 82*]: Tsichritzis, D. C. and Lochovsky, F. H.: Data Models. Prentice-Hall, Englewood Cliffs, N.J., 1982.

[*Ullman, 1982*]: Ullman, J. D.: Principles of Database Systems. Rockville, Md., Computer Science Press, 1982.

[*Zdonik, Wegner, 86*]: Zdonik, S. B. and Wegner, P.: "Language and Methodology for Object Oriented Database Environments". Proc. 19th. Annual Hawaii International Conference on System Sciences, January, 1986.

Object Oriented Programming and Computerised Shared Material

Pål Sørgaard*

Computer Science Department

Aarhus University

Abstract

Computer supported cooperative work currently receives much attention. There are many aspects of cooperative work. One of these is the use of shared material. Much cooperation is based on silent coordination mediated by the shared material used in the work process. The properties of the shared material are, however, often ignored when work is computerised. Instead the emphasis has been on automating frequent work procedures. This has resulted in very inflexible systems.

A fundamental idea in object oriented programming is to model the phenomena in the part of reality the system addresses. These modelling techniques can be used to implement shared material on computers. The result is a raw system providing the material and the essential primitive operations on this material. Such a system can be seen as a specialised programming environment which can be tailored to the needs of individual users or be modified for future needs.

This use of object oriented programming requires persistent and shared objects. Some objects may be active and execute as parallel processes. Incremental change of a running system will be needed to allow evolution.

It is non-trivial to decide which properties of the material to model. An example demonstrates that this decision may depend on the kind of technology being considered.

Key words and phrases: computer supported cooperative work, object oriented programming, shared material, persistent objects.

1 Introduction

In this paper an attempt will be made to bridge two research areas in informatics. These are object oriented programming, or actually object orientation in general, and computer supported cooperative work. This connection is based on the observation that shared material may play an

*Authors addresses: Computer Science Department, Ny Munkegade, 8000 Århus C, Denmark; E-mail paal@daimi.dk or from some countries paal@daimi.uucp; phone +45-6-127188

important role in cooperative work and that object orientation will be useful when implementing computerised shared material. This paper is based on the Mjølner project [5] and on the research programme on computer supported cooperative work [1] at Aarhus University.

The Mjølner project aims at developing an industrial prototype of a programming environment for object oriented programming. The project is deeply rooted in the "Scandinavian approach" to object oriented programming, a fact which also is reflected in the programming languages which the Mjølner environment is going to support: Simula [4] and Beta [12]. In this approach to object oriented programming a program execution, or actually a computer system, is seen as a model of an external system, the referent system: The phenomena in the referent system, the objects, go through their life processes, interacting with each other in various ways. The lives of these objects are reflected by the lives of corresponding objects in the computer system. Thus the ability to write programs where a set of objects execute in parallel is not an extravagant feature of some exotic programming languages; it is one of the most basic means of expression in programming. Correspondingly the semantics of a program is the interpretation of the program as a physical model. More in-depth expositions of this view on object oriented programming can be found elsewhere, see Nygaard [19], Knudsen and Madsen [11], and Madsen and Møller-Pedersen [17].

During the last couple of years computer supported cooperative work has started to receive attention [3], see also the special issues of the *ACM Transactions on Office Information Systems* (5(2), April 1987) and of *Office: Technology and People* (3(2), August 1987).

The term cooperative work is somewhat ill-defined. One tentative definition is given by this author in [24]. Cooperative work is nothing new, but now its existence has been recognised by people working with computers. It turns out that new technical possibilities open up if we may assume that the users are few in number and geographically close. In the transaction cost theory organisations are classified as markets, bureaucracies, and groups or clans [20]. Currently the state of the art of commercial computer systems is that they either support individual work or work divided in a formal manner as in bureaucracies and markets. Support for markets and bureaucracies has typically taken the shape of transaction processing systems or management information systems. The fact that people share tasks, work closely together in teams, know each other, exercise mutual solidarity, etc., is not only unreflected in current computer systems. It is often hampered by the way the computer systems change the conditions of work. The interest in computer supported cooperative work can thus be seen as an attempt to deal with some of the shortcomings of current computer systems.

Computer systems suitable for cooperative work can take many forms. Two main classes of systems are those supporting *explicit communication* and those providing — or supporting the existence of — *shared material*. Typical examples of systems supporting explicit communication are electronic mail, bulletin boards, and synchronous media like the UNIX commands write and talk. Support for shared material is more subtle, but the support for idea generation and shared writing provided by Cognoter is one example [6,26]. Shared material is the only aspect of computer supported cooperative work dealt with in this paper.

Jackson has written a system development method which is based on object orientation [9]. This method proposes an approach where the basis of the computer system is a model of the referent system. A crucial step in the development process is to identify which kinds of objects the system should model, and which objects are "outside model boundary." The method consists of several steps. After the kinds of objects that are to be modelled have been identified the model is constructed. Thereafter functions may be added to the model. Jackson argues that a model is more stable than the concrete functions, and that a system built in this way will be amendable to new kinds of functions. This requires that the model is sufficiently general, and that the new functions lie within the functionality the model has been made for. The ideas presented in this paper can be seen as an application of some of Jackson's ideas.

The paper proceeds as follows: The next section discusses the role of shared material in cooperative work. It also presents some examples of computer systems providing shared material. Section 3 presents several arguments for modelling material rather than frequent work procedures; these arguments are not restricted to the context of cooperative work. Section 4 presents some ideas for how the design of a computer system could be based on modelling the material used in the work process. Finally, section 5 identifies some of the requirements of the programming languages and run time systems which are incurred by the use of object oriented programming to model shared material.

2 Cooperative work and shared material

Two ways of coordinating cooperative work can be identified. One is by explicit communication about how the work is to be performed, another is less explicit, mediated by the *shared material* used in the work process. A simple example is the way two people carry a table. A part of the coordination may take place as explicit communication, for example in a discussion about how to get the table through a door. When the table is carried, however, the two people can follow each others' actions because the actions get mediated through the shared material. This coordination is not very explicit, it does by no means involve an explicit exchange of messages. Also, it has been learned. There is a big difference between two persons' first attempt at carrying something together, and the way people with experience do it. The learning is both on part of the individual, and on part of the team. It is crucial to this coordination that the actions of the other actor can be immediately observed or felt, so that appropriate corrective or supplementary actions may be taken. The pattern of cooperation is not fixed, it is often defined by the actors. The material and the situation in general make a wide range of patterns of cooperation possible. For people with computer support in mind it should be noted that the mediation of actions taking place through a table is a mediation with a very high bandwidth. Another example is the way a manual file is used as a shared material. A record in such a file can only be in one place at a time. When a document — a record — is gone, it may mean that somebody is already dealing with the problem in question. It may also mean that some other person is using out-of-date information in a potentially dangerous way. This could be the case with a medical journal. The meaning of an absent record, or some other sign attached to the material, depends on the context, but it is mediated through the material.

It is a central hypothesis in this paper that the properties of the shared material often determine or strongly structure the pattern of cooperation in a work process. If we look at how shared material is used we will get design ideas which are appropriate for computer supported cooperative work. In this way the concept of shared material is used as a design metaphor, a technique proposed by, among others, Madsen [15,16].

When work is computerised the material people work with is normally changed. As a result some properties of the material which were crucial to cooperation may get lost or changed. One example of such a change is that a record is no longer only in one place at a time. Many of the above mentioned "nice" properties of a manual file may therefore disappear. Another example is the transition to computerised text processing. On the surface the typewriter is replaced by a text processor, a "modern typewriter" which has the convenient capacity of storing the text so that corrections can be made without having to retype the whole document. In reality paper is replaced by magnetic storage. Printouts are just snapshots of a document. The document is *in* the computer. This has many consequences, one of these is that the concept of original is more blurred than ever before. There can be many originals since a copy of a file is just as good as the original, and an original can be perceived to be in many places at the same time. People are able to produce new versions of documents more often, and with smaller changes, than before.

Computerisation of material does not, however, have to have negative consequences. There is also a potential for giving the material new properties suitable for coordination. It is the interpretation of this author that the examples presented below implement shared material.

In Colab at Xerox PARC a number of experimental computer systems for cooperative use have been made [26]. Many of these systems are based on the WYSIWIS (What You See Is What I See) metaphor. The users are in other words given the impression of working on a shared surface. Actions of the other users are visible and pieces of material grabbed by somebody else is typically shown to be reserved or removed by being greyed out. For a discussion of the WYSIWIS metaphor see especially Stefik et al. [25].

Kaiser, Kaplan, and Micallef [10] present an experimental programming environment which allows concurrent update of different well defined parts, modules, of the same program. If inconsistencies arise, i.e. if the definition of a module is changed so that its use becomes invalid, the involved programmers will be informed. The technique used in this environment is attribute grammars, like in the Cornell Program Synthesizer [21,28]. The computations are performed in parallel, however, with one process for each module. Information about changed attributes will only in a few cases need to travel from one module, and process, to another. The points where attributes may enter or exit are well defined, they are explicit import and export statements in the programming language. This makes it possible to distinguish between attribute changes that may affect other modules and changes that are certain to be local.

In UNIX and in many other programming environments facilities have been made to support controlled access to different versions of a file (a program). Two examples running on UNIX are the Source Code Control System (SCCS) [22] and the Revision Control System (RCS) [29]. These two systems perform a multitude of functions. Many of these functions are specific to programming and to a specific way of representing programs (as text files), but they do both implement a material which has the same property as a paper-based document: It can only be in one place at a time. This is implemented through commands like check-out and check-in. In addition to version numbering of the different modules symbolic naming is supported. This can be used to name configurations and make it easier to retrieve these at later points in time. Different versions of the same module may be updated concurrently. Different branches in a version tree may later be joined or merged, provided the updates of the common ancestor do not textually overlap.

3 Model material, not work procedures

Newman [18] and Suchman and Wynn [27] describe office work as a mixture of problem solving and work according to known procedures. There is not a sharp distinction between these two kinds of work. Newman states [18, p. 55]:

> Existing procedures are used extensively in solving problems. They suggest manageable subgoals and thus simplify the development of problem-solving strategies. In some cases, only a minor modification is needed to a basically adequate procedure; in other cases, an ad hoc solution is constructed from a number of procedural elements.

In many administrative systems, however, the emphasis has been on automating frequent work procedures. This approach can be motivated by its simplicity, and by the fact that much time is spent on these seemingly trivial work procedures. Hence large and immediate savings appear to be obtainable. There are several arguments against this approach:

- The computerisation of work procedures ignores the learning aspects of the supported work. Intimate knowledge of the composition of the procedures is needed in order to perceive the procedure as what it is: a work procedure which sometimes is used "as is", sometimes is combined with other procedures, and often needs slight modification to be applied to a deviant case. This need not cause trouble for users who know the work procedures from before the system was introduced. New users, however, will have little chance for learning this, and will often perceive their work as that of an operator.

- The view of an office as a collection of machines executing procedures is false. The work can be *approximated* by a number of procedures, but this will not cover the work in the office. The strategy of starting with a few, recurrent procedures will turn into a process of implementing an apparently infinite number of procedures having less and less volume. As the procedures get less typical, they will be harder to describe, making the development process more difficult. Also, since the volume is low, the gains from rationalisation, if any, do not outweigh the development costs. Thus a strategy which start with large potential savings as the aim, may end up as a money sink.

- Work is not static. It is hard and expensive to change tailor-made systems to new or modified procedures. The degree of reuse of the old pieces of software is often low and conversion can cause large difficulties.

- The view of work as the execution of procedures ignores or underemphasises the cooperative aspects of the work. Cooperation which took place in the manual system is reduced to issues of shared access to databases. In other words, only the formalisable, bureaucratic, aspects of cooperation are supported.

The picture given above is not pure speculation. We get confronted with the consequences of the procedure-automating approach in many situations. Some authentic cases:

- A family reserves berths in the night-train from Oslo to Trondheim. The reservation is made for seven persons, but it is explicitly stated that only six are certain to come, the seventh berth should therefore in some way be independent of the other six.

  When they go to the railway station to pick up the tickets, the seventh person has decided not to go. It turns out, however, that one single reservation has been made (probably the only way to get seven berths next to each other). It also turns out that one cannot selectively "dereserve" one berth from a reservation. The clever clerk in the ticket office decides to cancel the seven berths and thereafter reserve six berths. The cancellation works fine, but the new reservation fails because the train is full! (The seven released berths had probably been taken by a waiting list.) Luckily there was another train to Trondheim with free berths that night.[1]

- As a student I used to have a teaching assistantship (TA) and a research assistantship (RA) at Aarhus University. At the beginning of a new term I decided to drop the TA since time was shorter than money. For some obscure reason the university paid me salary as a TA also in the new term. As a consequence more tax was deduced from my RA salary since my deductions had been "spent" on the TA salary. I made the salary department aware of the error and I also stated that I at any time was willing to pay the money I had received in excess of what I would have received only as an RA. This amount was less than what I had received as TA because of progressive tax deductions.

  This was the start of a long series of complicated transactions. It took three months before it all was settled. Although I complained the matter was settled in such a way that I had to pay back all the money I had received as a TA before my tax deduction as an RA was corrected.

- In a Danish bank a typing error led to the erroneous deposit of an enormous amount, say 100 million kroner, on a customer's account. The error was immediately discovered, and the amount was withdrawn only a few instances after it was deposited. So far so good.

---

[1] Leikny Øgrim, Institute of Informatics, University of Oslo, told me about this case.

Due to the way interest is computed (deposits take place from the next day, withdrawals from the same day) it was computed that the customer had to pay interest for the erroneous amount for one day, approximately 15,000 kroner, roughly equivalent to a months salary. Consequently the bank informs the customer that his account is grossly overdrawn. The customer, of course, complained, and after some dispute the case was settled.

The problem in these cases is not only bad design of the tasks performed by the system, it is the failure to observe the intimate relationship between procedures and problem solving. The alternative proposed here is that instead of automating the procedures one should model, or implement, the *material* used in the work process. Material, or substance, can in a natural way be modelled as objects. Modelling of substance is one of the defining characteristics of object oriented programming. These objects should have value sets corresponding to the different states of the material. There should be operations on the objects corresponding to the *primitive operations* performed on the material. It is in other words proposed that object oriented programming should be applied to implement computerised shared material.

Primitive operations are the most primitive meaningful single operations modifying the material we can identify. Thus deposit-money-on-an-account is not such an operation because it involves several primitive operations, like change balance, record information for the computation of interest, etc. To write a single digit in a bank-book is too primitive. It does only have a meaning using a specific kind of technology. Care must be taken in identifying the material and the primitive operations to model. Much inspiration can be obtained from the different ways the corresponding traditional material is being manipulated, but in general a thorough analysis of the work in question, its purposes, etc., is needed. In the terminology of Jackson System Development [9] this corresponds to the entity-action step and some of the activities which have to precede the application of JSD.

There is a conceptual difference between primitive operations and work procedures, not only a difference in level of detail. A work procedure often reflects a normative view of the work, often as seen by management. Procedures are aggregates of simpler operations. Also, procedures are subject to change, and there may be deviations from a procedure in the actual performance of a work task. Primitive operations are stable, they have not changed, and are not expected to change in the future. Also they cannot meaningfully be divided in yet simpler operations.

The distinction between procedures and primitive operations is different from the psychological distinction between actions and operations which has been used by Bødker in [2]. The latter distinction applies to how a person conceives a task. What is a composite action to one person can be an operation to another, typically a professional. The distinction between procedures and primitive operations presented here is not individual. The distinction is given by the system and is in principle common to all users.

The raw model coming out of a development process focussing on material and primitive operations can be used, but it will be very clumsy for many practical purposes. The raw model should be seen as, and designed as, a specialised programming environment with a user-oriented or profession-oriented programming language. Machine implementation of many entire procedures should be made in this environment, and these programs or functions should be made available to the users for inspection and modification. Users who are inclined to do so can construct their own programs. The programs can, of course, contain calls to other programs as well as to primitive operations. Besides being convenient in programming, it allows some procedures to be seen as simple operations by some users. This implementation of procedures allows a user to reuse, modify, and combine machine implemented procedures in much the same way as can be done with manual procedures. When needed the user may resort to "manual" execution of primitive operations. The new or modified functions can be seen as incremental changes to the existing system. The system will also be tailorable to the needs of the individual users.

Facilities like this are often implemented by mechanisms like accelerators or macros. The possibility to write shell scripts in UNIX is a typical example. This way of doing it works, but is far from ideal because of the poor syntax and semantics of most macro-facilities.

4 Computerised material: an example

This section will discuss how some important aspects of material can be retained or created when the material is computerised. A train seat reservation system is used as example. The example illustrates how many improvements can be made by carefully implementing a new material. Some aspects of cooperation are also illustrated.

Manual seat reservation systems, as the author remembers them, were based on a number of sheets illustrating the cars in the train; see figure 1. The seats could be checked or the part of the

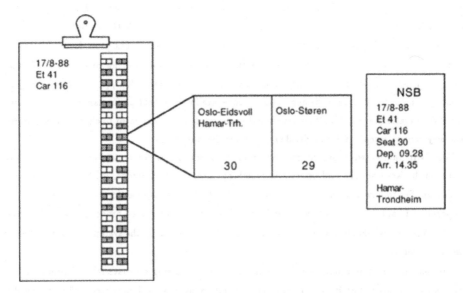

Figure 1: The manual seat reservation system

trip where the seat was occupied could be indicated. Travellers were given small tickets indicating the details of their reservation. This scheme was very flexible in terms of the kinds of services which could be supplied. For example, the reservation of a seat next to another specific seat, and the reservation of a seat close to the door (my grandmother always wanted that), etc., were feasible. Entries were of course written with pencil to allow easy update. The reservation sheets were kept at the train's station of departure so that they could be sent with the conductor.

This system obviously had many drawbacks making it impossible to retain: (The reasons given here are speculation; the author never worked with the railroad.)

- It could not handle high volumes of transactions incurred by compulsory reservation in some trains.

- Updates could only take place at the station of departure. Reservation from other railroad stations and travel agencies had to be made by phone calls to this station.

- The system was inadequate for a new policy where charges were made for seat reservation.

- Reservations from and to abroad were not well integrated in the system.

The system which was implemented solved some of the above-mentioned obvious problems with the manual system. At the same time almost all of the flexibility of the sheet-based system disappeared. It appears that the only functions which were implemented were **make-reservation** and **cancel-reservation**. In fact the only way to figure out whether there was room on a train was to make a reservation.

Many of the necessary improvements could have been made while retaining much of the flexibility of the sheet based system. This could be done by appropriate object oriented modelling of trains with cars and seats, reservations and reservation agents, and by making this model visible to the user. The manual system was actually based on a model of the trains and their seats. A major part of a computerisation of the system should have been to make a computerised model of the trains using the reservation sheets as inspiration for the design.

The approach taken here is different from the approach taken in most database systems. In database systems much effort is made to make it appear as if each user is working alone, unaffected by all others, on the whole database. In the approach taken here the sharing of the material, the reservation sheets, should be made explicit. The users should have access to them in a way which clearly distinguishes between looking at the current state of a piece of information which can be modified by others, and having a unique sheet at disposal for update. Also the process of, in competition with others, obtaining a sheet for unique manipulation must be made explicit. In this way patterns of cooperation can be retained and developed further. In the following a brief sketch will be given of how such a system might look to the user.

Users could have access to shared information by obtaining displays of trains satisfying some criterion, see figure 2. The trains listed in the figure are the trains bringing passengers from Oslo

Trains Oslo S - Hamar   17/8-88

Et 41   0800 0928 43/28/3/2
Ht 351   1010 1157 50/32
Ht 341   1140 1327 25/10
Ht 343   1430 1622 44/35
Ht 375   1542 1733 54/22
Ht 307   1640 1840 11/7
Ht 345   1900 2050 37/22
Ht 405   2300 0049 20/7

Figure 2: List of trains

to Hamar a specific day. The trains are listed with train number, time of departure and arrival, and the number of free seats in different categories. All but one train have only second class. The information displayed here is volatile, and this should be made clear to the users. The best way to do this would be to update the displayed numbers of free seats as reservations and cancellations are made. In this way the users would get an idea of the update activity on the trains, and they could develop their work practices on the basis of that. If a cheaper solution is needed one could display information subject to change in a special font or in a special colour.

Access to "reservation sheets" for single cars can be obtained in a shared display of the train. See figure 3. The figure indicates that car 117 is sold out and that somebody else is updating car

Et 41   Oslo S - Trondheim o/Dovre   17/8-88

Figure 3: Shared train display

115. The users working with this display compete with each other about obtaining unique access to single reservation sheets. One way to do this would be to implement this display in the WYSIWIS style [25], where the pointing and selection devices of the other users would be visible. In such a setting it could happen that two users try to get the same sheet at the same time. The system does not prevent such collisions, but it must of course detect them. Collision prevention is up to the users, they can see all cursors and can therefore "keep away" from each other. A WYSIWIS solution requires very high bandwidth in the communication network since updates and cursor movements should be propagated in real time between the users. Today such a solution can perhaps only be implemented using a high-speed local area network. This makes it hard to build a seat reservation system that way, but the principle could be applied in a case where the users are geographically close. This would typically be the case in cooperative work. A cheaper solution is to accept that the users do not have any means of seeing each other, and that they therefore will run into collisions more often. The system could also have a built-in function which gets hold of some free car in the train. It will not, however, be acceptable to force users to use this function. It would prevent

services like reserving seats in a car near the restaurant car.

Assume our user successfully gets hold of car 116, see figure 4. The car is now in a private

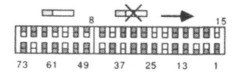

Figure 4: Car display

display, and is displayed to others in the same way as car 115 in figure 3. The user may return the car unmodified to the shared display. This could be the case if the user was searching for a seat near the door and could not find one in this car. The typical mode of operation, however, will be to "enlarge" a part of the car and make a reservation, see figure 5. When the user checks the wanted

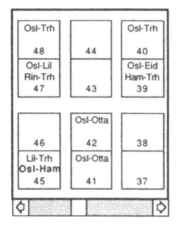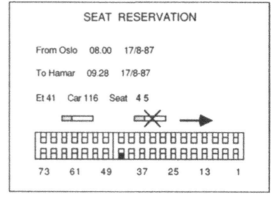

Figure 5: Making a reservation

seats, the corresponding seat reservation is built up in the display. When necessary, seats can be released by updating an old reservation. Standard procedures like make-reservation should be available as programmed functions.

It would be too much to say that the example is a system which supports cooperative work. But it does not prevent cooperation either. If cooperation is needed it may evolve because the users to some extent can "see" each other's actions, and because the work organisation is made visible by defining three types of display: a shared information display, a shared selection display, and a private update display.

The grain size of the system, the size of the object allocated to the single user for update, should of course be small enough to allow the needed degree of concurrency. At the same time it should be as large as possible to allow maximal flexibility. In this example the car was made the "grain", it could also have been the compartment or the whole train. This paper focuses on support for cooperative work, a kind of work where the number of users is limited. A large grain size does not necessarily cause too many collisions in a small group. In addition to this we can interpret collisions as a natural property of the material. It may simply reflect the fact that several users need to coordinate their actions. In computer supported cooperative work the "grain" should therefore be chosen to correspond to a natural piece of material.

Cooperation between users can also be supported by letting two users look at the same reservation, perhaps even work on it concurrently. This can be useful when problems show up or when a case needs to be taken over by another user.

The example shows that the properties of the phenomena which need to be modelled go beyond those which are modelled in traditional computer systems. Especially the "layout" of the train, for example the position of the restaurant car, needs to be reflected in the model. The system needs to keep track of different car types and also their physical layout. This should be done although there is no functionality in the system which depends on this information. Similarly search systems in libraries should contain information about the colour and physical shape of books, although such information do not immediately provide extra functionality.

The example also illustrates that new technology, in this case especially output media which can draw pictures of train cars, result in changes of which parts of the referent system that should be modelled in the computer system. Clearly traditional computer systems have been designed with slow devices with few capabilities and with high communication costs in mind.

It should be asked whether the kind of design proposed here is prohibitively expensive or otherwise unrealistic. Clearly it requires better terminals than those typically found in clerical workplaces, but media with graphic capabilities are getting cheaper. The communication costs involved in transferring drawings of trains, cars, etc., could be high, but they can be reduced by having the different graphics elements distributed once, and later only transfer what is needed to build the drawing in question. In addition, communication costs are also falling.

5 Requirements of the environment

The use of object oriented programming in this paper puts some requirements on the programming languages and the run time environments of the systems.

The objects in question need to be *permanent* or *persistent* across individual program executions. Thus a language with persistent objects is needed, and certainly persistence is more central than object orientation. Reasonable programming languages with no or little support for permanent, in practice disc-based, data structures will therefore have no chance in the competition with 4th generation tools, which can be characterised as data base management systems augmented with primitive programming facilities. We must hope that work on persistent languages and on object oriented databases will change this situation.

Objects used as shared material clearly need to be *sharable* in some well-defined way. In some cases the sharing will take place as a series of exclusive accesses, in other cases there will be concurrent use of the same objects by several users who make updates and see the effects of their own and others' updates. This issue is discussed by Greif and Sarin in [8] and by Stefik et al. in [26].

In making object oriented models of the referent system *parallelism* arises as a natural part of the model. In a seat reservation system the objects executing in parallel will be the objects modelling the users and, if the system is to be able to handle the reservations while the train is travelling, the objects modelling the trains. Reservation objects can be passive, they only change because some other object sends messages to them. It is therefore important that the language used not only facilitates classes as a kind of abstract datatypes (like Smalltalk [7]), but that the classes have some sort of action part, like in Simula and Beta. Furthermore the objects, i.e. the instances of the classes, must be able to execute in parallel. Parallel execution in Beta is described by Kristensen et al. in [12,14].

Finally these objects may reside at different places in a network of computers. One such place could be an object server. Work is going on to allow the cooperation of several operating system processes to constitute a Beta ensemble [13].

In this paper it has been assumed that it is possible to make a stable model of the shared material. In the seat reservation example it was shown, however, that changes in the model may be needed because new technology makes it relevant to model new aspects of the referent system. Jackson System Development is also based on the stability of the model. New functions are expected to be within the predicted functionality. This will not always be the case. We also know that no description, and hence no model, of reality is complete. If we have a system consisting of many long-lived objects we may therefore need to be able to modify the model while the system is running. In other words: we need *incremental execution*. This may lead to the concurrent presence of instances of different versions of the same class, some mechanisms for handling this is discussed by Skarra and Zdonik in [23]. It is still an open question, however, how we should understand a system which is a changing model of a changing referent system.

In this paper it has been argued that the modelling aspects of object oriented programming can be useful in the implementation of computer based shared material. There is, however, no guarantee that object oriented programming will lead to better computer support for cooperative work. Conversely object oriented programming is not a necessary requirement for making this kind of computer support. The only claim made is that object oriented programming will be *useful* for this purpose.

Acknowledgements

I am indebted to Riitta Hellman and Ole Lehrmann Madsen for their constructive comments.

References

[1] Peter Bøgh Andersen et al. *Research Programme on Computer Support in Cooperative Design and Communication.* IR 70, Computer Science Department, Aarhus University, Århus, 1987.

[2] Susanne Bødker. *Through the Interface — A Human Activity Approach to User Interface Design.* PB 224, Computer Science Department, Aarhus University, Århus, April 1987.

[3] *Conference on Computer Supported Cooperative Work*, MCC Software Technology Program, Austin, Texas, December 1986. Proceedings.

[4] Ole-Johan Dahl, Bjørn Myhrhaug, and Kristen Nygaard. *SIMULA 67 Common Base Language.* Pub. S-2, Norwegian Computing Center, Oslo, 1967.

[5] Hans Petter Dahle, Mats Löfgren, Ole Lehrmann Madsen, and Boris Magnusson. The MJØL-NER project. In *Software Tools: Improving Applications: Proceedings of the Conference held at Software Tools 87*, pages 81–87, Online Publications, London, 1987.

[6] Gregg Foster and Mark Stefik. Cognoter, theory and practice of a colab-orative tool. In *Proceedings from the Conference on Computer Supported Cooperative Work*, MCC Software Technology Program, Austin, Texas, December 1986.

[7] Adele Goldberg and David Robson. *Smalltalk-80, The Language and its Implementation.* Addison-Wesley, 1983.

[8] Irene Greif and Sunil Sarin. Data sharing in group work. *ACM Transactions on Office Information Systems*, 5(2):187–211, April 1987.

[9] Michael Jackson. *System Development.* Prentice-Hall, Englewood Cliffs, 1983.

[10] Gail E. Kaiser, Simon M. Kaplan, and Josephine Micallef. Multiuser, distributed language-based environments. *IEEE Software*, 4(6):58–67, November 1987.

[11] Jørgen Lindskov Knudsen and Ole Lehrmann Madsen. Teaching object-oriented programming is more than teaching object-oriented programming languages. In Stein Gjessing and Kristen Nygaard, editors, *Proceedings, Second European Conference on Object Oriented Programming (ECOOP'88), Oslo, Norway, August 1988*, Springer Verlag, Heidelberg, 1988.

[12] Bent Bruun Kristensen, Ole Lehrmann Madsen, Birger Møller-Pedersen, and Kristen Nygaard. The BETA programming language. In Bruce Shriver and Peter Wegner, editors, *Research Directions in Object-Oriented Programming*, pages 7–48, MIT Press, Cambridge, Massachusetts, 1987.

[13] Bent Bruun Kristensen, Ole Lehrmann Madsen, Birger Møller-Pedersen, and Kristen Nygaard. Dynamic exchange of BETA systems. January 1985. Unpublished manuscript.

[14] Bent Bruun Kristensen, Ole Lehrmann Madsen, Birger Møller-Pedersen, and Kristen Nygaard. Multi-sequential execution in the BETA programming language. *Sigplan Notices*, 20(4), April 1985.

[15] Kim Halskov Madsen. Breakthrough by breakdown. In Heinz K. Klein and Kuldeep Kumar, editors, *Proceedings of the IFIP WG8.2 Working Conference on Information Systems Development for Human Progress in Organisation, Atlanta, 29–31 May 1987*, North-Holland, Amsterdam, 1988 (forthcoming). Also available as PB 243, Computer Science Department, Aarhus University, Århus, March 1988.

[16] Kim Halskov Madsen. *Sprogbrug og Design — sammenfattende redegørelse*. PB 245, Computer Science Department, Aarhus University, Århus, November 1987.

[17] Ole Lehrmann Madsen and Birger Møller-Pedersen. What object oriented programming may be — and what it does not have to be. In Stein Gjessing and Kristen Nygaard, editors, *Proceedings, Second European Conference on Object Oriented Programming (ECOOP'88), Oslo, Norway, August 1988*, Springer-Verlag, Heidelberg, 1988.

[18] William M. Newman. *Designing Integrated Systems for the Office Environment*. McGraw-Hill, Singapore, 1986.

[19] Kristen Nygaard. Basic concepts in object oriented programming. *SIGPLAN Notices*, 21(10), October 1986.

[20] William G. Ouchi. Markets, bureaucracies, and clans. *Administrative Science Quaterly*, 25:129–141, March 1980.

[21] Thomas Reps and Tim Teitelbaum. The synthesizer generator. In Peter Henderson, editor, *Proceedings of the ACM SIGSOFT/SIGPLAN Software Engineering Symposium on Parctical Software Development Environments*, pages 42–48, May 1984. Published as ACM Software Engineering Notes 9(3) and ACM SIGPLAN Notices 19(5).

[22] Marc J. Rochkind. The source code control system. *IEEE Transactions on Software Engineering*, SE-1(4):363–370, December 1975.

[23] Andrea Skarra and Stanley Zdonik. Type evolution in an object-oriented database. In Bruce Shriver and Peter Wegner, editors, *Research Directions in Object-Oriented Programming*, pages 393–415, MIT Press, Cambridge, Massachusetts, 1987.

[24] Pål Sørgaard. A cooperative work perspective on use and development of computer artifacts. In Pertti Järvinen, editor, *The Report of the 10th IRIS (Information Research seminar In Scandinavia) Seminar*, pages 719–734, University of Tampere, Tampere, 1987. Also available as PB 234, Computer Science Department, Aarhus University, Århus, November 1987.

[25] Mark Stefik, Daniel G. Bobrow, Gregg Foster, Stan Lanning, and Deborah Tatar. WYSIWIS revised: early experiences with multiuser interfaces. *ACM Transactions on Office Information Systems*, 5(2):147–167, April 1987.

[26] Mark Stefik, Gregg Foster, Daniel G. Bobrow, Kenneth Kahn, Stan Lanning, and Lucy Suchman. Beyond the chalkboard: computer support for collaboration and problem solving in meetings. *Communications of the ACM*, 30(1):32–47, January 1987.

[27] Lucy Suchman and Eleanor Wynn. Procedures and problems in the office. *Office: Technology and People*, 2(2):133–154, January 1984.

[28] Tim Teitelbaum and Thomas Reps. The Cornell program synthesizer: a syntax-directed programming environment. *Communications of the ACM*, 24(9):563–573, September 1981. Also in David R. Barstow, Howard E. Shrobe, and Erik Sandewall, editors. *Interactive Programming Environments*. McGraw-Hill, New York, 1984.

[29] Walter F. Tichy. RCS: a revision control system. In Pierpaolo Degano and Erik Sandewall, editors, *Integrated Interactive Computing Systems*, pages 345–361, North-Holland, Amsterdam, 1983.

Asynchronous Data Retrieval
from an Object-Oriented Database

Jonathan P. Gilbert
Lubomir Bic

Department of Information and Computer Science,
University of California, Irvine, CA 92717, USA.

Abstract

We present an object-oriented semantic database model which, similar to other object-oriented systems, combines the virtues of four concepts: the functional data model, a property inheritance hierarchy, abstract data types and message-driven computation. The main emphasis is on the last of these four concepts. We describe generic procedures that permit queries to be processed in a purely message-driven manner. A database is represented as a network of nodes and directed arcs, in which each node is a logical processing element, capable of communicating with other nodes by exchanging messages. This eliminates the need for shared memory and for centralized control during query processing. Hence, the model is suitable for implementation on a multiprocessor computer architecture, consisting of large numbers of loosely coupled processing elements.

1. Introduction

The overall goal of the semantic data modeling project at UCI is to develop a semantic database system suitable for highly parallel processing. We believe that this can be accomplished if the underlying model is completely message-driven, i.e., without any centralized control and centralized memory. First, however, the semantics of the model and its operations must be defined. Based on these definitions, procedures that govern the propagation of messages during processing can be derived.

The present paper is a first step toward such a model. It describes the basic philosophy of our approach, the components of the model, and the semantics of queries. We also outline the *generic* procedures that permit queries to be executed in a purely message-driven manner.

The model has all the desirable features of a conceptual modeling system. These features are well known and have been presented many times before: see, for example, [BRODIE80, BORGIDA87]. In particular, the model combines the virtues of four concepts: the functional data model

[SHIPMAN81], a property inheritance hierarchy (common to most semantics networks [FINDLER79] and some frame based languages like KRL [BOBROW77]), the principles of message-driven computation [ARVIND78, AGHA85], and the data hiding/abstract data types of object-oriented programming systems [STEFIK86].

The paper is organized as follows: In section 2, we describe the representation and organization of base and derived data within our paradigm. We also sketch the syntax of queries and specify their semantics. Section 3 shows how requests can be processed asynchronously by propagating messages through the database hierarchy. Finally, section 4 contains some concluding remarks and points out the relationship of our model to some other approaches.

2. Components of an Object-Oriented Model

In this section, we begin by describing the representation, components and organization of data in our model. After the basics have been described, we present the message passing strategy.

2.1. Data Representation

A database is represented by a network of nodes and directed edges. Each node represents an independent database object. We adopt the philosophy found in many semantic data models (see, for example, [CODD79, HAMMER81, BANERJEE87]): higher-level (molecular) objects are recursively constructed from simpler database objects. Nodes of the network represent objects within the database enterprise (for example, people, colors, automobiles, or engines) and arcs represent various associations among these objects. There are two basic kinds of association: the IS-A relationship and the ROLE relationship. The first is used to construct an *inheritance* hierarchy (see, for example, [DAYAL84]) while the second is the functional "glue" that binds together molecular structures. These associations and the overall structure of a database is similar to those in an Omega knowledge base [ATTARDI86]. Data are organized in an *incremental* fashion, with more refined data descriptions beneath their more general ancestors' descriptions in the IS-A hierarchy. Figure 1 shows a single branch of a "modes-of-transportation" hierarchy. It is used to illustrate various aspects of the two hierarchies found in our model.

We distinguish two types of nodes: ellipses which represent sets of non-decomposable *atomic* objects and rectangular boxes which represent sets of compound *molecular* objects. The IS-A hierarchy (in which nodes are connected by the *unnamed* arcs) facilitates *inheritance* of properties and relationships, represented by ROLE associations. The arrows of the IS-A hierarchy show the direction in which inheritance takes place. We chose the name "role", rather than property, function, or relationship, to stress the fact that molecular objects are recursive compositions of simpler objects and each of the simpler objects plays a certain role in the "super" object. A database user may *choose* those roles he perceives as inherent (attributes) parts of an object and those which are more like relationships between independent objects. The former are displayed

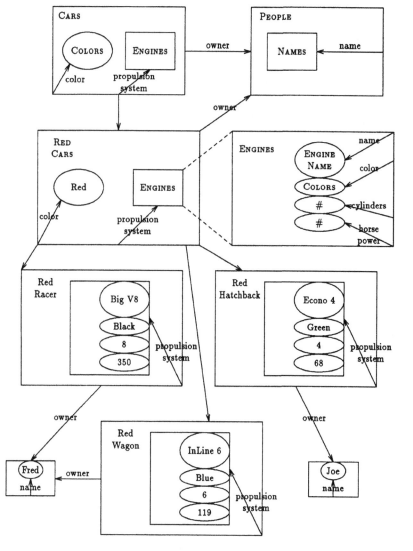

Figure 1

Single Branch "Modes-of-Transportation"

inside the objects description while the latter are displayed outside of the object. For example, in figure 1 the roles *color* and *engine* are perceived as part of an automobile while the role *owner* is identified as a relationship between an automobile and a person. Notice that these choices and many other choices related to the users' perception of the data are subjective. Although meaningful to the user, whether a role is displayed inside or outside of an object is irrelevant to the semantics of the database itself. On the other hand, there may be some roles (displayed inside or outside the node) which are absolutely essential to the description of an object. We call this type of role a *key role*; all other roles are *ordinary roles* (which may or may not be instantiated in all leaves).

For example, in figure 1 it the owner role (from cars to people) is key because (in this very simple world) all cars must be owned by people. However, if we were to look at that same relationship in the other direction we would find that it is *not* key because some, not all people own cars.

There is no explicit distinction made between sets of objects and individual elements in our model. Conceptually, each node contains a *generic description* of an object so that leaf nodes of the IS-A hierarchy are sets containing exactly *one* object. However, since there are relatively few *internal* (non-leaf) nodes it is desirable to store the bulk of the description and semantics at this level thereby minimizing the amount of redundant information at the "element" or *leaf level*. Furthermore, internal nodes serve a dual purpose: they represent the *set* of leaves reachable by following outgoing IS-A arcs and they serve as a *type* for those leaves. One major advantage of this uniform view of sets and elements can be illustrated by the following simple example: If we assume that CARS is a multi-set, then, if Fred owned a fleet of identical Red_Racers, instead of just one, it would not be necessary to repeat the Red_Racer's description for each car. Conceptually, the current Red_Racer node would become a generic description and *empty* children nodes would be inserted to represent the individual automobiles.

2.2. The External Schema

The global external schema contains only non-leaf nodes (set description objects). It describes for the user the entire database enterprise in a single connected graph. Even though leaf nodes (object instances) are not included in the global schema, the schema is often too large to display as a single graph; therefore, the user may view the global schema as several graphs rather than a single graph. The system provides an interactive graphics browser that permits users to explore the schema. An object is selected as the current point of interest. This node and the nodes which are *directly* connected to it by a single IS-A or *role* arc are displayed in a window for the user. For example, when displaying the PEOPLE object's node in figure 1 nodes representing CARS and NAMES would also be shown (without any further detail). The user can navigate through the schema (change the point of interest) by moving a mouse pointer to an object and pressing the appropriate mouse button. The new node's object then becomes the point of interest. Many objects and arcs in a schema are not *base* but *derived* (shown as dashed boxes and arrows). Base objects have a concrete representation *stored* in the database while derived (or virtual) data (described in more detail later) are calculated by applying rules when a user tries to "retrieve" that data. In the day to day interactions with the database, there is no visible difference between virtual and stored data for the user except that virtual data cannot be directly updated.

2.3. Derived Data

Much of the semantic richness of this model comes from its support of a variety of derived data. There are two types of derived data: sets and roles, which are represented by rules that are part of an object description. The syntax of these rules is beyond the scope of this short paper but we do discuss the derived data available and, in the next section, the data retrieval algorithms including the instructions necessary for retrieving data from virtual objects and arcs. To better illustrate the three kinds of *union-subset* and *aggregate data*, we present a non-trivial example (shown in figure 2) which is based on examples in [McLEOD78]. Note that dashed nodes and arcs represent derived data.

2.3.1. Derived Sets

Union-subset nodes are a grouping mechanism which allow the formation of heterogeneous sets. All union-subset nodes contain pointers to the base sets that are the basis for a set abstraction. There are three types of union-subset abstraction called *category, collection* and *power sets*. To define a derived set a user must specify: its name, its type, the sets whose union are the basis for the (maximal) derived set, restrictions on each set's roles (if any), and any new roles which are associated with objects in the virtual set.

Collection sets "automatically" include all leaves in all base sets which are in the union and whose descriptions are consistent with any restrictions placed on that set's roles. In figure 2 oil tankers is a collection because its members are all military and merchant ships whose class is "oil tanker". Unlike collection sets, a *category set's* node contains *explicit* pointers to its members which have been specifically inserted into that category. Banned Ships (see figure 2) are an example of a category. There is no rule associated with the banned ships object. Any ship may be banned but a user must explicitly ban it. *Power sets* can be thought of as a generalization of the category. The major difference between them is that the power set is based on the *power set of the union* of some base sets instead of their union — each *element* of a power set is a *category*. In figure 2 convoys are modeled as a power set because each convoy is a set of ships and not a single ship. Notice that the roles (location and max-speed) are associated with the convoy and not the individual ships in that convoy.

2.3.2. Derived Roles

Virtual role abstractions are classified by the action taken by the system when it instantiates them. Actions correspond to substituting a subquery for the virtual role, spawning the new query which is reprocessed by the node and "creating" a virtual arc or a virtual node. A *VR-arc* rule causes a virtual arc to be "created" while a *VR-node* causes a virtual node to be "created".

To create a VR rule a user must specify: the name of the role, the set on which it is defined, the domain of the operation (where the rule is mapped to) and the operation itself (which may be anything from a simple "restriction list" to a general purpose (external) procedure or both). In addition, the user must determine whether the rule will be evaluated at the set or instance level of the IS-A lattice.

An example of a VR-node abstraction is *aggregate data*. Aggregate data are defined by aggregate operators which abstract a single object from a set of objects. Examples of aggregate operations are: calculating the maximum speed of a convoy (see figure 2) or determining the average length of an oil tanker (not shown in the figure).

VR-arc abstractions are *inference rules*, so called because the relationship which they make explicit can be inferred from the structure of schema anyway. Information is retrieved by substituting a role request subquery for a VR-arc "role" thereby "creating" the virtual arc. For example, consider the *grandfather* relationship between people. This could be represented explicitly as a role (arc) from an individual to his parents' fathers or it could be represented *implicitly* by including a rule which states: "To find a person's grandfather, first find his parents and then find their fathers."

To the user, derived data of both kinds can be used to retrieve information in exactly the same way as any base role.

3. Message-Driven Processing

In an object-oriented environment, each object is an *abstract data type* which includes a description of the data it represents and a set of operations (*methods*) for manipulating that data. These methods are triggered when messages are received from other objects. The data representation is not visible to the outside world; the user "sees" a "black box" and the actions (which may vary from one abstraction to another) for the manipulation data inside the box. In our model, a similar situation exists except that communication between objects is achieved by a small number of *generic* methods.

The object-oriented paradigm with its abstract data types and message passing semantics make our model suitable for implementation on a highly parallel loosely coupled multiprocessor. The ideal architecture has no centralized control or memory and each node may be mapped onto a different processing element (PE) as long as there are physical communication paths for each logical arc. There are many architectures that satisfy this requirement.

3.1. Internal Representation of Arcs and Objects

Objects are data structures that are mapped onto the local memory of a processor (PE). The description of an object contains information about all data within that object. It must include components that represent arcs, derived data and operations (or *methods*) that are triggered by incoming messages. In addition, the description contains information about individual roles: i.e. which of them are *key* and where they are to be displayed. We have shown that roles' nodes may be *displayed* inside and outside of their "super" object's node. The semantics of these differences are in some sense "external". This means that, although the placement of a role node may make a difference to the way in which a user perceives a concept, placement makes no difference to the

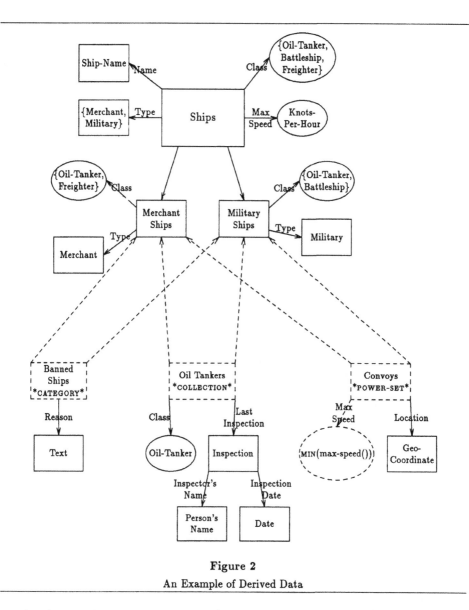

Figure 2

An Example of Derived Data

way that the system processes a query on an object. On the other hand, the difference between key and non-key roles are internal since they are absolutely essential to the description of an object.

Arcs represent either IS-A or ROLE relationships between objects; they are implemented by using pointers where each pointer identifies a PE and an address within that PE's local memory space. All arcs are bi-directional which means that each arc is actually represented by *two* pointers, one at each of its ends. Atomic roles are not represented by independent objects. Since atomic objects are simple values, it would be wasteful to have independent objects that just return a value. Instead, we store *singleton* roles locally so that they can be retrieved from an object's local memory without flooding the system with unnecessary messages.

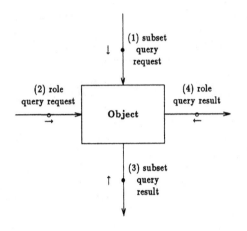

Figure 3

The Four Message Types

3.2. Information Retrieval

Queries are formulated and processed against the external schema. There are two kinds of information retrieval queries. The first variety of retrieval request refers to an object as a *set* while the second refers to it as a *type*. A user may want to retrieve all elements of a set which have particular properties (we call this kind of request a *subset query*) or a user may want information about the objects associated with a particular role. (This second type of query is called a *role query*.) The basic strategy is for the user to send a message to the injection point node which either replies to the request directly or propagates the query to other objects and waits for their response. When all objects have responded, the node can combine the results and return the result to the sender. This query processing strategy and the two query types are implemented using four types of generic message. These messages are called: (1) the subset query request message, (2) the role query request message, (3) the subset query result message and (4) the role query result message. The four message types are illustrated in figure 3. Note that the arcs at the top and bottom of the object represent IS-A relationships and arcs on the sides represent role relationships.

By examining the message, an object can determine which action it should take (there is exactly one action for each message type). We now describe the general strategy and show high-level descriptions of the procedures used to process user requests.

Conceptually, a request for information either points to a set of objects and retrieves the subset of those that satisfy some list of restrictions on their outgoing roles or retrieves information about some of an object's roles. Restrictions are recursively decomposed and applied to objects reached via role arcs starting at the original object, until the entire restriction is satisfied or fails. First we give the basic syntax of queries. Each query can be thought of as a four-tuple:

⟨⟨set⟩; ⟨query-type⟩; ⟨query-restriction⟩; ⟨query-output⟩⟩ where:

⟨set⟩ $\overset{\text{def}}{=}$ the name of the injection point node.

English "equivalents" of the queries are shown in *italics*; comments are shown in roman font.

1. *List all Red Cars Owned by a Person Named Fred*: The key word here is "list" the system produces a list of cars.
   ⟨RED CARS; SUBSET-REQUEST; owner.name = "Fred"; LIST(VALUE(ALL))⟩.

2. *Are there any Red Cars Owned by a Person Named Fred*: This time a "yes" or "no" answer will be produced.
   ⟨RED CARS; SUBSET-REQUEST; owner.name = "Fred"; EXISTS(ALL)⟩.

3. *Is it possible that a Person Named Fred could be the owner of a Red Car*: This is a query about the owner role and not the set of Red Cars.
   ⟨RED CARS; ROLE-REQUEST; owner.name = "Fred"; EXISTS(ALL)⟩.

Figure 4

Sample Queries

⟨query-type⟩ $\overset{\text{def}}{=}$ identifies the query as a role request or a subset request.

⟨query-restriction⟩ $\overset{\text{def}}{=}$ a set of paths which define the restrictions on roles involved in the query. Its format is comparable to the body of the *is-there?* query in Omega [ATTARDI86]. The processing, however, is not the same.

⟨query-output⟩ $\overset{\text{def}}{=}$ decribes roles and format of the output of the query.

To illustrate the expressive power of these queries and to provide a set of concrete examples for subsequent discussions, consider the list of queries in figure 4.

When processing any query, the system must differentiate between *key* and *non-key* roles. The reason for this is obvious: If a role is key to a set's object then it definitely exists for *all* instances of that set; if it is non-key then it may exist in some of a set's instances. Notice that this definition of *key* is quite different from a key attribute in many traditional database models since uniqueness is not necessary.

There are two kinds of question that can be asked about a role: (1) does the role exist and (2) if it exists, does it map to a particular set of objects or values. The semantics of a role request query are captured by the two procedures shown in figure 5. A query names an injection point r and lists the roles (and restrictions on those roles) which are the focus of the query. A status value is calculated for all roles named in the query by sending a role request (sub)query message along each of the named arcs. Each role object processes it's subquery independently of all other role objects and the strategy is exactly the same as that followed at the injection point. The overall strategy is that the query is *dynamicly* recursively decomposed for parallel processing. Eventually, for each role path a *terminal* node is reached. A terminal node is a node which can determine a status (and a value) for a particular (sub)role; it is *not* necessarily a leaf node. Once the status is

```
Procedure Role-Query-Request (Triggered by a message of type 2)
  create activity record for pending query &
  for each path R in query-restriction
    if head(R) is a base role then
      if it is a singleton or node is terminal
        then send a Role-Query-Result message to self
        otherwise remove R from the restriction list &
          send a Role-Query-Request message
            containing tail(R) along arcs that match head(R)
      otherwise (R is a virtual role)
        if node is not a leaf and R is a "set-level" rule
          OR if node is a leaf and R is a "instance-level" rule
            spawn appropriate subquery


Procedure Role-Query-Result (Triggered by a message of type 4)
  store result
  if last result for corresponding activity
    determine status of query
    Case 1: the original query was a Role-Query-Request
      SubCase 1.1: the object is a base set
        send a Role-Query-Result message to sender & destroy activity record
      SubCase 1.2: the object is union-subset node
        & its base sets have not been visited
        & the query has NOT definitely succeeded or failed
          for each base set in union
            send a Role-Query-Request message containing only unfound roles
          adjust activity record to reflect change in query
      SubCase 1.3: the object is union-subset node
        & its base sets have not been visited
        & the query has definitely succeeded or failed
          send a Role-Query-Result to sender & destroy activity record
      SubCase 1.4: the object is union-subset node
        & the result comes from a base set
          store result & destroy activity record
          if it also is the last result for original activity record
            then determine status of original query (minimum status found) &
              send Role-Query-Result to sender & destroy activity record
    Case 2: the original query was a Subset-Query-Request
      if node is not a leaf & status is not 5
        then for each non-leaf child send Subset-Query-Request message to child
          if status is 1, 2 or 3
            then for each leaf child send Subset-Query-Request message to child
          adjust original activity record to reflect change in query
      otherwise (the node is a leaf)
        send a Subset-Query-Result to sender
        destroy activity record
```

Figure 5
Role Request Procedures

known it is returned (on a role query *result* message) along the arc on which the original request arrived. When a non-terminal node has collected results from all its subqueries, they are used to determine its own status which is then sent back to the sender of the request. Note that because of the distributed structure of the database and the absence of centralized control in this strategy, the subqueries are distributed and the results collected in an asynchronous manner.

There are five possible status values for individual roles; their most general meanings are listed below. Note that, although all five status values are not necessary for processing role request queries, they are all necessary when processing subset requests.

1. This role was found and (the restrictions on it) satisfied for all possible instances of the set rooted at this node (for key roles only).
2. This role definitely exists for all possible instances, however, the restriction on this role may not be satisfied (once again key roles only).
3. This role was found and exists for some instances of the rooted set (for non-key roles only).
4. This role was not found.
5. This role was found and is definitely not satisfiable for any instance of the rooted set.

The *maximum* value of the individual roles' status values is taken as the status of the query for the entire object. The basic meanings of the object status values (used by all query types) are listed below:

1. All restrictions (on roles) were satisfied.
2. All restricted roles definitely exist but some *may* not be satisfied.
3. Some restricted roles may exist for some instances and not others.
4. Some resticted roles were not found.
5. Some restricted roles are definitely not satisfiable.

The semantics of subset query request processing is slightly more complicated because subset queries spawn role queries. Figure 6 shows sketches of the two procedures executed by a database object when it receives a subset query message. The processing strategy depends on the propagation of messages from the injection point down through the IS-A hierarchy possibly all the way to the leaves. At each node visited, subset query requests spawn role request subqueries to determine whether individual restrictions have been satisfied. There are four basic assumptions about what happens to object descriptions as the IS-A hierarchy is traversed towards the leaves: (1) more role descriptions may be added, (2) *any* role's definition may become more restricted, (3) *non-key* roles may become *key* or so restricted that they "disappear" and (4) *virtual* roles are treated like *non-key* roles.

The semantics of a subset request query are captured by the two recursive procedures shown in figure 6. They are applied as follows: the query names a node s as the target set, from which elements are to be retrieved; S represents the set of nodes reachable from s by following IS-A arcs and L is a subset of S containing only leaf nodes (elements). Each element of L is an object which may be retrieved by the query, if it satisfies the specified restrictions.

In each element of S, the status of all roles named in the query is determined by sending role request queries along all role arcs listed on the query restriction list. In each node of the set S–L (i.e., non-leaf nodes), a status is determined for each role by the role request query which is compared with the status obtained by the node's parent. This is necessary because some non-key

```
Procedure Subset-Query-Request (Triggered by a message of type 1)
  if the node is a leaf & query originated from a category node &
    object is not directly connected to that category node
      then (report failure) send a Subset-Query-Result to sender
      otherwise create activity record for pending query
        for each path R in query-restriction
          if head(R) is a base role then
            if it is a singleton or node is terminal
              then send a Role-Query-Result message to self
              otherwise remove R from the restriction list &
                send a Role-Query-Request message
                  containing tail(R) along arcs that match head(R)
            otherwise (R is a virtual role)
              if node is not a leaf and R is a "set-level" rule
                OR if node is a leaf and R is a "instance-level" rule
                spawn appropriate subquery


Procedure Subset-Query-Result (Triggered by a message of type 3)
  store result &
  if last result for corresponding activity
    then determine status of query & send Subset-Query-Result to sender &
    destroy activity record
```

Figure 6

Subset Request Procedures

roles "disappear"; if the previous status was 3 and the current status is 4 then the current status must be changed to 5. The object's status is then calculated and if it is *not* 5 then the query (including the status values) is passed to its descendants. Nodes in the set L determine the status in a similar way. This final value determines whether the object satisfies the given query; if it does, the data specified in the query's output field are retrieved and output.

Notice that all non-singleton role status values are calculated independently and that an *object* must wait for all of its roles to report their status before it continues processing a query. The first observation suggests a potentially high degree of parallelism if the system is implemented on a loosely coupled multiprocessor architecture. The second observation seems to imply that any benefit from this parallelism is lost because objects spend much of their time waiting for results from other objects. This conclusion is incorrect for several reasons: First, the fact that objects spend much of their time waiting does *not* imply that PEs are *busy waiting* or even idle. When a *PE* receives a request message, it creates an activity record for the request and when all the necessary subqueries have been spawned, it stores the activity record until it receives result messages for that request. When a result message is received, the PE determines whether it is the last result for the query; if it is not, the message is stored with the activity record. Otherwise, it is combined with the other results in order to calculate the object's status. This strategy allows for true asynchronous processing of queries and enables a high degree of parallelism without using a database management system query optimizer.

3.2.1. An Example — Processing a Simple Query

To clarify our asynchronous query processing strategy, we will describe the processing of the first sample query shown in figure 4. In order to satisfy that request, it is propagated through the schema shown in figure 1. We assume that all roles are key and that, initially, the status of the query and all of its roles are 4 (not found). Since the user requested all information about red cars the system will add all RED_CARS' roles that are not explicitly mentioned in the query to the (query-restriction) (in this case there are just two: propulsion-system and color). Note that these new roles can be assigned a status of 1 and, therefore, do not add significantly to the processing time. When the RED_CARS object receives the subset request, it decomposes the (query-restriction), stores the status of propulsion-system and color, and sends a role request message to PEOPLE. The (query-restriction) of this new message contains *name = "Fred"* and since name is a singleton the PEOPLE object determines that "Fred" is a (not the) valid name and, therefore, returns a status of 2 to RED_CARS. The RED_CARS object then calculates the status of the query by taking the maximum of the roles' status values: 2. From this status RED_CARS determines that any of its children may satisfy the query and it sends subset request messages to the Red_Racer, the Red_Hatchback and the Red_Wagon. If each of these objects is mapped to a different PE then each will be able to look up its singleton roles and send role request messages to its non-singleton roles independently and in parallel with the other objects. Eventually, Red_Hatchback determines that its status is 5 and, therefore, it returns its status but no data. At the same time Red_Racer and Red_Wagon determine that their status values are 1 and they, therefore, do return data. When RED_CARS has received subset results from each of its children it combines the successful results and returns the objects' descriptions to the user.

4. Conclusions

Similar to other semantic and object-oriented database models, our approach has a clear advantage over the classical database models. The classical models are relatively low-level and capture little of the semantics of the application domain.

There have been many research efforts directed towards improving the semantics of database modeling and several surveys have been published on the subject — see, for example, [BIC86, HULL86]. Some research has produced significant enhancements to the relational model. For example, J. Smith and D. Smith added aggregation and generalization abstractions to the relational model (both of which are integral parts of our model) to produce their hierarchical semantic model [SMITH77]. Codd also introduced an enhancement to the relational model [CODD79] (known as the Tasmania relational model) which includes many forms of abstraction (including aggregation and generalization). Another approach has been to develop new semantic models which replace the relational data model; Hammer and McLeod's SDM [HAMMER81] is a good example of this. A major drawback of both the latter models is their *extreme* complexity — only the most sophisticated users may find them useful modeling tools. By comparison, object-oriented models like ORION [BANERJEE87] and this model are very simple to use.

We believe that object-oriented models have some advantages over each of the semantic models. In particular, in object-oriented approaches, objects include the procedures (methods) for manipulating the data which they contain. Because objects communicate by sending each other messages and their methods are independent local procedures, there is an excellent *potential* for parallel processing. Finally, because of the *generic* methods which are built into its objects, our model provides a general framework for the development of database applications.

REFERENCES

[AGHA85]   AGHA. G.A. Actors: A Model of Concurrent Computation In Distributed Systems.
              Tech. Rep. No. 844. MIT Artificial Intelligence Lab., MIT, Cambridge, Mass..

[ARVIND78]   ARVIND, GOSTELOW, K.P. AND PLOUFFE, W. An Asynchronous Programming and
              Computing Machine. Tech. Rep. No. 114a. Univ. of CA., Irvine, Dept. of Info. and
              Comp. Sci..

[ATTARDI86]   ATTARDI, G. AND SIMI M. A Description-Oriented Logic for Building Knowledge
              Bases. *Proc. of the IEEE 74*, 10 (Oct., 1986), 1335–1344.

[BANERJEE87] BANERJEE, J. ET AL. Data Model Issues for Object-Oriented Applications. *ACM
              Trans. on Office Information Systems 5*, 1 (Jan., 1987), 3–26.

[BIC86]   BIC L. AND GILBERT J.P. Learning from AI: New Trends in Database Technology.
              Computer 19, 3 (Mar., 1986), 44–54.

[BOBROW77]   BOBROW D.G. AND WINOGRAD T. An Overview of KRL. *Cognitive Science 1* (1977),
              3–36.

[BORGIDA87]   BORGIDA, A. Conceptual Modeling of Information Systems. In *On Knowledge Base
              Management Systems*, Brodie, M.L. and Mylopoulos, J., Ed., Springer-Verlag, 1987.

[BRODIE80]   *Proceedings of the Workshop on Data Abstraction, Databases and Conceptual Model-
              ling*, Brodie, M.L. and Zilles, S.N, Ed., Sponsored by the Nat'l. Bureau of Standards,
              ACM SIGART, SIGMOD and SIGPLAN, Pingree Park, Colordo, 1980.

[CODD79]   CODD, E.F. Extending the Database Relational Model to Capture More Meaning.
              ACM Trans. on Database Systems 4, 4 (Dec., 1979), 397–434.

[DAYAL84]   DAYAL, U. AND HWANG, H.-Y. View Definition and Generalization for Database
              Integration in a Multibase System. *IEEE Trans. on Software Engineering SE-10*, 6
              (Nov., 1984), 628–645.

[FINDLER79]   *ASSOCIATIVE NETWORKS Representation and Use of Knowledge by Computers*,
              Findler, N., Ed., Academic Press, 1979.

[HAMMER81]   HAMMER M. AND MCLEOD D.J. Database Description with SDM: A Semantic Data
              Model. *ACM Trans. on Database Systems 6*, 3 (Sept., 1981), 351–386.

[HULL86]   HULL, R. AND KING R. Semantic Database Modeling: Survey, Applications, and
              research Issues. Tech. Rep. No. TR-86-201. U.S.C., Comp. Sci. Dept..

[MCLEOD78]   MCLEOD, D. A Semantic Data Base Model and its Associated User Interface. Rep.
              No. MIT/LCS/TR-214. Lab. for Computer Sci., MIT, Cambridge.

[SHIPMAN81]   SHIPMAN, D.W. The Functional Data Model and the Data Language DAPLEX.
              ACM Trans. on Database Systems 6, 1 (Mar., 1981), 140–173.

[SMITH77]   SMITH, J.M. AND SMITH D.C.P. Database Abstractions: Aggregation and General-
              ization. *ACM Trans. on Database Systems 2*, 2 (June, 1977), 105–133.

[STEFIK86]   STEFIK, M. AND BOBROW D.G. Object-Oriented Programming: Themes and Varia-
              tions. *The AI Magazine 6*, 4 (Jan., 1986), 40–62.

AN OVERVIEW OF OOPS+,

AN OBJECT-ORIENTED

DATABASE PROGRAMMING LANGUAGE

Els Laenens - Dirk Vermeir

Philips International B.V.
Corp. ISA / AIT
Building VN3
P.O. BOX 218
5600 MD Eindhoven, The Netherlands

Dept. of Math. and Computer Science
University of Antwerp, U.I.A.
Universiteitsplein 1
B2610 Wilrijk, Belgium

Abstract

This paper provides a brief introduction to the OOPS+ knowledge-representation language. While basically object-oriented, OOPS+ integrates database concepts as well as classical knowledge-representation techniques such as rule-based inference and demons. In addition, the language supports types as first-class objects, inheritance, imperative function definition, and query facilities based on logic programming.

1. Situation and motivation

OOPS+ has been designed within the framework of the Esprit KIWI project[1] . The aim of the project is to design and develop a knowledge-based user-friendly system for the utilization of information bases. The KIWI system consists of four software layers (Figure 1): the user interface (UI), the knowledge handler (KH), the advanced database environment (ADE) and the information base interface (IBI). In order to support information retrieval both directly from the KIWI knowledge base and indirectly, through the IBI, from the existing external databases, the ADE uses the relational model to represent knowledge. At present the relational model is widely accepted to represent data. However, the fact that it is restricted to flat relations has proven to be a problem in many applications. Therefore, the KH is introduced within the general architecture of the KIWI system. Its main function is to provide an environment that supports the definition and also the manipulation of knowledge using a semantically rich formalism.

This research was supported in part by the EEC Esprit program under contract P1117.

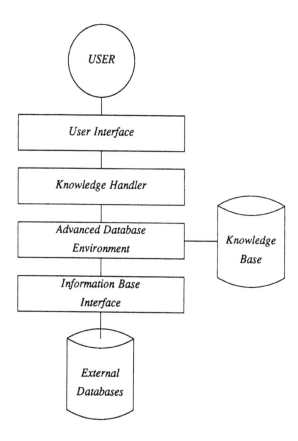

Figure 1. KIWI Architecture

The KH makes use of the facilities of the ADE for the storage and retrieval of knowledge. It is the responsibility of the ADE to perform the appropriate translations between the KH's complex objects and the ADE's underlying 'flat' relational structures. On the other hand, the KH interfaces to the User interface (UI) module which will provide a user-friendly view of the knowledge in the system using state-of-the-art graphical techniques. In this paper, we will discuss some features of the knowledge handler.

2. Requirements

It follows from the above that the KH is characterized mainly by the chosen knowledge-representation formalism, which in turn will be influenced by the intended area of application: intelligent access on information stored in large conventional databases, enriched with a local knowledge base. As a consequence of the functionality of the KH within the KIWI system, we get the following requirements for the knowledge-representation language (OOPS+) that realizes the formalism[2] .

R1   Concerning the *semantic modeling* concepts we expect organizational facilities such as classification, aggregation, generalization and specialization for structuring the knowledge base. OOPS+ will need to manipulate complex objects.

R2   Regarding *database* concepts, we want powerful query facilities as well as easy translation between the complex objects of the KH and the flat relational structures of the ADE.

R3   Another requirement is the provision of tools for *knowledge-based application programming*, which means that we want to be able to apply some knowledge representation technology to general data processing problems.

R4   Finally, we also want to keep *simplicity* by introducing all features in their minimal, essential form (e.g. as was done in the Amber language)[3] .

3. The resulting language: OOPS+

In order to meet all of the requirements, OOPS+ is a multi-paradigm knowledge-representation language[4] .

The query facilities (R2) are based on *logic programming* while knowledge-based application programming (R3) is supported through the facilities of *rule-based* and *access programming*: a *trigger* mechanism is available that can be used to enforce integrity constraints. As far as the knowledge-representation aspect (R1) is concerned, we will look upon the knowledge base as a collection of objects and relations defined over them. Modifications to the knowledge base occur through the insertion and deletion of objects and the manipulation of relations.

In the sequel, we will discuss OOPS+ in two layers.

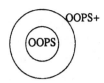

Figure 2.

The innermost layer deals with all aspects concerning knowledge representation. The purpose of this layer is to provide an environment suitable for the definition and manipulation of knowledge. In other words, this layer should meet both the semantic modeling (R1) and the simplicity requirements (R4). As this is the most fundamental part of OOPS+ we will be referring to this layer as the kernel of OOPS+, called OOPS. The other layer provides the different programming paradigms. As shown in Figure 2, this layer is build on top of OOPS as each of the paradigms makes use of the environment established by the kernel. So OOPS+ consists of a kernel (OOPS) enriched with some powerful paradigms in order to satisfy the variety of requirements (i.e. R1-R4).

4. OOPS: the OOPS+ kernel

4.1. Introduction

Different ways of structuring data have generated distinct classes of programming languages and induced different programming styles. Data is said to be taxonomically organized if it constitutes a hierarchy of classes and subclasses, and if data at any level of the hierarchy inherits all the attributes of data higher up in the hierarchy. Programming with taxonomically organized data is often called object-oriented programming, and has been advocated as an effective way of structuring programming environments, databases and large systems in general. In more conventional languages, data is organized as cartesian products, disjoint sums and function spaces i.e. the basic data type constructors in denotational semantics. Our semantic modeling requirement (R1) for knowledge representation suggests to use a taxonomically way of structuring data whereas the conventional way seems more appropriate if we want to introduce as few different concepts as possible to meet the requirement of simplicity (R4). For this reason, both the taxonomically and conventional ways of structuring data are merged in OOPS.

4.2. Objects and values

The notion of **object** is considered to be primitive in OOPS. We use O to denote the set of all objects. Each object has an internal state, called its **value.** An object retains its identity through arbitrary changes in its own state. One can look upon an object as a kind of identifier with which a value, the data represented by the object, is associated. Thus the **state** of an OOPS program is defined by a partial function

$$s : O \rightarrow V$$

where V is the domain of possible object values. Let us write $State$ to denote the set of all such functions.

We distinguish primitive values (integers, strings, etc.) and complex values made up of primitive values and constructors[5,6,7] . Complex (primitive) objects are objects having a complex (primitive) value. Because of the existence of complex objects, a single entity can be modeled as a single object: entity features need not be simple data values, but can be entities of arbitrary complexity. On the contrary, in the traditional relational database approach an entity is represented as multiple tuples spread amongst several relations.

In the next sections, we will start off with three kinds of complex values. Records (cartesian products), finite sets and functions will be presented subsequently. Sets are supported explicitly, without the encoding required in the relational model, and can have arbitrary objects as elements, they need not be homogeneous. Furthermore, functions can be used to protect the integrity of the knowledge base.

As will be shown, the subset

$$Prim + Rec + Set + Fun$$

of V already supports some aspects of aggregation, classification, generalization and specialization.

Some considerations concerning the object hierarchy (inheritance) along with the introduction of three additional complex objects (meet, join and power objects) will round off our discussion on OOPS.

Recapitulating, we get the following semantic domain of object values in OOPS:

$$V = Prim + Rec + Set + Fun + Meet + Join + Power$$

4.2.1. Primitive values

Oops supports the following primitive values:

• predefined domains:

 - `Int`,

 - `Str`,

 - `Log`

which represent the sets of all integers, strings and boolean values respectively;

• atomic values:

 - integers (e.g. -1, 0, 1) having type `Int`,

 - strings (e.g. "Fred", "") having type `Str`,

 - the boolean values `True` and `False` having type `Log`,

i.e. all elements of the predefined domains;

• special values:

 - an error value `error`,

 - an undefined value `*` (or any),

 - a nil value `nil`

With each of the primitive values, there corresponds a constant expression which denotes the unique predefined object in the state that has this value, i.e. we assume the existence of a subset O_{Prim} of O which is isomorphic with $Prim$. Thus

$$O \supset \{ o_p \mid p \in Prim \} = O_{Prim}$$

such that

$$\forall \; s \in State, p \in Prim \quad s(o_p) = p$$

O_{Prim} is the set of primitive objects. Hence, in this case, we can identify the objects with the values. For example, the expression

```
"Fred"
```

will evaluate to the object $o_{"Fred"}$ which has as value the string constant "Fred".

4.2.2. Record values

A record value is a finite labeled set of (references to) objects. Labels are denoted by identifiers. Formally,

$$Rec = (L \rightarrow O)$$

where L is the set of labels which is disjoint from V. Note that they are not identifiers nor strings and can not be the result of the evaluation of an expression.

Record valued objects are constructed using the '( )' operator, e.g. evaluation of

```
(name = "Fred"; birthdate = 1960)
```

will create an object with a record value containing two fields, one labeled *name* pointing to $o_{"Fred"}$, the other labeled *birthdate* referring to o_{1960}.

Using ':=' instead of '=' in a record field definition indicates that the field at hand may be *updated*, otherwise fields retain their creation time value. For example, the object created by the expression

```
(name = "Fred"; age := 3)
```

has a constant *name* field while the object pointed to by *age* can be updated.

OOPS supports *aggregation* through records. A collection of concepts can be treated as a single concept having several components. For example, an address could be thought of as an aggregation of its street, city and country.

```
(
name = "Fred";
address = (
            street = "Fifth avenue";
            city = "New York";
            country = "USA"
            )
)
```

A record object may be declared *extendible* by preceding it with a ':'. One can add/delete label-value associations to/from an extendible record value. E.g.

```
(
name = "Fred";
address = : (street=...)
)
```

creates a record object having two fields. The *address* field points to an updatable record object r. However, the association between the label *address* and r is fixed. Thus, extendability is a feature of the record and not of the reference to it.

Records also play the role of record types:

$$(name = "Fred"; age = 25) : (name = Str; age = Int)$$

The latter record is a record type for the former record because "Fred" and 25 are of type Str and Int respectively.

4.2.3. Set values

A set value is a finite set of (references to) objects

$$Set = Fin(O)$$

Set valued objects are constructed using the '{ }' operator. E.g. evaluation of

```
{"Jane", 64, FALSE}
```

will create an object with a set value

$$\{ O"Jane", O\ 64, OFALSE \}$$

Note that different objects may have identical values i.e.

```
{ (i=3) , (i=3) }
```

will create an object with a set value containing two objects. This illustrates one of the principles of object-oriented programming: objects - in contrast to tuples in a relation - are not identified by their intrinsic properties. Thus two objects can be distinct even if all their defined components have the same values; objects are distinguished by reference.

Sets will be the basic constructs for providing OOPS with the *classification abstraction* method.

A set object may be declared *updatable* by preceding it with a ':'. E.g.

```
(
name = "Fred";
kids = : {(name="Pebbles";age:=1)}
)
```

The set valued object s which *kids* points to is updatable. However, the association between *kids* and s is fixed.

4.2.4. Functions

A function is an object that, when activated, takes a number of parameters and returns an object. Functions may have side effects, so

$$Fun = State * O \rightarrow State * O$$

A function definition consists of a parameter environment object (used as a record type), a function body and an optional type for the return value. The parameter environment object is a record type for the parameter object the function may be called with.

E.g.

```
(x=Int;y=Int)
    {
    if (x>y)
            return(x);
    else
            return(y);
    }
Int
```

Here, valid function calls have a parameter object containing two integer valued parameters x and y.

4.3. Inheritance

As is the case in Amber[3] , the OOPS type system exploits inheritance, introduced in Taxis and Galileo, and extends this to work on higher-order types. Unlike Amber, where inheritance is based on the notion of *type inclusion*, our basic notion is the *instance-of* relation between objects. This relation (denoted ':') captures both the notions of *having a type* and *explicit set membership*. In other words, it will support OOPS in both *type checking* and *referential integrity*. For example

$$1 : Int$$

since each integer valued object is by definition of type (the object) Int. Also

$$1 : \{1,"Fred"\}$$

because the instances of a set valued object are exactly its members.

Let us define the instance-of relation for primitive objects and set objects in a more formal way. In a given state s, an object, say o_i with value $v_i = s(o_i)$ is an instance of an object o_m with value $v_m = s(o_m)$ if neither v_i nor v_m is the error value and if one of the following conditions hold.

- v_m is a predefined domain and v_i is one of its elements. In other words, v_m is Int, Str or Log and v_i is an integer, a string or a boolean value respectively.

- v_m is the undefined value *.

- v_i is the nil value.

- v_m is a set and v_i is one of its members.

v_m is also called a meta object of v_i.

As mentioned before, a record is an instance of another record if the instance-of relation holds for equally labeled fields of these records. For example

$$(name="Fred";sex="Male") : (name=Str;sex=\{"Male","Female"\})$$

because "Fred" and "Male" are instances of Str and {"Male","Female"} respectively.

Let us now consider the following definitions.

```
person = (name=Str; age=Int);
student = (name=Str; age=Int; number=Int);
fred = : (name="Fred"; age:=19)
```

It follows from the above that fred is an instance of person. However it is not an instance of student because of the missing number field. Suppose that we wish to extend fred so that it becomes a student:

```
fred = : (name="Fred"; age:=19; number=878)
```

We have added some information to fred to create a more informative object. Intuitively, we still want fred to be an instance of person.

In general, this gives rise to the definition of the instance-of relation between record objects. Using the same notation as above, the additional condition is defined recursively

- v_i and v_m are record values $(l_1 = o_1,..,l_n = o_n, l_{n+1} = o_{n+1},..,l_k = o_k)$ and $(l_1 = O_1,..,l_n = O_n)$ respectively such that equally labeled fields satisfy the instance-of relation, i.e. $o_1:O_1$ and .. and $o_n:O_n$.

The fields of person match (both in label and value) fields that are present in student. From this it is inferred that any instance of student is also an instance of person. Therefore, student is called a *subobject* (or subtype) of person.

This leads to our definition of the *subobject relation* in OOPS. We say that an object o_s is a *subobject* of an object o_m (denoted $o_s < o_m$) if any instance of o_s is also an instance of o_m. Hence this relation is both reflexive and transitive.

Note that the instance-of relationship (and consequently the subobject relationship) is inferred from the intrinsic properties (i.e. the structure) of the objects and need not be declared explicitly. As an effect of this interpretation, OOPS is a *polymorphic language* which means that an object may have more than one type.

Now consider the function age

```
age = (p=person)
  {
  return(p.age);
  } Int
```

It takes a person as argument and returns an integer valued object. We say that this function is of type (i.e. an instance of) (p=person) → Int. As fred is an instance of person, we can compute age(p=fred) which will result in an integer (i.e. an instance of Int).

So far we can write in general, if f is a function of type $\alpha \to \beta$ then for each instance, say i of α (i.e. i:α) f(i) is meaningful and of type β.

Since any object that is an instance of student is also an instance of person, age is also of type (or an instance of) (p=student) → Int.

Moreover, if we think of the parameter environment object as a whole (i.e. the record object (p=person)) we see that (p=person,s=Str) → Int is another type of age because any instance of (p=person,s=Str) is also an instance of (p=person) and the latter is exactly the domain on which the function age operates.

From this we can infer the following property. When f is a function of type $\alpha \to \beta$ then it is also of type $\alpha' \to \beta$ for all the subobjects α' of α. Note that this requires α and α' to be record objects as they are supposed to be parameter environments. This is an example of what is called *horizontal polymorphism*, i.e. polymorphism based on inheritance.

On the other hand, suppose that the following function picks the best out of a number of students.

```
bestStudent = (class=studentSet)
  {
  ..
  } student
```

Apparently, this function is of type (class=studentSet) → student. As any instance of student is necessarily an instance of person, bestStudent is also of type (class=studentSet) → person.

Thus a next property appears. When a function is of type $\alpha \to \beta$ then it is also of type $\alpha \to \beta'$ for all the *superobjects* β' of β (i.e. $\beta < \beta'$).

As a result, we can meaningfully compute

```
age(p=bestStudent(class=s))
```

where s is a set of students.

Recapitulating we get, if f:$\alpha \rightarrow \beta$ and $\alpha' < \alpha$ and $\beta < \beta'$ then f:$\alpha' \rightarrow \beta'$.

4.3.1. Power objects

Up to now we have types for primitive objects and for record objects as well as for function objects. How about set objects?

We would like to specify that the set object {1,2,3} is a set of some integer valued objects. One way of doing so is through the subobject relation defined in a previous section.

$$\{1,2,3\} < Int$$

However, we do not feel like introducing a direct representation of this subobject relation as it is based on the more primitive instance-of relation which is already explicitly available in OOPS. To this end, a new form of objects is introduced, called power objects. A power object is a complex object constructed by using the power operator, denoted '[ ]'. Now we can write

$$\{1,2,3\} : [Int]$$

Similarly, a possible definition of the set of students studentSet (used in an earlier example) is

```
studentSet = [student]
```

because this requires that each instance of studentSet is a set of instances of student.

In general we can write: the power operation takes an object, say x and constructs a new object [x] which has the property that

$$y : [x] \quad iff \quad y < x$$

Hence the power object of x has all x's subobjects as instances.

4.3.2. Meet

As OOPS supports the notion of multiple inheritance, one may require to specify that an object is an instance of several other objects simultaneously. Therefore, we introduce another constructor, namely *meet* (or intersection), denoted '&'.

For example, reconsider the example where fred was a person with an additional integer valued field labeled number. Hence

$$fred : person \ \& \ (number = Int)$$

Note that the earlier definition of person and student are equivalent to the following.

```
person = (name=Str; age=Int);
student = person & (number=Int);
```

The & construct arranges two declarations to be simultaneously effective, and consequently allows the additional fields to be introduced simultaneously with those 'inherited' from the 'superobject'. This is equivalent to *specialization*.

However, the meet operator also operates on other objects. A more complicated example is the following. Let men be the set of all male persons, i.e. {fred,bill,..}. Then

$$student \ \& \ men$$

which is the meet between a record object (student) and a set object (men) will have all male students as instances. In other words, its instances are exactly the members of men that are of type student.

Generally we have: the meet operation takes 2 objects, say x and y and constructs a new object x&y which has the property that

$$z : x \& y \quad iff \quad z : x \ and \ z : y$$

4.3.3. Join

It is often desirable to specify that an object is an instance of at least one of a number of other objects. To this aim, we introduce a new object constructor called *join* (or union), denoted '|'.

For example, fred is an instance of the join between student and employee (whatever this may be) since fred is an instance of student.

$$fred : student \ | \ employee$$

In general: the join operation takes 2 objects, say x and y and constructs a new object x|y which has the property that

$$z : x | y \quad iff \quad z : x \ or \ z : y$$

4.3.4. Extents

We have discussed the instance-of relation with respect to programming languages. E.g. person denotes the set of all possible objects that are instances of (i.e. of type) person. However this set is not actually available to the programmer as an object. Let us now have a look at the instance-of relation from a different angle. In database programming, person not only describes a type (a set of possible persons) but it is also used to describe the set of all persons that are currently in the database. This set is called the *extent* associated with the 'type' person.

Similarly, in OOPS we can think of an object as its extent being the set of all its instances (excluding nil) that are currently present in the knowledge base. Note that this means that OOPS merges the notions of 'type' and 'class'. We distinguish implicit and explicit extents. Explicit extents or user defined extents are set objects constructed using '{ }', whereas implicit extents are system created extents (i.e. through a scan of the knowledge base).

It is worth noting that OOPS is a persistent language which means that any value may persist. In other words persistence is not determined by type. Rather, any object that can be referenced directly or indirectly from the program object is persistent. Objects which cannot be refered are automatically deleted.

Some consequences of the definition of extent are:

1. The extent of a set valued object is the object itself;

2. The extent of a meet object is the intersection of the extents of its component objects;

3. The extent of a join object is the union of the extents of its component objects;

4. The extent of an atomic object is the empty set;

5. The extent of * (i.e. any) is the set of all objects in the knowledge base.

Some examples of the use of extents will be shown later on.

4.3.5. Typed environments

As OOPS contains a number of object constructors, it is mostly desirable that it is strongly typed.

A record object can be thought of as an environment in which both constant and variable (field) declarations are allowed. Hence, to our feeling, it is obvious to allow an instance-of specification for each field, as this makes it possible to specify the type of a constant-field value or to restrict the range of possible variable-field values. In order to get such a *typed environment* we need to extend the definition of a record value. For example,

```
(name="Fred"; Int age:=19)
```

defines a record object of which the age field is updatable, but restricted to refer to an instance of Int (i.e. an integer valued object). Note that types in OOPS are 'first class' objects[8].

It is also possible to specify *referential integrity* rules in this way. E.g.

```
person = : {fred,jane,bill};
anEmployee = (
                employee = jane;
                Int salary := 1000;
                person boss := fred
                )
```

At any time, the boss field of anEmployee should refer to an instance of the set person i.e. to an existing person. Hence an error may occur upon deletion of an element of this set.

5. OOPS+

As mentioned before, besides the OOPS layer of OOPS+ - which is mainly responsible for the definition of knowledge - there is a second layer that should provide the programmer with a number of tools for querying the various available databases and in general for knowledge manipulation and application programming. We will now introduce these tools in a rather informal way.

5.1. Logic programming

OOPS+ uses predicates to incorporate the logic programming paradigm. In most object-oriented languages, it is rather difficult and tedious to reference objects by contents: the programmer has to "hand code" support for such references explicitly by providing methods. In OOPS+, predicates can be used to query the object space. A predicate object consists of a set of defining clauses (in the logic programming sense).

A predicate can be activated in two ways: as a logical function that verifies that the predicate holds for the parameter object or as a set containing all tuples satisfying this predicate. It is then possible to further qualify this set using a *where* condition (see below).

$$Pred = State * O \rightarrow State * O$$

The following defines a predicate called parent. Let person be a set of persons.

```
* parent =
  {
  ? parent(child=_c, parent=_p) :-
    person(father=_p, self=_c)
    ;
    person(mother=_p, self=_c)
  };
```

E.g.

```
parent(child=fred)
```

will return a set object containing all record objects that satisfy the parent predicate. Such record objects are equipped with properties as specified in the predicate parameters, i.e. child and parent.

```
parent(child=fred,parent=jane)
```

will return true if jane is a parent of fred.

Predicates can refer to both objects with instances (like person in the previous example) and other predicates. The following predicate refers to the earlier defined parent predicate.

```
* ancestor =
  {
  ? ancestor(older=_x,younger=_y) :-
      parent(child=_y,parent=_x)
      ;
      parent(child=_y,parent=_z)
      ancestor(older=_x,younger=_z)
  }
```

An object may also be queried using *set restriction*, i.e. the *where* condition. For example

```
parent() where (child.age<1)
```

will only select those record objects that satisfy the parent predicate and of which the child field refers to a baby. The object to be queried by the where condition need not be a set object since its extent is used in the selection process. E.g.

```
student where (age=fred.age)
```

will make a set object consisting of all existing students which are of fred's age.

Using both predicates and set restrictions makes it possible to easily formulate rather complex queries.

5.2. Access programming

A powerful programming paradigm is 'access-oriented programming' where actions are triggered by the occurrence of events. Originally conceived as part of AI and knowledge-representation languages[9] , similar mechanisms have recently been proposed also for database programming[10] where they can be used e.g. to support constraint checking. OOPS+ supports two forms of access-oriented programming namely triggers and laws. Triggers are demons that get activated whenever a particular event occurs. Laws are similar except that their execution may be postponed which, together with a transaction mechanism, makes them suitable to implement integrity rules.

Like functions, triggers take a parameter environment upon activation and may have side effects. However, triggers can only be activated indirectly e.g. through the occurrence of an assignment.

$$Trig = State \times O \rightarrow State$$

A trigger consists of a specification of the events that will cause its activation and a body describing what actions are to be executed upon activation.

The next example defines a trigger called updatePosition.

```
movingObject = (position = point);
updatePosition = trigger on position
        (x = movingObject)
        {
        updateDisplay(x);
        }
```

The function updateDisplay is called whenever the position field of an instance of movingObject is updated.

Laws are used to provide support for consistency rules.

```
CheckSalaryChange = law on salary
        (e = employee)
        {
        check (e.salary > e..salary)
        }
```

The main difference between laws and triggers is that the activation of a law can be postponed using a transaction mechanism. Also, the activation of a law results in either the commitment or the rejection of a transaction, depending on the value of the boolean expression to be checked. Note that access to the previous field binding in a record is possible through the '. .' operator.

5.3. Rule-based programming

Rule-based programming is convenient for describing flexible responses to a wide range of events characterized by the structure of the data. This helps to explain the wide popularity of this paradigm, e.g. to construct expert systems. In OOPS+, rule-based programming is supported through functions: the function body can be either a statement, supporting procedural programming, or a ruleset, supporting rule-based programming.

A ruleset consists of a set of production rules, each specifying a condition-action pair. A condition is a logical expression, while an action is a statement.

```
thermostat = (r = room)
        {
        rule tooHot:
            (r.temp > MAXTEMP) ->
                lowerTherm(r);
        rule tooCold:
            (r.temp < MINTEMP) ->
                higherTherm(r);
        }
```

The call thermostat(r=A212) causes rules to be fired as long as the temperature is not comfortable in room A212.

6. A test case

6.1. The problem

In this subsection we will introduce four tasks - that Atkinson and Buneman[11] believe are characteristic of database programming - together with their solutions in OOPS+.

The example they use is an illustrative fragment of a manufactering company's parts database. The database represents among other things the inventory of a manufacturing company. In particular, it presents the way certain parts are manufactured out of other parts: the subparts that are involved in the manufacture of a part, the cost of manufacturing a part from its subparts, the mass increment or decrement that occurs when the subparts are assembled. A manufactured part may themselves be subpart in a further manufacturing process. The relationship between parts is therefore hierarchical, but it is a directed acyclic graph rather than a tree, for part D may be used in the manufacture of parts B and C, which are both used in the manufacture of part A. In addition, certain information must be held on the parts themselves: their name and, if they are imported, (i.e. manufactured externally) the supplier and purchase cost.

The first task is

> Task 1: Describe the database.

The next task is simple:

> Task 2: Print names, cost and mass
> of all imported parts that cost
> more than $100.

Task 3 is somewhat more complicated and defeats many query languages:

> Task 3: Print the total mass and
> total cost of a composite part.

The last task requires an update by adding some information to the database in order to examine where in the program or type system integrity constraints are implemented.

> Task 4: Record a new manufacturing
> step in the database, i.e. how a new
> composite part is manufactured from
> subparts.

In their paper [11] Atkinson and Buneman show different approaches to meet these tasks. Pascal as well as SQL mainly failed because of inadequate persistence and lack of computational power respectively. They also discuss several other languages and in particular database programming languages with respect to these tasks.

6.2. The OOPS+ solution

In this section, the OOPS+ solutions to the tasks are presented together with some remarks.

Task 1

Figure 3 shows the declarations corresponding to task 1.

```
dollars = Int;
grams = Int;
partType = (name=Str);
basePartType = partType &
                (
                cost=dollars;
                mass=grams;
                suppliedBy=[suppliers]
                );
compositePartType = partType &
                    (
                    assemblyCost=dollars;
                    massIncrement=grams;
                    components=[useType]
                    );
useType = (
        subPart=parts;
        quantity=Int
        );

[basePartType] baseParts =:{};
[compositePartType] compositeParts =:{};
parts = baseParts | compositeParts;
```

Figure 3: Task 1 in OOPS+

We define among other objects, four objects that will be used for typing: partType, basePartType, compositePartType and useType.

PartType isolates the common features of both kinds of parts.

This partType is then used in the definition of basePartType and compositePartType which are two meet objects. It follows from the definition of meet objects that each instance of basePartType must be an instance of partType and of the record (cost=dollars;mass=grams;..). Hence basePartType is a subobject or a specialization of partType.

Similarly, compositePartType is a subobject of partType.

As a basepart may be supplied by several suppliers we define the suppliedBy field in basePartType as a power object: suppliedBy=[supplier]. This means that the suppliedBy field of any instance of basePartType must be a set of (instances of) suppliers (note that we do not examine suppliers in order to concentrate on the rest).

BaseParts and compositeParts are sets that will contain instances of basePartType and of compositePartType respectively. This is specified by the power objects. Both sets are declared updatable (:{}) and initialized to the empty set.

Note the difference between basePartType and baseParts. basePartType is used as a type in the classical sense whereas baseParts plays the role of a class as in other object-oriented languages. In other words, basePartType is an intentional type whereas baseParts is an extensional type.

Parts is defined as the join between baseParts and compositeParts. Thus, at each moment an object is an instance of parts if it is an instance of either baseParts or compositeParts. In other words, each base and composite part is also a part.

UseType represents the use of a part in a composite part. Writing (subPart=parts;..) instead of (subPart=partType;..) requires that a compositePart is made of existing parts.

It follows from the definition of useType that the components field of compositePart also keeps track of the quantity used of a subPart.

Task 2

Figure 4 shows an OOPS+ solution to task 2.

The let statement allows the declaration of local variables. E.g. 'let useType use' declares use as a local variable of type useType.

(parts where cost>100 and suppliers!={}) selects among the instances of parts those objects that cost more than $100 and have a nonempty set of suppliers (i.e. they are imported). The expression returns a set containing these selected objects. Then the print statement is executed for each element in this resulting set.

```
expensiveParts =
        {
        let parts part;
        foreach part in (parts where cost>100 and suppliedBy!={} ) do
                print(part.name,part.cost,part.mass);
        }();
```

Figure 4: Task 2 in OOPS+

Task 3

Figure 5 presents an OOPS+ solution to task 3.

```
totalMassAndCost =
        (p=part)
        {
        if (p:basePart)
                return((totalMass=p.mass;totalCost=p.cost));
        else
                {
                let useType use;
                let grams resultMass:=p.massIncrement;
                let dollars resultCost:=p.assemblyCost;
                foreach use in p.components do
                        {
                        let tmpMandC := totalMassAndCost(p=use.subPart);
                        resultMass += tmpMandC.totalMass*use.quantity;
                        resultCost += tmpMandC.totalCost*use.quantity;
                        }
                return((totalMass=resultMass;totalCost=resultCost));
                }
        } (totalMass=grams;totalCost=dollars);
```

Figure 5: Task 3 in OOPS+

(p=part) is the parameter environment object or the formal parameter object of which the actual parameter must be an instance.

372

The ':' denotes the instance-of operator. Thus the condition p:basePart is true if p is an instance of basePart and false otherwise.

Note that local variables may be initialized: 'let Int resultMass := p.massIncrement'.

return((totalMass=..;..)) creates a record object which is an instance of the formal return object (totalMass=grams;totalCost=dollars).

Task4

Figure 6 sketches an OOPS+ solution to task 4.

```
newCompositePart =
        (name=Str;assemblyCost=dollars;
                massIncrement=grams;components=[useType]);
        {
        compositeParts += param;
        }
```

Figure 6: Task 4 in OOPS+

Param is a reference to the actual parameter object.

Note that the definition of useType along with the parameter type checking (i.e the actual parameter object must be an instance of the formal parameter object) forces the components field to refer to existing objects.

'+=' denotes a set update operator: 'compositeParts += param' inserts the actual parameter object (param) in the set object compositeParts.

The constraint that the relationship between parts is acyclic i.e. no part can be directly or indirectly a subpart of itself is also expressed by this definition of the function object newCompositePart. Param always refers to the actual parameter object which is newly created at the moment of the function call. Thus param cannot refer to a component of another object.

7. Conclusion

OOPS+ is object-based and multi-paradigm, supporting procedural as well as logic, rule-based and access (demon) programming. In addition, it unifies the notions of type (types are first-class objects) and collection in a general instance-of relationship. Also, the concept of name space (environment) and tuple (record) object are confused (i.e. unified), both being a set of typed and labeled references to other objects. Persistence is defined by preserving only those objects that can be referred to, using either names or (operations on) collections from a set of designated initial objects, making an explicit delete operation unnecessary and motivating an approach in which a query results in the creation of new objects which are automatically discarded unless action is taken by storing a reference to the collection in some referable object.

8. Acknowledgement

Thanks to Prof. S. Ceri for his constructive criticism which led to the writing of this paper.

References

1.   D. Sacca, D. Vermeir, A. D'Atri, J. Snijders, G. Pedersen, and N. Spyratos, "Description of the overall architecture of the KIWI system," in *Proceedings of the Esprit Technical Week*, Elsevier Publ. Co., 1985.

2.   D. Vermeir and E. Laenens, *Requirements document of the knowledge handler (main features)*, 1986. B3 report, Esprit project P1117 - KIWI

3.   L. Cardelli, "Amber," in *Proceedings of the Treizieme Ecole de Printemps d'Informatique Theorique*, 1985.

4.   D. Vermeir and E. Laenens, *Formal description of the OOPS language*, 1987. B2 report, Esprit project P1117 - KIWI

5.   F. Bancilhon and S. Khoshafian, "A Calculus for Complex Objects," in *Proceedings of the fifth ACM Symposium on Principles of Database Systems*, 1986.

6.   L. Cardelli, "A Semantics of Multiple Inheritance," in *Lecture Notes in Computer Science*, vol. 173, pp. 51-67, Springer, 1984.

7.   D. Maier, J. Stein, A. Otis, and A. Purdy, "Development of an Object-Oriented DBMS," in *OOPSLA'86 conference proceedings*, pp. 472-482, 1986.

8.   K. J. Lang and B. A. Pearlmutter, "Oaklisp: an Object-Oriented Scheme with First Class Types," in *Proceedings of the OOPSLA'86 conference*, 1986.

9.   D. G. Bobrow and Stefik, *The LOOPS Manual*, Tech. Report Xerox Park, 1981.

10.   M. Stonebraker and L. A. Rowe, "The design of POSTGRES," in *Proceedings of the ACM Sigmod International Conference on Management of Data*, ed. C. Zaniolo, 1986.

11.   M. P. Atkinson and O. P. Buneman, "Types and Persistence in Database Programming Languages," *ACM Computing Surveys* , to be published in 1988.

PCLOS: A Flexible Implementation of CLOS Persistence

Andreas Paepcke

Hewlett-Packard Laboratories

1501 Page Mill Rd.

Palo Alto, California 94304

paepcke@hplabs.hp.com

Abstract

We describe the design of a prototype which makes objects persistent. Our target language is the CommonLisp Object System (CLOS), although we pay attention to the eventual sharing of data with other languages. Our design is very flexible, in that it allows the simultaneous use of multiple, different databases. This is accomplished by defining a virtual database layer which consists of a core protocol that is expected to be implemented on all databases, and of protocol adapters which accommodate features offered by some databases, but not by others. This virtual database has been implemented for a simple, single-user, in-core data store, and for Iris, a multi-user, object-oriented database management system. We outline the advantages of the CLOS Metaclass Protocol for implementing object persistence or other low-level modifications to the CLOS implementation.

Keywords and phrases: Object persistence, CLOS.

1 Introduction

The attraction of making data structures persistent has been recognized for some time. PS-algol [ea82, ABC*83, AM86] has, for instance, attempted to introduce persistence into an Algol-like language. The spread of object-oriented programming has been followed by several more efforts in this direction. Coral3 [ML87] and GemStone [CM84, MSOP86] are examples of systems that provide object persistence for Smalltalk-80 [Gol84]. Other projects, like *Altaïr O₂* [Ban87, BBDV87, BLRV87, LRV87, AH87], attempt to add objects and persistence to existing languages, like C, Lisp or Basic. Some efforts [BKKK87, GK87, KBC*87, Rei85] attempt to fuse data definition, manipulation and persistence into one system, using a single language. All of these systems use one particular database or object server to provide persistence.

We plan to experiment with object models in general and with object persistence in particular and have therefore decided to modify an object-oriented system so that its objects may reside on persistent storage. We do not, however, want to be limited to particular databases, and therefore need a design that is flexible enough to allow the use of significantly different databases and that does not exclude the use of multiple languages sharing data. A prototype of such a system has now been completed. It is called Persistent CommonLisp Object System (PCLOS) and this paper presents its design and our initial experiences with it.

2 Design Goals and Early Decisions

The following sections explain our design goals and some of our early design decisions.

2.1 Support for Multiple Databases

Functionality must usually be paid for by decreased performance, and database management systems differ in their position on this functionality versus performance spectrum. The ability to accommodate concurrency is, for instance, a piece of functionality which requires sophisticated mechanisms that impact performance. Systems which are particularly tolerant of hardware failure will also generally incur some run-time cost. We expect that applications will have varying needs which will determine how fast object storage or retrieval must be, and which pieces of functionality are indispensable. Application developers will therefore want to pick among various storage facilities offering different tradeoffs to handle object persistence. Such a choice might also be made on the grounds that data is to be shared with other parts of an organization which already use one particular database management system.

We therefore decided that our system is to support multiple data storage mechanisms, and that we want the ability to store objects using any of these mechanisms. The interface to object storage must be defined well enough that an adapter module may be written for any database or storage scheme that meets some minimal requirements.

2.2 Support for Multiple Languages

If a system is to benefit a wide range of users, it should not unnecessarily limit the choice of languages that can access it. Programming languages differ in their expression of data structure, and earlier experiments have taught us [BK86] that we should not expect data persistence to make these differences transparent. Even among object-oriented languages, immediate, perfectly matched sharing of data *structure* might not be possible, for example because some object models allow multiple inheritance, while others provide single inheritance. We believe, however, that data sharing may be useful, even if arrangements must be made to accommodate different capabilities of expressing the structure of data.

Another design goal is therefore the support of multiple programming languages sharing persistent objects. We would like the sharing of data *structure* to degrade gracefully, as producers and consumers of data are made progressively different. We expect stored data to be organized in such a way that objects in one object-oriented language may be mapped to analogous objects in another object-oriented language through the storage medium. For non-object-oriented languages we can still utilize the fact that the stored state of an object can be viewed as a logical unit. Our work does, however, focus on an object-oriented view of data, and we will use the term "objects" to refer to units of storage.

2.3 Integration of Persistence

Wherever possible we would like to introduce longevity as a transparent property of objects in an existing language. Programs are to run on transient and on persistent objects alike, and we want to avoid syntactic changes to languages. We have found that two conditions may necessitate a deviation from this guiding principal: when database concepts prove to be so useful that they should be introduced into the realm of programming, appropriate means must be found to reflect these concepts in the programming language. A second reason for compromising on complete transparency of persistence is the desire to bring performance to acceptable levels. Whenever we do introduce persistence-related operations, we try to incorporate them naturally into the programming language.

2.4 Our First Supported Language

Our initial target and implementation language is the CommonLisp Object System (CLOS) [ea87b]. This is an object model which is in the process of being standardized and is integrated into the CommonLisp language. The language is particularly well-suited because it offers the *metaclass* concept which is a well-defined protocol for extending its implementation. CLOS is defined in [ea87b], and we merely summarize the aspects relevant to making CLOS objects persistent: CLOS supports the concept of *classes* which define *slots* that hold pieces of state. In general, slots are instance-allocated which means that every instance of a class controls a private set of values for its slots. The value of a class-allocated slot, on the other hand, is shared by all instances of the class. The combined values of all the slots represent the state of an object. Classes may inherit from multiple parents. Inheritance means that instances of a subclass will have the union of all slots of the subclass and its ancestors as their state. A *method* is conceptually a set of programs, from which one is selected to run when a method is invoked. This selection is based on the types of the arguments passed to the method.

2.5 Our First Supported Databases

Since we want to avoid making hidden assumptions about the properties of one particular database in our implementation, we decided to support two very different storage mechanisms from the start.

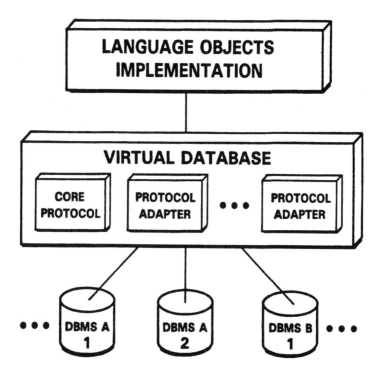

Figure 1: Design Model

The first is a very simple in-core database serving one user at a time and allowing data to be written to a file on request. The second database is Iris, a multi-user, object-oriented database offering transaction management and sophisticated query processing [ea87a].

3 System Model

Figure 1 shows the current model of our system as seen by the implementor of object persistence for one language. In order to support multiple databases, we need to provide an interface between the language implementation at the top of the figure and the set of data storage mechanisms at the bottom. We call this conceptual interface a *virtual database*. Every physical database appears to the language implementation as one such virtual database.

The virtual database consists of a *core protocol* and multiple *protocol adapters*. All operations of the core protocol must be implemented for each physical database. It is therefore kept as simple as possible, while still providing the capability to store and retrieve data comprising the state of objects. A virtual database, of which only the core protocol is implemented, would therefore be sufficiently powerful to provide persistence 'without frills'. While the core protocol is sufficient to store and recover objects and associated administrative information from most databases, an unfortunate drawback of using the core protocol alone is the fact that it would prevent the use of any advanced features which some databases might offer. One protocol adapter is therefore added to the virtual database for each *concept* which is useful but which may not be supported by all physical databases. Examples of such concepts are transactions, support for the storage of particular language datatypes, special query capabilities or unusual data locks. We use the protocol adapter idea to help us organize the various database features and to define unified programmatic interfaces to them. This usage could conceivably be expanded towards complete formal specifications, but we have taken no steps in this direction.

The *implementation* of a protocol adapter is called a *protocol converter*. One such converter is provided for each of the physical databases. If the associated adapter's concept is directly supported by a converter's database, the converter is straight-forward. Otherwise, it may choose to either implement the concept, or to return a failure indication when used.

The virtual database guarantees that we may make use of very simple database managers, while not restricting the use of sophisticated features supported by advanced database management systems.

When implementing the model of Figure 1, we generally expect that we must make changes in the *implementation* of a language, in order to have object state accesses go through the virtual database to physical storage. The interface ensures, however, that we may subsequently add database management systems without having to change the language implementation further. In addition, the definition of the virtual database is a very concise specification of the capabilities that may be relied upon when a language layer is modified to use object persistence.

3.1 The Core Protocol

The core protocol uses as its basic data structure a table model, similar to relational databases. The following assumptions are made about tables:

- Tables have unlimited height and width.

- They are uniquely named.

- Rows of tables are uniquely identified by a row number.

- Tables, and rows within them, may be dynamically created and destroyed.

- Field values are not required to have type restrictions.

Operations are defined to perform these functions. The permissive typing for values of fields is required for object-oriented languages that do not restrict the type of values assigned to slots.

3.2 Protocol Adapters

No universally valid implementation rule can, of course, be stated for the realization of protocol adapters. But we outline two adapters in this section to illustrate how we use the scheme.

Some databases allow the use of built-in types, such as `integer` when declaring data items in a schema. The use of such built-in types is desireable because of efficiency reasons or to facilitate data access through other languages. PCLOS handles the mapping of language datatypes to datatypes understood by various databases through a protocol adapter. It is implemented through a class hierarchy of *type mappers*. Instances of the root of this hierarchy convert all language items to strings which are assumed to be a supported datatype on all databases. A subclass of this root type mapper is written for each database to be supported. Such a subclass will shadow the parent mapper's conversion for the types that are directly supported by the associated database. Each protocol converter for the type mapping adapter is thus an instance of an appropriate type mapper.

Transactions are also represented in the virtual database as a protocol adapter. When the message `begin-transaction` is sent to the implementation of the virtual database for Iris – which directly supports transactions – an appropriate command is passed to the database: the protocol converter is trivial in this case. When the same operation is invoked on the virtual database implementation of the single-user in-core database, the concurrency control aspect of the transaction concept is ignored. But all subsequent operations on the database are logged as they pass through the implementation of the virtual database. The protocol converter then does implement transaction rollback.

4 Implementation Strategy

Implementing the system model described above implies that two mappings must be provided: language objects must be mapped onto the virtual database, and the virtual database must be mapped onto the physical database management systems. The first of these two is currently done as shown in Figure 2. We reserve one table in each database for information about the classes of objects stored in the database. This table, called the MasterClassTable, can be used to check for class consistency, or to communicate a class structure to other users.

All instances of one class are then stored in one table, with rows representing instances, and columns holding values of instance variables. The concept of 'table' therefore models the concept of 'class' in its role as the keeper of a set of instances related by a common structure.

The mapping of the core protocol onto physical databases varies with the characteristics of the database. For object-oriented databases, this mapping is particularly easy, since a database type can be created for each virtual table. An instance of such a type models one row, in that its state contains all the information contained in the columns of a row. A set of database functions is defined for each type, where each function takes an instance of the type and returns one piece of the object's state. This is used to model the retrieval of one virtual field. Since functions are made updatable,

Master Class Table:

Row ID	Class Name	Inheritance	Slot Names	...
0	"Ship"	NIL	"Captain Tonnage Flag"	...
1	"Book"	NIL	"Title Author Publ. Price"	...
⋮	⋮	⋮	⋮	⋮

Ship Instances:

Row ID	Slot 0	Slot 1	Slot 2
0	"Cook"	3	"US"
⋮	⋮	⋮	⋮

Book Instances:

Row ID	Slot 0	Slot 1	Slot 2	Slot3
0	"Oil Spills"	<corp-obj>	"Glib, Inc."	32.50
1	"Barter Economy"	R. Blank	"Nostalgia Publ."	<pig-obj>
⋮	⋮	⋮	⋮	⋮

Figure 2: Mapping Language Objects onto the Virtual Database

the same function may be used to retrieve and to set one virtual column of one virtual row.

If we combine the mapping of programming language objects onto the core protocol with the mapping of the core protocol onto an object-oriented physical database, we find the following, very natural relationship between language objects and their storage:

$$\begin{aligned}
\text{Language classes} &\longrightarrow \text{Database types} \\
\text{Language instances} &\longrightarrow \text{Database instances} \\
\text{Language slots} &\longrightarrow \text{Database functions}
\end{aligned}$$

The virtual database and the implementation strategy described here do not make sharing of data among multiple languages trivially easy. But several aspects do help us along towards this goal. The simplicity of the core protocol ensures that there is a clear way to describe the structure of the data as it is stored in the database. In the case of an object-oriented database the structures used to implement the core protocol will be more elegant than tables and rows, but we know that the core protocol represents an equivalent representation.

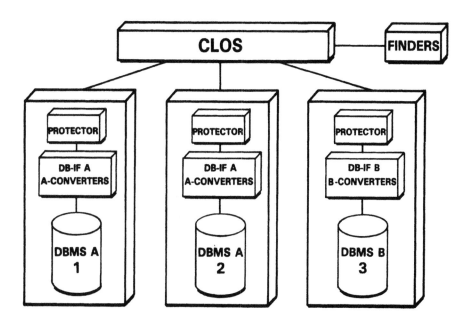

Figure 3: Modules of the PCLOS Implementation

In addition, the database access implementation of a language can use the MasterClassTable to recover data structure that might have been lost in the storage implementation.

Protocol adapters represent a further benefit of the virtual database abstraction in the context of multi-language data access because they force a separation of functionality description and implementation. The type mapping adapter, for example, does not solve the problem of translating an elaborate built-in type of one language to an equivalent type in another language, but it isolates the problem which is helpful.

5 Some Implementation Details

The following sections on selected implementation details begin with a description of the current architecture of the prototype which implements the design model. We then describe our attempts at handling performance issues and the advantages of CLOS in the context of modifications to the language implementation.

5.1 Current PCLOS Architecture

Figure 3 restates the conceptual system view of Figure 1 in terms of system modules of our prototype. The box at the top represents the CLOS implementation. Each protector and database interface (DB-IF) pair represents the implementation of the virtual database for one session on one of the databases.

```
(find-all iris-protector
          'ship
          '(and
               (> tonnage 2)
               (or
                  (= captain ''Cook'')
                  (= flag ''Kuwait'')))))
```

Figure 4: A Sample Query

Figure 3 thus shows the modules active in a session where CLOS objects are located on either of two physical databases of kind **A**, or on one database of kind **B**.

The database interfaces above each DBMS in the figure are instances of database interface classes of which one is defined for each kind of DBMS. Each instance has the responsibility to translate core protocol operations into actions on the associated physical database, and to implement various protocol adapters.

Above each database interface, we place a *protector* object. Each of these protectors is an instance of a single system class, and it represents one database session to the CLOS implementation and directly to users. Protectors perform administrative functions, such as object caching, and they are responsible for presenting the unified virtual database interface to the language layer.

Finders in Figure 3 represent the query protocol adapter. Queries are expressed by users or programs in terms of CLOS classes, slots and objects. They are then converted by a finder object into queries on the virtual database, and are passed on through the protector of one of the sessions to the associated database interface object. The 'virtual query' is re-formulated there to be recognized and solved by the underlying DBMS. The same virtual query could be handed to more than one of the open sessions.

The query in Figure 4 is an example of our associative search capabilities at the core-protocol level. The system converts the form shown in the figure into an equivalent intermediate form that replaces the type and slot references with table and column specifications. The database interface module for the Iris database, for example, then converts this intermediate form to a query involving objects, types and functions. A "complex-query" protocol adapter supported only by the Iris database allows much more sophisticated queries that include multi-valued results and existential variables and may range over multiple classes.

5.2 Performance

Access concurrency is maximized, if no objects are cached into memory but reside on a database only. Every atomic slot manipulation then accesses the database without automatically caching the affected object into memory. Such a scheme is feasible for slots and objects that are accessed rarely, and it is the default in PCLOS. This mechanism is, however, too slow when object slots are accessed frequently. In addition to this mode of operation, PCLOS therefore provides a large degree of flexibility for determining locality of information. All but the last of the following facilities has been implemented:

- Individual and recursive object caching/snapshots.

- Slot-level caching/snapshots.

- Transient slots.

- Transient objects.

- Caching within the scope of a method (method auto-cache).

Objects may be brought into memory within a transaction, which is in effect a caching operation because it write-locks the database copy. If this operation is not done under the protection of a transaction, it represents a snapshot of the object and will not prevent the database copy from being modified. For convenience, these operations may be done recursively for all objects reachable through references starting with some root object.

Granularity of locality control is further increased by the ability to cache or snapshot individual slots, while leaving the rest of the object exclusively on the database.

Slots whose state is not important across sessions may be declared transient. Values for such slots are only kept in memory, and will potentially differ among users sharing the object. If none of the slots of an object need to persist, the object can be made transient. In this case the database does not hold a copy of the object.

We plan to facilitate object caching further in the context of method execution. This will include the option to have objects that are passed as parameters cached automatically, and to have them uncached when the method terminates. We also plan to provide a macro which causes specified objects to be cached while operations within its scope execute.

Caching, while necessary for reasons of performance, can destroy the important advantage of transaction rollback, if the rollback does not also restore the state of cached, that is in-memory objects. The PCLOS prototype therefore optionally takes a snapshot of cached objects when a transaction is begun. Any subsequent rollback then includes these objects.

5.3   The Use of CLOS

The addition of persistence to an object system implementation is generally intrusive, in that permanent modifications must be made at a low level. Here are some examples of requirements that call for low-level changes:`

- Every slot (instance variable) access must be intercepted so that the appropriate slot manipulation is done on the database, if the proper data structures are not cached.

- It must be possible to inspect most low-level aspects of class hierarchies so that appropriate virtual database schemas may be produced.

- It is advantageous to add room for administrative information to the memory representation of instances.

- Access to the entire state representation of instances must be provided to support rollback for cached objects.

CLOS is exceptionally well suited to solving these problems for two reasons:

- All important data structures of the CLOS implementation are themselves objects. This includes, in particular, the representation of classes.

- The CLOS Metaclass Protocol is flexible enough to augment virtually any aspect of the CLOS implementation.

The Metaclass Protocol defines interfaces to class definition and re-definition, instance allocation, slot access mechanisms, the inheritance scheme and many other areas. Every user-defined class is conceptually an instance of some metaclass. Different metaclasses therefore implement sets of classes which are potentially radically different. All of these classes may coexist during one session. CLOS objects were made persistent without a single change to the standard language implementation. Everything was accomplished by writing a metaclass that inherits from the default metaclass and adds all persistence-related behavior. One advantage of this is that persistence capabilities may be added to a system by simply loading appropriate modules. The CLOS session is not disrupted, and modules whose classes are not of the PCLOS metaclass operate properly without change or recompilation. The remainder of this section outlines how our metaclass works.

PCLOS objects which are created during a Lisp session or which are retrieved from a database through associative search must somehow be represented in memory to provide a proper destination for messages sent to them and to preserve Lisp *eqness* for objects. This is done through the use of so-called *husk* objects which are regular objects that do not, however, contain slot values unless they are cached or transient. In this latter case they are very similar to standard CLOS objects. The ability to create and manipulate husk objects is embodied in the `pclos-class` metaclass which inherits from the standard metaclass `class` and uses four mechanisms to accomplish its goals:

- Appropriate methods inherited from `class` are shadowed.

- Some slots are added to class objects to hold persistence-related information.

- The class precedence list of user classes is modified during the processing of `defclass` forms to include a persistence-related class called `persistent-class-parent-class`.

- The physical storage of instances is modified to make room for some administrative information needed on a per-instance basis.

Standard metaclass methods which must behave differently for persistent classes are, of course, all methods involved with slot accesses, most notably those that implement `slot-value`. In addition, all slot access optimization is suppressed by the `pclos-class` metaclass, because instances may be transient or persistent at run time. Different action must be taken to access slots in these cases and compile-time decisions are therefore not possible.

Other modifications of standard metaclass behavior involve the implementation of the new slot option `:transient`. This requires shadowing of methods involved during the processing of `defclass` forms to accept this option as legal and to perform appropriate bookkeeping.

The class from which all user classes inherit automatically through the mechanism described above introduces no slots to those classes but makes relevant methods available to their instances. These include, for example, methods to `cache` and `uncache` instances, to make them persistent or transient, or to obtain persistence-related information from them.

Information needed for each instance includes a pointer to the database interface object currently responsible for it and an indication of whether the instance is currently cached. This is kept in fields of an array that is normally used by CLOS to contain slot values only. Methods which index into this array simply use a fixed offset to index past this administrative information.

In addition to these mechanisms, the `pclos-class` metaclass features methods needed to cache class-allocated slots, to map slots onto virtual columns by using their position in the class definition and to take care of other necessary functions related to persistence.

6 Status and Plans

Persistence for CLOS objects has just been implemented and is now being tested. Figure 5 shows an example configuration of our system. Workstations are connected through a local area network. One or more Iris server processes provide database service, and some workstations use the in-core database.

Initial experiments confirm the obvious suspicion that object caching is important for performance and must be optimized. Test users have quickly expressed their desire to specify arbitrary groups of objects which should be cached into memory together whenever one of the objects is referenced. A similar mechanism has been proposed in [Row86].

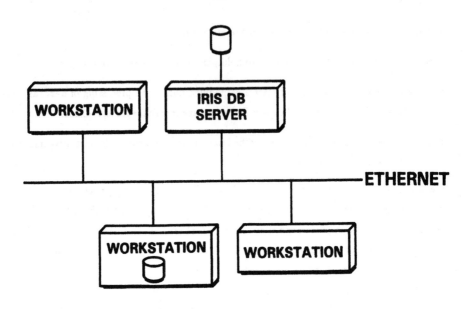

Figure 5: Example System Configuration

Notification locks were another mechanism that was requested quickly, when programmers began thinking about their use of the system. These would allow reading and writing of an item in the database, but would cause 'interested parties' to be notified whenever the item was modified. PCLOS does not currently support such a facility.

We need to investigate the problems and opportunities arising from the simultaneous use of multiple databases in one session. Deadlocks are one potential problem to be dealt with. The inclusion of many databases in a query is one of the opportunities we plan to take advantage of. This means merely that the system will submit the same query to several databases or that we might specify some filtering operations on the data from the various scans.

It is too early to tell what system designers will do with the new ability to search associatively over the object space. Through use of our prototype we hope to gain some insight into searching needs of systems and applications programmers, and to expand PCLOS query capabilities accordingly.

We have found that the rollback support for cached objects is very useful. Since all state information of an object-oriented system resides in objects, transaction rollback may be used as a convenient *undo* facility.

The PCLOS experiment is proving to be very valuable in increasing our understanding of object persistence, and its usefulness in this respect is by no means exhausted. As we learn more, we will assess the suitability of our virtual database model and the usefulness of protocol adapters.

7 Acknowledgments

Brian Beach, Jim Kempf and Joe Mohan laid extensive groundwork for PCLOS with their DOOM project. Nancy Kendzierski has improved the system model and its implementation through her suggestions. Alan Snyder, Dennis Freeze and Craig Zarmer have helped to clarify many aspects of previous versions of this paper, and initial test users have cheerfully reported the painstakingly collected details of their system crashes to me – with a smile.

References

[ABC*83] M.P. Atkinson, P.J. Bailey, K.J. Chisholm, W.P. Cockshott, and R. Morrison. An approach to persistent programming. *The Computer Journal*, 26(4):360–365, 1983.

[AH87] Timothy Andrews and Craig Harris. Combining language and database advances in an object-oriented development environment. In Norman Meyrowitz, editor, *Proceedings of the Conference on Object-Oriented Programming Systems, Languages and Applications.*, Association of Computing Machinery, 1987.

[AM86] M.P. Atkinson and R. Morrison. Integrated persistent programming systems. In B.D. Shriver, editor, *Proceedings of the 19th Annual Hawaii Conference on System Sciences*, pages 842–854, , , 1986. Vol. IIA, Software.

[Ban87] François Bancilhon. *Object Oriented Multilanguage Systems: the Answer to Old and New Database Problems?* Technical Report, Altaïr, BP 105; 78153 Le Chesnay Cedex; France, October 1987.

[BBDV87] François Bancilhon, Véronique Benzaken, Claude Delobel, and Fernando Velez. *The O₂, V0 Object Manager Interface.* Technical Report Altaïr 11-87, Altaïr, BP 105; 78153 Le Chesnay Cedex; France, September 1987.

[BK86] Brian Beach and James Kempf. *DOOM: Permanent Objects for Common Lisp.* Technical Report STL-TM-86-09, HP Labs, September 1986.

[BKKK87] Jay Banerjee, Won Kim, Hyoung-Joo Kim, and Henry F. Korth. Semantics and implementation of schema evolution in object-oriented databases. In Umeshwar Dayal and Irv Traiger, editors, *Proceedings of the ACM Special Interest Group on Management of Data*, Association of Computing Machinery, 1987.

[BLRV87] Gilles Barbedette, Christophe Lécluse, Philippe Richard, and Fernando Velez. *The O₂ Programming Environment, Version V0.* Technical Report, Altaïr, BP 105; 78153 Le Chesnay Cedex; France, October 1987.

[CM84] G. Copeland and D. Maier. Making Smalltalk a database system. In *Proceedings of the ACM/SIGMOD International Conference on the Management of Data*, 1984.

[ea82] Malcom Atkinson et al. PS-Algol: an Algol with a persistent heap. *Sigplan Notices*, 24–30, July 1982.

[ea87a] D. Fishman et al. Iris: an object-oriented database management system. *ACM Transactions on Office Information Systems*, 5(1):48–69, April 1987.

[ea87b] Daniel G. Bobrow et al. *Common Lisp Object System Specification.* Technical Report 87-001, ANSI, September 1987.

[GK87] Jorge F. Garza and Won Kim. *Transaction Management in an Object-Oriented Database System.* Technical Report ACA-ST-292-87, MCC, September 1987.

[Gol84] Adele Goldberg. *Smalltalk-80: The Interactive Programming Environment.* Addison Wesley, 1984.

[KBC*87] Won Kim, Jay Banerjee, Hong-Tai Chou, Jorge F. Garza, and Darrell Woelk. Composite object support in an object-oriented database system. In Norman Meyrowitz, editor, *Proceedings of the Conference on Object-Oriented Programming Systems, Languages and Applications.*, Association of Computing Machinery, 1987.

[LRV87]   Christopher Lécluse, Philippe Richard, and Fernando Velez. *O2, an Object Oriented Data Model.* Technical Report Altaïr 10-87, Altaïr, BP 105; 78153 Le Chesnay Cedex; France, September 1987.

[ML87]   Thomas Merrow and Jane Laursen. A pragmatic system for shared persistent objects. In Norman Meyrowitz, editor, *Proceedings of the Conference on Object-Oriented Programming Systems, Languages and Applications.*, Association of Computing Machinery, 1987.

[MSOP86] David Maier, Jacob Stein, Allen Otis, and Alan Purdy. Development of an object-oriented DBMS. In Norman Meyrowitz, editor, *Proceedings of the Conference on Object-Oriented Programming Systems, Languages and Applications.*, Association of Computing Machinery, 1986.

[Rei85]   Stephen P. Reiss. *GARDEN: An Environment for Graphical Programming.* Brown University, October 1985. Reference and Programmers Manual.

[Row86]   Lawrence A. Rowe. A shared object hierarchy. In Klaus Dittrich and Umeshwar Dayal, editors, *Proceedings of the International Workshop on Object-Oriented Database Systems,* Association of Computing Machinery, 1986.

A Shared, Persistent Object Store

Colin Low

Department of Computer Science,
Queen Mary College,
Mile End Rd, London E1 4NS.

ABSTRACT

Smalltalk-80 is presented as a useful testbed for prototyping applications involving shared, persistent objects, and a detailed design of a shared persistent object store is discussed. The store is a set of named containers for object state, and it provides low-cost atomic transactions using an optimistic synchronisation technique. The standard Smalltalk-80 virtual machine is modified to support a new object class, the **Transaction**, and an example of a Smalltalk program using nested sub-transactions is given. Immutability of object state is identified both as an important property of objects, and a basis for producing an efficient implementation within a distributed system environment.

Keywords and phrases: object-oriented programming, distributed systems, atomic transactions, Smalltalk-80, persistence, immutable objects.

Many die too late, and some die too soon. Still the doctrine sounds strange: "Die at the right time".

Nietzsche

1. Introduction

Programmers have been using one particular type of shared, persistent object for years: this object type has many guises, but it is normally called a **file**. Files are the donkeys of persistent programming; anything and everything that will fit is perched on the backs of these poor animals. The shared, persistent object is a natural generalisation of a concept of which a file is the most common example.

The terms are loosely defined as follows. An object is a container for a value of some kind. The object is associated with a set of operations for observing, and possibly changing that value. It is possible to create new objects with an initial value. It must be possible to reference a particular object in order to apply one of it's operations, and so objects are named. Objects may cease to exist; it is possible to have referential failure.

An object is persistent if it "dies at the right time"; persistence is an observation about the lifetime of an object, namely that it exists for precisely as long as intended, and then it disappears. There are various ways of deciding the right time, but generally speaking a programmer or user demands that an object disappears according to some predetermined criteria, and only then. It should not cease to exist as a result of a processor crash, or a bad block on a disc drive, or some other predictable failure mode. An object can be persistent without being shared, and shared without being persistent; the problems are orthogonal.

The work described in this paper grew out of the author's interest in the programming of distributed applications. A large proportion of such applications involve objects which are shared and which are persistent. One approach is to use shared files held on public file servers such as the SUN NFS file servers [24] used at Queen Mary College. There are two problems however. The first is that files are the *only* persistent objects provided in the author's environment. This is a nuisance but it can be circumvented by using files to contain the representation of different types of object using ad-hoc conventions. The second problem is that the commonly available programming languages, such as C, Pascal, Modula 2 etc. do not provide the programmer with persistent objects. The data types that represent objects are held in volatile memory, and the state of an object disappears along with the process that created it. The programmer has almost no control over the lifetimes of objects created within the programming language environment. The normal solution is to devise a scheme for saving the state of an object in a file, and to read it back into volatile memory again at an appropriate time. A substantial amount of programming effort is required to map between volatile and non-volatile representations of an object's state. This is clumsy and a poor use of a programmer's time; the advantages of integrating persistence into programming languages have been discussed by a number of authors, for example Atkinson *et al* [2].

If sharing of objects is involved then the problems are compounded. A realistic solution to the problem of sharing is complex if several objects are involved, integrity constraints exist, and concurrency is to be maximised; reviews of work in this area have been carried out by Kohler [15], and Bernstein and Goodman [5], and we return to the problem later in this paper. It is unreasonable for every programmer to have to re-invent a solution to the dual problems of persistency and sharing, and one answer is to use a programming language that supports the use of shared, persistent objects as naturally and transparently as possible.

The choice of Smalltalk-80 [14] as the host language for experimenting with distributed applications was motivated primarily by the simplicity and expressive power of the language. It is an excellent testbed for trying out new ideas, and the language is embedded in an interactive programming environment that provides support for sophisticated user interface design. The addition of shared, persistent objects to Smalltalk makes it possible to build on the existing and extensive class libraries in the construction of working prototypes; it is possible to experiment with user interfaces to shared objects without the systems programming expertise that is usually needed for this type of work. Much of the language is defined at a level where it is easy to modify; for example, in the programming of distributed applications the ability to handle many kinds of exception can be important, and Smalltalk provides the building blocks for implementing exception handling.

There were other reasons for choosing Smalltalk. It is portable and runs on most of the workstations in the author's environment. The virtual machine that provides run-time support is described in detail [14], and an efficient high-level language implementation is available locally [21].

It is important that any modifications to the Smalltalk-80 system preserve the reactive quality of existing Smalltalk-80 implementations; this is a severe constraint that has guided much of the design.

The system described is tightly coupled to Smalltalk-80, but many of the techniques used have wide applicability. It can be regarded as an instance of the application of distributed system techniques to a particular problem; multiple Smalltalk-80 client virtual machines sharing a persistent object storage service.

Sections 3 of this paper discusses how Smalltalk-80 objects are represented in object memory. It shows how it is possible to remove part of the object memory to a remote server, the persistent object store how persistent objects are loaded into volatile object memory on demand. Section 4 describes the optimistic concurrency control algorithm. Section 5 outlines the importance of immutable objects for improving performance. Section 6 discusses new object classes and extensions to the Object protocol. Section 7 is a description of a Smalltalk program in Appendix 1. that shows the use of nested transactions for modifying a shared, persistent object.

2. The Environment

The environment assumed is a typical distributed system such as described in [12]: a number of client graphics workstations interconnected by a fast local area network. Shared services such as filing, naming, authentication etc. are provided by dedicated server systems, and a fast, lightweight remote procedure calling mechanism is used for client-server communication. Users wishing to use Smalltalk will run it on a client workstation; a number of dedicated servers are used to implement the shared persistent object store. This is shown in Figure 1. below. Each server has a significant amount of volatile semiconductor storage and a large amount of slower non-volatile storage (e.g. disc drives).

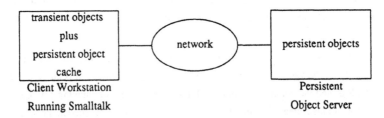

| transient objects plus persistent object cache | network | persistent objects |
| Client Workstation Running Smalltalk | | Persistent Object Server |

Figure 1: The Client-Server Relationship

There can be several client systems using persistent objects, and the objects themselves can be spread across several persistent object stores.

3. Object Memory

Smalltalk-80 objects must be represented in memory of some kind. Normally this is the volatile semiconductor memory in a workstation. This section describes the abstract relationships of objects and shows that it is possible to spilt the object space between different physical memories, some volatile, and some non-volatile. This gives rise to two types of object: transient objects whose death is unimportant, and persistent objects which satisfy guarantees about their lifetimes.

3.1. The Object Space

A Smalltalk-80 system contains a large number of objects. Objects reference oneanother. Even though all objects reference a class object, they can be split into two major categories: some objects *only* reference a class object and correspond to primitive data types, such as symbols or floating point numbers. The other category of objects directly reference one or more other objects in addition to the class reference. The first type of object I will refer to as *primitive*, while the second type will be referred to as *composite*.

A composite object *directly depends* on those objects it references (including its class). A primitive object directly depends only on its class. An object *indirectly depends* on any object that can be reached through the chain of direct dependencies. The direct dependency relationship between objects structures the object space into a directed graph. This is shown in Figure 2. below; the object at the root of the graph directly depends on every object to which it is directly connected with an arrow, and indirectly depends on every object beneath it.

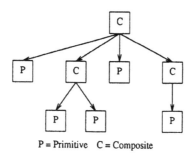

P = Primitive   C = Composite

Figure 2: Object Dependencies

A chain of direct dependencies is terminated as follows:

- small integers directly depend on nothing.
- primitive objects directly depend only on their class.
- the **nil** object depends only on its class.
- the class **Object** (from which all classes are derived) depends on nothing.

Indirect dependency is important; an object of type **Class** directly depends on its method dictionary, and indirectly depends on method texts and bytecodes. Change a method and the behaviour of a class changes. Indirect dependency, and the consequent set of objects so related to a given object, reflects the way in which the programmer builds complex objects and behaviour out of simple components; it is an important concept when considering persistent objects.

An object is a container for a value; for example, a tuple of references to other objects, and in general that value may change in time as the operations associated with an object modify it. The object behaves like a finite state machine, beginning in an initial state, and moving through a succession of new states. The *total* state of an object is the collective state of every object on which it indirectly depends. Like indirect dependency, the total state of an object is important because objects are not normally used in isolation; they are meaningfully related by the programmer into complex structures.

Object names in Smalltalk-80 have only local significance within a given instance of the Smalltalk-80 virtual machine. They can be 16 or 32 bit quantities depending on the implementation.

3.2. The Smalltalk-80 Object Memory

The Smalltalk-80 object memory is the repository for all of the object state in a conventional Smalltalk system. It provides the following functionality:

- a set of named containers for object state.
- a set of operations on containers. The normal operations on containers are **create** a new container, **put** a value in a container, **get** a value from a container, and **delete** a container.

The actual interface differs in detail. There is no delete; object deletion is handled implicitly. The object state is divided into parts, with explicit operations for putting or getting individual components of the object state. Everything in Smalltalk is an object, so with minor exceptions (e.g. small integers) the representation of objects is simple and uniform. The visible components of object state are the **class**, the **size**, and the **instance variables**. The size refers to the number of instance variables. The instance variables can either be used as a tuple of object references, corresponding to a composite object, or as numeric values representing, for example, floating point numbers, long integers, strings etc, which correspond to a primitive object.

Object names, or references, can either be memory addresses, or more commonly, integer indices into an Object Table; these names have no significance outside of the Smalltalk-80 virtual machine. Smalltalk implementations provide persistent objects by saving and loading much of the state of the Object Memory, but because object names retain their local significance sharing is not possible. This procedure is manual; a snapshot of the object memory is written to disc and so the object memory can be reconstructed at a later time, but it is an all-or-nothing procedure. It is not possible to select one object and access it from another different virtual machine.

3.3. Persistent Objects

The first step in providing shareable, persistent objects is to provide objects with global, persistent names so that the same object can be referenced from anywhere at any time using the same name. However, not all objects need to be persistent (for example, an instance of the browser), and a large proportion of the objects in a running Smalltalk system could be transient even if persistence was available. Fast creation and manipulation of objects is too important to be compromised. A second consideration is that the number of global persistent objects that a user could reference is potentially enormous; persistent object names could be larger than ordinary object names, particularly if they are viewed as capabilities and sparsely embedded within a range of values. This is a strong argument in favour of having a two-level naming scheme. The existing Smalltalk Object Memory interface is retained, along with the existing object names, and the conversion between local and persistent object names is carried out transparently within the Object Memory. This approach is modular; a large proportion of the virtual machine remains unchanged.

If an arbitrary Smalltalk object is to be made persistent then every object on which it indirectly depends must also be persistent, otherwise there will be dangling references in the persistent store. This is a strong requirement; every object references a class, so this implies that every class referenced by a persistent object must also be global and persistent. In turn this implies that the class methods are persistent, as are all the literal symbols referenced by the methods.

There must be at least one object from which the persistency of every other object depends, and the solution chosen was to make the global directory **Smalltalk** a global persistent object. Any object reachable from **Smalltalk** is global to every virtual machine. Any object not reachable from **Smalltalk** is transient.

A two-level naming scheme, with the local Object Memory translating between local and non-local references is like the LOOM system [16]. The LOOM system was designed to overcome address space restrictions in workstations at the time, and provided an object virtual address space very much larger than that possible in a memory resident system. It was essentially an "object paging" system, with non memory resident objects being paged in whenever an "object fault" occured. The LOOM system did not separate out transient and persistent objects, nor was sharing of objects possible, but some of the algorithms used have been adopted, as described in 3.5.

3.4. The Persistent Object Store

The persistent object store, like the object memory, is a collection of named containers for object state. An object name is referred to as a *persistent object identifier* or PID. A PID can be decomposed into two parts: a partition of the total object space, and an object within a partition. A partition is synonymous with a server; persistent objects can be spread amongst several servers. The basic operations provided are:

CreatePSObject: ( ) → PID
Create a new persistent object in the null state and return its PID.

GetPSObject: PID → ObjectState
Get the current state of the specified object.

PutPSObject: PID × ObjectState
Assign a new state to the specified object.

SwapPid: PID × PID
Swaps the state associated with a PID so that the state associated with the first PID becomes the state associated with the second PID and vice-versa.

SetPSNil: PID
Store the PID of the object that represents the undefined object **nil**.

GetPSNil: ( ) → PID

Get the PID of the **nil** object.

SetPSRoot: PID

Store the PID of the root object.

GetPSRoot: ( ) → PID

Get the PID of the root object.

An object is created in a null state because of objects that refer to themselves; unless the PID of an object is known, the value of the object (which contains a reference to itself) cannot be assigned. **SwapPid** is necessary to support the primitive method **become**. One of the important, uses of **become** is to grow collection classes. There are two special objects that must be individually accessible. **SetPSRoot** makes public the PID of the root object via **SetPSRoot**. **SetPSNil** and **GetPSNil** do the same for the **nil** object. Every persistent object is a component of the total state of the root object and is indirectly related to it. There is no difficulty in having several root objects, but in this implementation there is only one, the global directory **Smalltalk**.

The object state can be decomposed into components. These are

class: the PID of the object that is the class of this object.

hash: a unique value permanently associated with this object.

timestamp: the (pseudo)time of the last write to the object.

size: the number of components of a given type making up the instance variables (see **type**).

type: indicates whether the instance variables should be interpreted as object names (PID's), or as values of a simple type. (e.g. words, bytes, bits. See [14]).

immutable: a flag to indicate whether an object's instance variables are frozen (see below).

instanceVariables: there are **size** items of a type indicated by **type**.

All object references within the persistent object store are PID's.

3.5. Object Loading

Persistent objects are loaded into Smalltalk object memory on demand. The algorithm is very similar to, but simpler than, the algorithm used in LOOM [16]. The LOOM system was designed to overcome address space limitations, and some of its complexity is a result of trying to obtain satisfactory performance in the face of these limits. There are broad similarities to the persistent heap used with S-Algol [3] and the subsequent PS-Algol [8]. The following description assumes a familiarity with the implementation of the Smalltalk object memory described in [14].

When a persistent object is loaded into local object memory it must look like any other local object. Persistent objects only reference other persistent objects, so all the PID's that are part of the state of a persistent object must be transformed into local references. A reference to a persistent object (PID) is resolved in three parts:

1. check to see if it is already loaded. A dictionary called **LoadedPIDs** is used to store an association between a PID and a local object pointer. If it is already loaded then the PID can be replaced by a local reference. If the object is not loaded then......

2. a vacant object table slot is allocated; the index in the table is the local object reference or identifier. The persistent object now has a local reference that can be stored in the instance variables of other local objects. The association between the PID and the object identifier is added to **LoadedPIDs**. A flag is set in the object table entry to indicate that the object state has not been loaded. The state of the object is not loaded until......

3. it receives a message. When an object receives a message its state is loaded. The **class** component, and the **instanceVariables** component of composite objects, contain PID's which must be transformed into local references by going back to step 1.

The partly loaded object in step 2 is identical to what in LOOM is called a *leaf*. Object table slots were at a premium in LOOM, and this was behind the decision to implement *lambdas*, unresolved PID's in local Object Memory, rather than turning every PID in an object into a leaf. Object table slots in 32 bit architectures are no longer at a premium, and *lambdas* were not implemented. The resulting algorithm is simple. Almost all the changes required are confined to the object memory. The object memory maps object pointers to object state; it must check that the state is loaded while doing this. This test must be made for every object access, but it is typically a couple of machine instructions to do this.

Adding support for persistent objects increases the overall size of object memory. The object memory keeps more administrative information about each persistent object than a non-persistent object, but careful design can minimise the increase.

An unexpected cost of persistence is the need to explicitly and permanently associate with every object a random value. All objects respond to the method **hash**, which is used, for example, in the class **Dictionary**. A **Dictionary** uses the hash value for a hash table lookup, and as it *could* be persistent, the hash values it uses must remain constant over its lifetime; this implies that objects, whether persistent or not, must return a hash key which is not calculated from transient information.

There are two separate garbage collectors. Persistent objects are garbage collected if they are inaccessible from the root object **Smalltalk**. At present this is an off-line activity. The local Object Memory maintains its own garbage collector and discards objects which are not currently referenced within the Object Memory; discarding a persistent object has no global effect. Persistent objects may be discarded from local memory at any time as part of LRU space recovery.

If the state of a persistent object is changed, then its new state should be reflected in the persistent object store at some point. Even when the object store is not shared this is not a trivial problem. The simplest method is to carry out an immediate write-through whenever a change to a persistent object is made. This is not usually what is wanted. If a message is sent to an object, it may send further messages to objects on which it is directly dependent, as may the objects which receive the messages. It is the change to the *total state* of the object which is of interest, not just the individual changes to each object. We want *all of the* changes to be done or *none* of the changes to be done.

As an example, I may write a program to iterate over a persistent **Set** of numbers, adding 1 to each number in the set. I either want all of the numbers to be incremented, or none of them; if there is a crash I do not want a situation where I am unable to tell how many of the elements have been incremented.

It is reasonable to expect the total state of an object to change consistently to new states. This all-or-nothing property of changes in the face of failure and concurrency is familiar in database systems where it is referred to as *atomicity*. Atomicity is composed of two properties:

> **recoverability**: integrity constraints on data are preserved in the face of crashes and other unexpected events.

> **serialisability**: integrity constraints on data are preserved in the face of multiple, concurrent sets of changes, and the effect of concurrents sets of changes is the same as if they had been executed consecutively.

The requirement that the total state of an object changes consistently to a new total state means that even in the absense of sharing we need recoverability. The next section shows how both serialisability and recoverability are provided using atomic transactions.

4. Concurrent Access

The persistent object store is designed to be shared by many users. Sharing introduces the traditional problems of concurrency control encountered in shared databases. It is important to make some assumptions about the environment in which objects are shared, for without assumptions one is forced to make pessimistic design decisions.

1. The persistent object store is not intended to be used as an object-oriented database. It *is* a database, but lacks the sophistication and optimisation of special purpose database systems. An examination of what additional facilities might be needed has been made by Bloom and Zdonik [7].

2. It is possible to make the *optimistic* assumption that, because the number of objects is very large, the probability that any two sets of concurrent changes intersect, is small.

It is the author's belief that such a system is useful. There are many applications in a distributed system in which sharing is required, but not at the level of commercial database systems. At QMC I use various shared databases for such things as telephone numbers, reports, student applications, mail and electronic bulletin boards, purchases, authentication and authorisation, and so on. I also share documents, programs and program sources. This sharing is implemented using UNIX files and simple file locking. It works because the optimistic assumption holds. It is unreasonable to implement every instance of sharing as if it was a no-holds-barred state-of-the-art transaction processing system. The system described makes no compromises; it provides serialisability and recoverability, but uses the optimistic assumption to provide them cheaply and efficiently.

Each Smalltalk object memory can function as a large cache of persistent objects. Reading the state of an object is cheap, because it is local, and it was decided to use a concurrency control algorithm that takes advantage of this cache. Bernstein and Goodman [5] have analysed a large number of algorithms and concluded that all are variations on two basic techniques: 2-phase locking (2PL), and time-stamp ordering (TSO). 2PL with time-out based deadlock prevention, such as used in Violet [13], is essentially a pessimistic technique. It is valuable in an environment where conflicts between transactions are common. Under such circumstances it is better to use an algorithm that enforces serialisation through waiting rather than restarting. A number of TSO schemes were considered and two which seem most appropriate are optimistic algorithms proposed by Thomas [25], and Kung and Robinson [17]. The two algorithms use the same principles, but differ in detail; Thomas's algorithm associates a timestamp with each item of data, while Kung and Robinson do not, at the expense of a more complex validating procedure. Both algorithms can be used, but Thomas's variation was chosen because it allows each Smalltalk virtual machine to use its cache of persistent objects very efficiently.

The solution adopted is a conventional one in which the programmer brackets access to persistent objects inside a **Transaction**. Transactions are atomic, and may span several persistent object stores. Transactions may also be nested inside oneanother. A transaction may involve reading and writing many persistent objects; this is done locally using the persistent object cache in the client workstation, and assuming the objects are present, does not require any external communication. A transaction adds very little processing overhead to the interactive cost of reading and writing persistent objects; the cost is incurred at commit time, when the read and write set of the transaction is *validated* by the persistent object stores involved. The outcome of validation is that the transaction is accepted at every store, and committed, or it is rejected at one or more, and aborted; the coordination of multi-server commits or aborts is carried out by an new external service called the **Transaction Manager** (TM).

Each persistent object is associated with a timestamp which is updated when a new value is assigned to the object. If an object is loaded into an object memory, it carries with it the timestamp current at the time it was loaded. A transaction records the sequence of persistent object references carried out in local object memory; for each reference it records the object, the operation (read or write) and the timestamp. This information is used by the validating algorithm in the persistent object store to decide whether to accept or reject a transaction. Each transaction consists of at least three, possibly four phases:

1. the read phase. This can last an indefinite period. A transaction which has entered its read phase is invisible. No external communication between the Smalltalk virtual machine and any other external agent are necessary during the read phase, unless of course, an object fault occurs and it is necessary to load a persistent object from a store. Any changes made to persistent objects are local and hence invisible.

2. the validation phase. The transaction becomes visible at the end of the read phase, when the Smalltalk virtual machine requests a new transaction number from the TM. The virtual machine then communicates with every object store involved, passing it the transaction number, the set of names (PID's) of objects read from that store, the set of objects written to in that store, and for each updated object, a new value. Each component of the transaction is separately validated by the relevant object store, and is either accepted or rejected. Each component of the transaction enters the *pending* stage. The Smalltalk virtual machine then sends to the TM the transaction number, and a list of the object stores involved in this transaction, and awaits the total result. The TM is

responsible for handling the usual 2-phase commitment protocol (see, for example Ceri and Pelagatti [11]). The TM communicates with each object store asking whether the component of the transaction handled in that store was accepted or rejected. If all are accepted, then the transaction as a whole is committed, if any are rejected then the transaction as a whole is aborted. The TM logs the result in stable store, and sets about informing every object store of the result. If the result is commit, then the component at each object store enters the write phase, otherwise the pending write is deleted.

3. the write phase. During the pending phase the new values for each object in the write set were written to secondary storage, *but the objects themselves were not changed.* For a period of time both the original object, and a new "shadow object" exist. When the object store is informed by the TM that it can complete the transaction each object is swapped with the "shadow object" that has been created, and when all the swaps have been carried out, the write phase is marked as complete. The write phase is irrevocable and non-interruptable. Only one transaction can be in the write phase in any object store. If a crash occurs, only one write phase could have been affected, and it is restarted at the beginning, over and over, until it is done. If a transaction is aborted, the object store is informed eventually and the shadow objects are garbage collected in the normal way.

4. the pre-write phase. This phase does not occur in every case. If several transactions are pending, and their write sets intersect, then they must enter the write phase in the order defined by their transaction numbers, otherwise inconsistencies will result. In this case the write phase of a transaction will be delayed, although it is guaranteed to be done eventually. A transaction in the pre-write phase has been committed, but it cannot enter the write phase until all earlier pending transactions have been resolved. If the write set of a transaction is a subset of the read set, then the transaction will never enter the pre-write phase, because the read set is known not to intersect the write set of any pending transaction.

It is important that the 2-phase commit mechanism runs separately from client workstations; 2-phase commit is an essential part of the algorithm for ensuring that updates preserve integrity constraints, and for this reason an independent service, the TM, is used to carry it out. It could have been built into the persistent object store, but it has nothing to do with object storage as such, and separating it out produces a clean division of responsibilities with the object stores in a subordinate relationship to the TM.

The validation of a transaction works as follows:

1. if the timestamp on each object in the read set is still current, the the transaction is accepted.

2. if the read set intersects the write set of any pending transaction then the transaction can either be rejected or the decision can be deferred. The transaction will be rejected if the transaction number of the pending transaction is more recent than that of the transaction being validated, otherwise the decision is deferred.

The purpose behind deferring a decision is the hope that a pending transaction which conflicts with the transaction being validated might be aborted, in which case there will no longer be a conflict. A transaction is split into components, one in each object store involved, and it is possible that a component of transaction T1 might be deferring on the outcome of transaction T2 in one object store, and the other way around in a second object store. The result is a deadlock, and the way to avoid deadlock is to use the ordering on transaction numbers. For this reason an older transaction is not permitted to defer on the result of a younger transaction.

Despite the optimism of the concurrency control mechanism described above, the processing and communication overheads are not neglible. It is not as bad as it looks; only mutable persistent objects require concurrency control, and as will be described below, large numbers of persistent objects can be immutable. A major advantage of this method is that almost no overhead is incurred by the Smalltalk virtual machine until a transaction attempts to commit, and then all the overheads are incurred in one go. This is different from 2PL for example, where locks are acquired progressively. This means that it is possible to produce the same kind of reactive Smalltalk program as before; if a user is interacting with a persistent data-base, there are no unpredictable gaps in the interaction where the program fails to respond because slow concurrency control measures are being applied.

One problem, and a major one, is the problem of restarts. A transaction will be rejected and restarted if its read set has been invalidated, and the probability increases as the read set increases in both size and age. The bigger the read set, the more likely a conflict will occur. The older the read set, the more likely that a conflict has already occurred. If the size of a read set is a problem, then the algorithm of Kung and Robinson offers a finer grain of concurrency, so reducing the chance of conflict. The problem of age can be solved by always reloading an object prior to reading it , but this is pessimistic. If the network supports broadcasting, or better still, multicasting, then each object store can broadcast a list of PID's that have been invalidated by the last transaction. Each Smalltalk virtual machine will examine the list, and delete (turn into a *leaf*) any of the objects that are currently resident. If the object is referenced again it will be reloaded in its most modern version.

A second method of controlling restarts is to use sub-transactions. The room-booking program in Appendix 1. uses this method. When a sub-transaction commits and is accepted, nothing actually changes in the persistent store, because the sub-transaction is dependent on the overall outcome of its enclosing transaction; it is guaranteed to commit or abort if necessary, but it remains a shadow pending the outcome of the 2 phase commit. In this state it behaves like a lock on all the objects it has changed; no more recent transaction can invalidate it. The behaviour is not precisely the same as 2PL, as subsequent transactions are not serialized through waiting, but it is possible to obtain some of the advantages of 2PL without sacrificing any of the advantages of the optimistic technique.

5. Immutable Objects

The term "immutable object" as used by, for example Liskov and Guttag [18], refers to objects whose state cannot change, because there are no operations which "mutate" that state. New objects are created in an initial state, which remains fixed for all time. It is useful to relax this idea of immutability to include objects which were mutable up to a certain time and thereafter have a fixed state. The crucial difference is that the second type of object *does* have mutating operations, which suddenly become invalid. A common example of this can be found in containers which have a "read-only" switch which can be toggled. The configuration RAM in my music synthesiser has such a switch; UNIX files have such a switch. When the switch is set to read-only, attempting to write causes an error.

It is possible to distinguish three types of immutability:

1.   objects which have no mutating operations, and are immutable by declaration e.g. integers.

2.   objects which become "read-only", but whose state at the time it is frozen may contain references to objects which are mutable.

3.   objects which become "read-only", and whose state at the time it is frozen contains only references to other immutable objects of type 3 or type 1.; e.g. the *total state* of the object is immutable.

Type 1. will be called *intrinsic immutability*, type 2. will be called *read-only*, and type 3. will be called *recursive read-only*. Instrinsic immutability is a property which can be declared at the language level; Emerald [6] provides the ability to declare immutable objects of this type. The other two types of immutability are dynamic; a mutable object becomes immutable, and in this system, remains immutable.

Immutable objects are the keystone to building an efficient, shared, persistent object store. Immutable objects do not have to participate in concurrency control. Once loaded into local object memory they can be read at no further cost. There are enormous benefits in trying to make as many objects as possible immutable. In particular, it is proposed that Smalltalk classes can be made recursive read-only.

One of the problems of sharing classes is that if someone modifies a shared class, everyone else will be affected by the modification. It is not in the common interest to permit important system-wide classes to be mutable. The rule adopted is that every shared class is made recursive read-only. When a Smalltalk program is compiled, the compiler binds textual references to classes into object references by using the global dictionary **Smalltalk**. The behaviour of a class can be modified by binding a new class to the class name in the dictionary, rather than by directly modifying the class object. The procedure for modifying a class is

-    copy the current immutable shared class.

-    modify the copy.

-    make the copy recursive read-only.

-    bind the copy to the name of the class in **Smalltalk**.

New instances of the class will conform to the new behaviour. As instances of classes refer to their class via an object reference, old instances will still refer to the original class, which is now inaccessible but will remain in the persistent object store for as long as there is an instance to reference it. This behaviour is exactly what is needed: an object always refers to the immutable class object of which it is an instance, and this relationship exists until the last object of that class is removed. Classes can be modified, but not at the expense of existing instances. The installation of modified shared classes still needs to be controlled, but change control is necessary in any shared environment, and is beyond the scope of this paper.

A Smalltalk programmer usually develops new classes; these are frequently changed during development and debugging, and it would be counter-productive to slow down program development by insisting that all classes are shared, persistent and immutable. It is possible to have mutable private classes; to do this requires a minor modification to the Smalltalk compiler so that it resolves global names by searching a local name dictionary in addition to **Smalltalk**.

Smalltalk in its current practice does not favour the use of immutable objects; the ability to change anything at any time is one of its attractions for rapid prototyping. The functionality of the persistent object store does not depend on having immutable objects, but its performance does. Aside from the performance issues, immutability is an important concept that is deeply connected with persistence and sharing. An object is a container for a value; values are intrinsically immutable. By distinguishing between immutable and mutable objects we are really distinguishing between containers and the values they contain. Sometimes we wish to share a value, at other times a container. It is important that a language can distinguish between these two very different ideas.

The introduction of immutable objects into the Smalltalk system is not a trivial change and requires changes to be made to many of the system classes, such as the **Browser**.

6. New Classes and Modifications to the Object Protocol

This section introduces a new class and changes to the Smalltalk **Object** protocol. Additions to the object protocol are:

> **persist** - the receiver and every object indirectly dependent on it will be made persistent. Note that unless the object is inserted into the tree of persistent objects it will not be accessible, and hence garbage collected.

> **reload** - the receiver and every object on which it indirectly depends is reloaded into the cache in its most recent version, on demand. The current version in the cache is invalidated.

> **shallowReadOnly** - returns a copy of the receiver that is read-only. The object is copied in the same way as **shallowCopy**.

> **deepReadOnly** - returns a copy of the receiver that is recursive read-only. The object is copied in the same way as **deepCopy**.

The new object class is **Transaction**. A **Transaction** is a primitive object type that acts as an anchor for the read and write sets that must be accumulated as part of the concurrency control algorithm. Sending the message **commit** to a transaction object results in the read and write sets being sent to the appropriate object stores; the result of committing a transaction is either **True** or **False**, depending on whether it completed or aborted. Creating a transaction modifies the global state of the virtual machine on a per-process basis; an internal boolean flag **IsTransaction** is set to true, and the transaction object id. is added to a push-down stack of active transactions. Any reads or writes on the state of a mutable persistent object are logged by adding an entry to the transaction at the top of the stack. This provides nested sub-transactions, and permits a single large transaction to be split into several parts. The ability to handle nested sub-transactions does add to the complexity of the implementation, but as the TM service already has the ability to manage a transaction split into components, the addition of components which are themselves transactions is not a significant change.

The protocol for **Transaction** is:

new - returns a new transaction.

restart - discards any previous state associated with a transaction and begins again. This is necessary to support nested sub-transactions. Restarting the same sub-transaction until it succeeds is not the same as creating one new transaction after another until one succeeds.

commit - commit a transaction; returns **True** or **False**. The meaning of true varies; if the transaction is a sub-transaction then true means that the effects of the transaction will be visible if the parent commits, but until then the changes are pending, and are guaranteed to be able to commit if necessary. If a transaction is at the top-level then **True** means that changes to the persistent object store are visible to everyone.

abort - abort the transaction.

7. An Example

The following example is taken from a real problem at QMC: we have a seminar room which is used for meetings, seminars, tutorials etc. The seminar room is booked by obtaining a particular notebook, looking up the date, and booking the time slot on the appropriate day. This concurrent access problem has been solved in Smalltalk[1] in a very simple (and unsophisticated) way, to illustrate how transactions and persistence can be used. The booking sheet is an array of 12 entries, one for each month. Each month is a dictionary, with a possible association between a day number and a booking page. A dictionary is used because the room is booked on only a few days each month. Each booking page consists of nine slots corresponding to the hours between 9.00 and 17.00 inclusive. The relationship of these objects is illustrated in Figure 3. below.

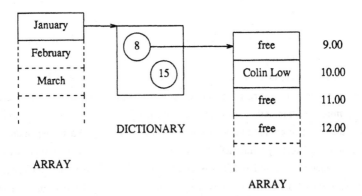

Figure 3: A Persistent Booking Sheet

---

[1] Actually Little Smalltalk version 2 [9]. The advantage is that the example is compact and completely self contained, including the user interaction. The additions to the protocol for **Object** and the **Transaction** class are not part of standard Little Smalltalk.

The user follows a simple interactive dialogue to pick the month, the day, and then the hour in the day. If the slot is free, the user inserts his or her name, and may then attempt to book more slots. The full text of the program can be found in Appendix 1.

The booking sheet is initialised as an array of 12 dictionaries, made persistent, and then made **shallowReadOnly** because the array will never be changed after initialisation. It is placed in a global dictionary, which can be assumed to be the root for all persistent objects. It is found in this example by using the class **Finder** to map a pathname onto a persistent object.

The **Booker** class uses a main enclosing transaction t1 and a number of subtransactions represented by t2 and t3. It is possible to book several slots as individual sub-transactions, and then to commit or abort all the bookings in one go. t2 is an example of a transaction that restarts automatically and is invisible to the user; if there is no booking page for a particular day in the month, one is inserted, but this only has to be done once. If there is a concurrency conflict and the transaction fails, it is guaranteed to succeed next time, because it will not be necessary to insert a page; it will see the new insertion. It is guaranteed to see the new insertion because the dictionary is reloaded. t3 is restarted until either the user gives up or books a slot. Thus both t2 and t3 can be completed, and so the enclosing transaction t1 can be completed. Sub-transactions are an important method for avoiding lengthy restarts.

8. Ownership

The example above works because the **Booker** class is trusted to be civilised. It would be simple to write another version of **Booker** that rips out pages, discards previous bookings, and so provides preferential access to the owner of the modified class. There are many places in a persistent object store where different classes of ownership and protection may be needed to prevent the malicious manipulation of object state; such facilities are normal in filing systems.

Providing a protection scheme poses many practical difficulties, and is a significant problem in its own right. This design for persistent object storage has no protection, a serious weakness that requires further work.

9. Status

A prototype of the persistent object store has been integrated and tested with Little Smalltalk version 2, but it does not contain the concurrency control algorithm described. The object store is currently being completely re-implemented to run under UNIX 4.2 BSD and will have the functionality described in this paper. Modifications are being made to the Smalltalk-80 virtual machine implementation available at QMC [21].

10. Related Work

The integration of languages and persistence overlaps conventional database technology, and the two directions of research are converging rapidly. The following selection of related work does not attempt to be comprehensive, and concentrates on a few closely related projects.

The work described in this paper was influenced by the PISA (Persistent Information Space Architecture) project [4] at the Universities of Glasgow and St. Andrews. As has already been mentioned, the work is based around the language PS-Algol, which uses a shared, persistent heap. The persistent object space is split into databases which provide the unit of granularity for lock based concurrency control.

The Telesophy system described by Caplinger [10] provides an object-oriented framework for the management of *Information Units* (IU). An IU can be any kind of information; IU's can be stored, viewed and modified within a distributed system environment. Concurrency control is based on times-tamping.

The GemStone database system [19] combines an object-oriented programming language, OPAL, and persistent objects. The system as described appears to be centralised, with remote user agents to provide a user interface. Concurrency control is achieved with an optimistic technique, and a hybrid approach incorporating locking has been proposed [23].

The VBASE system [1] is based on a pair of languages, TDL and COP, which together constitute a strongly typed object-oriented programming language something like CLU [18]. Database features such as data clustering, inverse relationships and triggers are provided, as are synchronisation and recovery, although the latter are not described.

Jasmine [26] is an object-oriented system for manipulating persistent objects that describe the structure and versions of software. This system is interesting because the objects are immutable and can be repli-cated at low cost, a factor of great importance in a distributed application.

Merrow and Laursen [20] describe a system called Coral3 which extends Smalltalk to include shared per-sistent objects. An object class **SharedObjectHolder** is used to encapsulate a representation of the state of an arbitrary object. Sharing is coordinated by locking but transactions are not supported. Behaviour is not shared, that is, classes do not seem to be persistent, leading to possible inconsistencies between dif-ferent uses of the same object.

The Amoeba File System [22] supports a tree-structured hierarchy of files, and uses file versions and both optimistic concurrency control and locking to provide atomic transactions. In so far as a persistent object store is normally built on top of a filing system, the Amoeba Filing System solves almost all the problems at an underlying level; a persistent object store needs very little extra functionality. It is clear that the difference between a filing system and an object store is largely one of nomenclature, and it will be interesting to see whether a general filing system can match the performance of a persistent store optim-ised to match the characteristics of a particular object-oriented programming environment.

11. Conclusion

A number of issues have arisen in this work that are of general interest. The first is the identification of the *total state* of an object, which in Smalltalk includes all of the classes from which the object inherits behaviour. The need to store behaviour as well as object state is paramount; there is no actual distinction, and it is a credit to Smalltalk that it supports such a fundamental idea so well.

The second issue is whether the behaviour of an object should be mutable or immutable; Smalltalk permits behaviour to be modified dynamically, but in a shared environment this is a recipe for chaos. The system described adopts the convention that shared behaviour is immutable. The binding between a name, for example, the name of the class **Set**, and the behaviour associated with the class, is mutable; the behaviour itself is immutable and will persist for as long as it is referenced.

The third issue is how the use of immutable objects can be extended as far as possible; the resources wasted in maintaining many immutable objects can be solved by the decreasing cost of hardware, but the difficulty in providing efficient implementations of atomic transactions and the related problem of replication in a distributed system are unlikely to be solved so easily.

Lastly, the uniformity of Smalltalk's object model and the way in which a powerful programming notation can be developed out of simple elements make it useful for low cost experimentation in an area which normally requires a high investment in languages, systems and expertise. It should be an invaluable tool for developing prototypes of highly interactive applications that use shared persistent information.

12. Acknowledgements

Thanks to Eliot Miranda for helping with Smalltalk-80 esoterica and to Takis Anastassopoulos for conversations about concurrency. Thanks also to the referees for their comment and advice.

13. References

[1] Andrews, T., & Harris, C., *Combining Language and Database Advances in an Object-Oriented Development Environment*, ACM OOPSLA Proceedings 1987.

[2] Atkinson, M.P., Bailey, P.J., Cockshott, W.P., Chisholm, K.J., & Morrison, R., *An Approach to Persistent Programming*, Computer Journal 26(4) 360-365 (1983).

[3] Atkinson M.P., Chisholm K., Cockshott, P., & Marshall R., *Algorithms for a Persistent Heap*, Software Practise and Experience, 13(3), 259-271 (1983).

[4] Atkinson M.P., Morrison, R., & Pratten, G.D., *A Persistent Information Space Architecture*, Persistent Programming Research Report 21, Persistent Programming Research Group, Dept. of Computing Science, University of Glasgow.

[5] Bernstein, P.A. & Goodman, N., *Concurrency Control in Distributed Database Systems*, Computing Surveys, 13(2), 185-219 (1981).

[6] Black, A., Hutchinson, N., Jul, E., & Levy, H., *Object Structure in the Emerald System* ACM OOPSLA Proceedings (1986).

[7] Bloom, T., & Zdonik, S.B., *Issues in the Design of Object-Oriented Database Programming Languages*, ACM OOPSLA Proceedings (1987).

[8] Brown, A.J., & Cockshott, W.P., *The CPOMS Persistent Object Management System*, Persistent Programming Research Report 13, Persistent Programming Research Group, Dept. of Computing Science, University of Glasgow.

[9]   Budd, T., *A Little Smalltalk,* Addison-Wesley, 1987.

[10] Caplinger, M., *An Information System Based on Distributed Objects,* ACM OOPSLA 1987 Proceedings.

[11] Ceri, S., & Pelagatti, G., *Distributed Databases* McGraw-Hill, 1985.

[12] Coulouris, G., & Dollimore, J., *Distributed Systems: Concepts and Principles,* (to be published) Addison-Wesley 1988.

[13] Gifford, D., *Violet: an Experimental Decentralised System,* Operating Systems Review, **13**(5) 1979.

[14] Goldberg, A., & Robson, D., *Smalltalk-80, The Language and its Implementation,* Addison-Wesley, 1983

[15] Kohler W.H., *A Survey of Techniques for Synchronisation and Recovery in Decentralised Computer Systems,* Computing Surveys, **13**(2), 149-183(1981).

[16] Krasner, G., *Smalltalk-80, Bits of History, Words of Advice,* Addison-Wesley 1983.

[17] Kung, H.T. & Robinson, J.T., *On Optimistic Methods for Concurrency Control,* ACM-TODS, **6**(2), 1981.

[18] Liskov B., & Guttag, J., *Abstraction and Specification in Program Development,* MIT Press, 1986.

[19] Maier, D., Stein, J., Otis, A., & Purdy, A., *Development of an Object-Oriented DBMS,* ACM OOPSLA Proceedings 1986.

[20] Merrow T., & Laursen J., *A Pragmatic System for Shared Persistent Objects,* ACM OOPSLA Proceedings 1987.

[21] Miranda, E., *Brouhaha: A Portable Smalltalk Implementation,* ACM OOPSLA 1987 Proceedings.

[22] Mullender, S.J., & Tanenbaum, A.S., *A Distributed File Server based on Optimistic Concurrency Control,* Proc. of the 10th. Symp. on Operating Systems, ACM. N.Y. P51-62.

[23] Penny, D.J., & Stein, J., *Class Modification in the GemStone Object-Oriented DBMS,* ACM OOPSLA 1987 Proceedings.

[24] SUN Microsystems Inc. *The Network Services Guide,*

[25] Thomas, R.W., *A Majority Consensus Approach to Concurrency Control for Multiple Copy Databases,* ACM-TODS, **4**(2), 1979.

[26] Wiebe D., *A Distributed Repository for Immutable Persistent Objects,* ACM OOPSLA Proceedings 1986.

Appendix 1

```
*  Room Booking Class
Declare Booker Object
Class Booker
      bookRoom   |aBookingSheet month day more t1|
         aBookingSheet <- (Finder new) lookup: 'a pathname'.
         more <- true.
         t1 <- Transaction new.
         [more] whileTrue:
            [month<- self selectMonth: aBookingSheet.
             day <-  self selectDay: month.
             self selectSlot: day.
             (slot isNil)
                ifFalse:
                   [day at: slot put: self createEntry].
               more <- self continue
             ]
         'you may now commit any bookings you have made' print.
         (self continue) ifTrue:.[t1 commit] ifFalse: [t1 abort]

      selectMonth: aBookingSheet
         ^aBookingSheet at: (self prompt: 'select month [1..12]: ') asInteger

      selectDay: aMonth    |day t2|
         day <- ((self prompt: 'select day [1..31]: ' ) asInteger).
         t2 <- Transaction new.
         [true] whileTrue:
            [ (aMonth includesKey: day)
                 ifFalse: [aMonth at: day put: (Array new: 9)].
              (t2 commit)
                 ifTrue: [^(aMonth at: day)]
                 ifFalse:[t2 restart. aMonth reload]
             ]

      selectSlot: aDay     |hour more t3|
         "iterate until a free slot is chosen or user gives up"
         t3 <- Transaction new.
         more <- true.
         [more] whileTrue:
         [  t3 restart.
            self showBookings: aDay.
            hour <- (self prompt: 'select slot beginning hour [9..17]: ') asInteger.
            hour <- (hour - 8).       "convert hour to array index"
            (aDay at: hour) isNil
               ifTrue: [ aDay at: hour put: self createEntry.
                            (t3 commit)
                               ifTrue:
                                  ['done' print. ^nil]
                               ifFalse:
                                  ['just booked - try again' print]
                          ]
                  ifFalse: ['this slot is already booked' print].
             aDay reload.
             more <- self continue.
         ].
         t3 abort

      showBookings: aDay
         (1 to: 9) do:
            [:slot| ((slot + 8) printString , ' ') printNoReturn.
             ((aDay at: slot) isNil)
                ifTrue:  ['free' print]
                ifFalse: [(aDay at:slot) print]
             ]

      createEntry
         ^self prompt: 'enter your name: '

      continue
         ('y' = (self prompt: 'continue y/n: '))
             ifTrue: [ ^true]
             ifFalse: [ ^false]

      prompt: promptString   |aString|
```

```
                    [ promptString printNoReturn.
                      aString <- smalltalk getString. aString size = 0] whileTrue: [ nil ].
                      ^aString
]
Declare BookingSheet Array
Class BookingSheet
        initialise
            (1 to: 12) do: [:month| self at: month put: Dictionary new].
            self persist; shallowReadOnly
]
Declare Finder Object
Class Finder
        "turns a pathname into a persistent object - here it is a stub"
        lookup: aPath
            ^globalNames at:£bookingSheet
]
Declare SetUp Object
Class SetUp
        new
            globalNames at:£bookingSheet put: (BookingSheet new:12) initialise
]
```

Vol. 270: E. Börger (Ed.), Computation Theory and Logic. IX, 442 pages. 1987.

Vol. 271: D. Snyers, A. Thayse, From Logic Design to Logic Programming. IV, 125 pages. 1987.

Vol. 272: P. Treleaven, M. Vanneschi (Eds.), Future Parallel Computers. Proceedings, 1986. V, 492 pages. 1987.

Vol. 273: J.S. Royer, A Connotational Theory of Program Structure. V, 186 pages. 1987.

Vol. 274: G. Kahn (Ed.), Functional Programming Languages and Computer Architecture. Proceedings. VI, 470 pages. 1987.

Vol. 275: A.N. Habermann, U. Montanari (Eds.), System Development and Ada. Proceedings, 1986. V, 305 pages. 1987.

Vol. 276: J. Bézivin, J.-M. Hullot, P. Cointe, H. Lieberman (Eds.), ECOOP '87. European Conference on Object-Oriented Programming. Proceedings. VI, 273 pages. 1987.

Vol. 277: B. Benninghofen, S. Kemmerich, M.M. Richter, Systems of Reductions. X, 265 pages. 1987.

Vol. 278: L. Budach, R.G. Bukharajev, O.B. Lupanov (Eds.), Fundamentals of Computation Theory. Proceedings, 1987. XIV, 505 pages. 1987.

Vol. 279: J.H. Fasel, R.M. Keller (Eds.), Graph Reduction. Proceedings, 1986. XVI, 450 pages. 1987.

Vol. 280: M. Venturini Zilli (Ed.), Mathematical Models for the Semantics of Parallelism. Proceedings, 1986. V, 231 pages. 1987.

Vol. 281: A. Kelemenová, J. Kelemen (Eds.), Trends, Techniques, and Problems in Theoretical Computer Science. Proceedings, 1986. VI, 213 pages. 1987.

Vol. 282: P. Gorny, M.J. Tauber (Eds.), Visualization in Programming. Proceedings, 1986. VII, 210 pages. 1987.

Vol. 283: D.H. Pitt, A. Poigné, D.E. Rydeheard (Eds.), Category Theory and Computer Science. Proceedings, 1987. V, 300 pages. 1987.

Vol. 284: A. Kündig, R.E. Bührer, J. Dähler (Eds.), Embedded Systems. Proceedings, 1986. V, 207 pages. 1987.

Vol. 285: C. Delgado Kloos, Semantics of Digital Circuits. IX, 124 pages. 1987.

Vol. 286: B. Bouchon, R.R. Yager (Eds.), Uncertainty in Knowledge-Based Systems. Proceedings, 1986. VII, 405 pages. 1987.

Vol. 287: K.V. Nori (Ed.), Foundations of Software Technology and Theoretical Computer Science. Proceedings, 1987. IX, 540 pages. 1987.

Vol. 288: A. Blikle, MetaSoft Primer. XIII, 140 pages. 1987.

Vol. 289: H.K. Nichols, D. Simpson (Eds.), ESEC '87. 1st European Software Engineering Conference. Proceedings, 1987. XII, 404 pages. 1987.

Vol. 290: T.X. Bui, Co-oP A Group Decision Support System for Cooperative Multiple Criteria Group Decision Making. XIII, 250 pages. 1987.

Vol. 291: H. Ehrig, M. Nagl, G. Rozenberg, A. Rosenfeld (Eds.), Graph-Grammars and Their Application to Computer Science. VIII, 609 pages. 1987.

Vol. 292: The Munich Project CIP. Volume II: The Program Transformation System CIP-S. By the CIP System Group. VIII, 522 pages. 1987.

Vol. 293: C. Pomerance (Ed.), Advances in Cryptology — CRYPTO '87. Proceedings. X, 463 pages. 1988.

Vol. 294: R. Cori, M. Wirsing (Eds.), STACS 88. Proceedings, 1988. IX, 404 pages. 1988.

Vol. 295: R. Dierstein, D. Müller-Wichards, H.-M. Wacker (Eds.), Parallel Computing in Science and Engineering. Proceedings, 1987. V, 185 pages. 1988.

Vol. 296: R. Janßen (Ed.), Trends in Computer Algebra. Proceedings, 1987. V, 197 pages. 1988.

Vol. 297: E.N. Houstis, T.S. Papatheodorou, C.D. Polychronopoulos (Eds.), Supercomputing. Proceedings, 1987. X, 1093 pages. 1988.

Vol. 298: M. Main, A. Melton, M. Mislove, D. Schmidt (Eds.), Mathematical Foundations of Programming Language Semantics. Proceedings, 1987. VIII, 637 pages. 1988.

Vol. 299: M. Dauchet, M. Nivat (Eds.), CAAP '88. Proceedings, 1988. VI, 304 pages. 1988.

Vol. 300: H. Ganzinger (Ed.), ESOP '88. Proceedings, 1988. VI, 381 pages. 1988.

Vol. 301: J. Kittler (Ed.), Pattern Recognition. Proceedings, 1988. VII, 668 pages. 1988.

Vol. 302: D.M. Yellin, Attribute Grammar Inversion and Source-to-source Translation. VIII, 176 pages. 1988.

Vol. 303: J.W. Schmidt, S. Ceri, M. Missikoff (Eds.), Advances in Database Technology – EDBT '88. X, 620 pages. 1988.

Vol. 304: W.L. Price, D. Chaum (Eds.), Advances in Cryptology – EUROCRYPT '87. Proceedings, 1987. VII, 314 pages. 1988.

Vol. 305: J. Biskup, J. Demetrovics, J. Paredaens, B. Thalheim (Eds.), MFDBS 87. Proceedings, 1987. V, 247 pages. 1988.

Vol. 306: M. Boscarol, L. Carlucci Aiello, G. Levi (Eds.), Foundations of Logic and Functional Programming. Proceedings, 1986. V, 218 pages. 1988.

Vol. 307: Th. Beth, M. Clausen (Eds.), Applicable Algebra, Error-Correcting Codes, Combinatorics and Computer Algebra. Proceedings, 1986. VI, 215 pages. 1988.

Vol. 308: S. Kaplan, J.-P. Jouannaud (Eds.), Conditional Term Rewriting Systems. Proceedings, 1987. VI, 278 pages. 1988.

Vol. 309: J. Nehmer (Ed.), Experiences with Distributed Systems. Proceedings, 1987. VI, 292 pages. 1988.

Vol. 310: E. Lusk, R. Overbeek (Eds.), 9th International Conference on Automated Deduction. Proceedings, 1988. X, 775 pages. 1988.

Vol. 311: G. Cohen, P. Godlewski (Eds.), Coding Theory and Applications 1986. Proceedings, 1986. XIV, 196 pages. 1988.

Vol. 312: J. van Leeuwen (Ed.), Distributed Algorithms 1987. Proceedings, 1987. VII, 430 pages. 1988.

Vol. 313: B. Bouchon, L. Saitta, R.R. Yager (Eds.), Uncertainty and Intelligent Systems. IPMU '88. Proceedings, 1988. VIII, 408 pages. 1988.

Vol. 314: H. Göttler, H.J. Schneider (Eds.), Graph-Theoretic Concepts in Computer Science. Proceedings, 1987. VI, 254 pages. 1988.

Vol. 315: K. Furukawa, H. Tanaka, T. Fujisaki (Eds.), Logic Programming '87. Proceedings, 1987. VI, 327 pages. 1988.

Vol. 316: C. Choffrut (Ed.), Automata Networks. Proceedings, 1986. VII, 125 pages. 1988.

Vol. 317: T. Lepistö, A. Salomaa (Eds.), Automata, Languages and Programming. Proceedings, 1988. XI, 741 pages. 1988.

Vol. 318: R. Karlsson, A. Lingas (Eds.), SWAT 88. Proceedings, 1988. VI, 262 pages. 1988.

Vol. 319: J.H. Reif (Ed.), VLSI Algorithms and Architectures – AWOC 88. Proceedings, 1988. X, 476 pages. 1988.

Vol. 320: A. Blaser (Ed.), Natural Language at the Computer. Proceedings, 1988. III, 176 pages. 1988.

Vol. 322: S. Gjessing, K. Nygaard (Eds.), ECOOP '88. European Conference on Object-Oriented Programming. Proceedings, 1988. VI, 410 pages. 1988.